Proceedings of the

21st European Conference on Cyber Warfare and Security

ECCWS 2022

A Conference Hosted By

University of Chester UK

16-17 June 2022

Review Process

Papers submitted to this conference have been double-blind peer reviewed before final acceptance to the conference. Initially, abstracts were reviewed for relevance and accessibility and successful authors were invited to submit full papers. Many thanks to the reviewers who helped ensure the quality of all the submissions.

Ethics and Publication Malpractice Policy

ACIL adheres to a strict ethics and publication malpractice policy for all publications – details of which can be found here:
http://www.academic-conferences.org/policies/ethics-policy-for-publishing-in-the-conference-proceedings-of-academic-conferences-and-publishing-international-limited/

Self-Archiving and Paper Repositories

We actively encourage authors of papers in ACIL conference proceedings and journals to upload their published papers to university repositories and research bodies such as ResearchGate and Academic.edu. Full reference to the original publication should be provided.

Conference Proceedings

The Conference Proceedings is a book published with an ISBN and ISSN. The proceedings have been submitted to a number of accreditation, citation and indexing bodies including Thomson ISI Web of Science and Elsevier Scopus.

Author affiliation details in these proceedings have been reproduced as supplied by the authors themselves.

The Electronic version of the Conference Proceedings is available to download from https://papers.academic-conferences.org/

The Conference Proceedings for this year and previous years can be purchased from http://academic-bookshop.com

E-Book ISBN: 978-1-914587-41-2
E-Book ISSN: 2048-8610
Book version ISBN: 978-1-914587-40-5
Book Version ISSN: 2048-8602

Published by Academic Conferences International Limited
Reading, UK
+44 (0) 118 324 6938
www.academic-conferences.org
info@academic-conferences.org

Contents

Paper Title	Author(s)	Page No
Preface		vi
Committee		vii
Biographies		ix
Keynote Outlines		
Research papers		
Desired Cybersecurity Skills and Skills Acquisition Methods in the Organizations	Kirsi Aaltola, Harri Ruoslahti and Jarmo Heinonen	1
Identification of Violence in Twitter Using a Custom Lexicon and NLP	Jonathan Adkins	10
The U.S. Cyber Threat Landscape	Elie Alhajjar and Kevin Lee	18
Impact of Moral Disengagement on Counterproductive Work Behaviours in IT Sector, Pakistan	QaziMuhamamd Ali	25
Supporting Situational Awareness in VANET Attack Scenarios	Dimah Almani, Steven Furnell and Tim Muller	37
The Emergence of IIoT and its Cyber Security Issues in Critical Information Infrastructure	Humairaa Yacoob Bhaiyat and Siphesihle Philezwini Sithungu	46
Including Human Behaviors Into IA Training Assessment: A Better Way Forward!	Henry Collier	52
Automatic Construction of Hardware Traffic Validators	Jason Dahlstrom, Brandon Guzman, Ellie Baker and Stephen Taylor	60
Operationalizing Cyber: Recommendations for Future Research	Baylor Franck and Mark Reith	70
Obstacles on the Path to the Internet of Things: The Digital Divide	John Gray	78
Societal Impacts of Cyber Security in Academic Literature: Systematic Literature Review	Eveliina Hytönen, Amir Trent and Harri Ruoslahti	86
Planning the Building a SOC: A Conceptual Process Model	Pierre Jacobs, Sebastiaan von Solms and S. van der Walt	94
Pedagogical and Self-Reflecting Approach to Improving the Learning Within a Cyber Exercise	Anni Karinsalo, Karo Saharinen, Jani Päijänen and Jarno Salonen	105
Strategies for Internet of Things Data Privacy and Security Using Systematic Review	Sithembiso Khumalo, Amanda Sibiya and Teballo A. Kekana	115
Public Authorities as a Target of Disinformation	Pekka Koistinen, Milla Alaraatikka, Teija Sederholm, Dominic Savolainen, Aki-Mauri Huhtinen and Miina Kaarkoski	123
An Ontological Model for a National Cyber-Attack Response in South Africa	Aphile Kondlo, Louise Leenen and Joey Jansen van Vuuren	130
Combining System Integrity Verification With Identity and Access Management	Markku Kylänpää and Jarno Salonen	140

Paper Title	Author(s)	Page No
SIEM4GS: Security Information and Event Management for a Virtual Ground Station Testbed	Yee Wei Law and Jill Slay	150
Physical Layer Security: About Humans, Machines and the Transmission Channel	Christoph Lipps and Hans Dieter Schotten	160
On the Road to Designing Responsible AI Systems in Military Cyber Operations	Clara Maathuis	170
Responsible Digital Security Behaviour: Definition and Assessment Model	Clara Maathuis and Sabarathinam Chockalingam	178
A Model for State Cyber Power: Case Study of Russian Behaviour	Juha Kai Mattila	188
Building Software Applications Securely With DevSecOps: A Socio-Technical Perspective	Rennie Naidoo and Nicolaas Möller	198
Effective Cyber Threat Hunting: Where and how does it fit?	Nombeko Ntingi, Petrus Duvenage, Jaco du Toit and Sebastian von Solms	206
Two Novel Use-Cases for Non-Fungible Tokens (NFTs)	Alexander Pfeiffer, Natalie Denk, Thomas Wernbacher, Stephen Bezzina, Vince Vella and Alexiei Dingli	214
Cybersecurity Risk Assessment Subjects in Information Flows	Jouni Pöyhönen, Aarne Hummelholm and Martti Lehto	222
Exploring Care Robots' Cybersecurity Threats From Care Robotics Specialists' Point of View	Jyri Rajamäki and Marina Järvinen	231
Cyberterritory: An Exploration of the Concept	Jori-Pekka Rautava and Mari Ristolainen	239
Researching Graduated Cyber Security Students: Reflecting Employment and job Responsibilities Through NICE Framework	Karo Saharinen, Jarmo Viinikanoja and Jouni Huotari	247
ZTA: Never Trust, Always Verify	Char Sample, Cragin Shelton, Sin Ming Loo, Connie Justice, Lynette Hornung and Ian Poynter	256
A Collaborative Design Method for Safety and Security Engineers	Taito Sasaki, Takashi Hamaguchi and Yoshihiro Hashimoto	263
Siamese Neural Network and Machine Learning for DGA Classification	Lander Segurola-Gil, Telmo Egues, Francesco Zola and Raúl Orduna-Urrutia	271
Probability of Data Leakage and its Impacts on Confidentiality	Paul Simon and Scott Graham	280
An Analysis of the Prevalence of Game Consoles in Criminal Investigations in the United Kingdom.	Iain Sutherland, Huw Read and Konstantinos Xynos	289
How are Hybrid Terms Discussed in the Recent Scholarly Literature?	Ilkka Tikanmäki and Harri Ruoslahti	296
Application of Geospatial Data in Cyber Security	Namosha Veerasamy, Yaseen Moolla and Zubeida Dawood	305
Layer 8 Tarpits: Overwhelming Malicious Actors With Distracting Information	Toni Virtanen and Petteri Simola	314
Cybersecurity Threats to and Cyberattacks on Critical Infrastructure: A Legal Perspective	Murdoch Watney	319

Paper Title	Author(s)	Page No
Cyber Security Norms: Trust and Cooperation	Allison Wylde	328
A Managerial Review and Guidelines for Industry 4.0 Factories on Cybersecurity	Najam Ul Zia, Ladislav Burita, Aydan Huseynova and Victor Kwarteng Owusu	336
A Cyber-Diplomacy and Cybersecurity Awareness Framework (CDAF) for Developing Countries	Hendrik Zwarts, Jaco Du Toit and Basie Von Solms	341
PHD Papers		350
Assessing Information Security Continuous Monitoring in the Federal Government	Tina AlSadhan and Joon Park	351
Expectations and Mindsets Related To GDPR	Pauliina Hirvonen	360
A Cyber Counterintelligence Competence Framework	Thenjiwe Sithole and Jaco Du Toit	368
The Cyber era`s Character of War	Maija Turunen	378
Enhancing the STIX Representation of MITRE ATT&CK for Group Filtering and Technique Prioritization	Mateusz Zych and Vasileios Mavroeidis	385
Masters Papers		392
Cyber Concerns With Cloud Computing	Jacob Chan and Mark Reith	393
Impact of Information Security Threats on Small Businesses During the Covid-19 Pandemic	Inga Mzileni and Tabisa Ncubukezi	401
Analysis of Sexual Abuse of Children Online and CAM Investigations in Europe	Johanna Parviainen and Jyri Rajamäki	411
Forensic Trails Obfuscation and Preservation via Hard Drive Firmware	Paul Underhill, Toyosi Oyinloye, Lee Speakman, and Thaddeus Eze	419
WIP Papers		429
DRAM-Based Physically Unclonable Functions and the Need for Proper Evaluation	Pascal Ahr, Christoph Lipps and Hans Dieter Schotten	430
ECHO Cyber-Skills Framework as a Cyber-Skills Education and Training Tool in Health and Medical Tourism	Eleonora Beltempo, Jussi Karvonen and Jyri Rajamäki	434
How to Utilize E-EWS as a Tool in Healthcare	Janne Lahdenperä, Joonas Muhonen and Jyri Rajamäki	438

Preface

These proceedings represent the work of contributors to the 21st European Conference on Cyber Warfare and Security (ECCWS 2022), supported by University of Chester, UK on 16-17 June 2022. The Conference Co-chairs are Dr Thaddeus Eze and Dr Nabeel Khan, University of Chester and the Programme Chair is Dr Cyril Onwubiko from IEEE and Director, Cyber Security Intelligence at Research Series Limited.

ECCWS is a well-established event on the academic research calendar and now in its 21st year the key aim remains the opportunity for participants to network and share ideas. The scope of papers will ensure an interesting conference. The subjects covered illustrate the wide range of topics that fall into this important and ever-growing area of research.

The opening keynote presentation is given by Dr Charles Clarke, Senior Lecturer in Cyber Security and Programme Convenor, University of Roehampton, UK on the topic of *Employability by Design: A Strategy for Career Ownership in Cyber.* The second day of the conference will open with an address by Dr Brett Van Niekerk, Noëlle Van der Waag-Cowling and Dr Trishana Ramluckan speaking on *Multidisciplinarity and Multistakeholderism for Cyber Resilience of Emerging Economies: Lessons from Cyber Challenges.*

With an initial submission of 105 abstracts, after the double blind, peer review process there are 41 Academic research papers, 5 PhD research papers, 4 Masters research paper and 3 work-in-progress papers published in these Conference Proceedings. These papers represent research from Australia, Austria, Czech Republic, Finland, Germany, India, Japan, Lithuania, Netherlands, New Zealand, Norway, Pakistan, South Africa, Spain, Netherlands, UK and USA.

We hope you enjoy the conference.

Dr Thaddeus Eze
University of Chester
UK
June 2022

ECCWS Conference Committee

Rege, Temple University, USA; Dr Mark Reith, Full time academic, Air Force Institute of Technology; Dr Neil Rowe, US Naval Postgraduate School, Monterey, USA; Prof Vitor Sa, Catholic University of Portugal, Portugal; Dr Char Sample, Carnegie Mellon University/CERT, USA; Prof Henrique Santos, University of Minho, Portugal; Prof Leonel Santos, Polytechnic of Leiria, Portugal; Dr Keith Scott, De Montfort University, UK; Prof Dr Richard Sethmann, University of Applied Sciences Bremen, Germany; Dr Yilun Shang, Northumbria University, UK; Prof Paulo Simoes, University of Coimbra, Portugal; Prof Jill Slay, University of South Australia, Australia; Dr Lee Speakman, University of Chester, UK; Dr Joseph Spring, University of Hertfordshire, UK; Prof Iain Sutherland, Noroff University College , Norway; Dr Hamed Taherdoost, Hamta Group, Hamta Business Corp, Vancouver, Canada; Unal Tatar, University at Albany - SUNY, USA; Dr Selma Tekir, Izmir Institute of Technology, Turkey; Prof Dr Peter Trommler, Georg Simon Ohm University Nuremberg, Germany; Dr Brett van Niekerk, Durban University of Technology, South Africa; Richard Vaughan, General Dynamics UK Ltd, UK; Dr Namosha Veerasamy, Council for Scientific and Industrial Research, South Africa; Dr Sangapu Venkata Appaji, KKR & KSR Institute of Technology and Sciences, India; Stilianos Vidalis, School of Computer Science, University of Hertfordshire, UK; Prof Trish Williams, Flinders University, Australia; Prof Richard Wilson, Towson University, USA; Dr Justyna Żywiołek, Czestochowa University of Technology, Polska.

Biographies

Conference and Programme Chairs

Dr Thaddeus Eze is a Senior Lecturer in Cyber Security at the University of Chester. He is the founder and convener of the IEEE UK & Ireland YP Postgraduate STEM Research Symposium, Vice Chair, IEEE UK & Ireland Young Professionals, a technical committee member for a number of international conferences (e.g., ICAS, CYBERWORLDS, EMERGING etc.), and currently runs the Computer Science departmental research seminar series. His research interests include Trustworthy Autonomics, MANET and Cyber Security (specifically, Return Oriented Programming, Policing the Cyber Threat and Cyber Education) and he has a number of publications in these areas.

Dr Nabeel Khan is a Senior Lecturer and Programme Lead for BSc Cybersecurity at the University of Chester. Nabeel has been the lead researcher for numerous national and international research projects, funded by the European Commission (e.g., CONCERTO), U.K. research councils (e.g., EPSRC), U.K. Technology Strategy Board/Innovate-UK., and international industries (DoCoMo Eurolabs Munich). He gained his PhD in Resource Allocation in Wireless Networks from Kingston University, London, in 2013. He joined the University of Chester in 2021 as the Program Leader to further develop and deliver Cybersecurity courses. His research interests are in Network Security, Machine Learning techniques for threat hunting, and Cloud computing.

Dr Cyril Onwubiko is the Secretary, IEEE UK & Ireland, Chair, IEEE UK & Ireland Blockchain Group, and Director, Cyber Security Intelligence at Research Series Limited, where he is responsible for directing strategy, IA governance and cyber security. Prior to Research Series, he had worked in the Financial Services, Telecommunication, Health sector and Government and Public services Sectors. He is a leading scholar in Cyber Situational Awareness (Cyber SA), Cyber Security, Security Information and Event Management (SIEM), Data Fusion & SOC; and interests in Blockchain and Machine Learning. He is the founder of the Centre for Multidisciplinary Research, Innovation & Collaboration (C-MRiC) https://www.c-mric.com Detailed profile for Cyril can be found on https://www.c-mric.com/cyril

Keynote Speakers

Dr Charles Clarke is a Senior Lecturer in Cyber Security and Programme Convenor (Course Leader) for the Undergraduate Computer Science Programme at the University of Roehampton. He has extensive experience of delivering creative, engaging, and innovative learning experiences at both undergraduate and postgraduate levels. Charles is also a co-founder of CISSE UK, a national network of Cyber Security Education (CSE) strategists, practitioners, and learners. CISSE UK works closely with government, industry, and academia, to establish a culture of innovative cyber security education across the UK. Prior to academia, Charles worked extensively in the IT industry at all levels, including senior management and directorships. He has many technical interests which include cloud computing, data visualisation for cyber security and steganographic messaging protocols.

Mini Track Chairs

Dr Edwin "Leigh" Armistead is the President of Peregrine Technical Solutions, a certified 8(a) small business that specializes in Cyber Security. A retired United States Naval Officer, he has significant Information Operations academic credentials having written his PhD on the conduct of Cyber Warfare by the federal government and has published three books, in an unclassified format in 2004, 2007 and 2010, all focusing on full Information Warfare. He is also the Chief Editor of the Journal of Information Warfare (JIW) https://www.jinfowar.com/ ; the Program Director of the International Conference of Cyber Warfare and Security and the Vice-Chair Working Group 9.10, ICT Uses in Peace and War. Shown below are the books on full spectrum cyber warfare and the JIW:

Dr Chris Flaherty has had a varied career withtin the cyber domain. He is now retired, but a regular contributor. on various topics on in "Space and Defence" https://spaceanddefense.io/?s=flaherty

Dr Brett van Niekerk is a senior lecturer at the Durban University of Technology, chairs the International Federation of Information Processing Working Group on ICT in Peace and War, and is co-Editor-in-Chief of the International Journal of Cyber Warfare and Terrorism. He actively participates in international cybersecurity forums (Global Commission on the Stability of Cyberspace, Paris Call working groups, Carnegie Endowment for International Peace's project on countering influence operations). He is CISM certified, with over 50 academic publications and 20 presentations at industry events.

Dr Trishana Ramluckan is the research manager at Educor Holdings, prior to which she was a Postdoctoral Researcher in the School of Law and an Adjunct Lecturer in the Graduate School of Business at the University of KwaZulu-Natal. She is a member of the IFIP working group on ICT Uses in Peace and War and is an Academic Advocate for ISACA. She graduated with a Doctor of Administration specialising in IT and Public Governance (2017) and was listed as in the Top 50 Women in Cybersecurity in Africa (2020). Her current research areas include Cyber Law and Information Technology Governance.

Dr. Char Sample is the Chief Cybersecurity Research Scientist for the Cybercore division at Idaho National Laboratory. Dr. Sample is a visiting academic at the University of Warwick, Coventry, UK and a guest lecturer at Bournemouth University, Rensselaer Polytechnic University and Royal Holloway University. Dr. Sample has over 20 years' experience in the information security industry. Dr. Sample's research focuses on deception, and the role of cultural values in cybersecurity events. More recently she has begun researching the relationship between human cognition and machines. Presently Dr. Sample is continuing research on modeling cyber behaviors by culture, other areas of research are data resilience, cyber-physical systems and industrial control systems.

Noëlle Van der Waag-Cowling is the Cyber Programme Lead at the Security Institute for Governance and Leadership and teaches cyberwarfare in the Department of Strategic Studies, Stellenbosch University. She is a member of the International Journal of Cyber Warfare and Terrorism review board, the Tana High-Level Forum on Security in Africa, the International Committee of the Red Cross Group of Global Experts on protecting civilians during cyber-attacks, the Carnegie Endowment for International Peace's Project for Cybersecurity, Capacity-Building, and Financial Inclusion. She is an affiliate member of the University of Canberra's National Security Hub, and was voted one of the Top 50 Women in Cyber in Africa.

Workshop Facilitator

Dr Edwin "Leigh" Armistead is the Principal of ArmisteadTec LLC, a Veteran-Owned Small Business (VOSB) that is certified by the Small Business Administration (SBA) as a Historically Underutilized Business Zone (HBZ), https://www.armisteadtec.com/. This company primarily focuses on the academic functions of Information Warfare (IW), to include curriculum development, teaching as well as other related tasks to include lecturing and writing. A retired United States Naval Officer, Dr Armistead has significant Information Operations academic credentials having written his PhD on the conduct of Cyber Warfare by the federal government and has published three books, in an unclassified format in 2004, 2007 and 2010, all focusing on full Information Warfare. He is also the Chief Editor of the Journal of Information Warfare (JIW) https://www.jinfowar.com/; the Program Director of the International Conference of Cyber Warfare and Security and the Vice-Chair Working Group 9.10, ICT Uses in Peace and War.

Biographies of Contributing Authors

BKirsi Aaltola is director of development at the Finnish Institute of Public Management, with background in educational science and information technology (University of Jyväskylä) focusing on modern learning environments and skills acquisition, with experience of coordinating European Commission funded development projects, leadership and management, and training programs in peacebuilding, international security leadership and cybersecurity.

Jonathan Adkins has a PhD in Information Assurance from Nova Southeastern University in Fort Lauderdale, Florida in the United States. Currently he teaches Digital Forensics and Topics in Information Assurance at Norwich University in Vermont. His research interests include deepfakes, sentiment analysis and machine learning applied in the field of disinformation.

Pascal Ahr, B.Sc., is a Master of Science student of Electrical and Computer Engineering in Embedded Systems at the University of Kaiserslautern Germany. He works as Junior Researcher at German Research Center for Artificial Intelligence and researches in the field of Hardware Physical Layer Security. The main focus of his work is the Static Random Access Memory Physical Unclonable Function and its usage in security applications.

Milla Alaraatikka MA is a researcher at the Finnish National Defence University Department of Leadership and Military Pedagogy. Alaraatikka's main focus area is communication studies in which she received her master's degree. In addition, she has studied municipal and regional management and political science.

Elie Alhajjar is a senior research scientist at the US Army Cyber Institute. His research interests include mathematical modelling, machine learning and network analysis, from a cybersecurity viewpoint. He is a recipient of the Civilian Service Achievement Medal and the NSF Trusted Cybersecurity Fellowship. He holds a PhD in mathematics from George Mason University.

Dimah Almani is a Computer Science lecturer at Shaqra University in Saudi Arabia and is currently undertaking a PhD in cyber security at the University of Nottingham. She holds a Master's degree with honours in Computer Science from Nova Southeastern University, Florida and IBM Mastery Award as an Artificial Intelligence Analyst.

Tina AlSadhan has over 25 years of experience working in Information Technology and Cyber Security within the United States Department of Defense. She holds the following professional certifications: CISSP, CISA, and CISM. She is presently working towards a doctorate degree with her research focused on Cyber Security, specifically, Information Security Continuous Monitoring, in the United States Federal government.

Franck Baylor B.S. degree in Electrical Engineering (2019), Masters in (2021). Now working on PhD all from University of Dayton, Dayton, OH, USA Phd has an expected graduation date in 2024. Currently works for NASIC (National Air and Space Intelligence Center) at WPAFB (Wright-Patterson Air Force Base) as electronics engineer. Research interests revolve around electrical hardware cybersecurity with focus micro/nanoelectronics with defense and medical industries, also cybersecurity theory and policy.

Micki Boland is a global cybersecurity warrior and evangelist with Check Point Software Technologies Office of the CTO. A practitioner with 20 years in IT, cybersecurity, emerging technology innovation, Micki holds ISC2 CISSP, Master of Science in Technology Commercialization from the University of Texas at Austin, MBA with Global Security concentration from East Carolina University.

Jacob Chan is an officer in the U.S. Space Force and a master's student at the Air Force Institute of Technology, OH. His area of study is in Cyber Operations, with a thesis focused on the value proposition of digital badges to the Air Force.

Henry Collier is an associate professor and the Director for the MS in Cybersecurity, BS in Cybersecurity and BS in Computer Science & Information Systems programs at Norwich University. He received his Ph.D. in engineering security from the University of Colorado Colorado Springs. His main research areas are human factors in cybersecurity and networking.

Jaco du Toit is working as a senior lecturer at the Academy of Computer Science and Software Engineering at the University of Johannesburg. He responsible for lecturing courses in Computer Science and Information Security. He is also the deputy director at the Centre for Cyber Security at the University of Johannesburg.

Petrus ('Beer') Duvenage has been in the field of information security for more than 30 years. He holds a PhD (Political Science) from the University of Pretoria and a PhD (Computer Science) from the University of Johannesburg. His extensive academic research in Counterintelligence and Cyber Counterintelligence has been published in various journals and proceedings of conferences.

John Gray received his Ph.D. in Information Systems Security from Nova Southeastern University and is employed in the U.S. defense industry as an Information Systems Scurity Manager. Areas of research interest include the effects of human factors on information assurance and cybersecurity, insider threats, and information systems ethics.

Pauliina Hirvonen is a PhD student in the Information Technology Faculty of University of Jyväskylä, in Finland. Her on-going dissertation relates to Information Systems and more specifically on information and/or cyber security field. The research considers the overall impacts and situational awareness related to GDPR from the organizational perspectives.

Bhaiyat Yacoob Humairaa is from Johannesburg and holds a Bachelor of Science Honours in Information Technology (University of Johannesburg). She is currently a software developer at Business Systems Group (Africa).

Aarne Hummelholm, PhD in Information Technology (University of Jyväskylä, 2019). He has over 30 years' experience in the design, development of architectures` of authorities` telecommunications networks and information systems. Key themes in his work have been critical service availability, usability, cyber security and preparedness issues.

Aydan Huseynova Tomas Bata University in Zlin Faculty of Management and Economics: Univerzita Tomase Bati ve Zline Fakulta managementu a ekonomiky Zlin, CZECH REPUBLIC.

Her main research interest areas are employer branding, marketing, HR, social media and digitalization.

Eveliina Hytönen is a Senior lecturer of Security and risk management at Laurea University of Applied Sciences, a project expert in related projects, and Dissemination Manager in SAFETY4RAILS (Data-based analysis for SAFETY and security protection, FOR detection, prevention, mitigation and response in trans-modal metro and RAILway networkS).

Pierre Jacobs has over 20 years consultation experience in the IT Security industry. He is currently with CyberAntix where he serves as Head of Security Operations and Compliance. He holds a MSc with Specialisation in Information Security from Rhodes University, a Ph.D at the University of Johannesburg and various Industry certifications.

Dr. Connie Justice has over 30 years' experience in cybersecurity, computer, and systems engineering. She designed courses in cybersecurity curriculum to NSA/DHS Center of Academic Excellence and NIST National Initiative for Cybersecurity Education standards. Research areas include: misinformation, industrial controls risk, experiential learning, information and security risk management, digital forensics.

Miina Kaarkoski (Dr) is a researcher at the Finnish National Defence University Department of Leadership and Military Pedagogy. Her areas of expertise are crisis talk and influence on public opinion, societal beliefs and political decision-making processes.

Anni Karinsalo is working as a Research Scientist in the applied cryptography team at VTT Technical Research Centre of Finland. She has background in various fields of cybersecurity, such as distributed ledgers, privacy and post-quantum cryptography.

Sithembiso Khumalo is Lecturer and a Deputy HoD of Information and Knowledge Management IKM at the University of Johannesburg, SA. He received his Master's in IKM in 2016. His main research areas are Knowledge Portals, Strategic Information Management, entrepreneurial institutions, commercialisation dynamics. Miss Amanda Sibiya and Miss Teballo A. Kekana are IKM Honours graduates.

Jan Kleiner is a PhD student of Political Science at Masaryk University in the Czech Republic. He focuses primarily on cybersecurity, the relationship between a state and citizens in cyberspace (e.g., how states secure their citizens in cyberspace), and propaganda and information warfare. He mainly employs quantitative (statistical) and mixed methods research designs.

Pekka Koistinen Msc is a researcher and a PhD student at the Finnish National Defence University Department of Leadership and Military Pedagogy. His main research areas are national security and cooperation between authorities.

Aphile Kondlo is a part-time master's student at University of the Western Cape majoring in Cybersecurity. He is currently a full-time employee at Woolworths Holdings Limited as an intermediate software developer. He is also, a member of the Golden Key International Society and UWC-CPUT Space Association.

Markku Kylänpää is a senior scientist in the applied cybersecurity research group at VTT Technical Research Centre of Finland. He received M.Sc. and Lic.Tech. from Information Technology in Helsinki University of Technology in 1985 and 1989, respectively. His main research area is platform security.

Janne Lahdenperä is a 3rd year Business Information Technology student at Laurea University of Applied Sciences, Finland. His interests include cyber security and ethical hacking.

Dr Yee Wei Law is a Senior Lecturer at UniSA STEM, University of South Australia. He is also a founder of the start-up Mesh In Space, which specalises in secure mesh networking for space vehicles. His research is currently focusing on space system security, adversarial machine learning and machine diagnostics.

Christoph Lipps, M.Sc., graduated in Electrical and Computer Engineering at the University of Kaiserslautern where he meanwhile lectures as well. He is the Lead of the *Cyber Resilience & Security* Team and Ph.D. candidate at the German Research Center for Artificial Intelligence (DFKI) in Kaiserslautern. His research focuses on Physical Layer Security (PhySec), Physically Unclonable Functions (PUFs), Artificial Intelligence (AI), entity authentication and all aspects of network and cyber security.

Clara Maathuis is Assistant Professor in AI and Cyber Security at Open University in the Netherlands. With a PhD in AI and Military Cyber Operations from Delft University of Technology, she is involved in teaching different AI and cyber security courses and conducts research in AI, cyber/information operations, military technologies, and social manipulation.

Juha Kai Mattila is a freelance researcher and a postdoc at Aalto University, Helsinki, Finland. He received his D.Sc. (Tech) from Aalto University in 2020 and G.S. Officer (C4ISTAR) from National Defence University, Finland in 1993. He is consulting Armed Forces in their digitalization or digital transformations. His main research areas are organizational evolution, enterprise architecture in modelling complex open systems, use of power in the cyber domain, and implementing digital transformations.

Qazi Muhammad Ali is a a PhD Scholar at Superior University Lahore, Pakistan. I am perusing my PhD in Business Administration. I had attended many conferences in my research life. My main research area is on Behaviours and Human Resource Management.

Rennie Naidoo is an Associate Professor of Informatics at the University of Pretoria (South Africa). His research focuses on the interplay between information systems, people and organizations. His papers have appeared in Journal of Strategic Information Systems, European Journal of Information Systems, Information Technology & People, and The Information Society Journal.

Zia Najam ul Zia is an early career researcher. He is PhD scholar at Tomas Bata University in Zlin, Czech Republic. His broader research interests include Industry 4.0, knowledge management and big data management.

Olga Navickienė is associate professor of the Research Group on Logistics and Defense Technology Management at General Jonas Žemaitis Military Academy of Lithuania. Research interests are related to defining and analysing economic and social phenomena by applying different methods of mathematical statistics. Area of Expertise: Operations Research, Structural Equation Modelling, Statistical Analysis, Structured Judgement.

Tabisa Ncubukezi is a communication networks lecturer at the Cape Peninsula University of Technology. She is a doctoral candidate at the same institution. Her primary research areas are communication networks, cloud computing, cyber security, and e-Learning.

Narayan Nepal is a Faculty Lead of the School of Technology at Yoobee College, Christchurch, New Zealand. He received his PhD in Electrical and computer engineering from University of Canterbury in 2018. His main research areas are general areas of wireless communications and networking, cybersecurity, cognitive radio, artificial intelligence, deep learning, 5G and beyond networks.

Teija Norri-Sederholm (Dr) is an adjunct professor at the Finnish National Defence University's Department of Leadership and Military Pedagogy. Her main research areas are situational awareness, inter-organisational communication in command centres and hybrid environments, national security, and the dark side of social media.

Nombeko Ntingi holds a BSc (Computer Science) University of Transkei, Master's degree in IT Carnegie Mellon University, BCom (Financial Management) University of South Africa. She holds certificates in SAP ABAP, Cyber Security from UJ, CPUT and F'SAT; Managing Risk Harvard Online. She is currently pursuing PhD studies from the University of Johannesburg.

Jacob Oakley is a cybersecurity expert and author with over 15 years of experience improving strategic, enterprise-level security architectures for threat and risk management of globally distributed Department of Defence (DOD) and Fortune 500 networks. Additionally, he has served as the technical principal for multiple cybersecurity R&D efforts.

Dr. Fredrick Ochieng' Omogah (h.c) is a head of department of Information Systems, Sci. & Technology and a lecturer in I.T & Medical Informatics at Uzima University, Kenya. He is currently finalising Msc. I.T Security and Audit from Jaramogi Oginga Odinga University, Kenya. Received Bachelor of I.T from Australia, 2009. His main research areas are in I.T and Cyber security in electronic healthcare

Johanna Parviainen, student of Security Management at Laurea University, works as a Detective Inspector at local police. She has years' experience of leading different crime investigations including crimes against children, last 2.5 years she has worked with economic crimes. Her Master Thesis explores sexual abuse of children online and CAM investigations in Finland.

Alexander Pfeiffer is recipient of a Max Kade Fellowship awarded by the Austrian Academy of Science to work at the Massachusetts Institute of Technology (MIT), Department for Comparative Media Studies / Writing in 2019 and 2020. In 2021 he returned to Donau-Universität Krems and is now head of the Emergent Technologies Lab. He is currently approaching his second PhD at the department of AI at the University of Malta. . https://www.alexpfeiffer.at

Jouni Pöyhönen, Col (ret.) is a postdoctoral researcher of cyber security programs in University of Jyväskylä. He received his PhD in information technology from University of

Jyväskylä in 2020. He has over 30 years' experience of C4ISR systems in Finnish Air Forces. He has more than twenty cyber security research reports or articles.

Jyri Rajamäki is Principal Lecturer in Information Technology at Laurea University of Applied Sciences and Adjunct Professor of Critical Infrastructure Protection and Cyber Security at the University of Jyväskylä, Finland. He holds D.Sc. degrees in electrical and communications engineering from Helsinki University of Technology, and a PhD in mathematical information technology from University of Jyväskylä.

Jori-Pekka Rautava M.Sc. has a Master's degree in wireless communications engineering (2019). He is currently a postgraduate student at the University of Oulu and writing his doctoral thesis on the concept and implementation of cyberterritory.

Huw O.L. Read is Professor of Digital Forensics and the Director of the Centre of Cybersecurity and Forensics Education and Research (CyFER) at Norwich University, Vermont, USA. For over 15 years, he has taught in several countries, authored over 20 peer-reviewed publications in the field and has been awarded numerous competitive grants.

Mari Ristolainen Researcher at Finnish Defence Research Agency. Studied psychology at Moscow State University and earned a doctorate in Russian Language and Cultural Studies from University of Joensuu in 2008. Been conducting postdoctoral research in the field of Russian and Border Studies in several Academy of Finland- and EU-funded projects at the University of Eastern Finland and at the University of Tromso, Norway. Her current research interests include cyber warfare as a phenomenon, Russian digital sovereignty, and the governance of cyber/information space.

Harri Ruoslahti is a Senior lecturer of Security and risk management at Laurea University of Applied Sciences, a researcher in related projects, and Laurea's point of contact in ECHO (the European network of Cybersecurity centres and competence Hub for innovation and Operations).

Karo Saharinen (M.Eng) is working as a Senior Lecturer in IT and handling the responsibility of degree programme coordinator of the master's degree programme in IT, Cyber Security at JAMK University of Applied Sciences. He is currently working on his PhD related to Cyber Security Education.

Taito Sasaki received his Bachelor of Engineering degree, Nagoya Institute of Technology, Japan, in 2021. He is now a graduate school student at Nagoya Institute of Technology. His research interest includes the Security and Safety of the Internet of Things.

DR KEITH SCOTT Is Programme Leader for English Language at De Montfort University in Leicester. His research operates at the intersection of communication, culture and cyber, with particular interests in influence, information warfare, and simulations and serious gaming as a training. teaching, and research tool.

Lander Segurola is a researcher on data science applied to cybersecurity in the research centre Vicomtech, where he has been working for three years. He finished a master on

Mathematics and applications in 2018, and recently, he has started his PhD thesis focusing his research line on Bayesian approaches for Machine Learning applied to cybersecurity.

Paul M. Simon is an electrical engineering PhD candidate at Air Force Institute of Technology (AFIT) scheduled to graduate June 2022. He is a Research Engineer for Air Force Research Lab and Principle Hardware Engineer for Huntington Ingalls Industries. His main research areas are reverse engineering, hardware system design, and cyber security of embedded devices.

Thenjiwe Sithole is a PhD student at the University of Johannesburg. She holds a Masters in Information Technology (Information Systems) from the University of Pretoria and a Master of Engineering Sciences in Electronics (Telecommunications) from the University of Stellenbosch. She also has Certificate in Cyber Security from the University of Johannesburg.

Siphesihle Sithungu is from Johannesburg, South Africa and holds a MSc in Computer Science (University of Johannesburg). He is a lecturer at the University of Johannesburg and his research interests are artificial intelligence and critical information infrastructure protection. Mr. Sithungu is a technical committee member for the International Conference on Computational Intelligence and Intelligent Systems.

Iain Sutherland is Professor of Digital Forensics and Head of Research at Noroff University College in Kristiansand, Norway. He is a recognised expert in the area of computer forensics. He has authored articles ranging from forensics practice and procedure to network security. His current research interests lie in the areas of computer forensics and computer security.

Stephen Taylor is a Professor of Computer Engineering at Dartmouth College, and a co-owner in Web Sensing LLC, an R&D company specializing in network security solutions. His academic research focusses on systems security using System-on-Chip and FPGA devices. He is a former DARPA Program Manager and member of the US Air Force Scientific Advisory Board.

Ilkka Tikanmäki is a researcher at Laurea University of Applied Sciences and a doctoral student of Operational art and tactics at the Finnish National Defence University. He holds an MBA degree in Information Systems and BSc degree in Information Technology.

Maija Turunen is a Military Science student in the Finnish National Defense University. Her main research areas consist of cyber warfare, cyber deterrence and Russia

Paul Underhill is a lecturer and programme leader for MSc Data Science at the University of Chester and has a strong research interest in cyber security and related fields.

Namosha Veerasamy has obtained a BSc:IT Computer Science Degree, and both a BSc: Computer Science (Honours Degree) and MSc: Computer Science with distinction from the University of Pretoria. She also holds a PhD from the University of Johannesburg. She is currently employed as a senior researcher at the Council for Scientific and Industrial Research (CSIR) in Pretoria. Namosha is also qualified as a Certified Information System Security Professional (CISSP) and Certified Information Security Manager (CISM). She has been involved in cyber security research and governance for over 15 years.

Toni Virtanen has a PhD. in Psychology. He is currently working as a researcher at the Human Performance Division in the Research Institute of the Finnish Defence Forces. His topics of interests are Human Factor issues in cyber warfare, military psychology, social hacking and information security.

Basie von Solms is a Research Professor in the Academy for Computer Science and Software Engineering at the University of Johannesburg. He specialises in Cyber Security. He is a Past President of the International Federation for Information Processing (IFIP).

Murdoch Watney is a professor at the University of Johannesburg, South Africa. She teaches criminal law. Her field of interest is cybercrime and cybersecurity. She has published in textbooks on criminal law and has delivered peer-reviewed papers at national and international conferences.

Richard L. Wilson is a Professor in Philosophy at Towson University in Towson, MD. Professor Wilson Teaches Applied Ethics in the Philosophy and Computer and Information Sciences departments at Towson while also serving as Senior Research Scholar in the Hoffberger Center for Professional Ethics at the University of Baltimore. Professor Wilson's interest's are directed towards applying phenomenology to issues with emerging and innovative technologies.

Allison Wylde is a cyber security researcher with the Cardiff University Centre for Cyber Security Research. Expertise includes- artificial intelligence and innovation, trust and zero trust, and norms in cyber security with the UN Internet Governance Forum on Cyber Security Norms. Allison was formerly an international commissioner on security standards with ASIS International.

Hendrik Zwarts is a master's student at the Centre for Cyber Security at the University of Johannesburg where his main research areas are cyber-diplomacy and cybersecurity. He works within the government sector as a security consultant.

Mateusz Zych is a PhD Research Fellow at the University of Oslo, Norway. He received his MSc in computer sciences from the University of Oslo in 2018. In addition, he has four years of experience in programming from the industry. His min research areas are cyber threat intelligence, standardizations, interoperability and automation.

Desired Cybersecurity Skills and Skills Acquisition Methods in the Organizations

Kirsi Aaltola[1], Harri Ruoslahti[2] and Jarmo Heinonen[2]
[1]Finnish Institute of Public Management andUniversity of Jyväksylä, Finland
[2]Laurea University of Applied Sciences, Finland
kirsi.aaltola@haus.fi
harri.ruoslahti@laurea.fi
jarmo.heinonen@laurea.fi

Abstract: Key personnel and their competences play important roles in continuity management and improving resilience of cybersecurity in organizations. Researchers have addressed many topics and studies in the cybersecurity domain. However, relevant cybersecurity skills and acquisition of them in expertise development, have only been partially touched. If designed systematically and properly, cybersecurity training can improve cybersecurity expertise to ensure better performance in complex cybersecurity situations. More through study on the acquisition of cybersecurity skills, and work-life needs are needed. The research three questions of this study are: How do work-life representatives see cybersecurity? How do work-life representatives see cybersecurity related skills? How do work-life representatives see methods for skills acquisition in the organizations? The work is multi-method, as it builds on both a literature review on skills acquisition in cybersecurity, and on empirical findings of a questionnaire study on cybersecurity skills desired by the work-life representatives. The findings show that cybersecurity is seen important in the organizations. The demanded skills from the employees focus especially on communication and situational awareness. There is a specific need for training with Cyber Ranges (CR) to ensure skills acquisition on cybersecurity. These results can be used to plan and design training and education for future professionals. This study aims to promote constructive discussion on skills and their acquisition in the cybersecurity domain.

Keywords: information technology, resilience, cybersecurity, cybersecurity skills, skills acquisition, training, cyber ranges (CRs)

1. Introduction

There is a need to increase cyber resilience in organizations (Aaltola & Taitto, 2019; Ruoslahti 2020), and to find solutions for the lack of skilled cybersecurity professionals in organizations (Dawson & Thomson, 2018). Recently, there have been interests to understand cybersecurity skills more comprehensively, and to build taxonomies that support the design of training of work-life needs (Furnell et al., 2017; Carlton et al., 2019). This is partly because there is a high need for professionals in the field of Information Technology (IT) (Crumpler & Lewis, 2019), and to design efficient solutions to secure digital technologies (Soni & Bhushan, 2019).

There are several research, development, and innovation (RDI) initiatives on cybersecurity on the European level, such as European Commission funded projects. This paper is based on the study conducted as part of project ECHO (European Network of Cybersecurity Centres for Innovation and Operations). Project ECHO consists of 30 partners from different sectors including health, transport, manufacturing, ICT, education, research, telecom, energy, space, healthcare, defence & civil protection. The ECHO project promotes European-wide network building, methods, and models that, within regulatory requirements, promote information sharing among network partners (Mengidis et al., 2021; Rajamäki & Katos, 2020), with a future governance model that aims to bring academia, industry, cybersecurity practitioners and end-users together (Yanakiev, 2020). This survey was conducted to provide direct input to the development Federated Cyber Ranges (FCR) concept in the ECHO project, and to serve as a potential case study to deepen understanding of skills acquisition and training in relation to the societal impacts of project ECHO.

Firstly, this paper reviews some previous literature on cybersecurity in organizations, cybersecurity skills and methods for skills acquisition. Secondly, the findings of survey responses by organizational work-life representatives on these topics are presented. Thirdly, previous findings by scholars and our findings to discuss the future directions in cybersecurity training are combined. The overall aim of the study is to improve the understanding of cybersecurity skills and their acquisition to support the work-life gaps in terms of cybersecurity expertise.

2. Theoretical framework

Cybersecurity combines several academic disciplines, and may therefore be seen as being a multi-disciplinary domain that joins mathematics, psychology, engineering, law and computer science (Dawson & Thomson, 2018). New technologies have radically changed the human dimension in organizations (Aaltola, 2021). Dalai et al. (2021) have emphasized increasing focus on cybersecurity work in organizations. The field has developed rapidly, which has challenged academic works in understanding what skills make a good cyber expert, and how should organizations recruit these professionals (Dawson & Thomson, 2018).

Consequences of cyberattacks may greatly vary, and new techniques to improve cyber resilience within organizations are required (Aaltola & Taitto, 2019; Ruoslahti, 2020). This literature review section of this paper focuses on viewpoints that scholars have on cybersecurity in organizations, cybersecurity skills and methods for skills acquisition. These scholarly papers mainly focus on improving cyber resilience within organizations by identifying risks and identifying different methods of cybersecurity counter measures or training.

2.1 Cybersecurity workforce in the organizations

Organizations must secure every critical element of their infrastructure to be well prepared to withstand threats that can compromise the security and continuity of their operations (Topham et al., 2016), also, users are often deemed a weak link due to not being educated in cyber threats concepts and not having the experience to mitigate cyber threats that may arise; e.g., social engineering and phishing are some common attacks that everyday users may encounter. Not having relevant cybersecurity training may leave them with little possibility to distinguish between a legitimate request and a cyber-attack.

Ruoslahti (2020) finds that to improve resilience and managing continuity, it is important that the network organization consider their key personnel through the possible event management stages; suggesting to 1) identify key people and develop and exercise their (cyber) skills, while in the planning phase, 2) have needed skills available during the absorb phase, and 3) broaden involved people and their skills in the recovery phase, and 4) revise the list of key people and (cyber) skills once the organization has reached the adopt phase, which would then become a new plan phase in expectation of the next unexpected cyber event.

Neal, Facteau & O'Connell (2021) offer that the growing demand for cybersecurity professionals could be solved by recruitment from "occupations with similar profiles to cybersecurity jobs include: electrical and electronics repairers, telecommunications and equipment installers and repairers, geographers, purchasing managers, personal financial advisers, sociologists and budget analysts." (p. 2/4). Modern cybersecurity specialists mainly need flexibility, as due to the available IT tools technical skills are relatively easy to master (Skorenkyy et al, 2021). Neal et al. (2021) note a shortage in cybersecurity professionals, and identify the most promising new source of future cybersecurity experts being people possessing an aptitude to acquire the new skills that make them likely to succeed in a cybersecurity career; identifying what occupations have similar work profiles and identifying the individuals who could "have the greatest potential to acquire new technical skills and succeed in cybersecurity" (p. N/A). Dawson & Thomson (2018) suggest the use of Big Five Personality traits to better understand the fit between cyber professionals and organizations.

According to Tomić et al. (2020) cybersecurity experts often begin as IT professionals, and therefore the authors suggest that tomorrow's cybersecurity experts be recruited from today's IT professionals. The study by Tomić et al., (2020) perceive among cybersecurity professionals, a lack of skills for solving different cybersecurity issues, together with poor capabilities or even unwillingness for continuously learning and skills improvement. Modern cybersecurity specialists mainly need flexibility, as due to the available IT tools technical skills are relatively easy to master (Skorenkyy et al, 2021). Nevmerzhitskaya et al. (2019) see that cybersecurity skills are needed to continuously be advanced at all levels to achieve preparedness and resilience. They suggest following "a constant learning process, to address complex demands of individual and organizational level capacity building through trainings and exercises." (Nevmerzhitskaya et al., 2019, p. 311).

2.2 Cybersecurity skills

Technical and engineering skills tend to become emphasized in the cybersecurity domain (Gates et al., 2014), ignoring the important social and organizational aspects needed to perform successfully in everyday work-life settings (Dawson & Thomson, 2018). Cyber professionals who work in operations maintaining security, need significant skills and knowledge about computer systems and how to use analytical tools, such as vulnerability analysis or network scanning (Dawson & Thomson, 2018). Jajodia et al. (2010) identify a need for strong situational awareness skills, which include continuing risk assessment skills, for network professionals.

Non-technical knowledge, skills and abilities (KSA), such as problem-solving, communication and collaboration can be useful. Higher education and professional development training should integrate these non-technical KSAs into their programs that train cybersecurity professionals. (Sussman, 2021). Cyber professionals are asked to be able to communicate the technical information for the people with no technical background (Dawson & Thomson, 2018). Due to complexity of cyber domain, it has been acknowledged that cyber professionals are required to have teamwork abilities (Mathieu et al., 2000). Tomić et al. (2020) find that the most helpful IT skills for successful cybersecurity activities are use of technologies and applications, networking technologies and infrastructure, coupled with knowledge and skills of IT operations.

The skills that Neal et al. (2021) find relevant for cybersecurity professionals are critical thinking, complex problem-solving, monitoring, systems analysis, and coordination, which they would combine with technical knowledge of computers and electronics, telecommunications, customer and personal service, and administration and management. They also recommend as a suitable background experience of work activities in interacting with computers, gathering and evaluating information, updating and using relevant knowledge, making decisions and solving problems, documenting and recording information. Aaltola & Taitto (2019) raise the importance of decision-making models among cybersecurity professionals.

Dawson & Thomson (2018) discuss identifying relevant cybersecurity skills, and whether the ideal cyber workforce is required more cognitive ability than personality traits or values; the authors identify the need to determine cognitive underpinnings of the expertise that can ensure that organizations, work roles and individual skills are successfully aligned with each other.

2.3 Skills acquisition methods

According to Adams and Makramalla (2015) the main obstacle, which affects personnel from learning how to apply security measures and establishing cybersecurity skills, are the type of instruction they receive from cybersecurity education programs. Most programs teach security concepts with a traditional approach, where it is difficult to retain information, or to put it into practice.

Technical training is currently the prevalent form of cybersecurity education, and study programs often fail to include development cycle requirements, professional standards or regulations (Skorenkyy et al, 2021). Ghafir et al. (2018) identify challenges in implementing cybersecurity training in organizations, and promote knowing how to properly provide training that a) effectively engages non-ICT personnel to practice security awareness and to b) develop their cyber skills, and c) facilitate ICT professionals become more proficient in analysing and managing the constantly evolving cyber threats.

Aaltola and Taitto (2019) find that experiential learning principles can deepen the level of cyber learning. By supplementing theoretical knowledge with experiential learning and interactive training (e.g., games, puzzles, scenarios) for general employees could provide a more practical hands-on training that looks at real situational threats (cyber-ranges). The cognitive learning can bring the foundations to the discussions and practical solutions for acquisition of skills, but also to design of training and education for cybersecurity professionals.

Augmented reality interfaces and specialized scenarios with content that reflects the context are in use in gaming, which may be very useful in forming required competences, skills and abilities; games may be the appropriate method of implementing skills frameworks into study programs (Skorenkyy et al, 2021). Developing skills and competences, which are needed to navigate within the cyber domain are constructive processes, which use and recognize previously adapted competences of learners (Aaltola & Taitto, 2019). Simulation environments can be used to assess preparedness against cyber crises, technology failures or incidents against critical information infrastructure (Nevmerzhitskaya, Norvanto & Virag, 2019). Davis & Magrath (2013) identify skills, such as, penetration testing, hardening critical infrastructure, defending networks and responding to attacks that can be practiced in Cyber Range (CR) environments. CRs are complex IT environments where organizations can practice handling real-world cyber scenarios, and train users on the latest cyber threats. The capabilities of CRs may include simulations of real-world network environments and electronic warfare

(Priyadarshini, 2018), addressing exercises on network forensics, social engineering, reverse engineering and penetration testing, which exercises are supported by self-directed and problem-based learning (Raybourn et al., 2018). Simulations and game theories are also acknowledged methods of CRs (Wang 2010), and agent-based simulation platforms focus on simulating the effects of attacks and analyses their impacts (Grunewald et al. 2011). Practical trainings, including network simulated exercises, can be beneficial when developing relevant cyber skills (Topham et al., 2016). CRs and their features can be used as a method in cybersecurity skills acquisition, and with CRs, staff can practice their skills, identify vulnerabilities, run attack simulations and benefit CR capabilities for human improvement (Aaltola, 2021). More generally, the use of reality-virtuality technologies for learning purposes can allow more dynamic and autonomous roles in the creation of learning experiences (Ostrom et al., 2015) that may lead to higher perceptions of value (Patrício et al. 2011).

Critical discussion about the use of digital technologies in human skills acquisition often addresses challenges to transfer skills in different actual life contexts. Experts that exercise their skills in digital training platforms with the purpose to apply those skills in actual-life situations, require at least understanding of the context-specific nature of some skills, and ideally understanding of the methods that support this skills transfer from one context to another (Aaltola, 2021). Skills assessment should be built on a validated skills taxonomy, with regard to the level of the required skills, and their alignment against the job task requirements (Nevmerzhitskaya et al., 2019). Also, the use of learning outcomes in the training design ensures more thorough training needs analysis and metrics to assess learning (Aaltola & Taitto, 2019).

3. Research methods

To understand desired cybersecurity skills of professionals, this study collected empirical data as part of the ECHO project, and related to the development of its CR capabilities and FCR concept development. Guided by the research questions, survey questions were prepared for work-life representatives from different organizations. These survey questions were framed based on cybersecurity and cybersecurity skills, combining both, multiple-choice and open-ended questions. The multiple choice questions were framed with potential responses and an open ended “something else” alternative to choose. The survey questionnaire was sent to specified respondents selected to ensure cybersecurity expertise and experience of the respondents. The survey analysis was conducted with the sample of 43 respondents. Data validation ensured that the questionnaire was fully completed and presented the consistent data. The background data included organization related information and position in the organization. This paper presents the analysis and findings of CR cybersecurity skills study. Quantitative questions of the survey addressed mainly cybersecurity competences of the personnel in the respondents’ organizations. At first the data was analysed with factor analysis to find out groups (factors) in which loadings are connected together. The most meaningful (the highest loadings of the same factor) of the quantitative questions were analysed with correspondence analysis utilizing the Euclidean distance in two dimensional figures, where two variables and their connection to each other are presented two-dimensionally on X and Y axis. Correspondence analysis is an exploratory multivariate technique that converts a data matrix into a particular type of graphical display in which the rows and columns are depicted as points (Yelland, 2010; Greenacre & Hastie, 1987). The open ended survey responses were analysed with qualitative content analysis (Denzin & Lincoln, 1994). There were forty-three respondents (n=43) from forty European organizations. The respondents’ roles were developers, architects or engineers (n=9), managers (n=7), directors (n=5), experts, researchers or analysts (n=5), security officers (n=4), coordinators or experts (n=4) advisors or consultants (n=2), and professors (n=2). The respondents mainly represented private organizations (n = 32), while 11 of the respondents were from public organizations. By size of organization respondents represented mainly medium, over 50 employees, and large, over 250 employee, organizations, with only a few small organizations of less than five employees represented in this survey. Though, the number of responses are not sufficient to be representative of Europe, the study results are valid as a case study to provide a basis for ECHO CR development and in part to provide understanding of skills acquisition.

4. Results

4.1 Cybersecurity in the organizations

The majority of the respondents (40 of 43) described the importance of cyber security as an “important” or a “very important” thematic topic to their organization. The respondents elaborated on the importance by answering: "We have dedicated in-house competence and capability building programmes" (n=28), "We conduct regular in-house vulnerability assessments" (n=27) and "We use security as an enabler (maintaining proper cyber hygiene and security measures positively affects business processes)" (n=24) as the most crucial reasons, while "We conduct regular vulnerability assessments by external providers" and "We have out-sourced partners for handling cyberattacks" were also mentioned as relevant reasons. However, six respondents noted that cyber security is not an important factor for their organizations. One open ended response notes that the missing parts in relation to cybersecurity are comprehensive: "The management is changing, no time for analysing the vulnerabilities, too few people, small professional staff. Too old colleagues. Not real and exact definitions, lack of knowledge of good practices." One respondent described that they are behind the technology developments, and it is hard for the employees to keep up with the latest trends and developments. The rest of the responses focused on solutions and tools for increased data protection, and issues such as the lack of compliancy policies and standards, and challenges. For example, the age of the personnel was seen to effect organizational cybersecurity capabilities, such as preparing against vulnerabilities, and setting up Security Operation Centres (SOC) or Early Warning Systems.

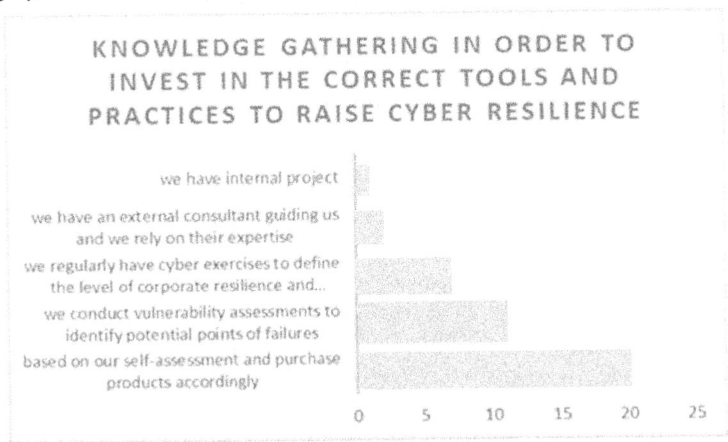

Figure 1: Knowledge gathering for tools to raise cyber resilience

Figure 1 shows how respondents gather the knowledge that they need to invest in the correct tools and practices that raise cyber resilience. Most respondents (n = 20) base their purchases on self-assessment, while many other conduct vulnerability assessments or conduct regular cyber exercises to determine their needs.

4.2 Cybersecurity skills

The respondents defined situational awareness and communication as the key cybersecurity skills that they demand from their employees. Moreover, being collaborative and approachable, but also analytical and having a hacker mind-set were seen as some key 'skills or behavioural attributes for their employees working on cybersecurity. Based on the responses technical skills, such as understanding of the IT domain, programming, and architecture skills, are also needed, while leadership and writing skills were not seen as demanded (Figure 2).

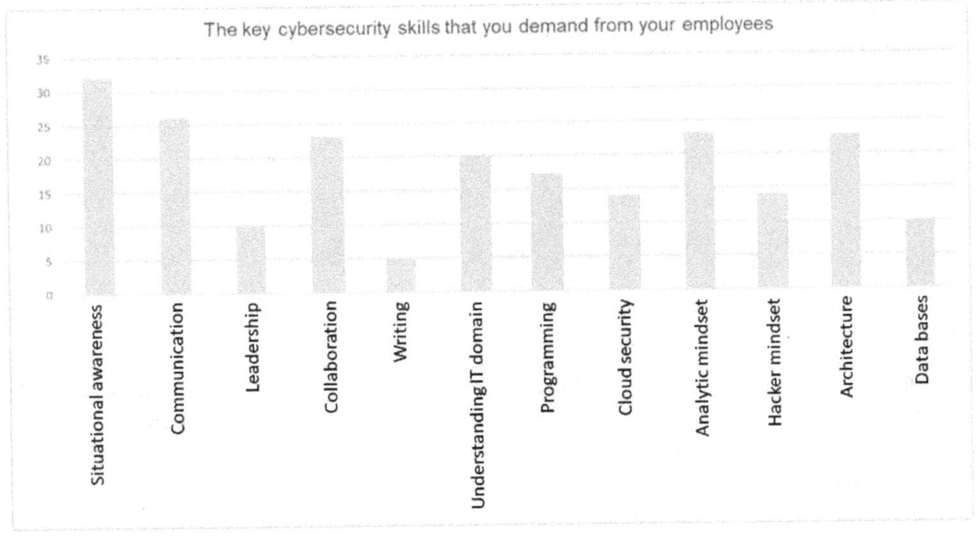

Figure 2: Key cybersecurity skills demanded of employees

As seen in Figure 2, Situation awareness was seen as the single most important skill for people working in cybersecurity, while in addition an appropriate array of technical skills and collaboration are needed.

4.3 Desired methods for skills acquisition

Seven respondents see cyber exercises as a key metric to identify correct tools and practices ("we regularly have cyber exercises to define the level of corporate resilience and identify key milestones for improvement"). The open-ended responses show seven out of twenty respondents addressing a specific training need or cyber range, while general awareness was the most needed or missing from the organization for some.

When asking about preferred usage of Cyber Range (CR), 25 of the respondents valued in-house training than external training service. The correspondence analysis conducted in the study and especially the in-house training was favoured by the organizations which had employees 25-50 or 50-250 (Figure 3).

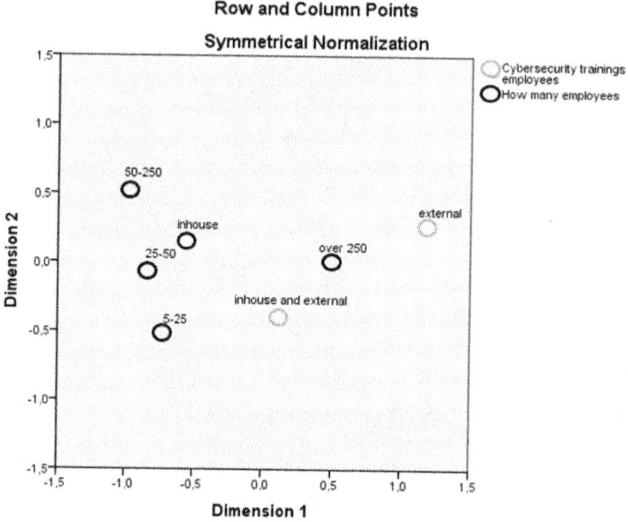

Figure 3: Cybersecurity trainings by type and number of employees

The technical capabilities of the CR used in the trainings focused on attack and defence simulations, learning platform and real-time monitoring. Also, traffic simulations and performance evaluation were mentioned (Figure 4).

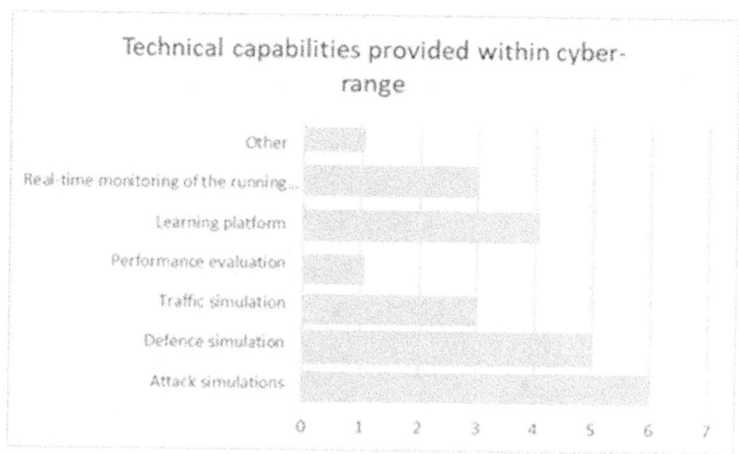

Figure 4: Technical capabilities of cyber ranges

On how training debriefing and performance evaluation are performed in their cyber range, three respondents noted that they have evaluation activities such as hot-wash debriefing, providing feedback that is based on the collection of data during trainings, or debriefing by company personnel. The respondents would like to further develop their sector specific capabilities (e.g., healthcare, space, transportation, energy) and provide better automation for quicker development and deployment and add automatic performance evaluation. The domain specific approach was seen necessity and one respondent described "we always draw domain-specific expertise when create a cyber-exercise".

5. Discussion

Different employees have different competences, while several vulnerabilities may occur due to the comprehensiveness of cybersecurity. Therefore, the better understanding of how important organizational representatives see cybersecurity that this study brings, supports earlier findings that cybersecurity is seen important. Different in-house capability improvements, and cybersecurity topics are seen as being complex. The respondents raise challenges, such as a lack of cybersecurity knowledge in management, lack of compliancy policies and standards, and lack of time to keep up with technology developments.

According to this study, the most desired skills for employees are situational awareness, communication, and technical skills (e.g., programming and architectural issues). In addition, having an analytical mind-set was also mentioned. The findings partly support the results of other scholars, providing in those definitions of 'skills' should become more un-ambiguous, and there is a need for more focus on better analysing what is meant by cybersecurity skills and by e-skills, and if these are domain-specific and their relevance to professionals in this domain.

The methods on skills acquisition were studied from the responses of both multiple choice and open-ended questions. Organizational representatives addressed the need for training, and related use of CRs. Cyber exercises are clearly a way to "define the level of corporate resilience and identify key milestones for improvement". Many respondents value in-house training over the purchase of external training services. The technical capabilities of CRs should include attack and defence simulations, a learning platform, real-time monitoring, and capabilities to run performance evaluations. The ability to create sector-specific scenarios in CRs was seen as a positive asset.

6. Conclusions

The expanded use of technology has increased the research focus on cybersecurity issues that are seen being multidisciplinary. Scholars from different disciplines, who focus on studying human factors in cybersecurity raise multiple viewpoints: organisational needs, challenges and competence gaps. There seems to be a continuing need to further deepen the understanding of the specific nature of cybersecurity in organisations, with related cybersecurity skills and relevant methods of skills acquisition. Some authors have e.g., recommended solutions to narrow the current gaps in workforce within the cybersecurity field (see for example Dawson & Thomson 2018; Tomić et al., 2020; Aaltola & Taitto, 2019; Ruolahti, 2020). The findings presented in this paper help in part to describe the current situation of organisational cybersecurity, understand cybersecurity skills and different skills acquisition methods. The respondents' (n=43) backgrounds varied, and they represented organisations with different sizes and from different European countries. Wider study will be needed to provide results that can be generalized throughout Europe, and to achieve this, the questionnaire and results of are used in the development the ECHO Societal Impact Assessment Toolkit questionnaire.

Most respondents value cybersecurity as either very important or important. They also identify several gaps in the capabilities, awareness, and employee skills within their organisations. All of these are needed to implement cybersecurity and to address related issues in everyday work life. Organizational representatives can use organisational self-assessments to gain knowledge and information on how to improve resilience, cybersecurity, mitigation and to base the purchase of new tools and solutions.

Figure 5 highlights the findings of this study. Organizations require and look for a range of skills, knowledge and attributes from their employees. In this broader context these skills can be called 'e-skills', as they directly relate to using ICT technology, being aware of the risks against the used ICT environment, solutions and information, and being able to address, solve or appropriately escalate possibly emerging problems and issues.

The demand for employee cybersecurity skills, as shown in the results of this study, vary from situational awareness, communication and collaboration to technical (e.g., architecture and programming) skills. In addition, attitudinal needs, such as being analytical and having a hacker mind-set were mentioned. Literature raises skills relevant to cybersecurity (e.g., Sussman, 2021; Tomić et al., 2020; Neal et al., 2021). Further study and development of organisational skills acquisition could be addressed through three relevant categories:

technical, situation awareness, and problem solving e-skills. Structuring cyber range training features, such as tailored in-house scenarios based attack simulation training, with roleplaying exercise cycles, around these three categories of e-skills may provide needed focus for impactful cybersecurity training in organizations. E.g., the ECHO E-skills and Training Toolkit will be based on these three e-skills categories.

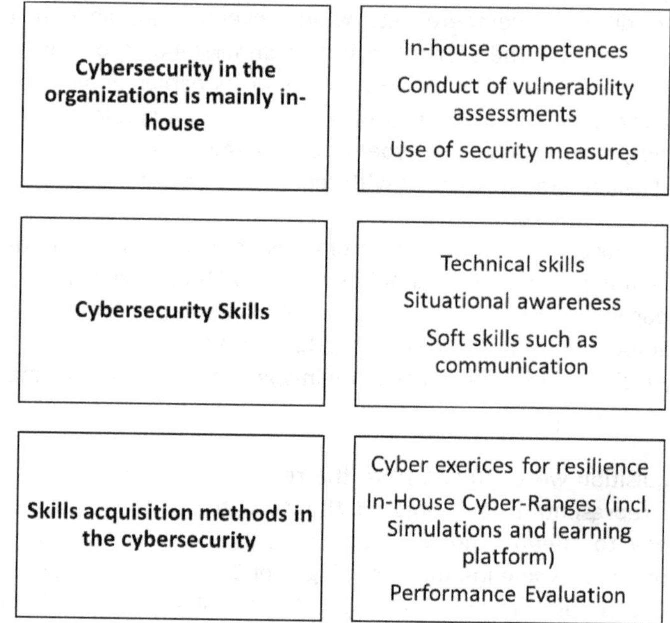

Figure 5: Cybersecurity in the organizations, skills and skills acquisition methods

In terms of skills acquisition, organizational representatives value cybersecurity training with CRs as a method. Relevant CRs should be uniquely designed and adapted for organisational purposes. While, modern learning environments, such as CRs are expected to promote attack and defence simulations and even real-time monitoring of the cybersecurity domain, they can be used to support skills acquisition through relevant learning platforms, and when possible, using organisational in-house capabilities. CRs cannot, however, encompass all cyber related training while, especially small and medium size organizations would prefer in-house training capabilities for experiential learning and interactive training. CRs may be too complicated to encompass all cyber related training in-house, and CR capacity can be bought from CR service providers, such as the ECHO FCR. Organisations are looking for ways to attract and develop relevant skills that take them to success. One important element in today's digitalized work and communication environments is continuity in the face of possible cyber issues and incidents. Thus, organizations are looking for better methods to identify and develop their ICT and cyber related e-skills needed to prevent, detect and address any occurring cyber incidents in a timely manner. Modern learning environments e.g., cyber ranges, provide increasing opportunities for focused skills acquisition. The model provided in Figure 5 provides a way to structure training approaches that address these three categories of e-skills in both recruitment and training. Further research and development is recommended to provide scientifically rigorous, but practical methods to identify the skills needed to successfully implement appropriate counter measures against cyber threats and incidents. Cyber ranges have so far, been mostly used to train cyber security experts. It is recommended to expand the usage of cyber ranges also for a wider range of ITC users and persons working in information intensive positions, as their careless actions may provide the needed access for a cybersecurity wrongdoer. Work-life representatives identify especially communication and situational awareness as the skills most demanded for their employees. Specified training with cyber ranges help ensure cybersecurity skills acquisition. These results can be used to plan and design practical training and education for future professionals. This study aims to promote scientific theory with its constructive discussion on cyber and e-skills and their acquisition in the cybersecurity domain.

References

Aaltola, K. (2021). Empirical Study on Cyber Range Capabilities, Interactions and Learning. *Digital Transformation, Cyber Security and Resilience of Modern Societies*, 84, 413.

Aaltola, K., & Taitto, P. (2019). Utilising Experiential and Organizational Learning Theories to Improve Human Performance in Cyber Training. g. Information & Security: An International Journal 43, no. 2 (2019): 123-133.

Adams, M., & Makramalla, M. (2015). Cybersecurity Skills Training: An Attacker-Centric Gamified Approach. *Technology Innovation Management Review*, 5-14.

Carlton, M., Levy, Y., & Ramim, M. (2019). Mitigating cyber attacks through the measurement of non-IT professionals' cybersecurity skills. *Information & Computer Security*.

Crumpler, W., & Lewis, J. A. (2019). *The cybersecurity workforce gap*. Washington, DC, USA: Center for Strategic and International Studies (CSIS).

Davis, J. & Magrath, S. (2013). A Survey of Cyber Ranges and Testbeds. Cyber Electronic Warfare Division. Commonwealth of Australia 2013. October 2013.

Dawson, J., & Thomson, R. (2018). The future cybersecurity workforce: going beyond technical skills for successful cyber performance. *Frontiers in psychology*, 9, 744.

Denzin N. K. & Lincoln Y. S. (1994). Handbook of Qualitative Research (Sage Publications,
Thousand Oaks, USA).

Furnell, S., Fischer, P., & Finch, A. (2017). Can't get the staff? The growing need for cyber-security skills. *Computer Fraud & Security, 2017*(2), 5-10.

Gates, A. Q., Salamah, S., & Longpre, L. (2014). Roadmap for Graduating Students with Expertise in the Analysis and Development of Secure Cyber-Systems.

Ghafir, I., Saleem., Hammoudeh, M., Faour, H., Prenosil, V., Jaf, S., Jabbar, S. & Baker, T (2018). Security threats to critical infrastructure: the human factor. The Journal of Supercomputing 74, 4986-5002.

Greenacre M. and Hastie T. (1987). The Geometric Interpretation of Correspondence Analysis. Journal of the American Statistical Association, Vol. 82, No. 398 (Jun., 1987), 437-447.

Grunewald, D., Lützenberger, M, Chinnow, J. (2011). Agent-based network security simulation. In: proceedings of the 10th international conference on autonomous agents and multiagent systems. Taipei, Taiwan, 2–6 May 2011, pp.1325–1326. Richland, SC: International Foundation for Autonomous Agents and Multi-agent Systems.

Jajodia, S., & Noel, S. (2010). Topological vulnerability analysis. In *Cyber situational awareness* (pp. 139-154). Springer, Boston, MA.

Mengidis, N., Spanopoulos-Karalexidis, M., Voulgaridis, A., Merialdo, M., Raisr, I., Hanson, K., . . . Votis, K. (2021). ECHO federated cyber range: Towards next-generation scalable cyber ranges. Piscataway: The Institute of Electrical and Electronics Engineers, Inc. (IEEE). doi: http://dx.doi.org/10.1109/CSR51186.2021.9527985

Milgram, P., and Kishino, F. (1994). A taxonomy of mixed reality visual displays. IEICE Trans. Inform. Syst. E77-D, 1321–1329.

Neal, J. Facteau, J. & O'Connell, B. (2021). To find cybersecurity talent, poach from other fields. Nextgov.com (Online). Availabe: https://www.proquest.com/magazines/find-cybersecurity-talent-poach-other-fields-xa0/docview/2555709888/se-2?accountid=12003.

Nevmerzhitskaya, J., Norvanto, E., & Virag, C. (2019). High impact cybersecurity capacity building. Bucharest: "Carol I" National Defence University.

Ostrom, A., Parasuraman, A., Bowen, D., Patricio, L., Voss, A. (2015). Service research priorities in a rapidly changing context. Journal of Service Re-search, 18 (2) (2015), pp. 127-159

Patrício, L., Fisk, R. Falcão e Cunha, J. Constantine, L. (2011). Designing multi-interface service experiences: The service experience blueprint. Journal of Service Research, 10 (4) (2008), pp. 318-334

Priyadarshini, I. (2018). Features and Architecture of the Modern Cyber Range: A qualitative analysis and survey. University of Delaware. Available: http://udspace.udel.edu/handle/19716/23789 © 2018 Ishaani Priyadarshini

Raybourn E., Kunz M., Fritz D., Urias V. (2018) A Zero-Entry Cyber Range Environment for Future Learning Ecosystems. In: Koç Ç. (eds) Cyber-Physical Systems Security. Springer, Cham.

Rajamäki, J. and Katos, V., (2020). Information Sharing Models for Early Warning Systems of Cybersecurity Intelligence. Information & Security: An International Journal 46, no. 2 (2020): 198-214.

Ruoslahti, H. (2020). Business Continuity for Critical Infrastructure Operators. Annals of Disaster Risk Sciences: ADRS, 3(1), 0-0.

Ruoslahti, H., Coburn, J., Trent, A. & Tikanmäki, I. (2020). Cyber Skills Gaps – a Systematic Literature Review of Academic Literature. Submitted to a peer-reviewed journal.

Skorenkyy, Y., Kozak, R., Zagorodna, N., Kramar, O., & Baran, I. (2021). Use of augmented reality-enabled prototyping of cyber-physical systems for improving cyber-security education. Journal of Physics: Conference Series, 1840(1)

Soni, S., & Bhushan, B. (2019). Use of Machine Learning algorithms for designing efficient cyber security solutions. In *2019 2nd International Conference on Intelligent Computing, Instrumentation and Control Technologies (ICICICT)* (Vol. 1, pp. 1496-1501). IEEE.

Sussman, L. (2021). Exploring the value of non-technical knowledge, skills, and abilities (KSAs) to cybersecurity hiring managers. Journal of Higher Education Theory and Practice, 21(6), 99-117.

Tomić R. & Komnenić, V. (2020). Cybersecurity Talent Shortage. Annals of Disaster Risk Sciences, 3(2).

Topham, L., Kifayat, K., Younis, Y., Shi, Q., & Askwith, B. (2016). Cyber security teaching and learning laboratories: A survey. Information & Security: An International Journal, 35, 51-80.

Yanakiev, Y. (2020). A governance model of a collaborative networked organization for cybersecurity research. *Information & Security, 46*(1), 79-98.

Yelland, P. M. (2010). An introduction to correspondence analysis. The Mathematical Journal, 12(1), 86-109.

Wang, B, Cai, J, Zhang, S. (2019). A network security assessment model based on attack defense game theory. In: proceedings of the IEEE 2010 international conference on computer application and system modeling (ICCASM), Tai-yuan, China, 22–24 October 2010, pp. V3–639. Piscataway, NJ: IEEE.

Identification of Violence in Twitter Using a Custom Lexicon and NLP

Jonathan Adkins
Norwich University, Northfield, USA
jadkins@norwich.edu

Abstract: Information warfare is no longer a denizen purely of the political domain. It is a phenomenon that permeates other domains, especially those of mass communications and cybersecurity. Deepfakes, sock puppets, and microtargeted political advertising on social media are some examples of techniques that have been employed by threat actors to exert influence over consumers of mass media. Social Network Analysis (SNA) is an aggregation of tools and techniques used to research and analyze the nature of relationships between entities. SNA makes use of such tools as text mining, sentiment analysis, and machine learning algorithms to identify and measure aspects of human behavior in certain defined conditions. One area of interest in SNA is the ability to identify and measure levels of strong emotions in groups of people. In particular, we have developed a technique in which the potential for increased violence within a community can be identified and measured using a combination of text mining, sentiment analysis, and graph theory. We have compiled a custom lexicon of terms used commonly in discussions relating to acts of violence. Each term in the lexicon has a numerical weight associated with it, indicating how violent the term is. We will take samples of online community discussions from Twitter and use the R and Python programming languages to cross-reference the samples with our lexicon. The results will be displayed in a Twitter discussion graph where the user nodes are color-coded according to the overall level of violence that is inherent in the Tweet. This methodology will demonstrate which communities within an online social network discussion are more at risk for potentially violent behavior. We assert that when this approach is used in association with other NLP techniques such as word embeddings and sentiment analysis, it will provide cybersecurity and homeland security analysts with actionable threat intelligence.

Keywords: social network analysis, natural language processing, sentiment analysis, text mining

1. Introduction

The Internet has brought with it several immeasurable benefits, such as the ability to instantly share information between multiple entities over great distances. It has also ushered in several negative variables that appeal to some of the basest aspects of the human psyche. These include anonymity, aggression, criminal opportunism, and appeals to engage with others in violent acts. The most ubiquitous venue on the internet where these antisocial behaviors are explored is social media. Twitter has been used by terrorists to recruit as well as disseminate information (Oh, Agrawal and Rao, 2011). Facebook accounts have been used for countless cases of cyberbullying (Sharif & Hoque, 2021). A niche microblogging site called *Parler* served as a staging platform for the January 6 rioters in Washington DC (Prabhu et al, 2021). There have been many studies in the academic literature that seek to use Natural Language Processing (NLP) techniques to identify and measure aspects of aggression and violence in social media (Lytos et al, 2019). Ni et al (2020) used interview data with analytics to predict a student's risk of violence in a school setting. Bigrams and trigrams were used with an unsupervised classifier to identify hate speech in a polarized political environment in Nigeria (Udanor and Anyanwu, 2019). Studies such as these only capture a small piece of the puzzle. Bigrams and trigrams by themselves do not provide a full context of what we seek to know from an online population. Based on empirical data, many sentiment lexicons used in the literature do not provide the granularity that is needed to isolate the variables that are needed to identify people who are at risk for violence (Rekik, Jamoussi and Hamadou, 2019). Lexicons such as the Bing and NRC are either too narrow or too broad to succinctly identify the elements that precipitate violence in a community. These limitations have created a gap in the literature concerning studies of online aggression and violence. To this end, we propose a custom niche lexicon containing terms specifically related to violence in the English Language (Beigi and Moattar, 2021). Our lexicon is not confined to specific domains of politics or deviant psychology. Our lexicon will contain terms from politics, popular culture slang, hacking terms, and criminal justice terms. Each term is weighted for severity, ranging from one to three. Our custom lexicon will identify terms from Twitter and will aggregate the weighted values to form a *violence score*. Our Violence Score will be represented in graphs on the x-axis compared with sentiment, subjectivity, and emotion. When the violence score is evaluated against these three variables, we will provide more insight into the context of violence that is inherent within an online community (Akhtar, Ekbal and Cambria, 2020). The remainder of this paper will be divided into eight sections. Sections two and three will discuss the violence lexicon and the Twitter dataset. Sections four, five, and six will discuss the sentiment, subjectivity, and emotion graphs juxtaposed to the violence score. Section seven will discuss our color-coded social network graph. In the final section, we will discuss our conclusions and future work.

2. Violence lexicon

Beigi and Moattar (2021) developed a framework for creating a domain-specific lexicon for positive and negative sentiment. Our approach differs from their technique in two fundamental ways. First, we include multiple domains under the larger umbrella of violence. Under our paradigm, terms from different domains are relevant if they in some way suggest or directly invoke the construct of violence. Second, we don't assign *positive* or *negative* valence to our terms. We assign a numerical weight to a term depending upon its perceived severity. In its current proof-of-concept version, a word such as warlord suggests the potential for violence but does not directly invoke it, therefore it gets a score of one. A term such as eviscerate directly describes a severely violent act, therefore it gets a score of three (the highest). In later research, we plan to expand the number of terms. We also will re-evaluate the scoring metric to see if a larger range of values is warranted. A weight of two is assigned to words that are perceived as violent but not as severe as a level three term. The weights assigned to individual terms will also be re-evaluated in later research since manual labeling is a subjective task and requires additional consideration. Table 1 below displays a small sample from the lexicon. The term associated with violence is displayed in the left column. The weight, ranging from 1 to 3, is displayed in the right column.

Table 1: Sample from the Violence lexicon

word	weight
abduct	2
abuse	2
aggressor	2
aggression	2
agitated	2
agitator	2
airstrike	3
ambush	2
anarchism	1
anarchy	1
anguish	2
annihilate	3

3. Twitter dataset

Khader, Awajan and Al-Naymat (2018) stated the value of using social media data for sentiment analysis. According to their study, platforms such as Twitter provide large volumes of "high-velocity data" that contains valuable information that can be extracted for analysis. Kausar, Soosaimanickam and Nasar (2021) used Twitter data from the most infected countries in the world during the height of the COVID-19 pandemic to understand how the countries dealt with the crisis. Instead of applying sentiment analysis tools and social media data toward studies of public health, we intend to apply the "high-velocity" social media data from the Twittersphere toward the identification and metastasizing of violence in a given population. This is a proof-of-concept study and our custom lexicon currently only hosts 400 terms with associated weights. For the sake of reference, the National Research Council Canada (NRC) lexicon has approximately 20,000 terms to evaluate eight different emotions in text (Khoo & Johnkhan, 2018). Based on prior empirical studies, we decided upon a Twitter dataset of 30,000 tweets to evaluate for *violence*, *sentiment*, *subjectivity*, and *emotion*. We collected the dataset using the R programming language's "twitteR" library, which contains a plethora of tools for querying Twitter's public-facing Application Programming Interface (API). We queried the Twitter API for any tweets currently in circulation based on the keyword search "assault." We saved the query results to a comma-separated value (csv) file. For our evaluation, we used three columns from the csv file. These were *screenName*, *text*, and *isRetweet*. In the next section, we will discuss assessing the *violence score*.

3.1 Assessing the violence score

The first step in our study was to identify any tweets that contained terms that were listed in the violence lexicon and discard any tweets which did not contain any violent terms. A script was used in R which looped through all 30,000 rows of the dataset and compared the terms in the text field to the terms that were in the violence lexicon. Out of the original 30,000 rows of tweets from the dataset, approximately 15,000 contained one or more violent terms that were listed in our lexicon. For the tweets in the dataset that remained, the R script looped through each one row by row and identified one or more violent terms in the tweet. The weights for each term were added together and the resulting score was written to a new column called *ViolenceScore*. Depending on how many violent terms were in a tweet, the resulting score could range from a one to approximately thirteen. A score of *one* suggests a solitary word with mild violent content. A score of *five* or more suggests that the tweet was significantly more violent in content. The values in the ViolenceScore column were placed on the x-axis for the sentiment, subjectivity, and emotion graphs. By approaching the graphs in this manner, we were able to see the distribution of violence scores relative to the responses by users in the population.

4. Sentiment and violence

To evaluate the sentiment analysis of our assault dataset, we used an R library called *word2vec*. Word2vec uses word vectors (otherwise known as word embeddings) to mathematically define words and evaluate their context (Giatsoglou et al, 2017). A machine learning neural network is used to train a sample of tokenized words taken from the original tweets. Word2vec assigns a probability for each tweet as to *how negative* or *how positive* the tweet is. Each dot in this graph represents a tweet. The closer to zero a tweet is, the more negative it is. The closer to 1.0 the tweet is, the more positive it is. As seen in Figure 1 below, the ViolenceScore column was placed on the x-axis, so that the distribution of tweets can be seen juxtaposed to their sentiment valence (positive or negative). We made the following observations with regard to the sentiment analysis of the *assault* dataset. The majority of the tweets had a violence score of 3 (still mild). In this level of violence, there is a full range of sentiments, from extremely negative to extremely positive. Another observation we made was that the more violent the content became, the fewer in number the tweets were for those violence scores. The tweets with the highest violence scores were outliers and were highly negative.

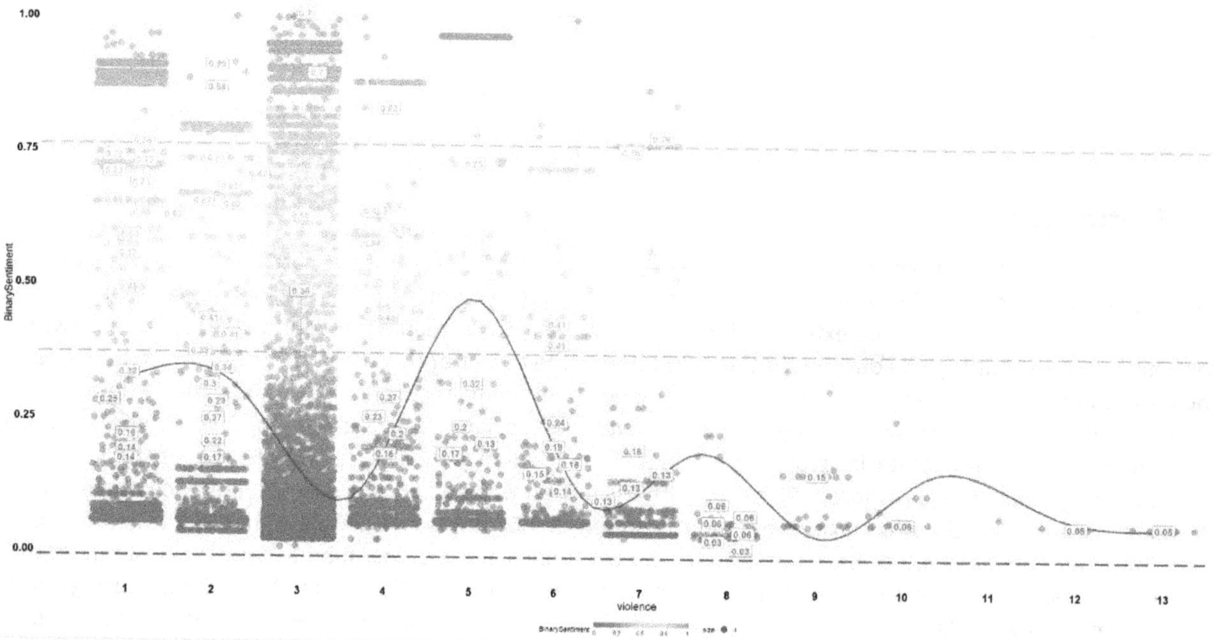

Figure 1: Sentiment analysis and violence

5. Subjectivity and violence

Subjectivity is a metric that is often included in the same class as *sentiment* and *emotion*. There are notable differences between these three constructs. *Sentiment* (which we discussed in the previous section) is a valence that exists between positive and negative. By itself, it does not articulate a specific emotion (Neogi et al, 2021). *Subjectivity* is a metric that seeks to evaluate how opinionated or factual a person's statements are (Yaqub et al,

2018). By combining the sentiment and subjectivity metrics, there is a more vivid context for a person's intent (Akhtar, Ekbal and Cambria, 2020). For example, a person can speak at length and have his speech qualify as highly negative based on sentiment analysis. However, if it was demonstrated that his subjectivity measured very low, it could suggest that the person was speaking academically and objectively about negative subject matter.

We used word2vec with the assault Twitter dataset in order to evaluate the level of subjectivity in the text. Even though this is the same approach we used for sentiment analysis, the technique differed in that we had to calculate the subjectivity scores for the assault dataset. We accomplished this task by using the Python programming language and a library called textblob. The textblob library assesses input text for words expressing opinion and feelings toward a topic (Saha, Yadav and Ranjan, 2017). For example, if a speaker says, "I think" or "this should be," he is conveying *opinion*. If statements lack qualifiers such as these, there is a higher probability that the statements are more *objective* or *neutral*. The textblob library was used in Python to score the assault Twitter dataset for subjectivity. Word2vec was then used to evaluate the overall subjectivity of the tweets with the violence score values on the x-axis. The results can be seen in the graph below in Figure 2.

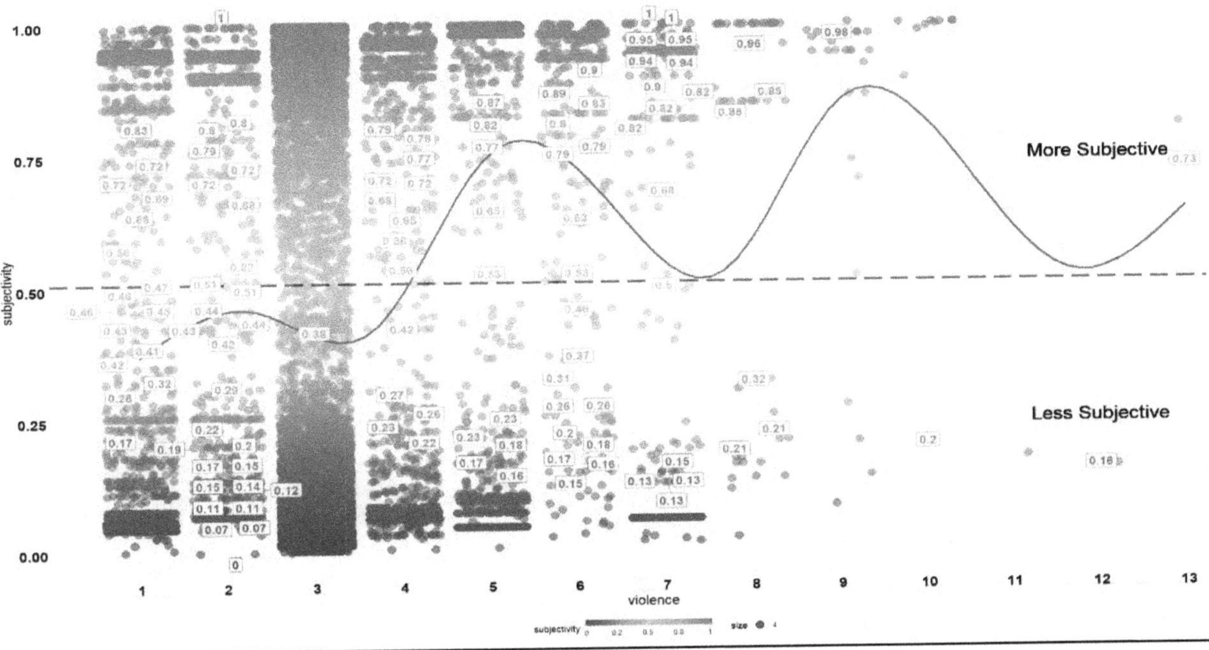

Figure 2: Subjectivity analysis and violence

Based on our findings, there was one significant difference between the subjectivity and sentiment graphs. We observed that as the violent content of the tweets increased (higher violence score), subjectivity also increased. This observation by itself if inconclusive. Several more samples would be needed to assess whether this was a recurring pattern. There is one variable in this approach that can be tweaked. Word2vec works on a probability scale. In both sentiment and subjectivity, the tweets were assessed as to how likely they were to be positive or negative, subjective or not subjective. For subjectivity, we set the threshold to .25 or 25%. This means that if a tweet was scored as 25% subjective, it was overall more subjective. The .25 value itself may be too high. The more words of personal opinion we add to a statement, the more subjective it becomes. We felt for this proof-of-concept experiment that 25% was a valid threshold.

6. Emotion and violence

In sections 4 and 5 we compared our Twitter sentiment and subjectivity distributions to their relative violence scores (Peng et al, 2021). We have observed thus far that the most violent content tends to be outliers that are highly negative and highly subjective. The integration of sentiment and subjectivity provides us with a sharper, more enhanced perspective into a speaker's intent. What we were missing up until this point was an emotional metric to provide a more complete answer to the question: *why* is the speaker more violent, more subjective, and more negative (Akhtar, Ekbal and Cambria, 2020)?

The National Research Council Canada (NRC) lexicon measures the amount of emotion in text. The NRC lexicon is actually an aggregation of ten different lexicons. Eight of the ten lexicons are emotions: anger, fear, anticipation, trust, surprise, sadness, joy, and disgust. It also includes lexicons for positive and negative sentiments (Agarwal and Toshniwal, 2018). Based on empirical findings, some of the emotion lexicons do not provide useful feedback. For example, there are more words in the anticipation lexicon than in the other lexicons. This frequently causes a skewed result where the anticipation observations are higher than any other emotion (Zad, Jimenez and Finlayson, 2021). Based on this empirical observation, we removed the anticipation, trust, and disgust lexicons when we run assessments. We assessed the assault Twitter dataset using a truncated version of the NRC lexicon: anger, fear, joy, sadness, and surprise. The results can be seen in the graph below in Figure 3. Violence scores were placed on the x-axis and the percentage of associated emotion words on the y-axis. We tried several different types of plots, but most did not provide us with insightful observations. We decided to create our emotion graph using violin plots. The violin plots allowed us to adequately measure the levels of violence juxtaposed to the percentage of emotion-related words from the Twitter dataset (Sinha et al, 2021).

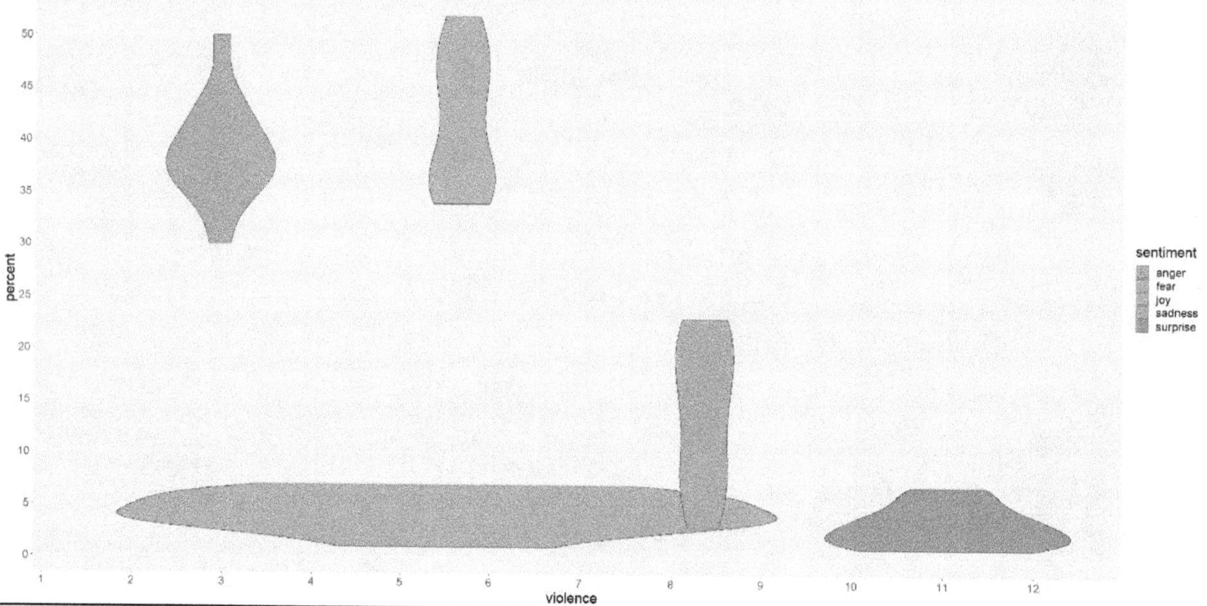

Figure 3: NRC emotion lexicon graph

After reviewing the violin plots of anger, fear, joy, sadness, and surprise we made the following observations. As violence scores increased, the prevailing emotion was surprise. Tweets with violence scores ranging from 8 to 9 tended to convey sadness. Violence scores that ranged between 5 and 6 conveyed fear. The majority of tweets in the assault dataset scored in the 3 range, which suggested that they were predominantly angry. From these scores it could be extrapolated that the average Twitter user from this dataset was only mildly violent and angry. The sentiment scores for this level of violence were evenly distributed between positive and negative. The subjectivity score for 3 range was also more objective. Overall, these observations suggest that the Twitter users were debating more objective facts from ideologically opposing viewpoints. In section 7 we will further demonstrate the utility of our violence lexicon by creating a social network graph using the violence scores as attributes.

7. Social network graph using the violence scores as attributes

Social Network Analysis (SNA) is an interdisciplinary domain of research that has been around since the early 1990's (Min et al, 2021). One of the many tools used by SNA researchers is the social network graph. One of the primary purposes of a SNA graph is to model relationships between entities in a population. SNA graphs are composed by two principal features called edges and nodes. Nodes represent specific entities and edges represent relationships between entities (Tabassum et al, 2018). There are several implementations of SNA graphs that range from public health and law enforcement to search engine optimization. The application of the graph depends on which aspect of the graph researchers wish to focus (nodes or edges). Homeland Security and other law enforcement agencies use SNA graphs to model the relationships between criminal organizations such as terrorist networks or organized crime (Min et al, 2021). Graphs have also been used by epidemiology researchers to model the spread of COVID-19 in a population (So et al, 2020). Another implementation of SNA graph analysis is called community detection. A graph of an online discussion taken from Twitter is represented by a composite set of nodes and edges. Within this larger composite collective of points and lines, there are smaller subnetworks where the connections between certain nodes are denser. These subnetworks that exist within a larger online discussion are referred to as *communities*. The area of SNA research that endeavors to identify graph subcommunities is called *community detection* (Du et al., 2007). Community detection can provide researchers with an abundance of information concerning group identities, ideological differences, and the flow of information (Kanavos et al, 2018). A good example of community detection being implemented is Campan, Cuzzocrea and Truta (2017). In their study, community detection was used to detect fake news as it spread through online communities.

For our proof-of-concept, we created a SNA graph (seen in Figure 4) which displays the retweet relationships between Twitter users in an online conversation concerning assault. The composite violence score for this dataset ranged from one to thirteen. In order to efficiently depict this information, we scaled the sizes of each node in the graph to correspond to their violence score. The larger a node is, the higher its violence score. In addition, we color-coded the nodes using thirteen shades of red. The nodes with the highest violence scores have the deepest shade of red. Nodes with lower violence scores are smaller and have lighter shades of red. If we look at the graph below in Figure 4, we see that several subnetworks or communities have formed. In each community, there is at least one node or Twitter user that is more violent than the others. The benefit of this type of modeling is that researchers can isolate individual subnetworks in this discussion graph and identify which communities have a larger number of larger scaled nodes that are deeper red in color.

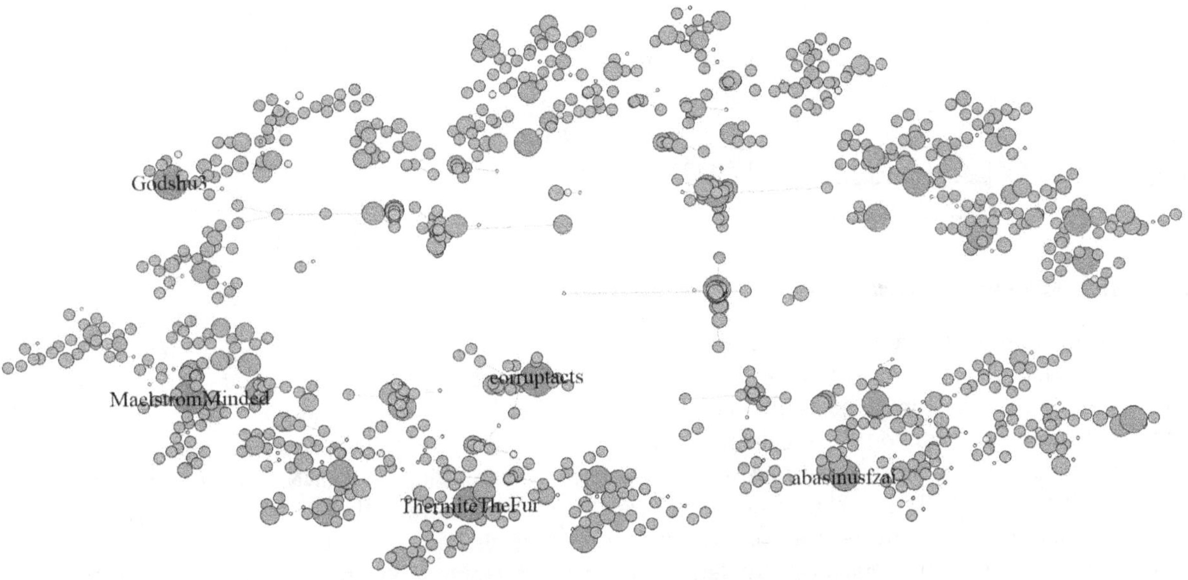

Figure 4: Social network graph of Twitter *assault* online discussion

7.1 Bigrams and trigrams found in the Twitter discussion network

In many studies using SNA and NLP, unigrams (keywords), bigrams (two-word phrases), and trigrams (three-word phrases) are used to capture principal themes that exist in a body of text (Arts, Hou and Gomez, 2021). We extracted the top 10 bigrams and trigrams from the assault Twitter dataset. All three n-gram combinations are necessary for extracting themes. Through empirical analysis, we have found that some themes will not become apparent until larger n-gram combinations are extracted. Table 2 and Table 3 display the themes that were most prevalent in the network seen in Figure 4. From our list of bigrams, the most vivid themes we found were "assault weapons" and "background checks." If we extrapolate the larger discussion based on the phrases in this table, the evidence suggests that Twitter users were discussing themes relating to assault weapons and firearm legislation.

The most insightful trigrams from Table 3 are "high capacity magazines," "ban assault weapons," and "universal background checks." These themes further validate what was found using bigrams. Based on the frequency of the bigrams and trigrams, topics relating to automatic firearms and gun control legislation are the most frequently discussed themes among Twitter users in this online conversation.

Table 2: Top 10 bigrams from *assault* Twitter dataset

bigram	n
assault weapons	1996
ban assault	1822
high capacity	1820
capacity magazines	1800
background checks	1541
universal background	1537
checks ban	1532
pass universal	1531
magazines repeal	1474
occupied jerusalem	1189

Table 3: Top 10 trigrams from *assault* Twitter dataset

trigram	n
high capacity magazines	1800
ban assault weapons	1790
universal background checks	1538
background checks ban	1532
pass universal background	1531
checks ban assault	1530
capacity magazines repeal	1474
sheikh jarrah neighborhood	849
u.s rightly condemns	849
also under threat	846

8. Conclusion and future work

In this study, we proposed a proof-of-concept algorithm for the identification and evaluation of violence within an online social network. For this paper we used Twitter as our case study, however, the framework can be implemented to suit any online social media platform. The novel contribution that we offer in this study is the creation of a violence lexicon. The lexicon is highly focused on areas that articulate concepts of violence, therefore it is more unique than existing lexicons such as the NRC, Bing, or AFINN lexicons. To maximize the utility of this lexicon, we have included numerical weights for each term in the lexicon. Currently, the weights range from scores of one to three. We plan to revisit the scoring mechanism and perhaps expand the range after further consideration. In this study, we also demonstrated an ensemble of SNA metrics that were used in concert with our violence score. Specifically, we used word embedding-based sentiment analysis, subjectivity analysis, and emotion analysis in plots with violence scores on the x-axis. By plotting SNA graphs with violence along the x-axis, we could quickly identify the distribution of sentiment, subjectivity, and emotions per unit of aggregated violence. This technique allows us to identify trends concerning human responses to certain inds of subject matter in online discussions. The aggregation of sentiment, subjectivity, and emotion allowed us to put user responses into a more vivid context. For future research, we plan to integrate sentiment, subjectivity, emotion, and violence into a single composite score that incorporates time series into its algorithm. By integrating these features with a probabilistic function, we will be able to predict *how violent* an online community will get in the future.

References

Agarwal, A. and Toshniwal, D., 2018, June. Application of lexicon based approach in sentiment analysis for short tweets. In *2018 International Conference on Advances in Computing and Communication Engineering (ICACCE)* (pp. 189-193). IEEE.

Akhtar, M.S., Ekbal, A. and Cambria, E., 2020. How intense are you? Predicting intensities of emotions and sentiments using stacked ensemble [application notes]. *IEEE Computational Intelligence Magazine, 15*(1), pp.64-75.

Arts, S., Hou, J. and Gomez, J.C., 2021. Natural language processing to identify the creation and impact of new technologies in patent text: Code, data, and new measures. *Research Policy, 50*(2), p.104144.

Beigi, O.M. and Moattar, M.H., 2021. Automatic construction of domain-specific sentiment lexicon for unsupervised domain adaptation and sentiment classification. *Knowledge-Based Systems, 213*, p.106423.

Campan, A., Cuzzocrea, A. and Truta, T.M., 2017, December. Fighting fake news spread in online social networks: Actual trends and future research directions. In *2017 IEEE International Conference on Big Data (Big Data)* (pp. 4453-4457). IEEE.

Du, N., Wu, B., Pei, X., Wang, B. and Xu, L., 2007, August. Community detection in large-scale social networks. In *Proceedings of the 9th WebKDD and 1st SNA-KDD 2007 workshop on Web mining and social network analysis* (pp. 16-25).

Giatsoglou, M., Vozalis, M.G., Diamantaras, K., Vakali, A., Sarigiannidis, G. and Chatzisavvas, K.C., 2017. Sentiment analysis leveraging emotions and word embeddings. *Expert Systems with Applications, 69*, pp.214-224.

Kanavos, A., Perikos, I., Hatzilygeroudis, I. and Tsakalidis, A., 2018. Emotional community detection in social networks. *Computers & Electrical Engineering, 65*, pp.449-460.

Kausar, M.A., Soosaimanickam, A. and Nasar, M., 2021. Public sentiment analysis on Twitter data during COVID-19 outbreak. *Int. J. Adv. Comput. Sci. Appl, 12*(2), pp.415-422.

Khader, M., Awajan, A. and Al-Naymat, G., 2018, November. The effects of natural language processing on big data analysis: Sentiment analysis case study. In *2018 International Arab Conference on Information Technology (ACIT)* (pp. 1-7). IEEE.

Khoo, C.S. and Johnkhan, S.B., 2018. Lexicon-based sentiment analysis: Comparative evaluation of six sentiment lexicons. *Journal of Information Science, 44*(4), pp.491-511.

Lytos, A., Lagkas, T., Sarigiannidis, P. and Bontcheva, K., 2019. The evolution of argumentation mining: From models to social media and emerging tools. *Information Processing & Management, 56*(6), p.102055.

Min, S., Gao, Z., Peng, J., Wang, L., Qin, K. and Fang, B., 2021. STGSN—A Spatial–Temporal Graph Neural Network framework for time-evolving social networks. *Knowledge-Based Systems, 214*, p.106746.

Neogi, A.S., Garg, K.A., Mishra, R.K. and Dwivedi, Y.K., 2021. Sentiment analysis and classification of Indian farmers' protest using twitter data. *International Journal of Information Management Data Insights, 1*(2), p.100019.

Ni, Y., Barzman, D., Bachtel, A., Griffey, M., Osborn, A. and Sorter, M., 2020. Finding warning markers: leveraging natural language processing and machine learning technologies to detect risk of school violence. *International journal of medical informatics, 139*, p.104137.

Obembe, D., Kolade, O., Obembe, F., Owoseni, A. and Mafimisebi, O., 2021. Covid-19 and the tourism industry: An early stage sentiment analysis of the impact of social media and stakeholder communication. *International Journal of Information Management Data Insights, 1*(2), p.100040.

Oh, O., Agrawal, M. and Rao, H.R., 2011. Information control and terrorism: Tracking the Mumbai terrorist attack through twitter. *Information Systems Frontiers, 13*(1), pp.33-43.

Peng, S., Cao, L., Zhou, Y., Ouyang, Z., Yang, A., Li, X., Jia, W. and Yu, S., 2021. A survey on deep learning for textual emotion analysis in social networks. *Digital Communications and Networks.*

Prabhu, A., Guhathakurta, D., Subramanian, M., Reddy, M., Sehgal, S., Karandikar, T., Gulati, A., Arora, U., Shah, R.R. and Kumaraguru, P., 2021. Capitol (Pat) riots: A comparative study of Twitter and Parler. *arXiv preprint arXiv:2101.06914.*

Rekik, A., Jamoussi, S. and Hamadou, A.B., 2019, September. Violent vocabulary extraction methodology: Application to the radicalism detection on social media. In *International Conference on Computational Collective Intelligence* (pp. 97-109). Springer, Cham.

Rezaeinia, S.M., Rahmani, R., Ghodsi, A. and Veisi, H., 2019. Sentiment analysis based on improved pre-trained word embeddings. *Expert Systems with Applications, 117*, pp.139-147.

Saha, S., Yadav, J. and Ranjan, P., 2017. Proposed approach for sarcasm detection in twitter. *Indian Journal of Science and Technology, 10*(25), pp.1-8.

Sharif, O. and Hoque, M.M., 2021. Tackling Cyber-Aggression: Identification and Fine-Grained Categorization of Aggressive Texts on Social Media using Weighted Ensemble of Transformers. *Neurocomputing.*

Sinha, S., Chen, H., Sekhon, A., Ji, Y. and Qi, Y., 2021. Perturbing inputs for fragile interpretations in deep natural language processing. *arXiv preprint arXiv:2108.04990.*

So, M.K., Tiwari, A., Chu, A.M., Tsang, J.T. and Chan, J.N., 2020. Visualizing COVID-19 pandemic risk through network connectedness. *International Journal of Infectious Diseases, 96*, pp.558-561.

Sullivan, Bebe. 2022. Violence Lexicon [Unpublished lexicon]. Peace and War Center, Norwich University.

Tabassum, S., Pereira, F.S., Fernandes, S. and Gama, J., 2018. Social network analysis: An overview. *Wiley Interdisciplinary Reviews: Data Mining and Knowledge Discovery, 8*(5), p.e1256.

Udanor, C. and Anyanwu, C.C., 2019. Combating the challenges of social media hate speech in a polarized society: A Twitter ego lexalytics approach. *Data Technologies and Applications.*

Yaqub, U., Sharma, N., Pabreja, R., Chun, S.A., Atluri, V. and Vaidya, J., 2018, May. Analysis and visualization of subjectivity and polarity of Twitter location data. In *Proceedings of the 19th annual international conference on digital government research: governance in the data age* (pp. 1-10).

Zad, S., Jimenez, J. and Finlayson, M., 2021, August. Hell hath no fury? correcting bias in the nrc emotion lexicon. In *Proceedings of the 5th Workshop on Online Abuse and Harms (WOAH 2021)* (pp. 102-113).

The U.S. Cyber Threat Landscape

Elie Alhajjar and Kevin Lee
Army Cyber Institute, West Point, USA
elie.alhajjar@westpoint.edu
kevin.lee@westpoint.edu

Abstract: Cybersecurity is concerned with protecting information, hardware, and software on the internet from unauthorized use, intrusions, sabotage, and natural disasters. It is the body of technologies, processes and practices designed to protect networks, computers, programs and data from attack, damage, or unauthorized access. The numerous ways in which computer systems and data can be compromised and the dramatic increase in cybercrimes have made cybersecurity a growing field. One of the most problematic elements of cybersecurity is the quick and constant evolving nature of security risks in critical infrastructure and major businesses all around the world. In this paper, we sketch a general frame for the cyber threat landscape in the United States of America by focusing on five major categories: ransomware, social engineering, third party software, deep fakes, and insider threats. We elaborate on each of these pillars by providing case studies from the past decade, as well as discussing ways to move forward.

Keywords: cyber threat, ransomware, insider threat, malware, social engineering, deepfakes

1. Introduction

Advancements in information technology (IT) have raised concerns about the risks to data associated with weak IT security, including vulnerability to viruses, malware, attacks and compromise of network systems and services. Inadequate IT security may result in compromised confidentiality, integrity, and availability of the data due to unauthorized access. Nowadays, a general rule of thumb is the following: All networks are vulnerable to cybersecurity threats!

According to the Practical Law Company (Farhat, McCarthy and Raysman, 2011), a cyberattack is an attack initiated from a computer against a website, computer system or individual computer that compromises the confidentiality, integrity or availability of the computer or information stored on it. Cyberattacks may take different forms, while having many objectives such as gaining unauthorized access to a computer system or its data, disrupting normal activities including the take down of entire web sites, installing viruses or malicious code on a computer system, changing the characteristics of a computer system's hardware, firmware or software without the owner's knowledge, instruction, or consent, inappropriately using computer systems by current and/or former employees, etc.

A cyber threat is an activity intended to compromise the security of an information system by altering the availability, integrity, or confidentiality of a system or the information it contains. Cyber threat actors are states, groups, or individuals who, with malicious intent, aim to take advantage of vulnerabilities, low cyber security awareness, and technological developments to gain unauthorized access to information systems in order to access or otherwise affect victims' data, devices, systems, and networks. The globalized nature of the Internet allows these threat actors to be physically located anywhere in the world and still affect the security of information systems. Cyber threat actors can be categorized by their motivations and, to a degree, by their sophistication. Threat actors value access to devices, processing power, computing resources, and information for different reasons. In general, each type of cyber threat actor has a primary motivation.

Cyber threat actors are not equal in terms of capability and sophistication, and have a range of resources, training, and support for their activities. Cyber threat actors may operate on their own or as part of a larger organization. In some cases, even sophisticated actors use less sophisticated and readily available tools and techniques because these can still be effective for a given task and make it difficult for defenders to attribute the activity. Nation-states are frequently the most sophisticated threat actors, with dedicated resources and personnel, and extensive planning and coordination. Cybercriminals are generally understood to have moderate sophistication in comparison to nation-states. Nonetheless, they still have planning and support functions in addition to specialized technical capabilities that affect a large number of victims. Hacktivists, terrorist groups, and thrill-seekers are typically at the lowest level of sophistication as they often rely on widely available tools that require little technical skill to deploy. Their actions, more often than not, have no lasting effect on their targets beyond reputation. Finally, insider threats are individuals working within their organization who are particularly dangerous because of their access to internal networks that are protected by security perimeters. Access is a key component for malicious threat actors and having privileged access eliminates the need to employ other remote means. Insider threats may be associated with any of the other listed types of threat actors but can also include disgruntled employees with motive.

Of particular interest is the notion of cyber-terrorism (Denning, 2000): "Cyber-terrorism is the convergence of terrorism and cyberspace. It is generally understood to mean unlawful attacks and threats of attack against computers, networks, and the information stored therein when done to intimidate or coerce a government or its people in furtherance of political or social objectives. Further, to qualify as cyber-terrorism, an attack should result in violence against persons or property, or at least cause enough harm to generate fear. Attacks that lead to death or bodily injury, explosions, plane crashes, water contamination, or severe economic loss would be examples. Serious attacks against critical infrastructures could be acts of cyber-terrorism, depending on their impact." Terrorists' activities via cyberspace include creating websites/blogs, communication via email, discussion via chat rooms, e-transactions (e-commerce/ e-banking), using search engines to collect data and information, phishing/ hacking, viruses, malicious code, etc. (Alhajjar, Fameli and Warren, 2020).

There are various threats to an organization's information system; understanding the vast array of threats is the first step in ensuring adequate protection of sensitive data. A holistic approach to data security begins with understanding the network, its architecture, user population, and mission requirements. In this paper, we aim at providing a "big picture" view of the cyber threat landscape in the United States of America. To this end, the current paper is organized as follows. After this brief introduction, we give an overview of the main five cybersecurity threats, namely ransomware, social engineering, malware, deep fakes, and insider threats in sections 2-6, respectively. We discuss many case studies and highlight their details. We conclude our work in section 7 and suggest some ways to mitigate such threats in the future.

2. Ransomware

Ransomware is a type of malware that involves an attacker locking the victim's computer system files typically through encryption and demanding a payment to decrypt and unlock them. It is a rising threat that utilizes malicious software that prevents a user from accessing computer files, systems, or networks demanding a ransom for their return. If the ransom is not paid, the victim's data remains unavailable. Cyber criminals may also pressure victims to pay the ransom by threatening to destroy the victim's data or to release it to the public.

Generally speaking, one can distinguish ransomware attacks into three categories (Loman 2019). First, cryptoworm is a standalone ransomware that replicates itself to other computers for maximum reach and impact. Second, Ransomware-as-a-Service (RaaS) is sold on the dark web as a distribution kit to anyone who can afford it. Such RaaS packages allow people with little technical skill to attack with relative ease. They are typically deployed via malicious spam emails, via exploit kits as a drive-by download, or semi-manually by automated active adversaries. Third, automated active adversary is a ransomware attack deployed by actors who use tools to automatically scan the internet for IT systems with weak protection. When such systems are found, the attackers establish a foothold and from there carefully plan the ransomware attack for maximum damage.

Although cyber criminals use a variety of techniques to infect victims with ransomware, the most common means of infection are email phishing campaigns, Remote Desktop Protocol (RDP), and software vulnerabilities. In the first case, the cybercriminal sends an email containing a malicious file or link, which deploys malware when clicked by a recipient. Cyber criminals historically have used generic, broad-based spamming strategies to deploy their malware, though recent ransomware campaigns have been more targeted and sophisticated. Criminals may also compromise a victim's email account by using precursor malware, which enables the cybercriminal to use a victim's email account to further spread the infection. In the second case, RDP is a proprietary network protocol that allows individuals to control the resources and data of a computer over the internet. Cybercriminals have used both brute-force methods, a technique using trial-and-error to obtain user credentials, and credentials purchased on dark web market. Once they have RDP access, criminals can deploy a range of ransomware attacks to victim systems. In the third case, cybercriminals can take advantage of security weaknesses in widely used software programs to gain control of victim systems and deploy ransomware.

In May of 2021, Colonial Pipeline announced that it had fallen victim to a devastating ransomware attack that shut down operations and cut off fuel supplies to millions, causing massive economic disruption across the

Eastern United States. The FBI confirmed the attacks were executed by the DarkSide ransomware gang, a relatively new threat actor that Cybereason has been tracking since August 2020. They have a mature Ransomware as a Service (RaaS) business model and affiliate program. The group has a phone number and even a help desk to facilitate negotiations with and collect information about its victims, not just technical information regarding their environment but also more general details relating to the company itself like the organization's size and estimated revenue. DarkSide follows the double extortion trend, where the threat actors first exfiltrates sensitive information stored on a victim's systems before launching the encryption routine. After the ransomware encrypts the target's data and issues the ransom demand for payment in exchange for the decryption key, the threat actors make the additional threat of publishing the exfiltrated data online should the target refuse to make the ransom payment.

On or about 18 Feb 2018, a threat actor gained access to the Colorado Department of Transportation (CDOT) network via a virtual server and installed the SamSam ransomware malware variant. Two days later, the ransomware became active and infected approximately 150 servers and 2000 workstations. The Remote Desktop protocol is how this attack was initiated: the attacker discovered the system available on the internet, broke into the Administrator account using approximately 40,000 password guesses until the account was compromised. From there, the attacker was able to access CDOT's environment as the domain administrator, installing and activating the ransomware attack. The alleged attackers behind the SamSam ransomware, who operated from Iran, have been identified and are wanted by the FBI.

3. Social engineering

Social engineering is one of the most prolific and effective means of gaining access to secure systems and obtaining sensitive information yet requires minimal technical knowledge. It refers to the manipulation of individuals in order to induce them to carry out specific actions or to divulge information that can be of use to an attacker. Attacks vary from bulk phishing emails with little sophistication through to highly targeted, multi-layered attacks which use a range of social engineering techniques. Social engineering works by manipulating normal human behavioural traits and as such there are only limited technical solutions to guard against it.

Social engineering techniques are commonly used to deliver malicious software but, in some cases, only form part of an attack, as an enabler to gain additional information, commit fraud or obtain access to secure systems. Social engineering techniques range from indiscriminate wide scale attacks, which are crude and can normally be easily identified, through to sophisticated multi-layered tailored attacks which can be almost indistinguishable from genuine interactions. Social engineers are creative, and their tactics can be expected to evolve to take advantage of new technologies and situations.

Although social engineering attacks differ from each other, they have a common pattern that involves four phases: (i) collect information about the target; (ii) develop relationship with the target; (iii) exploit the available information and execute the attack; and (iv) exit with no traces (Mouton, Leenen and Venter 2016). In the research phase, also called information gathering, the attacker selects a victim based on some requirements. In the relationship phase, the attacker starts to gain the trust of the victim through direct contact or email communication. In the exploitation phase, the attacker influences the victim emotionally to provide sensitive information or perform security mistakes. In the exit phase, the attacker quits without leaving any proof (Salahdine and Kaabouch 2019).

Social engineering attacks can be classified into two categories: human-based or computer-based. In human-based attacks, the attacker executes the attack in person by interacting with the target to gather desired information. Thus, they can influence a limited number of victims. The software-based attacks are performed using devices such as computers or mobile phones to get information from the targets. In a different perspective, social engineering attacks can be classified into three categories according to how the attack is conducted: social, technical, and physical-based attacks (Kalnins, Purins and Alksnis 2017). Social-based attacks are performed through relationships with the victims to play on their psychology and emotion. These attacks are the most dangerous and successful attacks as they involve human interactions. Technical-based attacks are conducted through internet via social networks and online services websites, and they gather desired information such as passwords, credit card details, and security questions. Physical-based attacks refer to physical actions performed by the attacker to collect information about the target. An example of such attacks is searching in dumpsters for valuable documents.

4. Malware

Malware is an abbreviation of the words malicious and software. The term refers to software that is deployed with malicious intent. Malware is easy to deploy remotely and tracking the source of malware is hard. In the past decade, malware has emerged as a major threat for cybersecurity as software companies have to constantly work to write newer versions and security patches and release them so that the existing and in-use software system can be upgraded to confront new forms of threats. The financial services industry is a primary target for malware-enabled cyber-attacks because financial institutions operate software that tracks ownership of monetary assets. Cybercriminals also directly target customers of such institutions and business partners using malware-enabled attacks.

Malware may take as many forms as software. It may be deployed on desktops, servers, mobile phones, printers, and programmable electronic circuits. Sophisticated attacks have confirmed data can be stolen through well written malware residing only in system memory without leaving any footprint in the form of persistent data. Malware has been known to disable information security protection mechanisms such as desktop firewalls and anti-virus programs. Some even have the ability to subvert authentication, authorization, and audit functions. It has configured initialization files to maintain persistence even after an infected system is rebooted. Upon execution, sophisticated malware may self-replicate and/or lie dormant until summoned via its command features to extract data or erase files.

In general, a single piece of malware is generally described by four attributes of its operation - PISC (Zeltser 2010). First, propagation is the mechanism that enables malware to be distributed to multiple systems. Second, infection is the installation routine used by the malware, as well as its ability to remain installed despite disinfection attempts. Third, self-defence is the method used to conceal its presence and resist analysis, these techniques may also be called anti-reversing capabilities. Last, capabilities refer to software functionality available to the malware operator.

There are six steps a malware criminal tends to follow in order to accomplish a typical cybercrime: reconnaissance, assembly, delivery, compromise, and command (Cloppert 2010). In the beginning, the attacker surveys the target to identify points of vulnerability as well as an attack-planning phase. Then, he/she creates, customizes, or otherwise obtains malware to satisfy attack requirements. At the delivery step, the malware propagation occurs followed by the infection at the compromise level which leads to unleashing the malware capabilities at the command phase. Finally, the malware delivers data to malware operator, i.e., exfiltration, or otherwise accomplishes attack objective.

Many remarkable incidents took place in the past couple of years, and they vary in magnitude and damage level. In early 2017, the multinational credit reporting company Equifax got hacked in early 2017 and more than 800 million consumer financial data was leaked online. The hackers launched their attack by exploiting the Apache Struts vulnerability, then accessing dozens of databases and creating more than 30 backdoors into Equifax's systems. In March 2018, the US Department of Justice indicted nine Iranian hackers over an alleged spree of attacks on more than 300 universities in the United States and abroad. The hackers stole 31 terabytes of data, estimated to be worth $3 billion in intellectual property. In May 2021, the Alaska Department of Health and Social Services (DHSS) warned that a highly sophisticated cyber-attack has exposed residents' personal data, including financial information. Before systems were shut down attackers potentially had access to full names, dates of birth, Social Security numbers, addresses, phone numbers, driver's license numbers, health information, and financial information. Internal identifying numbers such as for Medicaid or case reports, and historical information concerning individuals' interaction with DHSS were also potentially exposed.

5. Deepfakes

Developments in Artificial Intelligence (AI) have led to the emergence of deepfake technologies, which pose a significant threat to American and global institutions. Deepfake is defined as an AI-based technology that can manipulate images, sounds, and video content to represent an event that did not occur (Wojewidka 2020). Deepfake technology has taken a new level of sophistication as cybercriminals now can manipulate sounds, images, and videos to defraud and misinform individuals and businesses. As a common example, there have been multiple stances where the faces of politicians are being edited onto other individuals' bodies who appear to say things that they never did.

The term deepfake is a combination of deep learning and fake. Deepfake technology is powered by a special deep learning technique, namely Generative Adversarial Networks (GANs). GANs employ two artificial neural networks working against each other in a game-like setting. A typical GAN is generally composed of two deep neural network models: a generator and a discriminator. The generator's task is to learn from the discriminator's output and thus train so that its output may deceive the discriminator. The discriminator attempts to discern if its inputs are from the genuine data set or from the adversarial data set. As described in the original work (Goodfellow et al. 2014), the competition between the two components continues until the discriminator is unable to detect media forgery, thereby improving the overall quality of the deepfake before it can be deployed.

Deepfake technology poses many threats to global public and private entities in general, and to US institutions in particular, in many ways such as defrauding businesses, conducting misinformation campaigns in politics, and raising cybersecurity concerns in enterprises. At the judicial level, evidence tampering is one of the major threats posed by deepfakes in the judicial system. Evidence in the court of law can be manipulated with the use of deepfake technology to sway a case one way or the other (Pfefferkorn 2020). At the political level, deepfakes can be quickly created and easily circulated to a wide audience to spread disinformation among the general public. Deepfake technology can be used knowingly or unknowingly to misinform the public for political advantage (Goss and Burkell 2020). At the economic level, deepfake technology has an adverse budgetary impact on businesses as it can be used to defraud them and ruin their reputations. In the same realm, face scanners that grant access to restricted areas can be breached with the use of deepfakes, which can lead to unauthorised access to sensitive information and intellectual property. Such an attack could lead to monetary loss due to the costs incurred from containing the breach, compensating customers, and heightening security costs (Rosati et al. 2017).

Several incidents took place in the past couple of years that showed the effect of deepfake technology in the political realm. One notable example was the circulation of an altered video of an American politician, Nancy Pelosi on social media. In the video, she appeared intoxicated while mispronouncing her words; the video had been viewed and shared over 2.5 million times on Facebook (Greengard 2019). On the international theater, deepfakes can have a damaging impact on geopolitics and relationships between countries. Recently, the Australian Prime Minister, Scott Morrison demanded an apology after Zhao Lijian, a spokesperson for China's foreign ministry posted a fake image on Twitter that depicted an Australian soldier holding a knife to the throat of an Afghan child (BBC 2020).

6. Insider threat

Insider threats are malicious events from people within the organization, which usually involve intentional fraud, the theft of confidential or commercially valuable information, or the sabotage of computer systems. The subtle and dynamic nature of insider threats makes detection extremely difficult. The2018 U.S. State of Cybercrime Survey indicates that 25% of the cyber-attacks are committed by insiders, and 30% of respondents indicate incidents caused by insider attacks are more costly or damaging than outsider attacks (CERT 2018).

A malicious insider is defined as "a current or former employee, contractor, or business partner who has or had authorized access to an organization's network, system, or data, and has intentionally exceeded or intentionally used that access in a manner that negatively affected the confidentiality, integrity, or availability of the organization's information or information systems." (Costa, Albrethsen and Collins 2016). Compared to the external attacks whose footprints are difficult to hide, the attacks from insiders are hard to detect because malicious insiders already have the authorized power to access the internal information systems. In general, there are three types of insiders: (i) traitors who misuse their privileges to commit malicious activities, (ii) masqueraders who conduct illegal actions on behalf of legitimate employees of an institute, and (iii) unintentional perpetrators who innocently make mistakes. Based on the malicious activities conducted by the insiders, the insider threats can also be categorized into three types: (i) IT sabotage which directly uses IT to make harm to an institute, (ii) theft of intellectual property which steals information from the institute, and (iii) fraud which indicates unauthorized modification, addition, or deletion of data (Homoliak et al. 2019).

Insider threat research constitutes one of the facets of the new and emerging field of social cybersecurity (Carley 2020). Social cybersecurity is a computational social science with a large foot in the area of applied research. It uses computational social science techniques to identify, counter, and measure (or assess) the impact of communication objectives. The methods and findings in this area are critical, and advance industry-accepted practices for communication, journalism, and marketing research.

Multiple noticeable insider threat incidents occurred in the last decade, whose level of sophistication ranged all over the spectrum. In Indiana in mid-2011, a foreign national with permanent residence status was convicted in the state's first ever case of insider theft. Working for two companies, he stole formulas for pesticides and food additives with the intent to establish a company in his home country with co-conspirators. The technology and the site of the theft were deliberately targeted. In Louisiana, a naturalized U.S. citizen with 27 years of service working for a U.S. chemical company conspired with former colleagues and overseas partners to steal a specific formula and market that technology to companies in his native country. In Colorado, insiders stole plans for a chip that controlled sound quality in cell phones. The insiders created a joint venture with a foreign university with the intent to mass produce the technology and market it to commercial entities.

7. Conclusion

In this paper, we provide a general overview of the main five cyber threats in the United States of America. The reader is warned that this list is by no means exhaustive, it is rather correlated with the author's expertise and fields of research. We conclude below by providing a handful of suggestions to mitigate the risks of such attacks in the near future, as well as highlight some future directions in this wide realm.

First, there are many best practices to minimize ransomware risks such as backing up data, images, and configurations in an offline pipeline, using multi-factor authentication, updating, and patching systems, making sure security solutions are up to date, reviewing the incident response plan, etc. In the United States, The FBI does not encourage paying a ransom to criminal actors. Paying a ransom may embolden adversaries to target additional organizations, encourage other criminal actors to engage in the distribution of ransomware, and/ or fund illicit activities. Paying the ransom also does not guarantee that a victim's files will be recovered. Moreover, the FBI urges people to report ransomware incidents to your local field office or the FBI's Internet Crime Complaint Center (IC3). Doing so provides investigators with the critical information they need to track ransomware attackers, hold them accountable under US law, and prevent future attacks. As a side note, it is valuable to understand and measure the impact of a ransomware attack on the victim entity, so that adequate risk management plans can be put in place (Alhajjar and Olchowoj 2022).

Second, defense procedures for social engineering attacks include: encouraging security education and training, increasing social awareness of social engineering attacks, providing the required tools to detect and avoid these attacks, learning how to keep confidential information safe, reporting any suspected activity to the security service, organizing security orientations for new employees, and advertising attacks' risks to all employees by forwarding sensitization emails and known fraudulent emails.

Third, when planning an approach to malware prevention, organizations should be mindful of the attack vectors that are most likely to be used currently and in the near future. Malware prevention efforts such as user and IT staff awareness, vulnerability mitigation, threat mitigation, and defensive architecture are essential for the mitigation of malware attacks. An effective awareness program explains proper rules of behaviour for use of an organization's IT hosts and information. Also, a vulnerability can usually be mitigated by one or more methods, such as applying patches to update the software or reconfiguring the software.

Forth, digital forensics can provide an effective solution for detecting deepfakes. Using computational techniques, forensics experts can observe whether image pixels have been altered, by isolating anomalies, such as shadows and reflections. They can also inspect the metadata of the file to check if it has been altered, by checking the edit history and how many times the file has been compressed.

Finally, many recent works have studied ways to detect insider threats as well as find factors that lead to such incidents (Alhajjar and Bradley 2021). However, it remains a very difficult task to identify such threats, mainly because there is no straightforward way to diagnose anomalous human behavior with high certainty. A single point of anomalous behavior may not stand out, but anomalies in several areas may indicate an insider threat issue. Examples of this behavior might look like changes in job performance or work habits, downloading large volumes of information from the network, entering or exiting organization facilities at odd hours, making changes to the IT system, etc.

Acknowledgements

The second author would like to thank the Intelligent Cyber-Systems and Analytics Research Lab (ICSARL) at the Army Cyber Institute (ACI) in West Point, NY for their hospitality. Most of the work was done during his summer research stay at the institute.

References

Alhajjar, E. and Bradley, T. (2021). "Survival analysis for insider threat", *Computational and Mathematical Organization Theory*, 1-17.

Alhajjar, E. and Olchowoj, O. (2022). "The Impact of a Ransomware Attack", *Proceedings of the Sixth Annual Workshop on Naval Applications of Machine Learning, San Diego, USA.*

Alhajjar, E., Fameli, R. and Warren, S. (2021). "Are Terrorist Networks Just Glorified Criminal Cells?" *Northeast Journal of Complex Systems (NEJCS), Vol. 3: No. 1, Article 1.*

BBC (2020). "Australia demands China apologize for posting 'repugnant' fake image," [Online]. https://www.bbc.co.uk/news/world-australia-55126569. Accessed 31 October 2021.

Carley, K. M. (2020). "Social cybersecurity: an emerging science". *Computational and Mathematical Organization Theory*, pp. 365–381.

CERT (2018). "U.S. State of Cybercrime." Tech. rep. CERT Division of SRI-CMU and Force Point.

Cloppert, M. (2010). "Evolution of APT State of the ART and Intelligence-Driven Response," *US Digital Forensic and Incident Response Summit, SANS*. http://computer-forensics.sans.org.

Costa, D. L., Albrethsen, M. J. and Collins, M. L. (2016). "Insider threat indicator ontology." Tech. rep. Carnegie Mellon University, Pittsburg, PA.

Denning, D. (2000). "Cyber-terrorism". *Testimony before the Special Oversight Panel of Terrorism Committee on Armed Services*. Washington, DC: US House of Representatives.

Farhat, V., McCarthy, B. and Raysman, R. (2011). "Cyber Attacks: Prevention and Proactive Responses", Practical Law Company.

Goodfellow, I., Pouget-Abadie, J., Mirza, M., Xu, B., Warde-Farley, D., Ozair, S., Courville, A. and Bengio, Y. (2014). Generative Adversarial Nets. *Advances in neural information processing systems*, 2672–2680.

Gosse, C. and Burkell, J. (2020). "Politics and porn: how news media characterizes problems presented by deepfakes," *Critical Studies in Media Communication*, vol. 37, no. 5, pp. 497-511.

Greengard, S. (2019). "Will deepfakes do deep damage?" *Communications of the ACM*, vol. 63, no. 1, pp. 17-19.

Homoliak, I., Toffalini, F., Guarnizo, J., Elovici, Y. and Ochoa, M. (2019). "Insight into insiders and IT: a survey of insider threat taxonomies, analysis, modeling, and countermeasures". *ACM Computing Surveys (CSUR)* 52.2, pp. 1–40.

Kalnins, R., Purins, J. and Alksnis, G. (2017). "Security evaluation of wireless network access points." *Appl. Comput. Syst.*, 21, 38-45.

Loman, M. (2019). "How Ransomware Attacks: what defenders should know about the most prevalent and persistent malware families," Sophos Labs white paper.

Mouton, F., Leenen, L. and Venter, H. (2016). "Social engineering attack examples, templates and scenarios." *Comput. Secur.* 59, 186–209.

Pfefferkorn, R. (2020). "Deepfakes in the courtroom," *Boston University Law Journal*, vol. 29, no. 2, pp. 245-276.

Rosati, P., Cummins, M., Deeney, P., Gogolin, F., der Werff, L., Lynn, T. (2017). "The effect of data breach announcements beyond the stock price: Empirical evidence on market activity," *International Review of Financial Analysis*, vol. 49, pp. 147-148.

Salahdine, F. and Kaabouch, N. (2019). "Social Engineering Attacks: A Survey." *Future Internet*. 11(4):89.

Wojewidka, J. (2020). "The deepfake threat to face biometrics," *Biometric Technology Today*, vol. 2020, no. 2, pp. 5-7.

Zeltser, L. (2010). "Analysing Malicious Software in Cyberforensics," *J. Bayuk, Editor, Springer.*

Impact of Moral Disengagement on Counterproductive Work Behaviours in IT Sector, Pakistan

QaziMuhamamd Ali
Business and Management Sciences, Superior University, Lahore, Pakistan
Phdba2@gmail.com

Abstract: This research examines the role of moral disengagement towards counterproductive work behaviour in the information technology sector of Pakistan. Furthermore, research is also focused on the mediating effect of information security awareness (Attitude & knowledge) and information security awareness behaviours. The target population consisted of public sector I.T. departments of Punjab, Pakistan. A convenience sampling technique is utilized. Data collection has been done through a survey questionnaire from technical and non-technical staff currently employed in the Public sector I.T. departments of province Punjab. Statistical software PLS-SEM is used for analysis. This study highlights the role of the information technology sector staffing level of engagement that affects the employee's counterproductive work behaviour and information security awareness behaviour. Moreover, the study proposes that management should take the initiative for the implementation of strategies that may be helpful to get awareness about information security amongst employees.

Keywords: moral disengagement (M.D.), information security awareness (ISA), information security awareness knowledge (ISAK), counterproductive work behaviour (CWB), and information security awareness behaviour (ISAB)

1. Introduction

Cyber security is gaining importance worldwide because of the excessive usage of computers in all spheres of life (Chang and Coppel 2020). The potential outcome of information security breaches can have a broader effect, which includes disrepute of firms, competitive advantage, efficiency, and bankruptcy in a minor case (Jeong et al. 2019), (Schatz and Bashroush 2016). In the United Kingdom, primarily breaches of information security are found because of human error like irresponsible behaviour of workers in the firm and fraudulent emails responded by the employees (Hadlington 2021).

Cyber security is becoming the most important element for developing countries due to having future threats and weak organizational procedures and processes (Chang and Coppel 2020). Pakistan is an example of cyber security laws that have been established/applied by the state to combat cyber-attacks. Yet the threat remains significant because of gaps in applying these practices for various reasons (Khan and Anwar 2020).

In the last few years, there have been exponential advances in a study examining the role of the human component as an indicator of global information security awareness (Egelman and Peer 2015), (Parsons et al. 2014), (McCormac et al. 2018), (Janicke et al. 2018). The main focus of the current study is on individual differences that relate to gender and personality characteristics like amicability and carefulness (Butavicius et al. 2017), (McCormac et al. 2017).

However, the little investigation found in the literature shows how ISA is influenced by individual differences like; little research has been done to examine the complex relationship between adherence to information security awareness and a willingness to be morally detached, a key factor considered in a dysfunctional context or unproductive firm engagement. Literature has shown that M.D. is seen as a likely coping process to cope with the pressure of workplace safety necessities (D'Arcy et al. 2014). CWB is associated with a general disregard for firm procedure and safety (Spector et al. 2006), (Spector and Fox 2010). Furthermore, the main focus of information security awareness studies is on banking sectors compared to other information technology sectors (Dharmawansa and Madhuwanthi 2020), (Nasser et al. 2020), (Akinbowale et al. 2020).

Thus, this research aims to identify the connection between counterproductive work behaviour, M.D, and ISA. The second goal is to investigate how cognitive-emotional characteristics of ISA, like knowledge and attitude of ISA, can act as mechanisms that are underlying the connection between M.D and ISA behaviour and CWB. Although a previous researcher has shown that moral withdrawal predicts higher counterproductive work behaviour (Moore et al. 2012), the procedures after this connection are unclear. Likewise, studies showing a link between M.D. and ISA are very limited, especially in Asian countries (Chen et al. 2019), for later results in a Chinese perspective. There is no organized study of possible methods that combine the two.

This study contributes towards the information security in the Information Technology sector a neglected sector. Further, it enriches the literature by investigating a double mediation mechanism between moral disengagement and counter productive work behaviour by examining the role of information security awareness attitude, knowledge and behaviour of IT departmental staff in improving the counterproductive behaviour. The study contributes to the limited literature available on information security in the Asian context in different public or private information technology sectors of Punjab.

2. Hypothesis development

2.1 Relationship between moral disengagement and information security awareness

The researcher defines moral disengagement (Bandura 1986) as controlling individual actions that self-regulate in nature but can be selectively motivated (Hystad et al. 2014). MD includes eight types of interrelated cognitive processes that enable an employee to abandon intrinsic moral values to behave morally questionable (Moore 2015). These cognitive practices enable a person to reduce the distress feeling (Moore et al. 2012).

The author highlights that few studies examine that M.D can play for an immoral act in firms (Moore et al. 2012). Researchers connected the tendency of M.D. to behaviour like theft and dishonesty (Detert et al. 2008). The further author highlights that moral disengagement is considered a strong predictor for planned violation of information security awareness. It is noted that the main focus of previous studies is on a small set of violations that are linked with information security awareness while neglecting the other aspects that are related to disengagement in information security awareness like knowledge related to ISA processes and procedures as well as attitude of employees towards the ISA (D'Arcy et al. 2014). Few established studies show the connection between Moral disengagement and ISA & ISA behaviour (Hadlington et al. 2021).

> ***H1:*** *Moral disengagement has a significant positive relationship with Information Security awareness attitude*
>
> ***H2:****Moral disengagement has a significant positive relationship with Information Security awareness Knowledge*
>
> ***H3:*** *Moral disengagement has a significant positive relationship with Information Security awareness behaviour*
>
> ***H4:*** *Moral disengagement has a significant positive relationship with counterproductive work behaviour*

2.2 Relationship between information security and information security awareness:

The researchers from different perspectives have defined the term information security awareness (ISA), and most of the studies change that term to cyber security. A researcher argued with another concept and meaning and we do not consider it the same entity (Von Solms and Van Niekerk 2013). The main focus of information security is the information protection and system to save, transmit and utilize that information (Whitman and Mattord 2012).

Information Security is usually directed by a set of regulations that are established for employees by the organization to explain in terms of protocols and procedures that they must follow and do not share the credentials, report unusual activity (Parsons et al. 2014). Researchers said that unethical motives and deviancy affect the workers to oppose the protocols and procedures or to utilize information communication technology improperly (Wilks 2011). Such non-compliance of employees with firm standards may be seen as a part of firm citizenship. However, it is also mainly a problem with a solid moral aspect (Wilks 2011). So, employees who have a high tendency involve more counterwork behaviour. The people who are most involved in the counterwork behaviour and who have a strong propensity for M.D can also be the ones who have a more deficient commitment level with ISA.

> ***H5:*** *Information Security awareness attitude has a significant positive impact on Information Security awareness knowledge*
>
> ***H6:*** *Information Security awareness attitude has a significant positive impact on Information Security awareness behaviour*

H7: Information Security awareness knowledge has a significant positive impact on Information Security awareness behaviour

2.3 Relationship between information security awareness and counterproductive work behaviours

Researchers said that workplace deviance includes any conduct that breaches the firm standards wilfully threatens the well-being of employees and the firm itself (Robinson and Bennett 1995). The study proposed that CWB are volitional and excluded from those activities that may be considered directly instructed (Fox et al. 2012). Furthermore, researchers highlight that most of the studies in this perspective mainly focused on predicting why employees are more committed to counterwork behaviour and how to prevent them from doing it (Robinson 2008). Whereas, the researcher suggests that counterproductive work behaviour can be considered as a kind of complaining behaviour; for example, Employees or a cluster of individuals can attentively participate in the CWB to correct perceived inequality or firm indisposition (Kelloway et al. 2010)

As per a few studies, misuse of information communication technology at the organization is usually neglected due to counterwork behaviour (Weatherbee 2010). A previous study confirmed that cyberloafing (non-professional use of ICT in the workplace) is significantly associated with poor information security awareness, with more frequent participation in cyber-loafing associated with lower participation in information security awareness (Hadlington and Parsons 2017). Disengagement in ISA might be one of the potential features of counterproductive work behaviour (Carpenter and Berry 2017). Researchers point out the connection between the withdrawal of employees and counterproductive work behaviour.

H8: Information Security awareness attitude has a significant positive impact on counterproductive work behaviour.

H9: Information Security awareness knowledge has a significant positive impact on counterproductive work behaviour

H10: Information Security awareness behaviour has a significant and positive relationship with counterproductive work behaviour

H11: Information Security awareness attitude mediates the relationship between moral disengagement and counterproductive work behaviour

H12: Information Security awareness knowledge mediates the relationship between moral disengagement and counterproductive work behaviour

H13: Information Security awareness behaviour mediates the relationship between moral disengagement and counterproductive work behaviour

H14: Information Security awareness attitude, Information Security awareness knowledge, and Information Security awareness behaviour mediates the relationship between moral disengagement and counterproductive work behaviour

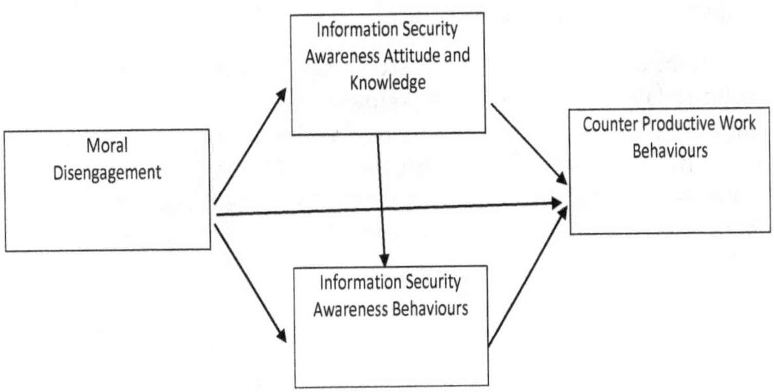

Figure 1: Theoretical framework

3. Methodology

3.1 Questionnaire

The measurement items of all variables were adopted from previous literature. A 5-point Likert scale was used from (1) Strongly Disagree to (5) strongly agree Current research aims to determine the connection between Moral Disengagement, ISA attitude and Knowledge, ISA Behaviours, and Counterproductive work behaviour based on technical staff's perception. The scale of 24 items was adopted from the literature of (Moore et al. 2016) to measure M.D. The researcher adopted ten items from the literature (Kaur and Mustafa 2013) to measure Information Security awareness. To measure the ISA behaviour, a scale of 06 items was adopted from the literature of (Kaur and Mustafa 2013). Moreover, the scale of 17 items was adapted from the literature of (Lee et al. 2005) to measure CWB in the information technology department.

3.2 Sample design and data collection

A quantitative survey method was used to collect the data from the public sector employees in the I.T. department of Pakistan, and a convenience sampling technique was utilized. Employees from several public sectors I.T. departments of Pakistan participated in the current study. The unit of analysis for this study is the employees of the I.T. department. The respondents were management staff, and technical staff responses were taken from two groups because the non-management staff is engaged in making the strategies and policies. In contrast, the technical staff is responsible for implementing those strategies. About 322 questionnaires were distributed among different workers employed in the I.T. department in Punjab after providing detailed information about the need for the current study, from which 223 were being returned out of which 16 were imperfect, so they were discarded and thus it was found that 207 were valid responses. The response rate of the current study was 64.28%.

4. Empirical findings

In this chapter, statistical data analysis has been done by using the smart PLS software as it is considered the most advanced technique for data analysis. Furthermore, PLS-SEM is utilized because of lesser needed data and data normality (Hair Jr et al. 2016). They keep going with this study by using Smart PLS-3 for analysis of data and evaluation of the hypothesis. The two-step procedures employed in this study highlighted the results recommended by (Henseler et al. 2009) and were considered most suitable in the field of social science research (Hair Jr et al. 2016).

4.1 Normality of data

In PLS-SEM, normal distribution of data is not required. It is critical to evaluate the normality distribution of data before applying inferential statistics (Hair et al. 2007). As per the researcher's recommendations (Munro 2005), Skewness and Kurtosis, and histogram charts are used to check the normality of data in PLS-SEM. The threshold for skewness and kurtosis is -2 to +2 for checking the normality of data. Results show that all variables are in between the threshold value, which shows that data are normally distributed.

4.2 Assessment of reflective measurement model

For assessing the measurement model in the current study, scholars confirmed both the validity and reliability of the data set. Composite reliability is used to assess data reliability whereas convergent, and discriminant validity measures the data validity. These results show the validity of measurements. Average Variance Extract (AVE) was used to assess convergent validity. The threshold for AVE is 0.500, and as shown in the table given below, 4.4.1. The AVE value of all items was more significant than the threshold value in the range of 0.504 to 0.936. Further, it also shows that all the measures of the 05 constructs were valid. Therefore, the model has sufficient convergent validity.

Table 1: Convergent validity

Variables	Items	Loadings	Chronbach Alpha	Alpha	C.R.
Moral Disengagement	MD1	0.569	0.96	0.968	0.965
	MD10	0.622			
	MD11	0.715			

Variables	Items	Loadings	Chronbach Alpha	Alpha	C.R.
	MD12	0.742			
	MD13	0.902			
	MD14	0.813			
	MD15	0.875			
	MD16	0.656			
	MD17	0.759			
	MD18	0.564			
	MD19	0.892			
	MD20	0.801			
	MD21	0.652			
	MD22	0.936			
	MD23	0.817			
	MD24	0.918			
	MD7	0.617			
	MD8	0.792			
	MD9	0.85			
Information Security Awareness Attitude	ISAA1	0.739	0.834	0.846	0.883
	ISAA2	0.827			
	ISAA3	0.773			
	ISAA4	0.835			
	ISAA5	0.697			
Information Security Awareness Knowledge	ISAK1	0.775	0.828	0.877	0.882
	ISAK2	0.898			
	ISAK3	0.811			
	ISAK4	0.74			
Information Security Awareness Behaviour	ISAB1	0.834	0.886	0.933	0.919
	ISAB2	0.878			
	ISAB3	0.916			
	ISAB4	0.808			
Counterproductive Work Behaviour	CWB1	0.763	0.875	0.901	0.902
	CWB14	0.641			
	CWB16	0.779			
	CWB2	0.912			
	CWB3	0.909			
	CWB5	0.504			
	CWB6	0.692			
	CWB7	0.605			
	MD1	0.569	0.96	0.968	0.965
	MD10	0.622			
	MD11	0.715			

Source: Author's Design by using Smart PLS-3

4.3 Discriminant validity

HTMT ratio is a new criterion that is introduced to check the discriminant validity for variance-based SEM. (Henseler et al. 2015). The threshold value for HTMT ratios is less than 0.90, and if the value is greater than the threshold value, then the problem of discernment validity occurs. Table 4.3.1 shows the HTMT ratios of the 1st order construct, which shows each value is less than the threshold value, whereas table 4.3.2 shows the ratios. It also shows the values are less than 0.90 which means discriminant validity for constructs is established.

Table 2: HTMT ratio

Items	CWB	ISAA	ISAB	ISAK	MD
CWB					
ISAA	0.591				
ISAB	0.247	0.348			
ISAK	0.564	0.874	0.306		
MD	0.751	0.704	0.364	0.893	

Source: Author's Design by using Smart PLS-3

Structure Equation Modelling (SEM) Path Analysis

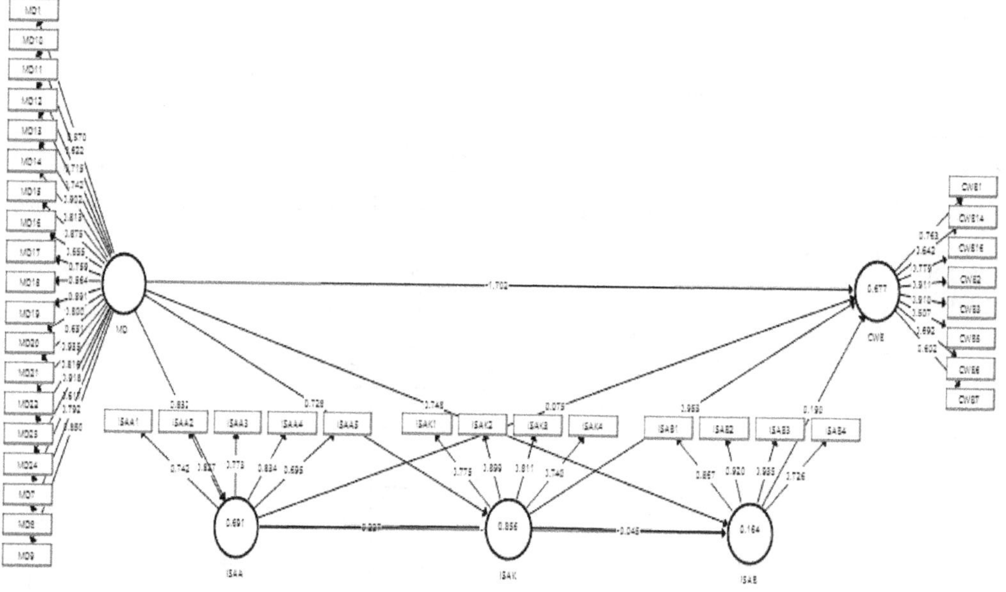

Figure 2: Measurement model assessment (Confirmatory Factor Analysis)

4.4 Assessment of Structural Model (SEM)

Hypothesis testing has been done using bootstrapping and PLS-SEM which shows the significant, positive, and negative relationships of variables. The indirect effect is used for mediation analysis. Table 4.4.1 shows the bootstrapping results. PLS-SEM and bootstrapping show the significant and positive relationship between moral disengagement and Information Security Awareness attitude (S.D. = 0.016, t = 5.801, p = 0.00). That means the 1st hypothesis is supported. Furthermore, a significant and positive connection between moral disengagement and Information Security Awareness knowledge shows β = 0.045 t = 16.216, p = 0.00). Therefore the 2nd hypothesis is also supported. A positive and significant relationship was found between moral disengagement and Information Security Awareness behaviour as the t value is greater than the threshold value. The p-value is less than 0.05 (β = 0.113, t = 6.597, P= 0.00) means hypothesis 3 is supported. However, results show the significant and positive relationship between moral disengagement and counterproductive work behaviour (β = 0.11, t = 15.526, P= 0.00), which means hypothesis 4 is supported. A significant and positive connection between Information Security Awareness and Information Security Awareness shows (β = 0.046 t = 4.904, p = 0.00). Therefore the 5th hypothesis is also supported. Further, a significant and positive connection has been found between Information Security Awareness attitude and Information Security awareness behaviour which shows β = 0.136 t = 3.736, p = 0.00). Therefore the 6th hypothesis is also supported. An insignificant connection between

Information Security awareness knowledge and Information Security awareness behaviour shows β = 0.142 t = 0.316, p = 0.752). Therefore the 7th hypothesis is also not supported. Besides this, the p-value of 0.437 was greater than the cut-off value of 0.05 which shows the insignificant relationships between Information Security awareness attitude and counterproductive work behaviour (β = 0.96, t = 0.778, p = 0.437) therefore hypothesis 8th is not supported and rejected. Moreover, a Significant and positive connection between Information Security awareness knowledge and counterproductive work behaviour shows β = 0.117, t = 8.129, p = 0.000). Based on the results hypothesis, 9th is supported. A positive and significant relationship was found between Information Security awareness behaviour and counterproductive work behaviour as the t value is greater than the threshold value. The p-value is less than 0.05 (β = 0.056, t = 3.407, P= 0.001); therefore it means hypothesis 10th is supported. As the t-value 0.776 and p-value = 0.438, which was lower than the threshold values, this study found an insignificant relationship between moral disengagement, Information Security awareness attitude, and counterproductive work behaviour (β = 0.08, t = 0.776, p =0.438). Based on the results hypothesis, 11th is rejected. A positive and significant relationship was found between moral disengagement, Information Security awareness knowledge, and counterproductive work behaviour as the t value is greater than the threshold value. The p-value is less than 0.05 (β = 0.117, t = 5.936, P= 0.000) therefore, it means hypothesis 12th is supported. However, results show the significant and positive relationship between moral disengagement, Information Security awareness behaviour, and counterproductive work behaviour (β = 0.029, t = 4.915, P= 0.000), which means hypothesis 13th is supported. Results of mediation analysis shows the insignificant relationship (β = 0.005, t = 0.338, P= 0.735) which means hypothesis 14th is not supported.

Table 3: Results of hypothesis (direct, indirect, mediation and moderation)

		SD	T	P	LLCI	ULCI	Decision
H1	MD -> ISAA	0.016	5.801	0.000	0.799	0.86	Supported
H2	MD -> ISAK	0.045	16.216	0.000	0.64	0.814	Supported
H3	MD -> ISAB	0.113	6.597	0.000	0.5	0.938	Supported
H4	MD -> CWB	0.11	15.526	0.000	1.949	1.532	Supported
H5	ISAA -> ISAK	0.046	4.904	0.000	0.135	0.314	Supported
H6	ISAA -> ISAB	0.136	3.736	0.000	0.783	0.255	Supported
H7	ISAK -> ISAB	0.142	0.316	0.752	0.341	0.224	Not Supported
H8	ISAA -> CWB	0.096	0.778	0.437	0.123	0.244	Not Supported
H9	ISAK -> CWB	0.117	8.129	0.000	0.748	1.225	Supported
H10	ISAB -> CWB	0.056	3.407	0.001	0.085	0.303	Supported
H11	MD -> ISAA -> CWB	0.08	0.776	0.438	0.09	0.211	Not Supported
H12	MD -> ISAK -> CWB	0.117	5.936	0.000	0.502	0.949	Supported
H13	MD -> ISAB -> CWB	0.029	4.915	0.000	0.078	0.193	Supported
H14	MD -> ISAA -> ISAK-> ISAB -> CWB	0.005	0.338	0.735	0.011	0.008	Not Supported

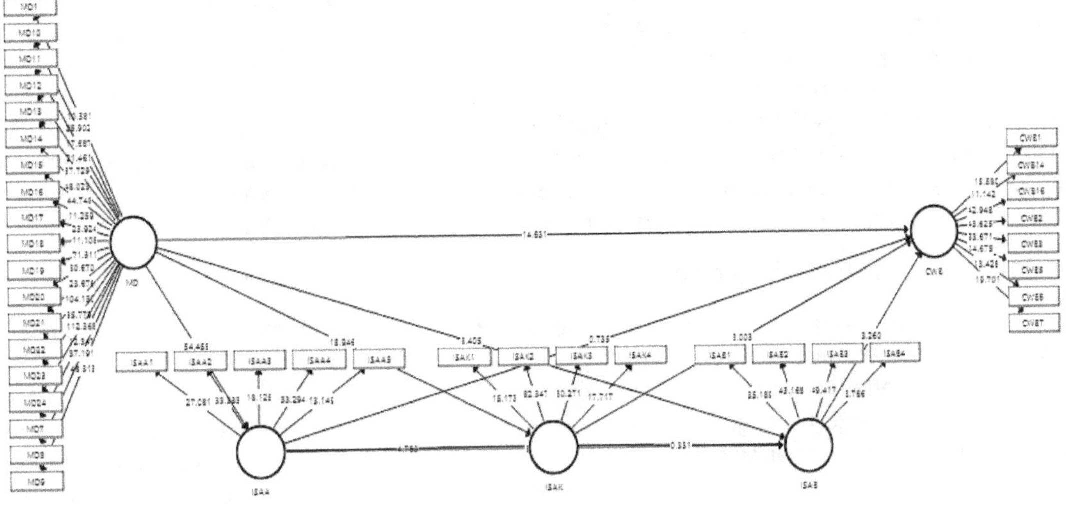

Figure 3: Structural model assessment

5. Discussion

The main objective of the present study was to inspect the way through which individual differences in M.D. and CWB are related to ISA. Whereas literature has pointed out the general connection between moral disengagement and counterproductive work behaviour related to the firm's information technology security protocol, more comprehensive and reliable models were lacking, allowing for the expansion of productive training and no other interventions. Based on various I.T. workloads, it has been theorized that ISA will at least partially explain the relationship between M.D. and CWB. A framework was projected in which Information Security attitude, knowledge, and behaviour mediates the relationship between moral disengagement and counterproductive work behaviour.

Overall, the findings revealed numerous fascinating trends among the tendency for M.D. and ISA. All the hypothesis of moral disengagement is significantly positively related to Information Security awareness and behaviour. Vigorous correlations among moral disengagement and Information Security awareness were those that imply a separation of concerns if all subscales are checked. From the Information Security perspective, this can indicate a significant obstacle to active protocol compliance. It can be effortless for people to bypass their Information Security awareness in several firms and depend on others to burden it. Literature has shown that many employees are often unaware of their role in the active firm cyber security of an organization and instead rely on that this is something that firm administration should be held accountable for (Hadlington 2017).

It should be noted that Information Security awareness attitude and knowledge were as closely related to counterproductive work behaviour as the behavioural aspect. In particular, this may reflect some conceptual overlap between counterproductive work behaviour and Information Security awareness behaviour. Research participants may understand to include a task related to information technology. As mentioned earlier, I.T. activities in many professions make up a significant portion of all work-related activities.

It is not surprising that counterproductive work behaviour and Information Security awareness negative behaviour go hand in hand. A second related point that needs to be emphasized is that Information Security awareness attitude and knowledge can, therefore, as our many mediation models postulate, influence counterproductive work behaviour and Information Security awareness behaviour. In addition, the counterproductive work behaviour was positively associated with the moral disengagement subscales and provided an accurate mirroring of the relationship between Information Security awareness and moral disengagement. The further endorses previous evidence of the moral disengagement association with immoral behaviour and general disrespect for basic security policies (Cohen et al. 2013), (Spector and Fox 2010).

The results of multivariate causal modeling are consistent with our theoretical model. As predicted, Information Security knowledge and behaviour reflect the relationship between M.D. and behavioural outcomes. Essentially, the behavioural results evaluated in the current study were mixed. Firstly, the process of mediation that leads to CWB instantly seems believable. It also leads to a lesser progressive attitude about ISA guidelines and protocols. Knowledge attitudes are essential ancestors of behaviour that impact the Information Security of a single user and the firm. However, the model's second mediation method expands the role of communicating information security awareness specific views and knowledge to the broader domain of counterproductive work behaviour. That may specify that the counterproductive work behaviour and the Information Security awareness behavioural element overlap at a theoretical or operational level, but the two constructs were defined in diverse theoretical contexts. The methods were used in current research change significantly in terms of a particular behaviour. Additional clarification could be such that Information Security awareness attitude and Information Security awareness knowledge follow a more general path from moral disengagement to counterproductive work behaviour. It characterizes more important motivators for compliance behaviour.

6. Limitation and future directions

Chances of further research always exist because of theoretical and methodological limitations. Several information technology departments in Pakistan are also making efforts to implement strategies that are helping to secure the information. Future research work should be sufficient enough to study the impact of moral disengagement and counterproductive work in Information Security awareness in-depth to understand I.T. sectors in a better way. Moreover, this study should be applied in different countries while considering the cross-cultural setting. It will be helpful to measure Information Security awareness in the I.T. sectors globally. The current study is based on quantitative; the data were collected based on adopted questionnaires from previous studies based on respondents' perception with limited information. Therefore, upcoming studies may focus on a mixed-method approach to examine the above-said relationships. It will be helpful for better understanding in analysis. Other Independent and dependent variables may be used to explore just like organizational abuse and discipline.

Appendix 1

Data Normality

Histogram Charts

- 1. Moral Disengagement

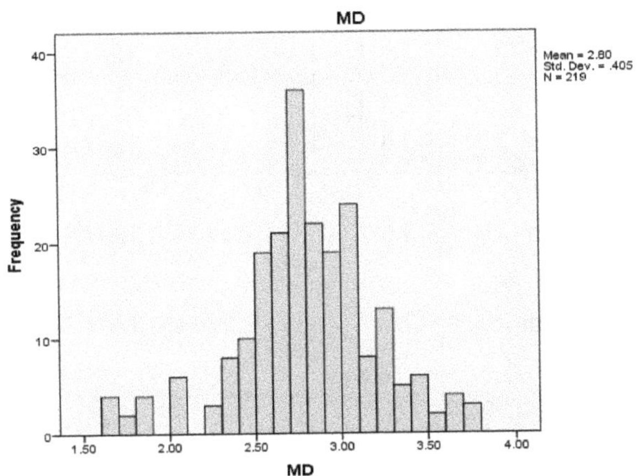

Source: Designed by using IBM SPSS-23

- 2. Information Security Awareness Attitude

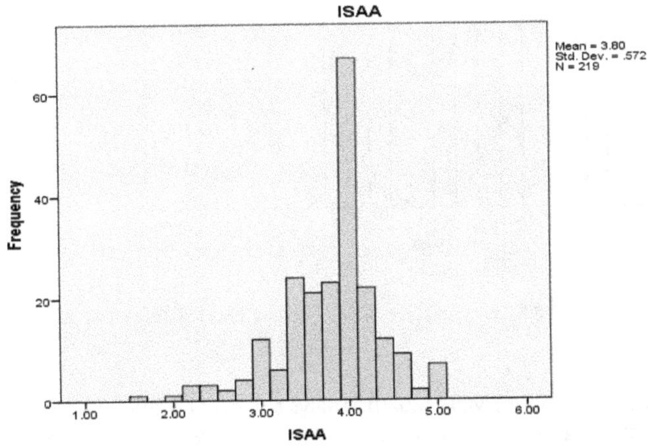

Source: Designed by using IBM SPSS-23

- 3. Information Security Awareness Knowledge

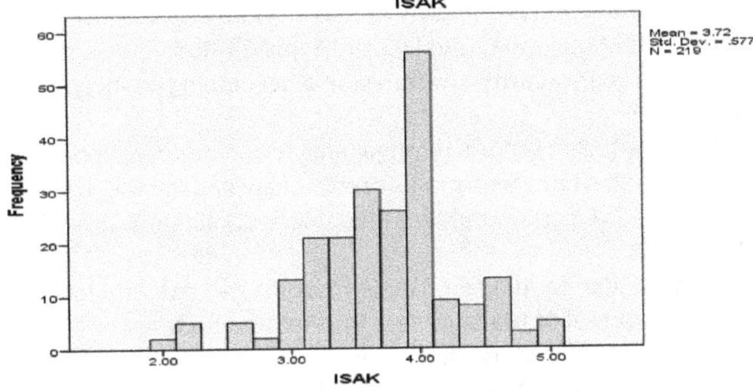

Source: Designed by using IBM SPSS-23

- 4. Information Security Awareness Behaviour

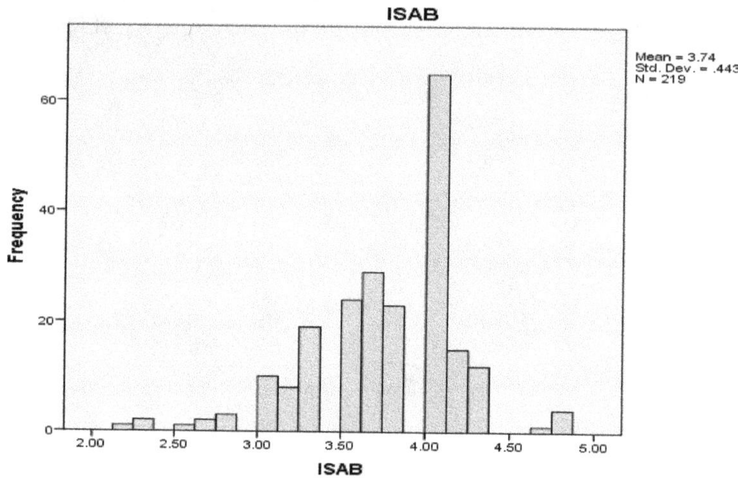

Source: Designed by using IBM SPSS-23

- 5. Counterproductive Work Behaviour

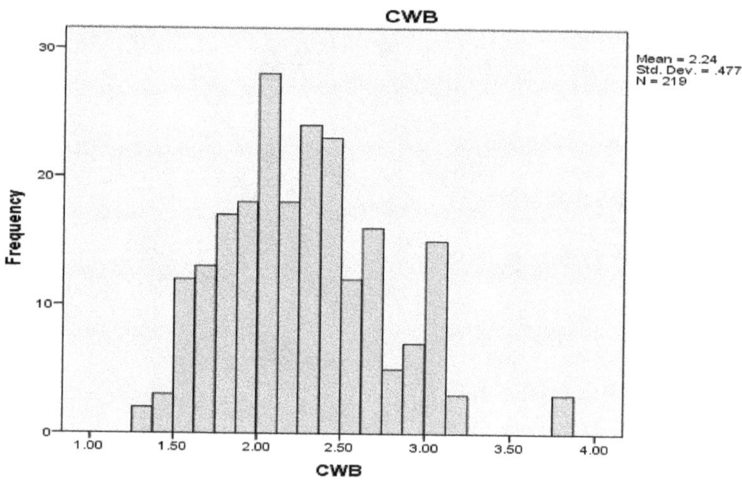

Source: Designed by using IBM SPSS-23

References

Akinbowale, O. E., Klingelhöfer, H. E. andZerihun, M. F. (2020). Analysis of cyber-crime effects on the banking sector using balance score card: a survey of literature. Journal of Financial Crime,Vol 27, No. 3, pp 945-958.

Bandura, A., & National Inst of Mental Health. (1986). Social foundations of thought and action: A social cognitive theory. Prentice-Hall, Inc.

Butavicius, M. A., Parsons, K., Pattinson, M. R., Mccormac, A., Calic, D. and Lillie, M. (2017). Understanding susceptibility to phishing emails: Assessing the impact of individual differences and culture,pp 12-23.

Carpenter, N. C. and Berry, C. M. (2017). Are counterproductive work behaviour and withdrawal empirically distinct? A meta-analytic investigation. Journal of Management, Vol 43 No. 3, pp834-863.

Chang, L. Y. andCoppel, N. (2020). Building cyber security awareness in a developing country: lessons from Myanmar. Computers & Security, Vol 97, 101959.

Chen, H., Chau, P. Y. and Li, W. (2019). The effects of moral disengagement and organizational ethical climate on insiders' information security policy violation behaviour. Information Technology and People,Vol 32 No. 4, pp 973-994.

Cohen, T. R., Panter, A. andTuran, N. (2013). Predicting counterproductive work behaviour from guilt proneness. Journal of Business Ethics, Vol114 No. 1, pp45-53.

D'arcy, J., Herath, T. &Shoss, M. K. (2014). Understanding employee responses to stressful information security requirements: A coping perspective. Journal of Management Information Systems, Vol31 No. 2, pp. 285-318.

Detert, J. R., Treviño, L. K. andSweitzer, V. L. (2008). Moral disengagement in ethical decision making: a study of antecedents and outcomes. Journal of Applied Psychology, Vol 93 No. 2, pp 374-391.

Dharmawansa, A. D. andMadhuwanthi, R. (2020). Evaluating the Information Security Awareness (ISA) of Employees in the Banking Sector: A Case Study.

Egelman, S. & Peer, E. Scaling the security wall: Developing a security behaviour intentions scale (sebis). Proceedings of the 33rd annual ACM conference on human factors in computing systems, 2015. Pp. 2873-2882.

Fox, S., Spector, P. E., Goh, A., Bruursema, K. & Kessler, S. R. (2012). The deviant citizen: Measuring potential positive relations between counterproductive work behaviour and organizational citizenship behaviour. Journal of Occupational and Organizational Psychology, Vol 85 No. 1, pp 199-220.

Hadlington, L. (2017). Human factors in cybersecurity; examining the link between Internet addiction, impulsivity, attitudes towards cybersecurity, and risky cybersecurity behaviours. Heliyon, Vol 3 No. 7, e00346.

Hadlington, L. (2021). The "human factor" in cybersecurity: Exploring the accidental insider. Research Anthology on Artificial Intelligence Applications in Security. IGI Global.

Hadlington, L., Binder, J. andStanulewicz, N. (2021). Exploring role of moral disengagement and counterproductive work behaviours in information security awareness. Computers in Human Behaviour, Vol. 114, 106557.

Hadlington, L. and Parsons, K. (2017). Can cyberloafing and Internet addiction affect organizational information security? Cyberpsychology, Behaviour, and Social Networking, Vol20 No. 9, pp567-571.

Hair, J. F., Money, A. H., Samouel, P. and Page, M. (2007). Research methods for business. Education+ Training.Vol 49 No. 4, pp 336-337.

Hair Jr, J. F., Hult, G. T. M., Ringle, C. andSarstedt, M. (2016). A primer on partial least squares structural equation modeling (PLS-SEM), Sage publications.

Henseler, J., Ringle, C. M. andSarstedt, M. (2015). A new criterion for assessing discriminant validity in variance-based structural equation modeling. Journal of the academy of marketing science, Vol 43 No. 1, pp 115-135.

Henseler, J., Ringle, C. M. andSinkovics, R. R. (2009). The use of partial least squares path modeling in international marketing. New challenges to international marketing. Emerald Group Publishing Limited,Vol 23, pp 277-319.

Hystad, S. W., Mearns, K. J. and Eid, J. (2014). Moral disengagement as a mechanism between perceptions of organisational injustice and deviant work behaviours. Safety Science, Vol68, pp138-145.

Janicke, H., Hadlington, L., Yevseyeva, I., Jones, K. andPopovac, M. (2018). Exploring the role of work identity and work locus of control in information security awareness,Vol 81, pp 41-48.

Jeong, C. Y., Lee, S.-Y. T. and Lim, J.-H. (2019). Information security breaches and IT security investments: Impacts on competitors. Information and Management, Vol56 No. 5, pp681-695.

Kaur, J. & Mustafa, N. Examiningthe effects of knowledge, attitude and behaviour on information security awareness: A case on SME. (2013). International Conference on Research and Innovation in Information Systems (ICRIIS), 2013. IEEE, pp. 286-290.

Kelloway, E. K., Francis, L., Prosser, M. and Cameron, J. E. (2010). Counterproductive work behaviour as protest. Human resource management review, Vol 20 No. 1, pp. 18-25.

Khan, U. P. and Anwar, M. W. (2020). Cybersecurity in Pakistan: Regulations, Gaps and a Way Forward. Cyberpolitik Journal, Vol5 No. 10, pp 205-218.

Lee, K., Ashton, M. C. and Shin, K. H. (2005). Personality correlates of workplace anti-social behaviour. Applied Psychology, Vol54 No. 1, pp 81-98.

Mccormac, A., Calic, D., Parsons, K.,Butavicius, M., Pattinson, M. and Lillie, M. (2018). The effect of resilience and job stress on information security awareness. Information & Computer Security, Vol 26 No. 3, pp 277-289.

Mccormac, A., Zwaans, T., Parsons, K., Calic, D., Butavicius, M. and Pattinson, M. (2017). Individual differences and information security awareness. Computers in Human Behaviour, Vol 69,pp151-156.

Moore, C. (2015). Moral disengagement. Current Opinion in Psychology, Vol 6,pp199-204.

Moore, C., Detert, J. R., KlebeTreviño, L., Baker, V. L. and Mayer, D. M. (2012). Why employees do bad things: Moral disengagement and unethical organizational behaviour. Personnel Psychology, Vol65 No.1.pp 1-48.

Moore, C., Detert, J. R., Treviño, L. K., Baker, V. L., & Mayer, D. M. (2016). "Why employees do bad things: Moral disengagement and unethical organizational behaviour": Corrigendum. Personnel Psychology, Vol 69 (1), pp 307.

Munro, B. H. (2005). Statistical methods for health care research, lippincottwilliams&wilkins.

Nasser, A. A., Al Ansi, N. K. A. and Al Sharabi, N. A. (2020). On The Standardization Practices of the Information Security Operations in Banking Sector: Evidence from Yemen. Int. J. Sci. Res. in Computer Science and Engineering, Vol 8 No. 6. pp 8-18.

Parsons, K., Mccormac, A.,Butavicius, M., Pattinson, M. andJerram, C. (2014). Determining employee awareness using the human aspects of information security questionnaire (HAIS-Q). Computers & security, Vol 42.pp 165-176.

Robinson, S. L. (2008). Dysfunctional workplace behaviour. The Sage handbook of organizational behaviour, Vol1, pp 141-159.

Robinson, S. L. and Bennett, R. J. (1995). A typology of deviant workplace behaviours: A multidimensional scaling study. Academy of management journal, Vol38 No. 2, pp555-572.

Schatz, D. andBashroush, R. (2016). The impact of repeated data breach events on organisations' market value. Information & Computer Security,Vol 24 No. 1, pp 73-92.

Spector, P. E. and Fox, S. (2010). Counterproductive work behaviour and organisational citizenship behaviour: Are they opposite forms of active behaviour? Applied Psychology, Vol 59 No. 1, pp 21-39.

Spector, P. E., Fox, S., Penney, L. M., Bruursema, K., Goh, A. and Kessler, S. (2006). The dimensionality of counterproductivity: Are all counterproductive behaviours created equal? Journal of vocational behaviour, Vol*68 No.* 3, pp 446–460.

Von Solms, R. and Van Niekerk, J. (2013). From information security to cyber security. computersand security, Vol 38. pp 97-102.

Weatherbee, T. G. (2010). Counterproductive use of technology at work: Information and communications technologies and cyberdeviancy. Human Resource Management Review, Vol20 No. 1, pp35-44.

Whitman, M. andMattord, H. (2012). Legal, ethical, and professional issues in information security. Principles of information security (4th ed.; pp. 133–147). Boston, MA: Course Technology, Cengage Learning. Retrieved from http://www. cengage. com/resource_uploads/downloads/1111138214_259148. pdf.

Wilks, D. C. (2011). Attitudes towards unethical behaviours in organizational settings: An empirical study. Ethics in Progress, Vol 2 No. 2 pp 9-22.

Supporting Situational Awareness in VANET Attack Scenarios

Dimah Almani, Steven Furnell and Tim Muller
University of Nottingham, UK
dimah.almani@nottingham.ac.uk
steven.furnell@nottingham.ac.uk
tim.muller@nottingham.ac.uk

Abstract: The integration of sensors and communication technologies is enabling vehicles to become increasingly intelligent and autonomous. The Internet of Vehicles (IoVs) is built from intelligent vehicles that work collaboratively and interact with the surrounding environment in real time. The underlying communications infrastructure is provided by Vehicular Ad-hoc Networks (VANETs), for vehicle to infrastructure (V2I) and vehicle to vehicle (V2V) communications. The volume of autonomous vehicles (AVs) increases, as well as the level of automation for vehicles. The potential for related incidents and attacks increases as a result. A particular concern is the ability to disseminate alerts and emergency messages effectively and securely via the V2V/V2I nodes, given the diminishing involvement of autonomous vehicle users with the operation of the autonomous vehicles. With this challenge in mind, this paper investigates the issue of situational awareness for occupants in autonomous vehicles. Building from the concept of VANETs and recognised classification of automation levels, the discussion considers a range of related attack scenarios that could be encountered, each of which illustrates also contexts in which occupants may need to be made aware and take decisions in response. Consideration is then given to resulting support for situational awareness that would be required, particularly highlighting the associated requirements for user responsibility at different levels of automation. The resulting discussion serves to articulate the challenge and serves as a basis for further research to inform the mechanisms to address the resulting requirements.

Keywords: autonomous vehicles, VANET, situational-awareness, attacks, messages, communications

1. Introduction

The Internet of Vehicles (IoVs) evolves towards ever-higher levels of vehicle autonomy. Autonomous Vehicles (AVs) are the most essential entities in Vehicular Ad-hoc Networks (VANETs) that facilitate wireless communication and exchange safety messages (e.g., attacks and congestions) through vehicle- to-vehicle (V2V) and vehicle-to-infrastructure (V2I) communication. Exchanging safety messages in VANETs has different scenarios based on vehicle automation. The current Society of Automation Engineers (Inagaki and Sheridan, 2019) automation level for vehicles provides a clear breakdown of environmental monitoring, autonomy control, fallback procedures, and system limitations. However, it does not expressly classify situations where the driver should resume control of the vehicle, which relates to situational awareness (environmental monitoring).

This study investigates users' situational awareness in VANET based on different levels of automation. In a partially autonomous system, users shift to a more supervisory position. However, they have to remain situation aware as they may be called upon to retake control of the vehicle or decide if the AV cannot deal with some incidents effectively. This discussion is divided into five sections: Section 2 discusses an overview of VANET technologies, as well as the different levels automation that are possible within vehicular contexts. Section 3 then proceeds to examine different attacks scenarios in VANETs and the different elements of security that may be targeted. Section 4 represents the main contribution of the paper and considers the resulting requirements for Situational Awareness, highlighting the types of messages involved and extent to which users need to be alerted to them at different levels of automation. Finally, section 5 concludes the discussion and highlights the resulting directions for future work.

2. Background

This section provides an overview of the supporting concepts that underpin the discussion in the paper, beginning with background on nature of VANETs and then examining the level of automation

2.1 An overview of Vehicular Ad-Hoc Networks (VANETs)

A VANET is a type of mobile ad-hoc network (MANET); it is capable of spontaneous creation of a network of mobile vehicles. In VANETs, vehicles are moving wireless access nodes, providing wireless connectivity to other vehicles and users in their surroundings. Expanding this concept is the Internet of Vehicles (IoV), which vehicles turn into intelligent nodes on the road, with their storage, compute, and networking capability (Qian and

Moayeri, 2008). From an engineering perspective, the study of VANETs focuses on network infrastructure, and the vehicle is mainly considered as a node that distributes different messages (i.e., V2V or V2I).

In the IoV, each vehicle is considered an intelligent entity equipped with an efficient multi-sensor platform, computation units, communications tools, and IP-based connectivity in V2V/V2X either directly or indirectly. Additionally, a vehicle in IoVs is envisioned as a multi-communication system that enables communications between intra-vehicle components, V2V, V2I, and V2X. As (Sateesh and Zavarsky, 2020) stated, VANETs consist of On-Board Units (OBUs) and Roadside Units (RSUs). The former is installed on the vehicle to provide wireless communication with other vehicles or RSUs, while RSUs themselves are communication units located aside the road. They are connected to the application server and trusted authority (TA). The main VANET components and their purposes can be summarised as shown in Table 1.

Table 1: Outlining the main components of a VANET

VANET Component	Definition	Purpose
On-Board Unit (OBU)	A GPS-based tracking system embedded in each AV that allow the vehicles to communicate with each other and with RSU.	Retrieving the vital information. Supporting many electronic components such as resource command processor (RCP), sensor devices and user interfaces. Communicating between different RSUs and OBUs via a wireless link.
Roadside Unit (RSU)	A computation unit fixed at specific location on roads, intersections, and parking areas.	Providing the V2I connectivity. Supporting vehicle's localization. Connecting vehicle with other RSUs using different network topologies. Calculate vehicles' trajectories to avoid threats.
Trusted Authority (TA)	Controls the entire VANET processes. Only legitimate RSUs and vehicle OBUs can be registered and communicate.	Affording protection by checking the OBU ID. Detecting malicious nodes or suspicious behaviour.

In terms of data broadcasting, VANET is a tool for controlling messages. It regulates message broadcasting between vehicles to support the timely delivery of safety messages. Furthermore, VANET aims to support security on the roads during potentially dangerous scenarios such as congestion and accidents. Specifically, vehicles communicate with their neighbours to share safety messages (Mariani, 2018). It is this context that is of particular interest in this paper, as it leads to the role of the AV occupant as the target of messages.

2.2 Levels of automation

The VANET environment in which a vehicle operates is complex. For example, a vehicle or user may have to respond to confused traffic situations, such as accidents and congestion. Therefore, the ability to understand AV system and respond to any events in VANETs (even with partial information) is critical. To communicate effectively in VANET (L3-L5), users will need to share personal information over V2V or V2I. However, sending/ receiving, collecting, and storing data pose a risk to users during this process, leading to attacks and exploitation. To illustrate, AV users might need to reveal personal information to RSU or other AV users. It is similar to using internet services. However, in VANETs, the collected data will be more accurate such as user identity, IP address, video, emotional state, etc. Therefore, the large-scale collection of data makes exploiting personal information more accessible and more lucrative; hence, the attacker will find it an attractive environment to launch attacks.

VANETs have increasingly exhibited advantages and made numerous benefits for AV users. However, AV users tend to have low acceptance of autonomous systems. Forecasting the acceptance and the understanding of autonomous driving is a new topic. Furthermore, the established understanding of automated driving is constantly being updated and now automatically recognises the need for occasional user's control, even at high levels (Mutzenich et al, 2021).

As shown in Table 2, the Society of Automation Engineers (SAE) identified six automation levels (Inagaki, and Sheridan, 2019). At higher levels of taxonomy, vehicles become fully autonomous, and the passenger is not normally required to take action. However, even fully autonomous vehicles of this type sometimes require user intervention. Therefore, providing secure vehicular communication to guarantee the user-safety is the main goal in implementing VANETs where vehicles can send and receive safety messages to each other to ensure user safety (Muhammad and Al Hussein, 2021).

Table 2: Level of automation in autonomous vehicles

SAE Level		Description	Engagement level	Occupant role
0	Zero Autonomy	The driver must perform all the driving tasks.	No Automation	Driver
1	Driver assistance	An advanced driver assistance system (ADAS) assists the driver with either steering or breaking/ accelerating, but not both simultaneously.	Hands on	
2	Partial Automation	ADAS on the vehicle controls both steering and accelerating simultaneously under some circumstances. The driver must continue to control the tasks (monitor the driving environment) and performs the rest of driving task.	Hands off	
3	Conditional Automation	An Automated Driving System (ADS) performs all the driving task under some circumstances. The driver must be attention to take back control at any time when the ADS requests.	Eyes off	Passenger
4	High Automation	ADS on the vehicle performs all driving tasks and monitor the driving environment. Do all the driving- in certain circumstances. The driver needs not to pay attention in those circumstances.	Mind off	
5	Full Automation	ADS performs all the driving tasks under all circumstances, even when there is no occupant in the vehicle.	Body off	

Consideration needs to be given to the implications of the different automation levels in the event of an attack. This issue is highlighted because the user in L0-L2 is often expected to be aware of the situation and be fully responsible, whereas, in L3-L5, the situation will be different. Therefore, the user cannot respond correctly, for example, in a sudden emergency in L3-L5, where users may be responsible for having adequate situational awareness, they may have performed other tasks (social media/sleep) and therefore may not be fully conscious. However, prior to exploring this aspect further, it is firstly relevant to consider the types of attack that may occur in the VANET context.

3. Attacks scenarios in VANETs

As VANETs are open-access and self-organised networks, they are prone to potential attack. Some attacks aim to disseminate fake messages to disrupt safety-related services or misuse the VANET's communication systems, leading to various types of damage such as Denial of Service (DoS) (Poongodi et al, 2019). In this study, a focus on the attackers and their behaviour of launching attacks on VANET will be stated. Attacks are considered the most severe threats in VANETs that compromise V2V and V2I communications messages. Therefore, broadcasting emergency messages in VANETs to prevent attacks is a significant concern. For example, some VANET users need to broadly announce specific messages in real-time (e.g., emergency messages). Selecting a single trusted node to store and disseminate critical information will also be challenging. Unfortunately, users might not trust the automated driving system, preventing handing over the driving task or entirely focusing on the other task. We assert that enhancing situational awareness can increase a user's trust in Automation and lead to better decision-making. In addition, it is essential to mention VANET security requirements, as failure to meet these will lead to various vulnerabilities. At a high level, the requirements in VANETs requirements are broadly common to other areas and have been categorised into five main classes: Authenticity, confidentiality, availability, integrity, and nonrepudiation (Zubairu, 2018; Poongodi et al, 2019).

To illustrate how the security requirements may be compromised by particular attacks, Figure 1 depicts a range of scenarios that may be launched in VANET at L3-L4 automation. The scenarios are described as follows:

- **Scenario A:** This scenario depicts a Sybil attack (Figure 1A), in which the attacker (red vehicle) will have different fake identities to disrupt the standard mode of VANET operation. First, the malicious vehicle broadcasts multiple counterfeit messages. Then, the attack manipulates other vehicles' directions. For example, the attacker will broadcast congestion ahead; if the victim vehicle acts upon this, it is forced to alter its paths and exit. In this case, AV users should react quickly to confirm the received safety messages with RSU to thwart such attacks.
- **Scenario B:** In broadcast tampering, attackers' issue false safety messages in the VANET. These sometimes hide traffic threats, which can lead to situations like accidents and road congestion. As shown in Figure 1B,

the malicious vehicle will broadcast fake messages "there is no congestion ahead, and the road is clear" to mislead other vehicles to continue straight. The white front vehicle, for example, will continue proceeding then will encounter congestion. By monitoring the AV sensors, users can identify the attack, ignore the fake message, and exit the road. Moreover, users can safeguard the VANET by alerting other vehicles on the road by broadcasting a corrective message.

- **Scenario C**: This attack occurs in the middle of V2V communications (Figure 1C). The attacker checks the target vehicles closely then alters the messages between them. In this case, the attacker manipulates the V2V communications while they think they communicate privately. Under this attack, AV users must be aware of their surroundings and authenticate the source of the received messages by using cryptography techniques such as PKI.
- **Scenario D**: A masquerading attack occurs when the attacker logins into the VANET system using a stolen ID and passwords then attempts to broadcast false messages which appear to come from the registered vehicle (Bagga et al., 2020). For example, in Figure 1D, the red vehicle pretends to be police, then forces the yellow front vehicle to expose information such as an ID or a social number. In this case, AV users need to be aware of this situation and know how to react to deny revealing information to the untrusted vehicle; they also need to check the accuracy of received messages with the Trusted Authority before starting the communication.
- **Scenario E**: In this scenario, two or more vehicles share the same key. The two vehicles will not be distinguished from each other, so their actions can be repudiated. For example, as shown in Figure 1E, after the malicious vehicle (A) did the road accident, the attacker (A) will send malicious messages in VANET. As stated in Figure 1E, the same (A) yellow vehicle that caused the accident continued proceed as if nothing happened and denying the fact of sending the message in case of any dispute.
- **Scenario F:** This depicts a replay attack, in which the malicious vehicle replays the previous message's transmission to exploit its contents at the moment of transmission. As shown in Figure 1F, a malicious vehicle alters a duplicate of the received message then resends it again to the neighbouring vehicles causing further VANET incidents.

While the notion of AVs is fundamentally aimed towards reducing or removing the need for human involvement, it is still crucial to define what responsibilities users may still be required to fulfil (such as roles and duties, control transfer, operational mode, and most importantly, decision making). Security, legal and ethical responsibilities need that occupant to remain aware of the VANET situation. AV drivers will shift to a more supervisor position in such a scenario. They have to remain situation-aware as they may be called upon to make decisions or take control of the vehicle as the vehicle will be confused, and the AV system will be unable to address this situation successfully. Under this concept, the taxonomy of AVs by the National Highway Traffic Safety Administration (NHTSA) specifies that the occupant in L2 must be situation-aware at all times (Mutzenich et al, 2021). In L3, the occupant must become situation-aware and control the vehicle after a brief period. In VANETs, since the attentional demand will be low, maintaining adequate alertness will be a challenge. It is therefore necessary to assess the level of situation awareness that occupants maintain in different scenarios in VANETs. Critical decisions in some attack scenarios depend on whether the occupants are responsive and aware of external conditions. For example, based on the Sybil attack scenario (shown in Scenario A in Figure 1), a different forged identity is launched in VANETs by sending numerous incorrect messages to the neighbouring vehicles. As a result, these vehicles will leave the road to let the attacker pass the road quickly. In this case, the real identities of the sender will be hidden, and the attackers create a deception to the other vehicles. Hence, the occupant needs to be aware of such an event to verify the receiving messages and authenticate these messages with the third agent on the roadside.

As the safety messages are broadcast in an open-access system, the whole communication in VANETs disturbs if the attacker injects, alters, or blocks the messages in VANETs. Moreover, it will make VANETs more vulnerable to other attacks that may fill bogus information in the transmitted message. These challenges expose VANETs to various scenarios of dangerous attacks.

Interaction mechanisms in AVs need an appropriate design for exchanging information and function. The functions in AVs have to meet the cognitive characteristics of users and guarantee the efficiency and safety of any communication during autonomous driving (Ghazi et al, 2020). Generally, since machines and humans differ significantly in terms of their competences and capacities any design in AV must enable the user and the automation to function over a high-level guarantee of quality system performance (Petersen et al, 2019).

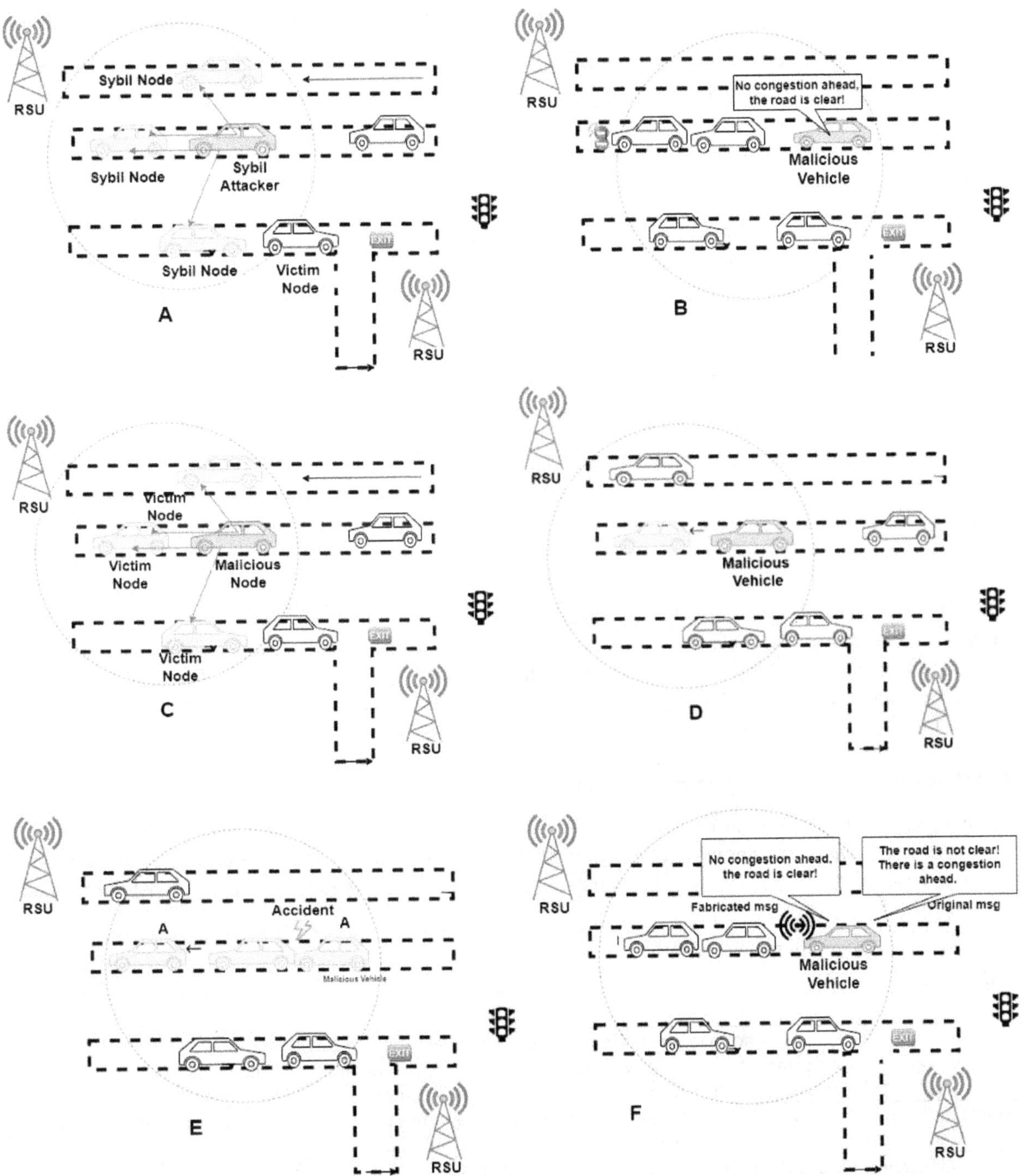

Figure 1: Scenarios where VANET vehicle is susceptible to different malicious attacks

4. Recognising situational awareness requirements in VANETs

Situations mean understanding the perceived data, starting by taking the activity, then predicting the following appropriate action based on the taken activities. It has been used widely in the aviation industry for pilots. Recently, it has started gaining notoriety in the automotive system. Situational awareness can be classified in three stages (Hashem et al, 2015) as follows:

- Perception: Extracting information as features from VANETs.
- Comprehension: Collecting all the extracted information and providing an understanding to them. This stage may take actions based on the understanding of the features.
- Projection: Predicting the future state based on the action taken in the comprehension stage.

The issue of real situations identification plays a significant role in avoiding road incidents and attacks. In V2V communications, it is suggested that 60% of the dangerous incidents could be avoided if the warning messages had been disseminated to alert neighbouring vehicles on time (Moharrum and Al-Darkish, 2012). Therefore, a key role of AV users is verifying the receiving warning message and selecting the correct next reception of warning messages in the vehicle's neighbourhood as soon as an emergency occurs. The collaboration of AV users in V2V will reduce the message delivery latency for nearby vehicles and achieve higher awareness for vehicles in the vicinity.

Studies that considered role of Situational Awareness in AV have been limited to automation levels 0-3 in which users have to take over the driving of an AV that can operate autonomously in a specific time. These have included calculating the response times of users taking control after receiving an emergency message has been studied (Banks et al, 2018), and examining the issues of complacency when users are needed to be in charge of monitoring state for prolonged periods in AVs (Larsson et al, 2014). However, there is no clear guideline for L3-L5 users explaining their responsibilities and the right direction to deal with an attack. (e.g., communications with the safety messages).

With the above in mind and to ensure occupant safety from attacks over VANETs, it is essential to ascertain the following questions across the wider range of automation levels:

- How do AV users exchange emergency messages in different automation levels under attack?
- Will AV users be aware in the event of attacks?
- What will be the information and functions required to support such activities?

To date, relatively few studies have attempted to answer the above questions. However, recent works (Liu and Parkinson, 2020) have suggested the need to analyse this concept in-depth to design an efficient framework that supports AV user awareness under any attacks over VANETs. As such, the implications of the questions are discussed in the sub-sections that follow.

4.1 AV user's role in exchanging messages in VANETs

Little research has been done in respect of clearly illustrates the AV users' role in emergencies. Therefore, understanding the types of emergency messaging in VANETs is the key to comprehending the communication to improve road safety. Generally, the communication in VANETs is classified into six types of safety messages, as shown in Table 3.

Table 3: Types of safety messages in VANETs

Message Type	Description	Comm. type	Priority
Group Communication	AV users who share the same or some vehicles features can participate in this communication. E.g., vehicles, that have the same models, or vehicles sharing the exact location in the time interval.	V2V	Low
Road Condition Warning	Nearby Vehicles exchange safety messages about the condition of the road (e.g., congestion, maintenance, closed road, etc.)	V2V	Medium
Low Connection Warning	The exchange messages contain information about the VANET connection conditions in some areas (e. g. type of wireless and the communication speed. etc.)	I2V V2V	Medium
Collision Warning	In different collision situations, safety messages are needed to be sent to a nearby vehicles to avoid further incidents and increase safety. (e.g., post and pre-crash warning.)	V2V	High
VANET Warning	The warning messages alert all the nearby vehicles about the event of any incidents (disruption, attacks) affecting the VANET. These messages can contain the security incidents features and behaviour (e.g., Vehicle colour, model, speed, velocity etc.).	V2V V2I	High
I2V Warning	The infrastructure broadcast messages via RSUs to all vehicles within its surrounding area about environmental weather and safety issues when an issue is detected.	I2V	High

Osika et al. (2017) emphasised that AV users will be unaware of their surroundings; they used a simulated AV to prove their view and explore user activity situations. Their experiment analysed some non-driving activities, including smart device usage, reading a book, and sleeping. As they found, the unsecured non-driving activities are open to all risks, which means that any corresponding driver-vehicle interaction will be subject to high levels of attacks. In addition, users have high expectations for AVs to relieve them from driving responsibilities to engage in other activities. In such a case, AV users will be unaware of their surroundings and will be less able to take the right action to face such an attack. This leads to the need to ensure user awareness at key points when it becomes relevant.

4.2 AV user's awareness in the event of attack

Despite the advantages obtained from AV in high levels of automation, the sole source of communication in VANETs is through wireless links, which are sensitive to a variety of attacks. Depending upon the implementation, VANET entities may be susceptible to third parties injecting faked messages (altering and repeating old messages). Because these messages are urgent and usually life critical. While they may no longer be performing the traditional driving task, AV users need to suitably aware the technology they use as well as how to take the appropriate action in the right time.

Warning messages in VANET are sent in broadcast mode where all the vehicles inside the coverage area of the sender should receive the message. However, as vehicles with the same area can receive the emergency messages, some vehicles from outside the area are unable to receive any alert or receive it late, which results in undesirable consequences. Therefore, research in this area suggests AV users engage in V2V/V2I communications to boost VANET safety. Under this area, many vehicle concepts have been devoted to proving how driving might change when it reaches the highest level of automation, where a user is neither needed nor, in some circumstances, even recommended to monitor the vehicle. However, based on SAE's six levels, any system short of full automation will still need driver control in some situations, and some fully automated vehicles will still recommend driver monitoring under some conditions.

While SAE Level 0 means no automation and Level 5 means full autonomy, the middle levels are rather more blurred. The SAE is clear that the first three levels (L0, L1, and L2) must be referred to as "Driver Support Systems", while L3 to L5 are actual "Automated Driving Systems". A survey conducted by Tang et al. (2020) noted that users in high automation levels (L3-L5) will primarily attend non-driving activities, such as sleep and social media. However, unlike low-level automation (L0 to L2) where there is no vehicle connectivity (no possibilities of communication attack), the high-level automation contexts require users to be aware of their surroundings to protect themselves from any attacks, as explained in Table 4.

Table 4: Levels of user awareness based on the automation levels

Automation Level	Level 3 Conditional	Level 4 High	Level 5 Full
Features	Has its own internal connectivity (i.e. connected services inside the vehicle only)	In-vehicle experience of a broad range of online services that can be ported inside and outside the vehicle	In-vehicle experience of a broad range of online services that can be ported inside and outside the vehicle
Vehicle Connectivity	Vehicles are connected partially in V2V/V2I	Vehicles are directly connected in V2V/ V2I	Vehicles are connected to everything (i.e. V2X)
User Awareness Tasks	Check the vehicle connections. Monitor the system (non-driving activities). Control the communication (V2V/V2I) Make decisions in emergency situations.	Monitor the system (non-driving activities). Control the communication (V2V/V2I) Make decisions in emergency situations.	A 'driver' is not expected to be present in the vehicle. User needs to remotely monitor the system. Control the communication (V2X) Make decisions in emergency situations.
Primary occupant role	Monitor/Passenger	Monitor/Passenger	Passenger/ Remote driver

4.3 The required information that supports the AV user's awareness

AV users need to understand AV technology to support the decision-making under any VANET threat. Furthermore, an attacker may compose these messages then compromise the user's privacy by obtaining his location. Thus, the privacy of the user must be protected from unauthorised access. Various even new types of attacks might be launched on VANET differently. The impact of these attacks depends on the intention of the attacker and the way used to perform the attack.

The different awareness and intervention requirements at different levels of automation suggest different assessments and responsibilities, depending on how the AV user is involved in the driving task. However, vehicle automation is relatively new and constantly changes, especially when automation is high (L4, L5). Therefore, there is no clear timetable for control transitions and no unified responsibility model for different levels of vehicle automation. The current model will continue to apply as long as most daily traffic remains at 0-3 levels, but AV users may need to review and reassess their roles for future accountability.

In order to raise SA in AV users, it is necessary to develop an effective human control mechanism with a well-designed user interface so that users can efficiently monitor and control AVs in when circumstances require it. However, understanding of users' activities during problem scenarios is still limited. The published data corresponding to the AV users and their responsibilities under VANET attack is insignificant. Recent fatal AV accident investigations have revealed issues such as over-reliance on systems and a low SA, resulting in poor driving handling in an emergency (Litman T, 2021). Nonetheless, it is recognised that user presence is a critical factor in driving safety. Hence, ignoring the user responsibility in the event of attacks and other events will result in undesired consequences. Moreover, the ability to take the correct actions in AV will serve to both protect the occupants and support the security of communications over the VANET more generally.

5. Conclusion and future work

Attempting to understand the issue of vehicular automation is a long-term process, and the implications and challenges arising from the different levels of automation are yet to be fully understood and resolved. Our study offers insights into the requirements to support situational awareness for users in AVs context to safeguard VANETs in the event of attacks that seek to compromise V2V/V2I communications. Until vehicles are fully autonomous and sufficiently trustworthy, a level of user engagement and responsibility is likely to required, which necessitates making them appropriately aware of the current situation. While the paper has illustrated some contexts that are likely to drive the need for awareness and identified the challenges in achieving this as the automation levels change, work is still required to establish the actual mechanisms that would support this in practice. This will therefore form one of the areas of further attention, and the authors intend to build upon this conceptual foundation as part of ongoing research.

References

Bagga, P., Das, A. K., Wazid, M., Rodrigues, J. J. P. C., and Park, Y. (2020). Authentication Protocols in Internet of Vehicles: Taxonomy, Analysis, and Challenges. IEEE Access, 8, 54314–54344. https://doi.org/10.1109/access.2020.2981397

Ghazi, M., Khan Khattak, M., Shabir, B., Malik, A. and Sher Ramzan, M., 2020. Emergency Message Dissemination in Vehicular Networks: A Review. IEEE Access, 8, pp.38606-38621

Hashem Eiza, M., Owens, T., Qiang Ni, and Qi Shi. (2015). Situation-Aware QoS Routing Algorithm for Vehicular Ad Hoc Networks. IEEE Transactions on Vehicular Technology, 64(12), 5520–5535. https://doi.org/10.1109/tvt.2015.2485305

Inagaki, T., and Sheridan, T. B. (2019). A critique of the SAE conditional driving automation definition, and analyses of options for improvement. Cognition, technology & work, 21(4), 569-578.

Larsson, A. F. L., Kircher, K., and Hultgren, J. A. (2014). Learning from experience: Familiarity with ACC and responding to a cut-in situation in automated driving. Transport. Res. F Traffic Psychol. Behav. 27, 229–237. https://doi.org/10.1016/j.trf.2014.05.008

Lee, J. M., Park, S. W., and Ju, D. Y. (2020). Drivers' user-interface information prioritization in manual and autonomous vehicles. International Journal of Automotive Technology, 21(6), 1355-1367.

Liu, N., Nikitas, A., and Parkinson, S. (2020). Exploring expert perceptions about the cyber security and privacy of Connected and Autonomous Vehicles: A thematic analysis approach. Transportation research part F: traffic psychology and behaviour, 75, 66-86.

Litman.T. (2021), Autonomous Vehicle Implementation Predictions: Implications for Transport Planning, Victoria Transport Policy Institute, 17 December 2021.

Mariani, R. (2018, March). An overview of autonomous vehicles safety. In 2018 IEEE International Reliability Physics Symposium (IRPS) (pp. 6A-1). IEEE.

Moharrum. M and Al-Daraiseh. A, Toward Secure Vehicular Ad-hoc Networks: A Survey, IETE Technical Review (Medknow Publications & Media Pvt. Ltd.), vol. 29, no. 1, pp. 80-89, Jan/Feb 2012.

Muhammad, G., and Alhussein, M. (2021). Security, Trust, and Privacy for the Internet of Vehicles: A Deep Learning Approach. IEEE Consumer Electronics Magazine, 1. https://doi.org/10.1109/mce.2021.3089880

Mutzenich, C., Durant, S., Helman, S., and Dalton, P. (2021). Updating our understanding of situation awareness in relation to remote operators of autonomous vehicles. Cognitive Research: Principles and Implications, 6(1). https://doi.org/10.1186/s41235-021-00271-8

Petersen, L., Robert, L., Yang, J., and Tilbury, D. (2019). Situational Awareness, Driver's Trust in Automated Driving Systems and Secondary Task Performance. SSRN Electronic Journal. https://doi.org/10.2139/ssrn.3345543

Poongodi, M., Vijayakumar, V., Al-Turjman, F., Hamdi, M., and Ma, M. (2019). Intrusion prevention system for DDoS attack on VANET with reCAPTCHA controller using information-based metrics. IEEE Access, 7, 158481-158491.

Qian, Y., and Moayeri, N. (2008, May). Design of secure and application oriented VANETs. In VTC Spring 2008-IEEE Vehicular Technology Conference (pp. 2794-2799). IEEE.

Rashid, S. A., Hamdi, M. M., and Alani, S. (2020, June). An overview on quality of service and data dissemination in VANETs. In 2020 International Congress on Human-Computer Interaction, Optimization and Robotic Applications (HORA) (pp. 1-5). IEEE.

Sateesh, H., and Zavarsky, P. (2020, November). State-of-the-Art VANET Trust Models: Challenges and Recommendations. In 2020 11th IEEE Annual Information Technology, Electronics and Mobile Communication Conference (IEMCON) (pp. 0757-0764). IEEE.

Sun, X., Cao, S., and Tang, P. (2021). Shaping driver-vehicle interaction in autonomous vehicles: How the new in-vehicle systems match the human needs. Applied Ergonomics, 90, 103238. https://doi.org/10.1016/j.apergo.2020.103238

Zavvos, E. Gerding, V. Yazdanpanah, C. Maple, S. Stein and m. Schraefel, Privacy and Trust in the Internet of Vehicles, IEEE Transactions on Intelligent Transportation Systems, pp. 1-16, 2021. https://doi.org/10.1109/tits.2021.3121125

Zubairu, B. (2018). Novel approach of spoofing attack in VANET location verification for non-line-of-sight (NLOS). In Innovations in Computational Intelligence (pp. 45-59). Springer, Singapore.

The Emergence of IIoT and its Cyber Security Issues in Critical Information Infrastructure

Humairaa Yacoob Bhaiyat and Siphesihle Philezwini Sithungu
Academy of Computer Science and Software Engineering, Faculty of Science, University of Johannesburg, South Africa
humairaabhaiyat@gmail.com
siphesihles@uj.ac.za

Abstract: The emergence of the Industrial Internet of Things (IIoT) can transform and improve industrial domain processes. This is achieved by IIoT's ability to collect and process vast amounts of data using technology such as sensors. IIoT capabilities can improve the manufacturing processes of these sectors and contribute to the improved functioning of critical information infrastructure. In addition, current trends - such as the Fourth Industrial Revolution (4IR) - use IIoT to realise specific goals. While the emergence of IIoT systems does introduce many benefits, such as improved efficiency and sustainability, it can also introduce security concerns. These security concerns pose a significant threat to the industrial domain, including critical information infrastructures. The resulting threats emphasise the need to implement solutions to secure IIoT systems. The paper aims to discuss the emergence of IIoT and its cyber security issues within the context of critical information infrastructure. The research paper follows a theoretical research methodology to provide an improved understanding of the emergence of IIoT and its cyber security issues in critical information infrastructure. The paper contains an exhaustive discussion of what is IIoT. A discussion on where IIoT fits within the context of critical information infrastructure and its impact on 4IR is also highlighted in the paper. Due to the many vulnerabilities that IIoT systems can contain, the paper also discusses security concerns surrounding the emergence of IIoT. The security concerns make IIoT systems attractive targets for cyberattacks. Therefore, different approaches that can be applied to secure IIoT systems is also provided. Since IIoT capabilities can impact the critical information infrastructure of businesses and nations, the authors' stance on how IIoT systems could transform the current understanding of critical information infrastructure is also discussed.

Keywords: internet of things, industrial internet of things, critical information infrastructure, fourth industrial revolution, security concerns

1. Introduction

The Industrial Internet of Things is an emerging field that can assist in improving the industrial sector by expanding manufacturing, agriculture, and military processes (Jaidka, Sharma & Singh, 2020). The emergence of IIoT systems can also impact both critical information infrastructure (CII) and the 4IR. The role that IIoT plays within CII and the 4IR can benefit these two domains through its ability to collect and process vast amounts of data (Serpanos & Wolf, 2017). However, due to the increased use of IIoT systems, there are security concerns surrounding the emergence of IIoT systems. These security concerns emphasise that security approaches that protect the integrity of IIoT devices must be designed and implemented (Sadeghi, Wachsmann & Waidner, 2015). The abilities and security concerns surrounding IIoT can transform the ability of CIIs to manage and control CI.

This research paper aims to discuss the emergence of the IIoT and its cyber security issues within the context of CII. The paper is structured in the following manner to achieve this objective: Section 2 discusses IIoT. Section 3 provides a discussion on the role IIoT can have on CII. Section 4 highlights the impact that the IIoT can have on the Fourth Industrial Revolution. Section 5 discusses security concerns surrounding the emergence of IIoT. Section 6 discusses different approaches that can help secure IIoT systems. Section 7 discusses the authors' view on how IIoT could transform CII. The last section, Section 8, concludes the paper.

2. What is the industrial internet of things?

The concept of the Internet of Things (IoT) and IIoT are closely related. However, there are differences between them (Sisinni, Saifullah, Han, Jennehag, & Gidlund, 2018). To understand IIoT, one should first understand the difference between IoT and IIoT. The basic concepts between IoT and IIoT are similar: interconnected smart devices that perform data collection, remote sensing, processing, monitoring and control (Serpanos & Wolf, 2017). However, the focus of IoT is human-centered (i.e., smart consumer devices that are interconnected with one another to make people's lives easier in terms of saving time and money) (Sisinni et al., 2018). In comparison, IIoT's focus is safety and operation, which includes the operational technology used in the industrial sector (Serpanos & Wolf, 2017). IIoT combines information and Communication Technology (ICT) trends with industrial production systems (Arnold, Kiel & Voigt, 2016). IIoT can also be seen as a subset of IoT that operates in the industrial sector (Serpanos & Wolf, 2017). The main goal of IIoT is to integrate industrial control systems, analytics, business processes, and enterprise systems (Bajramovic, Gupta, Guo, Waedt & Bajramovic, 2019).

IIoT can also be described as communication between machines. These machines communicate and interact with other objects and machines (Simon, 2017). This communication can lead to optimal industrial operations because IIoT can detect failures. IIoT systems can activate the manufacturing process, obtain information from different objects and sensors and send readings to the cloud-based centres (Sisinni et al., 2018; Wan, Tang, Shu, Li, Wang, Imran & Vasilakos, 2016). Sensors are key IIoT technologies as they produce different kinds of data, which must be precise and is usually predictive. Other key technologies are Big Data and advanced analytics for predictions, historical analysis, and insight on machines and processes (Gilchrist, 2016). IIoT can also transform businesses by improving worker safety, worker productivity, sustainability, customer experience, reducing operational costs, and creating new revenue streams (Simon, 2017). These capabilities have resulted in the emergence of IIoT in CII. The following section discusses where IIoT fits within CII.

3. Where does Industrial IoT fit within critical information infrastructure?

IIoT can be applied to different areas, but one of the main areas that can benefit from the applications of IIoT is critical infrastructure (Mcginthy & Michaels, 2019). Critical infrastructures are managed and controlled through CIIs. CIIs manage and control other systems that provide essential economic services, such as gas utilities, electrical power grids, air transportation, and many other crucial systems (Lopez, Setola & Wolthusen, 2012). Furthermore, CII systems are controlled by remote systems, also known as Industrial Control Systems (ICS), such as supervisory control and data acquisition (SCADA) systems. Therefore, IIoT can be integrated within CII to improve efficiency in critical areas (Mcginthy & Michaels, 2019).

IIoT can lead to efficient management by intelligently processing and analysing the huge amounts of data obtained from communications between the different machines in critical areas such as transportation, energy, medical, etc. (Simon, 2017). IIoT can do this because wireless sensor networks and IIoT allow for the sensors to be placed in remote locations that provide data to the central system. This can be done across many CII sectors (Mcginthy & Michaels, 2019). However, considering the application of IIoT to CIIs, a failure in IIoT can lead to life-threatening situations (Magomadov, 2020). Examples of IIoT applied in CIIs would be automated vehicles that use sensors to reduce the exposure of noise, hazardous gases, and chemicals to workers involved in industries such as gas and oil (Agenda, 2015).

Nations such as the United Kingdom have started integrating IIoT in their CII. Some providers of wastewater services and drinking water in the United Kingdom use a combination of IIoT sensors, real-time data, and analytics to determine and anticipate equipment failures and respond quickly to emergencies such as water leakage (Simon, 2017). These examples show that IIoT can augment CII to provide automation to optimise operations. The optimisation of IIoT can also impact 4IR. The following section discusses the impact of IIoT on 4IR.

4. Impact of Industrial IoT on the fourth industrial revolution

IoT and IIoT can assist in understanding the impact of the Fourth Industrial Revolution (4IR), also known as Industry 4.0 (Mcginthy & Michaels, 2019). The Fourth Industrial Revolution can be described as digitalising the manufacturing sector by embedding sensors in virtually all cyber-physical systems, manufacturing equipment and product components. 4IR also includes analysing the data generated by sensors for effective decision-making (Lampropoulos, Siakas & Anastasiadis, 2019). For this reason, it is believed that the Fourth Industrial Revolution began with the inception of IIoT (Jaidka, Sharma & Singh, 2020). IIoT achieves key features of 4IR, such as horizontal integration. IIoT achieves horizontal integration through value networks for supporting companies' business strategies. Other key features include vertical integration and integrating the digital with the real world in the value chain, thus achieving end-to-end integration (Pivoto, de Almeida, da Rosa Righi, Rodrigues, Lugli, & Alberti, 2021).

IIoT is also seen as one of the drivers of 4IR. IIoT, automation, and the digitalising of the industrial manufacturing sector are believed to be drivers that initiated 4IR. IIoT in 4IR can also improve and change current industries by improving productivity, reducing costs and wastage, digitalising production, and creating production systems that are flexible, adaptable, agile, and interoperable (Lampropoulos et al., 2019). Although IIoT can positively impact 4IR, there are significant challenges, especially in terms of security. The following section looks at some of the major security concerns related to the emergence of IIoT.

5. Security concerns surrounding the emergence of IIoT

Since IIoT applications connect sensors, machines, and actuators in critical industries such as power grids, there are concerns that a security breach in IIoT applications can lead to devastating effects (Mumtaz, Alsohaily, Pang, Rayes, Tsang & Rodriguez, 2017). The emergence of IIoT security concerns is discussed in the following sub-sections.

5.1 Increased connectivity

The increased connectivity of IIoT systems creates several attack surfaces (Sadeghi, Wachsmann & Waidner, 2015). This is due to many connected technologies such as sensors deployed in the industrial sector, creating many access points for attackers to exploit. Therefore, IIoT systems increase the risk of exposure to cyberattacks for critical infrastructure (Paez & Tobitsch, 2017). IIoT systems are subject to physical attacks, such as hardware attacks, reverse engineering attacks, and side-channel attacks. IIoT software can also be compromised by runtime attacks, viruses, and Trojans. While communication protocols also have a risk of protocol attacks such as Distributed Denial of Service attacks (DDOS) and man-in-middle attacks (Koushanfar, Sadeghi & Seudie, 2012). IIoT applications are challenging to monitor daily, which is one of the reasons why there are several security vulnerabilities. Additionally, it is a challenge to notify users when there is a security breach, which can cause the breach to continue for an extended period without being detected (Paez & Tobitsch, 2018).

The interoperability capabilities of IIoT technologies are another security concern. Interoperability increases the scope of a data breach because the open and connected IIoT technologies make it easier for the damage from cyber-attacks to spread to other devices connected to the same network (Paez & Tobitsch, 2018). Many attacks against CI and CII have occurred in the past. An example is a cyberattack against a German Steel Mill. The attackers gained access to the CII systems by targeting the ICS devices through a phishing email. They further manipulated other systems on the network, which resulted in a system failure. Furthermore, the attackers caused physical destruction on the systems and prevented a safe shutdown on a blast furnace (Bajramovic et al., 2019).

5.2 Huge amounts of data

Another security IIoT security concern is the enormous amounts of generated data. IIoT applications connected to critical industries generate vast amounts of data processed in real-time. While this is an impressive ability of IIoT, there is a concern with how this data is stored (Yu & Guo, 2019). This data is often stored on cloud platforms, and with decreasing cost of cloud platforms, there is a broader number of IIoT applications storing their data on these cloud platforms. Storing this confidential data makes it attractive for cybercriminals to breach and get access to this sensitive data (Paez & Tobitsch, 2018). A data breach in these IIoT applications could cause devastating effects in the CIIs such as electrical power systems. Additionally, this data has to be often transferred from sensors to cloud platforms, introducing new cyber risks since attackers can also breach the data during transmission (Yu & Guo, 2019).

5.3 Legacy systems

Legacy systems are a serious concern for security in IIoT. Companies still use older systems known as Legacy systems because replacing them with new systems is disruptive and costly (Paez & Tobitsch, 2018). These companies then add a layer of IIoT applications on these Legacy Systems they use. Implementing IIoT devices on Legacy systems is a concern because attackers can find new ways to attack them (Bajramovic et al., 2019). Legacy systems are not designed for connectivity, and they are harder to secure (Paez & Tobitsch, 2018).

5.4 ICS security

Before IIoT devices were integrated into CII, most ICSs were isolated from the enterprise IT infrastructure (Yu & Guo, 2019). Merging CII with IIoT allowed ICSs to handle sophisticated environments (Bajramovic et al., 2019). The role of ICSs expanded from control and safety to providing processed information or responding to instructions from enterprise systems such as enterprise resource planning (ERP). However, the increased IIoT integration with ICSs creates cybersecurity risk concerns. IIoT exposes ICSs to cyber risks because of its highly interconnected networks (Yu & Guo, 2019). These risks make finding appropriate and secure ways to integrate IIoT and ICS challenging (Bajramovic et al., 2019). As such, the following section discusses some approaches to secure IIoT systems.

6. Approaches for securing IIoT systems

Due to the number of security concerns of IIoT devices, several different approaches can be implemented to secure IIoT applications. Such approaches are discussed in the following sub-sections.

6.1 Data confidentiality protection

As mentioned in Section 5.2, the huge amounts of generated data can be breached during transmission and storage. Therefore, a solution to this issue would be data encryption, where the data is converted into a ciphertext for transmission and storage (Yu & Guo, 2019). However, traditional encryption algorithms cannot be applied because this would require downloading and decrypting the entire dataset (Spathoulas & Katsikas, 2019). As a result, more flexible encryption approaches have been proposed in the literature. He, Ma, Zeadally, Kumar, and Liang (2017) proposed a certificateless public key authenticated encryption with keyword search (CLPAEKS) for the Industrial Internet of Things where users can search for keywords in the ciphertext while also preserving privacy. The proposed CLPAEKS scheme specifically created for IIoT allows the data owner to encrypt a keyword and authenticate it. This means that an attacker cannot encrypt the keyword without the owner's private key, thus protecting the privacy and integrity of the data (He et al., 2017).

Traditional encryption approaches also suffer from a single point of failure. If the secret key is compromised, data confidentiality will also be compromised (Yu & Guo, 2019). A proposed approach by Mahalle, Prasad and Prasad (2014) solves this problem. The authors proposed threshold cryptography in IIoT systems where the key is divided into parts stored at different locations. This eliminates the single point of failure with the keys.

6.2 Network segmentation

Due to the connectivity of IIoT systems as mentioned in section 5.1. attackers could use other devices on the network to gain access to the industrial settings. A solution to prevent this would be to ensure that the IIoT systems are segregated in the network. This can be done where devices and sensors that manage other SCADA devices are on a separate network and not the same network as the IT infrastructure (Bajramovic et al., 2019).

6.3 Detection of attacks

It is important to implement approaches for detecting attacks to ensure the security of IIoT systems. Spathoulas and Katsikas (2019) describe using CNNs, Convolutional Neural Networks. CNNs can process huge amounts of data to detect anomalies in the functioning of industrial systems.

Another solution is the use of the Squeezed Convolutional Variational AutoEncoder (SCVAE) model. This solution makes use of low processing power, and it can detect the abnormal states of industrial systems (Kim, Yang, Chung, Cho, Kim, Kim, Kim, Kim, 2018). Additionally, SCVAE is a useful solution because it can also detect cyberattacks without using cloud solutions. Instead, the detection process takes place locally on the IIoT/ IoT network (Spathoulas and Katsikas, 2019).

6.4 Cyber-physical systems integrity

Since most IIoT systems are Cyber-Physical systems, the IIoT systems need to support the integrity verification of the Cyber-physical systems. A solution to this would be the use of attestation (Yu & Guo, 2019). Attestation is where the device that needs to be verified – called the prover – sends a status report of its software configuration to another device – called the verifier (Sadeghi, Wachsmann & Waidner, 2015). The prover does this to prove that it is in a trustworthy state. Although malicious software could forge the status report, the authenticity can be assured by trusted software and secure hardware (Yu & Guo, 2019).

6.5 Training employees

Most cyber-attacks are caused by human error. Therefore, to ensure that IIoT systems are secure, it is vital for the employees using IIoT systems to be trained in cybersecurity before gaining access to the IIoT system

(Bajramovic et al., 2019). The following section discusses how IIoT could transform our current understanding of CII.

7. How the Industrial IoT could transform critical information infrastructure

IIoT could change the way we understand Critical Information Infrastructures. The role of CIIs will no longer focus on managing and controlling CI. Instead, this role can expand where CII can also provide processed information. Additionally, IIoT could provide new ways for CIIs to control and manage CIs. For example, IIoT combined with CIIs could allow unmanned aerial vehicles to check oil pipelines to minimise workers' exposure to harmful chemicals in industries such as gas and oil (Agenda, 2015).

IIoT could also provide significant benefits for CII in the long run. From what was discussed in Section 3 paragraph 3, IIoT's use of real-time data can benefit CII. The data generated by IIoT devices can provide insights into the different CIs, such as power plants. Additionally, IIoT could transform how we think about CII processes, such as cost reductions, safety improvements, and increased efficiency. These potential benefits of IIoT could improve service delivery within nations.

Finally, IIoT could also transform the way we implement security for CII. Before the inception of IIoT, CII systems such as ICSs were thought of as isolated systems, and the connectivity brought by IIoT introduces new cyber security risks (Yu & Guo, 2019). This means that security measures previously applied to CII must change to accommodate the integration of IIoT. However, integrating ICSs with IIoT could also provide other challenges, such as increased research and development costs (Simon, 2017). This means that IIoT and CII integration could be challenging to implement due to the need for large investments to help initiate the integration.

8. Conclusion

Several topics were discussed in the aim to achieve the objective of this paper. The main critical points discussed were (1) understanding what IIoT means, (2) discussing where does IIoT fit within CII, (3) highlighting the impact of IIoT on 4IR, (4) discussing security concerns of IIoT in CII, (5) providing a few approaches to securing IIoT systems and (6) discussing how IIoT could transform CII.

While addressing the first critical point, the paper noted that IIoT can be seen as a subset of IoT that focuses on the industrial sector. Additionally, its use of sensors and other devices to obtain real-time operational data has shown that it can transform businesses. The second critical point could be seen in IIoT's ability to be integrated with CII. IIoT provides several benefits to ensure that CII can manage and control CI. IIoT improves CII efficiency, and cases of the integration of CII with IIoT have already been observed in nations such as the United Kingdom (Simon, 2017). To understand the third critical point, the paper noted that IIoT can be seen as one of the main drivers behind 4IR, and it meets the key 4IR requirements since the main aim of 4IR is digitalising the manufacturing industry.

It is important to note that security concerns have also been raised with the emergence of IIoT. The paper discussed this critical point by highlighting that the connectivity brought by IIoT will bring about new cyber risks. Additionally, the data collection and processing that comes with IIoT could create major concerns if the data were to be breached since it could contain vital information about a nation/business. The fifth critical point discussed in this paper highlighted a few approaches for addressing the noted security concerns. The approaches discussed were network segmentation, cyber-physical systems integrity and encryption. Finally, the paper discussed important aspects regarding how IIoT could transform our current understanding of CII. IIoT's ability to change the scope of CII, its security approaches, and providing additional benefits has shown that IIoT can positively transform CII.

References

Agenda, I. (2015) Industrial internet of things: unleashing the potential of connected products and services, World Economic Forum.

Arnold, C., Kiel, D. and Voigt, K.I. (2016) "How the industrial internet of things changes business models in different manufacturing industries", International Journal of Innovation Management, Vol 20, No. 8, pp 1-20.

Bajramovic, E., Gupta, D., Guo, Y., Waedt, K. and Bajramovic, A. (2019) "Security challenges and best practices for IIoT", INFORMATIK 2019 Workshops, Lecture Notes in Informatics (LNI), Gesellschaft für Informatik, Bonn.

Gilchrist, A. (2016). "Introduction to the industrial internet". Industry 4.0 , Apress, Berkeley, CA.

He, D., Ma, M., Zeadally, S., Kumar, N. and Liang, K. (2017) "Certificateless public key authenticated encryption with keyword search for industrial internet of things", IEEE Transactions on Industrial Informatics, Vol. 14, No. 8, 9 November, pp 3618-3627.

Jaidka, H., Sharma, N. and Singh, R. (2020) "Evolution of IoT to IIoT: Applications & challenges". Proceedings of the International Conference on Innovative Computing & Communications (ICICC),

Kim, D., Yang, H., Chung, M., Cho, S., Kim, H., Kim, M. and Kim, E. (2018). Squeezed convolutional variational autoencoder for unsupervised anomaly detection in edge device industrial internet of things. In 2018 international conference on information and computer technologies (icict), Dekalb, IL, USA.

Koushanfar, F., Sadeghi, A. R. and Seudie, H. (2012) "Eda for secure and dependable cybercars: Challenges and opportunities", Proceedings of the 49th Annual Design Automation Conference, June.

Lampropoulos, G., Siakas, K. and Anastasiadis, T. (2019) "Internet of Things in the context of Industry 4.0: An Overview". International Journal of Entreprenuerial Knowledge, Vol. 1, No. 7, June, pp 4-19.

Lopez, J., Setola, R. and Wolthusen, S. D. (2012) "Overview of critical information infrastructure protection", Critical Infrastructure Protection, Springer, Berlin, Heidelberg.

Magomadov, V. S. (2020) "The Industrial Internet of Things as one of the main drivers of Industry 4.0", IOP Conference Series: Materials Science and Engineering, Vol. 62, pp 1-4.

Mahalle, P. N., Prasad, N. R. and Prasad, R. (2014) "Threshold cryptography-based group authentication (TCGA) scheme for the Internet of Things (IoT)", 2014 4th International Conference on Wireless Communications, Vehicular Technology, Information Theory and Aerospace & Electronic Systems (VITAE), Aalborg, Denmark, October.

Mcginthy, J.M. and Michaels, A.J. (2019) "Secure industrial Internet of Things critical infrastructure node design", IEEE Internet of Things Journal, Vol. 6, No. 5, 5 March, pp 8021-8037.

Mumtaz, S., Alsohaily, A., Pang, Z., Rayes, A., Tsang, K. F. and Rodriguez, J. (2017) "Massive Internet of Things for industrial applications: Addressing wireless IIoT connectivity challenges and ecosystem fragmentation", IEEE Industrial Electronics Magazine, Vol. 11, No. 1, 21 March, pp 28-33.

Paez, M. and Tobitsch, K. (2017) "The Industrial Internet of Things: Risks, Liabilities, and Emerging Legal Issues", Exploring the Things in the internet of Things: Implications For Business, Consumers, and the Law, Vol. 62, No. 2, pp 217-247.

Pivoto, D. G., de Almeida, L. F., da Rosa Righi, R., Rodrigues, J. J., Lugli, A. B. and Alberti, A. M. (2021) "Cyber-physical systems architectures for industrial internet of things applications in Industry 4.0: A literature review", Journal of Manufacturing Systems, Vol. 58, January, pp 176-192.

Sadeghi, A. R., Wachsmann, C. and Waidner, M. (2015) "Security and privacy challenges in industrial internet of things", 2015 52nd ACM/EDAC/IEEE Design Automation Conference (DAC), IEEE, San Francisco, CA, USA, July.

Serpanos, D. and Wolf, M. (2017) "Industrial Internet of Thing", Internet of Things (IoT) Systems, Springer, Cham.

Simon, T. (2017) "Chapter seven: Critical infrastructure and the internet of things", Chapter seven: Critical infrastructure and the internet of things, Cybersecurity in a volatile world.

Sisinni, E., Saifullah, A., Han, S., Jennehag, U. and Gidlund, M. (2018) "Industrial internet of things: Challenges, opportunities, and directions", IEEE transactions on industrial informatics, Vol. 14, No. 11, 2 July, pp 4724-4734.

Spathoulas, G. and Katsikas, S. (2019). "Towards a Secure Industrial Internet of Things", Security and Privacy Trends in the Industrial Internet of Things, Springer, Cham.

Wan, J., Tang, S., Shu, Z., Li, D., Wang, S., Imran, M. and Vasilakos, A. V. (2016) "Software-defined industrial internet of things in the context of industry 4.0", IEEE Sensors Journal, Vol. 16, No. 20, 10 May, pp 7373-7380.

Yu, X., and Guo, H. (2019) "A survey on IIoT security", 2019 IEEE VTS Asia Pacific Wireless Communications Symposium (APWCS), Singapore, September.

Including Human Behaviors Into IA Training Assessment: A Better Way Forward!

Henry Collier
Norwich University, Northfield, USA
hcollier@norwich.edu

Abstract: Few can argue against the reality that humans are the weakest link in cybersecurity, and Social Engineers work very hard to take advantage of this human weakness. Many cybersecurity practitioners believe the only way to solve this problem is through a technical solution; however, this solution is elusive because humans are still in control and can circumvent these technical measures. In cybersecurity, the human is the critical component of the human firewall, and it is going to take a multi-disciplinary approach to solve the human problem. The human firewall is the first line of defense for cybersecurity. Historically, the primary solution to the human problem has been the Information Awareness training program, designed to teach the end-user about the risks and assess their risk. The biggest problem with the information awareness training program is that it does not modify behavior. Cybersecurity practitioners need to understand better the human firewall and how it can be strengthened. It is necessary to understand how the human makes security-minded decisions, how these decisions affect the cybersecurity decision-making process, and if there is a way to assess a person's susceptibility level more precisely when working to strengthen the human firewall. Humans are multifaceted, complex beings influenced by both internal and external factors. The most significant internal factor that affects a person's decision-making process is behavior, while Social Media is one of the most significant external factors that impact a person's decision-making capacity. This study presents a new method of assessing a person's susceptibility to cybercrime by including behavioral and social media usage factors into a Dynamic/Adaptable information awareness training assessment tool. This study shows that including human behaviors and social media usage behaviors into an Information Awareness (IA) training assessment tool produces a more precise measure of a person's accurate susceptibility level.

Keywords: cybersecurity, human behaviors, susceptibility, social engineering, information awareness

1. Introduction

Cybersecurity incidents can be classified as either technical or non-technical attacks (McIlwraith, 2006) (Schroeder, 2017). The critical difference between a technical and a non-technical attack is how the threat actor works to access the data on the computer or network system. The technical attack targets a weakness in a system, whereas the non-technical attack targets the human using the system (Ariu et al., 2017). Both forms of cyberattack commonly have one of two desired outcomes, either gaining the attacker profit or causing the target to suffer a service disruption (McIlwraith, 2006) (Schroeder, 2017) (Hallas, 2018). Technical attacks are conducted by malicious outsider threats seeking to take advantage of a technical weakness within a system. The non-technical attacks, also known as social engineering attacks, typically target the individual and work to get the individual to allow the malicious threat actor into their system, making the user a non-malicious insider threat (Ariu et al., 2017). Most social engineering attacks have a technical component to them, but their primary attack surface is the person, not the equipment (Ariu et al., 2017)(Hallas, 2018) (Collier, 2020)(Hadnagy, 2018). Many individuals in the cybersecurity industry want to find a technical means of preventing social engineering attacks. Unfortunately, technical defences can only go so far in preventing social engineering attacks because the attack surface is the human, and humans are complex, multifaceted beings that behave in ways that cannot always be predicted (McIlwraith, 2006)(Schroeder, 2017) (Hadnagy, 2018)(Ayates & Conolly, 2003). For a technical measure to be effective, it needs to apply logic to a situation consistently and manipulate the outcome based on this logic; humans muddle this logic by being non-logical beings.

There is no technical measure that can stop a human from making a poor information security decision, and therefore more needs to be done to mitigate the risk the human creates (McIlwraith, 2006)(Schroeder, 2017 (Wilcox & Bhattacharya, 2019) (American Psychological Association, 2021). The only way to mitigate this problem is to conduct research around the human firewall and better understand the decision-making process, cognition, and what influences a person's decision-making process.

If humans were entirely logical beings and decisions were made logically, without emotion, the risk posed would be significantly lower than it currently is. The human decision-making process is not entirely logical; rather, it is affected by various behaviours that impact the person's mindset and successively bear upon the decision's effectiveness (Ayates & Conolly, 2003) (Bada et al., 2019). The current method of defending against social engineering attacks is the information awareness (IA) training and assessment program. The IA program is designed to provide the end-user information about cyber risks and then assess the person's risk to an organization (Schroeder, 2017). The problem with this method is that the current information awareness training and assessment model does not consider the behavioural aspect of the person being assessed and therefore does not calculate an accurate measure of a person's risk (McIlwraith, 2006)(Ayates & Conolly, 2003). Current IA training assessments consist of a multiple-question quiz at the end of training. Unfortunately, many of the answers to these questions are readily available on the Internet. Many individuals will seek out these answers to help ensure they meet the minimum grade requirement of 70% (Schroeder, 2017). This process shows that most end users will have a similar susceptibility level calculated for them (Collier, 2021).

This paper presents the results of a mixed qualitative and quantitative study that investigated how human behavioural traits and social media usage behaviours could be used to measure a person's susceptibility level more accurately. This study consisted of a comparative analysis that compared the KnowBe4 information awareness training and assessment program results and a new Dynamic/Adaptable IA training assessment tool. The Dynamic/Adaptable IA training assessment tool consisted of information security questions and behavioral/social media usage questions making the tool's approach new and unique. The KnowBe4 tool is one of the top IA training and assessment programs currently available, and as of the first quarter of 2021, KnowBe4 has over 39,000 customers worldwide (KnowBe4, 2021). In no way is this paper meant to denigrate the KnowBe4 tool; instead, this paper intends to show a different way to assess a person's susceptibility that could prove fruitful in the cybersecurity industry.

2. Humans and their behaviours

Every end-user is a complex and multifaceted human being that holds the key to the success of an organization's information security posture. End users make decisions every day to help protect an organization from cyber threats or open the door to vulnerability and compromise (McIlwraith, 2006). On the surface, the decisions themselves appear to be simple; however, each decision is almost as complex as the person making the decision (Mackinnon & Wearing, 1980). A person's knowledge and logic certainly impact decisions, but they are also influenced by the person's behaviours which significantly increase the intricacy of the decision (Edwards, 1954)(Henderson & Nutt, 1980). Behaviour is the response of an individual to stimuli that can be external or internal (American Psychological Association, 2020). Research has shown that human behaviour results from various interconnected factors that work alone and together, to influence how a person responds to stimuli (UK Parliment, 2020). Understanding which behaviours impact the information security decision-making process is vital as efforts to improve the human firewall continue.

This study was based on the premise that behaviours could be used to improve information awareness training assessment. For this to occur, it was necessary to identify which behaviours have a link to susceptibility. It was essential to identify a list that could be used as a starting point to make this identification. A list of behavioural traits was obtained from an earlier study conducted by MIT's "List Man," Peter Gunkel called The Human Kaleidoscope. Gunkel (Gunkel, 1998) developed a list of 638 traits for his study, where he was looking to see if there were limits on human variability, diversity, and psychogenesis. The evaluation of the behaviours was purely qualitative and did leave room for improvement. The intent of this study was not to identify the behaviours related to susceptibility with absolute mathematical certainty but rather to determine if it is possible to better assess a person's susceptibility by including behavioural factors in the evaluation process. The extensive list of traits was reduced to a shortlist that could be more practically incorporated into an information security assessment tool. A small focus group consisting of a cybersecurity professional, two undergraduate psychology students, and a graduate psychology student was created for this study to assess the traits and specify which traits likely impact a person's susceptibility level. The focus group started with 234 positive traits, 292 negative traits, and 112 neutral traits. Tables 1, 2, and 3 are examples of the three groups of traits from the study.

Table 1: Positive trait example (Collier, 2021) (Collier, 2020)

Accessible	Calm	Dedicated	Fair	Gallant

Table 2: Negative trait example (Collier, 2021) (Collier, 2020)

Aimless	Desperate	Egocentric	Faithless	Gullible

Table 3: Neutral trait example (Collier, 2021) (Collier, 2020)

Absentminded	Busy	Competitive	Emotional	Emotional

The trait assessment process resulted in a revised list of 128 behavioral traits that more closely represented traits that impact a person's decision-making process based on the trait's characteristics. The 128 traits were chosen based on the likelihood that the trait would not only influence the decision-making process but more significantly result in a poor information security decision.

The 128 traits were then grouped based on how similar the traits were. An example is grouping traits like arrogant, conceited, and egocentric, which represent a state of mind where a person is likely overconfident in their skills. Someone who is overconfident is at risk of becoming a victim of social engineering because they do not believe it will ever happen. Not being aware of the risk of becoming a victim of social engineering could lead to the person becoming a victim of social engineering by making a poor information security decision.

After grouping the list of behavioural traits, the next step was to determine a set of questions that would reflect the underlying behaviour that could lead to a cybersecurity incident. The focus group again took on this qualitative task and developed forty questions related to the eighteen groups of traits. An example of the questions and the associated traits is in Table 4. As can be seen, if a person answers yes to the question, then they likely meet the definition of at least one of the traits. For example, the trait Trusting implies that the person exhibiting the trait tends to believe in someone's honesty or sincerity. Social engineers work very hard to take advantage of their victim's trusting nature (Luo et al., 2011) (Mitnick & Simon, 2002). From an information security perspective, trust needs to be earned, not fully awarded, when you first meet someone. The concept of "trust but verify" promotes the idea of believing what you have been told is accurate but also questioning what you were told and working to verify its truthfulness, rather than simply accepting it blindly as the truth. Information security professionals would rather have an end-user question something than merely take it at face value. In contrast, social engineers would rather have the end-user take it at face value and become their victim.

Table 4: Behavioural question and trait example (Collier, 2021) (Collier, 2020)

Question: Do you believe that people are generally good and trustworthy?
Traits: Gullible, Impressionable, Imitative, Trusting, Dependent, Guileless, Naïve, Soft.

Social engineers develop attacks that target a person's emotions or trigger a particular behavior. For example, social engineers will build a level of urgency in their attacks to trigger an immediate emotional response rather than a thought-out response. A form of attack known as CEO-Fraud is an example whereby an attacker pretends to be an executive of an organization and sends an email to a subordinate ordering them to send money straight away to cover an overdue invoice (Kemp, 2016). The sense of urgency that the attacker imposes in their email prompts the employee to send money to the account information in the phishing email, rather than simply checking with the executive, even when the executive may be in the next room (Kemp, 2016). If the concept of "trust but verify" is applied to this type of attack, the attack's success rate would be significantly reduced, if not eliminated.

The 21st century adds a new level of complexity to the end user's behaviour that has not existed before—Social Media. Social Media was developed to allow individuals to remain connected with others in their familial or social circles, especially when physically displaced. The concept of Social Media supports the social nature of humanity. Unfortunately, social engineers have embraced Social Media and work to use the data held within its digital walls as a goldmine of information (Hallas, 2018) (Collier, 2020). Social influences impact an individual's behaviours, which further impact the security-minded decision (UK Parliment, 2020). There is a constant barrage of social forces bombarding people through Social Media. The continuous connection to Social Media does not allow a person to digest what is happening and then make a conscious decision but instead promotes the instant, knee-jerk reaction that is frequently influenced by emotions and behavioral traits. Add to this how Social Media allows people to connect with others worldwide without ever actually meeting them. Before Social Media, the idea of a group of friends might include 10-20 individuals. Still, with the introduction of Social Media, this group of people called "friends" increases and can do so exponentially. Now, it is possible to have hundreds, thousands, and even millions of so-called friends following you on Social Media (Vishwanath, 2014) (Kayes & Iamnitchi, 2017). This behaviour opens the door to social engineers who will connect with you on social media to work to develop more directed social engineering attacks (Hallas, 2018) (Vishwanath, 2014) (Kayes & Iamnitchi, 2017) (Tsikerdekis & Zeadally, 2014). Before the invention of Social Media and Social Media scraping tools, the reconnaissance process was very labour-intensive, so a social engineer would be limited to targeting a small number of people at a time (Postnikoff & Goldberg, 2018). Now social engineers can gather information on hundreds of potential targets and then develop an increased number of attacks, which increases their likelihood of being successful (Wilcox & Bhattacharya, 2019) (Gharibi & Shaabi, 2012) (Weir et al., 2011).

The only way to decrease the success rate of the social engineer is by improving the end user's information security decision-making process. Improving how end-users make information security decisions requires changing how end users are trained and assessed. The current information awareness (IA) training models are based on the idea of simply presenting the end-user with the information, not modifying their behaviour (Schroeder, 2017). Identifying behavioural traits related to susceptibility led to developing a new Dynamic/Adaptable IA Training Assessment Tool for this study. The development of this tool is the first step in validating the concept of including behavioural and social media factors in determining a person's accurate susceptibility level.

3. Dynamic/adaptable assessment tool

The Dynamic/Adaptable IA Training Assessment Tool created for this study was developed to incorporate behavioural and social media usage factors in determining a person's susceptibility level. The tool was designed to improve current information awareness training assessment tools in two distinct ways. The first way is by including human behaviours and social media usage questions which gauge a person's behavioural risk of becoming a victim of social engineering. The second way is by incorporating an adaptive assessment approach to evaluating a person's information security knowledge level. The results of both parts of the tool are then fed into a susceptibility algorithm designed for this study. The tool results are then compared to the results of the KnowBe4 IA Training program completed by the same individuals to determine if there is a statistically significant difference between the two tools.

The Dynamic/Adaptable IA Training Assessment Tool was built around three unique but connected components. The first is the participant registration section, where the participant inputs their demographic information and KnowBe4 score. A unique identification number is assigned to each user to ensure anonymity. For the data from the behavioural and social media section to be valid and usable, the participants needed to answer the questions honestly; therefore, participant anonymity was a must. If participants felt their answers would come back and haunt them, they might answer the questions with a less truthful answer. Anonymity was guaranteed, and not even the researchers could correlate a participant and a specific data set.

The second section of the Dynamic/Adaptable IA Training Assessment Tool is the behavioural questionnaire. The user is presented with forty behavioural questions related to the 128 behavioural traits identified during this study's qualitative portion. The behavioural questions used a Likert scale, and each response was assigned a value from 10 for least risk to 50 for most risk. Upon completing the behavioural questionnaire, the combined total value of the responses is then put into variable Hb for use in the susceptibility algorithm.

Upon completing the behavioural trait-based questions, the individuals are presented with 30 questions related to social media behaviours, representing an enlarged social media footprint. A more prominent social media footprint increases a person's susceptibility to social engineering by providing access to data that could be used to develop a targeted social engineering attack. A person's social media footprint is calculated using two factors—how much data is available and how many people have access to the data. Large amounts of data a person posts to social media increase the amount of information a social engineer can use to create a targeted attack. In addition, an increased number of "friends" raises the probability that a social engineer has gained access to the data. The responses to the social media questions are assigned a value between 50 and 100. The resulting social media score is then placed into the variable SM in the susceptibility algorithm.

Upon completion of the tool's behavioural and social media usage sections, the participants then progress to the information security portion of the tool. Unlike other IA training assessment tools, which only ask between 10 and 20 questions, the Dynamic/Adaptable IA Training Assessment Tool takes a learning approach to question the participants. This learning approach is where the Dynamic/Adaptable IA Training Assessment Tool gets its dynamic/adaptable nature. Furthermore, this approach shows the tool's value in better assessing a person's information security knowledge because the tool does not require the person to "pass it." Instead, the tool measures a person's actual knowledge of information security. Other IA training assessment tools, and their coordinated training programs, provide the end-user with information security information and assess that the person can regurgitate the basic knowledge in a simple quiz (McIlwraith, 2006).

The Dynamic/Adaptable IA Training Assessment Tool is built around seven information security topics: Safe Surfing & Human Firewall, Identity Theft & Privacy, Passwords, Social Engineering, Malware, Physical Security & Policy, and Incident Response & Backup and Recovery. Information security questions, spanning four difficulty levels, were developed for each topic. The individuals who participated in this study were first presented with four level 1 questions related to topic 1. If the participant answered all the questions correctly, the difficulty level increased by 1, and a new set of questions were presented from the same topic. This process continued so long as the participant answered all four questions correctly. If the participant did not answer all four questions correctly or answered the four questions related to the fourth level of difficulty, then the difficulty value reset to 1, and the topic number increased by 1. The process then starts over with the new topic.

A value is calculated using the formula (Qr/T)C as each topic is completed. Qr represents the percent of correctly answered questions, divided by the average time (T) it took the participant to answer the questions, which is then multiplied by the highest difficulty level (C) reached by the participant. When all seven topics are completed, each topic's (Qr/T)C values are added together for a cumulative total. The sum value is then put into the variable QrTotal of the algorithm. The last part of the susceptibility algorithm is to multiply the results by 1000, then divide by 2, which converts the results into a numeric value between 0 and 100. The complete susceptibility algorithm is expressed as:

$$S=(QrTotal/(Hb+SM)*1000)/2$$

4. Experiment setup

The experiment for this study was structured around three phases—Tool Development, Beta Testing, and Implementation. Since this study included human subjects, it was necessary to obtain Institutional Review Board (IRB) approval before moving from the prototype phase to implementation. IRB approval was obtained on February 25th, 2021.

This study was limited to participants who were 18 years old or older, with varying levels of information security knowledge. Participants were required to complete the 2021 KnowBe4 IA training and assessment and the Dynamic/Adaptable IA Training Assessment Tool. Participants for this study were recruited from the faculty, staff, and students at Norwich University. The population at Norwich University provided this study with an appropriate level of diversity reflective of the broader United States population. For this study, only three aspects of demographic data were collected and analysed—gender, age, and education level.

5. Data analysis

Forty-seven individuals volunteered to participate in this study. The demographic breakdown for this study is shown in tables 5, 6 and 7.

Table 5: Gender (Collier, 2021)

Male	Female	Transgender
25	21	1

Table 6: Age (Collier, 2021)

18-24	25-34	35-44	45-54	55-64	65-74
7	4	11	11	10	4

Table 7: Education level (Collier, 2021)

High School	College Certificate	Associate Degree	Bachelor Degree	Master Degree	Doctorate/Term
7	4	11	11	10	4

One of the significant problems that current models of IA training assessment present are that the assessment tool tends to group users closely together, implying that all users have approximately the same level of susceptibility. A comparative analysis was conducted comparing the results of the KnowBe4 assessment and the Dynamic/Adaptable IA Assessment tool to show that the Dynamic/Adaptable IA Training Assessment Tool provided a more precise measure of a person's susceptibility. Figure 1 shows a visual comparison between the KnowBe4 results and the Dynamic/Adaptable IA Training Assessments Tool's results. It is easy to see there is a difference between how each of the tools assesses susceptibility. The results for each of the demographic categories are similar where the KnowBe4 tool groups the results together, while the Dynamic/Adaptable IA Training Assessment Tool spreads them apart.

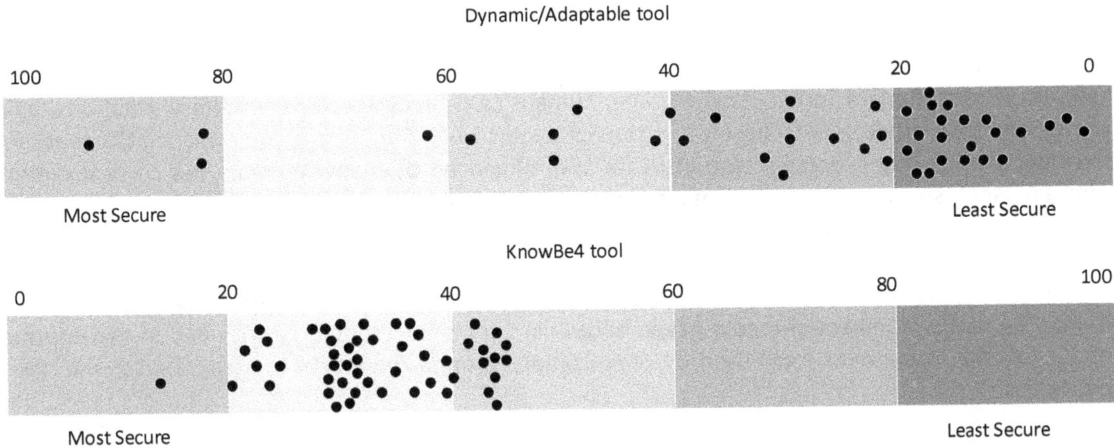

Figure 1: KnowBe4 Compared to Dynamic/Adaptable IA training assessment tool (Collier, 2021)

Comparing specific KnowBe4 scores to their associated Dynamic/Adaptable scores makes it clear that the Dynamic/Adaptable tool measures susceptibility more precisely. Table 8 is an example of three users who score virtually the same when assessed by the KnowBe4 assessment tool. In contrast, the Dynamic/Adaptable IA Training Assessment Tool assesses their risk level significantly differently.

Table 8: Similar KnowBe4 Scores example 1 (Collier, 2021)

KnowBe4	Dynamic/Adaptable
25.9	65.5
25.9	19.5
26	6.4

Although the KnowBe4 assessment tool scored the three users in Table 8 as having virtually the same score, the Dynamic/Adaptable IA Training Assessment Tool evaluated those same three users with significantly different levels of susceptibility. The Dynamic/Adaptable IA Training Assessment Tool assessed the user with the higher KnowBe4 score with the lowest susceptibility rating. The results of this comparison show the disparity that exists between the results of the two tools and further show that the current method of assessing someone's susceptibility is not as precise as it could be

6. Future work

This study consisted of both qualitative and quantitative aspects. To solidify the value of the qualitative approach to identifying the behaviors related to susceptibility, a more quantitative approach to assessing the behavioral traits is an area to conduct future research. The Covid-19 pandemic placed limitations on this study by not allowing the researchers to observe body language responses to the behavioral questions or include biometric data related to the behavioral or social media usage questions. Both an observational approach and a data collection approach would have increased the viability of the qualitative methodology of identifying the validity of the behavioral factors and their relationship to susceptibility.

Taking this research to the next level will include identifying what role culture plays in the information security decision-making process. Do people from western countries perceive the threat of social engineering differently than those from eastern countries? Is there an impact from different cultures within a single country where multiple sub-cultures exist? Does culture support or block improved information security responses? These are but a few questions that need to be answered to understand better how culture impacts a person's perception of cyber threats.

Another area where this study opens the door is finding ways to alter the current model of information awareness training, so the training modifies a person's behavior, rather than simply teaching them about risks. Taking what has been learned from this study and working with an interdisciplinary team of information security professionals and behavioral modification professionals, it could be possible to develop a new form of IA training that incorporates behavioral modification techniques like Cognitive Behavioral Therapy(CBT). CBT works to change the person's behavior, which would improve their response to a cyber-attack.

7. Conclusion

Humans are certainly the most vulnerable part of any information security chain. Although technical solutions are great at preventing many forms of cybercrime, humans can still bypass those technical solutions, opening themselves and their employers to become victims of cybercrime. People are complex, multifaceted beings, and their decision-making process is also intricate and influenced by many factors. The current method of assessing a person's susceptibility to becoming a victim of cybercrime does not include the behavioural or social media usage factors that influence a person's ability to make a good information security decision. Since necessary behavioural data is not used in the current IA training assessment tools, these tools artificially group users, giving the impression that users all have approximately the same level of susceptibility. This study showed that human behaviours and social media usage behaviours could and should be included in the information awareness training assessment tools used by organizations around the world to assess better the risk their employees pose.

Acknowledgements

I would like to thank the following people for their support of my research. Dr. Edward Chow, Dr. Yanyan Zhuang, and Dr. Rick White—University of Colorado, Colorado Springs, Dr. Huw Read, Dr. Matthew Thomas and Dr. Mich Kabay—Norwich University. Alexandra Collier, Southern New Hampshire University, Morgan Woods, Elizabeth Gregory, Aiden Cruz, Charles Grunert, Nicole Folib—Norwich University.

References

American Psychological Association, 2020. *APA Dictionary of Psychology.* [Online] Available at: https://dictionary.apa.org/behavior [Accessed July 28th, 2021].

American Psychological Association, 2021. *Personality.* [Online] Available at: https://www.apa.org/topics/personality [Accessed July 27th, 2021].

Ariu, D., Frumento, E. & Fumera, G., 2017. *Social Engineering 2.0: A Foundational Work.* Sienna, s.n.

Ayates, K. & Conolly, T., 2003. *A Research Model for Investigating Human Behavior Related to Computer Security.* Tampa, s.n.

Bada, M., Sasse, A. M. & Nurse, J. R., 2019. Cyber Security Awareness Campaigns: Why do they fail to change behaviour?. *CoRR,* Volume abs/1901.02672.

Collier, H. D., 2020. *Social Media: A Social Engineer's Goldmine.* Larnaca, s.n.

Collier, H. D., 2021. *Enhancing Information Security By Identifying and Embracing Executive Functioning and the Human Behaviors Related to Susceptibility.* Colorado Springs: s.n.

Edwards, W., 1954. The Theory of Decision Making. *Psychological Bulletin,* pp. 380-417.

Gharibi, W. & Shaabi, M., 2012. Cyber Threats in Social Networking. *International Journal of Distributed and Parallel Systems,* Volume 3, pp. 119-126.

Gunkel, P., 1998. *638 Primary Personality Traits - Ideonomy.* [Online] Available at: http://ideonomy.mit.edu/essays/traits.html [Accessed 2019].

Hadnagy, C., 2018. *Social Engineering: The Science of Human Hacking.* Indianapolis: Wiley.

Hallas, B., 2018. *Rethinking the Human Factor.* s.l.: The Hallas Institute.

Henderson, J. C. & Nutt, P. C., 1980. The Influence of Decision Style on Decision Making Behavior. *Management Science,* pp. 119-126.

Kayes, I. & Iamnitchi, A., 2017. Privacy and security in online social networks: A survey. *Online Social Networks and Media,* 3(4), pp. 1-21.

Kemp, T., 2016. Social Engineering Fraud: A Case Study. *Risk Management,* 63(6), pp. 8-9.

KnowBe4, 2021. *KnowBe4 Human Error Conquered.* [Online] Available at: https://www.knowbe4.com/knowbe4-timeline/#:~:text=Now%20647%20employees%2C%20and%2023%2C000,is%20going%20stronger%20than%20ever.[Accessed 19 04 2021].

Luo, X., Brody, R., Seazzu, A. & Burd, S., 2011. Social Engineering: The Neglected Human Factor for Information Security Management. *Information Resources Management Journal,* pp. 1-8.

Mackinnon, A. J. & Wearing, A. J., 1980. Complexity and Decision Making. *Behavioral Science,* pp. 285-296.

McIlwraith, A., 2006. *Information Security and Employee Behaviour: How to Reduce Risk Through Employee Education, Training and Awareness.* Burlington: Gower Publishing Company.

Mitnick, K. D. & Simon, W. L., 2002. *The Art of Deception.* Indianapolis: Wiley Publishing Inc.

Postnikoff, B. & Goldberg, I., 2018. *Robot Social Engineering Attacking Human Factors with Non-Human Actors.* Chicago, s.n.

Schroeder, J., 2017. *Advanced Persistent Training: Take your Security Awareness Program to the Next Level.* Edinburgh: Apress.

Tsikerdekis, M. & Zeadally, S., 2014. Online Deception in Social Media. *Communications of the ACM,* 57(9), pp. 72-80.
UK Parliament, 2020. *Science and Technology Committee - Second Report Behaviour Change.* [Online] Available at: https://publications.parliament.uk/pa/ld201012/ldselect/ldsctech/179/17902.htm [Accessed July 27th, 2021].
Vishwanath, A., 2014. Habitual Facebook Use and its Impact on Getting Deceived on Social Media. *Journal of Computer-Mediated Communication,* 20(1), pp. 83-98.
Weir, G. R., Toolan, F. & Smeed, D., 2011. The Threats of Social Networking: Old Wine in New Bottles?. *The Herald*, August 11th, p. 3.
Wilcox, H. & Bhattacharya, M., 2019. *A Human Dimension of Hacking: Social Engineering through Social Media.* Guangzhou, s.n.

Automatic Construction of Hardware Traffic Validators

Jason Dahlstrom[1], Brandon Guzman[2], Ellie Baker[2] and Stephen Taylor[2]
[1]Web Sensing LLC, Lebanon NH, USA
[2]Thayer School of Engineering, Dartmouth College, Hanover NH, USA
Jason.Dahlstrom@websensing.com
Ellie.F.Baker.22@dartmouth.edu
Brandon.J.Guzman.22@dartmouth.edu
Stephen.Taylor@dartmouth.edu

Abstract: This paper describes a fully automated process that creates a custom hardware traffic validator directly from a formal grammar and deploys it within a specialized network security appliance. The appliance appears as a hidden, all-hardware "bump-in-the-wire" that can be inserted within any network segment; it stores and validates messages on-the-fly, and either forwards or drops individual packets in real-time. Consequently, it serves to disrupt and mitigate stealthy remote attacks that leverage zero-day exploits and persistent implants. Allowed traffic, files, and mission payload formats are specified formally using a standard Look-Ahead, Left-to-Right (LALR) grammar that operates on ASCII and/or binary data. The grammars can be expressed either in Backus-Naur Form (BNF), used by industry standard tools such as Bison, or through state-of-the-art combinators, such as Hammer, under development within the DARPA SafeDocs program. Bison and Hammer compiler tools are used to generate standard shift/reduce parsing tables. These tables are post-processed to improve their compactness and practical viability. The optimized tables are then combined with a generic push-down automaton to form a complete parser. The parser is then automatically transformed into a hardware circuit using High-Level Synthesis (HLS). The result is a composable block of circuitry that can be directly inserted into a generic communications harness embedded within a Field Programmable Gate Array (FPGA) on the network appliance.

Keywords: parsing, LALR grammar, traffic validation, FPGA, Bison, Hammer

1. Distribution

This research is supported by the Defense Advanced Research Projects Agency (DARPA) under contract W31P4Q-20-C-0033. The views, opinions and/or findings expressed are those of the authors and should not be interpreted as representing the official views of the Department of Defense or the U.S. Government. The paper is released with the following distribution determination:

Distribution A: (Approved for Public Release, Distribution Unlimited).

2. Introduction

Many organizations handle sensitive data and files relating to military missions, trade secrets, intellectual property, private personal data, and/or classified projects (NBAR, 2021). Traditionally, these organizations have been well advised to implement an *airgap* that physically isolates computers containing sensitive information from the Internet, to protect against theft. Unfortunately, airgaps come with a substantial cost in productivity and assume that profession staff, with the expertise to handle sensitive information, is co-located. Airgaps are increasingly impractical given the need to connect critical systems, such as industrial plant, to cloud-based analytic platforms (e.g., Google Analytics) or support condition-based vehicle maintenance (Adams 2012, Carter 2013, Shanthakumaran 2010). Similarly, individual air, space, and ground vehicles increasingly rely upon standard networking technologies to link embedded control systems with sensors, actuators, and human machine interfaces through industry standard buses (e.g., CAN, J1939, MIL-STD-1553, USB) and communications interfaces (e.g., Gigabit Ethernet (GigE), OpenVPX, and PCIe). Network connected personal devices – phones, tablets, and laptops – are increasingly being used to manage and interact with these systems. Though conceptually air gapped, these systems are intermittently connected with military installations to provide mission parameters, or to effect maintenance and upgrades. Though network boundary protections are used during these activities, there are many threat vectors that circumvent these protections, and airgaps in general; these include unintended network connections, insiders, zero-day exploits, supply chain interdiction, and persistent implants (Kushner 2013).

This paper describes an automated process to validate network traffic employing Field Programmable Gate Array (FPGA) technology. This technology is employed at the core of an all-hardware *network appliance* that continuously validates the integrity of network traffic. This combines the isolation of an airgap, with the convenience of an Internet connection, at considerably less risk than an open connection: it provides a base-of-trust in hardware that both disrupts, and is impervious to, stealthy remote attacks perpetuated through zero-day exploits and persistent implants. Two such appliances are shown in Figure 1 – an Ethernet appliance and a Smart Network Interface Card (Smart NIC). Both consist, by design, of a *single* FPGA chip lying between industry standard buses and network connections. Each appliance thereby forms a "bump-in the-wire" with the FPGA acting as a communication bridge. Consequently, the FPGA can monitor and interact with all systems attached to its interfaces, and may act to store, validate, drop, or forward traffic. Both appliances can use a variety of pin-compatible FPGA sizes from the low-cost Spartan-7 devices pictured on the Ethernet device, to the Artix-7 pictured on the Smart NIC; however, only the Artix devices allow partial reconfiguration.

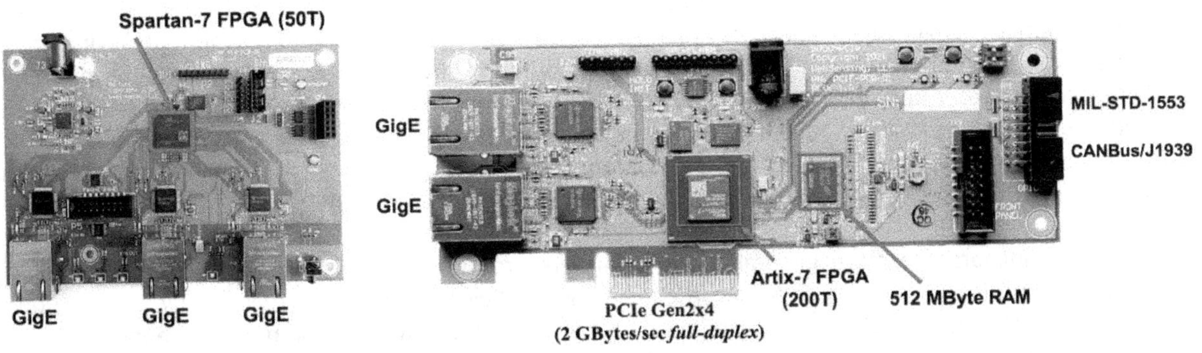

Figure 1: Ethernet (left) and Smart-NIC (right) appliances

The appliances offer several key security advantages: All sensitive data -- encryption keys, buses, and algorithms -- is strictly hidden within the security perimeter afforded by the FPGA chip-boundary (Dahlstrom and Taylor 2018, Aug 2019, Oct 2019, 2020, 2021) thereby mitigating reverse engineering if an appliance is lost or stolen; No software is present in the device, mitigating malicious implants and zero-day attacks (Kushner 2013); any connection can be used as an out-of-band channel to adapt the device to alternate mission profiles, augment its internal functions, or upgrade the device; Extensive anti-tamper and circuit destruction techniques enhance resilience; Large files and data repositories can be held within RAM attached to the FPGA, without violating the strict visibility constraints imposed by the FPGA chip-boundary, by encryption and decryption algorithms embedded inside the FPGA. For versatility, all circuits resident in the appliances are developed using a rapid prototyping technology called High Level Synthesis (HLS). This process allows algorithmic specifications to be designed and tested in C, C++, or System-C. The working code is then automatically transformed into a standalone, reusable, *hardware block*. These reusable circuit plugins can be directly integrated into a static circuit design in the FPGA for production deployments. Alternatively, the block can be treated as an Open Container Initiative (OCI) compliant *container*; Using a technique known as *partial reconfiguration*, the FPGA can then be partitioned into segments and containers can be dynamically loaded into a partition linking it into the overall function of the device on-the-fly. To manage this process, we have developed a thin, hypervisor-like hardware layer termed a *Nanomarshal* (Dahlstrom et al, Nov 2019).

3. Parser plugins

Recall that each network appliance is concerned with monitoring the flow of traffic across its interfaces and *validating* both that messages adhere to an industry standard protocol and that message content is valid in the context of some tactical mission. Generally, its action on detecting a valid message is to allow it to pass; conversely, its action on detecting invalid data is to drop the message -- mitigating potential exploitation -- and/or generate an alert. To achieve message validation, the TSNIC uses custom *parsing engines* inserted across its communication paths. Parsing is the general process of taking an input stream of symbols and understanding their format (syntax) and meaning (semantics). For example, compilers such as GCC use a parser to validate that a computer program, written in some programming language such as C/C++/Java/Fortran is written correctly (i.e., is syntactically valid), and to understand its structure (i.e., its semantics) for the purpose of machine code generation and optimization. Parsers are tools that apply a collection of formal *grammar rules*, defining some input language, to determine if the input adheres to the rules. For example, the following 3-rule grammar G defines a language in which a stream of symbols is valid only if it begins with the character 'a', ends with 'c', and contains one or more intervening 'b' characters:

G : 'a' Bs 'c' ;
Bs: 'b' | Bs 'b' ;

The "or" symbol | designates an alternative definition for the rule defining "Bs". This grammar accepts as valid the input streams *abc, abbc* and *abbbc* etc, but rejects any other stream, e.g., *a, ac, aaa, ccc, adx, abbbx*, etc. Individual characters such as 'a' are *terminal symbols* that must be present in the input stream; all other symbols are non-terminals representing intermediate structural elements. For binary grammars, hexadecimal terminal values can also be used, for example, the value '\xFF' represents a single byte value corresponding to 255 in decimal.

Parser generators are tools that take a grammar as input and automatically generate a program that implements the associated parser. The most mature and widely used generator is Bison which accepts two primary classes of grammar: Generalized Look-Ahead (GLR) and the more restrictive Left-to-Right Look Ahead (LALR). Both classes of grammar are expressed in Backus-Naur Form (BNF), used above to define the grammar G. Under the DARPA SafeDocs program, new tools are being developed based formal methods. One of the more mature is the Hammer *combinator* library (Bratus, et al. 2016) which provides a collection of well-defined base parsers and methods to combine them to build more complex parsers – all expressed in the C-programming language. The resulting parsers are provably correct by construction. The Hammer library provides a collection of backends that allow different classes of grammar to be implemented, including GLR and LALR.

Though GLR grammars are more general, LALR grammars are sufficient for validating a wide variety of communication protocols and file formats. Their simplicity allows them to be realized by a *push-down automaton* – a finite state machine employing a single stack to store symbols while parsing the input stream. The state machine relies on two core operations *shift* – involving saving a symbol from the input onto the stack -- and *reduce* – involving the application of a grammar rule to detect a structure in the input and reduce the symbols on the stack. The state-machine is generic and common to all grammars, however, the order in which shift and reduce operations are applied to the input is based on a collection of *parsing tables* derived from the grammar, usually referred to as Action/Goto tables (Aho, Sethi, and Ullman, 1988). These tables map the current state of the automaton, to a next state based on the symbol read from the input stream.

The network appliances in Figure 1 can parse any LALR grammar expressed in either Bison or Hammer. Figure 2 shows the automation process used to transform a Grammar into a hardware parser-plug. Using Bison, the input grammar, defined in BNF, is fed directly into the standard Bison parser generator (leftmost path). This produces a set of parsing tables, expressed in a machine-readable XML format. For Hammer, an equivalent parser can be defined using pre-existing combinators in C linked with the Hammer library (alternate path). We have augmented the Hammer library to output the same XML format as Bison. A conversion tool – *xml2h* – is then used to convert the XML tables into a C-header file (pda.h) containing a large two-dimensional C-array. Rows in the array correspond to *states* in the automaton, while columns designate terminal and non-terminal symbols encountered when reading the input stream. Entries in the array designate shift or reduce actions applied in each state. Consequently, the C-array provides a complete definition of how the push-down automaton should operate to validate any particular grammar. The C-array (pda.h) is combined with a generic LALR *automaton* (pda.c) (Aho, Sethi, and Ullman, 1988) and *testbench* code (main.c) to produce a runnable C-program implementing the parser. This software parser is validated using a set of representative test vectors that are successively loaded from files by the testbench (main.c) to ensure that the parser operates correctly.

The parser code is carefully constructed to operate on streaming-data and feed directly into the Xilinx High-Level Synthesis (HLS) tools. These produce a hardware circuit block that implements the software parser derived from the pda.[ch] files. The HLS hardware-software co-simulation tool is then used to validate the hardware parser. It uses the same software testbench and test vectors, used to validate the software version of the parser, to ensure the hardware version operates correctly. The validated hardware is subsequently output as a *parser-plugin* component that can be loaded directly into one of the network appliances shown in Figure 1. The plugin is organized to connect with the communication interfaces on the board through industry standard AXI-streams. These are either statically connected through the Xilinx Vivado tool chain, or dynamically managed through partial reconfiguration by the Xilinx Vitis tool chain.

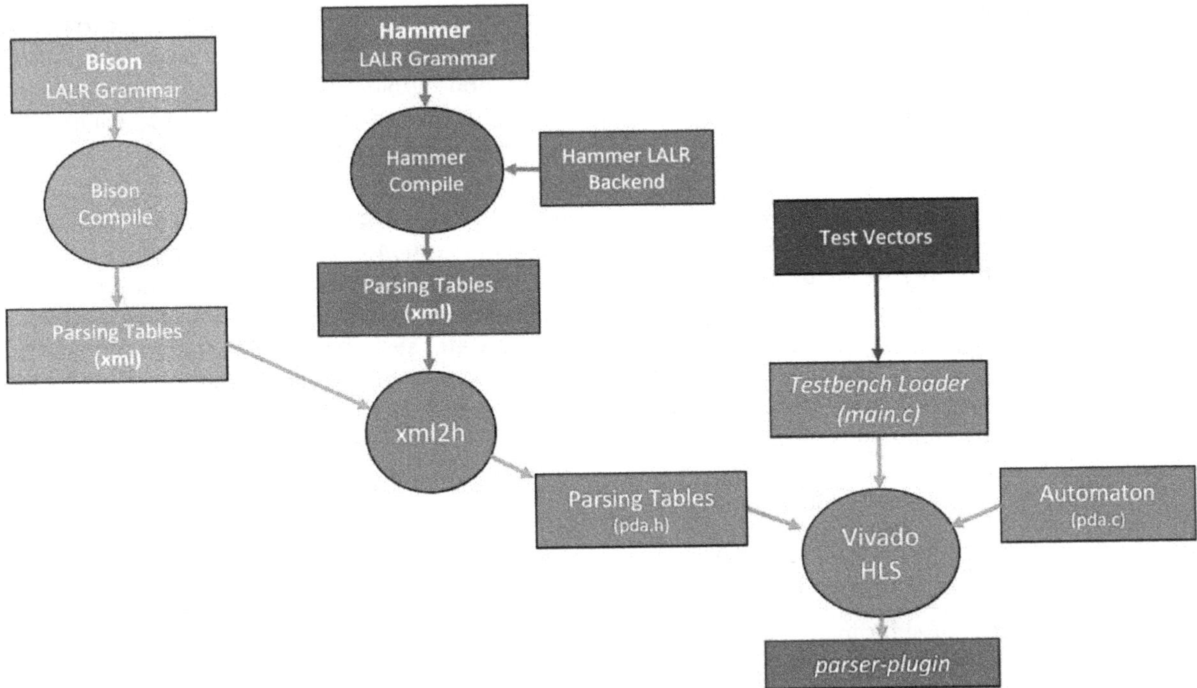

Figure 2: Automated Parser generation process

4. Table Optimization

Unfortunately, though a conventional two-dimensional C-array (pda.h) is a convenient conceptual framework for organizing the automation process, it is impractical since it contains many states that cannot be reached in practice. Consequently, a highly optimized alternative representation is used that removes much of the empty structure from the array. These optimizations follow similar techniques to those employed internally by Bison and are described in detail online (Popuri, 20006). For example, consider the following Hammer Grammar:

```
HParser *init_parser() {
        HParser *G;
        H_RULE(chara, h_ch('a'));
        H_RULE(obrac, h_ch('('));
        H_RULE(cbrac, h_ch(')'));
        HParser *P = h_indirect();
        HParser *M = h_indirect();
        h_bind_indirect(P, h_choice(chara, h_sequence(obrac,M,cbrac,NULL), NULL) );
        h_bind_indirect(M, h_optional(P) );
        G = h_sequence(P,h_end_p(),NULL);
        return G;
}
```

The resulting parsing table for this grammar, involves 10 states:

```
static int16_t table[10][9] = {      /* line 1 */
 {0,5,0,3,2,0,6,4,32767},            /* state 0 */
 {-7,-7,-7,-7,0,0,0,0,0},
 {-6,-6,-6,-6,0,0,0,0,0},
 {-5,-5,-5,-5,0,0,0,0,0},
 {0,5,-3,3,2,8,9,0,0},
 {7,0,0,0,0,0,0,0,0},
 {-4,-4,-4,-4,0,0,0,0,0},
 {0,0,10,0,0,0,0,0,0},
 {-2,-2,-2,-2,0,0,0,0,0},
 {-1,-1,-1,-1,0,0,0,0,0} };
```

Each line in the table represents a state in the push-down automaton. The first 4 entries in each line correspond to the standard ***action*** table associated with detection of a terminal symbol in the grammar; while the last 5 entries correspond to the standard ***goto*** table associated with non-terminal symbols. Positive values in the action table represent shift operations, negative values are reductions, zeros represent parse failures and unreachable states, and 32767 is a reserved value representing the *accept* symbol.

A first optimization (O1) arises from inspection of the data in the tables: notice the entries in lines 3,4,5,8,10 and 11 (bold-faced) of the parsing table. Each designates a *default reduction* (-ve) operation applied for *every* terminal symbol, with unreachable (0) *goto* table entries. Consequently, it is possible to represent each of these lines by a single number designating the default reduction to be applied in that state (7,6,5,4,2,1 respectively). Since the state numbering is arbitrary, it is possible to reorder the states such that all of these transitions fall, in order, to the end of the table, the *goto* segment can then be represented independently, and all of the default lines can be removed and represented by a single array (*deftable*). This array is accessed only when a shift operation exceeds the number of remaining states in the action table. The resulting parsing tables are shown below:

```
static uint16_t deftable[6] = {
 -7,-6,-5,-4,-2,-1
};

static int16_t actions[4][4] = {
 {2,0,6,0},
 {2,-3,6,0},
 {0,0,0,8},
 {0,10,0,0}
};

static int16_t gotos[4][5] = {
 {5,0,3,7,32767},
 {5,4,9,0,0},
 {0,0,0,0,0},
 {0,0,0,0,0}
};
```

A second optimization (O2) reduces the *gotos* table by adding a default goto table (*defgotos*) then generating a remaining *gotos* table with the defaults removed. Positive values in the *defgotos* table indicate unique defaults; negative values (e.g. -1) indicate where to find a non-default element (i.e. column = (-default)+1) :

```
static int16_t defgotos[5] = {5,4,-1,7,32767};

static int16_t gotos[4][1] = {{3}, {9}, {0}, {0} };
```

Obviously, to utilize these smaller tables requires changes to the operation of the automaton, since it must now detect and use indirection rather than directly index into a two-dimensional array. Table 1 shows the resulting table size, in bytes, for a variety of parsers taken from the Thayer Parser Experimentation repository online at *https://github.com/lvln/thayer_parsers*: O0 represents unoptimized tables, O1 uses the first optimization, and O2 the second. The last two columns show the percentage reductions adding O1 only (O0-O1) and subsequently adding O2 (O1-O2). Only a single ASCII parser – *json* (without Unicode) – is reported in Table 1 for both Bison and Hammer to provide a point of comparison. The others all combine both binary and ASCII within a single parser and are more representative of expected use cases – on these grammars we see a dramatic cost saving: more than 80% across the board. The second round of optimization produces only a small improvement. Though a third level of optimization is possible (removing the zeros in the final gotos table), the 01-02 results indicate a diminishing return for added complexity in the automaton.

Table 1: Parser optimization results

Parser	Version	O0	O1	O2	O0-O1 %	O1-O2 %
json	Bison	51712	14080	12278	72.8	12.8

Parser	Version	O0	O1	O2	O0-O1 %	O1-O2 %
json	Hammer	74098	17534	13254	76.3	24.4
command	Hammer	480564	74128	69498	84.6	6.2
response	Hammer	151140	6426	6288	95.7	2.1
json.unicode	Hammer	406944	36006	30320	91.2	15.8

5. Notational conveniences

Our experiments with parser compactness showed that Hammer provides a variety of notational conveniences not available in Bison BNF. These primarily revolve around the ability to express *ranges*, *enumerations*, and *sequences* of terminal symbols. For example, in BNF, to specify a hexadecimal digit we might use 25 rules:

```
HexDigit: Digit | LowerAF | UpperAF ;
Digit: ‘0’ | ‘1’ | ‘2’ | ‘3’| ‘4’| ‘5’| ‘6’| ‘7’| ‘8’| ‘9’ ;
LowerAF: ‘a’ | ‘b’ | ‘c’| ‘d’| ‘e’| ‘f’ ;
UpperAF: ‘A’ | ‘B’ | ‘C’| ‘D’| ‘E’| ‘F’ ;
```

Similarly, to recognize any bracketing symbol we might use 8 rules:

```
Bracket: ‘(‘ | ‘)’ | ‘[‘ | ‘]’ | ‘{‘ | ‘}’ | ‘<’ | ‘>’ ;
```

Finally, to represent the values true and false we would use 2, rather ugly, rules:

```
True: ‘t’ ‘r’ ‘u’ ‘e’ ;
False: ‘f’ ‘a’ ‘l’ ‘s’ ‘e’ ;
```

Bison generally assumes that such rules would be handled by a separate lexical analysis program that produces *tokens* to be consumed by the parser. Hammer integrates such operations as elementary combinators, implemented efficiently through comparisons, and consequently builds them formally into a single cohesive and general system. Integration of Lex, which generates lexical analysers, into the automaton framework shown in Figure 2 would of necessity require the added complexity of calling Lex functions fed into the HLS process. Instead, we choose to dispense with Lex completely, and simply pass all input bytes directly into Bison, with each grammar incorporating its own rules for lexical analysis, consistent with Hammers treatment. To simplify these lexical steps, we have developed a pre-processor (implemented in Bison) that unambiguously accepts an extended version of BNF, termed xBNF. The pre-processor accepts the following notations:

[t1-t2] -- Accepts any terminal value in the range **t1** to **t2**, where **t2>t1**.
[t1,t2,...,tn] -- Accepts an enumerated range of terminal values **t1** to **tn**; **n>1**.
“string” -- Accepts any string of terminal characters

Consequently, in xBNF the above examples would be rendered:

```
HexDigit : [‘0’-‘9’] | [‘a’-‘f’] | [‘A’-‘F’] ;          /* ranges */
Bracket:   [‘(‘, ‘)’, ‘[‘, ‘]’, ‘{‘, ‘}’, ‘<’, ‘>’] ;     /* enumeration */
True: “true” ;                                        /* sequences */
False: “false” ;
```

Pre-processing would result in an equivalent set of rules to the original, however, with less non-terminals for intermediate rules (c.f. HexDigit), since these exist purely for human readability. Bison's BNF input language allows the parsing of binary formats by virtue of the ability to express terminal values in an equivalent hexadecimal format. For example, the rule for LowerAF could have been expressed:

```
LowerAF: ‘\x61’ | ‘\x62’ | ‘\x63’ | ‘\x64’ | ‘\x65’ | ‘\x66’ ;
```

Unfortunately, Bison expects Lex to return the value zero (0) to signify the end of input. Consequently, it is not possible to parse the full range of binary values without special treatment of zero. To resolve this issue, we replace lex with a simple function that distinguishes the special case of zero in the input stream and returns a special token in its place. Using this convention, it is then possible to add two new notations to xBNF that simplify the parsing of binary formats:

'\x00' -- Accepts the terminal symbol for zero in hexadecimal.
* -- Accepts any byte (i.e. in the range **'\x00'** to **'\xFF'**)

Table 2 compares the expressiveness of a variety of parsers written in xBNF, Bison BNF and Hammer using the parsers in Table 1. As expected, xBNF consistently improves upon BNF. Hammer is consistently the most complex, this is due to its use of the comparatively complex syntactic structure of the C programming language and the need to handle forward references, as exemplified by lines 6-9 of the example Hammer grammar shown previously in Section 4.

Table 2: Source lines of code

Parser	xBNF	Bison BNF	Hammer
json	27	44	72
command	22	41	25
response	7	32	20
json.unicode	32	58	76

Though the source code for xBNF and BNF is more compact, this does not necessarily translate into a smaller parsing table by virtue of the optimization strategies, such as used for combining equivalent states, used internally by Bison and Hammer. To gain an appreciation for this underlying internal compactness, it is useful to consider the number of *terminals, non-terminals*, and *states* used by the associated automaton *prior* to any post-processing that optimizes the tables using the *same* algorithms. Recall that, to a first order, the size of the parsing table (*Tsize*) is governed by the number of *states, terminals* (action table) and *non-terminals* (goto table) i.e. Tsize = States x (Terminals+Non-terminals). Table 3 shows the results for *json.unicode.*

Table 3: Internal compactness

	xBNF	BNF	HMR
Terminals	228	228	224
Non-terminals	49	45	69
States	346	342	689
Tsize	17,182	15,618	47,765

These results are consistent with a complete study of all the parsers in the repository. The results indicate that the xBNF pre-processing enhancements cause Bison to generate up to 10% larger parsing tables; obviously, the impact of this increase is largely absorbed by optimizations. Hammer tables, in comparison, are substantially larger than the others reflecting its current lack of maturity compared to Bison. A considerable saving was observed for ASCII parsers by virtue of the constrained set of characters (i.e. terminal symbols) employed in a grammar; Binary parsers can employ all 256-bit combinations available in a single byte, plus additional terminals to represent the value of zero (in Bison) and the end-of-input.

6. FPGA resource optimization

The FPGA resources -- Block RAM (BRAM), Flip Flops (FF) and Lookup Tables (LUT) -- needed for a parser is ultimately used to represent its *space-optimized parsing tables*, *pushdown automaton*, and the associated *stack* used during parsing. Table 4 shows the resource utilization associated with an Artix 200T -- for the *json.unicode* parser with a 2Kbyte stack – large enough to parse all of our most complex test vectors. The parser is compared with an *empty* parser that contains no parsing tables. Four versions of the empty parser are shown which vary only by stack size associated with the parser in increments of 2 Kbytes. Values in the table are actual number of resources with the percentage of the overall resources on the chip. Where the overall resource usage is less than 1% it is signified by the notation (~1%).

Table 4: FPGA resource utilization

Parser	BRAM	FF	LUT
json.unicode	16 (2%)	306 (~1%)	942 (~1%)
empty0	0 (0%)	167 (~1%)	533 (~1%)
empty2	2 (~1%)	263 (~1%)	588 (~1%)
empty4	4 (~1%)	263 (~1%)	589 (~1%)
empty8	8 (1%)	263 (~1%)	590 (~1%)

The results demonstrate that the most complex parser tested to date consumes only 2% of the available BRAM resources and an insignificant percentage of the other resources. The automaton alone consumes only 167 Flip Flops and 533 Lookup tables -- less than 1% of each resource. When present, the size of the stack affects only the BRAM resources used and is generally a small percentage of the overall chip resources. This study indicates that for complex parsers, only the smallest and least expensive Spartan-7 FPGA will ultimately be required for production parsers.

7. Latency and Bandwidth

Recall the most complex parser in the repository accepts valid JSON inputs with Unicode (*json.unicode*). The time to store, parse, and forward an Ethernet Frame is proportional to the length of the input that must be traversed by the parser state-machine, to detect either a valid or invalid input. This value could be immediate for short failure cases. Instead, we base analysis on randomly selected *valid* inputs, of various sizes, to force the parser to traverse the entire input and detect success through a variety of alternative paths in the state-machine. This provides an upper bound on latency and lower bound on bandwidth. The performance numbers are obtained directly by counting hardware clock cycles under co-simulation, where the actual circuit, running at 125MHz, is driven from test-vectors provided via a software testbench in High-Level Synthesis – only the cycles used by the actual circuit are registered and have proved to be highly accurate in a variety of prior experiments. Figures 3 and 4 show the measured Latency and Bandwidth curves for a variety of test inputs of differing length. Note that the largest parse is 1600 bytes -- the size of the largest Ethernet frame; messages longer than a single message were shown to be dominated by the byte-to-byte sequential nature of parsing and add linearly to the graphs.

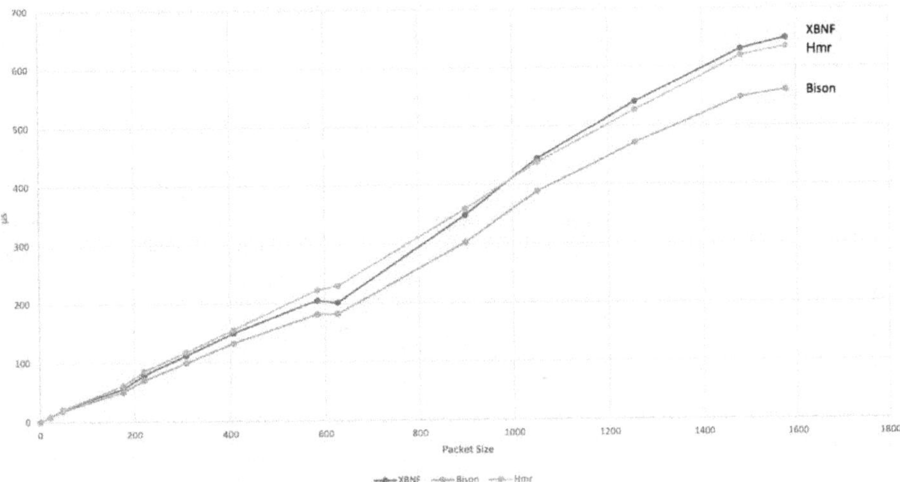

Figure 3: Parser Latency (microseconds)

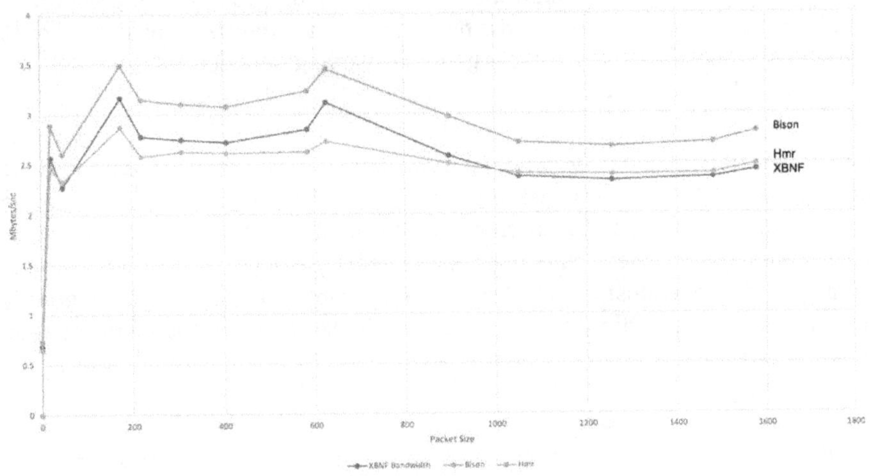

Figure 4: Parser Bandwidth (Mbytes/sec)

Figures 3 and 4 are useful for comparing the relative performance of the automation processes. Note that the random nature of the input drives, what is a relatively complex state machine, through a broad variety of differing state-sequences. In consequence, variability in the results, for example between 200 and 600 bytes is likely dependent on the sequences used in practice in the experiments. It is more important to observe that Bison, by virtue of the maturity of its internal optimization, consistently outperforms Hammer and XBNF in Latency and Bandwidth, but there are no dramatic unexpected deviations in performance inherent in the alternative formulations. The improvements in readability and compactness provided by XBNF, illustrated in Table 2, clearly comes at a cost of up to 15% in latency and bandwidth; its overall performance is comparable to Hammer. It is not yet clear if this overhead can be eliminated through alternative implementation techniques and represents the subject of ongoing research. Using linear approximations, general formulae characterizing the expected overall performance can be constructed as shown in Table 5, where *x* is the packet size.

Table 5: Linear approximations

Grammar	Latency L (microseconds)	Bandwidth BW (Megabytes/sec)
xBNF	L = 0.4225x - 14.022	BW = 0.0005x + 1.9878
Bison	L = 0.3662x - 10.607	BW = 0.0007x + 2.2239
Hammer	L = 0.4131x - 6.4267	BW = 0.0005x + 1.9467

Obviously, these approximations are inaccurate for very small messages, however, they are helpful in gauging practical expectations. The MAVLink network protocol is an unclassified protocol in common use for command and control of drones. It uses control messages that range in size between 12 and 280 bytes in length. Two of the most frequently used messages -- SETMODE and STATUS -- are 18 and 66 octets in length respectively. Table 6 shows the expected latency when, during normal operations, every valid MAVLink message is parsed to the end; additionally, the latency for a maximum sized message of 1600 bytes and the expected number of clock cycles per octet are calculated.

Table 6: Typical latency

Parser	12-byte SETMODE Latency	66-byte STATUS Latency	280-byte MAVLink Latency	Large Packet Latency (1600bytes)	Estimated cycles/octet (1600 bytes)
xBNF	< 1 μs	14 μs	104 μs	662 μs	51
Bison	< 1 μs	13.6 μs	92 μs	575 μs	45
Hammer	1 μs	21 μs	109 μs	655 μs	51

In summary, as one would expect, deep packet inspection requires significantly more time than simple checks on a packet header, such as those required to validate the presence or absence of particular values (e.g., protocol flags, IP addresses, etc.). Latency currently accrues at approximately *50 clock cycles per octet*. At 125Mhz, typical latencies for frequently used, valid MAVLink messages are between 1 and 20μs, with a maximum of 110μs; obviously, invalid messages would be detected more quickly. Though the link may operate at Gigabit rates, parsing causes bandwidth to drop -- by virtue of the need to store, sequentially consider every byte in turn, and subsequently forward individual Ethernet frames -- generally varying between 2.25 and 3.5 Mb/sec.

8. In conclusion

The technology described here occupies a middle ground between the fully air-gapped and fully connected systems and trades a marginal increase in risk for the opportunity to observe, optimize, and interact with networked systems remotely. This paper has discussed an automated process for developing custom hardware parsers using formal grammars. In the extreme, this allows every byte of every message – including payload data -- to be inspected and validated against a formal specification. Obviously, parsing is inherently a byte-by-byte sequential process that drives detection through a potentially large state-machine – consequently, it is not surprising that latency varies linearly with message length and that bandwidth is limited by the store-and-forward nature of Ethernet frames.

The performance figures presented here represent the baseline for an initial prototype and utilize only the resource optimizations described. Many additional sources of performance enhancement exist that can be expected to improve the technology as it matures: running parsers at higher clock rates; using extra resources to parse multiple frames concurrently; widening input and output data streams; using data-flow optimizations within the parsing pipeline rather than packet-by-packet store and forward etc. Moreover, the technology need not be used in isolation, or, in an all-or-nothing effort to validate every byte: only a subset of message traffic considered "unsafe" might be considered in conjunction with other, higher-performance tests. For example, by quickly checking an IP address or port, the decision might be made to perform whole packet analysis on only a subset of traffic from specific locations. The low resource requirements achieved here offer not only the ability to produce further optimizations, but also to include other algorithms concurrently with parsing. For example, they have already been combined with AES encryption and decryption blocks, as well as IPSec packet encapsulation techniques, both of which are able to operate at GigE line rates.

References

Adams,C. (2012) "HUMS Technology", Aviation Today (aviationtoday.com/2012/05/01/hums-technology).

Aho, A.V., Sethi, R., Ullman,. J.D. (1988) *Compilers*, Addison Wesley.

Barham P, et al (2003) "Xen and the Art of Virtualization", In: Proceedings of the nineteenth ACM symposium on Operating systems principles, pp 164–177.

Bratus,S. et al., (2016) "Implementing a vertically hardened DNP3 control stack for power applications", ICSS'16 ACM Proceedings of the 2nd Annual Industrial Control System Security Workshop, pp 1-9. (c.f. https://gitlab.special-circumstanc.es/hammer/hammer).

Carter, M.C. (2003) "Post Implementation CBM Benefit Analysis", Annual Conference of the Prognostics and Health Management Society.

Dahlstrom, J., and Taylor, S. (2018) "System-on-Chip Data Security Appliance and Methods of Operating the Same", U.S. Patent 10,148,761.

Dahlstrom, J., and Taylor, S. (Aug 2019) "System-on-Chip Data Security Appliance and Methods of Operating the Same", U.S. Patent 10,389,817.

Dahlstrom, J., and Taylor, S. (Oct 2019) "Endpoints for Performing Distributed Sensing and Control and Methods of Operating the Same", US Patent 10,440,121.

Dahlstrom, J., et al, (Nov 2019) "Hardening Containers for Cross-Domain Applications", In Proceeding of MILCOM 2019, Norfolk, VA, USA, pp. 1-6.

Dahlstrom, J., and Taylor, S. (2020) "System-On-Chip Data Security Appliance Encryption Device and Methods of Operating the Same" U.S. Patent 10,616,344.

Dahlstrom, J., and Taylor, S. (2021) "Hardware Turnstile", U.S. Patent 10,938,913.

Kushner, D. (2013) "The Real Story of STUXNET", IEEE Spectrum.

NBAR (2021) -- The National Bureau of Asian Research, The IP Commission Report, 2013, updated 2021.

Popuri, S.K., "Understanding C parsers generated by GNU Bison", Sept 2006 (cs.uic.edu/~spopuri/cparser.htm).

Shanthakumaran, P. (2010) "Usage Based Fatigue Damage Calculation for AH-64 Apache Dynamic Components", The American Helicopter Society 66th Annual Forum, Phoenix, Arizona.

Operationalizing Cyber: Recommendations for Future Research

Baylor Franck and Mark Reith
Department of Electrical & Computer Engineering, Air Force Institute of Technology, WPAFB, USA
baylor.franck@us.af.mil
mark.reith@afit.edu

Abstract: The goal of this paper is to argue for the mandatory reporting of cyber-attacks on critical U.S. infrastructure, industries, and companies to the Department of Defense (DoD) for the DoD to improve national security through a clearer understanding of the threats and how to position the U.S. for better defense. The paper will first discuss who will be subject to mandatory reporting and propose a template for the requirements of reporting such as the turnaround time to report and the details needed from the attack. The paper will provide an argument showing the benefit to the DoD requiring reporting and why it should be concerned about external cyber-attacks on non-DoD systems. The paper will then look on the private sector viewpoints to discuss the benefits of mandatory reporting such as the bottom line and brand awareness. Additionally, the paper will also discuss how the consumer will benefit from mandatory reporting with a focus on both financial and privacy issues. Lastly, the paper will address some key points of dissent on the topic of mandatory reporting as well some evidence to push back or show how the negatives of not reporting outweighs the negative of reporting. After reading the paper, the reader will have a better picture of the current status of cyber-attacks on the private sector, how these attacks effect the DoD's mission, and why mandatory reporting can help enhance private sector cybersecurity. More research is needed to better understand the legal argument for requiring reporting on cyber-attacks as well as economic incentives for compliance, however this paper is not intending to answer that argument given the authors do not come from the legal or economic disciplines.

Keywords: cyber, DoD policy, business

1. Introduction

As the modern world continues to evolve, the DoD must also evolve itself in response to better fulfil its mission of defending the U.S. from internal and external threats. One key area of focus by the DoD is in the cyber domain as the rise of the Internet, accessible hacking techniques, and lucrative ransoms have allowed for a widespread increase in cyber-attacks both on the public and private sectors. Due to the large presence of contractors in infrastructure support and new technologies, the DoD needs to take critical steps to improve the cybersecurity of U.S. businesses and infrastructure to maintain the high performance of the U.S. military, not just its own systems. Therefore, the U.S. government needs to develop and enforce cybersecurity reporting standards for the DoD to preserve national security and its competitive technological advantages. The DoD needs to first assist companies and private entities in improving their cybersecurity by assisting companies when attacked as well as taking initiative to strengthen existing cybersecurity in other companies by evaluating systems and sharing information across the landscape to help companies better understand threats. Next, the U.S. government needs to push businesses to improve their cybersecurity by adding financial rewards for overachievers and sanctions for companies who do not meet minimum standards for cybersecurity. Lastly, the DoD needs to push the economic incentives such as better brand positivity, less ransom payments, and loss of intellectual property (IP) to help companies better realize the value of cybersecurity in terms of how it pertains to the bottom line. With the numerous benefits that exist for the DoD and private businesses themselves, key U.S. corporations and other key U.S. non-governmental organizations should report all cyber-attacks including the resolution of the attack to the DoD due to the interests of national security, interests of the attacked entity, and protecting individuals affected in these cyber-attacks. This paper will be split into four main parts; proposing the reporting requirements for U.S. entities, outlining the benefits of mandatory reporting, discussing the main counterarguments to mandatory reporting, and providing some concluding remarks that may temper counterarguments

2. Explanation of new reporting requirements

Before debating the importance of mandatory reporting, this section will be used to clearly layout who is required to report what information within certain time frames of the attack. Without clear rules and obligatory participation via the U.S. government, a combination of entities either deciding that they do not need to report when they do or giving information that is not helpful to determine how to stop the attack as well as what was stolen will lead to an ineffective cybersecurity response to the growing problem. However, the current number of attacks pared with U.S. entities where money is exchanging hands will be too much for the DoD to keep up with. This means that a huge backlog of incidents will develop leading to companies not receiving help during the attack to critical attacks not being analyzed in detail many months later. As such, U.S. entities that are required to report on cyber-attacks must be limited to key areas and are as follows (NIAC-2017);

- a) Critical Infrastructure Companies - Power, Water, Sewage, etc.
- b) DoD Contractors/Sub-Contractors - Those who work on military contracts regardless of what part of the contract they belong to
- c) Medium to Large-Scale Companies - U.S. businesses who have a market cap/size over a certain amount

The desired result is that the number of companies who will have to report would be manageable for the DoD to monitor and respond while covering the most critical entities that support the DoD and its interests. Now that this has been determined who must report cyber-attacks, clear rules and guidelines must be implemented to effectively stay on top of cyber incidents that are ongoing. The timeline of reporting what information is provided below (NIAC-2017);

- 1. Initial Notification (Within 24 hours)
- a. Explains the status of the attack, type of attack used, and target of the attack
- b. Ask for assistance of federal authorities if needed or required if DoD mandates it.
- 2. Midterm Notification (Within 72 hours)
- a. Status of Attack - Has it been completed and contained or is attack still occurring?
- b. Was a financial payment required to stop attack, how is it occurring if so.
- c. Critical DoD risks - Was any software or hardware affected directly used by the DoD. Simply understanding what systems may be at risk.
- 3. Full Comprehensive Report (Within 14 Days)
- a. Clear understanding of attack, how bad actor got in and steps taken to resolve this issue
- b. Clear list of everything that was affected, including stolen data/IP, financial payment, etc.
- c. Bad actors behind attack - specific country or individual hacker?
- d. Full impact on DoD such as were any systems/IP used by the DoD affected and if so, what is your analysis on the potential impacts that could be seen by the DoD.
- e. Impact on consumers - need to change passwords, personal data stolen, etc.

In terms of sanctions, there needs to be strict penalties to generate compliance of the rules as meager fines and reprimands by Congress have showed little to no change in previous scenarios. There also needs to be incentives for the companies as well as top executives to further push compliance whether it is in terms of adding tax credits, adjusting bonuses for executives based on cybersecurity/reporting, and requiring minimum cybersecurity standards to receive government contracts. Combined with the limited power of the DoD to lawfully impose certain rules on privates businesses, additional research on both the legal and economic routes must be examined to both force and encourage compliance. Finally, the DoD should also look at the European Union (EU) NIS directive and other practices as a template for their policies and reporting.

3. Comparison to EU Standards

The DoD should look to other large governmental organizations to better guide their mandatory reporting policy. One organization is the EU as they have developed their NIS Directive below (ENISA).

- *1. National capabilities:* EU Member States must have certain national cybersecurity capabilities of the individual EU countries, e.g. they must have a national CSIRT, perform cyber exercises, etc.
- *2. Cross-border collaboration:* Cross-border collaboration between EU countries, e.g. the operational EU CSIRT network, the strategic NIS cooperation group, etc.
- *3. National supervision of critical sectors:* EU Member states have to supervise the cybersecurity of critical market operators in their country: Ex-ante supervision in critical sectors (energy, transport, water, health, digital infrastructure and finance sector), ex-post supervision for critical digital service providers (online market places, cloud and online search engines)

There are a few key takeaways that can be used by the DoD to effectively implement a mandatory reporting system. First, the DoD needs to ensure that it works with other U.S governmental agencies such as the FBI, CIA, and DHS to ensure that all agencies are on board with the policies, willing to help out, and able to effectively contribute to the problem at hand. Communication and collaboration is important in solving all sorts of problems and given the complexity of cyber-attacks, it is needed now. The other main idea to take from the EU is to determine which sectors are of importance to the health of the U.S. and this includes economically, medically, and defensively among others. Given the complexity of just the needs of the DoD, it is imperative that all critical sectors are determined and analyzed for their cybersecurity. The recent pandemic how shown how relatively few component shortages across several industries can affect the U.S. as it has led to production delays and re-designs. Given that cyber-attacks can also disrupt with the supply chains of these same industries more severely and rapidly, the DoD needs to be aware of where its resources should be dictated to combat cyber-attacks thus preventing the most critical production setbacks. With the large amount of possible rules and regulations that can be enacted, the DoD needs to use the success/failures of the EU NIS Directive to better implement their own policies. The proposed rules along thoughts on increasing participation, the benefits of mandatory reporting for both the DoD and business will now be discussed.

4. Discussion of National Security

U.S. contractors pride themselves on their ability to work and support the DoD. Cyber-attacks that expose key secrets and cyber security flaws put the DoD's advantages and intelligence at risk, therefore jeopardizing the national security of the U.S. U.S. businesses often develop key technologies to advance DoD operations both in the cyber domain and with physical assets; allowing foreign adversaries to steal critical information on these targets both improves their defensive response to our attacks as well as accelerating their own development of similar capabilities. An estimated 18% of attacks were performed by state actors, leading to the point that valuable IP and data are of importance instead of money (Council of Economic Advisers-2018). With this study being done 3 years ago and cyber-attacks continuing to grow from state-backed entities, more and more secret information and IP will keep being stolen from U.S. companies. One example of such an attack can be seen in the loss of valuable contractor IP through F-35 hack. Lockheed Martin was hacked by Chinese nationals, aiding in the quick development of a similar plane, the J-31 using F-35 data as evidenced in the similar design and intelligence gathered by the DoD. This is a considerable cost in national security as the advantage of the F-35 is mitigated by the abilities of J-31 as well as better understanding to counter F-35 through other ground/air to air attacks (Council of Economic Advisers 2018).

U.S. businesses also often serve as software contractors to the DoD and other government entities, resulting in lots of interactions and sharing of information between the two. Foreign actors can target weaker private cyber security systems first before then targeting DoD systems, making their chance of success much higher than a direct attack on DoD internal systems. This is since cybersecurity is like a chain; it is only as strong as its weakest link. Considering the DoD relies heavily on contractor work in terms of personnel and software, the DoD needs to ensure that these companies not only meet the high standards of the DoD cybersecurity but also report these incidents quickly. One area where the DoD is vulnerable due to the work its contractors do are through supply chain attacks where software and firmware that are used from private businesses are attacked and used against the DoD. The recent SolarWinds attack was a key example of enemy state actors using vulnerabilities in contractors such as SolarWinds, Microsoft, and VMWare to get into DoD systems as well as the U.S. Treasury and allied organizations such as NATO (Financial Stability Board-2021). While the current damage is still not fully known, it is believed that key policy and financial information was stolen in the attack which could affect the ability of the DoD to communicate and develop new technologies without risk of state actors knowing about their development. If not for the cybersecurity firm FireEye voluntarily releasing the information to the U.S. government, the lack of mandatory reporting could have cost the U.S. months to realize the issue and resolve it, leading to more information being stolen and used by our enemies.

Another aspect of the risk that the DoD faces is how the U.S. government sets standards for key contractors including those in cybersecurity which often propagates throughout other U.S. businesses. With these contactors often relying on other large U.S. cyber-focused companies to provide software, firmware, and other technology that is used, these large businesses tend to set the guidelines and rules for most cybersecurity. Poor cyber security business standards for large businesses will propagate through the entire U.S. business sectors as smaller to medium size companies look to market leaders to ensure their products are similar in security. Currently, few standards exist at which cyber-attacks are reported, what must be reported, how long after the attack it must be reported with different jurisdictions having different requirements. Additionally, no standard for cyber terminology along with these previous factors makes cyber incident reports from companies widely vary (Cisco-2021). As per recent history, most companies will most likely meet the minimum standards of both cybersecurity and reporting which more than often have proven not to be enough. Until larger businesses get on board with better reporting standards and cybersecurity, both they and smaller companies will not make the change to be able to better block and mitigate cyber-attacks. This is especially true with small businesses as when they see larger businesses struggling with cyber issues and receiving no or minimal government sanctions, what is the incentive to upgrade their security. They also use the fact that they are a small business meaning that they may be able to fly under the radar as they cannot provide a large monetary ransom to bad actors meaning they are less likely to be attacked. This shows in the IT spending data from the Small Business Administration is less than $120 million for FY21. (Bluestein-2021) This number is a miniscule amount compared to the large spending done by bigger companies, only exacerbating the issue of cybersecurity. While many companies pride themselves on being supportive of the U.S. military and DoD, most companies will struggle to make serious changes in their reporting if it does not positively impact their bottom line.

5. Discussion on business aspects for corporations

Businesses often initiate change if the bottom line is affected, so the DoD needs to highlight the incentives to improve cyber collaboration between the U.S. government and private entities. This collaboration will allow for such as more efficient purchasing and deployment of cyber resources in addition to limiting costly attacks. Corporations must spend large amounts of financial and human resources to develop highly secure cyber systems and protocols with estimates believing around 3.8 billion was spent on IT in the 2020 fiscal year (Ziff Davis-2022). Even if one considers that year to be an aberration with transitions to working from home and heightened expenses for COVID, better understanding of current attacks on the industry and sharing of the cost of cyber security advances could cut down on those costs. The first example would be in purchasing and setting up internal systems that are more secure and less likely to be exposed to ransomware. Not only would this lead to more efficient purchasing of IT equipment instead of each company going at it alone but the savings in limiting ransomware would be significant as well. It is estimated that between $57 to $109 billion dollars was lost in the U.S. economy alone due to cyber-attacks in 2018 (Council of Economic Advisers 2018). Combined with the fact that this number is only from reported incidents to the DoD where the ransom could be confirmed, it is widely believed that the actual cost is much higher (Council of Economic Advisers 2018). Additionally, most firms believe cybersecurity and IT spending in general will continue to increase at a rate higher than that of their revenue (Ziff Davis-2022). Companies are already planning and having to spend large amounts on new cybersecurity, so efficient spending is a must to limit malware attacks while maintaining profit margins. By working with the DoD and other U.S. entities, companies can alleviate spending in addition to creating a more effective defense through mandatory reporting by determining current malware in existence, ensuring up-to-date patches from other software providers, and best security practices from the DoD to setup internal systems. By limiting the amount of money wasted on IT hardware and software that does not actually help against malware threats, companies will be able to see higher profit margins through lower costs.

In addition to the high costs of setting up cybersecurity or paying ransomwares, Corporations spend large amounts on research and development (R&D) to develop new technologies to gain advantages on competitors and increase profits. Poor cybersecurity can leave IP, communications, and other private company data vulnerable to bad actors of whom may be willing to sell to the highest bidders. Given the importance of, they can easily take the files they have stolen and sell them to rival companies with certain countries viewing corporate espionage as acceptable and even encouraging it. Company secrets can often be worth much more to competitors and a single breach could affect years of research and other built-up advantages, allowing competitors to catch up and reduce possible revenue from highly secretive projects. Looking at Apple and their development of the iPhone, one can realize how crucial secrecy really is due to how game changing the device really was. The iPhone became so dominant in part by how advanced it was compared to competitors at the time with features such as multi-touch and the compression of apps into such a small device. If a competitor had gotten iPhone IP sooner, one only must think how less dominant Apple would be today as instead of taking years for others to develop a phone of similar or better quality, it took years in which Apple carved out a large chunk of the smartphone industry. Another example can be seen in the recent SolarWorld (not to be confused with SolarWinds) attack in terms of theft of both key IP and long-term strategy planning. In the SolarWorld attack, critical IP and trade secrets were stolen by Chinese hackers that enabled Chinese companies to duplicate their technologies at a cheaper price leading to a loss in market cap and future business loss that exceeded $150 million. To add further damage, financial documents were stolen allowing competitors to determine their future strategy to attack SolarWorld on costs through pricing where SolarWorld cannot make a profit or determine suppliers/customers they rely on (Council of Economic Advisers 2018). This simple cyber-attack has not only cost the company millions in lost profit, but to gain an advantage they will have to go a new direction in the industry to stand out in the market or hope that their R&D can generate another product that is better than their competitors which could take several more years.

Another area where corporations often spend large amounts of money is on their public relations (PR) to reflect key company values and the importance of hot-button issues such as privacy and secured data. A breach could hurt consumer confidence in the brand and expose internal documents that they may not want the public to see. Corporations value their positive brand awareness and a simple hack has the potential to tarnish the image the company has worked so hard to build up. One example is Facebook as several privacy scandals such as the Cambridge Analytica scandal have caused irreversible damage to the company. Despite their large status, many people have no confidence in Facebook when it comes to key issues such as privacy, honesty, and transparency. The lack of privacy has severely hurt the brand and one can wonder if the name change reflects the desire to shed the Facebook name and replace it with Meta (Jun,Kostyuk-2021). All the previous work, time, and money spent on PR is now down the drain due to a simple cybersecurity incident.

Finally, many U.S. companies have contracts with the U.S. to provide all sorts of advanced goods and services with many of them being secret in nature due to the heightened importance of the work being done. Repeated issues with cyber-attacks and reporting them to the DoD could hurt future chances at contracts as the U.S. government may be concerned in the ability of the company to execute the contract or safely protect DoD assets when they are entrusted to them. The DoD values the cutting edge technology provided by key U.S. businesses, but if they cannot secure it then the advantage will be short-lived. If the pattern repeats, the DoD is unable to rely on the company to deliver the advantages as they need as they will constantly worry if our adversaries already have seen this asset and are able to successfully defend against it. This will cost the company key revenue and profit as most government contracts are highly profitable for the companies who receive them. While certainly many positives to the arguments can be made for businesses to report cyber-attacks to the DoD, significant drawbacks exist that might adversely result from this practice.

6. Counterarguments to mandatory reporting

While U.S. businesses want to prevent cyber-attacks, mandatory reporting to the DoD would lead to adverse effects that would do more harm than good for most U.S. businesses. Often as a group grows larger, it is hard to maintain secrecy and not divulge sensitive information. Since many businesses would be required to partake in mandatory reporting and thus would see all shared data, bad actors could leak information to hackers to better target attacks, determine which attacks are not useful to limit wasting resources on them, and report on which companies are the worst at preventing cyber-attacks. Even if we could determine these bad actors, as they could create new businesses with the sole intention of observing information passed in the group. It would also be nearly impossible to restrict or block businesses from joining the group where cybersecurity is shared upon due to U.S. belief in equality of competition, meaning it would be impossible to prevent bad actors from seeing information in this database such as attacks that have succeeded and patches that have guarded against other attacks. Additionally, current SEC reporting standards require public companies to disclose IT spending, successful/failed attacks, and description of insurance coverage. (Division of Corporation Finance - SEC 2013) This information can allow for hackers to better target companies with minimal resources for maximum success as they can gauge what attacks work best and what companies can pay the most for it. The DoD would also need to regulate how information would be better shared to ensure that businesses stay secure without giving better financial information to certain investors or foreign entities.

Another pressing issue is the cost of upgrading cybersecurity as customers don't want to have to pay extra for something that they don't pay for already and businesses don't want to want to cut their profit margins while having to compete with limited resources and other firms overseas. This can be better realized in smaller U.S. businesses as they may not be able to afford and scale up cyber defenses at a rate of bigger companies, making them for vulnerable to attacks as well as identifying these businesses to a wider group of bad actors. The biggest issue is that small businesses tend to have even thinner margins and working capital than larger operations, meaning that they often lack the funds to divert to cybersecurity as well as the inability to reduce profits through higher prices. They also may require a dedicated employee or hire a contractor/consultant to design/install their system as they may not have the cybersecurity knowledge necessary to implement an effective system given many businesses have a low amount of personnel. Another issue for smaller companies to join mandatory reporting is they may be stuck following federal rules on how to deal with a cyber-attack given their lack of political influence. Smaller businesses have less leverage and influence with the federal government meaning that if they do experience a cyberattack, they may be at the guidance of the federal agency instructions. This can be seen through the Kaseya investigation as the federal government as the FBI held the victim decryption key longer after paying the ransom to reverse engineer it to stop future attacks causing harm to the businesses affected by the attack but possibility benefitting their rivals (Jun,Kostyuk-2021). In addition, many of the businesses still did not invest in cybersecurity measures after the attack while other hackers could shift their resources away towards new attacks (Jun,Kostyuk-2021). With small companies at the mercy of federal agencies and their instructions on solving the cyber-attack then it might cause more trouble than simply going at it alone.

The last key argument against collaboration through mandatory reporting is that U.S. companies pride themselves for their innovations and technological advances, so obligatory sharing data or IP on cyber-attacks make companies more hesitant to invest and even share cyber incidents so that competitors cannot glean any insight into them (NIAC-2017). Many companies already take their cybersecurity more seriously and have spent the money and effort to develop tough systems that have been able to stand up to attacks, making them weary of sharing this info with competitors who can build the same system for cheap nullifying any competitive advantage. This will end up with some companies just doing the bare minimum and relying on the government and other companies to make improvements or spend their resources to implement effective cybersecurity before they make an effort to improve their own cybersecurity since now they can just copy. Competitors often try to glean any information about their competitors as evidenced by how outside company personnel read reports where the SEC requires all companies to disclose information what cybersecurity measures they have implemented and how they are working. (Division of Corporation Finance - SEC 2013) Competitors can easily look up the public information and the success to better determine where to invest in. This will allow them to catch-up to any cybersecurity advantages their competitors may have leading to a race to the bottom where a simple few companies or government efforts are used to build the cyber infrastructure for every other company. While some important points were brought up regrading cost and future risk to U.S. businesses which will need to be evaluated to develop mitigating solutions, many solutions can be found to solve or at least mitigate these problems as well as more benefits that outweigh the possible downsides.

7. Refuting counterarguments

The DoD will have challenges with getting U.S. companies on board with mandatory reporting but taking key steps and providing safeguards to companies will allow for an effective tool to maintain high national security and improving cybersecurity of private sector. The first key will be ensuring that outside actors cannot distinguish which companies are being discussed in mandatory reporting so as to not alert which companies may have weaker security measures. To maintain secrecy, companies will only be identifiable to the DoD and not to other companies. In addition, cyber-attacks will be brief to limit understand of how to produce these attacks with most of the shared information reporting being solutions to defend against similar versions of the attack. Secrecy requirements and rules already exist for cybersecurity attack reporting. In addition, current mandatory reporting of cybersecurity risks already occurs meaning that the infrastructure for companies to interact with the SEC and other agencies already exists. It would not be a burden on the companies and only strengthen both federal help and future systems (SEC-2013). In order to ensure that the reporting data is secure, the DoD can take note of other anonymous reporting systems exist such as HIPAA as they protect patient confidently through secure systems and compartmentalization of data to limit human access. Similar systems with DoD grade security can be used to secure the data properly and limit access both within and outside of the DoD. In addition, the companies will have to disclose cyber-attacks in SEC filings to investors, so the attack will be must public in-time, diminishing how important it is for companies to remain anonymous in these attacks.

Another issue is how to ensure small businesses build up their cyber capabilities and also deal with issues such as limited operating capital and small margins. Therefore, more effort needs to be put into the Small Business Administration (SBA) to assist these companies with some funding earmarked specifically for cybersecurity as well as technical support to get the system up and running. This shows in the IT spending data from the SBA as it is less than $120 million for FY22 (Bluestein-2021). They cannot afford to spend lavishly on this area so targeted spending will yield better outcomes in preventing key attacks such as ransomware. Additionally, small U.S. businesses will be able to see solutions proposed to defend against attacks based on their industry and their operations so they can better focus their limited resources on certain attacks through data gathered by mandatory reporting. Through data collected by mandatory reporting, U.S. officials can help companies of this size on the most common attacks being conducted against them and how to build up their systems at an affordable rate to stop or better mitigate these attacks. At a minimum, this will establish a solid baseline to make bad actors put in some serious effort to get around their defenses while also ensuring that other small business can learn from the attack and not suffer the same fate.

Lastly, cybersecurity is such a rapidly changing field where constant innovations are required, meaning constant investment is needed. The reward is that companies who invest and have tougher cybersecurity not only experience less successful attacks but are also able to reduce the amount/importance of information stolen as well. This will also push bad actors to target other entities driving down the total number of attacks as they will want to go after easier targets as the effort required to obtain the reward is not worth it compared to other companies. Based upon the key points refuting the opposing viewpoints of mandatory reporting of cyber-attacks along with the advantages discussed earlier, mandatory reporting of cyber-attacks and sharing key details is a win-win for the public-private sector alliance the U.S. employs today for the DoD.

8. Conclusion

U.S. corporations and other non-governmental organizations that fit within the criteria listed in the second paragraph should be mandated to report all cyber-attacks up to and including the resolution of the attack to the DoD due to the interests of national security, efficiently building cyber security to minimize costs, and protecting individuals affected in these cyber-attacks. With U.S. businesses often developing key technologies to advance DoD operations both in the cyber and physical domain, poor cybersecurity will allow foreign adversaries to steal critical information or compromise communications on these targets resulting in improved enemy defensive responses to our attacks as well as accelerating their own development of similar capabilities. Another issue is that since U.S. business often serve as contractors to the DoD and other government entities, large volumes of interactions and sharing of information and cyber assets between the two occur. Therefore, foreign actors can target weaker private cyber security systems first before targeting DoD systems, making their chance of success much higher than a direct attack on DoD systems. Finally, most U.S. businesses often rely on other U.S. cyber-focused companies to provide software, firmware, and other technology that is used. Poor cyber security business standards for large businesses will propagate through the entire U.S. business sector putting DoD and U.S. businesses at risk of being attacked. The DoD needs to continue to get more serious about the possibility of cyber-attacks and realize that internal DoD systems are not the only target for our many adversaries. If the DoD is exposed on the cyber domain, our ability to produce advanced technologies at an industrial scale, which has been in a key factor in winning previous global conflicts, will be easily surpassed by our adversaries threatening the ability to defend ourselves in future conflicts.

Disclaimer: The views expressed are those of the authors and do not necessarily reflect the official policy or position of the Air Force, the Department of Defense, or the U.S. Government.

References

Bluestein, Keith - U.S. Small Business Administration. Information Technology Agency Summary. (2021, Sept. 30) Extracted from U.S. Small Business Administration Website: https://itdashboard.gov/drupal/summary/028

Cisco. Data Privacy Benchmark Study. (2021) Extracted from Cisco Website: https://www.cisco.com/c/dam/en_us/about/doing_business/trust-center/docs/cisco-privacy-benchmark-study-2021.pdf

Council of Economic Advisers. The Cost of Malicious Cyber Activity to the U.S. Economy. (2018, Feb.) Extracted from the Department of Homeland Security Website through PDF download.

Davis, Ziff. 2022 State of IT. (2021, July) Extracted from Spiceworks Website: https://swzd.com/resources/state-of-it/#soit-2022

Division of Corporation Finance - SEC. CF Disclosure Guidance Topic No. 2. (2013, Oct. 11) Extracted from the SEC website: https://www.sec.gov/divisions/corpfin/guidance/cfguidance-topic2.htm

ENISA. NIS Directive. (14 Oct 2016). Extracted from EU Website: https://www.enisa.europa.eu/topics/nis-directive

Financial Stability Board. Cyber Incident Reporting: Existing Approaches and Next Steps for Broader Convergence. (2021, October 19) Extracted from the FSB Website: https://www.fsb.org/wp-content/uploads/P191021.pdf

Foret, Will – Forbes. Using Cybersecurity as a Competitive Advantage. (2019, Oct. 9) Extracted from Forbes Website: https://www.forbes.com/sites/forbesbusinesscouncil/2019/10/09/using-cyber-security-as-a-competitive-advantage/?sh=35d3ee1c7ff7

Jun, Jenny and Kostyuk, Nadiya - Lawfare. The Pros and Cons of Mandating Reporting From Ransomware Victims. (2021, Nov. 1) Extracted from the Lawfare Website: https://www.lawfareblog.com/pros-and-cons-mandating-reporting-ransomware-victims

NIAC. Securing Cyber Assets: Addressing Urgent Cyber Threats to Critical Infrastructure. (2017 August) Extracted from NIAC (National Infrastructure Advisory Council) Website: https://www.cisa.gov/sites/default/files/publications/niac-securing-cyber-assets-final-report-508.pdf

SEC. Commission Statement and Guidance on Public Company Cybersecurity Disclosures. (2018, Feb. 26) Extracted from the SEC website: https://www.sec.gov/rules/interp/2018/33-10459.pdf

Obstacles on the Path to the Internet of Things: The Digital Divide

John Gray
Nova Southeastern University, Fort Lauderdale, USA
jg1553@mynsu.nova.edu

Abstract: The Internet of Things holds has the potential to provide an array of technological benefits and online resources to individual users and society in general. However, the Digital Divide, the gap between information computing technology (ICT) and those who can effectively take advantage of it, presents challenges to the global implementation of the Internet of Things. Factors contributing to the Digital Divide include lack of broadband access, cost of ICT, user socioeconomic challenges, user security concerns, and political or governmental restrictions.

Keywords: internet of things, digital divide, internet access, online content restrictions, freedom of access

1. Introduction

There are varying definitions of the Internet of Things (IoT) in both the practitioner and academic communities. Gartner Research defines the IoT as "the network of physical objects that contain embedded technology to communicate and sense or interact with their internal states or the external environment". The International Telecommunications Union states that the IoT is "a global infrastructure for the information society, enabling advanced services by interconnecting physical and virtual things based on existing and evolving interoperable information and communication technologies"; and the Oxford Dictionaries defines the IoT as "a development of the Internet in which everyday objects have network connectivity allowing them to send and receive data" (Teppler, 2015). The Pew Research Center defines it as a global network of information computing devices, electronics, and sensors which will provide real time data and information that can positively enhance people's lives (Anderson and Rainie, 2014).

Others choose to break the term down – with the Internet being described as being the commercial, educational, and government information systems which form a single worldwide network which is interconnected by protocols that are determined by the Internet Architecture Board (IAB), and in which the Internet Corporation for Assigned Names and Numbers (ICANN) oversees the names and address spaces (CNSS Instruction 4009, 2015); or the Internet being termed as an internetwork that encompasses large geographical areas, "enabled and managed" by a set of common and accepted ports, protocols, services, and interconnected devices and technologies as defined by the IAB and ICANN (Oriwoh and Conrad, 2015). These commonly agreed on ports, protocols, and services facilitate communication and the exchange of information between interconnected entities and devices. Oriwoh and Conrad (2015) also clarify that "of" makes it unmistakable that the Internet is comprised of specific items or "things." Patel and Patel (2016) build on this by defining the IoT as an environment of a variety of objects that interact with each other through wired and wireless connections to create services and applications. They advocate that the IoT includes numerous types of items – including vehicles, appliances, medical and industrial systems, buildings, and even humans which communicate and interact using common protocols and addressing schemas to achieve a particular goal.

Whichever description is used for the IoT, what is agreed upon is that connected devices will impact and improve careers, educational opportunities, health services, and overall quality of life of those individuals who participate in it. The myriad of potential uses include real time tracking of health and fitness activities, control of residential appliances and utilities, and self-reporting of equipment/device maintenance and repair needs. Envisioned future uses include incorporation of large numbers of devices that generate and require information, such as robotics, self-driving automobiles, automated machinery, and a wide range of living beings – to include animals and plants (De Guglielmo, Anastasi, and Seghetti, 2014). As illustrated in Figure 1, the IoT will touch or influence most aspects of people's lives (Tech Team Tree, 2016).

It has been estimated that in 2020 there are currently four IoT devices for every person on earth - exceeding 30 billion connected devices worldwide, and as shown in Figure 2 the number is projected to increase to over 75 billion devices by 2025 (Greenouch and Camhi, 2016; Statista, 2020).

Figure 1: Internet of Things

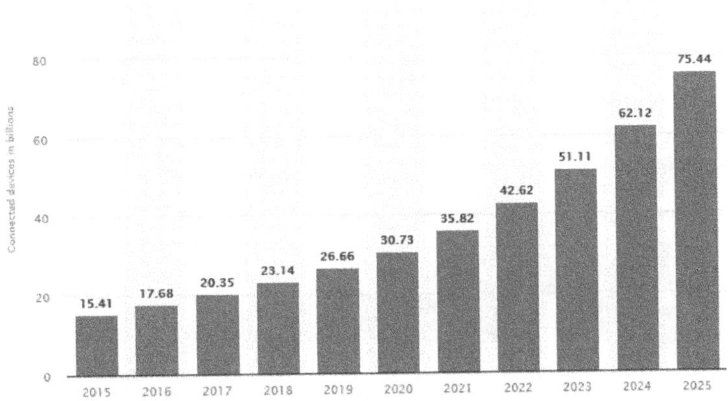

Figure 2: IoT connected devices in billions – 2015 to 2025

The benefits of the IoT will be to provide tools to effect positive changes to individual behaviors such as making healthier choices, safer decisions, and being more efficient in various activities (Anderson and Rainie, 2014). However, the growth, use, and effectiveness of the IoT also has the potential to contribute to and be affected by the so-called Digital Divide.

2. What is the Digital Divide?

The concept of what the Digital Divide is has changed over time. Previously it was characterized as the disparity between people who had access to Information Computing Technology (ICT) – the computing hardware, software, and access to the Internet, and those who did not (Goth, 2005). This early definition meant that the divide was based primarily on factors such as income, education, occupation, and geographical location (van Dijk and Hacker, 2003).

While access was described as having a computer connected to the Internet, van Dijk and Hacker (2003) further interpreted access to include a person's lack of experience with ICT due to not having an interest in being a user, users having a fear of the technology, insufficient user skills due to lack of education or social support, or individuals having few opportunities to use the technology.

Currently the definition of the Digital Divide includes that of users having poor quality ICT devices, not having an affordable connection to effectively use devices, or having dialup or restricted wireless connections versus high speed access as a person's connection to the Internet (Crawford, 2011). Soltan (2019) advocates that while information technology improvements and increased internet access have addressed many of the technology's earlier accessibility issues – a divide still exists based on the financial income levels of users, with "poor" people having less access to various digital resources – particularly those that are bandwidth intensive.

At present, ICT is described as consisting of information, resources, applications, and services; including computers, software, digital television, mobile phones, and telecommunication and broadband technologies (Selwyn, 2004). Emerging uses include devices for security controls, health monitors, sensors, traffic management controls, fitness trackers, and household appliance/device controls. While physical access to these devices may be available, the issue of sufficient user skills and knowledge to effectively use and take meaningful advantage of the available resources and information which users can access remains an issue. Additionally, the absence of a high-speed connection can limit how effectively digital resources can be used (Soltan, 2019).

Early social and political opinion was that the Digital Divide would be eliminated once every individual had a computer connected to the Internet. In 2016 approximately 88.5% of the United States population - representing 8.4% of the world's Internet users, had access to the Internet; however, only 40% of the world's population had an Internet connection (Internet Users, 2019); and a high-speed connection was not available to all of those users. By mid-2019, as detailed in Table I, during the years of 2000 to 2019 the percentage of world Internet users had increased by 1,157 percent. While the number of Internet users has increased to 58.8% of the world population, that still leaves over 40% without access to the IoT (Internet World Stats, 2020). Moreover, of the 58.8% who can get online many have limited or restricted access to digital resources.

These numbers indicate that a significant portion of the world's population did not have the opportunity to benefit from the emerging technology of the IoT. And events have shown that just because an individual has access - if they choose not to utilize it or if the bandwidth or access to digital resources is restricted, then the Digital Divide remains (van Dijk and Hacker, 2003). Consequently, the Digital Divide could more accurately be characterized as who can benefit from the IoT technology and who cannot.

Table 1: World internet usage 2019 mid-year estimates

World Region	% World Population	% of World Internet	% of Pop. Penetration Rate	% Growth 2000-2019
Africa	17.1	11.5	39.6	11,481
Asia	55.0	50.7	54.2	1,913
Europe	10.7	16.0	87.7	592
Latin America	8.5	10.0	68.9	2,411
Middle East	3.3	3.9	67.9	5,243
North America	4.7	7.2	89.4	203
Australia	.05	.06	68.4	276
World Total	100	100	58.8	1,157

3. Does the divide exist?

The existence of the Digital Divide has been well documented by researchers since the mid-1990s. Shortly after the Internet began being used by the public, the Digital Divide was recognized as an issue. As a result of the United States (US) technology sector leadership being challenged by then President Clinton to address Information Computing Technology disparities between those citizens who had access and those who did not, in 1999 the Digital Divide Network was established by the National Urban League and the Benton Foundation (Goth, 2005). While government policy has addressed many of those issues in the US, as noted by Goth (2005) they continue to be issues in many other areas of the world. Numerous nations have established policies and programs to ensure their citizens do not get "left behind" as a result of the implementation of ICT, particularly in the areas of access to technology and information. These policies and programs address Digital Divide issues between social groups within each specific country and in the global economy Selwyn, 2004). The United Nations ICT Task Force was established to address the Digital Divide problem worldwide; and while it was originally thought that the task force would not need to exist beyond 2004 (Goth, 2005), the task force is still currently in existence with the ongoing mission of offering policy advice to world governments and to assist in establishing partnerships between technology companies, nations, private industry, and other organizations in bridging the Digital Divide.

A high-speed Internet connection has become a key tool for participation in society. The expectation that job seekers, employees, students, patients, and consumers to use the Internet have evolved broadband from a luxury into a necessity. Institutions are increasingly assuming that their customers have online access, and they changing their service and business models accordingly (Anderson and Rainie, 2014; Crawford, 2011; Shapiro, 2016).

Numerous national governments around the world utilize technical, regulatory and censorship strategies to regulate access to online content. Politically imposed restrictions and obstacles to Internet freedom and information access create another category of users considered to be among those who are digitally divided.

4. Who are the digitally divided?

In 2000 the Digitally Divided were generally defined as the people who had access to the necessary ICT and associated connection to the Internet as opposed to those who did not; termed as the information haves and have-nots (Wresch, 1996). This has been further expanded to include the "information want-not's", those individuals who either have a fear of or a feeling of insecurity when interacting with information technology, or that they have no interest in its use (van Dijk and Hacker, 2003). It has been estimated that the digitally divided population numbers exceed four billion people (Smith, 2010).

Even within technologically developed countries such as those in Southeast Asia, the US, and in Western Europe there remain geographic or social groups where citizens would be classified as digitally divided (Selwyn, 2004). They are deprived of the benefits of meaningful access; usage that either can provide them an escape from poverty, that would empower them to improve their lives, help sustain the world's markets, or provide solutions to their problems and issues (Smith, 2010). Lack of access makes it harder for them to find work or to train for in-demand skills job skills that would qualify them for good-paying jobs.

Other characterizations include people who have access to the information, but who do not use it for meaningful benefit or do not understand how to effectively use it in order to create real benefit to themselves. These individuals comprise a significant portion of the population of Third World countries, but it also includes various social groups and geographic regions in technologically advanced countries such as the US.

There are four primary areas that contribute to the digital divide – listed in Table II.

Table 2: Digital Divide categories

Lack of Broadband Service
User Socioeconomic Challenges
Affordable Information Computing Technology
Political and Governmental Restrictions

Each category has specific issues that prevent or restrict access to IoT resources. Individuals that fall into one of the four categories of the digital divided are prevented from fully participating in and benefiting from the Internet of Things.

4.1 Lack of broadband

A key aspect of achieving digital inclusion is the availability of broadband service with the speed and reliability required by users to make the capability worthwhile. In many sparsely populated, rural, or low-income areas broadband connectivity is unavailable, unreliable, or the required infrastructure is underdeveloped (Crawford, 2011; Bates, Malakoff, and Kane, 2012).

In other instances, cost may not be a barrier to use, and users may be willing to pay, but a broadband service may not be available. Nations with rural or isolated areas are particularly prone to an uneven distribution of quality service. In some cases a quality broadband connection may be available but the benefits of the connection to a first time user are outweighed by the cost.

Previously connected users who have subsequently cancelled their broadband service cite the high cost of maintaining the connection, the increasing availability and opportunity to access the Internet elsewhere such as in a community library or other public locale, and the inadequacy of their ICT equipment or service as reasons for discontinuing their use of broadband. These types of users are termed "un-adopters" by Whitacre and Rhinesmith (2016).

The result is that a significant number of people are unable to benefit from the technology.

4.2 User socioeconomic challenges

Social and economic factors such as age, education, financial income, gender, occupation, and geographic location are demographic determinants in whether an individual is considered to be digitally divided. In many cases, the cost for a high-speed connection is prohibitive – even though the user may understand of the value of home broadband and the service is available, they simply cannot afford the price. Therefore, they either use a connection that is not broadband or do not have any type of connection at all (Crawford, 2011; Bates, Malakoff, and Kane, 2012; Rhinesmith, Reisdorf, and Bishop, 2019). Low-income households have historically had poor broadband adoption rates, and the number is even more pronounced along ethnic lines (Soltan, 2019; Shapiro, 2016).

Online education relies heavily on streaming videos and live feeds that require a high-speed connection to be effective as an educational tool. While broadband access may be available at their educational facility, many students from low-income households lack an adequate connection at home. This hampers their ability to participate in many leaning activities. Teachers of low-income students reported more obstacles to effectively using this technology as a teaching aid because of inadequate access (Crawford, 2011; Soltan, 2019; Shapiro, 2016). Ultimately, this places the students at a learning disadvantage.

Van Dijk and Hacker (2003) and Idiegbeyan-Ose et al. (2018) also point out that learned cultural and social skills play a role in processing the meaning and taking advantage of any information that is accessed. The meaning of available information and how to use it is lost on many people if they cannot relate to it or place it into context with their cultural background and experiences. Not having these skills is a contributing factor in who is termed as digitally divided.

4.3 Affordable ICT

Related to the cost of a broadband connection is the affordability of ICT. The cost for ICT equipment such as computers, internet modems, and software is out of reach for many users and subsequently contributes to the digital divide (Idiegbeyan-Ose et al., 2018). Additionally, the design model of the IoT and the wide range of devices that make up the IoT introduces security concerns at the physical, transport, and application layers (Patnaik, Padhy, and Raju, 2021).

Growing end-user concerns about data and privacy protections contribute to the digital divide. Reports of cybersecurity breaches and user data being lost, stolen, or compromised are frequently in the news. The cost of data protection tools, solutions or services, and the associated user skills required to provide effective security for their personal devices, data, and privacy are a challenge to many users (Lee and Ahmed 2021). The result is that many users choose not to fully engage in the benefits of the IoT because by doing so they feel that they may be placing their sensitive data or personal privacy at risk of being compromised.

The general consensus is that there is a growing inequity in ICT user skill levels because of the types of technology being produced (2003). This is exacerbated by the perception among technology companies that there is very little profit in selling products that are inexpensive and only have basic functionality. It is contended that producers of ICT make production decisions based on profit motives. This is based on the fact that 80% of technology profits are made from marketing products to the most affluent 20% of society (Smith, 2010. This results in the development of more advanced products for experienced users while the "have-nots" continue to be denied access to current technology.

The result is that high-end products are not affordable to a significant number of users, and that less experienced users lack the skillset required to use the advanced technology. Consequently, their use may be restricted to outdated, less capable technology or that they may not have access to any form of ICT. These decisions ultimately affect user access to digital technology.

4.4 Political and governmental restrictions

An interesting perspective is that some political groups and governments may actually promote the existence of the Digital Divide (van Dijk and Hacker, 2003). The claim is that the divide increases income, occupational, education, and social class differences which can be exploited for political gain. It is not uncommon for political groups or governments to promote the divide in order to advance their specific agendas by restricting free communication, religious and political participation, and economic activities (Shirazi, Ngwenyama, and Morawczynski, 2010) or other online content restrictions.

Politically motivated blocking of digital communication and knowledge acquisition occurs in numerous societies across the globe. Governments implement various tools and controls to censor speech and restrict access to information. Various groups and organizations from countries including Turkey, Saudi Arabia, Mongolia, Iran, and China struggle to access and post content on-line. The blocked or restricted information includes political information and content critical to the ruling political faction, content on embarrassing medical conditions, and controversial social issues (Nekrasov, Parks, and Belding, 2017). Freedom House is an independent, nongovernmental organization that conducts research on political freedom and human rights. In 2018 Freedom House ranked 65 nations for their degree of online freedom with zero being the most free and 100 with the most restrictions (Shahbaz, 2018). A subset of those nations are listed in Table III to illustrate that nations with authoritarian governments have the worst score for online freedoms. China scored the worst, having the most restrictions, and Iceland had the lowest score indicating they have the least restrictions to online resources.

In China government policies, businesses, and scientific institutions collaborate to control the development direction and management of the nation's IoT industry (Zhang et al. 2021) The People's Republic of China (PRC), as a sub-project of its Golden Shield Project, has instituted a combination of legislative actions, regulatory barriers, and technology solutions to institute Internet surveillance and control – commonly known as the "Great Firewall of China". This government controlled gateway provides censorship and control over the international connections to the global Internet and any information that is considered politically inconvenient or inappropriate to the ruling communist political party (Shahbaz, 2018; Lv and Luo, 2018). Local and foreign companies are required to cease transmission of what the government considers "banned" content as well as adapt to and abide by Chinese Internet regulations. The latest PRC directed effort is to ban all virtual private networks (VPNs) not under government control, which opponents state could erode Chinese scientists ability to stay connected with peers outside of the country (Shahbaz, 2018; Normile, 2017).

Additionally, PRC officials have worked with 36 of the 65 nations listed in the Freedom House survey in order to establish a network of countries that will "follow its lead on Internet policy" and laws. The result is that several nations with primarily authoritarian governments have introduced cybersecurity and cyber media laws that mimic those of the PRC (Shahbaz, 2018).

China's censorship system, and similar blocking/censorship systems of other nations, prevents their citizens from unencumbered access to digital resources – in effect creating a digital divide to scientific research, innovation, free thought, and commerce.

Table 3: Online freedom score chart

Nation	Score
Peoples Republic of China	88
Iran	87
Syria	87
Ethiopia	83
Cuba	79
Vietnam	76
Saudi Arabia	72
Russia	65
Turkey	61
India	43
Mexico	40

Nation	Score
United Kingdom	24
Japan	22
USA	19
Canada	16
Iceland	6

5. Addressing the Divide

Broadband and broadband enabled products and services are now the key to addressing the Digital Divide in order to shape the behavior of individuals and transform governments, businesses, education systems, and communities (Bates, Malakoff, and Kane, 2012). It is the opinion of digitaldivide.org that eliminating the divide will require that nations restructure their telecommunications infrastructure so that broadband is available to the majority of their population, not just the most affluent. Addressing user security concerns and ensuring that affordable Information Computing Technology and services are available to underserved or less economically well off populations are also important factors for closing the divide. While many nations have integrated technological training into their educational systems, others still struggle to provide basic educational services, which typically do not include instructing about or actually utilizing ICT. Consequently, the cost to obtain information is much more to impoverished peoples whose limited funds may be otherwise needed for day-to-day survival. Additionally, if they are also geographically isolated from access to ICT then their information isolation increases as the cost to travel or purchase technology to bridge the difference is more than they may be able to afford (Wresch, 1996). Another aspect to consider is that the uses of ICT technology must be meaningful to users or the Digital Divide could grow even wider (Smith, 2010). Digitaldivide.org contends that a significant number of users become caught up in the entertainment aspect of the technology and thereby waste time, cease their education efforts, and subsequently fail to contribute to society. This ultimately promotes continued poverty and ultimately results in increasing the divide. In short, inappropriate or un-meaningful access could be as damaging as no access at all. Moreover, while ICT may offer the benefits of automation and reduction of manpower requirements for businesses, it can result in the loss of jobs due to those reductions. If closing the Digital Divide is characterized as being benefits derived from access, then these losses could be considered a negative effect.

In countries that impose constraints, restrictions, and repression of access to IoT digital resources, affected citizens subsequently resort to various technical skills, digital applications, tools, and other circumvention techniques and methodologies in an attempt to access restricted content - particularly if those techniques provide anonymity. User workarounds to technological blocking of information flow is countered by new governmental blocking methods or regulatory enforcement, including punishment of violators. This results in a back and forth effort to block or gain access to content

6. Conclusion

The concept of a Digital Divide that consists of haves and have-nots is likely oversimplified. As can be seen by the changing definition and interpretation of the Digital Divide and the impact and consequences of the access to and the use of information technology on the quality of users' socioeconomic status - the problem is dynamic, multi-faceted, and complex. How that one defines the divide drives the definition of who is considered to be among the digitally divided. The numbers of the people considered as divided changes as well since the cost and approach to solving the issue depends on the definition used. What is not disputed is that a significant portion of the world population cannot or is not able to, or chooses to not take advantage of one of mankind's greatest achievements. The Internet of Things is poised to be a core component of personal, economic, and political life across the world. Successful implementation of the IoT is dependent on being able to bridge the Digital Divide by providing consumers with the required yet affordable broadband backbone necessary to support the myriad of connected devices; development of economically priced information computing technology which possesses ease of use qualities, compelling features, security, and benefits which promote user desirability; and education efforts which demonstrate the benefits of those devices to potential consumers. Additionally, the ability of world citizens to fully participate in digital information access and exchange in order to take advantage of the cultural, economic, educational, political and social opportunities the IoT affords remains vulnerable to the actions of political and governmental regulators. Certain aspects to enable reaching the digitally divided will need to be addressed through policy, regulations, and subsidies, international diplomacy, and public – private partnerships and cooperation.

References

Anderson, J. and Rainie, L. (2014) "The Internet of Things will Thrive by 2025", Pew Research Center, [online], http://www.pewinter net.org/2014/05/14/internet-of-things/.

Bates, K., Malakoff, L., and Kane, S. (2012) "Closing the Digital Divide: Promoting Broadband Adoption Among Underserved Populations", Port of Clarkston, [online], http://portofclarkston.com/ uploads/ Benefits %20of%20Broadband.pdf.

CNSS Instruction 4009 (2015) *National Information Assurance Glossary*, Committee on National Security Systems.

Crawford, S. P. (2011) "The New Digital Divide", *The New York Times*, Vol. 12, No. 03.

Goth, G. (2005) "Digital-Divide Efforts are Getting More Attention", *Internet Computing*, Vol. 9, No. 4, pp. 8-11.

De Guglielmo, D., Anastasi, G. and Seghetti, A. (2014) *A step towards the Internet of Things. In From IEEE 802.15. 4 to IEEE 802.15. 4e: Advances onto the Internet of Things*, Springer, Berlin.

Greenouch, J. and Camhi, J. (2016) "How the 'Internet of Things' will affect the world", Business Insider, [online], https://www.businessinsider.com/internet-of-things-2015-forecasts-of-the-industrial-iot-connected-home-and-more-2015-10.

Idiegbeyan-Ose, J., et al. (2018) "Digital Divide: Challenges for Library and Information Services Provision in Developing Countries," *Proceedings of 11th annual International Conference of Education, Research and Innovation, ICERI 2018, Seville, Spain*, pp. 0717 – 0722.

Internet Users. (2019) Internet Live Stats, [online], http://www. internetlivestats.com/ internet-users/.

Internet World Stats. (2020) Internet World Stats, [online], https://www.internetworldstats .com/ stats.htm.

Lee, C. and Ahmed, G. 2021. "Improving IoT Privacy, Data Protection and Security Concerns". *International Journal of Technology, Innovation and Management*, Vol 1, No. 1, pp.18-33.

Lv, A. and Luo, T. (2018) "Asymmetrical power between Internet giants and users in China", *International Journal of Communication*, Vol. 12, pp. 3877–3895.

Nekrasov, M., Parks, l. and Belding, E. (2017) "Limits to Internet Freedoms: Being Heard in an Increasingly Authoritarian World", *Proceedings of the Third Workshop on Computing Within Limits, ACM LIMITS 17,* pp. 119-128.

Normile, D. (2017) "Science suffers as China plugs holes in Great Firewall" *Science*, Vol. 357, No. 6354, pp. 856.

Oriwoh, E. and Conrad, M. (2015) "'Things' in the Internet of Things: Towards a Definition", *International Journal of Internet of Things*, Vol. 4, No. 1, pp. 1-5.

Patel, K. K. and Patel, S. M. (2016) "Internet of Things-IOT : Definition, Characteristics, Architecture, Enabling Technologies, Application & Future Challenges", *International Journal of Engineering, Science, and Computing,* Vol. 6, No. 5, pp. 6122-6131.

Patnaik, R., Padhy, N., and Raju, K.S. 2021. *A Systematic Survey on IoT Security Issues, Vulnerability and Open Challenges. In Intelligent System Design*, (pp. 723-730). Springer, Singapore.

Rhinesmith, C., Reisdorf, B. and Bishop, M. (2019) "The Ability to pay for Broadband", *Communication Research and Practice,* Vol. 5, No. 2, pp. 121-138.

Selwyn, N. (2004) "Reconsidering Political and Popular Understandings of the Digital Divide", *New Media & Society*, Vol. 6, No. 3, pp. 341- 362.

Shahbaz, A. (2018) "Freedom on the Net 2018. The rise of Digital Authoritarianism", Freedom House, [online], https://freedomhouse.org/report/freedom-net/2018/rise-digital-authoritarianism.

Shapiro, I. (2016) "FCC Broadband Initiative Could Reduce Barriers to Low-Income Americans' Advancement and Promote Opportunity", Center on Budget and Policy Priorities, [online], http://www.cbpp.org/sites/ default/files/atoms/files/fcc broadband initiative could reduce barriers to low-income americans advancement and promote opportunity.

Shirazi, F., Ngwenyama, O., and Morawczynski, O. (2010) "ICT Expansion and the Digital Divide in Democratic Freedoms: An Analysis of the Impact of ICT Expansion, Education and ICT Filtering on Democracy", *Telematics and Informatics*, Vol. 27, pp. 21-31.

Smith, C. W. (2010) "Digital Divide Defined (Hint: It's not About Access)," Digital Divide Institute, [online], http://www. digitaldivide.org/digital-divide/digitaldividedefined/digitaldivide/.

Soltan, L. (2019) "Digital Divide: The Technology Gap Between the Rich and the Poor," Digital Responsibility, [online], http://www.digitalresponsibility.org/digital-divide-the-technology-gap-between-rich-and-poor.

Statista. (2020) "Internet of Things (IoT) Connected Devices Installed Base Worldwide from 2015 to 2025", Statista Inc., [online], https://www.statista.com/ statistics/471264/iot-number-of-connected-devices-worldwide/.

Tech Team Tree, (2016) "IOT Devices to Touch 34 Billion By 2020", [online], https://www.tech tree.com/ index.php?q=content/ news/10889/iot-devices-touch-34-billion-2020.html.

Teppler, S. (2015) "The Internet of Things and Liability. Let the Lawsuits Begin...," *ISSA Journal*, Vol.13, No. 1, pp. 38-40.

van Dijk, J. and Hacker, K. (2003) "The Digital Divide as a Complex and Dynamic Phenomenon", *The Information Society*, Vol. 19, No. 4, pp. 315-326.

Whitacre, B. and Rhinesmith, C. (2016) "Broadband Un-adopters", *Telecommunications Policy*, Vol. 40, No. 1, pp. 1-13.

Wresch, W. (1996) *Disconnected: Haves and Have-nots in the Information Age*, Rutgers University Press, New Jersey.

Zhang, Z., Li, X., Xiong, J., Yan, J., Xu, L. and Wang, R. (2021) A Global Race to Dominate the Internet of Things: How China Caught Up. *Journal of Business Strategy*.

Societal Impacts of Cyber Security in Academic Literature: Systematic Literature Review

Eveliina Hytönen, Amir Trent and Harri Ruoslahti[1]
Security and Risk Management, Laurea University of Applied Sciences, Espoo, Finland
Eveliina.Hytonen@laurea.fi
Harri.Ruoslahti@laurea.fi

Abstract: The 2020 Allianz Risk Barometer, with 39% of responses, ranked cyber incidents as the number one risk threatening business continuity. Any organisation may face a number of challenges e.g. costly data breaches, ransomware incidents, and even litigation after an event. The Internet has, in many ways, changed society, transformed businesses, organisational communication and learning. People can now interact through social networking platforms. Modern society has become very technology driven, as ICT is now an integral component in peoples' lives. However, besides the many benefits that the Internet and other ICT technology bring, there are also threats, such as cyber-attacks looking to exploit vulnerabilities in ICT applications and systems. This study is a systematic literature review that explores how societal impacts of cyber security in modern society are discussed in academic literature. The Introduction discusses the overall importance of cyber security in today's society. The body of this paper presents the method in which the literature review was conducted, and a concise summary of the findings that answer the research question: How are societal impacts of cyber security discussed in academic literature? Six categories of investigation of societal impacts of cyber security are identified: 1) Impacts on Social and Societal Levels, 2) Detection of cyber-crime and incidents, 3) Critical infrastructures and services, 4) Impacts of incidents and individual technology, 5) Cybersecurity awareness, and 6) Cybersecurity and collaboration. Lastly, the conclusions, based on the research findings, address the feasibility, impact, strengths, weaknesses and possible ethical concerns of cybersecurity. This paper contributes to the overall understanding of current societal impacts of cyber security, and this understanding benefits the development of methods that assess societal impacts, as well as provides focus for future training and development of cyber and e-skills needed for better awareness of cyber threats, and to better address possible cyber incidents.

Keywords: society, cyber security, societal impacts

1. Introduction

In 2020, cyber incidents ranked as the most important business risk in the Allianz Risk Barometer (2020). Thus, businesses face a number of challenges such as large and costly data breaches, ransomware incidents and increasingly the prospect of litigation after an event. In 2013 cyber incidents had finished 15th, driven by companies' increasing reliance on data and IT systems, awareness of cyber threats have grown very quickly.

The Internet has changed many aspects of society, as it has transformed businesses, organisational learning, as it has enabled interaction between people through social networking platforms, so due to the emergence of the Internet, today's society today has evolved into a technological driven world (Chamie, 2020). ICT has essentially become an integral component to peoples' everyday lives, and besides the many benefits of the Internet and other ICT technology, there are unfortunately also threats, such as cyber attackers with malicious intent, who look to exploit vulnerabilities within these ICT applications (Singh, 2012).

Cyber risks continue to evolve, in e.g. significant increases in the numbers of ransomware incidents drive up the frequency of losses for companies. Overall, cyber-attacks are becoming more sophisticated and targeted as criminals seek higher rewards with multimillion-dollar extortion demands (Allianz Risk Barometer, 2020). Singh (2012) finds that e-skills are essential, due to the influence that ICT has on society, its organisations, and members, so investing in ICT skills provide needed possibilities to build competences to protect against cyber threats.

The purpose of this literature review is to identify how societal impacts of cyber security, and how the impacts of cyber security issues to individuals, communities, organizations or societies are discussed in academic literature. The goal of project ECHO (European Network of Cybersecurity Centres and Competence Hub for Innovation and Operations) is to organize a networked approach that by effective and efficient multi-sector collaboration aims at strengthening proactive cyber security in the European Union. The project ranges from 2019 to 2023 (ECHO, 2020; Yanakiev, 2020). This study adds in part to the practical body of knowledge that the

[1] https://orcid.org/ 0000-0001-9726-7956

project cumulates, while also extending current theoretical knowledge regarding the impacts that cyber security may have on society. The research question of this study is: How are societal impacts of cyber security discussed in academic literature?

2. Methodology

The method used in this research is a systematic literature review. This is a qualitative study. Systematic literature reviews are useful in identifying knowledge gaps in current literature, and bring new insights to the respective field for further investigation (Kitchenham, 2004).

2.1 Qualitative research design

According to Kitchenham (2004), a systematic literature review is a through process that can help present evidence that showcases the effects of selected events as they are described in research literature, and which may not be conveyed in traditional non-systematic literature reviews. Systematic literature reviews may thus, be more extensive than traditional ones. To conduct this literature review an academic search was conducted to provide answers to the research question. This study was conducted in a series of four steps: 1) search, 2) inclusion criteria, 3) DET analysis, and 4) writing of results and conclusions.

2.2 Search, inclusion criteria and DET-analysis

The search for the articles was performed in April 2020. The search was conducted using the Google Scholar, where a Boolean keyword search with the combination of "societal impacts of cyber-security + ICT + security + society + impact" were used as search parameters. The period for the search spanned literature published within the six-year period of 2014 - 2020.

The initial Google Scholar search returned a total of 265 peer reviewed articles. The abstract of these articles were examined against inclusion criteria: societal impact and cyber security keywords included in abstract, title and subject terms, Full Text and relevant to Research Question. The final sample included 33 papers that correspond to the inclusion criteria.

Following the identification of the appropriate peer reviewed articles for the literature review all 33 articles were read, and analysed by extracting relevant pieces of information to a data extraction table (DET) that was based on the research question. The next chapter of this paper discusses the findings of the sample articles.

3. Findings

Six streams of academic discussion emerge from the data. 1) Eleven papers discuss Impacts on social and societal levels, 2) four paper focus on Detection of cyber-crime and incidents, 3) seven papers relate to impacts on critical infrastructures, 4) five papers relate to impacts of individual incidents and technology, 5) seven papers discuss awareness building and training, 6) two papers look at creating common understanding of cyber security.

3.1 Impacts on social and societal levels

Eleven papers discuss Impacts on social and societal levels.

Schia and Gjesvik (2018) take a broader approach to cyber security than just national security, as they consider wider economical and societal impacts of digitization. 1) Cyber security has impacts on states and stakeholders. Digital technologies can promote freedom, as they provide digital arenas for expression of thought and debate free from governmental constraints; 2) Strengthen cyber security on an international level, because security on the online arena is only as strong as its weakest link, so focus on improving developing states to improve global cybersecurity structures; 3) Importance of digitalization and ICT in fostering economic and societal development is increasing, and there is a need to protect these benefits by cyber security.

Michel and King (2019) find that the Internet and connected technology platforms have enabled an increase of cyber influence and actions ranging from personal to national level security. Understanding how awareness (e.g. understanding what is real vs. fake) effects human decisions, may help avoid falling prey to cybercriminals cyber influence, and Internet fraud; appropriate technology can aid to increase cyber awareness by providing detection and support.

Kallberg & Burk (2014) promote defending national infrastructure from cyberattacks to protect information, network availability, and the global information grid, and to safeguard the lives of its citizens, protect their property, and preserve needed ecosystems and the ecosystem services. Attacks may cause environmental damages and have impact on societal stability.

Glisson and Choo (2017) find that the continued and increasing fusion of technology into wider dimensions of everyday life encourages cyber-crime. Hence, promoting evolution and diversification of cybersecurity, and responses that address growing concerns to highlight, investigate and address cyber-security vulnerabilities, especially in the context of cyber-of-things, the changing landscape demonstrates a need to develop innovative managerial, technological and strategic solutions.

La Torre, Dumay and Rea (2018) note that there is great societal power in Big Data. Stemming from cyber threats, the authors identify a need for corporate accountability, and call for a human-oriented approach to by understanding what detrimental implications the increasing usage of big data has for businesses and society, to promote equal society, transparency and a better decision-making big data. According to Bradshaw (2018) society is mostly aware of emerging technologies, and study to understand the impacts of novel innovations on society and the lives of its members is needed. Increased governmental oversight may be needed, and policy makers should prepare for a technology-driven disruption of society.

Afonasova et al. (2019) have studied the ways in which socio-economic background can affect how to allocate conditions for a transition to the digital economy. The authors find that focus should be put on management, information infrastructure, research and development, human resources and education, information security, smart city technology, and digital healthcare to boost to digitalization.

3.2 Detection of cyber-crime and incidents

Four paper focus on Detection of cyber-crime and incidents.

Gañán, Ciere and van Eeten (2017) find that cybercrime impacts are not limited to the direct consequences of a cyber-attack, but that there are significant costs to the well-functioning of the economy. The authors propose effective economic impact assessment with systematic data collection, guided by identified factors and indicators.

Tarafdar, Gupta, and Turel (2015) focus on, what they call the 'dark side' of information technology (IT) use. They find that cybercrime and other illicit use of information and communications technology (ICT) can seriously infringe the wellbeing of individuals, organizations, and society. Ibrahim et al. (2019) find that cyber-attacks have many impacts on national security, even putting the political, economic, and social welfare of the state in jeopardy, so a cyber warfare challenge on a national level, is rapidly detecting relevant threats as the scale of potential damages can be substantial if e.g. critical services are attacked.

Cuquet et al. (2017) see that data-driven innovations and business models, and data analytics of big data can improve event detection, situational awareness, and decision-making, and efficient allocation of resources; Interoperability is a key enabling factor, as is including data skills to general educational programs, and updating of legal frameworks to promote positive big data practices that positively address societal concerns.

3.3 Critical infrastructures and services

Seven papers relate to impacts on critical infrastructures.

Rajaonah (2017) notes that few studies on information systems combine trust and security, and those that do are mostly limited to two-agent interactions. The authors propose to introduce more holistic approaches of critical infrastructure protection (CIP), as information systems and the Internet are at the core of most modern businesses and services, and the information and knowledge stored and exchanged are of great value, which can attract cyber-attacks on information systems that can directly affect vital services; Critical infrastructures involve vital services, and their disruption have a significant impact on vital functions of society, so protecting society from cyber-attacs and ensuring people's well-being becomes a government level concern.

McLeod and Dolezel (2018) note unprecedented rates of health care data breaches. Elements associated with data breaches can be considered as: organizational factors, business process exposure factors, and technological security factors; Understanding these factors, and using computer security industry frameworks and healthcare standards, may help predict healthcare data breach weaknesses.

Steiger et al. (2018) propose that it is crucial to discuss hypothetical scenarios and analyse actual events, to achieve better understanding of cyber conflicts, and also: 1) Besides Western sources, also work with e.g. Russian and Chinese speaking coders; 2) Recognise the attribution made by actual conflict actors; 3) Rely on structured inclusion criteria to establish a dataset to evaluate cyber incidents; and 4) Build assessments on transparent indicators and develop criteria to gauge the intensity of cyberattacks that may represent the severity of cyber incidents.

Touhiduzzaman et al. (2019) view risk assessment important in understanding the potential consequences related to critical infrastructure. Different risk assessment methods vary in their goals, application domains, impacts, and consequences, and selection of appropriate method of risk assessment can be made based on the advantages and disadvantages associated with each respective method.

Rajamäki and Knuuttila (2015) show that public protection and disaster relief may form a complex software-intensive system that consists of several different sub-systems (e.g. 112-services, law enforcement, emergency medical services, firefighting and rescue services), which may be divided into many sub-sub-systems, calling for: 1) Proactive models of information security driven by awareness of vulnerabilities, threats, assets, potential attack impacts, the motives and targets of potential adversaries; 2) Effective tools and methods help cope with challenges of dynamic risk landscape with self-healing; 3) Integrate cyber security to every-day life, where efficient usage of tools and methods enable stakeholders co-operate, while protecting their privacy, and creating heightened public awareness and understanding.

Kotzanikolaou, Theoharidou and Gritzalis (2013) note the need to focus on assessing risks caused by the multi-order dependencies of critical infrastructures, and to resolve some major challenges: data accessibility, model development, model validation, and access to reliable real-time data to identify failures related dependencies. Methodologies should examine how threats and their impact may transfer from one infrastructure to another by identifying dependencies that may be based on existing security plans and service level agreements.

Adlakha et al. (2019) find that unawareness among the masses is surprisingly common, and while organizations need to maintain their security goals that in order to prevent the attacks, people should be aware that the more technology they use, the more vulnerable they become to possible attacks; Cyber criminals use evolving measures to get their hands on confidential information, and most cyber-attacks could be avoided by employing proper cyber-base hygiene, appropriately responding to cyber-attacks, and protecting confidential data.

3.4 Impacts of individual incidents and technology

Five papers relate to impacts of individual incidents and technology.

Henshaw and Van Barneveld (2016) conclude that cyber physical systems (CPS) are changing the nature of threats as they increase levels of complexity that individuals have to face; the effects of growing levels of autonomous machine decision-making in networks and added complexity in CPS bring new vulnerabilities and associated threats, with e.g. vast amounts of personal information stored in many databases.

Freeman (2018) discusses the future of drones and drone delivery, and concludes that to alleviate concerns, companies that are testing commercial drones should share with the public what information the on-board components capture and justify its need in improving drone delivery of packages. Legislation, such as who should be held accountable, and where is the line between public and private concern of the impacts of CPS may be difficult to determine (Henshaw & Van Barneveld, 2016).

Yan et al. (2018), note that ordinary users tend to be the weakest link in cybersecurity; understanding the weakest link phenomenon can help create and implement more effective cybersecurity awareness and intervention programs for ordinary users, who are the majority of society. Riek (2017) finds that, when users perceive the risk of cybercrime they become less willing to use unknown websites, or shop and bank online;

cyber policy makers can establish trust marks, standards, security certificates, and other such incentives that facilitate usage of online services.

Miftha, Conrad and Gibson (2019) call for initiatives that raise public awareness of communications being increasingly mediated via different technologies, which create possibilities of being cyber stalked. Preventative legal and technical support mechanisms recognizing possible secondary effects and negative impacts, should complement social reforms.

3.5 Cybersecurity awareness

Seven papers discuss awareness building and training.

Wilk (2019) proposes a curriculum for a course in computers, ethics, law, and public policy, which issues are very relevant to building and using intelligent systems; computer professionals and decision makers learn ethics and law fundamentals for increased professional responsibility, as future computing jobs will require technical knowledge coupled with legal and ethical awareness to cover relevant topics and challenges that tomorrow's computer professionals and decision makers will face.

Dewar (2018) notes that due to the development of digital technologies and ICT tools, cyber-attacks are prevalent, and calls for training in cyberdefense; early and clear goal definition and effective planning are key in implementing successful exercises, where scenarios, when used, should be realistic, and all exercises should serve clearly defined purposes; conducting exercises for the sake of conducting exercises becomes counterproductive.

Østby, Lovell, and Katt (2019) suggest a three-phase process that helps prepare for cyber exercises. First, the societal impact of the cyber crisis are identified; Second, roles and responsibilities of cyber crisis management are identified; Third, relevant training team roles are built; strategic approaches and excellent training skills are required to train and develop teams that are diverse in competence, measeured by maturity testing before and after exercises.

According to Aaltola and Taitto (2019) building resilience, preparedness, and responding to crisis require multidisciplinary approaches, because the cyber domain crosscuts all functions of today's organizations and society, and how humans behave and how they make decisions play a crucial role in building cyber security, and training and exercises should simulate this reality as accurately as possible; considering experiential learning principles can deepen the level of learning in cyber education and training, and developing the skills and competences needed to safely navigate the cyber domain, should be seen as a constructive process, utilising and are recognising learners' previously adapted competences.

Pouraimis et al. (2019) results indicate that awareness reduces risk, as does the nature of the organisation, e.g. in military organizations people are more likely to closely follow cybersecurity protocols; significant damage can be caused depending on what resources are compromised, so potential risks that could negatively affect end-user operations should be studied and an awareness approach can have significant improvements in long term lasting risk reduction.

Alotaibi (2019) sees training of users as essential to increase awareness about cybersecurity, and that serious games as a training method can be effective in providing user training and achieving a behavioural change.

3.6 Cybersecurity and collaboration

Two papers look at creating common understanding of cyber security.

Urgessa (2020) sees that international cooperation in cybersecurity has been difficult, because the subject has been defined and conceptualized differently, and that they is incompatibility in how the respective political systems of major cyber powers have been organized; in the west, cybersecurity is mostly seen as when the machine is secure, so are the societal functions that the machine runs, while China and Russia see cybersecurity as the security of the machine, of the state, as disseminating information can threaten their political systems.

Vishik, Matsubara and Plonk (2016) note that technology and services are rapidly changing and evolving, so cyberspace related policymaking must be innovative to support growth, security, trust, confidence, and stability in society; bringing all relevant stakeholders, from government, industry, academia, to civil society, to work together in ensuring that the benefits of cyberspace are accessible to citizens; governments develop policies, strategies, and regulate the development of cyber security, while industry deploy novel technologies.

4. Discussion and conclusions

The discussion on Social and societal impacts takes a broader approach to cyber security than just national security by considering wider economical and societal impacts of digitization. The Internet and connected technology platforms enable an increase of cyber influence, which has increased the importance of awareness, while technology can aid detection and support cyber awareness. Defending national infrastructure from cyberattacks, while protecting information, networks and grids, and safeguarding citizens, property, and ecosystem services becomes a focus.

Fusion of technology into our everyday lives encourages cyber-crime, which causes an increasing need to address cyber-security vulnerabilities, and corporate accountability and human-oriented approaches help understand detrimental implications for businesses and society. Governmental oversight aids in managing the impacts that new innovations may have on society and the on lives of its members.

Detection of cyber-crime and incidents calls for data-driven innovations and analytics of big data improve event detection, situational awareness, and decision-making so that resources can be allocated efficiently, and cybercrime and other illicit use of information and communications technology (ICT) can seriously infringe the wellbeing of individuals, organizations, and society, calling for innovative managerial, technological and strategic cyber security solutions. The direct consequences of cybercrime and cyber-attacks have significant impacts and costs to economic functions, thus data collection of agent-level costs and social impacts of cyber-crime are needed to understand the impacts that cyber-attacks may have on national security.

The literature on critical services and infrastructures note that attacks on information systems can affect vital services and critical infrastructures, and protection of society and people's well-being is a governmental concern. There have been unprecedented rates of health care data breaches on organizational, business process, and technological security levels. Challenges associated with developing economically quantifiable risk assessment methods or frameworks require understanding what potential consequences are related to critical infrastructure. Identifying relevant dependencies may be based on existing security plans and service level agreements. People seem to be unaware of the risk of being attacked, while organizations need to maintain their security goals to prevent the attacks proper cyber-base hygiene and protection of confidential and sensitive data to appropriately respond to cyber-attacks are needed.

Impacts of incidents and individual technology include cyber physical systems (CPS) that are changing the nature of safety and security threats, and bring new vulnerabilities and associated. Ordinary users tend to be the weakest links in cybersecurity, and understanding this phenomenon can help create, and implement more effective cybersecurity awareness and intervention programs for ordinary users, who are the majority of society. Perceiving the potential risk of cybercrime can decrease willingness to use unknown websites, to shop, or to bank online. Cyber policy makers establish digital literary campaigns, trust marks, standards, security certificates, and other incentives to facilitate usage of online services. These initiatives raise public awareness of communications, which are increasingly mediated via different technologies. Using these technologies creates the possibility of cyber crime, often facilitated through no action of the victim. Preventative legal and technical support mechanisms recognizing possible secondary effects and negative impacts, can complement social reforms.

Cyber security awareness includes understanding computers, ethics, law, and public policy, which are very relevant to building and using intelligent systems. Computer professionals and decision makers should learn to link technical knowledge with legal and ethical awareness to cover relevant topics and challenges. Understanding the context of unique cyber-exercises, where realistic scenarios can help serve clearly defined purposes. Strategic approaches and excellent training skills are required to train and develop diverse teams, where individuals vary in competence. Proficiently predicting social media security and privacy practices assist organizations to focus on improving end-user awareness, skill and ability for better security and privacy behavior. Building resilience, preparedness, and responding to crisis require multidisciplinary approaches, and training, where exercises can simulate reality as accurately as possible, and so better conceptualize cyber security training, education and exercises.. Awareness approaches can improve long-term lasting risk reduction. Training users is essential in increasing awareness about cyber security. Serious games can effectively provide user training to achieve behavioral change.

Regarding cyber security and collaboration, international cooperation in cybersecurity has been difficult, because there has not been common conceptualization of cybersecurity. Technology and services are rapidly changing and evolving, also cyberspace related policymaking must be innovative and support growth, security, trust, confidence, and stability in society. To achieve international harmonization, better coherence create a common context that supports multi-stakeholder interactions can enable a multi-disciplinary scientific view of cyber security that helps model needed chance in cyberspace.

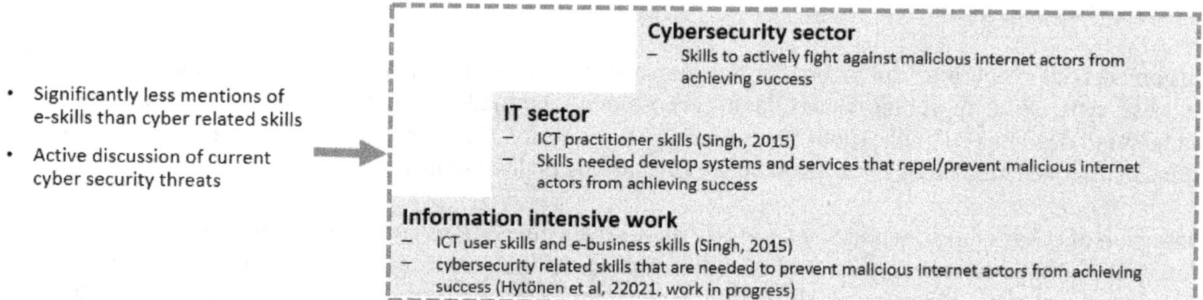

Figure 1: Cybersecurity literature focus on the workplace on levels of IT users, and IT and cyber experts

The sample literature discusses developing organizational cybersecurity focusing on the people working in information intensive jobs, where IT is used daily, and on the ITC and cybersecurity expert levels (Figure 1). ITC users may often have access rights that may put organizational information and systems in jeopardy. Distinguishing between these different user groups helps promote the skills development for successful cybersecurity.

Further study is recommended to better understand what cyber and IT-related e-skills are needed to detect, minimize and prevent possible cyberincidents. Limiting the effects of cyber incidents may directly and indirectly have wider impacts on society. Developing methods to further assess societal impacts, and e-skills and training gaps provide organizations with very practical tools to identify relevant training and recruitment needs, and understand the societal effects of cybersecurity. This could allow individuals to identify and develop their job relevant e-skills. This accumulation of practical data can contribute to science and deepen theoretical understanding.

References

Aaltola, K., & Taitto, P. (2019). Utilising Experiential and Organizational Learning Theories to Improve Human Performance in Cyber Training.

Adlakha, R., Sharma, S., Rawat, A., & Sharma, K. (2019, February). Cyber Security Goal's, Issue's, Categorization & Data Breaches. In 2019 International Conference on Machine Learning, Big Data, Cloud and Parallel Computing (COMITCon) (pp. 397-402). IEEE.

Afonasova, M. A., Panfilova, E. E., Galichkina, M. A., & Ślusarczyk, B. (2019). Digitalization in economy and innovation: The effect on social and economic processes. Polish Journal of Management Studies, 19.

Allianz Risk Barometer (2020). Available: https://www.agcs.allianz.com/news-and-insights/expert-risk-articles/allianz-risk-barometer-2020-cyber-incidents.html

Alotaibi, F. F. G. (2019). Evaluation and Enhancement of Public Cyber Security Awareness (Doctoral dissertation, University of Plymouth).

Bradshaw, D. J. (2018). Technology Disruption and Blockchain: Understanding Level of Awareness and the Potential Societal Impact (Doctoral dissertation, Dublin, National College of Ireland).

Chamie J. (2020) World population 2020: Overview. Yale Global Online, February 11, 2020. Available: https://yaleglobal.yale.edu/content/world-population-2020-overview.

Cuquet, M., Vega-Gorgojo, G., Lammerant, H., & Finn, R. (2017). Societal impacts of big data: challenges and opportunities in Europe. arXiv preprint arXiv:1704.03361.

Dewar, R. S. (2018). Cybersecurity and cyberdefense exercises. ETH Zurich.

ECHO (2020). the European network of Cybersecurity centres and competence Hub for innovation and Operations Webpage. Available: https://echonetwork.eu/

Freeman, E. (2018). The societal issues of drone delivery on public behaviour in the UK (Doctoral dissertation, Cardiff Metropolitan University).

Gañán, C. H., Ciere, M., & van Eeten, M. (2017, October). Beyond the pretty penny: the Economic Impact of Cybercrime. In Proceedings of the 2017 New Security Paradigms Workshop (pp. 35-45).

Glisson, W., & Choo, R. (2017). Introduction to Cyber-of-Things: Cyber-crimes and Cyber-Security Minitrack.

Henshaw, M., & Van Barneveld, J. (2016). CPS for community security and safety. Ethical Aspects of Cyber-Physical Systems. Brussels: European Parliamentary Research Services STOA, 64-74.

Ibrahim, A., Mahmud, N., Isnin, N., Dillah, D. H., & Dillah, D. N. F. (2019). Cyber Warfare Impact to National Security-Malaysia Experiences. KnE Social Sciences, 206-224.

Kallberg, J., & Burk, R. A. (2014). Failed Cyberdefense: The Environmental Consequences of Hostile Acts. Military Review, 94(3), 22.

Kitchenham B. (2004). Procedures for Performing Systematic Reviews. Keele University 33:1-26.

Akbari Koochaksaraee, A. (2019). End-User Security & Privacy Behaviour on Social Media: Exploring Posture, Proficiency & Practice (Doctoral dissertation, Université d'Ottawa/University of Ottawa).

Kotzanikolaou, P., Theoharidou, M., & Gritzalis, D. (2013). Risk assessment of multi-order dependencies between critical information and communication infrastructures. In Critical Information Infrastructure Protection and Resilience in the ICT Sector (pp. 153-172). IGI Global.

La Torre, M., Dumay, J., & Rea, M. A. (2018). Breaching intellectual capital: critical reflections on Big Data security. Meditari Accountancy Research.

McLeod, A., & Dolezel, D. (2018). Understanding Healthcare Data Breaches: Crafting Security Profiles.

Michel, M. C. K., & King, M. C. (2019, November). Cyber Influence of Human Behavior: Personal and National Security, Privacy, and Fraud Awareness to Prevent Harm. In 2019 IEEE International Symposium on Technology and Society (ISTAS) (pp. 1-7). IEEE.

Miftha, A., Conrad, M., & Gibson, M. (2019). Cyber stalking is a social evil: from the Indian women's perspective.

Pouraimis, G., Thanos, K. G., Grigoriadis, A., & Thomopoulos, S. C. (2019, May). Long lasting effects of awareness training methods on reducing overall cyber security risk. In Signal Processing, Sensor/Information Fusion, and Target Recognition XXVIII (Vol. 11018, p. 110180N). International Society for Optics and Photonics.

Rajamäki, J., & Knuuttila, J. S. (2015, November). Cyber Security and Trust. In KMIS (pp. 397-404).

Rajaonah, B. (2017). A view of trust and information system security under the perspective of critical infrastructure protection.

Riek, M. (2017). Towards a Robust Quantification of the Societal Impacts of Consumer-facing Cybercrime (Doctoral dissertation, Westfälische Wilhelms-Universität Münster).

Schia, N. N., & Gjesvik, L. (2018). Managing a Digital Revolution-Cyber Security Capacity Building in Myanmar.

Singh, S. (2012). Developing e-skills for competitiveness, growth and employment in the 21st century: The European Perspective. International Journal of Development Issues, Emerald Group Publishing, 11(1):37-59.

Steiger, S., Harnisch, S., Zettl, K., & Lohmann, J. (2018). Conceptualising conflicts in cyberspace. Journal of Cyber Policy, 3(1), 77-95.

Tarafdar, M., Gupta, A., & Turel, O. (2015). Special issue on'dark side of information technology use': an introduction and a framework for research. Information Systems Journal, 25(3), 161-170.

Touhiduzzaman, M., Gourisetti, S. N. G., Eppinger, C., & Somani, A. (2019, November). A Review of Cybersecurity Risk and Consequences for Critical Infrastructure. In 2019 Resilience Week (RWS) (Vol. 1, pp. 7-13). IEEE.

Urgessa, W. G. (2020). Multilateral cybersecurity governance: Divergent conceptualizations and its origin. Computer Law & Security Review, 36, 105368.

Vishik, C., Matsubara, M., & Plonk, A. (2016). Key Concepts in Cyber Security: Towards a Common Policy and Technology Context for Cyber Security Norms. NATO CCD COE Publications, Tallinn, 221-242.

Yan, Z., Robertson, T., Yan, R., Park, S. Y., Bordoff, S., Chen, Q., & Sprissler, E. (2018). Finding the weakest links in the weakest link: How well do undergraduate students make cybersecurity judgment?. Computers in Human Behavior, 84, 375-382.

Yanakiev, Y. (2020). A governance model of a collaborative networked organization for cybersecurity research. Information & Security, 46(1), 79-98.

Wilk, A. (2019). Teaching AI, Ethics, Law and Policy. arXiv preprint arXiv:1904.12470.

Østby, G., Lovell, K. N., & Katt, B. (2019, December). EXCON teams in cyber security training. In 2019 International Conference on Computational Science and Computational Intelligence (CSCI) (pp. 14-19). IEEE.

Planning the Building a SOC: A Conceptual Process Model

Pierre Jacobs[1], Sebastiaan von Solms[1] and van der Walt[2]
[1]University of Johannesburg, South Africa
[2]Sizwe IT Group – South Africa
pierrej06@gmail.com
basievs@uj.ac.za
svdwalt@gmail.com

Abstract: There are few frameworks available to consult when building Security Operation Centers (SOCs). (P. Jacobs, 2015). Jacobs proposed such a framework, and this paper builds on the "Planning" part of that framework. The authors could not find any existing conceptual process models where it comes to the planning phase when building SOCs. We propose a conceptual process model to follow during the planning phase of the SOC. Conceptual models are used to represent systems typically made up of the composition of concepts (Robinson; Arbez; Birta; Tolk; Wagner, 2015). The aim of our conceptual process model is to help SOC builders understand the proposed process to be followed during the SOC planning phase and is meant to guide the SOC builder's thinking during the planning phase. The conceptual process model will start by determining the services that the SOC in development will be offering, followed by deciding on a SOC model. After the determination of the SOC services and model we will identify the technologies and tools to facilitate the services, keeping in consideration the influence the SOC model has on the service. For each of the steps in our conceptual model we have identified existing, public frameworks, standards or best practices. Our conceptual process model will be mapped to these frameworks, standards or best practices with the intention to be used to augment our model.

Keywords: SOC planning, SOC conceptual process model, SOC services, SOC models, systems engineering, SOC builder, MSSP, EDR, MDR

1. Introduction

Cybersecurity is a topic being discussed globally by executives. In todays connected world, the protection of cyber assets is of extreme importance. There are many reasons for considering cybersecurity – some of these are the protection of organisational sensitive information, ensuring availability and integrity of cyber resources, as well as to comply with legislative and internal governance requirements. Information technology is a key enabler in today's business world, and it should be protected like any other asset in the organisation. Therefore, it is important for any company to have a Cyber Security Strategy.

A good, solid cyber security strategy will benefit business in the following ways as taken from (The British Standards Institution, 2014):

- Protect networks, computers and data from unauthorized access
- Improved information security and business continuity management
- Improved stakeholder confidence in your information security arrangements
- Improved company credentials with the correct security controls in place
- Faster recovery times in the event of disruption

Furthermore, important outcomes which can be expected when a Cyber Security Strategy is supported by an effective governance approach, is reduced cyber risk, and reducing potential business impact to an acceptable level. Such a governance approach allows for the strategic alignment of security with the enterprise strategy and organisational objectives. Market share could also potentially be preserved and increased due to the reputation for safeguarding of information, and lastly, security investments are efficiently utilised to support organisational objectives.

In achieving such governance goals, technical and administrative controls are used. The function of the deployed technical controls varies widely – some protect against unauthorised access, and loss of data, to intrusion and malware detection and prevention on the network, and endpoints. Furthermore, these technical controls need to be monitored to ensure that they function as intended, and that they are configured correctly.

Some of these technical controls such as firewalls, Intrusion Prevention Systems (IPS) and anti-malware software are used to monitor for events with the aim to detect possible attacks, breaches, and compromises, and to perform specified actions to react to these attacks such as containing or blocking them. These technical actions are supported by a solid incident response capability offered from the SOC. We will in Section 2 propose a definition for a SOC.

2. Background - What is a SOC?

Experience showed that a SOC can be defined as a single, central repository where logs from multiple sources are collected, aggregated, and correlated. The log events are monitored and investigated and elevated into incidents if needed. These incidents are then escalated internally or to clients and then responded to by the relevant teams depending on the incident at hand. Response could be in the form of remediation or containment. A SOC is made up of people, processes and technologies (Logsign, 2018)(McAfee, 2021).

When planning on consuming or using SOC services, there are two different choices open to organisations (Digital Hands, 2021); (Dave Young, 2020). The first choice is to build their own SOC (organisational SOC – also called an in-house SOC, or insourced SOC). The second choice is to consume SOC services from a Managed Security Service Provider (MSSP – also called an outsourced SOC). From a services perspective MSSPs cater for multiple clients from different industries and they are externally focussed in that they look after the interests of customers. Organisational SOCs looks after the interests of a single organisation, and they are internally focussed in that they look after the interested of the organisation they serve.

There are thus two distinct types of SOCs with different operational requirements and approaches, but from a technology perspective they would both use the same tools, The primary tools used to monitor for and respond to incidents are the Security Incident and Event Monitoring (SIEM) and Security Orchestration, Automation and Response (SOAR) tools. The two SOC types are shown in Figure 1. Our conceptual process model was developed to cater for both organisational SOCs and MSSP SOCs. Taking the above into consideration, we will use the term *"SOC"* as an umbrella term to describe both organisational SOCs and MSSPs.

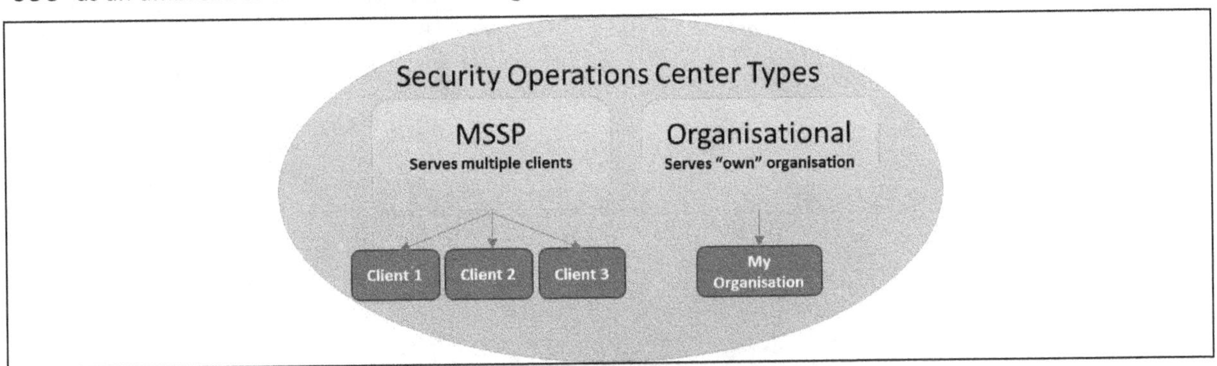

Figure 1: SOC types

The authors could not find any conceptual process model describing the higher-level planning steps when planning to build a SOC. In this paper we propose a SOC conceptual planning process model that describe the higher-level steps to follow when planning to build a SOC. This conceptual process model should be used first before a detailed "Build Framework" such as the one described by *Jacobs* (2015) (P. Jacobs, 2015) is used.

In Section 3 we introduce a methodology to identify SOC services. In Section 4 we identify SOC services and in Section 5 we introduce the different SOC models. In Section 6 we will map the SOC services to technologies and in Section 7 we follow a systems engineering approach to identify the most suitable SOC technologies. The planning steps in our model is mapped back to existing, public frameworks, standards and best practices in Section 8.

3. An introduction to our conceptual process model to plan for SOC building

When planning and building a SOC, experience showed that it is beneficial to follow a structured approach (such as a framework) to ensure that all people, processes and technology aspects are considered. An example of such a framework is the one proposed by Jacobs (P. Jacobs, 2015). Experience showed that the choices made during the planning phase influences the technology requirements as well as the number of resources (people) and the skills they need to have. At a high level, we propose the following six steps to guide and focus the SOC builder's planning and thinking.

- **Step 1:** Define the services to be offered by the SOC – services are supported by people, processes and technologies. Services definitions are normally defined during structured interviews with stakeholders.
- **Step 2:** Determine the SOC model – the SOC model further influences the number of resources needed. It also influences facility requirements as well as legal and regulatory requirements.
- **Step 3:** Identify and map technologies to facilitate and support SOC services.
- **Step 4:** Select technologies such as SIEM and SOAR platforms (see Paragraph 5.1). This should be done by soliciting user requirements, draft functional and technical specifications and issuing a Request for Proposal (RFP).
- **Step 5:** Plan Facility and Infrastructure. This is the building, the physical security and the screens, seats and desks and communications infrastructure needed to make the SOC work.
- **Step 6:** Identify people requirements (skills and experience) and develop SOC operational processes to support SOC services.

After the SOC services and SOC models are defined, the facilities and infrastructure are built, resources recruited, and technology procured. Once all this is done, processes, procedures, use cases and playbooks are developed. The six steps of the high-level conceptual process model are shown in Figure 2.

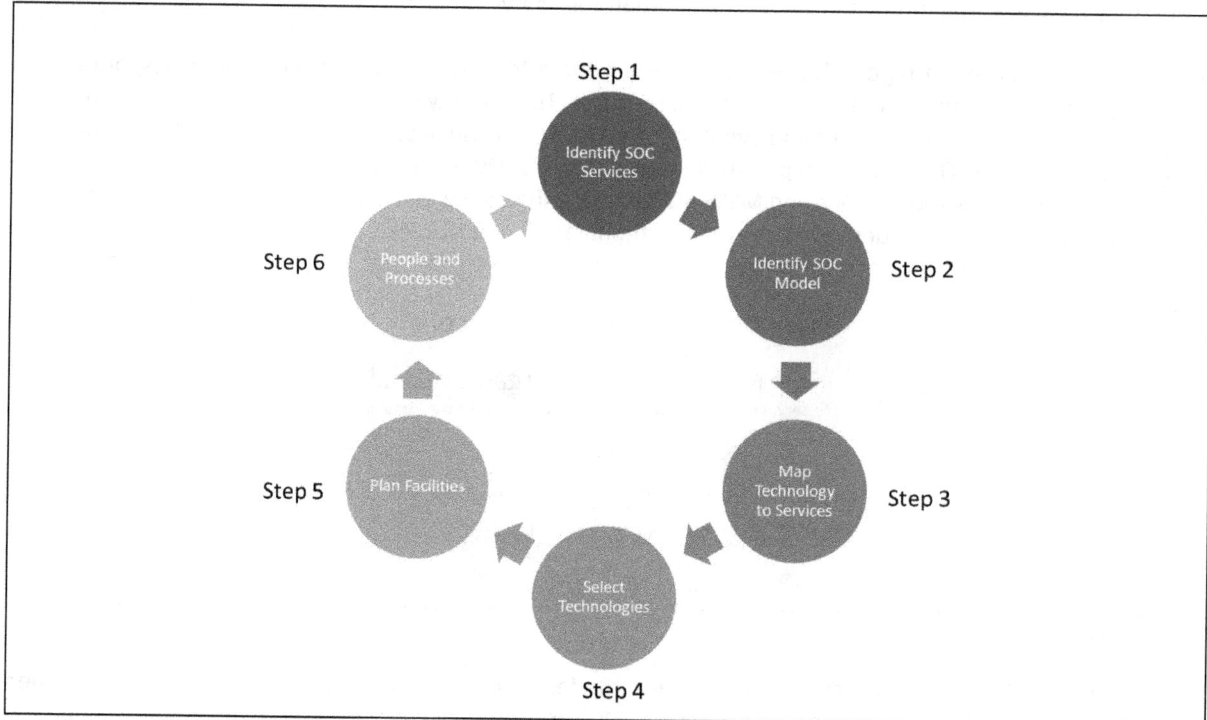

Figure 2: SOC builder conceptual process model

To illustrate the application of our conceptual process model we will describe Step 1 to Step 4. Steps 5 to 6, the planning for facilities and the planning for people and processes steps will be described in future work. We will now introduce Step 1 of our conceptual process model - the identification of SOC services, and a methodology to follow during their identification. Potential SOC services, and the methodology we have followed to discover them, are introduced in Section 4.

4. Step 1: Identify SOC services

It is our experience that some organisations follow a bottom-up approach in that they first procure tools and technology and then define the services around the technological capabilities. From an organisational SOC perspective this is the least desirable way to identify services as it will almost always be limited by technological constraints. MSSPs however may find this a useful way of identifying service offering to clients as they typically cater for multiple clients at a time.

For MSSPs functional capabilities such as multi-tenancy and role-based access may be important, while automated workflow and correlation capability may be more important in an organisational SOC. We will follow a top-down approach by identifying the SOC services and models first, and then build the enabling technology stack around it.

The SOC services selection will influence the SOC tools and technology as well as its processes. It is our experience, and supported by literature (SANS Institute, 2021) that the services offered by MSSPs can be grouped into primary and secondary services. The SOC Builder may use five different methodologies to determine SOC primary and secondary services. Primary services are the core services offered by SOCs such as monitoring, detection response and threat intelligence. These are the most basic services that should be offered from a SOC. Secondary services are services such as device management and user awareness training. They are not core to the SOC but augments the SOC service basket. Primary services can be determined by observing the following five-step methodology described next.

- 1. Through Market research

Market research is done to determine what services clients require from SOCs. The market research can be a simple desktop research exercise or using a third party to do the market research on the SOC's behalf.

- 2. By Identify existing pre-defined SOC services

It is our experience that existing MSSPs will already have a basket of pre-defined services on offer to clients. SOC builders could leverage off this knowledge. This pre-supposes a good relationship between the upcoming SOC and other industry players. Jacobs *et al* (Jacobs *et al.*, 2016) identified and pre-defined a basket of SOC services that may be used by the upcoming SOC to select services from.

- 3. Through Structured interviews with industry and stakeholders

SOC service requirements are solicited during structured interviews with stakeholders. The advantage of following this approach is that the establishing SOC starts building relationships with potential clients and will get a solid understanding of stakeholder's service requirements and performance metrics.

- 4. By Services implied from authoritative and normative sources.

Authoritative and normative sources are consulted to identify potential SOC services. This approach has the advantage that establishing SOCs consider national and international requirements and prescriptions expressed in these sources. Sources could be national acts and regulations, or standards and frameworks used by industry and stakeholders.

- 5. Through a combination of the listed methodologies

Following a combination of all these methodologies will yield the best results in that stakeholders as well as national and international prescripts and requirements are consulted. This leads to a relevant and market related services definition.

Following our proposed five-step methodology and backed by experience, and supported by literature, the three primary services that should be offered by SOCs are (SANS Institute, 2021) (*SOC-as-a-Service, MDR, and Managed SIEM. What's the Difference? - MSSP Alert*, no date)(C. Zimmerman and Operations Center, 2014):

- **Monitoring** - this service describes the near real time analysis of data from telemetry sources for potential threats. This is mostly achieved using a SIEM.
- **Detection** - this describes "the action or process of identifying the presence of something concealed." (*Detection | Meaning of Detection by Lexico*, no date). This is mostly achieved using a SIEM or SOAR platform.
- **Response** - could be in the form of incident handling or facilitating the actual remediation. Response could be automated using a SOAR tool.

Where the detection and response services are managed, it is known in the industry as Managed Detection and Response (MDR). The MDR service can be delivered using the clients own set of tools, or by tools provided by the service provider (Zhang, 2018).

To make sure the primary and secondary services are offered in an effective and measurable way, service metrics and key performance indicators need to be identified, and processes supporting the service need to be developed and their effectiveness and maturity managed.

Now that we have introduced the idea of SOC services, we move to the identification of SOC models in Section 5.

5. Step 2: Identify SOC models

It is important to identify the available SOC models since the model influences the services, and thus the technology requirements. We identified two SOC types in Section 2. The types are an organisational SOC, and MSSP SOC. Both types offer or provide a service or services to clients. The organisational SOC provides a service to an internal client (its organisation) and a MSSP to external clients. SOC models can thus then be viewed from two contexts.

The first context is the consumer context. The consumer context views the SOC services from the internal or external client's perspective. Our experience showed that the consumer has five different operational requirements when looking for SOC services.

The second context is the service provider context (organisational SOC or MSSP perspective). This context describes how organisational SOCs or MSSPs structure their services operationally. We identified two different service provider operational models. We will now discuss these in more detail:

5.1 Consumer context

From experience, we identified five SOC operational models from a consumer context. These are:

- Monitoring detection, and response are performed internally (organisational SOC).
- Outsourced monitoring and detection (MSSP) with internal response (organisational SOC).
- Outsourced monitoring and detection with outsourced response (MSSP).
- Co-sourced monitoring and detection with internal response (organisational SOC and MSSP).
- Co-sourced monitoring and detection (organisational SOC and MSSP) with outsourced incident response (MSSP).

5.2 Service provider (MSSP) context

From experience we have identified two primary MSSP operational models. These are:

- SOC as a Service (SOCaaS) - this model describes the traditional MSSP SOC that provides monitoring detection and response services to clients. These capabilities are developed internally and are inherent to the MSSP model.
- Co-Sourced SOC - This model describes an MSSP SOC that outsources the delivery of some of its services to external entities (partners or clients). In some instances, responsibility to deliver a specific service may be shared, and these are typically described as part of the SOC playbooks i.e. where the SOC outsources threat intelligence or remedial efforts to a third-party service provider.

The two contexts from where SOCs can be viewed from, and the operational models are summarised in Table 1.

Table 1: SOC contexts and operational models

Context 1: Consumer Context	Context 2: Service Provider (MSSP) Context
[1]Internal monitoring, detection and response [2]Outsourced monitoring and detection with internal incident response [3]Outsourced monitoring and detection with outsourced response (MSSP) [4]Co-sourced monitoring and detection with internal incident response [5]Co-sourced monitoring and detection with outsourced incident response	[1] MSSP/ MDR [2]Co-Sourced MSSP / MDR

Now that we have identified the SOC primary services as well as the SOC operational models, we need to map the services and SOC models to technology requirements in Section 6. For illustrative purposes we will focus on the MSSP context.

6. Step 3: Mapping services to technology

The three primary SOC services that we have identified for both organisational SOCs and MSSP are the monitoring detection and response services (see Section 3). A service is made up of people, processes and technologies (P. Jacobs, 2015). The focus of this section is on the technologies needed for these primary services, and in context of MSSPs operational model we identified in Section 4.

6.1 Monitoring service

Hewlett Packard defines security monitoring as the collection of logs and analysing the information therein to detect possible threats and attacks. Security monitoring includes the capability to trigger alerts and action triggered alerts. The primary technology used to monitor for security related alerts are a SIEM tool (HPE, 2018).

This service presupposes that the client has security technical controls installed and that the logs from the security controls can are sent to the SIEM tool. The purpose of a SIEM is to collect many logs from different telemetry or log sources. These telemetry or log sources may be from perimeter and network controls, to end-point protection logs. These logs are parsed, in some cases aggregated, often correlated and algorithms are applied to identify incidents and events of significance (J. Miller, 2019). The SIEM capabilities are often supported by a SOAR.

A SOAR platform's primary function is to collect and organise information and present it in such a way that cybersecurity staff can easily manage and process the information. The aim is to standardise and incorporate incident investigation into existing workflows. Information is collected from a range of telemetry sources and this information is delivered to a central hub. A SOAR platform helps to automates incident response through the categorisation and classification of alerts, with the end goal of reducing the amount of alerts seen on SIEM (J. Miller, 2019).

6.2 Detection service

In context of a MSSP, this service could be offered using the client's own tools, or tools provided by the MSSP. When combined with the Response service, and offered by a MSSP, it is also sometimes referred to as Managed Detection and Response (MDR). MDR service offered by service providers, such as MSSP SOCs, share the same characteristics (Zhang, 2018).

The first characteristic of a MDR service offered by a MSSP is that MSSP's MDR service focus is on threat detection, and not so much on compliance. In some cases, SIEM and SOAR are deployed to satisfy compliance requirements, and from that perspective a MSSP can provide compliance reporting.

A second characteristic is the coverage of the tools serving as telemetry or log sources that are provided by the MDR service provider. The monitoring and management service is limited to the tools provided by the MDR service provider. The tools could be deployed at the perimeter, network, endpoint, a combination, or all. MSSP' coverage differ in that they monitor multiple telemetry and log sources.

The third characteristic is that the MDR service usually involves humans, but some automation may be implemented. Analysis and triage are done by analysts, and this provide for high fidelity alerts. Clients typically interact with analysts rather than looking at dashboards and other tools.

A fourth characteristic is that incident validation and remote response is provided by the MDR service provider. This means that analysis, sandboxing and reverse engineering of malware is handled by the MDR service provider. Some MDR service providers provide remediation advice. The major difference between the MSSP monitoring service and the MSSP MDR service is that with the MSSP monitoring service, the MSSP monitors the clients existing tools and critical assets as telemetry or log sources, while with the MDR service, the MSSP will install and use their own tools as log sources.

MDR service can, and are often offered from MSSP SOCs (J. Godgart, 2019)(M.K. Hamilton, 2020)(Secuvant, 2019). It is our experience that in some cases, MSSPs evolve to offer the MDR service where the MSSP provide the tools as opposed to relying on the tools provided by the client. Some examples of tools needed for a MDR service are Intrusion Prevention Systems (IPS) and endpoint detection and response (EDR) tools. Differences between MSSP monitoring and MDR services characteristics (Zhang, 2018) are shown in Table 2.

Table 2: Difference in characteristics between MSSPs and MDR

	MSSP	MDR
Characteristic 1: Compliance Reporting	Provided	Usually not provided
Characteristic 2: Coverage	Work with multiple logs and contexts. Customer decides what logs from which telemetry sources to send	Ingests and monitors logs from own deployed security controls
Characteristic 3: Human Interaction	Normally portals and e-mail	Interact with analysts
Characteristic 4: Incident Response	Onsite and remote – need separate retainer	Onsite and remote – retainer normally included

6.3 Response service

Together with the monitoring and detection service, the incident response service is of paramount importance and a foundational service. The service work in tandem with the monitor and detect service. Events are monitored and possible malicious activity is detected. Deeper investigation is done, and eventually a single event, a couple of events or correlated events are classified as an incident. An incident could be a hots or hosts infected with malware, an ongoing attack, or data leakage.

The purpose of the incident response service is fourfold:

- Triage and investigate incidents - during this sub-task, additional research is done about the incident so that the incident can be:
- *Prioritise incidents.*
- *Severity of incident.*
- *Classify incidents.*
- *Determine impact of incident on client's environment.*
- Escalate incidents to clients.
- Response to incidents - Advise clients on how to handle the incident – this part of the service overlaps with the "Remediation Service" and includes containment, eradication, and recovery. Incident response can be automated using a SOAR.
- In the case of MDR service, orchestrate and automated response, or where the client selected a retainer from the MSSP, the MSSP engineers could provide the response on-site or remotely.

The response service should be defined, and performance metrics coupled to it. Different priorities and different severities normally have different response times associated with them. The escalation time should be agreed with the client and described in the SLA with the client. Incident response also includes remediation and containment. Remediation and containment can be performed by internal resources, co-sourced, or outsourced, and KPIs and metrics should be developed for the remediation service as well.

For guidelines on how to respond to cybersecurity incidents, consider the National Institute of Standards and Technology's (NIST) recommendations in its Computer Security Incident Handling Guide (Cichonski *et al.*, 2012). It is also a good idea to map attacks to the cyber kill chain. This shows value in that the better the SOC gets, the earlier in the kill chain attacks are detected and acted on. Consider using Lockheed Martin's "Cyber Kill Chain Methodology (Cichonski *et al.*, 2012) or the Mitre "ATT&CK Framework" (*Matrix - Enterprise | MITRE ATT&CK™*, 2022) to prove value.

In terms of an actual incident handling, frameworks such as those developed by NIST (Cichonski *et al.*, 2012) and the SysAdmin, Audit, Network and Security's (SANS) (Kral, 2012) may be considered. When choosing a framework, it is important to consider that the incident handling process will have to integrate with the client's incident handling process.

The primary services to technology mapping are done in Table 3. This mapping only considers the tools to be used to facilitate and enable the primary services. It does not consider tools for secondary services and neither does it consider tools or telemetry sources such as Intrusion Prevention Systems and anti-malware systems. SOC infrastructure tools such as jump boxes, Domain Naming Service (DNS), mail, firewalls and so on must also be considered. The infrastructure tools are outside the scope of this article.

Table 3: Primary service to technology mapping

Monitor	Detect	Response
SIEM – monitor Productivity suite	SIEM/SOAR – monitor and analysis Productivity suite	Helpdesk Incident management tool Collaboration tool Recovery tool Forensic tool Productivity suite SOAR

We will now, in Section 7 propose a Systems Engineering (SE) approach to select the correct technology. Following SE approach involves tasks such as determining user requirements and translating those to functional and technical specifications.

7. Step 4: Select technologies by applying systems engineering principles

We have in Section 6 identified the primary MSSP services and their corresponding technologies. The technology component is one of the costliest elements in a MSSP. Therefor it is crucial that a technology with the correct capability and capacity is selected. In this Section we will be viewing the SIEM and SOAR tools as systems.

The Pareto principle - also known as the 80/20 rule - states that for many events approximately 80% of the effects are from 20% of the causes (Harvey and Sotardi, 2018). Applying the Pareto principle to computing it shows that 80% of organisational risk could be addressed by 20% of the technical security controls (Symantec, 2020).

The SIEM and SOAR tools are very costly, and the Pareto principle will assist organisations to spend wisely. Experience showed that the Pareto principle could be extended to the procurement of technology. When organisations buy tools blindly off the shelf, they often utilise only 20% of the tool's capabilities, ignoring the remaining 80% of capabilities. This translates to a waste of money. A SE approach is needed to ensure that only relevant tool capabilities are identified, and that the organisation gets value for money in that at least 80% of the tool's capabilities are used.

The most important system requirement is the system functional requirement, and this will eventually lead to the technical specifications to be included in the Request For Proposal (RFP). It is recommended that the SE lifecycle is followed when selecting a SIEM / SOAR tool. The Systems Engineering Body of Knowledge (*System Requirements - SEBoK*, 2016) classifies system requirements as follows as taken from (*System Requirements - SEBoK*, 2016):

Table 4: Types of system requirement as taken from (system requirements - SEBoK, 2016)

Types of System Requirements	Description
Functional Requirements	Describe qualitatively the system functions or tasks to be performed in operation.
Performance Requirements	Define quantitatively the extent, or how well, and under what conditions a function or task is to be performed (e.g. rates, velocities). These are quantitative requirements of system performance and are verifiable individually. Note that there may be more than one performance requirement associated with a single function, functional requirement, or task.
Usability Requirements	Define the quality of system use (e.g. measurable effectiveness, efficiency, and satisfaction criteria).
Interface Requirements	Define how the system is required to interact or to exchange material, energy, or information with external systems (external interface), or how system elements within the system, including human elements, interact with each other (internal interface). Interface requirements include physical connections (physical interfaces) with external systems or internal system elements supporting interactions or exchanges.

Types of System Requirements	Description
Operational Requirements	Define the operational conditions or properties that are required for the system to operate or exist. This type of requirement includes: human factors, ergonomics, availability, maintainability, reliability, and security.
Modes and/or States Requirements	Define the various operational modes of the system in use and events conducting to transitions of modes.
Adaptability Requirements	Define potential extension, growth, or scalability during the life of the system.
Physical Constraints	Define constraints on weight, volume, and dimension applicable to the system elements that compose the system.
Design Constraints	Define the limits on the options that are available to a designer of a solution by imposing immovable boundaries and limits (e.g., the system shall incorporate a legacy or provided system element, or certain data shall be maintained in an online repository).
Environmental Conditions	Define the environmental conditions to be encountered by the system in its different operational modes. This should address the natural environment (e.g. wind, rain, temperature, fauna, salt, dust, radiation, etc.), induced and/or self-induced environmental effects (e.g. motion, shock, noise, electromagnetism, thermal, etc.), and threats to societal environment (e.g. legal, political, economic, social, business, etc.).
Logistical Requirements	Define the logistical conditions needed by the continuous utilization of the system. These requirements include sustainment (provision of facilities, level support, support personnel, spare parts, training, technical documentation, etc.), packaging, handling, shipping, transportation.
Policies and Regulations	Define relevant and applicable organizational policies or regulatory requirements that could affect the operation or performance of the system (e.g. labor policies, reports to regulatory agony, health or safety criteria, etc.).
Cost and Schedule Constraints	Define, for example, the cost of a single exemplar of the system, the expected delivery date of the first exemplar, etc.
Physical Constraints	Define constraints on weight, volume, and dimension applicable to the system elements that compose the system.

8. Mapping the conceptual process model to existing frameworks, standards and best practices

Our conceptual process model can be mapped to existing standards, frameworks and best practises. In Section 2 we stated that we will place the focus on SOC services, SOC models, mapping tools and technologies to services and SE the tools or technologies. We however include Facilities and People and Processes in Table 5 to illustrate the comprehensive coverage of our conceptual process from a framework, standards and best practice perspective.

Table 5: SOC planning steps mapped to existing frameworks, standards and best practices

Conceptual Process	Standards / Framework / Best Practice
Step1: Identify SOC services	For a list of SOC services - Towards a framework for building security operation centers (P. Jacobs, 2015). ITIL for service management (ITGovernanceUSA and IT Governance USA, 2011) Service Determination Methodology described in Section 3
Step 2: Identify SOC Models	Section 5 for methodology
Step 3: Map Technology to Services	Section 6 for methodology
Step 4: SE for Technology	SEBoK (SEBoK, 2016) ISO/IEC/IEEE 15288:2015 - Systems and software engineering — System life cycle processes
Step 5: Plan Facilities	ISO/IEC 27001:2013 Annex A11 (ISO/IEC, 2009) Occupational Health and Safety Acts (South African Government, 1993) NIST SP 800-53 (NIST, 2009)
Step 6: People and Processes	NICE Framework (NIST, 2013) SFIA (SFIA Foundation, 2015) For a list of recommended processes - For a list of SOC services - Towards a framework for building security operation centers (P. Jacobs, 2015).

9. Future work

In future studies we will identify SOC secondary services to augment its primary services. We will also catalogue the services according to SOC models. From a technology perspective, we will determine the most common functional requirements for SIEM and SOAR with the intention to map them according to industry. The intention is to develop a guide that will be useful for MSSPs when selecting their SIEM / SOAR tools. We will also develop SOC facility requirements. Further we will develop a guide to assist with SOC facility planning.

10. Conclusion

In this article we have proposed a conceptual process framework to be used when planning for the building of SOCs. The first step of our framework is to identify SOC services. We identified the three primary SOC services as monitoring, detection and response to cyber threats. We have also provided a methodology to identify the SOC services.

We then introduced the different SOC models. The SOC models will influence the SOC services, which in turn influences the SOC tools. We identified three contexts from where SOCs can be viewed. These are the consumer and service provider contexts. We the identified the different SOC models applicable to each context. The SOC models – together with the SOC services - will have an influence on the SOC tools.

In Section 6 we mapped the SOC services to its complimentary tools. And in Section 7 we proposed that a systems engineering approach be followed to identify technical specifications for the SOC tools. For each of the steps in our conceptual process there are existing frameworks, standards or best practices to guide its completion of the process, and these were introduced in Section 6.

During our research we could not find any conceptual process models that could be used to guide the SOC builders thinking when planning to build SOCs. SOC Builders can now use our conceptual process model to guide their thinking when planning for the building of their MSSP SOCs. The process will not differ much where it concerns organisational SOCs, and with small adjustments our conceptual process can also be applied during the building of SOCs with operational models different to the MSSP model.

References

C. Zimmerman and Operations Center, S. (2014) *Ten Strategies of a World-Class Cybersecurity Operations Center, Carson Zimmerman, MITRE*. The Mitre Corporation. Available at: www.mitre.org/sites/default/files/publications/pr-13-1028-mitre-10 (Accessed: 25 November 2018).

Cichonski, P. *et al.* (2012) *Computer Security Incident Handling Guide : Recommendations of the National Institute of Standards and Technology, NIST Special Publication 800-61*. Gaithersburg, MD. doi: 10.6028/NIST.SP.800-61r2.

Dave Young (2020) *Security Operations Center: Insource or Outsource to MSSP?* Available at: https://www.netnetweb.com/content/blog/security-operations-center-insource-or-outsource (Accessed: 13 August 2021).

Detection | Meaning of Detection by Lexico (no date). Available at: https://www.lexico.com/definition/detection (Accessed: 27 February 2020).

Digital Hands (2021) *IN-HOUSE SOC VS MSSP*. Available at: https://www.digitalhands.com/guides/in-house-soc-vs-mssp (Accessed: 13 August 2021).

Harvey, H. B. and Sotardi, S. T. (2018) 'The Pareto Principle', *Journal of the American College of Radiology*, 15(6), p. 931. doi: 10.1016/j.jacr.2018.02.026.

HPE (2018) *What is Security Monitoring? – HPE Definition Glossary | HPE™ EUROPE*. Available at: https://www.hpe.com/emea_europe/en/what-is/security-monitoring.html (Accessed: 15 March 2018).

ISO/IEC (2009) *An Introduction to ISO 27001, ISO 27002....ISO 27008*. Available at: http://www.27000.org/ (Accessed: 25 November 2018).

ITGovernanceUSA and IT Governance USA (2011) *ITIL®– IT Infrastructure Library® & IT Service Management*. Available at: http://www.itgovernanceusa.com/itil.aspx.

J. Godgart (2019) *Ditch Your MSSP: Making the Case for Managed Detection & Response*. Available at: https://blog.rapid7.com/2019/10/07/why-do-managed-detection-and-response-mdr-services-exist-in-a-world-dominated-by-mssps/ (Accessed: 20 February 2020).

J. Miller (2019) *MDR vs. SIEM vs. SOAR Acronyms Explained | BitLyft Cybersecurity*. Available at: https://www.bitlyft.com/mdr-vs-siem-vs-soar-acronyms-explained/ (Accessed: 20 February 2020).

Jacobs, P. *et al.* (2016) 'E-CMIRC – Towards a model for the integration of services between SOCs and CSIRTs', *European Conference on Information Warfare and Security, ECCWS*, 2016-Janua, p. 350. Available at: https://books.google.co.za/books?id=ijaeDAAAQBAJ&pg=PA350&lpg=PA350&dq=E-CMIRC+–

+Towards+a+model+for+the+integration+of+services+between+SOCs+and+CSIRTs+jacobs&source=bl&ots=50kTs5LXP_&sig=vQetp1xRqVtCPukMqSi1yZgpKFU&hl=en&sa=X&redir_esc=y#v=onepa (Accessed: 19 August 2016).

Kral, P. (2012) *SANS Institute: Reading Room - Incident Handling*. Available at: https://www.sans.org/reading-room/whitepapers/incident/paper/33901 (Accessed: 26 February 2020).

Logsign (2018) *What Makes SOC Effective? People, Process, and Technology*. Available at: https://www.logsign.com/blog/what-makes-soc-effective-people-process-and-technology/ (Accessed: 4 August 2021).

M.K. Hamilton (2020) *SOC-As-A-Service & How We Differentiate | Secuvant*. Available at: https://secuvant.com/soc-as-a-service/ (Accessed: 20 February 2020).

Matrix - Enterprise | MITRE ATT&CK™ (2022). Available at: https://attack.mitre.org/matrices/enterprise/ (Accessed: 26 February 2020).

McAfee (2021) *What Is a Security Operations Center (SOC)?* Available at: https://www.mcafee.com/enterprise/en-us/security-awareness/operations/what-is-soc.html (Accessed: 4 August 2021).

NIST (2009) *Recommended Security Controls for Federal Information Systems and Organizations*. Available at: http://csrc.nist.gov/publications/nistpubs/800-53-Rev3/sp800-53-rev3-final_updated-errata_05-01-2010.pdf.

NIST (2013) 'National Cybersecurity Workforce Framework', p. 127. Available at: http://csrc.nist.gov/nice/framework/ (Accessed: 17 February 2016).

P. Jacobs (2015) *Towards a framework for building security operation centers*. Available at: http://contentpro.seals.ac.za/iii/cpro/DigitalItemViewPage.external?lang=eng&sp=1017932&sp=T&suite=def (Accessed: 25 November 2018).

Robinson; Arbez; Birta; Tolk; Wagner (2015) 'CONCEPTUAL MODELING: DEFINITION, PURPOSE AND BENEFITS', in *Proceedings of the 2015 Winter Simulation Conference*. doi: 978-1-4673-9743-8/15/$31.00.

SANS Institute (2021) *Guide to Security Operations*.

SEBoK (2016) *Capability Engineering*. Available at: http://sebokwiki.org/wiki/Capability_Engineering (Accessed: 7 October 2016).

Secuvant (2019) *MDR vs. MSSP vs. SIEM – InfoSec Acronyms Explained | CI Security*. Available at: https://ci.security/resources/news/article/mdr-vs-mssp-vs-siem-infosec-acronyms-explained (Accessed: 20 February 2020).

SFIA Foundation (2015) *Why SFIA*. Available at: http://www.sfia-online.org/en (Accessed: 1 July 2016).

SOC-as-a-Service, MDR, and Managed SIEM. What's the Difference? - MSSP Alert (no date). Available at: https://www.msspalert.com/cybersecurity-guests/socaas-mdr-siem-defined/ (Accessed: 18 February 2020).

South African Government (1993) *Occupational Health and Safety Act (No. 85 of 1993)*. Available at: http://www.labour.gov.za/DOL/legislation/acts/occupational-health-and-safety/read-online/amended-occupational-health-and-safety-act/ (Accessed: 22 October 2017).

Symantec (2020) *Symantec Security Response - 80-20 Rule of Information Security*. Available at: https://www.symantec.com/avcenter/security/Content/security.articles/fundamentals.of.info.security.html (Accessed: 26 February 2020).

System Requirements - SEBoK (2016). Available at: https://www.sebokwiki.org/wiki/System_Requirements (Accessed: 26 February 2020).

The British Standards Institution (2014) *Cyber Security*. Available at: http://www.bsigroup.com/en-GB/Cyber-Security/.

Zhang, E. (2018) *What is Managed Detection and Response? Definition, Benefits, How to Choose a Vendor, and More | Digital Guardian*. Available at: https://digitalguardian.com/blog/what-managed-detection-and-response-definition-benefits-how-choose-vendor-and-more (Accessed: 20 February 2020).

Pedagogical and Self-Reflecting Approach to Improving the Learning Within a Cyber Exercise

Anni Karinsalo[1], Karo Saharinen[2], Jani Päijänen[2] and Jarno Salonen[3]
[1]VTT Technical Research Centre of Finland, Oulu, Finland
[2]JAMK University of Applied Sciences, Jyväskylä, Finland
[3]VTT Technical Research Centre of Finland, Tampere, Finland
anni.karinsalo@vtt.fi
jani.paijanen@jamk.fi
karo.saharinen@jamk.fi
jarno.salonen@vtt.fi

Abstract: In the digitalized world, there is a growing need not only to improve one's cybersecurity skills and knowledge, but also to find ways to optimize the learning process, for example by motivating the learners or optimising the learning facilities, material and the learners for the process. Cyber exercises ran within cyber ranges/arenas (CR) are an efficient way for the exercise participants to improve their cybersecurity skills and knowledge level. The pedagogical way of orienteering the participant to a learning situation is to have a preliminary survey, which prepares the participant for the upcoming event, adds self-reflection, and may even provide feedback and background information for the educator about the upcoming event. The objective of the survey is to improve the quality of the exercise by knowing the interest areas, preferences and other useful information about the participants that is then be used optimise the exercise accordingly. This study analyses the structure of one preliminary survey targeted for the cyber exercise event to be held in January 2022. The questions are justified according to existing frameworks. We have collected a set of structured questions presenting different topics related to the participants' professional background and expectations towards the exercise. In addition to the short-term goal of analysing the survey for one cyber exercise, this work benefits the long-term goal for improving the skills of cybersecurity professionals. Our further work will validate the results of our preliminary analysis and analyse its correspondence with the survey results, and the final analysis constructed after the cyber exercise.

Keywords: cyber range, cyber exercise, cybersecurity skills, cybersecurity, survey

1. Introduction

In the current research, there is an acknowledged need to improve the level of cybersecurity knowledge on European level. This includes both means of personal skill development for the cybersecurity professionals (European Commission. Joint Communication to the European Parliament and the Council, 2020), but also larger-scale, administrative policies such as developing a common European framework for monitoring and developing the skills of cybersecurity professionals (ENISA, 2019) (Nurse et al., 2021).

We perceive the motivating factors for this study from three dimensions. First, we want to extend the pedagogical knowledge of the learning process. The pedagogical aspect of cybersecurity learning has been studied for example in (Karjalainen, 2021) and (Le Compte, 2015). However, to the best of our knowledge, the concept of using a preliminary survey before the cyber exercise has not been employed in a very broad manner. Second, as we will be facilitating a cyber exercise ourselves, we study the ways to improve the cyber exercise practical arrangement with the pre-study from the organiser's perspective. Third, the knowledge we gain regarding learning within the cybersecurity exercise can affect other similar exercises. Thus, we hope our experience will add to the lessons-learned of such events, especially on European level, and where possible, also on the education framework development for security professionals.

The aim of this article is to describe the structure and benefits of, and theory behind the survey that is sent to the participants before the cyber exercise in January 2022. In this article, we argue, that by using a pre-survey to collect information about the participants' professional skills and areas of interest, and fine-tuning the exercise according to the responses, we can impact the development of the participants' professional skills as well as enhance the learning experience during the cyber exercise or other cyber event. The benefits of this study relate to resolving the following research questions:

- How can we better understand the needs and interests of cyber exercise participants (that can also be considered as "customers" in some sense) by using a pre-survey?
- What kind of questions should the pre-survey consist of?

- What kind of existing frameworks can we use to create our pre-survey?

The survey questions proposed in this article are tailored to the targeted exercise, namely Flagship #2, but we will generalise them in future research as well as provide the results from our pre-survey. We consider that this study lays groundwork for the benefits of increased learning about motivation of the participants, acquiring the necessary information for the cyber exercise, and increased general knowledge for the organisation of cyber exercises.

The article is structured as follows. In section two we provide the background to our research, namely describing the European and worldwide guidelines, taxonomies and other frameworks that we have used to create our pre-survey. In section three we introduce the pre-survey and justify the questions that we have decided to use in it. Finally, in section four we discuss the general justifications and lessons-learned for the construction of the study, before concluding the article in the last section.

2. Theory and framework background

2.1 Regulation and theory

The European Higher Education Area (EHEA) was adopted in May 2005 and it specified three cycles of qualification to which national frameworks were encourage to be made compatible with (European Higher Education Area, 2005). The cycles of qualification were updated by 2008 in a recommendation of the European parliament and of the council in establishment of the European Qualifications Framework (EQF) for lifelong learning. This update gave way for an eight level of qualifications; each of which were described by Knowledge and Skills to create Competence. Within the recommendation was also the requirement of mapping National Qualifications Frameworks (NQF) to the EQF from the Member States of the European Union (European Commission. Directorate-General for Education and Culture, 2008). Just before the 10th year anniversary of the EQF, the Council of the European Union refreshed their recommendation. These recommendations were divided into 18 different topics, e.g. to have member states ensure their consistency of national frameworks with the EQF periodically. (Council of the European Union, 2017) Within the European Union this background of guiding frameworks and recommendations give a good background in individual competence building and have established a common terminology within the EU (Brockmann, Clarke and Winch, 2009).

Bloom *et al.* (1956) introduced in their book a taxonomy to *"help (curriculum builders) to specify objectives so that it becomes easier to plan learning experiences and prepare evaluation devices"*. This taxonomy declared six major classes: Knowledge, Comprehension, Application, Analysis, Synthesis and Evaluation. Even though the learner could perform the major classes in different order than introduced in the book; it is still used as a tool of evaluation. Bloom's taxonomy has been revised by Anderson *et al.* (2001) to have a more dynamic conception of the classifications made earlier. Thus, the revised categories / cognitive processes are as follows; Remember, Understand, Apply, Analyze, Evaluate and Create. Curriculum developers use the taxonomy extensively in different universities.

2.2 Cybersecurity frameworks

Cybersecurity, as a paradigm of computing, has been a continuous topic of framework definition in multiple countries and international organisations. Several guiding frameworks have been introduced at the end of the last decade, with continuous work being done at the start of this decade. This chapter introduces the main cybersecurity frameworks related to this research paper.

Background of the *NICE Framework* came from the Comprehensive National Cybersecurity Initiative where one of the objectives was to expand cybersecurity education (Rollins and Henning, 2009). This Initiative was further emphasized into the formation of a National Initiative for Cybersecurity Education or NICE (The White House, 2010). The first available version of the NICE framework was published in 2017 (Newhouse et al., 2017). The framework described the cybersecurity work through tasks assigned to different work roles. These tasks required *Knowledge*, *Skills* and *Abilities* (KSA's) and the work roles themselves were defined into specialty areas and categories.

Association for Computing Machinery publishes their Curricula Recommendations on their web pages (Association for Computing Machinery, 2022). The overview report from 2005 on Curricula guidelines (CC2005

Task Force, 2005) had no section on cybersecurity. This was later published as *"Cybersecurity Curricula 2017"* guideline book in 2018 (Joint Task Force on Cybersecurity Education, 2018) next to the Computing Curricula recommendations of 2005. Finally in 2020 the updated work of ACM published the Computing Curricula 2020 (CC2020 Task Force, 2020) which declared cybersecurity as its own field of education.

In the European Union, several research and development projects had the goal of producing a cybersecurity framework to be used within the European Union. ECSO has published a European Cybersecurity Education and Training - Minimum Reference Curriculum (ECSO 2021) aimed at providing *"the guidelines relative to the competence & skills development framework along with pedagogical methodologies for the higher education programme requirements"*. SPARTA -project published its deliverable on cybersecurity skills framework (Piesarskas *et al.*, 2020) with stating *"This document serves as a basis for setting in motion a process of development of a comprehensive European cybersecurity skills framework"*. The framework analysed that European Cybersecurity Taxonomy (European Commission. Joint Research Centre, 2019) to be coupled with the NICE Framework would be a good starting point for a more comprehensive framework for the EU. CyberSec4Europe -project published its own Design of Education and Professional Framework (Karinsalo and Halunen, 2021) which combined a small part of the NICE framework with the ACM Cybersecurity Curricula 2017 Knowledge Areas. Other notable framework is The Cyber Security Body of Knowledge (Rashid *et al.*, 2021) in the United Kingdom, however it is not used in this research paper.

2.2.1 Flagship #1 cyber exercise

Flagship #1 was an online-only cyber exercise, organised in January 2021. The exercise platform used was a cyber-arena, a large-scale cyber range, as a technical platform. Participants used the prepared environment to perform their tasks. Flagship #1 was a reactive cyber exercise, showcasing real-world skills needed in every organisation that uses ICT-services. The task was to detect and investigate a successful cyber-attack that the exercise organisation had previously faced. Once the attack was detected and deemed successful, the participants started following the prepared (cyber) incident management documents and procedures, alerting organisations' staff and stakeholders, and various authorities. Flagship #1 showcased that the organisation benefits from using the existing documentation and procedures in a cyber exercise. When a cyber incident happens, there is some knowledge on the expected behaviour to mitigate and respond to the incident.

During registration to the exercise, the participants completed a short self-assessment questionnaire on their skills and knowledge in cybersecurity and previous experience related to cyber-exercises. This self-assessment was the basis for the preliminary survey covered in this paper. After the exercise, a comprehensive self-assessment questionnaire in skills improvement was filled-out. The post-exercise questionnaire was based on NICE framework KSA's. (CyberSec4Europe, 2021)

2.2.2 Flagship #2 cyber exercise

The forthcoming two-day Flagship #2 exercise showcases a simulated successful cyber-attack targeting a critical infrastructure operator, a train operator using a (simulated) next-generation Rail Traffic Management System. In the scenario, trains have smart devices installed that include Trusted Platform Modules (TPMs). The (simulated) technology is dependent on various ICT-infrastructure services and functionalities located in the train and alongside the railway. Attacking against such technology stack requires besides malicious objectives, also technological skills to avoid or bypass the security controls in a train or infrastructure.

The objective of Flagship #2 is to showcase that analyzing and investigating a sophisticated attack against complex technology requires broad and deep understanding of the technology, and that a (simulated) company, whilst having competent cybersecurity employees may still lack the skills needed. Given the scenario is successful from this point of view, the exercise participants receive support from a (simulated) cybersecurity analyst company that they have hired. The analyst company has a vast amount of workforce that focuses on analysing and investigating complex cyber-attacks. Due to the aforementioned needs, we aim to impact the development of the participants' professional skills as well as enhance the learning experience during the cyber exercise or other cyber event with our pre-survey.

2.3 Target groups

Flagship #2 exercise is targeted to the following target groups:

- Project group members
- Other personnel from project member organisations
- External stakeholders of the project (external cybersecurity analyst role)

In general, the exercise is targeted to any members of the aforementioned groups with interest towards attending the cyber exercise. In other words, one does not need to be a cybersecurity professional to participate even though professionals might benefit from the exercise more than non-professionals. The main difference to the previous Flagship #1 exercise is the inclusion of external cybersecurity analysts who participate in a separate capture-the-flag (CTF) exercise during Flagship #2 and analyse a simulated cyber-attack using real tools and applicable methodology in a dedicated environment. The cybersecurity analyst role has a prerequisite of having previous experience in using Linux command line tools and naturally the exercise benefits cybersecurity professionals more than non-professionals.

3. Survey design

In this section, we analyse each of the survey questions and their theoretical background in order to justify their use. By "survey", we mean the preliminary survey (or pre-survey) which is targeted to the forthcoming Flagship #2 exercise participants.

3.1 Survey design and process

The survey in question is an online survey sent to the registered participants of the forthcoming cyber exercise and it collects information about their competence levels and preferences prior to the exercise. The survey consists of eleven questions with eight single-choice, two multiple-choice and one open question. All but the last question (#11) are mandatory in order to get responses to all survey questions. However, we have included a specific "I prefer not to disclose this information" response to questions #1-#5 that collect information concerning the educational background, knowledge/skill levels, participant job roles and the organisation sector in case the respondent is concerned about the responses. All the other questions are collecting information about areas of interest, preferred exercise roles and opinions about suitable exercise group sizes and session times and therefore they do not have the aforementioned response option. Since these extra response options do not provide additional value to this article, they are not included in the figures nor covered in the next sub-sections. In addition to the survey questions covered in the following sub-sections, the survey also consists of an introductory/invitation text and a field to ask/verify the respondent email address. The email is used for connecting the right pre-survey with the post-survey that will be sent to the exercise participants after the event and used to match the expectations to the learning experience. Since these aforementioned survey parts do not have additional value to this article, we just mention them here.

3.2 Survey questions

The first survey question is shown in the figure below. The question is a single-choice one with four response options categorised according to the European Qualifications Framework (Council of the European Union, 2017). It also includes an "Other, please specify" option in case the respondent doesn't belong to any of the following groups or even has multiple degrees from different areas and would like to clarify.

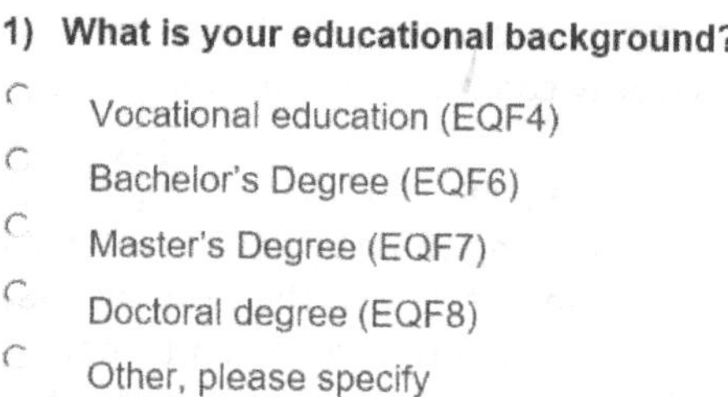

1) What is your educational background?
Vocational education (EQF4)
Bachelor's Degree (EQF6)
Master's Degree (EQF7)
Doctoral degree (EQF8)
Other, please specify

Figure 1: Survey question #1

The first survey question helps the exercise organisation to be more aware of the educational background and competence levels of the participants. With this gained awareness, the cybersecurity exercise could be adjusted or participant roles designed with more precision to match the capabilities of the participants.

The second survey question is shown below. The question is a single-choice one with 12 response options categorised according to the sectors specified in the European Cybersecurity Taxonomy (European Commission. Joint Research Centre, 2019). It also includes an "Other, please specify" option e.g. in case the respondent organisation doesn't belong or doesn't recognize him/herself to be in any of the groups.

2) What sector does your organisation primarily belong to?

- Audiovisual and media
- Defence
- Energy
- Financial
- Food and Drink industry
- Government (education)
- Health
- Manufacturing and Supply Chain
- Nuclear
- Public Safety
- Space
- Telecom Infrastructure
- Other, please specify

Figure 2: Survey question #2

Given the multipurpose cybersecurity exercises in development to day (Fischer-Hübner *et al.*, 2020) it would be of interest of the exercise conducting organization to get more familiar with the participants organization background. This gives way to customize the exercise towards a certain security of supply area.

The third survey question is shown below. The question is a single-choice one with three response options categorised according to Bloom's taxonomy (Bloom *et al.* (1956)). This gives a self-estimation of the participants' competence level in this particular area of expertise; of cybersecurity exercises in general.

3) In your opinion, what is your knowledge level (e.g. understanding of exercise concepts and types, etc.) regarding cybersecurity exercises/hackathons?

- Entry level (Remember/Understand)
- Intermediate (Apply/Analyze)
- Expert (Evaluate/Create)

Figure 3*:* Survey question #3

The objective of this question is to categorise participants according to their knowledge level and then, based on the exercise type and objectives, organise exercise groups accordingly. Generally, the groups are formed evenly, i.e. each group has members from each skill level, which makes it possible for the expert level members to assist the entry and intermediate level members during the exercise. However, in some exercise types it is also possible to assign members of the same level into one group, which among others helps the facilitation of the group. In practice this could mean e.g. that the entry level groups receive more comprehensive explanation than others do. The fourth survey question is shown. The question is a single-choice one with three response options categorised according to Bloom's taxonomy (Bloom *et al.,* 1956). As cybersecurity exercises usually are quite technical events, the participants are asked to self-evaluate their competence levels in technical skills.

4) In your opinion, what is your technical skill level (e.g. usage of operating systems and IT environments, etc.) regarding cybersecurity exercises/hackathons?

- Entry level (Remember/Understand)
- Intermediate (Apply/Analyze)
- Expert (Evaluate/Create)

Figure 4*:* Survey question #4

This question is very similar to the previous one, but focuses on the technical skill level of the exercise participants instead of the overall knowhow of the exercise types and processes. The objective of this question is to categorise participants according to their technical skill level and then, based on the exercise type and objectives, organise the exercise groups accordingly. For example, if the exercise supports multiple simultaneous tasks at different levels, then groups could be formed according to the participants' knowledge level and they would complete different tasks or "missions" during the exercise. In case the exercise consists of tasks or "missions" that every group must complete in the same order, then the groups would most likely be formed in such a way that each group has members from each knowledge level. In general, the advantage of having members of different technical skill level in one group may support the learning of those in the lower, i.e. entry and intermediate skill levels. However, there is a rather high probability that the members at expert level perform most of the exercise tasks, which may hinder the learning of the less advanced members. In most cases, it is the role of the group facilitator to monitor the progress and ensure that all members of the group understand the things done during the exercise despite their technical or other skill level.

The fifth survey question is shown below. The question is a single-choice one with seven response options categorised according to the sectors specified in the NIST - National Initiative for Cybersecurity Education (NICE) Cybersecurity Workforce Framework (Newhouse *et al.*, 2017). It also includes an "Other, please specify" option e.g. in case the respondent job role doesn't belong to any of the aforementioned groups.

5) In which category does your job role primarily belong to?

- Securely Provision (SP)
- Operate and Maintain (OM)
- Oversee and Govern (OV)
- Protect and Defend (PR)
- Analyse (AN)
- Collect and Operate (CO)
- Investigate (IN)
- Other, please specify

Figure 5*:* Survey question #5

The sixth survey question is shown below. The question is a multiple choice one with nine options that have been applied from the CyberSec4Europe deliverable "Design of Education and Professional Framework" (Karinsalo and Halunen, 2021). The respondents are instructed to choose from one to three options from the list.

6) Flagship 2 has defined goals. However, if you could choose, which knowledge area development/improvement are you most interested in? Choose 1-3 options.

- ☐ Data Security
- ☐ Software Security
- ☐ Component Security
- ☐ Connection Security
- ☐ System Security
- ☐ Human Security
- ☐ Organisational Security
- ☐ Societal Security
- ☐ Operate and maintain

Figure 6*:* Survey question #6

This question is very important one since it enables fine-grained exercise customisation according to the participants' areas of interest. In case the survey is conducted before or during the planning of the cyber exercise, it may enable quite radical customisation. However, as the question text in the previous figure specifies, the exercise may already have defined goals in which case the customisation could apply e.g. to spending more time in a desired type of session or include additional pieces of information to them in order to enhance the learning process. In case the exercise consists of different simultaneous tasks, then customisation could be done by grouping the members according to their desired interest areas and choosing suitable tasks for them.

The seventh survey question is shown below. The question is a single-choice one with seven response options categorised according to the sectors specified in the NIST - National Initiative for Cybersecurity Education (NICE) Cybersecurity Workforce Framework (Newhouse et al., 2017). It also includes an "Other, please specify" option e.g. in case the respondent job role doesn't belong to any of the aforementioned groups.

7) Which role do you want to primarily progress in the selected knowledge areas?

- Securely Provision (SP)
- Operate and Maintain (OM)
- Oversee and Govern (OV)
- Protect and Defend (PR)
- Analyse (AN)
- Collect and Operate (CO)
- Investigate (IN)
- Other, please specify

Figure 7: Survey question #7

The objective of this question is to assign suitable roles for each cyber exercise participant and where possible, target some tasks in order to support specifically the learning of specific roles. As an example, the Flagship #2 exercise consists of a parallel capture-the-flag (CTF) type of cybersecurity analyst exercise that is directed specifically to people interested in that role.

The eighth survey question is shown below. The question is a single-choice one with four response options with the objective of collecting the respondent's opinion about their preference regarding the ideal number of participants for the exercise teams.

8) In your opinion, what is the ideal number of participants for the exercise teams?

- 1-2
- 3-4
- 5-6
- more than 6

Figure 8*:* Survey question #8

The objective of this question is to assign the participants in groups that are pleasing in terms of the number of members and therefore enhance participation, learning and elements like peer teaching. According to the research by e.g. Koolos et al. (2011), the group-size effect is observed in favour or working in smaller groups (subgroups), i.e. students prefer smaller assignments and smaller groups that enable peer teaching.

The ninth survey question is shown below. The question is a single-choice one with six response options ranging from zero to more than 90 minutes.

9) In your opinion, how long should the average exercise sessions be (read: how often does the exercise/situation develop)?

- 0-15 minutes
- 16-30 minutes
- 31-45 minutes
- 46-60 minutes
- 61-90 minutes
- more than 90 minutes

Figure 9*:* Survey question #9

The question relates to the intensity of learning events in the cybersecurity exercise. The effective training length is a topic researched in education e.g. by Ericsson (2006) and Bunce et al. (2010). Since Flagship #2 lasts for two days, the individual sessions are bound to be quite lengthy. However, we are searching for possibilities to adjust the exercise intensity at least to some extent based on the responses to this question.

The tenth survey question is shown below. The question is a multiple choice one with nine options that have been applied from the Cybersecurity Curricula 2017 (Newhouse *et al.*, 2017). The respondent is instructed to choose from one to three options from the list.

10) What knowledge area development/improvement are you least interested in? Choose 1-3 options.

- ☐ Data Security
- ☐ Software Security
- ☐ Component Security
- ☐ Connection Security
- ☐ System Security
- ☐ Human Security
- ☐ Organisational Security
- ☐ Societal Security
- ☐ Operate and maintain

Figure 10*:* Survey question #10

Similarly to question six, this question enables customising the exercise contents in detail according to the responses. However since Flagship #2 exercise has already defined goals, customisation applies mainly e.g. to spending less time in the less desired knowledge areas or the related information can be provided as an extra.

The eleventh survey question is shown below. The question is an open one with the instructions to the respondent for giving any thoughts about the exercise or comments/greetings to the organisers.

11) Do you happen to have any thoughts about the forthcoming exercise or greetings to the organisers that you would like to say?

Figure 11*:* Survey question #11

The objective of this question is to allow participants express feelings and raise concerns about the forthcoming exercise, if any. The question is partially linked to the research by, e.g. Arbaugh and Benbunan-Fich (2007) that highlight the importance of participant interaction in online learning environments such as Flagship #2. In other words, the question also intends to motivate them by increasing their engagement to the exercise.

4. Discussion

In this study, we analysed how to use a pre-survey for understanding the needs and interests of the cyber exercise participants. We also analysed how to format the questions, and what frameworks to use when creating the survey. In this context, we constructed eleven questions using existing cybersecurity frameworks. We also provided related justifications based on the Flagship2 event requirements. One option could have been, that we would have used ranking scale for the question selections. However, if we rank the questions, we need to analyse the results with data analysis and come up with a weighted average, which does not serve our purpose with the survey end goal.

Regarding the general structuring of the survey, we concluded that since the audience consists of professionals and the event is voluntary for them (i.e. not a part of a student curriculum), the survey should not be too demanding or time-consuming. If the pre-survey has too complex or too many questions, there is a risk that the respondents do not care to answer. Instead, we will deepen our knowledge by sending another survey after the Flagship2 event. Thus, we optimized the questions to attain as much information as possible while trying to keep the number of the questions as low as possible. Regarding the question setting, we wanted to use the questions to improve the commitment of participants by increasing their motivation. Fishbach et al (2022) describe intrinsic (i.e. internally driven or rewarding) motivation to be "critical predictor of engagement". According to them, one approach for increasing intrinsic motivation is to factor "the positive experience while pursuing the activity, with choice." Questions formulated such as question 6, enabling participants feel they can affect or make choices of interest regarding the course content, potentially increase the intrinsic motivation of the participant towards the exercise. Further work will include analysing the pre-survey answers and reflecting them in the summary of the cyber exercise outcomes, lessons-learned and post-survey results.

5. Conclusions

This article presents the construction process and structure of a pre-survey targeted to the participants of a cyber exercise. We have constructed a survey consisting of eleven questions that are based on existing frameworks such as EQF, NICE, European and Cybersecurity Curricula. Based on our current analysis, the questions help us better understand the needs and interests of the Flagship #2 cyber exercise participants. The article also provides related justifications that are linked to the upcoming cyber exercise details.

References

Anderson, L. et al. (2001) A Taxonomy for Learning, Teaching, and Assessing: A Revision of Bloom's Taxonomy of Educational Objectives.

Arbaugh, J.B., Benbunan-Fich, R. (2007). The importance of participant interaction in online environments. Journal of Decision Support Systems, 43 (3), pp. 853-865. https://doi.org/10.1016/j.dss.2006.12.013.

Association for Computing Machinery (2022) Curricula Recommendations. Available at: https://www.acm.org/education/curricula-recommendations (Accessed: 10 January 2022).

Bloom, B.S. et al. (1956) Taxonomy of Educational Objectives - The Classification of Educational Goals. London: Longmans, Green and Co Ltd.

Brockmann, M., Clarke, L. and Winch, C. (2009) 'Competence and competency in the EQF and in European VET systems', Journal of European Industrial Training, 33, pp. 787–799. doi:10.1108/03090590910993634.

Bunce, D., Flens, E. and Neiles, K. (2010) How Long Can Students Pay Attention in Class? A Study of Student Attention Decline Using Clickers. Journal of Chemical Education 2010 87 (12), 1438-1443. doi: 10.1021/ed100409p.

CC2005 Task Force (2005) Computing Curricula 2005: The Overview Report. New York, NY, USA: Association for Computing Machinery. Available at: https://www.acm.org/binaries/content/assets/education/curricula-recommendations/cc2005-march06final.pdf (Accessed: 10 January 2022).

Council of the European Union (2017) Council Recommendation on European Qualifications Framework for lifelong learning. Available at: https://eur-lex.europa.eu/legal-content/EN/TXT/HTML/?uri=CELEX:32017H0615(01)&from=EN (Accessed: 5 January 2022).

CyberSec4Europe. (2021) "CyberSec4Europe Hosting Flagship 1: An Online Cybersecurity Exercise", [online], Cyber Security for Europe (CyberSec4Europe), https://cybersec4europe.eu/cybersec4europe-hosting-flagship-1-an-online-cybersecurity-exercise/ (Accessed: 15 January 2022).

ECSO (2021) European Cybersecurity Education and Professional Training: Minimum Reference Curriculum SWG 5.2 I Education & Professional Training. https://ecs-org.eu/documents/publications/61967913d3f81.pdf (Accessed: 5 January 2022).

ENISA (2019) Cybersecurity skills development in the EU. https://www.enisa.europa.eu/publications/the-status-of-cyber-security-education-in-the-european-union (Accessed: 15 January 2022).

European Commission. Joint Communication to the European Parliament and the Council (2020) The EU's Cybersecurity Strategy for the Digital Decade

European Commission. Joint Research Centre (2019) A proposal for a European cybersecurity taxonomy. LU: Publications Office. Available at: https://data.europa.eu/doi/10.2760/106002 (Accessed: 5 January 2022).

European Higher Education Area (2005) The Framework of Qualifications for the European Higher Education Area. Available at: http://www.ehea.info/media.ehea.info/file/WG_Frameworks_qualification/85/2/Framework_qualificationsforEHEA-May2005_587852.pdf (Accessed: 5 January 2022).

Ericsson, K.A. (2006) 'The Influence of Experience and Deliberate Practice on the Development of Superior Expert Performance', the Cambridge handbook of expertise and expert performance, p. 22.

Fischer-Hübner, S. et al. (2020) 'Quality Criteria for Cyber Security MOOCs', in Drevin, L., Von Solms, S., and Theocharidou, M. (eds) Information Security Education. Information Security in Action. Cham: Springer International Publishing (IFIP Advances in Information and Communication Technology), pp. 46–60. doi:10.1007/978-3-030-59291-2_4.

Fishbach, A. and Woolley, K. (2022). The Structure of Intrinsic Motivation. To be published in: Annual Review of Organizational Psychology and Organizational Behavior, Volume 9, number 1, doi:10.1146/annurev-orgpsych-012420-091122 (Accessed 17. January 2022)

Joint Task Force on Cybersecurity Education (2018) Cybersecurity Curricula 2017: Curriculum Guidelines for Post-Secondary Degree Programs in Cybersecurity. New York, NY, USA: Association for Computing Machinery.

Newhouse, W. et al. (2017) National Initiative for Cybersecurity Education (NICE) Cybersecurity Workforce Framework. NIST SP 800-181. Gaithersburg, MD: National Institute of Standards and Technology, p. NIST SP 800-181. doi:10.6028/NIST.SP.800-181.

Karinsalo, A. and Halunen, K. (2021) 'Design of Education and Professional Framework'. CyberSec4Europe. Available at: https://cybersec4europe.eu/wp-content/uploads/2021/06/D6_3_Design-of-Education-and-Professional-Framework_Final.pdf. (Accessed: 14 January 2022).

Karjalainen, M. (2021). Pedagogical Basis of Live Cybersecurity Exercises. https://jyx.jyu.fi/handle/123456789/76371 (Accessed: 12 January 2022).

Kooloos, J.G.M. *et al.* (2011) 'Collaborative group work: Effects of group size and assignment structure on learning gain, student satisfaction and perceived participation', *Medical Teacher*, 33(12), pp. 983–988. doi:10.3109/0142159X.2011.588733.

Le Compte, A., Elizondo, D. and Watson, T. "A renewed approach to serious games for cyber security," 2015 7th International Conference on Cyber Conflict: Architectures in Cyberspace, 2015, pp. 203-216, doi: 10.1109/CYCON.2015.7158478.

Nurse, J. R. C. and Adamos, K. and Grammatopoulos, A. and Di Franco, F. (2021) Addressing the EU Cybersecurity Skills Shortage and Gap Through Higher Education. European Union Agency for Cybersecurity (ENISA). Available at: https://www.enisa.europa.eu/publications/addressing-skills-shortage-and-gap-through-higher-education/@@download/fullReport (Accessed: 5 January 2022).

Piesarskas, E. et al. (2020) 'Cybersecurity skills framework'. Available at: https://sparta.eu/assets/deliverables/SPARTA-D9.1-Cybersecurity-skills-framework-PU-M12.pdf (Accessed: 10 January 2022).

Rashid, A. et al. (2021) 'The Cyber Security Body of Knowledge'. Available at: https://www.cybok.org/ (Accessed: 15 January 2022).

Rollins, J. and Henning, A.C. (2009) Comprehensive National Cybersecurity Initiative: Legal Authorities and Policy Considerations, UNT Digital Library. Library of Congress. Congressional Research Service. Available at: https://digital.library.unt.edu/ark:/67531/metadc743582/ (Accessed: 10 January 2022).

The White House (2010) Advancing Our Interests: Actions in Support of the President's National Security Strategy, whitehouse.gov. Available at: https://obamawhitehouse.archives.gov/the-press-office/advancing-our-interests-actions-support-presidents-national-security-strategy (Accessed: 10 January 2022).

Strategies for Internet of Things Data Privacy and Security Using Systematic Review

Sithembiso Khumalo, Amanda Sibiya, Teballo and A. Kekana
University of Johannesburg, South Africa
skhumalo@uj.ac.za
herexcell@gmail.com
tebantonny1@gmail.com

Abstract: The Internet of Things (IoT) now referend to as the Internet of Everything (IoE) has been in existence long before it was identified as a concept. It was introduced with the emergence of the Fourth Industrial Revolution and was aimed at improving people's lives and economies across the globe by connecting physical items to the internet so they can be able to deliver specific services implicitly. The nature of IoT requires that all the systems ensure data privacy and security because much of data that is uploaded into and used by the system is personal and private. Thus, the aim of this research was to identify the tools and strategies that can be used for IoT data privacy and security while also providing a brief but intensive understanding of the concept of IoT and data privacy and security challenges faced by IoT systems. This qualitative research study utilised a pragmatic paradigm and data was collected and analysed using text-based secondary data sources and a PRISMA protocol through systematic review. A PRISMA flow diagram was utilised to assess the eligibility of the sources used for this research. The findings showed that hacking is a major challenge that affects IoT systems and that there are strategies that can be used to protect data such as authentication, encryption technology, and anonymisation amongst many. Additional findings found that the strategies have not yet been found effective, but standards have been set upon the results expected from them. The conclusion is that for the identified strategies to be proven effective, they must be implemented and tested in IoT systems, so further investigation can be conducted if they prove to be ineffective.

Keywords: internet of things (IoT), strategies, tools, data privacy, data security

1. Introduction

The continuous developments in technology have led to several breakthroughs in the field of Information and Communication Technology (ICT) and they include the emergence of the Internet of Things (IoT) – a term discovered by Kevin Ashton in 1999. Rose, Eldridge, and Lyman (2015:5) define IoT as *"scenarios where network connectivity and computing capability extends to objects, sensors, and everyday items not normally considered computers, allowing these devices to generate, exchange, and consume data with minimal human intervention"*. IoT does not have one specific definition because people perceive and define it according to how it serves them and according to their own needs (Singh, 2014). Many economies across the world have improved using the IoT, making them more profitable while saving the costs of production and time consumption. Connecting devices and physical items to the internet can assist organisations, decrease data collection limitations by placing data exactly where it is needed, while also providing accurate information which improved reliability.

The concept of IoT has been around for a long time although it is said that it is a concept that came with Kevin Ashton where else he just came up with a more suitable word for the concept (Monther, 2020). Foote (2016) proves that this concept has been around for ages and can be seen in the 1980s where people were able to connect to a Coca Cola machine to check for available drinks before they can purchase. IoT has several advantages including coming with new ways of collecting data, connecting human beings with devices, and connecting devices to the internet, it also has its disadvantages (Weber, 2010). Every system requires some level of privacy and security especially in this type of technology that seems to be more personal and unique for every organisation or person. The many entry points in IoT system and data being shared daily, the system can be prone to cyber-attacks which makes privacy and security a big challenge (Rose *et al* 2015:6). The above mentioned makes it imperative for different IoT security and privacy tools and strategies to be continuously investigated and applied to ensure safety and integrity (Porras, Pänkäläinen, Knutas & Khakurel, 2018). Security in IoT is very important because users entrust the system with personal and sensitive information that when accessed by cyber attackers, they can perform illegal activities that may land users in trouble (Ning, Liu, & Yang, 2013).

This research investigated the tools and strategies of data privacy and security in IoT systems and understanding how these tools and strategies work to ensure the safety of the system. Furthermore, a basic understanding of what IoT is and the components within IoT were provided, as well as the main challenges that affect privacy and security in IoT (Zhao & Ge, 2013).

2. Research problem, aim and objective

The IoT is a data driven technology used by people and businesses to deliver services that are specific to their needs. This means that personal and private information is uploaded constantly which creates many entry points to the system. For this reason, an IoT system can be prone to cyber-attacks and hacking and people's private information may be used illegally. This makes data privacy and security in IoT a very important aspect to consider because with effective tools and strategies, data can be protected. The main aim of this research was to investigate data privacy and security strategies that can be used to protect data in IoT systems using the systematic review protocol. The main question: *What are the strategies and tools used for data privacy and security on the Internet of Things (IoT).*

3. Literature review

The IoT is an advanced form of technology that has allowed the sharing and trading of data between human being and machines, and/or between two devices without any interference from human beings (Majeed, Bhana, Ha, Kyaruzi, Pervaz, & Williams, 2016). Additionally, the many interconnected devices that are transmitting data and information to both symmetric and non-symmetric systems, an IoT system can be left vulnerable to privacy and security threats. Gulzar and Abbas (2018) argue that the IoT has proven to have more risks than benefits which makes it very challenging and risky for people to rely on it. The idea of introducing IoT was to assist businesses to increase profitability and productivity by having a technology that can interact with people and other devices and improve the value of life for people. However, the fact that data is the main driver of IoT, which means that data is forever being shared and transmitted between devices proves the system to be with many privacy and security faults that need to be addressed (Sfar, 2018). Gillis (2020) defines an IoT *"as a system of interrelated computing devices, mechanical and digital machines, objects, animals or people that are provided with unique identifiers (UIDs) and the ability to transfer data over a network without requiring human-to-computer interaction"*. It consists of four main components including sensors, data processing, connectivity, and user interface that work together seamlessly to provide the required service (Holdowsky, 2015).

The main security and privacy challenge facing IoT systems is hacking where illegal users find ways to get into a system by tracking user footprints through the enormous amounts of data shared and transmitted in the system Stoyanova (2020). Lally and Sgandurra, (2018) add that a lot of IoT system developers put their focus on the cost and usability and only implement measures of privacy and security once the system has been breached and exploited. Some of the strategies that can be used in IoT systems is an authentication system which deals with managing several users for a single device by including things such as biometrics, and digital certificates before access can be granted (Sultana & Gavrilova, 2014:416). Other strategies include encryption technology, anonymisation, routing, and Blockchain based Secure Data Aggregation strategy (BSDA) (Kirichek, Kulik & Koucheryavy, 2016:203). Moreover, none of the tools and strategies investigated have proven to be effective to the system, instead there are standards that have been set and it is hoped that these tools and strategies can meet them (Gionis & Tassa, 2009).

4. Research design and methods

This study used pragmatic paradigm to practically investigate the strategies and tools that are used for data privacy and security on the IoT. This was done by utilising the PRISMA protocol through systematic reviews to collect data on the grey literature and literally go through the data to analyse it to find out what strategies and tools are used in IoT for data privacy and security. To increase the validity of this research, deductive reasoning was used through the study. The study employed a qualitative research method which focused on the quality of textual data collected from databases and grey literatures using the PRISMA protocol. The study utilised a cross-sectional time horizon, where different academic journals are investigated, and results obtained at a single period. Selected journals that were scholarly and peer reviewed in the Information and Knowledge Management discipline and Information Technology were chosen through purposeful sampling. A PRISMA diagram was used for sampling to identify the eligibility criteria of articles that are used. Several articles were found on all the selected databases and grey literature when using the Boolean search string "strategies and IoT and data privacy and data security". The articles provided were screened to eliminate duplicates. The eligibility of the articles was assessed whether the documents are full text articles with the relevant textual information. The study used 37 of the sources found to be eligible to do the research this was completed using a data extraction form which consisted of an eligibility criterion for exclusion and inclusion of data sources (*cf* Appendix A).

Data was gathered from recommended databases (Emerald Insight, EBSCOhost, IEEE Xplore, SAJIM, ACMDL Digital Library and google scholar), these are text based secondary data sources, to help answer the research question. The study searches for academic journals using Boolean search strings: tools and strategies and IoT and data privacy, to gather relevant data and analyse it. The reliability and validity of the study was achieved through deductive reasoning by collecting enough data from different sources that were analysed to draw accurate conclusions. Some sources were excluded from the data analysis as they were irrelevance to the research aim and objective. This process of exclusion and inclusion was completed using a data extraction form which consisted of an eligibility criteria for exclusion and inclusion of data sources (*cf* Appendix A).

5. Results and discussion

This section is based on the articles that matched the eligibility criteria of this research paper. The chart below (figure 1) clearly articulates how the eligibility of the sources used was assessed.

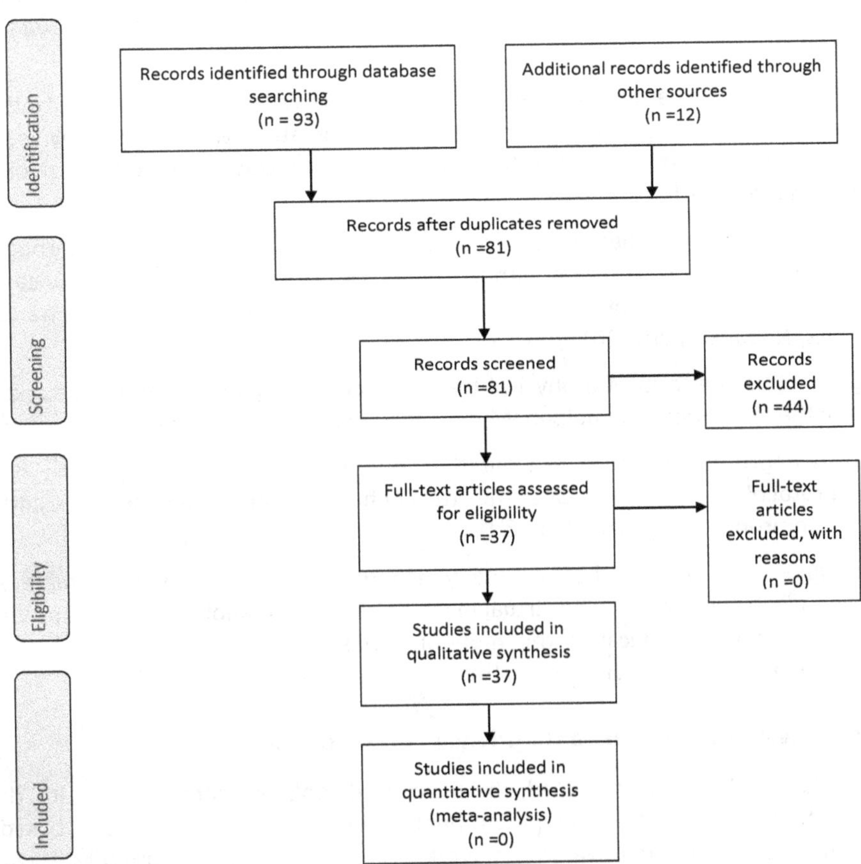

Figure 1: PRISMA flow diagram representation for the sources used for sampling

5.1 Defining the Internet of Things (IoT)

According to the research findings, the concept of IoT was coined by Kevin Ashton in 1999 who defined it as *"interconnected objects that can be uniquely identified with the radio frequency identification system (RFID) technology"*. The IoT is mainly the idea of having electronic systems integrated into physical objects to allow smooth, intelligent, and seamless sharing of information to create a new global infrastructure that can thrive in the fourth industrial revolution (Stankovic, 2014). Furthermore, the main reason why IoT does not have one specific definition is because of the involvement of a lot of research studies, businesses, stakeholders, etc. and all these parties want to define IoT in a manner that best suits their backgrounds, needs, and expectations. From the research findings, it was found that a more interesting idea of IoT, which seems more like a revelation currently, is the idea of Mark Weiser which he came up with in 1991 where he stated that, "The most profound technologies are those that disappear" which meant that the technologies that will thrive are those that form part of peoples' everyday lives, and this statement was supported by Gartner (2016) where he stated that in 2016, the internet will be connected with more than 60 billion objects which will be 30% more than the objects in 2015. He furthermore stated that the IoT will be one of the trends and key drivers of technology in the 21st century. It is significant to note both the positive and negative sides of IoT and investigate them further.

5.2 Data privacy and security challenges in IoT

According to the research findings, hacking is one of the most prominent privacy and security challenges in IoT systems and there are various forms of threats that can be used to attack (Majeed, Bhana, Ha, Kyaruzi, Pervaz & Williams, 2016). These threats will be discussed further below (Alwarafy, Thelaya, Abdallah, Schneider & Hamdi, 2021).

5.2.1 Hacking

The interconnection of many devices makes the system more prone to cyber-attacks or being hacked more especially if the device connected has poor security or when users leave data streams unprotected (Modi, Patel, Borisaniya, Patel & Rajarajan, 2013). There are several different types of attacks that hackers use to enter IoT Systems (Lu & Xu, 2019).

- *Denial of Service (DoS)* – the hacker may use a bug to attack the system, causing it to fail internally and cause damage to the hardware. The system is then infiltrated by the attacker, and they may use it however they see fit. IoT is most prone to this form of attack. This challenge includes flooding attack or one that simply denies access through a crash.
- *Man-in-the-middle (MITM)* – the hacker gets access and is allowed to listen and peep through a conversation between the sender and the receiver without their permission or knowledge. Once the hacker can eavesdrop, they can be able to replace the data shared between the two parties with fraudulent one (Ngu, Gutierrez, Metsis, Nepal & Sheng, 2017).
- *Node damaging* – this is a form of a physical threat which takes place when the attacker gains physical access to any of the IoT devices. (Tawalbeh, Muheidat, Tawalbeh & Quiwader, 2020).
- *Breakage of cryptographic protocols* – Limited resources can cause the systems; developer to use cryptographic protocols that are weak. Once the attacker hacks the encrypted messages and data, the whole system can be compromised (Roman, Zhou & Lopez, 2013).
- *Shared technology* – in IoT systems, there are many resources that are shared, for example, via virtualization (Rullo, Midi, Serra & Bertino, 2017). The virtual machine monitor of another user can be vulnerable and be penetrated by another user because the monitor can sometimes require the user to allow access (Makhdoom, Abolhasan, Lipman, Liu & Ni, 2019).

5.3 Tools and strategies used for IoT data privacy and security

According to the data that has been collected, the most used tools and strategies for IoT data privacy and security include, Anonymization, Encryption, Authentication, Routing, Blockchain based Secure Data Aggregation strategy (BSDA) and Data privacy frameworks (Ghaffari, Lagzian, Kazemi & Malekzade, 2019). Each of these tools and strategies will be fully explained, as to how they work or how they are used to ensure the security and privacy of data in IoT (Ari, Ngangmo, Titouna, Thiare, Mohamadou, Mohamadou & Gueroui, 2019).

5.3.1 Anonymization

According to the data collected, this is one of the strategies used which generally refers to hiding and modifying data that is related by rearranging some parts or all the original data so that attackers cannot combine and make sense of other information after stealing the anonymized data (Wang, Tong, Shancang, Geyong & Zhiwei, 2021). There is a location privacy protection scheme that has been proposed to utilise k-anonymity and trusted third party policies, where it protects the privacy of the location of the data when a sensing user uploads it and when a requester requests a server that could be untrustworthy (Kirichek, Kulik & Koucheryavy, 2016). In simple terms, k-anonymity hides the identification of the sender of data and upon request, it still does not show the source from which the data comes from. Although the scheme is expensive, it is immune to background attacks.

5.3.2 Encryption technology

According to the findings, encryption technology may be regarded as a process where the sender encrypts the first data using encryption technology, and therefore the receiver decrypts the info using the decryption algorithm.

Cryptographic hashes this is where small messages are read in correspondences to large messages as hashes and main aim of the method is ensure that the hash does not disclose anything related to the original data. Cryptographic hashes can simply be explained as a cryptography technique where data that is new is mixed in ciphertext form by making use of a secret key, transferred in public route and at a later stage decrypted by the receiver with a pre known secret key (Shammar & Zahary, 2019).

IoT encryption this includes encrypting data between IoT devices and back-end systems making use of standard cryptographic algorithms, to keep standard data integrity and to prevent sniffing.

Content encryption the data collected shows that this encryption is only related to data in the storage area, it assists to protect the confidentiality of data one adds into a database in case hackers or unauthorised persons gain access to the content.

5.3.3 IoT authentication

The findings state that IoT devices, begin with an easy static password to authentication mechanisms, through digital certificates and biometrics, nodes that store and/or communicate sensitive data. For instance, healthcare system may become valuable for hackers since they contain valuable information such as patient information and electronic forensic records. Therefore, the healthcare sector should employ means to guard the electronic records from being accessed by unauthorised persons.

A two-factor authentication (2FA) is one among the foremost recommended authentications which needs a user to use a combination of passwords and other authentication forms that do not depend on the knowledge of users, like randomly obtaining a code through Short Message Service (SMS) text. This is followed by a Context-Aware Authentication (CAA), which can be referred as an easily adaptive authentication, where related data and machine-learning algorithms is evaluated on a continuous basis to identify risk of malice without having to bother the end-user in demanding authentication (Jan, Nanda, He, Tan & Liu, 2014). Therefore, the hacker and/or subscriber will be requested to provide a multi-factor token to continue having access if the risk is found to be high. This may create a push or complicated authorisation which can make IoT devices to not have an easy-to-guess password/ authorisation code and keep the info within the devices secured and guarded (Khalid & Majeed, 2016).

5.3.4 Routing

Based on the data collected, routing refers to an unidentified technique used to anonymously sort out the source node to stop the situation information of the source node from leaking. This strategy involves having the power to route users, services and data to servers that are in possession of the user's data. These can use solutions like Amazon's Route service that allows routing based on the geographical location supported by the IP address of network requests done over the web (Tyagi & Goyal, 2020).

5.3.5 Blockchain based Secure Data Aggregation strategy (BSDA) for edge computing empowered IoT

In the data aggregation process, attention focused on the way to aggregate data without privacy disclosure for instance, people that release tasks to be aggregated cannot withstand the leakage of any sensitive information contained within the task (Wang, Garg, Kaddoum & Hossain, 2020). Therefore, they tend to settle on trustworthy workers to finish the task. On the data collected, it shows that the blockchain offers a distribution of ledger that is integrated, peer-to-peer networks, smart contracts and consensus mechanism which allows reliable access control, storage that is secure, and distributed computation, this suggests that applying blockchain to data aggregation will increase the safety and privacy of the information in those tasks (Sciancalepore, & Di Pietro, 2021). For example, purposefully modifying blockchain provide restrictions to workers on tasks that have certain restriction security levels and requirements of completion of requirements (Zanella, Bui, Castellani, Vangelista, & Zorzi, 2014). Furthermore, personal data is then best protected when data from a large group of users or devices are aggregated, mainly because the information on selected individual is concealed within a more general cohort (Wei, Wu, Long & Lin, 2019).

5.3.6 Data privacy frameworks

Amongst the framework proposed for IoT data security and privacy, Object Security Framework (OSCAR) for IoT is one among the foremost recommended frameworks to use in IoT (Irshad, 2016). This framework is predicated on the concept of object security that introduces security within the appliance payload (Xiong, Rong, Lei, Tian, Liu & Yao, 2020). Although it once considered separate confidentiality and authenticity trust domains, privacy is employed to supply capability-based access control and protection against eavesdropping during the communication. Meaning this security framework protects data from replay attacks by coupling the content encryption key with the duplicate detection (Ferracane, 2019).

6. The effectiveness of IoT tools and strategies in ensuring data privacy and security

Research has been done to find out how are the proposed tools and strategies effective in protecting data within the IoT. Based on the findings that has been collected, it is found that most researchers have not yet tested the tools or the strategies but have set standards as to what these tools and strategies can do in IoT data privacy and security (Cangea, 2019). The proposed standards, amongst others, include improved system performance and secured personalised data (Wang, Zhang, Liu, Bhuiyan & Jin, 2019). These standards or performance can be what is expected after implementing the proposed tools and strategies in securing data and its privacy.

The findings shows that data aggregation schemes ensure data confidentiality, integrity, and real timeliness.

Data confidentiality: data can be able to reach its destination without hackers accessing it or being leaked.

Data integrity: the users can be guaranteed that the data collected by the sensers is real and complete.

Data real timeliness: as soon as data is updated and uploaded by the user, it is received and becomes available in the platform.

7. Conclusions and recommendations

This paper has investigated the tools and strategies that are used in IoT for data privacy and security, while also providing an understanding of the concept of IoT and the security and privacy challenges within it. The use of the PRISMA protocol through systematic review protocol has been of great assistance because it allowed for the identification, evaluation, and summarisation of findings that are relevant to the study of the tools and strategies used in IoT data privacy and security. The findings of this study are not claimed to be universal but have found to be the most popular amongst different resources. The findings of this study have provided information about IoT as a concept and its historical information to shed some light on the technology that is being investigated for strategies that can be used to protects its data. Additionally, the challenges that affect data privacy and security have also been discussed to understand the complexity of the problem that needs the solutions being investigated. It is recommended that another study be conducted, of all the tools and strategies identified in this paper after they have been implemented and tested to find out if they are effective or not. And if they prove to be ineffective, other tools and strategies be investigated that are effective in ensuring data privacy and security in IoT.

In conclusion, the results of this study have been found from investigation the concept of IoT, the challenges facing data privacy and security in IoT, tools and strategies used in IoT data privacy and security, as well as the effectiveness of these tools and strategies using the PRISMA protocol through a systematic review. These findings may be of assistance to other researchers who are looking into IoT strategies and tools for data privacy and security and can be used to expand knowledge. It is however important to note that this study is limited to resources found in in the field of Information and Knowledge Management as well Information Technology. Developments in IoT data privacy and security tools and strategies should be seen as soon as possible considering that people entrust the system with private information. The developers of IoT systems must start considering security and privacy protocols more than they consider the cost, design, and the size. What is important in IoT is to protect the integrity of the data within the system.

Appendix A: Data extraction form and eligibility criteria

DATA EXTRACTION FORM

Topic: **Strategies for Internet of Things data privacy and security using systematic review.**

General Information

Data extractors	Sithembiso Khumalo, Amanda Sibiya and Teballo A. Kekana
Research Question	What are the strategies and tools used for data privacy and security in Internet of Things (IoT)
Setting	N/A
Time	Cross sectional time horizon (9 months)
Protocol and Registration	Systematic Review Protocol
Budget	R1000 for data to conduct the investigation

Eligibility Study

	INCLUDED	EXCLUDED
Publication Type	Article titles, abstracts, and findings on strategies for the Internet of Things data privacy and security.	Non-published material, Sources that are not peer reviewed.
Study Objective	Specific strategies used for data privacy and security on the Internet of Things.	Generic strategies used for data privacy and security.
Language	English	Other languages except English
Methodology	Qualitative	Quantitative, mixed methods
Results	Strategies for IoT data Privacy and security	Generic strategies used for data privacy and security.
Conclusions and Recommendations	Information including further study information related to IoT data privacy and security strategies	Information including further study information related to general data privacy and security strategies
References	Other studies related to IoT data privacy and security strategies	Studies related to general data privacy and security strategies.

References

Alwarafy, A., Al-Thelaya, A.K., Abdallah, M., Schneider, J. and Hamdi, M. 2021. Survey on Security and Privacy Issues in Edge-Computing-Assisted Internet of Things. *IEEE Internet of Things Journal*. 8(6):4004-4022.

Ari, A.A., Ngangmo, O.K., Titouna, C., Thiare, O., Mohamadou, K., Mohamadou, A., and Gueroui. A.M. 2019. *Enabling privacy and security in Cloud of Things: Architecture, applications, security & privacy challenges*. Emerald Publishing limited. Available at https://www.emerald.com/insight/2210-8327.htm.

Cangea, O. 2019. A Comparative Analysis of Internet of Things Security Strategies. *Seria Tehnica*. 71(1):1-10.

Corcoran, P. 2016. The Internet of Things: why now, and what's next? *IEEE Consumer Electronics Magazine*. 5(1):63-68.

Ferracane, M.F. 2019. Data flows and national security: a conceptual framework to assess restrictions on data flows under GATS security exception. *Information Management*. 21(1):44-70.

Foote, K.D. 2016. *A Brief History of the Internet of Things*. DataVarsity. Available form: https://www.dataversity.net/brief-history-internet-things/ Accessed (28 February 2022).

Ghaffari, K., Lagzian, M., Kazemi, M. and Malekzade, G. 2019. A comprehensive framework for Internet of Things development: A grounded theory study of requirements. *Journal of Enterprise Information. 38(1):26-30.*

Gillis, A. 2020. *What is Internet of Things (IoT)?* Tech Agenda. https://internetofthingsagenda.techtarget.com/definition/Internet-of-Things-IoT#:~:text=The%20internet%20of%20things%2C%20or,human%2Dto%2Dcomputer%20interaction.

Gionis, A. and Tassa, T. 2009. k-Anonymization with Minimal Loss of Information. *IEEE Transactions on Knowledge and Data Engineering*. 21(2):206-219.

Gulzar, M. and Abbas, G. 2018. *Internet of Things Security: A Survey and Taxonomy*. 2019 International Conference on Engineering and Emerging Technologies (ICEET). IEEE Xplore.

Holdowsky, J. 2015. *Inside the Internet of Things*. New York: Deloitte University Press.

Irshad, M. 2016. A Systematic Review of Information Security Frameworks on the Internet of Things. *Journal of Internet of Things*. 12(1):1270-1275.

Jan, M.A., Nanda, P., He, X., Tan, Z., and Liu. R. P. 2014. A robust authentication scheme for observing resources on the internet of things environment. *IEEE 13th International Conference on*. Conducted by IEEE.

Khalid, M. and Majeed, S. 2016. A smart visitors' notification system with automatic secure door lock using mobile communication technology. *International Journal of Computer Science and Information Security*. 16(4):97-101.

Kirichek R., Kulik, V. and Koucheryavy, A. 2016. *False Clouds for Internet of Things and Methods of Protection*. St. Petersburg State: St. Petersburg University.

Lally, D. and Sgandurra, G. 2018. Towards a Framework for Testing the Security of IoT Devices Consistently. *1st International Workshop on Emerging Technologies for Authorization and Authentication*. 11263(2018):88-102

Lu, Y., and Xu, L.D., 2019. Internet of Things (IoT) cybersecurity research: A review of current research topics. *IEEE Internet Things Journal*. 6(2):2103–2115.

Majeed, A., Bhana, R., Ha., A., Kyaruzi, I., Pervaz, S. and Williams, M. 2016. Internet of Everything (IoE): Analysing the Individual Concerns Over Privacy Enhancing Technologies (Pets*). International Journal of Advanced Computer Science and Applications (IJACSA*). 7(3):15-22.

Makhdoom, I., Abolhasan, M., Lipman, J., Liu, R. P. and Ni, W. 2019. Anatomy of Threats to the Internet of Things. *IEEE Communications Survey and Tutorial*. 21(2):1636-1675.

Modi, C., Patel, D., Borisaniya, B., Patel, A. and Rajarajan, M. 2013. A survey on security issues and solutions at different layers of Cloud computing. *The Journal of Supercomputing*. 63(2):561-592.

Monther, A.A. 2020. Security Techniques for intelligent spam sensing and anomaly detection in online social platforms. *Journal of Electronic Computer Engineering*. 2(10):143-145.

Ngu, A.H., Gutierrez, M., Metsis, V., Nepal, S. and Q. Z. Sheng. 2017. IoT middleware: A survey on issues and enabling technologies. *IEEE Internet of Things Journal*. 4(1):1-20.

Ning, H., Liu, H., & Yang, L. T. 2013. Cyberentity security on the Internet of Things. *Computer Journal*. 46(4):46-53.

Porras J., Pänkäläinen J., Knutas A. and Khakurel J. 2018. Security on The Internet of Things – A Systematic Mapping Study. *Proceedings of the 51st Hawaii International Conference on System Sciences*.

Roman, R., Zhou, J. and Lopez, J. 2013. On the features and challenges of security and privacy in distributed Internet of Things. *Computer Networks*. 57(10):2266-2279.

Rose, K., Eldridge, S. and Chapin, L. (2015). *The Internet of Things: an overview Understanding the Issues and Challenges of a More Connected World*. Reston, Virginia: The Internet Society (ISOC).

Rullo, A., Midi, D., Serra, E. and Bertino. E. 2017. Pareto Optimal Security Resource Allocation for Internet of Things. *ACM Trans. Privacy and Security*. 20(4)1-30.

Sciancalepore, S. and Di Pietro, R. 2021. PPRQ: Privacy-Preserving MAX/MIN Range Queries in IoT Networks. *IEEE The Internet of Things Journal*. 8(6)1-19.

Sfar, A.R., Natalizio, E., Challal, Y. and Chtourou, Z. 2018. A roadmap for security challenges on the Internet of Things. *Digital Communications and Networks*. 4(2):118-137.

Shammar, A. E. and Zahary, T. A. 2019. The Internet of Things (IoT): A survey of techniques, operating systems, and trends. *The Internet of Things*. 38(1):6-8.

Singh, J. 2014. Cyber-attacks in cloud computing: a case study. *International Journal of Electronics and Information Engineering*. 1(2):78-85.

Stankovic, A.J. 2014. Research directions for the Internet of Things. *IEEE Internet of Things Journal*. 1(1):7-9.

Stoyanova, M., Nikoloudakis, Y., Panagiotakis, S., Pallis, E. and Markakis, K.E., 2020. A Survey on the Internet of Things (IoT) Forensics: Challenges, Approaches, and Open Issues. *IEEE Communications Survey and Tutorial*. 22(2):1191-1221

Sultana, M. & Gavrilova, M. L. 2014. Face recognition using multiple content-based image features for biometric security applications. *International Journal of Biometrics*. 6(4): 414-434.

Tawalbeh, L., Muheidat, F., Tawalbeh, M. and Quiwader, M. 2020. IoT privacy and security: challenges and solutions. *Journal of Applied Science*. 10 (41): 1-17

Tyagi, A.K and Goyal, D. 2020. *A Survey of Privacy Leakage and Security Vulnerabilities on the Internet of Things. Emerald publishing limited*. Johannesburg: University of Johannesburg.

Wang, R., Tong, X., Shancang, I., Geyong, M. and Zhiwei, Z. 2021. *Privacy Enhancing Techniques on the Internet of Things Using Data Anonymisation*. Beijing: Peking University.

Wang X. Garg S. Kaddoum G. and Hossain H.S. 2020. A Secure Data Aggregation Strategy in Edge Computing and Blockchain empowered Internet of Things. *Journal of the Internet of Things*. 10(2):2327-4662.

Wang, T., Zhang, G., Liu, A., Bhuiyan, M.Z.A. and Jin, Q., 2019. A secure IoT service architecture with an efficient balance dynamic based on cloud and edge computing. *IEEE Internet Things Journal*. 6(3).

Weber, R. H. 2010. Internet of Things–New security and privacy challenges. *Computer law & security review*. 26(1):23-30.

Wei, L., Wu, J., Long, C. and Lin, Y.-B., 2019. The convergence of IoE and blockchain: security challenges. *IT* Professional. 21(5):26-32.

Xiong J., Rong M., Lei C., Tian Y., Liu X., and Yao Z. 2020. A Personalized Privacy Protection Framework for Mobile Crowdsensing in IIoT. *Journal of Industrial Informatics*. 16(6):4231-4241

Zanella, A., Bui, N., Castellani, A., Vangelista, L. and Zorzi, M., 2014. Internet of Things for smart cities. *IEEE Internet of Things Journal*. 1(2):22-128

Zhao, K. and Ge, L., 2013. A survey on the Internet of Things security. *Ninth International Conference on Computational Intelligence and Security*. Conducted by IEEE explore. New York: IEEE Explore.

Public Authorities as a Target of Disinformation

Pekka Koistinen, Milla Alaraatikka, Teija Sederholm, Dominic Savolainen, Aki-Mauri Huhtinen and Miina Kaarkoski
Department of Leadership and Military Pedagogy, National Defence University, Helsinki, Finland
pekka.koistinen@mil.fi
milla.alaraatikka@mil.fi
teija.sederholm@mil.fi
dominic.savolainen@mil.fi
aki.huhtinen@mil.fi
miina.kaarkoski@mil.fi

Abstract: Disinformation is a part of a modern digitalised society and thus affects public authorities´ daily work. Through disinformation, malicious actors can often erode the fundamentals of democratic societies. In practice, this can be achieved by influencing authorities' decision-making processes and creating distrust towards public organisations which can weaken authorities' ability to function. In Finland, public authorities have relatively transparent and open decision-making processes and communication practices compared to other democratic societies. This transparency and openness can be seen as a vulnerability, increasing the opportunities for malicious actors to use disinformation. The authorities of public services are also seen as producers of evidence-based official information. In general, Finns have very high trust in public authorities. Trust has a major impact on societies' psychological resilience and susceptibility to disinformation. The results of this article strengthen the idea that disinformation weakens authorities' ability to function. The producers of disinformation, aided by citizens' high confidence of public authorities, aim to utilise authorities' communication by misrepresenting the content according to their own agenda. In this study, our purpose is to describe public authorities' experiences relating to disinformation in their own organisation. This study follows a qualitative design framework by analysing data collected in September 2021 using inductive content analysis. The empirical data includes 16 government officials' interviews with themes exploring how disinformation affects their daily activities and why they are targets of disinformation. This article is part of a larger project relating to counterforces and detection of disinformation. The results contribute towards a broader understanding on how different types of public authorities, ranging from health to security organisations, communicate in complex social media environments.

Keywords: disinformation, communications, decision-making, public authorities, national security

1. Introduction

Public authorities face significant challenges in social media networks and the overall media structure. Their status as an information provider has been challenged by the changes in societies' communication and information avenues. The internet and social media have made it easier to form global networks and produce content in new ways, and thus, power has shifted to those who can influence information flows according to a specific agenda (Ikäheimo & Vahti, 2021). The algorithmic structures of popular social media sites are one of the factors that contributes to polarising public opinion and specifically, the intentional spread of inaccurate information can accelerate such developments (Pariser, 2011; Bozdag & Van Den Hoven, 2015; Woolley & Howard, 2016; Lazer et al., 2018; Cinelli et al., 2021). Even before the prevalence of social media, there were suggestions of social capital being strategically captured by political forces (Acemoglu, Reed & Robinson, 2014; Satyanath, Voigtlaender & Voth, 2017).

With the increasing speed of information exchange, the amount of disinformation circulating has been growing (Matasick, Alfonsi & Bellatoni, 2020). Disinformation should be understood as a separate term from fake news. Fake news frames information problems as unrelated incidents, whereas disinformation represents a more systematic and strategic approach when aiming to disrupt information flows (Bennet & Livingston, 2018). Disinformation can be used by political or governmental actors to manipulate public opinion online and reduce trust in existing institutions and authorities (Bradshaw & Howard, 2018). Distorting and questioning public authorities' messages is a significant challenge that threatens national security (Vasu et al., 2018). One way to counter the impacts of disinformation is efficient communication. The opportunity for institutions to communicate and engage with the public have expanded, but so have the challenges of providing sustained, timely, accurate and relevant information (Bennett & Segerberg, 2012; Sunstein, 2018; Tufekci, 2017).

In recent years, disinformation has been a popular research subject but there is little research about its relations to public authorities. It seems that western societies' public sectors are particularly vulnerable to digital propaganda, such as disinformation. Compared to other western democracies, Finland has relatively good resistance to digital propaganda. In Finland, there are coordinative responses to disinformation, public officials react quickly to false claims and citizens still have high trust in mass media (Bjola & Papadakis, 2020). Transparency and openness have been defined as the guiding principles of Finnish public authorities' communication, allowing citizens to evaluate institutional practices (Ministry of Justice, 2019). According to official instructions, in unusual circumstances, public authorities have the responsibility to communicate effectively to citizens and form a realistic picture of what is happening (Prime Minister's Office, 2019).

The security system in Finland is based on co-operation between state and civil society actors. This might be one element that helps to counter the effects of disinformation (Bjola & Papadakis, 2020). Furthermore, institutions represent a key pillar in the formation of society and trust is the foundation for the legitimacy of public institutions and a functioning democratic system. Public trust towards authorities in Finland is high and that has traditionally supported the workability of administrative and political models. This can be seen in Finland's successful response to the COVID-19 pandemic (OECD, 2021; Jallinoja & Väliverronen, 2021). Therefore, to maintain the legitimacy and sustainability of Finnish institutions, we should strive to understand the dynamics and dangers of information attacks that can reduce trust and increase polarisation.

This study aims to explore how public authorities view themselves as a target of disinformation and what aspects might cause vulnerability. Study focuses on how public authorities experience their position as a possible target to disinformation. The empirical data consists of 16 public authorities' interviews. This article argues that developing co-operation between authorities and coordinating counterforces are essential in countering disinformation.

2. Data and method

The data was collected in September 2021 by conducting semi-structured interviews with representatives from 15 public administrative institutions, which contained 21 questions exploring two key themes related to disinformation situational awareness. Specifically, the themes 'disinformation as a phenomenon from an organisational perspective' and 'current procedures' contained questions such as "What do you consider as disinformation?" and "Have there been any cases of disinformation in your organisation, and if so, how did you handle them?".

The interviewees represented 15 public administrative fields including national security, ministries, legal institutions and emergency services. Invitation letters were sent to organisations in coordinative positions in their respective administrative fields. The research conducted followed the ethical principles published by The Finnish Advisory Board on Research Integrity (TENK, 2012). Furthermore, the selected organisations had autonomy to decide who attended the interview. In total, 16 interviews were conducted, the majority of which were online. Furthermore, the average length of an interview was 50 minutes. The number of public authorities in Finland is relatively small and compared to this, 16 interviews provides adequate data sample. However, even tough interviewees spoke in behalf of their organisation, it is not possible to distinguish perfectly what is their personal opinion and what is the organisation policy.

To analyse the data, inductive content analysis techniques were used (Glaser & Strauss, 2017). This approach included open coding, categorisation and abstraction of data (Elo & Kyngäs, 2008). The text was coded using qualitative data analysis software Atlas.ti 9 (Atlas.ti, GmbH, Berlin Germany) and the data was organised by separating segments where the participants discussed being targets of disinformation. In total, there were 68 incidents. These findings were then further coded, focusing on different perspectives around being potential targets to disinformation. This coding process led to a further 11 groups, from which 8 themes were formed.

3. Results

The 8 themes identified describe how public authorities view themselves as targets of disinformation. The themes are ordered by incidence, with the most common occurring theme being presented first. The main results are summarised by Table 1. The first column presents the highlighted themes, the second column describes the content of these themes and the last column describes why the interviewees considered the themes challenging to public authorities.

Table 1: Summary of the results

Highlighted Theme	Theme Content	Challenge
Openess of Action by Public Authorities	Authorities communicate actively and openly; authorities' messages are truthful	Validating the information can be slow; disinformation spreads fast
Questioning Expertise	There are many sources of information; public authorities do not have monopoly status in information spreading	Questioning the expertise of public authorities aims to weaken trust
Public Authorities are not Potential Targets for Disinformation	Public authorities' task is to help citizens; public authorities' status protects them from disinformation	Public authorities do not prepare for disinformation
Duties of Public Authorities	Public authorities' duty is to guide citizens and to communicate what is proper behavior in society	Public authorities end up in confrontational positions in public discourse
Abuse of the Position of Public Authorities	Exploitation of the public authorities' trust	Public authorities' messages are used in disinformation
Resources for Combating Disinformation	There are not enough resources to repel disinformation or it is not the core duty of authorities	Detecting and countering disinformation is perceived as communication experts' duty
Disinformation is Used Internally	Agencies' employees post wrong information for example about the working conditions	Media publishes distorted information about the working conditions in public agencies
The Significance of Authorities' Co-operation Networks to Disinformation	Finnish public authorities are exposed to disinformation in their work-related networks	Disinformation can be part of larger influencing campaigns

Openness of Action by Public Authorities

The interviewees perceived the openness of actions, and communication, by public authorities as the most significant factor contributing to susceptibility to disinformation. Finnish public authorities must communicate openly and swiftly, although at the same time these represent vulnerabilities. Furthermore, the documents related to the actions of public authorities are primarily public. Even negative documents are made public. Some interviewees have perceived a change in how the media operates around these issues; even the largest media outlets no longer correct erroneous information like before. Some of the interviewees believe that reporters no longer function as gatekeepers for publishing information.

Most public authorities publish a considerable number of notices and announcements and have an extensive social media audience. Particularly, security authorities are an attractive target for news reporting. *"As we have an estimated, and I mean I'm just spitballing here, 50 000 news reports per year", "Yearly, we publish from 7 000 to 10 000 notices and announcements, and multiple times as many social media posts. Nowadays we are like a news agency, in a way"*. Public authorities are expected to instantly react to matters and to partake in public discourse.

Public authorities cannot defend themselves from disinformation in all cases because legislation prevents them from publicly disclosing certain matters as far as they are subject to confidentiality. Additionally, public authorities only publish validated, truthful information. In some cases, validating information takes time and creates an opportunity for disinformation to spread in the absence of official information.

According to the interviewees, relatively open publication of matters in the preparatory phase is a part of the openness of action by public authorities. Unfinished matters or decisions may be targeted by disinformation by framing them as value judgements and spreading deliberately misleading information. Based on the interviews, civil servants would like to simply prepare and present issues for particular questions of political decision-making without expressing their opinions on what is being prepared. *"I think that maybe those are more... I mean after all since political decision-making is about making value judgements and thus, like, particularly the value judgements and their correctness, they are much more easily influenced and that is why I believe that disinformation is certainly directed more towards that* [political decision-making] *and is probably more of a challenge there than in preparatory work by civil servants"*.

Questioning Expertise

Based on the interviews, civil servants perceive themselves to be targets of disinformation as representatives of an expert organisation. The product of an expert organisation is often information based on research. Disinformation seeks to call into question the expertise of public authorities or to make organisations seem untrustworthy. Questioning expertise aims to undermine the information produced by public authorities, alongside trust. Focusing on definitions, pointing out small errors of fact and polarising public discourse are methods used to question expertise. In many sectors, public authorities have an undisputed special position as the producers of official information.

Expertise in the communication of public authorities is challenged by converting the public discourse into one compatible with one's own world view by attempting to dispel the original topic of discussion. The interviewees say that communication has changed considerably in the past decade. Public authorities no longer have monopoly power to produce and share information. In today's news environment, citizens are able to source information from multiple different sources. *"Our greatest task is, in particular, that we are a part of what makes democracy work. And in order for it to work, someone must offer enough validated information so that people can make informed decisions. But it is not our job to affect what decision a person ends up making after assessing and weighing these different elements"*.

Public Authorities are not Potential Targets for Disinformation

According to the interviewees, some functions of public authorities are neutral and simply meant to help people, so there is no need to target them with disinformation. A certain kind of public authority status provides protection and, even in challenging circumstances, public authorities whose work is non-political are respected. The interviews showed that public authorities believe themselves to be experts in preparing against threats and that their staff is conscious of different threats. *"I mean citizens do have a lot of trust and a kind of neutrality perhaps brands* [public authorities]. *It is after all humanitarian action, broadly defined; it aims at helping"*. Many of the interviewees recognise disinformation as a phenomenon and the negative consequences it carries but believe that it has no significant effect on everyday activities.

The Finnish language is viewed as one of the factors that protects against supranational disinformation campaigns. From an international perspective, Finnish is not an appealing language for spreading disinformation.

Duties of Public Authorities

According to the interviewees, the legal duties of public authorities include citizen advice and guiding the public to follow common rules. Actions by public authorities help to set the limits of how one ought to act within the society and aim to influence citizens' attitudes and behaviour by means of effective communication.

Duties of public authorities also include issues that provoke passionate opinions, such as health, welfare and security. *"So precisely these kinds of topics that have strong interest groups that, from our point of view, maybe appear as spreaders of information, somewhat"*. The interviewees thought that public authorities as organisations are targeted by sensationalised news precisely based on their duties. Public authorities often have to strongly interfere with citizens' basic rights and make significant administrative decisions in the course of their duties. When it comes to research activities, the partial polarisation and antagonization of the research topics in public discourse was brought up in the interviews. Disinformation is used in attempts to affect how results of research are utilised in decision-making on emotive issues (e.g. equality and non-discrimination).

Individual actors may spread disinformation to take a position on the acts of public authorities. There might be frustration directed at the public authorities because a desired service or help has not been provided. In some cases, media may be used to present biased information, which considerably affects other citizens as users of public services.

Abuse of the Position of Public Authorities

The interviewees mentioned that the trust and reputation of public authorities is abused by means of disinformation. The notices and announcements made by public authorities are used to construct content that suits one's own needs, which is then published as the original. Often the motivation is to make public authorities party to a divisive debate involuntarily. *"I would also add to this that what I call the 'don't shoot the messenger' problem. So, this problem is born out of the fact that people think of reporting as an opinion"*. The interviewees said that the information produced by public authorities was used indirectly as a tool of disinformation, *"our news was then, in a way, used as a tool of disinformation to be used in another direction so that we kind of were, how would I put this, the ball was played through our court, that we were not directly the target"*. Some of the interviewees used the term 'framing'. Here, framing means emphasising certain issues in a way that, in a specific context, they become stronger than what their original meaning would entail.

Resources for Combating Disinformation

A significant number of the interviewees believe that there are not enough resources to combat disinformation or that combating disinformation is not a core duty of public authorities. Currently, disinformation is primarily combated using resources available to communications. *"Our resources are quite quickly used up if we are going to start chasing all those who share incorrect stuff and spin our publications by sharing them with incorrect accompanying info"*. To some extent, detecting and combating disinformation is made more difficult by the typical way in which civil servants only work during office hours. Combating the spread of incorrect information outside of office hours produces difficulties.

Disinformation is Used Internally

Based on the interviews, disinformation is also utilised within the operation of public authorities. Employees may publish incorrect or altered information because of a lack of resources or a bad working environment. The interviewees believe that the media is particularly interested in publishing news that involves some kind of antagonism between the employee and employer. *"In some cases, it can be a kind of 'cry for help' phenomenon, whereby it is thought that the solution to the problem, or for example the source of the problem is for instance the lack of resources, and in order to get more resources, it is thought that it is necessary to reveal defects"*.

Spreading incorrect information about the lack of resources of public authorities may, according to the interviewees, affect trustworthiness and, at worst, lead to citizens not utilising the services they are offered. Disinformation is used to erode the public image of the employee and to partially damage the duty of loyalty owed by the employee. *"I think that also the common duty of loyalty to the employer has probably suffered and been forgotten, and it is being broken, sometimes even severely, precisely owing to social media, because that is all that is needed"*.

The Significance of Authorities' Co-operation Networks to Disinformation

The interviews brought up the effects of disinformation on the network-like co-operation of public authorities domestically and abroad. A Finnish public authority may be targeted by disinformation, for example through the EU-community. For instance, in serious security-related situations, disinformation does not necessarily directly target a single public authority, but instead the effects reach all the actors involved through organs and networks of co-operation. *"It is really a kind of analysis of this phenomenon, like, more broadly as opposed to relating to just Finland, that we know how that method is used and how it then connects as a part to something bigger, and what its role is in this kind of broader campaign of influencing"*.

According to the interviewees, disinformation is used in attempts to influence the pillars of society by eroding trust. *"Yeah it is in a way the goal of taking a stand, or yes, in the broader frame of reference, it does also affect comprehensive security's ... co-operation model, which it seeks to erode in different ways, so that if I started to produce disinformation and, for example, attacking the Finnish state, then I would certainly start by eroding the pillars of society that have traditionally been sturdy, which effectively means that if I were to prepare this kind of broad operation of influencing then I would surely start by gnawing at the activities of public authorities or their co-operation"*.

4. Discussion

This study explored how public authorities view themselves as targets of disinformation and the aspects that might make them vulnerable to it. The main results pointed out that authorities in different organisations consider open decision-making processes, openness and transparency as potential causes of vulnerability to disinformation. Documents produced by Finnish authorities are often public and based on information that is considered official. In certain instances of communication, validating information takes time and allows disinformation to spread in the absence of official information. However, this openness is crucial from the viewpoint of the citizens, who expect open and clear communication from public authorities, especially during crisis situations (Mitu, 2021). The interviewees noted that especially security authorities are attractive targets for disinformation because they are active in their communication and have an extensive social media audience. In certain circumstances, the duties of public authorities include interfering with citizens' basic rights, as well as functions relating to health and security. The nature of these duties and negative experiences of authorities' activities may make authorities susceptible to being targeted by disinformation. On the other hand, respect for authorities, their neutrality and the assistance they provide to citizens are factors that protect authorities from disinformation.

Disinformation often aims to reduce citizens' trust towards public authorities and question the expertise of the public authorities. Disinformation may aim to point out that public authorities' communication only serves the state management and not citizens (Hillebradt, 2021; Mitu, 2021). According to the interviewees, for authorities' believability and their status as experts to be retained, they must enjoy citizens' trust. Disinformation aims to question authorities' expertise and thus sow distrust. The expert status of authorities can also be abused by disinformation by changing the original context to suit specific agendas. The interviewees also mentioned that authorities' trustworthiness and public image can be weakened within organisations by distorting the content of communication. The interviewees believed that the media is particularly interested in publishing news that involves some kind of antagonism between employer and employee. Furthermore, the interviewees noted that spreading false information about public authorities' lack of resources can negatively impact trustworthiness and, at worst, lead to citizens not utilising the services offered to them.

The results show that authorities' limited resources do not allow for efficient combating of disinformation in real-time. However, according to the interviewees it is not clear if the public authorities need to intervene or control social media content and correct erroneous information. They highlighted the increasing number of information channels in the current media environment as a challenge. This hinders the role of the public authorities as content providers. In the interviews it also became clear that public authorities don't have enough resources to detect and counter disinformation. Multiple organisations have been established, sanctions to spreaders of disinformation has been called out and underlined the responsibility of the media platforms to encounter this challenge (Alemanno, 2018).

It is notable while interpreting the results that public authorities' understanding about being targets to disinformation may be difficult because of the nature of the phenomenon. Public authorities can perceive fake news and misinformation as a similar kind of problem as actual disinformation. However, from the theoretical perspective, disinformation is systematic and strategic in action, but for public authorities, other kinds of misleading and incorrect communication may appear as disinformation. Questioning the authority's disposition, messages or expertise may not be an actual disinformation campaign (Filipec, 2019). This study indicates a contradiction between previous studies regarding how fast and coordinated Finnish public authorities react to disinformation (Biola & Papadakis, 2020). In this study, it seems that Finnish public authorities don't always react fast and coordinated to disinformation cases.

To conclude, it seems that public authorities think that disinformation is a single-case phenomenon and less of a systematic influencing campaign. Transparency, trust and neutrality can both expose to, and protect from, disinformation. Countering disinformation is challenging if public authorities feel that they are safe from disinformation campaigns based on their status, or if they feel that disinformation does not affect their daily routines. Based on the research results, we suggest that Finnish society's resilience to disinformation could be improved by promoting co-operation between authorities and by designating one actor to develop strategies to counter disinformation.

Acknowledgements

This study is part of a research project funded by the Academy of Finland.

References

Acemoglu, D., Reed, T. and Robinson, J. A. (2014). Chiefs: Economic development and elite control of civil society in Sierra Leone. Journal of Political Economy, 122(2), 319-368.

Alemanno, A. (2018). How to counter fake news? A taxonomy of anti-fake news approaches." European journal of risk regulation, 9(1), 1-5.

Bennett, W. L. and Livingston, S. (2018). The disinformation order: Disruptive communication and the decline of democratic institutions. European journal of communication, 33(2), 122-139.

Bennett, W. L and Segerberg, A (2012). The logic of connective action: Digital media and the personalization of contentious politics. Information, communication & society 15.5: 739–768.

Bjola, C., and Papadakis, K. (2020). Digital propaganda, counterpublics and the disruption of the public sphere: the Finnish approach to building digital resilience. Cambridge Review of International Affairs, 33(5), 638-666.

Bozdag, E and Van Den Hoven, J. (2015). Breaking the filter bubble: democracy and design. Ethics and information technology, 17(4), 249-265.

Bradshaw, S. and Howard, P. N. (2018). The global organization of social media disinformation campaigns. Journal of International Affairs, 71(1.5), 23-32.

Cinelli, M., Morales, G. D. F., Galeazzi, A., Quattrociocchi, W. and Starnini, M. (2021). The echo chamber effect on social media. Proceedings of the National Academy of Sciences, 118(9).

Filipec, O. (2019). Towards a disinformation resilient society? The experience of the Czech Republic. Cosmopolitan Civil Societies: An Interdisciplinary Journal, 11(1), 1-26.

Glaser, B. G., & Strauss, A. L. (2017). The discovery of grounded theory: Strategies for qualitative research. Routledge.

Elo, S. and Kyngäs, H. (2008) The qualitative content analysis process, Journal of Advanced Nursing, Vol. 62, No 1, 107-115.

Hillebrandt, M. (2021). The Communicative Model of Disinformation: A Literature Note. Helsinki Legal Studies Research Paper Forthcoming.

Ikäheimo, H and Vahti, J (2021). Mediavälitteinen yhteiskunnallinen vaikuttaminen. Murros ja tulevaisuus. Sitran selvityksiä 178 [Mediameditated social influencing. Turning point and the future]. Sitra's reports.

Jallinoja, P and Väliverronen, E (2021). Suomalaisten luottamus instituutioihin ja asiantuntijoihin COVID19 -pandemiassa. Media & Viestintä 44(2021): 1, 1-24. [Finns´ trust institutions and experts in the COVID19 -pandemic]. Media & Communication.

Lazer, D. M., Baum, M. A., Benkler, Y., Berinsky, A. J., Greenhill, K. M., Menczer, F. and Zittrain, J. L. (2018). The science of fake news. Science, 359(6380), 1094-1096.

Matasick, C., Alfonsi, C. and Bellantoni, A. (2020). Governance responses to disinformation: How open government principles can inform policy options.

Ministry of Justise (2019). Julkisuuslain soveltamisalan laajentaminen. Oikeusministeriön julkaisuja, 2019:31 [Extension of the application on the Act on the Openess of Government Activities] Ministry of Justice.

Mitu, N. E. (2021). Importance of Communication in Public Administration. Revista de Stiinte Politice, (69), 134-145.

OECD (2021), Drivers of Trust in Public Institutions in Finland, OECD Publishing, Paris, https://doi.org/10.1787/52600c9e-en.

Pariser, E. (2011). The filter bubble: What the Internet is hiding from you. Penguin UK.

Satyanath, S., Voigtländer, N. and Voth, H. J. (2017). Bowling for fascism: Social capital and the rise of the Nazi Party. Journal of Political Economy, 125(2), 478-526.

Sunstein, C. R. (2018). Is social media good or bad for democracy. SUR-Int'l J. on Hum Rts., 27, 83.

The Finnish Advisory Board on Research Integrity (TENK) (2012). Tutkimuseettinen Neuvottelukunta. Hyvä tieteellinen käytäntö ja sen loukkausepäilyjen käsitteleminen Suomessa. [Responsible conduct of research and procedures for handling allegations of misconduct in Finland. Guidelines of the Finnish Advisory Board on Research Integrity 2012]. Helsinki.

The Prime Minister's office (2019). Valtionhallinnon tehostetun viestinnän ohje. Viestintä normaaliloissa ja häiriötilanteissa. Valtioneuvoston kanslian julkaisuja 2019:23 [Guidelines for Enhanced Government Communications. Communications under Normal Conditions and During Incidents] Publications of The Prime Minister's office.

Tufekci, Z. (2017). Twitter and tear gas. Yale University Press.

Vasu, N., Ang, B., Teo, T. A., Jayakumar, S., Raizal, M. and Ahuja, J. (2018). Fake news: National security in the post-truth era. S. Rajaratnam School of International Studies.

Woolley, S. C. and Howard, P. N. (2016). Political communication, computational propaganda, and autonomous agents: Introduction. International Journal of Communication, 10.

An Ontological Model for a National Cyber-Attack Response in South Africa

Aphile Kondlo[1], Louise Leenen[1,2] and Joey Jansen van Vuuren[3]
[1]University of the Western Cape, South Africa
[2]Centre for AI Research, South Africa
[3]Tshwane University of Technology, South Africa
3579632@myuwc.ac.za
lleenen@uwc.ac.za
jansenvanvuurenjc@tut.ac.za

Abstract: South Africa is increasingly targeted by cyber criminals and is often ranked under the top five countries suffering the most cyber-attacks. In an initiative to counter these attacks, the South African government has initiated various measures such as a National Cybersecurity Policy Framework Policy (NCPF) and a Cybercrimes Act. However, the structures and policies that follow from these measures have not been fully implemented yet. Although the government published the NCPF in 2015 and enacted the Cybercrimes Act in May 2021, there is still a gap in terms of interoperability and shared understanding within the environment. In addition, numerous new structures have been established and others are still being planned. One example of a new structure is the Cybersecurity Hub, the national CSIRT, that is mandated to co-ordinate attack information and provide support for cyber incidents. In addition, the Hub must also implement a national Cybersecurity Awareness program. This paper presents a model for the Cybersecurity Hub in the event of a cyber incident in South Africa. The model is based on different attack scenarios and depicts the complex interoperability problem of the various roles, responsibilities, and interactions of role players when there is a cyber incident. One of the scenarios is an attack on critical infrastructure. The model is a prototype of a semantic knowledge base (an ontology) which will help with planning and decision making. Core queries that should be answered concern the critical role players during and after a cyber event; the communication activities that have to take place; and the response actions and the skills required to handle the event.

Keywords: cybersecurity governance, ontology, national critical infrastructure protection, national cybersecurity, cyber-attack

1. Introduction

The rapid development of information and communication technologies (ICT) has a major impact on the future, and digital transformation is one of the most significant societal changes. Many nations are driving initiatives to connect their citizens to the internet. The South African government, for example, introduced an initiative called SA Connect in 2013 as part of its National Broadband Policy (Alison, Onkokame, & Rademan, 2018). This initiative has two main goals; the first one was to deliver widespread broadband access to 90 percent of the country's population by 2020, and the second to deliver widespread broadband access to 100 percent of the country's population by 2030 (Alison, Onkokame, & Rademan, 2018). When comparing South Africa to the rest of Africa in terms of the percentage of the population that has access to the internet, South Africa is only ranked 13th in Africa with a percentage internet penetration of 59,8%, while countries such as Kenya and Libya have above 80% penetration (Statistica, n.d.). Although increased connectivity has many positive impacts ranging from economic growth to providing improved communication platforms, citizens and the private and government sectors also become increasingly vulnerable to cyber threats. Cybersecurity is a growing concern for governments across the world but especially in developing countries where governments are mostly not investing sufficiently in cybersecurity.

Many developing countries are struggling to provide national cybersecurity and are vulnerable to cyber-attacks on critical infrastructures. South Africa is one of the countries that suffer the most cyber-attacks and is thus very vulnerable; the Global Cyber Index placed SA sixth on its list of countries targeted most by cyber-attacks (Gavaza, 2021). Although South Africa has managed to make good progress in terms of cyber legislation and national cyber governance, the country still has a great deal of work to do in this regard. On 7 March 2012, the South African Cabinet approved the National Cybersecurity Policy Framework (NCPF) (State Security Agency, 2015) which had been developed over a number of years. The NCPF is a government policy document which recognises that the state is charged with implementing a government led, coherent and integrated cybersecurity approach. The NCPF was intended to provide a holistic approach to promote a cybersecurity implementation plan to be developed by the Justice, Crime Prevention and Security (JCPS) Cluster in consultation with relevant stakeholders, identifying roles and responsibilities, timeframes, specific performance indicators, and monitoring and evaluation mechanisms. The JCPS cluster consists of the Departments of Police, Defence, Home Affairs, and Justice and Correctional Services, and is responsible for the co-ordination of interdepartmental crime prevention initiatives across the criminal justice system. The JCPS works in consultation with other Government clusters to oversee the implementation of the policy framework, with the aim to ensure centralised coordination of cybersecurity issues. (State Security Agency, 2015).

The implementation of the NCPF led to various structures being developed and to the enactment of Cybercrimes Act 19 of 2020, in 2021 (RSA Government, 2021). An early draft of the Act was initially introduced to the National Assembly on 9 December 2016 as the Cybercrimes and Cybersecurity Bill (RSA Governmemnt, 2016). This Bill was revised to become the Cybercrimes Bill in 2017 (Parliamentary Montoring Group (PMG), n.d.) and then became Cybercrimes Act after going through some changes over the following four years until it was finalised in 2020, and enacted on 1 June 2021. The objective of the Cybercrimes Act is to consolidate legislation relating to cybercrime in a single act by addressing offences that are relevant to cybercrimes and to criminalise offences related to data, messages, computers and networks. It also regulates the investigation and prosecution of cybercrimes. (RSA Government, 2021).

It is interesting to note that the NCPF and the initial draft of the Cybersecurity Act, the Cybercrimes and Cybersecurity Bill, prescribed the establishment of a number of new structures required for national cybersecurity. These structures have been omitted from the Cybercrimes Act but are expected to be contained in a new Cybersecurity Bill which is apparently still to be drafted. In this paper, we adhere to the description of these structures contained in the NCPF, because the NCPF is still a valid policy. One of these structures is the Cybersecurity Hub[1], South Africa's National Computer Security Incident Response Team (CSIRT). The Hub was established in 2015 and strives to make Cyberspace an environment where all residents of South Africa can safely communicate, socialise, and transact in confidence[1]. It works with stakeholders from government, the private sector, civil society and the public to identify and counter cybersecurity threats. A few examples of other structures that have been established or are under development include a Point of Contact centre[2] operated by the South African Police Service (SAPS), a Cyber Security Centre operated by the South African National Defence Force (SANDF), a Cybersecurity Response Committee (CRC) operated by the State Security Agency, and a National Cyber Security Advisory Council (NCAC). (Sutherland, 2017).

This paper describes research to support the Cybersecurity Hub during its response to cyber incidents; specifically the development of an ontological model that will map the complexity of the interactions required between national cyber structures to respond during a cyber-attack. This first phase of the research focuses on a scenario where a National Critical Infrastructure (NCI) is attacked. The protection of National Critical Infrastructure (NCI) is discussed in Section 2. Section 3 provides a short introduction to ontologies and considers research on the use of ontologies in support of cyber governance. A scenario depicting a cyber-attack on an NCI in South Africa is presented in Section 4, as well as a model depicting the main role players and their interactions. The last two sections of the paper contain a brief evaluation of the initial model and the conclusion.

2. The protection of national critical infrastructure

The protection of NCIs against cyber-attacks is a critical aspect of national cybersecurity. Most developed and developing countries consider sectors ranging from telecommunications, energy, transport, water supply, health, finance, and other infrastructures that 'allow a nation to function' as their critical infrastructure, even though national security and military systems form significant components of critical national infrastructure (Cyber Infrastructure & Security Agency, 2019). Cyber-attacks on these infrastructures can cripple a country's economy and cause national emergencies. For example, on 23 December 2015, Ukraine Kyivoblenergo, a regional electricity distribution company experienced a cyber-attack that resulted in a power outage which left approximately 225 000 customers without power (Shehod, 2016). The attackers reportedly re-used a decade-old Trojan, BlackEnergy, to gain access to the company's network. Another incident occurred in December 2014 with a cyber-attack on the Korea Hydro & Nuclear Power Company, which operates nuclear power plants in the Republic of Korea (ROK) (Lee & Lim, 2016). The cyber-attack emerged as a national security threat and triggered

[1] https://www.cybersecurityhub.gov.za/

[2] Note that the establishment of the Point Of Contact centre is prescribed in the Cybercrimes Act. The NCFP prescribed two structures to be operated by SAPS but these have been replaced by the Point Of Contact.

a response from ROK society, which up until then had been sensitive to security incidents (Lee & Lim, 2016). Even though the cyber-attack did not result in loss of human life or the destruction of facilities, it showed that there is a need to pay attention to such incidents because it highlighted the inadequacies of the existing cybersecurity systems for critical infrastructure.

The protection of critical infrastructure and national crisis management is likely to increase cybersecurity's strategic significance for national security. Ensuring the safety of critical infrastructure is actually a matter of national security with or without a cyber aspect (Tatar, Calik, Celik, & Karabacak, 2014). One challenge highlighted by Tatar et al. is the fact that a significant part of critical infrastructures (CI) in liberal market economies are run by the private sector although they are deemed to be an integral part of national security. Consequently, governmental assistance is required for these service providers because most countries regard the protection of CIs to be a governmental responsibility. The risk of cyber-attacks on critical infrastructure and data fraud or theft is a constant concern for business leaders globally (ENISA, 2021). According to the World Economic Forum's Global Risks Report 2020 (World Economic Forum, 2021), the risk of cyber-attacks on critical infrastructure and data fraud or theft ranked in the top ten of risks most likely to occur, while the recent COVID-19 Risk Outlook identified cyber-attacks as the third greatest concern due to the current and sustained shift to digital work patterns.

A cyber-attack on entities that are part of critical infrastructure can have a debilitating impact on national security, and the risk increases significantly for electricity grids, nuclear installations, and telemetry or command and control networks of space assets (Samuel & Sharma, 2016). The US Department of Homeland Security identifies 16 critical infrastructure sectors whose assets, systems, and networks, whether physical or virtual, are considered so vital to the US that their incapacitation or destruction would have a debilitating effect on national security, economic security, public health or safety, or any combination thereof (Samuel & Sharma, 2016).

The financial sector has been a consistent target for hackers. The banking industry increasingly warn their customers of the risks of cyber fraud and has to prevent the possibility of criminals imitating their employees. An article released on the South African Banking Risk Information Centre (SABRIC) website, states that in the year 2017, 13 438 incidents occurred across banking apps, online banking and mobile banking, and cost the industry more than R250 000 000 in gross losses (SABRIC, 2017). Even though the financial sector may seem to be the most consistently targeted by hackers, power provision is a critical vulnerability for any nation, and the loss of power creates cascading damage in other critical sectors (Samuel & Sharma, 2016). In this paper, we describe an attack on a power utility as a use case.

According to Kyung-bok and Jong-in, malware attacks such as the Stuxnet and Dragonfly, which were capable of causing physical damage, were detected worldwide, and the cybersecurity of critical infrastructure resurfaced as a security issue among advanced countries. The U.S. issued Executive Order 13636 to improve and reinforce critical infrastructures cybersecurity, and Presidential Policy Directive 21 to address the role of government agencies in order to ensure the effective implementation of this order (Lee & Lim, 2016). The UK is establishing a "cybersecurity hub" for threat-information sharing with CI operators, and it has also formed a joint communique that regulates joint training and information sharing pertaining to CI cyber threats.

The consistent cyber-attacks on national critical infrastructure have become a major issue for various countries. It is clear that a cybersecurity strategy plan is a requirement for all countries that have a connection to the internet in their national critical infrastructures. This research builds on existing research on national cybersecurity in South Africa and addresses requirements to enable relevant new structures to become effective once they have been implemented.

3. The use of ontologies in cyber governance

An ontology is a semantic technology that provides a way to exchange semantic information between people and systems (Jansen van Vuuren, Leenen, & Zaaiman, 2014). A formal definition of an ontology is given by Gruber (1993): "...a formal, explicit specification of a shared conceptualisation". An ontology allows a formal, shared representation of the key concepts of a specific domain and it provides a way to attach meaning to the terms and relations used in describing the domain. The main advantages of an ontology is a semantic search ability, provision of a common shared vocabulary, the ability to share domain knowledge, and enablement of semantic integration and interoperability between various knowledge sources.

Although a number of ontologies have been developed for the cybersecurity domain, the use of ontologies to model the complexities of cyber governance in particular, has received limited attention in the research community. Greiman (2018) considered how ontological research methodologies contribute to the understanding of cyber governance at the global level. One of the numerous motivations she highlights for using an ontology to model the complexity of the cyber governance environment, is the ability of an ontology to formally define a vocabulary for the domain. Greiman found that although there have been a number of research endeavours on the development of cybersecurity ontologies (Obrst, Chase, & Markeloff, 2021), there are very few investigations on the use of ontological development to support the interoperability of cyber governance at the international level. Greiman explored an ontological approach to analyse cyber governance with the motivation that ontological models can provide insight into the entities, attributes and processes in this domain. She mentioned the following research publications in which ontologies were developed to support cyber governance:

1. Gcaza et al. (2015) developed an ontology to foster a national cybersecurity culture.
2. Jansen van Vuuren, et al. (2014) developed an ontological model for the implementation of the national cybersecurity policy framework for South Africa.

Talib et al. (2018) presented preliminary results of a project to develop an ontology that captures different national cybersecurity policies in Saudi Arabia. The aim of this ontology is to ensure an implementable national cybersecurity approach by mapping the different relevant policies with functions and responsibilities. In addition, the ontology will clarify the complexity of the necessary structures and provide a formal, shared description of the environment before implementation can succeed.

4. Critical infrastructure attack scenario and model

Veerasamay et al. highlight the use of scenario planning to deal with the numerous uncertainties that exist when strategic planning is done for cyber incident response (Veerasamy, Mashiane, & Pillay, 2019). Based on informal guidance received from the Cybersecurity Hub, results from Veerasamy et al.'s research and cybersecurity frameworks such as the popular NIST framework[3], the authors describe a high-level NCI cyber-attack scenario for South Africa. The next step was to develop an ontological model to map the numerous role players and their roles. Due to the fact that cyber measures and structures in South Africa are not yet mature, the coordination of national assistance will benefit from a model that depicts the relevant role players and their roles.

Our scenario depicts a cyber-attack on a South African NCI with national impact and describes the response from the Cybersecurity Hub's (the national CSIRT) perspective. Multiple entities must collaborate and communicate – the focus is on high-level role players, and not on the internal structures of any such entity. This first phase of modelling remain at a high-level of abstraction and will be expanded in the future. The scenario captures focal points with respect to the stakeholders, the response to an incident, the skills required and the communication amongst involved parties.

The government departments directly engaged in cybersecurity in South Africa (RSA Government, n.d.) are State Security, Justice and Constitutional Development, Defence, and Telecommunications and Postal Services - they all belong to the same cluster, the Justice, Crime Prevention and Security (JCPS) Cluster. Other departments will be engaged as required. The cyber structures that are involved in a response to an attack on NCI in South Africa are listed below (Phahlamohlaka & Hefer, 2019):

3. The JCPS's *Cybersecurity Response Committee* (CRC) has the responsibility to formulate strategy and decision-making in the national cyber domain and is chaired by the Director-General of State Security, and hosted by the SSA. The CRC is supported by the SSA's Cyber Security Centre.
4. The *Cybersecurity Hub*, or national CSIRT, will coordinate general cyber security activities in the private sector;
   inform Private Sector CIRTS, ISPs and other stakeholders, initiate cyber security awareness campaigns, support the development of Private Sector CSIRTs, and conduct cyber security audits, assessments and readiness exercises on request.
5. The *Cyber Security Centre* (CSC) is overseen by the State Security Agency (SSA) and is the operational entity of the CRC. The CSC is responsible for coordinating any NCI cyber incident. It has been mandated

[3] https://www.nist.gov/cyberframework

to analyse incidents, trends, vulnerabilities, share information, develop response protocols, to interact with stakeholders, to conduct exercises, and to perform security audits, assessments and readiness. In the event of a cyber incident the CSC will notify stakeholders, analyse information and systems involved in the incident, extract and pool data from various sources, monitor the incident, determine the assistance that is required, request status, reports, etc.

6. The *Cyber Command Centre* of the South African National Defence Force (SANDF) has the responsibility to act when national security is threatened. It has been mandated to facilitate the operational coordination of cyber security incident response activities regarding national defence and to support other activities relating to cyber incidents that involve national defence.
7. The *Point of Contact* centre of the South African Police Service (SAPS). Note that the NCFP refers to a Cybercrime Centre and a 24/7 centre, but as discussed before, the Cybercrimes Act refers to only a Point of Contact centre. This centre's main responsibility is to investigate any cybercrime incident. In addition, it has also been mandated to share information and technologies with law informant agencies, facilitate the analysis of cyber security incidents, and develop and maintain cross-border law enforcement cooperation in respect of cybercrime.

Figure 1 provides an illustration of the structure as described in the NCFP (Jansen van Vuuren J. C., 2016). Note that this figure still shows the two centres which are now replaced by the Point of Contact centre.

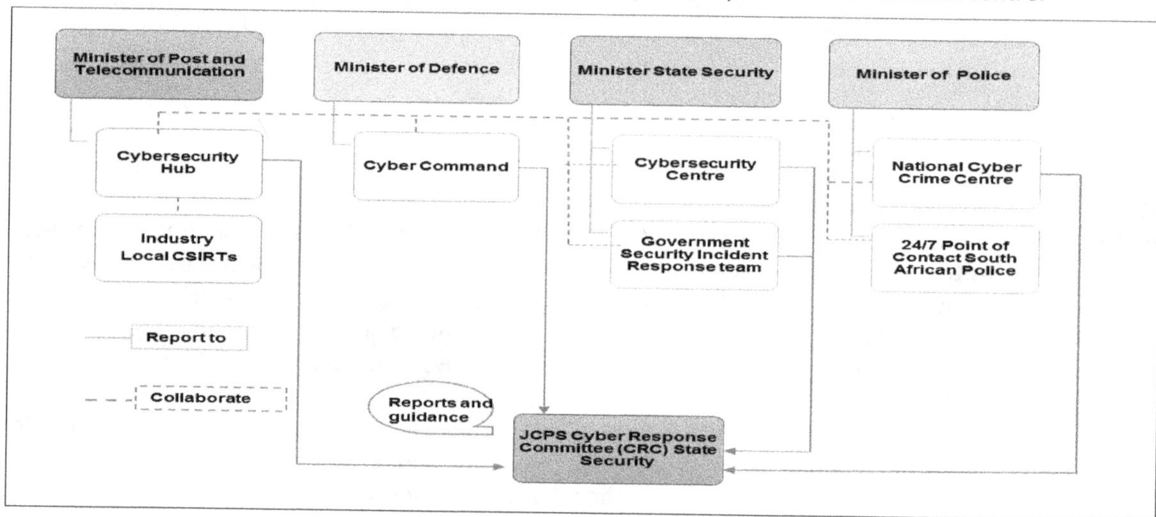

Figure 1: South African cyber structures according to the NCFP (Jansen van Vuuren J. C., 2016)

Our NCI attack scenario considers an attack on the main electricity provider in South Africa, Eskom. Eskom is a public utility and is required to monitor and analyse its systems for anomalies or cyber incidents. Eskom has to determine the severity of any incident, and process and document findings in an attempt to minimise effects and report any incident to the necessary structures. If Eskom finds an incident to be of a serious nature, it will report the attack to the Cyber Security Centre and request assistance from the CSC to co-ordinate the incident. The CSC will notify the Cybersecurity Hub, the Cyber Command Centre, and the National Contact Point centre, and make the first call to request assistance to respond and recover from the attack. Data gathering from various sources will commence under the CRC, and the CSC will announce the incident where necessary, analyse system data and continue to monitor the incident. They will also document the incident. A knowledge base will be created at the SSA (it hosts the CRC and the Cyber Security Centre) Eskom will begin the recovery process by requesting forensic investigation from the Point of Contact Centre for information updates, and then report to CRC. For this high-level scenario, an ontology was developed to map all the role players and relationships (depicting responsibilities and functions) that take place in an event of a cyber-attack incident. Note that the ontology does not aim to present the dynamic flow of events during and after an attack; it is a knowledge base that can be queried to gather information with respect to the various entities that are involved and the complex nature of their roles and responsibilities. The ontology captures the responsibilities of each structure which will assist Cybersecurity Hub in decision making. *Some* of the main concepts (classes) in the model are listed below:

Critical_Infrastructure

1. *WaterProvision*

2. *EmergencyServices*
3. *Energy*
    1. *PowerPlant*
4. *Healthcare*

Incidents

1. *DoS*
2. *Malware*
3. *Ransomware*

RolePlayers

1. *Body*
    1. *ECSP* *(Electronic Communication Service Providers)*
    2. *FIRST* *(The Forum of Incident Response and Security Teams)*
    3. *JCPS* *(Justice, Crime Prevention and Security Cluster)*
        1. *CyberResponseCommittee*
    4. *NCAC* *(National Cyber Security Advisory Council)*
2. *CSIRT*
    1. *CybersecurityHub*
    2. *ElectricitySectorCSIRT* *(Does not yet exist. Inserted for modelling purposes)*
    3. *ECS_CSIRT* *(Government CSIRT)*
    4. *SABRIC* *(Banking Sector CSIRT)*
3. *Private*
    1. *Company*
    2. *SABRIC*
4. *State*
    1. *ArmsofGovernment*
        1. *SANDF* *(South African National Defence Force)*
            1. *CyberCommandCentre*
        2. *SAPS* *(South African Police Service)*
            1. *PointofContact*
    2. *GovernmentAgency*
        1. *CyberCommandCentre*
        2. *CyberResponseCommittee*
        3. *PointofContactCentre*
        4. *SARB* *(South African Reserve Bank)*
        5. *SITA* *(State Information Technology Agency)*
        6. *SSA* *(State Security Agency)*
        7. *CyberSecurityCentre*
    3. *GovernmentDepartment*
        1. *DoJ&CD* *(Justice and Constitutional Development)*
        2. *Finance*
        3. *HomeAffairs*
        4. *Police*
        5. *PS&A* *(Public Service and Administration)*
            1. *SITA*
        6. *Security* *(Ministry of State Security)*
            1. *CyberResponseCommittee*
    4. *PublicEnterprise*
        1. *Telkom* *(Communication company)*
        2. *Eskom* *(Power utility)*

Sector

1. *Communications*

2. *Electricity*
3. *FinancialServices*
4. *GovernmentFacilities*
5. *Military*
6. *Water*
7. *Retail*
8. *Transportation*

The three main activities in a cyber-attack are to Monitor, Analyse and Respond. The Protect activity will not be considered as part of our model because the focus is on the mid-attack or post-attack phase. Each main activity contains sub-activities. A few examples of activities are listed below:

8. Monitoring requires the application of monitoring tools to raise alerts for detected anomalies.
9. Analysis activities entail analyses of the system data.
10. Share data from various sources.
11. Response activities consist of repairing systems and equipment, and minimising the effects of an attack for example the use the use of generators, water tanks or other means to deal with a critical infrastructure outage.
12. Processing and documentation of findings. This includes record keeping of detected threats, issues, and complications, and the preservation of log and system data for future reference or investigation of an incident.

Some of the relationships (object properties) between the main concepts (classes) are:

13. belongsToSector
14. documentFindings
15. hasCNI
16. hostsCNI
17. performsAttack
18. isInRegion
19. reportsTo

For example, the relationship *hasCNI* maps a member of the *Sector* class to a member of the *CriticalInfrastructure* Class.

Figure 2 shows a part of the ontological model for the CNI attack scenario.

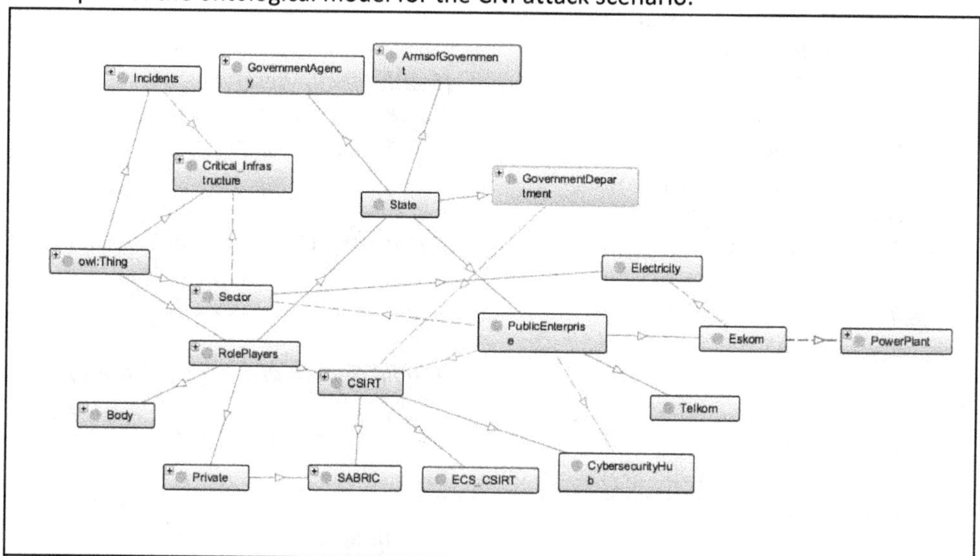

Figure 2: Part of the ontological model for the NCI scenario

The edges in the Figure 2 represents the following relationships:

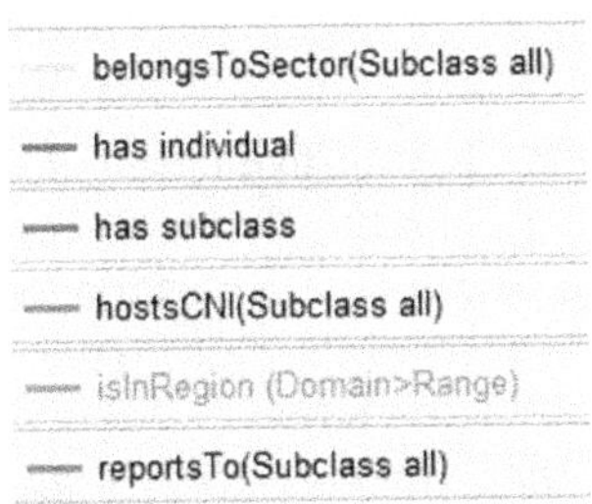

The solid lines represent subclass relationships, for example, the subclasses of the RolePlayer class are Body, CSIRT, Private and State. The dotted lines represent relationships. For example, the line from Eskom to Electricity indicates that Eskom hosts a CNI which is a powerplant, and the dotted line from Eskom to Electricity shows that Eskom belongs to the Electricity Sector. The dotted line from PublicEnterprise to CybersecurityHub shows that all public enterprises report (incidents) to the Cybersecurity Hub (this thus applies to Eskom because Eskom is a Public Enterprise). Note that the dotted line from PublicEnterprise to CSIRT shows that all public enterprises report incidents to CSIRTs, of which the Hub is just one in particular.

5. Evaluation

The ontological model reflects the first phase of the research project. It is still at a fairly high-level and needs to be expanded with more detail to provide a rich model. More scenarios will be developed in the future. A proper evaluation of the model will be done at a more advanced stage. We show the result of one query to the ontology in Figure 3.

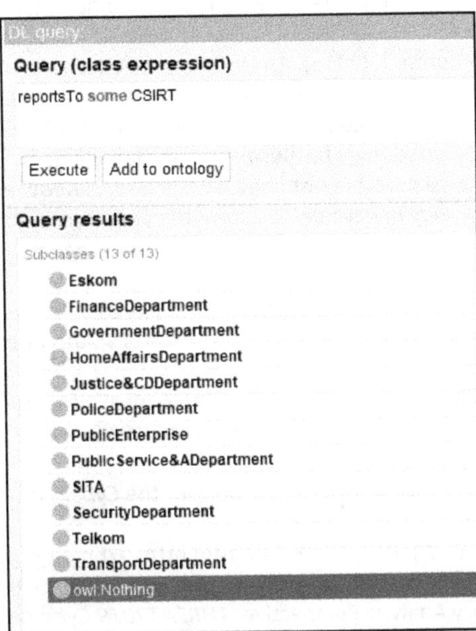

Figure 3: The query "Which entities have to report cyber incidents to a CSIRT?" posed to the ontology

Although this is a simple query, is it used to illustrate that the user can request an explanation for any query result. Figure 4 show the explanation provided for the result "SITA" (SITA is the State Information Technology Agency.) The explanation is that SITA belongs to the Department of Public Service and Administration and this department has to report to at least one CSIRT.

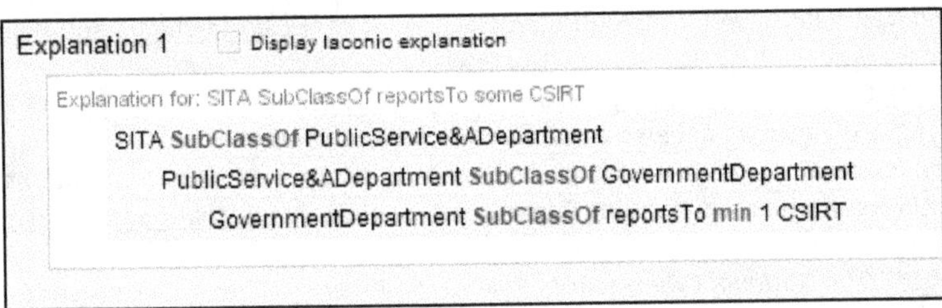

Figure 4: Explanation of the query result "SITA"

6. Conclusion and future work

South Africa has introduced several measures to improve national cybersecurity during the past decade. The core measure has been the approval of a National Cyber Policy Framework (NCFP) in 2012. While the implementation of the Cybercrimes Act and structures such as the Cybersecurity Hub (the national CSIRT) have been developed based on the NCPF, many aspects of the NCPF have yet to be addressed. This paper reflects on research to support the Cybersecurity Hub; a scenario of a cyber-attack on critical infrastructure and the development of an ontological model to map the role players and activities captured in the scenario are described. The next step of this research project is to extend the model and to add more scenarios. The aim of the ontological model is to provide a common and shareable view of the actions required to respond and recover from cyber-attacks. Cyber governance in South Africa is still being developed and this research will support a core national cyber structure.

References

Alison, G., Onkokame, M., & Rademan, B. (2018). *The State of ICT in South Africa.* Cape Town: Research ICT Africa. Retrieved from https://researchictafrica.net/wp/wp-content/uploads/2018/10/after-access-south-africa-state-of-ict-2017-south-africa-report_04.pdf

Cyber Infrastructure & Security Agency. (2019). *A Guide to Critical Infrastructure Security and Resilience*. Retrieved from https://www.cisa.gov/publication/guide-critical-infrastructure-security-and-resilience

ENISA. (2021). *ENISA Threat Landscape 2021: April 2020 to mid-July 2021.* Retrieved from https://www.enisa.europa.eu/publications/enisa-threat-landscape-2021

Gavaza, M. (2021, Sep 7). *Cyber threats are costing SA firms millions, Business Day 7 Sep 2021.* Retrieved from Business Day: https://www.businesslive.co.za/bd/companies/telecoms-and-technology/2021-09-07-cyber-threats-are-costing-sa-firms-millions/

Gcaza, N., von Solms, R., & Jansen van Vuuren, J. (2015). An Ontology for a National Cyber-security Culture Environment. *Proceedings of the Ninth International Symposium on Human Aspects of Information Security & Assurance (HAISA.*

Greiman, V. (2018). Reflecting on Cyber Governance for a new World Order: An Ontological Approach. Rome, italy: European Conference on Research Methodology for Business and Management.

Gruber, T. (1993). A translation approach to portable ontology specifications. *Knowledge Acquisition, 5,* 199-220.

Jansen van Vuuren, J. C. (2016). *Methodology and Model to Establish Cybersecurity for National Security in Africa using South Africa as a Case Study (Doctor of Philosophy (PhD) in Business Management).* Thohoyandou, South Africa: University of Venda.

Jansen van Vuuren, J., Leenen, L., & Zaaiman, J. (2014). Using an ontology as a model for the implementation of the National Cybersecurity Policy Framework for South Africa. (pp. 107-115). 9th International Conference on Cyber Warfare and Security (ICCWS).

Lee, K., & Lim, L. (2016). The reality and response of cyber threats to critical infrastructure: A case study of the cyber-terror attack on the Korea Hydro & Nuclear Power Co., Ltd. *KSII TRANSACTIONS ON INTERNET AND INFORMATION SYSTEMS, 10*(2).

Obrst, L., Chase, P., & Markeloff, R. (2021). *Developing an Ontology of the Cyber Security Domain.* McLean, Virginia: The MITRE Corporation,.

Parliamentary Montoring Group (PMG). (n.d.). *Bills.* Retrieved from https://pmg.org.za/bill/684/

Phahlamohlaka, J., & Hefer, J. (2019). The Impact of Cybercrimes and Cybersecurity Bill on South African National Cybersecurity: An Institutional Theory Analytic Perspective. *THREAT2019 Cybersecurity Summit.* Sandton, South Africa,: THREAT2019 Cybersecurity Summit.

RSA Governmemnt. (2016). *Cybercrimes and Cybersecurity Bill.* Retrieved from https://www.gov.za/documents/cybercrimes-and-cybersecurity-bill-b6-2017-21-feb-2017-0000

RSA Government. (n.d.). *Structure & Functions of the South African Government.* Retrieved from https://www.gov.za/node/537988

RSA Government. (2021). *The Cybercrimes Act 19 of 2020.* Retrieved from https://www.gov.za/documents/cybercrimes-act-19-2020-1-jun-2021-0000

SABRIC. (2017). *Media and News.* Retrieved from SABRIC: https://www.sabric.co.za/media-and-news/press-releases/digital-banking-crime-statistics/#:~:text=In%202017%2C%2013%20438%20incidents,the%20same%20period%20in%202017.

Samuel, C., & Sharma, M. (2016). *Securing Cyberspace: International and Asian Perspectives.* Pentagon Press.

Shehod, A. (2016). *Ukraine Power Grid Cyberattack and US Susceptibility: Cybersecurity Implications of Smart Grid Advancements in the US.* Retrieved from Working Paper CISL# 2016-22: https://web.mit.edu/smadnick/www/wp/2016-22.pdf

State Security Agency. (2015). *The National Cybersecurity Policy Framework (NCPF).* Government Gazette (39475). Retrieved from https://www.gov.za/sites/default/files/gcis_document/201512/39475gon609.pdf

Statistica. (n.d.). *Share of internet users in Africa as of December 2020, by country.* Retrieved from https://www.statista.com/statistics/1124283/internet-penetration-in-africa-by-country/

Sutherland, E. (2017). Governance of cybersecurity-the case of South Africa. *The African Journal of Information and Communication, 20*, 83-112.

Talib, A., Alomary, F., Alwadi, H., & Albusayli, R. (2018). Ontology-Based Cyber Security Policy Implementation in Saudi Arabia. *Journal of Information Security, 9*, 315-333. Retrieved from https://doi.org/10.4236/jis.2018.94021

Tatar, U., Calik, O., Celik, M., & Karabacak, B. (2014). A comparative analysis of the national cyber security strategies of leading nations. *In International Conference on Cyber Warfare and Security* (p. 211). 9th International Conference on Cyber Warfare and Security (ICCWS).

Veerasamy, N., Mashiane, T., & Pillay, K. (2019). Towards cyber incident response strategic planning. *THREAT 2019 Cybersecurity Summit.* Sandton, South Africa.

World Economic Forum. (2021). *The Global Risks Report 6th Edition*. Retrieved from https://www3.weforum.org/docs/WEF_The_Global_Risks_Report_2021.pdf

Combining System Integrity Verification With Identity and Access Management

Markku Kylänpää and Jarno Salonen
VTT Technical Research Centre of Finland Ltd, Espoo, Finland
markku.kylanpaa@vtt.fi
jarno.salonen@vtt.fi

Abstract: Digital transformation and the utilization of Industrial IoT (IIoT) introduces numerous interconnected devices to factories increasing among others the challenge of managing their software versions and giving attackers new possibilities to exploit various software vulnerabilities. Factory networks were earlier isolated from the Internet. However, this separation is no longer valid and there can be connections that allow intruders to penetrate into information systems of factories. Another issue is that although factories typically are physically isolated, it is not necessarily safe to assume that physical security is in good shape as the novel supply networks comprise subcontracted activities and temporary work force. Another threat can also arise from unauthorized monitoring of devices and the unauthorized replacement of existing ones. Based on the previous, it is crucial that IIoT security should be built into factories of the future (FoF) right from the design phase and even low-end devices need to be supported. Trusted computing concept called remote attestation should be used. Remote attestation allows remote parties to verify the integrity of each system component. System components should include trusted hardware components that can be used to measure executable software. The term measurement means calculating the cryptographic hash of the binary component before passing control to it. Trusted hardware components should also have a mechanism to protect the integrity of the measurement list and cryptographic keys that can be used to sign integrity assertions. The verifier part should have a storage of reference integrity metrics identifying the expected values of these measurements. Deploying trusted computing and remote attestation concepts to industrial automation is not straightforward. Even if it is possible to use remote attestation with suitable hardware components, it is not clear how remote attestation should be integrated with various operational technology (OT) industrial automation protocols. Approaches to use remote attestation with existing industrial automation protocols (e.g., OPC UA) is discussed. Advanced identity and access management (e.g., OAuth2, OpenID Connect) can be used to combine integrity measurements with device identity information so that the remote attestation process is triggered by authentication during the first transaction. The focus is on machine-to-machine (M2M) communications with immutable device identities and integrity evidence transfer.

Keywords: industrial IoT, remote attestation, OT protocols, factory of the future, IAM, trusted computing

1. Introduction

Typically, IIoT protocols are focused on network security without support for end-point integrity assessment. However, as the complexity of factory installations is growing, it becomes clear that there is a need to have software inventory of all installed software and mechanisms that can be used to verify the installation status and integrity of system components. Factory networks are no longer fully isolated as there are many connectivity requirements. As this separation is no longer valid, there can be connections that allow intruders to penetrate into factory information systems. If IIoT devices are left unpatched, there is a threat that these vulnerabilities can be utilized by attackers. There is a need to update software versions and quickly apply security patches. Physical security also needs rethinking. Typically, the FoF supply networks comprise subcontracted activities and temporary work force. Unauthorized monitoring of devices and the unauthorized replacement of existing devices can be a threat. Attackers may, e.g., seriously harm factory operations, join factory devices to a botnet, use factory devices for cryptocurrency mining, and install ransomware to them, just to name a few potential threats. In order to mitigate these threat scenarios, IIoT security needs more focus when developing factories of the future. The integrity of system components is a foundation for all security mechanisms and should be implemented already at design phase. Integrity protection should also include low-end devices that are often omitted when designing security mechanisms.

Remote attestation is a trusted computing concept that allows remote parties to verify the integrity of system components. Integrating trusted computing and remote attestation concepts to industrial automation is a challenging task as systems have a long lifespan and typically utilize legacy protocols. It is not realistic to assume that all components will quickly support recently added standardized features. The lack of trusted hardware, especially in low-end IIoT components, has been a challenge but the situation is gradually improving. However, even if it is possible to use remote attestation with suitable hardware components, it is not clear how remote attestation should be integrated with various OT industrial automation protocols since legacy protocol specifications do not include mechanisms for end-node integrity verification.

The convergence of IT and OT protocols with wider adaption of Internet-originated IT protocols in industrial automation is another IIoT trend. This will introduce more choices to combine integrity verification with advanced Identity and Access Management (IAM) approaches. Advanced IAM approaches for the factory of the future can be one approach to combine integrity measurements with device identity information so that the remote attestation process is triggered by authentication during the first transaction. Advances in confidential computing to build trust on remote systems can be considered as an analogue problem.

Basically, there are two approaches to endpoint integrity verification in IIoT systems. Either extend the existing legacy protocols or use IT originated IAM approaches to contain integrity verification of the endpoints. In this paper, these two approaches are discussed with examples from research literature. The purpose of this article is to collect information and clarify how this could be done by studying existing industrial automation protocols and figuring out what kind of approaches could be feasible. The focus is on M2M communications with immutable device identities and integrity evidence transfer. The rest of the paper is structured as follows. First, section 2 covers main remote attestation technology alternatives for IIoT devices. Section 3 describes IAM alternatives for IIoT devices including discussion and example use cases of remote attestation. Section 4 comprises some examples of extending legacy IIoT protocols to support remote attestation and section 5 contains the concluding remarks.

2. Remote attestation alternatives for IIoT devices

2.1 Background

The concept of remote attestation originates from the secure and measured boot research, where early prototypes, e.g., Tygar and Yee (1991), were using physically secure coprocessors as security hardware. Arbaugh et al. (1997) introduced the chained layered integrity checks model. These concepts inspired the Trusted Computing Group (TCG) industrial consortium to start the development of Trusted Platform Module (TPM) that is nowadays included in almost all PC hardware. TPM provides a root of trust that can be used to implement secure boot-time integrity measurements that can be extended to the operating system level. Attestation of network endpoints was one of the original goals of TCG (Berger 2005).

TCG defines attestation as the process by which an independent verifier can obtain cryptographic proof as to the identity of the device in question, evidence of the integrity of software loaded on that device when it started up, and then verify that what's there is what's supposed to be there (TCG 2019a). Coker et al. (2011) state that remote attestation is the activity of making a claim about properties of a target by supplying evidence to an appraiser over a network. They also identify some principles to guide the development of attestation systems (Coker et al. 2011). The first principle is that attestation must be able to deliver temporally fresh evidence, which is needed to prevent replay attacks. The second principle is that comprehensive information about the target should be accessible which requires that the attestation measurement mechanism should cover essential parts of the system. The third principle is that the target, or its owner, should be able to constrain disclosure of information about the target. This is sometimes hard to achieve if there are strict privacy requirements but that is not a case in IIoT systems. The fourth principle is that attestation claims should have explicit semantics to allow decisions to be derived from several claims. The verification process should be able to replay the measurement process and be aware of expected measurement values. The fifth principle is that the underlying attestation mechanism must be trustworthy. Typically, this requires that there is an isolated execution environment that can be used to protect the integrity of attestation measurements and to sign attestation evidence reports. (Coker et al. 2011)

In practice, remote attestation requires hardware support, as there must be trustworthy measurement mechanisms. The lack of hardware support has been an obstacle to the deployment of remote attestation in the IIoT environment. This is now changing as there are alternatives to implement isolated execution environments also to microcontroller class devices. There are multiple ways of implementing remote attestation and some examples are described in this section.

2.2 Standardization efforts

Nowadays, multiple organizations are standardizing trusted computing related security hardware and remote attestation. TCG was established already in 2003. Their main effort has been to specify the functionality of the TPM hardware component (TCG 2019b) focusing on Personal Computer (PC) platform but later they have expanded their scope. IETF has standardization efforts related to remote attestation. TCG has also developed a recommendation for the use of TCG technology in IIoT context (TCG 2022).

Additionally, IEEE, ENISA, and NIST have published many recommendations and specifications for IoT-related secure device identity and firmware update issues.

Classic remote attestation, described in Figure 1, is utilizing TCG TPM as a secure hardware. Measurement is a calculation of a cryptographic hash from a software component. One measurement mechanism example is Linux kernel Integrity Measurement Architecture (IMA) component, originally proposed by Sailer et al. (2004), which measures all loaded executables before execution. Measurements are stored as a measurement log that includes integrity protection using TPM Platform Configuration Register (PCR) values.

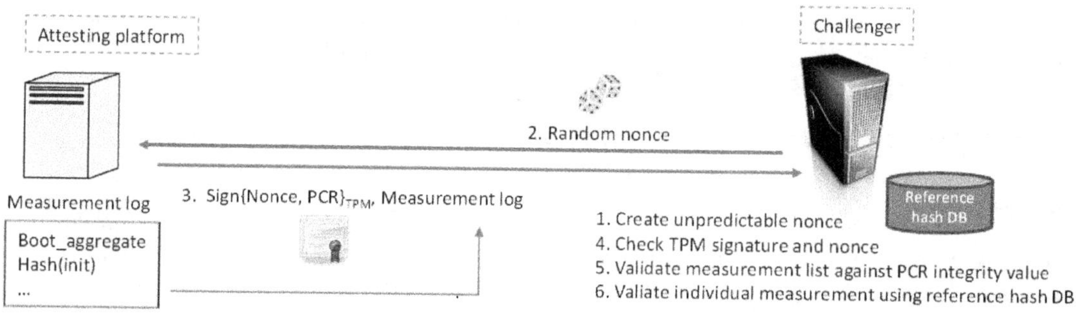

Figure 1: Classic remote attestation request/reply scenario utilizing TPM

TPM is not the only security hardware alternative. TCG Device Identifier Composition Engine (DICE) remote attestation (TCG 2020) expects that there is a statistically unique, device-specific, secret value called Unique Device Secret (UDS). The UDS must be stored in the non-volatile memory so that it is only accessible by the DICE layer. The DICE layer is using the UDS value to compute another secret called Compound Device Identifier (CDI) that is passed to the next layer.

IETF has a working group called Remote ATtestation procedureS (RATS) that covers remote attestation technologies. Their focus is networking infrastructure (e.g., routers) and not IIoT, but common topics and issues can still be found. The charter of the working group states that the entity (relying party) may require attestation evidence from a remote peer containing information about device identity, system component integrity, and device state.

The charter text also mentions that there are already domain-specific attestation mechanisms for TPM, FIDO Alliance attestation, and Android Keystore attestation. The goal of the working group is to standardize formats for describing assertions/claims about system components and their associated evidence. Another goal is to standardize procedures and protocols to convey these assertions/claims to relying parties. The working group has specified an Entity Attestation Token (EAT) that is used to transfer claims about device identity and state (Lundblade et al. 2021).

2.3 Industry initiatives

Platform Security Architecture (PSA) is an initiative of ARM that provides a security framework for IoT ecosystem. The framework is built to utilize ARM TrustZone for Cortex-M technology, also known as TrustZone-M, which provides isolated execution environment for microcontroller class devices (Pasham 2020). PSA also includes remote attestation.

Although attestation of cloud services is a different problem than the integrity verification of IIoT devices, both use quite similar technologies. There is a growing interest for confidential computing that requires remote attestation (Mulligan et al. 2021). Microsoft Azure Attestation (Baldwin et al. 2021) provides remote verification of the trustworthiness of a cloud platform and integrity of the binaries running in the cloud platform. Other cloud providers have similar offerings. For example, Amazon provides attestation support for their Nitro Enclaves (AWS 2022).

2.4 System level attestation

System level remote attestation has scalability problems when there are lots of nodes whose integrity status needs to be verified. The classic model describes only two-party communications. If the client needs to request attestation from all nodes this will easily become a bottleneck in the system. Attestation information is typically bound to low-level details, e.g., a cryptographic hash of firmware image or cryptographic hashes of certain application executable and library versions. The obvious idea is to split the system into subsystems with a gateway node that takes care of attestation of the subsystem nodes. Gateway nodes could hide low-level details from other system nodes and make higher level integrity claims of the system state. Property-based attestation is trying to define higher level properties that can be used outside of the subsystem.

Asokan et al. (2015) define a swarm-based attestation called Scalable Embedded Device Attestation (SEDA). The protocol includes two phases, offline and online. The offline phase includes device registration to a swarm. The online phase is the remote attestation operation. The verifier sends an attestation request to one swarm node. The node is then sending attestation requests to its neighbours. The operation is then repeated until all nodes have received the attestation request.

Remote attestation protocol is typically assumed to be a synchronous request-reply type of protocol. This is natural in simple cases as the process cannot proceed until the integrity status of the remote peer is verified. This can be a problem if there are lots of nodes as it requires that attestation operations must be done in a serial manner. Individual attestation operations to remote peers could be done in parallel. When replies from all remote peers have been received the process can continue. Dushku et al. (2020) describe asynchronous remote attestation.

3. IAM alternatives for IIoT devices

Modern authentication is an umbrella of mechanisms that can be used to replace traditional username/password-based mechanisms.

- Authentication methods: multi-factor authentication, smart cards, client certificates
- Authorization methods: OAuth2.0, OpenID
- Conditional access policies

Nowadays these are widely used in the office environment, e.g., Microsoft. Something similar can be used in factories also including M2M communications.

OAuth 2.0 developed by IETF Web Authorization (OAuth) WG is an industry-standard protocol for authorization. It is typically used in web-based services, but it is also increasingly used in IoT services (Hovsmith 2017, Sandoval 2017). IETF has a working group called Authentication and Authorization for Constrained Environments (ACE) that specifies a framework for authentication and authorization in IoT environments called ACE-OAuth (Seitz et al. 2021). The framework is based on a set of building blocks including OAuth 2.0 and the Constrained Application Protocol (CoAP).

OpenID is an open standard for authentication, promoted by the non-profit OpenID foundation. OpenID Connect is built on OAuth 2.0 protocol and is standardizing areas that OAuth 2.0 is not specifying, such as scopes and endpoint discovery. It is specifically focused on user authentication and is widely used to enable user logins on consumer websites and mobile apps.

Leicher et al. (2010) have described how remote attestation information can be merged into OpenID Connect transactions. They have added TPM-based integrity verification to OpenID Connect. The client is assumed to include a TPM and TPM PCRs are used to secure integrity measurements. The scenario begins in the same way as a normal OpenID Connect scenario when a client is accessing a service and the service provider is delegating authentication to the OpenID provider. The difference is that now the OpenID provider initiates the remote attestation challenge/response messaging with the client. OpenID provider then verifies the attestation response message and grants access to the service after the authentication and integrity is successfully verified. The scenario is presented in Figure 2.

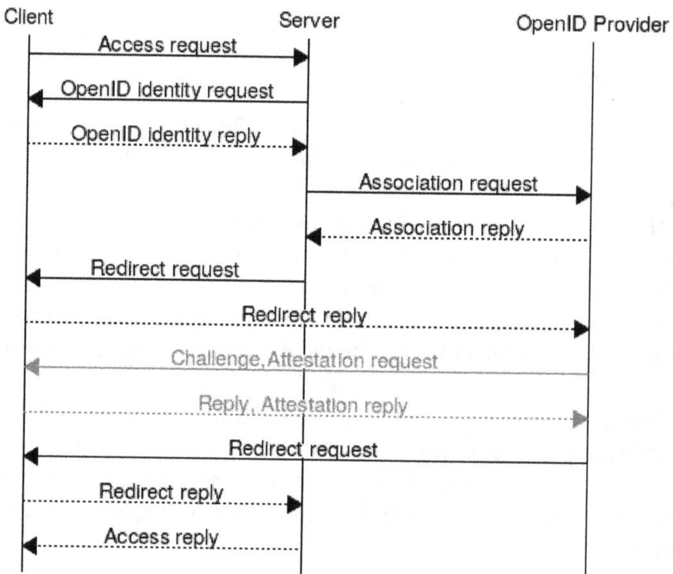

Figure 2: Integrate remote attestation with OpenID Connect transaction

Suomalainen et al. (2021) have extended OpenID Connect and OAuth 2.0-based authentication and authorization framework with DICE-based device integrity verification and remote attestation to verify the integrity of IoT devices.

SAML is an XML-oriented framework for transmitting user authentication, entitlement, and other attributes. The framework allows two federation partners to select and share identity attributes using a SAML assertion/message payload. SAML is using an XML-based syntax. SAML assumes three key roles in any transaction (Naik and Jenkins 2017). There is a trusted organization called Identity Provider. The Service Provider provides access-controlled services and User accesses these services. The SAML 2.0 specification describes assertions, protocols, bindings, and profiles (OASIS 2005). Ali et al. (2010) provide an example of the use of SAML-based federated identity management system with remote attestation.

4. Combining remote attestation with legacy IoT protocols

In this section we study a couple of widely used IoT protocols and discuss about possible ways to include remote attestation information. The goal is to extend existing protocols to carry remote attestation information. The convergence of IT and OT technology can introduce widely used Internet messaging protocols to industrial automation. Older protocols typically have fixed message structures and optimized wire protocols making it difficult to extend these protocols to support remote attestation.

4.1 OPC UA

OPC UA is a platform independent service-oriented M2M communications architecture targeting industrial automation. The specification is very large and complex consisting of more than 15 documents and over 3000 pages. The specification consists of multiple profiles (OPC 2017). Each profile includes a functionality set called a conformance unit. Despite the large size of the specification, the OPC UA security model (OPC 2018) does not yet include concepts that could be used to validate the integrity of connected devices. Recent study shows that there are large number of publicly accessible insecure OPC UA deployments (Dahlmanns et al. 2020).

Birnstill et al. (2017) have studied how OPC UA protocol can be integrated with remote attestation. Their approach is to include Trusted Platform Module (TPM) and use it as a hardware-based root of trust. Remote attestation is added as a separate conformance unit. Bienhaus et al. (2021) are also using TPM with OPC UA. Their focus is on securing a gateway node, but they also discuss about ideas of using sealed keys bound to certain

TPM PCR values. Another example of using OPC UA with remote attestation is by Matischek and Bara (2019). Instead of TPM they propose the use of a Secure Element (SE).

IETF draft by Birkholz et al. (2021a) proposes passing attestation information as part of existing security information exchange mechanism (e.g., X.509 certificates) that are already used in each protocol. OPC UA is using X.509 certificates and remote attestation evidence information is proposed to be embedded into an X.509 certificate as a certificate extension. Thus, remote attestation related information could be natively encoded in X.509 certificate extensions or could be encoded in some other format, which in turn is then encoded in an X.509 certificate extension. Certificates are normally considered to have relatively long lifetime, but remote attestation information is queried frequently. The approach may require the use of certificate chains, device acting as a certificate authority, and short-lived certificates.

Other possible approaches are to extend node endpoint information that is normally queried before a communications channel is created to contain attestation evidence information or use specific resource names that can be used to access attestation information.

4.2 DNP3

Distributed Network Protocol 3 (DNP3) is a protocol that is used, e.g., in SCADA industrial automation systems and electric power systems. DNP3 is also standardized as IEEE1815 (2012). DNP3 has had many protocol level vulnerabilities. Some of these are discussed by Lu and Feng (2018). Although the authors do not use remote attestation terminology, the measurement mechanism is described and remote attestation related modifications to DNP3 are discussed. DNP3 security framework has evolved to fix vulnerabilities, removing complex features, and adding new functionalities (e.g., features adapted from TLS) (Self 2020). TLS is often mentioned as a mechanism to secure DNP3 communications when communicating using Internet connections (PJM 2020). DNP3 seems to have many shortcomings and recent survey article by Pliatsios et al. (2020), covering proposed enhancements to SCADA with DNP3, does not refer to remote attestation at all.

4.3 OMG DDS

An article by Pardo-Castello and Lang (2013) discusses adding TPM and remote attestation to protect embedded systems. Their approach is to add a trusted hypervisor that can be used to run guest operating systems whose integrity can be measured by the hypervisor. The OMG DDS protocol is used to transfer integrity measurements to the attestation verifier using OMG DDS-based publish-subscribe protocol (OMG 2010).

There is another OMG DDS-based remote attestation example that describes hardening of nodes that are using OMG DDS based communications with Intel SGX Enclave based Trusted Execution Environment (TEE). The approach is described in three blog articles by Upchurch (2019). Intel SGX Enclaves use SCONE that can also be used to attest applications running in SGX Enclaves (Arnautov et al. 2016). The blog articles are more focused on providing confidentiality than discussing the integrity aspects and remote attestation.

Both examples are used in the Intel x86-based server room environment and not directly in the factory. OMG DDS is used just as a publish-subscribe protocol to carry integrity evidence information and not as an application protocol that will also include attestation information.

4.4 CoAP

The Constrained Application Protocol (CoAP) (Shelby et al. 2014) is a lightweight protocol that can be used as a replacement for HTTP in resource constrained environments. CoAP is typically used as a transport for higher-level protocols and not as an application protocol itself.

CHARRA is a proof-of-concept implementation of the "Challenge/Response Remote Attestation" interaction model of the IETF RATS Reference Interaction Models for Remote Attestation Procedures using TPM 2.0 (Eckel 2021). The mechanism has been applied to CoAP protocol.

IETF RATS WG work in progress draft “Reference Interaction Models for Remote Attestation Procedures” contains Appendix A with a title “CDDL Specification for a simple CoAP Challenge/Response Interaction”

(Birkholz et al. 2021b). The example describes the embedding of attestation information to CoAP FETCH request and response messages. Some of the authors of the draft are also working on CHARRA.

IETF RATS WG provides interaction models for remote attestation, but one must note that these are not transparent. If there are protocols on top of CoAP, changes are needed to include remote attestation support.

4.5 HTTPS

The original HTTP protocol was a simple request-reply protocol without transport layer confidentiality and integrity. Soon there was a need to utilize HTTP transport also for transactions that contain sensitive information. Secure Socket Layer (SSL) implementation by Netscape solved the problem by adding encryption and signing to provide confidentiality and integrity to communications. The approach was later standardized as Transport Layer Security (TLS). TLS is nowadays widely used in Internet traffic (all URLs starting with the prefix https). TLS also supports client certificates for (mutual) authentication.

Although TLS solves the security issues of communications, it does not provide integrity evidence of connected nodes. In a generic setup this is out of scope, as there is no way to get reference integrity metrics for heterogeneous end nodes. It is also a privacy issue as attestation evidence can reveal unwanted details. Factory environment is a more restricted environment. We can assume that the network node configuration is known and as there is no personal data in factory floor M2M communications, there are no privacy issues either. This means that it is feasible to add remote attestation to TLS.

Trusted Computing Group has a protocol called Trusted Network Connect (TNC) which includes remote attestation support using TPM as a security hardware. TNC never gained wide popularity but inspired developers to add remote attestation functionality to TLS. For example, Yu et al. (2013) are using TPM and the handshake protocol extensions of TLS to implement remote attestation functionality.

There are multiple examples of adding remote attestation to TLS with Intel SGX used as security hardware. Knauth et al. (2019) describe how they seamlessly combine Intel SGX remote attestation with the establishment of a standard TLS connection. Remote attestation is performed during the connection setup. There are no changes to the TLS protocol and not even modifications to existing protocol implementations. Intel has also SGX examples that include remote attestation support (Mechalas 2018).

TLS is often proposed as a mechanism to secure any legacy protocol by running it on top of TLS connection. As it is also possible to extend TLS to include remote attestation, this provides a mechanism to deploy remote attestation as well. However, there are still requirements for a measurement mechanism using isolated execution environments and reference integrity metrics. Error reporting also requires attention if remote attestation should take place transparently without the legacy protocol being aware of any integrity verification issues.

5. Conclusions

The factories of the future will interconnect and bring together OT and IT environments. The existing literature states that increased connectivity will also expose earlier isolated subsystems to attackers, which our research also supports. Network security provides confidentiality and integrity protection to communications preventing message eavesdropping, tampering, and replay attacks. However, according to our research, network security alone is not enough to protect IIoT systems. There is a need to have immutable device identities that are secured by security hardware and remote attestation mechanisms that can be used to verify the integrity of the connected device to detect that the device is in expected state. Our research shows that this is especially important in critical infrastructure systems but is also expected to be more widespread as hardware support for isolated execution environments in IoT devices becomes more common.

This paper first briefly described remote attestation mechanism alternatives for IoT. Based on our research work, remote attestation could be used in IIoT systems either as part of the shift towards IT protocols or there can be extensions to legacy protocols that are used in industrial automation. The literature shows that remote attestation is typically not used in identity and access management systems because of lacking security hardware and deployment complexity. However, we have noticed that cloud service providers have started to use remote attestation as part of confidential computing initiatives, giving examples of how it can be used in IIoT systems as well.

Legacy industrial automation protocols are difficult to extend because of their optimized fixed message structure. Some legacy protocols also have shortcomings in integrity and confidentiality protection. Even modern protocols like OPC UA do not yet contain options to support remote attestation. Based on our research, we argue that the current specification is not enough, but instead further specification work is needed. The IETF RATS working group has been active in specifying formats and reference architectures to remote attestation deployment in the IoT environment. We feel that especially their Entity Attestation token specification will be widely adopted as a model how to encode attestation evidence.

Acknowledgements

The work presented here has been carried out in the ITEA4 project 17032 CyberFactory#1.

References

Ali, T., Nauman, M., Amin, M. and Alam, M. (2010) "Scalable, Privacy-Preserving Remote Attestation in and through Federated Identity Management Frameworks", *2010 International Conference on Information Science and Applications*, 2010, pp. 1-8, doi: 10.1109/ICISA.2010.5480294.

Arbaugh, W. A., Farber, D. J. and J. M. Smith (1997) "A Secure and Reliable Bootstrap Architecture", *in IEEE Computer Society Conference on Security and Privacy*. IEEE, 1997, pp. 65–71.

Arnautov, S., Trach, B., Gregor, F., Knauth, T., Martin, A., Priebe, C., Lind, J. Muthukumaran, D., O'Keeffe, D., Stillwell, M.L., Goltzsche, D., Eyers, D., Kapitza, R., Pietzuch, P. and Fetzer, C. (2016) "SCONE: Secure Linux Containers with Intel SGX", *12th USENIX Symposium on Operating Systems Design and Implementation (OSDI 16)*, 978-1-931971-33-1, pp. 689-703, USENIX Association.

Asokan, N., Brasser, F., Ibrahim, A., Sadeghi, A.-R., Schunter, M., Tsudik, G. and Wachsmann, C. (2015) "SEDA: Scalable Embedded Device Attestation", *Proceedings of the 22nd ACM SIGSAC Conference on Computer and Communications Security (CCS '15)*. Association for Computing Machinery, New York, NY, USA, pp. 964–975, doi: 10.1145/2810103.2813670

AWS (2022) "AWS Nitro Enclaves User Guide", Available at: https://docs.aws.amazon.com/enclaves/latest/user/enclaves-user.pdf (Accessed: 25 Jan 2022).

Baldwin, P., Lyon, R., Sindhuri, D., Coulter, D. and Vilaysim, S. (2021) "Microsoft Azure Attestation", Microsoft, Available at: https://docs.microsoft.com/en-us/azure/attestation/overview (Accessed: 25 Jan 2022).

Berger, B. (2005) "Trusted computing group history", Information Security Technical Report, Volume 10, Issue 2, 2005, pp. 59-62, doi: 10.1016/j.istr.2005.05.007.

Bienhaus, D., Ebner, A., Jäger, L., Rieke, R. and Krauß, C. (2021) "Secure gate: Secure gateways and wireless sensors as enablers for sustainability in production plants", *Simulation Modelling Practice and Theory*, Volume 109, 2021, doi: 10.1016/j.simpat.2021.102282

Birkholz, H., Thaler, D., Richardson, M., Smith, N. and Pan, W. (2021a) "Remote Attestation Procedures Architecture", work in progress, IETF RATS draft, Available at: https://datatracker.ietf.org/doc/draft-ietf-rats-architecture/ (Accessed: 25 Jan 2022).

Birkholz, H., Eckel, M., Pan, W. and Voit, E. (2021b) "Reference Interaction Models for Remote Attestation Procedures", work in progress, IETF RATS draft, Available at: https://datatracker.ietf.org/doc/draft-ietf-rats-reference-interaction-models/ (Accessed: 25 Jan 2022).

Birnstill, P., Haas, C., Hassler, D. and Beyerer, J. (2017). "Introducing remote attestation and hardware-based cryptography to OPC UA", *22nd IEEE International Conference on Emerging Technologies and Factory Automation (ETFA)*, pp. 1-8, doi: 10.1109/ETFA.2017.8247591.

Coker, G., Guttman, J., Loscocco, P., Herzog, A., Millen, J., O'Hanlon, B., Ramsdell, J., Segall, A., Sheehy, J. and Sniffem, B. (2011) "Principles of remote attestation", *International Journal of Information Security* 10, pp. 63–81, doi: 10.1007/s10207-011-0124-7.

Dahlmanns, M., Lohmöller, J., Fink, I. B., Pennekamp, J., Wehrle, K. and Henze M. (2020) "Easing the conscience with OPC UA: An internet-wide study on insecure deployments," in Proc. ACM Internet Meas. Conf. New York, NY, USA: ACM, 2020, pp. 101–110, doi: 10.1145/3419394.3423666

Dushku, E., Rabbani, M. M., Conti, M., Mancini, L. V. and Ranise, S. (2020) "SARA: Secure Asynchronous Remote Attestation for IoT systems", *in IEEE Transactions on Information Forensics and Security*, vol 15, pp. 3123-3126, doi: 10.1109/TIFS.2020.2983282.

Eckel, M. (2021) "CHARRA: Challenge-Response based Remote Attestation with TPM 2.0", Sep. 2021, Available at: https://github.com/Fraunhofer-SIT/charra (Accessed: 25 Jan 2022).

Hovsmith, S. (2017) "Adapting OAuth2 for Internet of Things (IoT) API Security", blog article, Approov, Available at: https://blog.approov.io/adapting-oauth2-for-internet-of-things-iot-api-security, (Accessed: 25 Jan 2022).

IEEE1815 (2012) "IEEE Standard for Electric Power Systems Communications-Distributed Network Protocol (DNP3)," *in IEEE Std 1815-2012 (Revision of IEEE Std 1815-2010)*, pp.1-821, doi: 10.1109/IEEESTD.2012.6327578.

Knauth, T., Steiner, M., Chakrabarti, S., Lei, L., Xing, C. and Vij, M. (2019) "Integrating Remote Attestation with Transport Layer Security", arXiv:1801.05863.

Leicher, A., Schmidt, A. U., Shah, Y. and Cha, I. (2010) "Trusted Computing enhanced OpenID," *2010 International Conference for Internet Technology and Secured Transactions*, pp. 1-8.

Lu, Y. and Feng, T. (2018) "Research on trusted DNP3-BAE protocol based on hash chain", *EURASIP Journal on Wireless Communications and Networking*, 108, doi: 10.1186/s13638-018-1129-y.

Lundblade, L., Mandyam, G. and O'Donoghue, J. (2021) "The Entity Attestation Token (EAT)", work in progress, IETF RATS draft, Available at: https://datatracker.ietf.org/doc/draft-ietf-rats-eat/ (Accessed: 25 Jan 2022).

Matischek, R. and Bara, B. (2019) "Application Study of Hardware-Based Security for Future Industrial IoT," *22nd Euromicro Conference on Digital System Design (DSD)*, pp. 246-252, doi: 10.1109/DSD.2019.00044.

Mechalas, J.P. (2018) "Software Guard Extensions Remote Attestation End-to-End Example", Intel code sample, Available at: https://www.intel.com/content/www/us/en/developer/articles/code-sample/software-guard-extensions-remote-attestation-end-to-end-example.html (Accessed: 25 Jan 2022).

Mulligan, D.P., Petri, G., Spinale, N., Stockwell, G. and Vincent, H.J.M. (2021) "Confidential Computing—a brave new world," *2021 International Symposium on Secure and Private Execution Environment Design (SEED)*, pp. 132-138, doi: 10.1109/SEED51797.2021.00025.

Naik, N. and Jenkins, P. (2017) "Securing digital identities in the cloud by selecting an apposite Federated Identity Management from SAML, OAuth and OpenID Connect," *11th International Conference on Research Challenges in Information Science (RCIS)*, pp. 163-174, doi: 10.1109/RCIS.2017.7956534.

OASIS (2005) "Assertions and Protocols for the OASIS Security Assertion Markup Language (SAML) V2.0", OASIS Standard, March 2005, Available at: http://docs.oasis-open.org/security/saml/v2.0/saml-core-2.0-os.pdf (Accessed: 25 Jan 2022).

OMG (2010), "The Real-time Publish-Subscribe Wire Protocol DDS Interoperability Wire Protocol Specification (DDS-RTPS)", Version 2.1, Available at: https://www.omg.org/spec/DDSI-RTPS/2.1/PDF (Accessed: 25 Jan 2022).

OPC (2017) "OPC UA - Part 7: Profiles", Available at: https://reference.opcfoundation.org/Core/docs/Part7/ (Accessed: 25 Jan 2022).

OPC (2018) "OPC UA - Part 2: Security Model", Available at: https://reference.opcfoundation.org/Core/docs/Part2/ (Accessed: 25 Jan 2022).

Pardo-Castello G. and Lang, U. (2013) "Trusted remote attestation for secure embedded systems", Embedded.com, Available at: https://www.embedded.com/trusted-remote-attestation-for-secure-embedded-systems/ (Accessed: 25 Jan 2022).

Pasham, N. (2020) "Demystifying ARM TrustZone for Microcontrollers (and a Note on Rust Support)", blog article, Medium, Available at: https://medium.com/swlh/demystifying-arm-trustzone-for-microcontrollers-and-a-note-on-rust-support-54efc62c290 (Accessed: 25 Jan 2022).

PJM (2020) "DNP SCADA over Internet with TLS Security", Jetstream Guide, December 4, 2020, Available at: https://www.pjm.com/~/media/etools/jetstream/jetstream-guide.ashx (Accessed: 25 Jan 2022).

Pliatsios, D., Sarigiannidis, P., Lagkas, T. and Sarigiannidis, A. G. (2020) "A Survey on SCADA Systems: Secure Protocols, Incidents, Threats and Tactics", *in IEEE Communications Surveys & Tutorials*, vol. 22, no. 3, pp. 1942-1976, doi: 10.1109/COMST.2020.2987688.

Sailer, R., Zhang, X., Jaeger, T. and van Doorn., L. (2004) "Design and implementation of a TCG-based integrity measurement architecture", *Proceedings of the 13th conference on USENIX Security Symposium*, Volume 13 (SSYM'04), USENIX Association, USA, 16.

Sandoval, K. (2017) "OAuth 2.0 – Why It's Vital to IoT Security", blog article, Nordic APIs, Available at: https://nordicapis.com/why-oauth-2-0-is-vital-to-iot-security/ (Accessed: 25 Jan 2022).

Seitz, L., Selander, G., Wahlstroem, E., Erdtman, S. and Tschofeng, H. (2021) "Authentication and Authorization for Constrained Environments (ACE) using the OAuth 2.0 Framework (ACE-OAuth)", work in progress, IETF ACE draft, Available at: https://datatracker.ietf.org/doc/html/draft-ietf-ace-oauth-authz (Accessed: 25 Jan 2022).

Self, H. (2020) "Overview of DNP3 Security Version 6", Available at: https://www.dnp.org/Resources/Public-Documents (Accessed: 25 Jan 2022).

Shelby, Z., Hartke, K. and Bormann, C. (2014), "The Constrained Application Protocol (CoAP)", IETF RFC7252, Available at: https://datatracker.ietf.org/doc/html/rfc7252 (Accessed: 25 Jan 2022).

Upchurch, J., (2019) "DDS Security the Hard(ware) Way - SGX", a three-part blog article, RTI, Available at: https://www.rti.com/blog/author/jason-upchurch (Accessed: 25 Jan 2022).

Suomalainen, J., Julku, J., Vehkaperä, M. and Posti, H. (2021) "Securing Public Safety Communications on Commercial and Tactical 5G Networks: A Survey and Future Research Directions", *in IEEE Open Journal of the Communications Society*, vol. 2, pp. 1590-1615, doi: 10.1109/OJCOMS.2021.3093529.

TCG (2019a) "TCG Remote Integrity Verification: Network Equipment Remote Attestation System", Version 1.0, Revision 9b, June 15, 2019, Available at: https://trustedcomputinggroup.org/wp-content/uploads/TCG-NetEq-Attestation-Workflow-Outline_v1r9b_pubrev.pdf (Accessed: 2 Feb 2022).

TCG (2019b) "Trusted Platform Module Library Specification", Family "2.0", Level 00, Revision 01.59 – November 2019, Available at: https://trustedcomputinggroup.org/resource/tpm-library-specification/ (Accessed: 25 Jan 2022).

TCG (2020) "DICE Layering Architecture", Version 1.0, Revision 0,19, July 2020, Available at: https://trustedcomputinggroup.org/wp-content/uploads/DICE-Layering-Architecture-r19_pub.pdf (Accessed: 25 Jan 2022).

TCG (2022) "TCG Guidance for Securing Industrial Control Systems Using TCG Technology", Version 1.0, Revision 109, January 10, 2022, Available at: https://trustedcomputinggroup.org/wp-content/uploads/TCG_Guidance_for_Securing_Industrial_Control_Systems_v1_r109_pub10jan2022.pdf (Accessed: 2 Feb 2022).

Tygar, J. and Yee, B. (1991) "Dyad: A system for using physically secure coprocessor", Technical Report CMU-CS-91-140R, Carnegie Mellon University, May 1991.

Yu, Y., Wang, H., Liu, B. and Yin, G. (2013) "A Trusted Remote Attestation Model Based on Trusted Computing," *12th IEEE International Conference on Trust, Security and Privacy in Computing and Communications*, pp. 1504-1509, doi: 10.1109/TrustCom.2013.183.

SIEM4GS: Security Information and Event Management for a Virtual Ground Station Testbed

Yee Wei Law[1] and Jill Slay[1,2]
[1]UniSA STEM, University of South Australia, Mawson Lakes, Australia
[2]SmartSat Cooperative Research Centre, Adelaide, Australia
YeeWei.Law@unisa.edu.au
Jill.Slay@unisa.edu.au

Abstract: As the space sector continues to grow, so do the cybersecurity risks. As large as the attack surface of a space system is, the ground segment remains an attractive source of intrusion points, not only because of its relative accessibility but also because the ground system is often viewed as little more than a conventional IT system. Thus, a representative security assessment of a space system cannot avoid addressing the vulnerabilities of the associated ground system and the relevant threats. This motivates the construction of a virtual ground station testbed, as part of larger reference platform, to support our ongoing research on the cybersecurity of space systems. Presented here is a discussion of the preliminary work being undertaken at the University of South Australia node of the SmartSat Cooperative Research Centre on such a testbed. A distinguishing feature of the testbed is the integration of a security information and event management (SIEM) system justifying the name of the testbed, "SIEM4GS". Based on the latest literature on ground stations, a logical architecture and an implementation plan involving only open-source software building blocks for SIEM4GS are proposed. Features of the ground station and SIEM services are discussed. A plan is provided on how to extend the SIEM system from a primarily "detect" role in the NIST Cybersecurity Framework to a "detect and respond" role.

Keywords: ground station, ground system, mission operations service framework, security information and event management, extended detection and response, Elastic Stack

1. Introduction

The space sector has been expanding rapidly in recent years. For example, between 2019 and 2020, the number of spacecrafts launched per year more than doubled, and 100,000 satellites are expected to be in orbit by the end of the decade (McDonald et al., 2021). Accompanying this growth is the mounting recognition of the pivotal role of space capabilities in national security. As such, securing space assets is a national priority. However, the inherently large attack surface of space systems poses immense challenges. Attacks — whether cyber, physical or both — can come from a wide variety of threat actors, through the supply chain or from the user end, to wreak havoc on the space segment, ground segment or user segment, before the aftermath spills over to other segments (Pavur and Martinovic, 2020). The research reported here is concerned with the *ground system*.

Definition 1. A *ground system* is an integrated set of ground functions used to support the preparation activities leading up to mission operations or the conduct of mission operations itself (ESA, 2008).

A ground system consists of ground stations, ground networks, control centres and remote terminal as its primary elements (Weston, 2020, Table 12-1). In the literature, the term "ground station" is used to refer to both the physical infrastructure and the ground system; it is for this reason "ground station" and "ground system" are used interchangeably in the ensuing discussion. In the terminology of NASA's Advanced Multi-Mission Operations System (AMMOS, see Benecken, 2020), a ground system is composed of a Mission Operations System (MO System) and associated earth-bound communications and data acquisition infrastructure (see Figure 1). A standard and precise definition of a MO System does not exist, but it can be understood as a set of implementation components that include a flight team and a *Ground Data System* (GDS) for ensuring the correct operations of a space mission. A GDS is a set of software, hardware and facilities as well as support services (e.g., system administration support) for collecting and distributing mission data (Benecken, 2020). A rigorous discussion of the definitions of a MO System is deferred to Sec. 2.2.

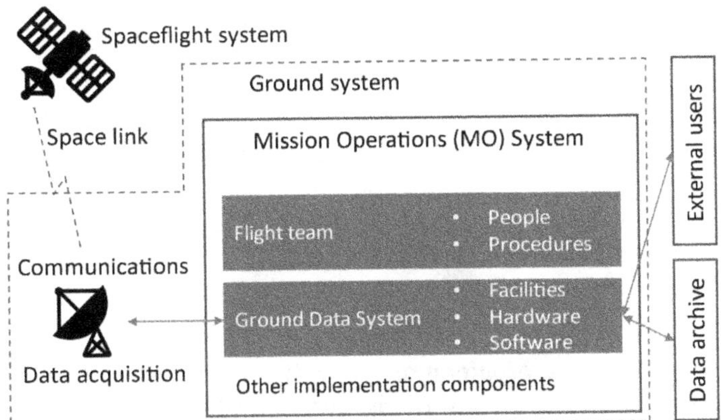

Figure 1: The components of a ground system in NASA's AMMOS terminology (Benecken, 2020)

Our focus on the security of ground systems through this work is based on the following rationales (Pavur and Martinovic, 2020, Sec. VI): (i) Coming from the ground-based Internet, ground systems represent an obvious point of entry into space systems. (ii) Ground stations are usually located in remote areas with limited physical security and are barely staffed, necessitating remote access to ground stations by operations centre staff located elsewhere, so ground stations can hardly be "air-gapped". (iii) A ground system typically serves multiple missions and multiple space agencies, so a pathway into a ground system can lead to compromise of multiple missions and agencies. In this scenario, a ground system represents a single point of failure. (iv) Ground system security has traditionally been treated as an extension of mainstream IT security, and as such, there has *not* been active research targeting the specific hardware, software, networks and architectures of ground systems.

The work presented here is concerned with a virtual ground station testbed, expanding our existing low-Earth-orbit satellite digital twin (Ormrod et al., 2021), to support our ongoing research into the cyberworthiness and cyberresilience of space systems. As our first step towards securing this testbed, a *security information and event management* (SIEM) system is used to monitor the network traffic of the testbed. SIEM is (i) the analysis of event data in real time for early detection of targeted attacks and data breaches, and (ii) the collection, storage, investigation and reporting on log data for incident response, forensics and regulatory compliance (Gartner, 2021). Our direction reflects industry best practice (Vera, 2016; Miller, 2021), and real-world needs to meet compliance requirements (Exabeam, 2021) such as the Sarbanes-Oxley Act.

Our contributions are as follows: (i) Our multivocal but rigorous literature review in Sec. 2 on ground systems, not only provides the theoretical basis for our virtual ground station testbed, but also clarifies space-related concepts and definitions that are not necessarily familiar to the security community, the primary audience of this paper. (ii) The proposed SIEM4GS architecture can serve as a reference architecture for future ground station security research. Once the source code for SIEM4GS is released, SIEM4GS can serve as a reference implementation too. The absence of a testbed similar to SIEM4GS testifies to the novelty of our work.

2. Literature review on ground systems

The primary ground system terminology is sourced from NASA, European Space Agency (ESA), and Consultative Committee for Space Data Systems (CCSDS). There is a large body of literature on ground system architectures and designs that can be sourced not only from the space agencies, but also from academia and industry. Traditional ground systems used a "chimney" or "stovepipe" architecture, catering to a single user and a single mission over a period of time. Current trends include:

- **Software-defined front-end processing**: A *front-end processor* (FEP) is a programmed-logic or stored-program device that interfaces data communication equipment with an input/output bus or memory of a data processing computer (Telecommunications Industry Association, 2021), and whose main functions are signal processing and encryption/decryption (Lowdermilk and Sethumadhavan, 2021). Fischer and Scholtz (2010) treat a "front end" as an assembly of antennas and rotators; and delegate signal processing functionality to a terminal node controller (TNC). However, since TNCs are only capable of VHF and UHF communications (Ahmad et al., 2016), *software-defined radios* (SDRs, i.e., radios in which the physical-layer functions are software-defined) are superseding TNCs to support communications in multiple frequency bands, especially the S band and X band (Weston, 2020). In other words, front-end processing is increasingly software-defined.

- **Adoption of SDR standard VITA 49:** The advent of SDRs created the need to ensure interoperability between diverse SDR components (Normoyle and Mesibov, 2008), especially within the signals intelligence community. To address this need, members of the defence industry and the VMEbus International Trade Association (VITA) initiated the analogue RF-digital standard called VITA 49 or VITA Radio Transport (Cooklev et al., 2012). This standard specifies a packet-based transport protocol for representing (i) digitised signal data, and (ii) metadata or context data about the radio (e.g., frequency, gain), the location of the radio and the processing done on the signal prior to the generation of the signal data packet. Besides interoperability, this standard enables accurate alignment of signal data and discrete events between multiple receivers that are either in the same location or separated by large distances. The many advantages of VITA 49 led to its quick adoption by the space sector.
- **Supporting multiple missions**: The proliferation of components ranging from field-programmable gate arrays (FPGAs) to SDRs has put deployment of low-cost satellites within the reach of organisations with modest budgets. However, barring educational projects like https://nyan-sat.com, the cost of licensing and setting up a dedicated ground station remains prohibitive for these organisations. By catering to multiple missions through technologies such as virtualisation and cloud computing, a ground station can offer its services to more organisations at a lower cost. After all, two consecutive passes of a satellite are typically separated by abundant idle time (Fischer and Scholtz, 2010).
- **Virtualisation**: This, in the context of ground stations, means reusing the same physical infrastructure for multiple simultaneous missions. To facilitate virtualisation, CCSDS (2010b) has defined a *Mission Operations Service Framework* (more details in Sec. 2.2) consisting of two interface layers that provide the patterns or templates of interaction between a consumer and a provider of a service. Through the standardised interfaces, any framework-aware mission operator can access any service that supports the framework. This allows any framework-compliant ground system to serve multiple missions. Besides standardised service interfaces, *scheduling* and *arbitration* are key to virtualisation, because a limited amount of hardware resources (e.g., antenna pool) means only a bounded number of operators can be given access to the resources at any given time. CCSDS (2018) has defined the XML-based Simple Schedule Format (SSF) for specifying scheduling information related to apertures at ground stations and/or relay satellites between operating entities. This allows mission operators to plan ahead to ensure there are sufficient aperture resources for their planned missions, and the ground station operator to ensure its resources are not overbooked. One way to achieve arbitration of resources is *software-defined networking* (Liu et al., 2020; Riffel and Gould, 2016). This mechanism enables a ground station operator to dynamically switch mission operators to their allocated antenna system at the scheduled time.
- **Cloud-based access and cloud-native implementation:** To be able to provide access and services that scale elastically, ground station operators increasingly resort to cloud computing technologies. As services and data are mirrored at redundant sites, cloud-based ground systems can achieve high reliability. An example of a "Ground Station as a Service" (GSaaS) offering is AWS Ground Station (see Sec. 2.3).
- **Network of ground stations:** Communications between a satellite and a ground station are limited to the portion of the satellite's orbit during which the satellite is within line of sight from the ground station, i.e., less than ten minutes in a single pass. This means a latency of several hours to several days before the user receives satellite's data. Furthermore, even when a satellite is in radio contact with a ground station, weather effects can cause more than 80% of packet loss (Vasisht and Chandra, 2020). Besides space-based relays, a network of geographically distributed ground stations can be used to provide all-time coverage and continuous access to communication and tracking services. *Virtualisation*, as discussed earlier, is one way to facilitate inclusion of a ground station into a ground station network. Examples of global ground station networks include Leaf Space, Satellite Network Open Ground Station (SatNOGS, see Sec. 2.1).

Below, discussion of ground system architectures is divided into (i) academia; (ii) space agencies; (iii) GSaaS.

2.1 Academia

The best-known architectures in academia include Standard University's Mercury (Cutler and Fox, 2006), and SatNOGS. The former is defunct, so we focus on SaNOGS.

SatNOGS is an open-source software and open hardware project initiated by Hackerspace.gr and embraced by academia. SatNOGS aims to provide participants worldwide with crowd-sourced resources for building a global network of satellite ground stations (Surligas et al., 2021). SatNOGS consists of four major subprojects:

- **SatNOGS Ground Station** is a collection of hardware designs and specifications for antennas, rotators with 3D-printed parts for directional antennas, and front-end processors.
- **SatNOGS Client** is a software running on a computer controlling Ground Station hardware. The Client regularly polls the Network (discussed later) for observation jobs scheduled for the local Ground Station. Client functions include scheduling, rotator control, GNU Radio-enabled Doppler tuning and signal demodulation. Demodulated signal data, logs and other reports are queued for upload to the Network.
- **SatNOGS Network** is a web application hosted at https://network.satnogs.org, to which users can submit scheduled observation requests. Based on the requests, the Network calculates the observation windows from available Ground Stations. Once an observer accepts an observation job proposed by the Network, the job is inserted into the job queue of the observer's Ground Station. Likewise, a Ground Station can query the Network for its list of scheduled jobs and uploads its data to the Network. When calculating possible observations, SatNOGS Network extracts the relevant information from the SatNOGS Database (discussed next).
- **SatNOGS Database** is a web application hosted at https://db.satnogs.org, providing access to crowd-sourced satellite information and collected telemetry data. Satellites are identifiable by their North American Aerospace Defense Command (NORAD) space object catalogue number and their common name. Transponder records for each satellite are extracted from https://CelesTrak.com.

2.2 Space agencies

In NASA's literature (Weston, 2020), a group station comprises (i) a ground station terminal, (ii) a MO Centre, (iii) a Science Operations Centre, and (iv) data storage and network. While the MO Centre is responsible for ensuring the success of a mission, the Science Operations Centre is responsible for instructing the MO Centre what science operations to perform, besides generating and disseminating science data products. In NASA's AMMOS architecture depicted in Figure 1, the MO System serves as the system supporting the MO Centre, while the "External users" block plays the role of the Science Operations Centre.

ESA's (2008) ground system architecture in the standard ECSS-E-ST-70C differs from NASA's, but it also features a MO System. A ground segment in ESA's ECSS-E-ST-70C architecture has four top-level systems, namely (i) a ground station system, (ii) a ground communications system, (iii) a MO System, and (iv) a payload operations and data system. The ground communications system provides the interconnections between systems, to enable data distribution, voice and video communications, and system maintenance. In terms of their functions, ESA's version of MO System does not completely overlap with NASA's version of MO System.

Instead of the MO System, MO Services receive the focus in CCSDS' standards. CCSDS' *MO Service Framework* (CCSDS, 2010b) is based on the principles of a *service-oriented architecture*, and the framework defines (i) a model for interaction between two entities, and (ii) a model for common services providing functionality common to most uses of the service framework. From the perspective of the open systems interconnection model, the framework defines two layers between the application layer and the transport layer:

- The **MO Services Layer** sits right below the application layer. Two types of services are provided through this interface: *Functional Services* and *Common Services*. Functional Services are MO-specific services, such as monitoring and control (M&C), scheduling, etc., while Common Services are general services that even Functional Services might need, such as service directory, user login/authentication, etc. Both Functional Services and Common Services are defined as specialisations or extensions to the Common Object Model (COM). The COM provides a common information model for all MO Service objects. In this model, any change in the attributes of an MO Service object triggers an automation event. For each MO Service, the operations are defined as specialisations or extensions to the generic interaction patterns defined in the Message Abstraction Layer, to be discussed next.
- The **Messaging Abstraction Layer** (MAL, see CCSDS, 2013) sits below the MO Services Layer and above the transport layer. For ground systems to be interoperable, i.e., a user to be able to access any service that is compliant with the MO Service Framework, the consumer-provider communication protocol must be standardised. The MAL meets this standardisation need by specifying data types and structures (which the COM inherits), message format, and message exchange patterns (e.g., SEND, SUBMIT).

2.3 Ground Station as a Service (GSaaS)

The separation of the MO System from the ground station system by the ground communications system in ESA's architecture is mirrored by cloud-based ground system architectures. AWS Ground Station (Amazon Web Services, 2021) and Azure Orbital (Microsoft, 2021) are two examples of these architectures, and a significant commonality between them is observable. The representative cloud-based architecture supports (i) radio communications in the UHF, S and X bands; (ii) digitising RF signals and serving the digitised streams through the standardised packet-based protocol VITA 49; (iii) process automation and security. We note this architecture classifies *telemetry, tracking and command/control* (TT&C, see Definition 2) as a function of the customer's MO System, in a departure from ESA's ECSS-E-ST-70C architecture.

Definition 2. In TT&C (Guest, 2017), (i) *telemetry* is the collection of on-board measurements and instrument readings required to deduce the health and status of all subsystems of a satellite (e.g., propellant supply) and the transmission of this data to the command segment on the ground; (ii) *tracking* or more precisely, *carrier tracking*, refers to the process of locating and locking onto a satellite from a ground station; (iii) *command/control* refers to the execution of action sequence (e.g., toggling a relay) on the satellite and its payloads to meet mission objectives.

Considering the increasing popularity of GSaaS, our testbed SIEM4GS is based on this architecture, while the core services of SIEM4GS are informed by CCSDS' MO Service Framework. Sec. 3 has more details.

3. Architecting and implementing SIEM4GS

The preceding review provides the basis for the ground station part of SIEM4GS. The following subsection covers some preliminaries on the SIEM part of SIEM4GS, while Sec. 3.2 describes the architectures.

3.1 Preliminaries on SIEM

SIEM, as defined in Sec. 1, is an integral part of a modern IT infrastructure. A core component of a SIEM system is its *rule engine* (Exabeam, 2021), which contains rules that are run periodically. A SIEM rule is a set of conditions (including thresholds) expressed in Boolean logic, schedules and actions that enable notifications (Elasticsearch, 2021). SIEM rules are often called *correlation rules* because they are processed by means of *correlation*. An *alarm/alert* is triggered when the result of correlation is significant. The mechanism of *alarm correlation* or *alert correlation* is to ensure the frequency of alarms is manageable, and it refers to the conceptual interpretation of multiple alarms such that a new meaning is assigned to these alarms (Jakobson and Weissman, 1993). When an interpretation is consistent with an attack scenario, then the original multiple alarms can be merged into one; otherwise, no alarm should be triggered so that the number of false alarms is minimised. It is unclear how many of the latest advances in alert correlation (Salah et al., 2013), including those for detecting multi-stage attacks (Shin et al., 2019) have been implemented in commercial SIEM systems. By popular assessment (Barros, 2017; Manor, 2021), real-world SIEM rules remain simplistic to the degree that threat coverage and false positive rate remain undesirable. Thus, additional technologies are needed to complement the functionality of a traditional SIEM system, including:

- **User and Entity Behaviour Analytics (UEBA):** Also known as User Behaviour Analytics, this refers to application of data analytics to the discovery of abnormal/risky behaviour by users or entities such as endpoints, servers and routers, often in conjunction with a SIEM (Fortinet, 2021). UEBA can help detect anomalies that SIEM rules cannot, using for example specialised neural networks (Sharma et al., 2020), but it requires determination of user and entity baseline behaviour and painstaking feature engineering.
- **Security Orchestration, Automation and Response (SOAR):** This is the planning, integration, coordination and cooperation of the activities of security tools and experts to produce and automate required remediations in response to security incidents across multiple technology paradigms (Islam et al., 2019). Whereas SIEM automates the "detect" function defined in NIST's (2018) Cybersecurity Framework, SOAR aims to automate both the "detect" and "respond" functions of the framework, alleviating the "alert fatigue" problem of traditional SIEM.
- **Extended Detection and Response (XDR):** This refers to the extension of traditional "endpoint detection and response" to threat detection, investigation and response solutions that work across all threat vectors in an organisation's infrastructure, including endpoints, servers, networks and cloud, rather than just one piece thereof (Palo Alto Networks, 2022). Whereas SOAR serves as an upgrade to SIEM, XDR is often seen

as comparable to SOAR but with a keener focus on threat detection and incident response (Miller, 2021). XDR started out as an industrial initiative for large vendors to integrate their security products and present a single pane-of-glass view of security information ("native XDR") but has evolved to embrace openness ("open XDR" or "hybrid XDR"). XDR aims to maximise situational awareness and coverage of the attack surface through enhanced data/intelligence ingestion and contextualisation. The success of both SOAR and XDR rests heavily on the synergistic leveraging of the latest software and hardware advances in big data analytics and artificial intelligence.

To implement the SIEM part of SIEM4GS, we use Elastic Stack because (i) its open source provides flexibility for customisation and experimentation; (ii) it has strong community support; (iii) it is widely used for cybersecurity research. Elastic Stack comprises four main components using the same Elastic Common Schema data format, namely (i) the distributed, RESTful search and analytics engine, Elasticsearch; (ii) the visualisation app, Kibana; (iii) the network and host data integrator consisting of endpoint kernel-level antimalware called Elastic Endpoint, distributed data shippers called Beats and server-side data processing pipeline called Logstash; and finally (iv) security content created by Elasticsearch and its user community. Elastic Stack supports open/hybrid XDR through its "Limitless XDR" feature, serving our research needs.

Our long-term plan is to push the envelope of XDR by exploring deep *semi-supervised*, *weakly-supervised* and *self-supervised* learning approaches for threat detection (Pang et al., 2022) because (i) manual data labelling for supervised learning is not scalable; (ii) unsupervised learning suffers from high false positive rates. Furthermore, learning will be done on *attributed networks*. An attributed network is a graph whose vertices have attributes (Liu et al., 2021). Their ability to model flows and multi-stage attacks, combined with a data-efficient learning paradigm with a low false positive rate, should pave way for fulfilling the promise of XDR and extending the "detect" function of SIEM's system to "detect and respond".

3.2 Architectures and implementation plan

Figure 2 and Figure 3 show the logical and physical architectures for SIEM4GS. For each of the non-SIEM logical blocks in Figure 2, a main reference implementation is identified with commentary, while alternative reference implementations are listed in Table 1 without commentary.

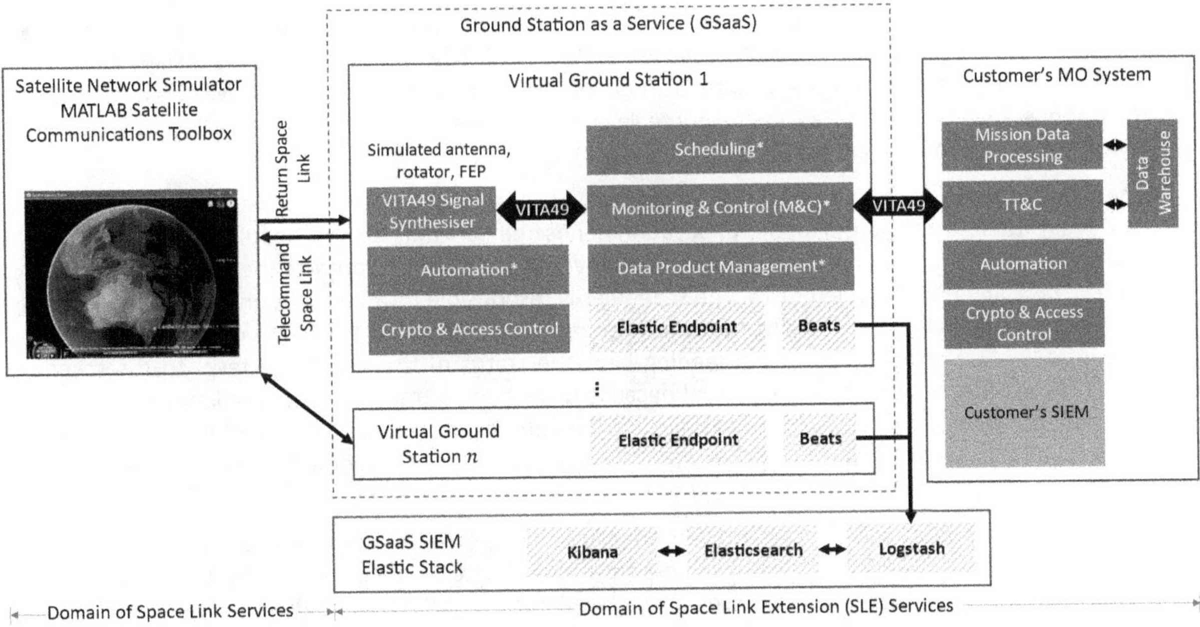

Figure 2: Logical architecture of the proposed virtual ground station testbed, SIEM4GS. Block names with an asterisk are associated with Functional Services identified by CCSDS (2010b). The hatch-filled orange blocks are part of the Elastic Stack SIEM

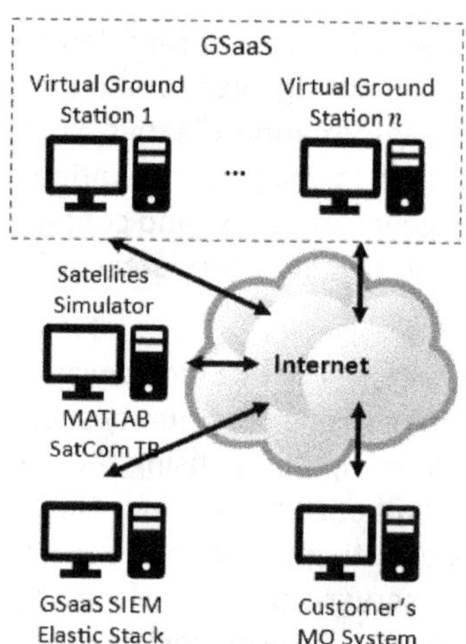

Figure 3: Physical architecture of SIEM4GS

The logical blocks are discussed below:

- **Satellite Network Simulator**: This simulates satellite orbits and space-ground links; and interact with one or more Virtual Ground Stations that are part of the GSaaS serving the Customer's MO System. The simulator can be implemented using MATLAB's Satellite Communications Toolbox and Aerospace Blockset.
- **VITA 49 Signal Synthesiser:** As SIEM4GS is by design virtual, radio signals are synthesised and furthermore in the VITA 49 format to comply with the current standard. The VITA 49 signal stream serves as input to the Monitoring & Control block. This synthesiser is implemented using the open-source C++-based REDHAWK SDR framework (Robert et al., 2015). Building on the Common Object Request Broker Architecture (CORBA), REDHAWK can interoperate with software components written in any other language provided an object request broker exists for that language (e.g., Java).
- **Monitoring & Control (M&C):** Real-time M&C of a spacecraft (both pre-launch and post-launch) includes downlink telemetry processing and visualisation, as well as formulation and initiation of the transmission of spacecraft (tele)commands (Benecken, 2020). Telemetry data is formatted in the CCSDS-prescribed XML Telemetric and Command Exchange (XTCE) schema (OMG, 2019). By design, operationally significant events or anomalies trigger alerts to be sent to subscribers of this service. XTCE-conformant M&C is built on Space Applications Services' Yamcs, an open-source Java-based software framework for M&C of spacecrafts, payloads and ground equipment (Schmitt et al., 2018).
- **Scheduling:** Upon generating a conflict-free schedule to automate MO (e.g., ground-space communications, data acquisition), a mission planning application sends the schedule to a scheduler via this Scheduling service. A schedule is a container for (i) predicted events, (ii) planned contacts (periods of ground-space connectivity), and (iii) scheduled tasks. The scheduler is responsible for (i) distributing the schedules to the applications responsible for executing the scheduled tasks, (ii) monitoring the status of the scheduled tasks, (iii) controlling the schedules, and (iv) feeding back the status of the scheduled tasks to the schedule-generating application. Scheduling is crucial because flight system tracking hours are limited by total user demand, as well as internal engineering and maintenance. There is currently no open-source implementation of the SSF (see Sec. 2), so SIEM4GS does not support SSF initially. The scheduler is built on that of the SatNOGS Client.
- **Automation:** Responses to events of interest are routinely automated. For example, AWS Ground Station enables automation of AWS services in response to satellite contact status via the mechanism of CloudWatch Events (Amazon Web Services, 2021). Implementation of this service can readily leverage the event handling mechanism or publisher-subscriber design pattern of a modern programming language, e.g., Java provides the Observable and Observer interfaces to support the observer design pattern.
- **Data Product Management:** This is the generation of scientific instrument data product, including processing, display and delivery, for use by instrument engineers, science planners and operators, as well as for public information releases. This service supports the transfer of data products in both directions, and alerts service subscribers about changes to the data product store, such as new product events. An archive pipeline typically includes metadata/label design, data format conversion, validation and delivery to archive storage, e.g., Open Archival Information System (CCSDS, 2012), Planetary Data System (Benecken, 2020).

This service can be implemented on top of the CCSDS File Discovery Protocol (CFDP, see CCSDS, 2020), an implementation of which exists in Yamcs.

- **Cryptography & Access Control:** One of the six Common Services in CCSDS' MO reference model is Login (CCSDS, 2010a). Here, the Login service is generalised to cryptographic and access control services. As per CCSDS' (2019a) mandate, SHA-256 is used for collision-resistant hashing; AES-GCM with 256-bit keys is used for authenticated encryption with associated data; DSA/RSA/ECDSA is used for digital signatures. Additionally, as per CCSDS' (2019b) report, Space Data Link Security (SDLS) is used for link security; TLS is used for end-to-end security. SDLS is implemented using NASA's open-source C-based CryptoLib, while the rest is implemented using OpenSSL. A Java API for OpenSSL is available from Wildfly. A simple implementation of access control is available from Yamcs.
- **TT&C**: This in the representative GSaaS architecture (see Sec. 2.3) is the responsibility of the customer's MO system. The tracking part can be technically demanding because it involves the use of tones/pseudocode or sensing of Doppler frequency shifts to achieve ranging (Kinman, 2021). For the purpose of our virtual testbed, tracking is assumed to be unnecessary and hence not implemented.

Table 1: Alternative reference implementations for select logical blocks in Figure 2.

Logical block	Project name, main programming language, URLs
Satellite Network Simulator	SatSim, Python, https://gitlab.com/librecube/prototypes/python-satsim
	SNS3, C++, https://github.com/sns3/sns3-satellite
	OS3, C++, https://github.com/inet-framework/os3
M&C	ReatMetric, Java, https://github.com/dariol83/reatmetric, https://github.com/dariol83/ccsds
	Only for MAL (see Sec. 2.2): CCSDS MO services – ESA's Java implementation, Java, https://github.com/esa/mo-services-java
	Only for monitoring: SatNOGS Client (see Sec. 2.1), Python, https://gitlab.com/librespacefoundation/satnogs/satnogs-client
	Only for telemetry visualisation: OpenMCT, JavaScript, https://github.com/nasa/openmct
Data Product Management	CCSDS File Delivery Protocol, Python, https://gitlab.com/librecube/lib/python-cfdp
	core Flight System (cFS) CFDP Application (CF), C, https://github.com/nasa/CF
Access Control	CCSDS MO services – ESA's Java impl., Java, https://github.com/esa/mo-services-java

4. Conclusion and future work

Even while space assets continue to grow, the ground segment remains crucial. For our space cybersecurity research, a virtual ground station testbed called SIEM4GS has been under construction. A multivocal but rigorous review of ground station designs was performed, and an abridged version of the review is presented in Sec. 2, informing the design decisions underlying our reference architectures in Sec. 3. The proposed implementation plan can guide any testbed-building effort similar to ours as it involves mostly open-source software building blocks. An alternative clean-slate approach based on model-based systems engineering could have been used for our testbed design, but ours is a security-focused engineering trade-off that builds on open-source contributions. Our mission is to explore new ideas of cyberattacks and countermeasures for space systems starting with the ground segment and within a SIEM framework. Besides pushing the envelope of XDR to achieve both the "detect" and "respond" functions of NIST's Cybersecurity Framework, our long-term plan includes growing this testbed into a *digital twin* where a two-way communication link synchronises the states of a small-scale physical testbed with those of its virtual replica. A journal version of this paper containing an extended version of our literature reviews on ground stations and SIEM (up to UEBA, SOAR, XDR) as well as implementation results, is under development.

Acknowledgements

The authors thank David Culpin (Raytheon Australia), Dr Ronald Mulinde (UniSA), Dr David Ormrod (UniSA) for their initial input; and Brandon Klar (UniSA) for undertaking the implementation.

References

Ahmad, Y.A., Nazim, N.J. and Yuhaniz, S.S. (2016) "Design of a terminal node controller hardware for CubeSat tracking applications", *AEROTECH VI*, IOP Publishing, p. 012031, DOI: 10.1088/1757-899x/152/1/012031.

Amazon Web Services (2021) "AWS Ground Station User Guide", [online], accessed 27 December 2021, https://docs.aws.amazon.com/ground-station/latest/ug/groundstation-ug.pdf.

Barros, A. (2017) "SIEM Correlation Is Overrated", [online], Gartner Information Technology Blog, http://blogs.gartner.com/augusto-barros/2017/03/31/siem-correlation-is-overrated/.

Benecken, Z. (2020) "AMMOS Catalog Version 5.3", Multimission Ground System and Services (MGSS) Program, Interplanetary Network Directorate (IND) Office, NASA, https://ammos.nasa.gov.

CCSDS (2010a) "Mission Operations Reference Model", Recommended Practice 520.1-M-1.

CCSDS (2010b) "Mission Operations Services Concept", Informational Report 520.0-G-3.

CCSDS (2012) "Reference Model for an Open Archival Information System (OAIS)", Recommended Practice 650.0-M-2.

CCSDS (2013) "Mission Operations Message Abstraction Layer", Recommended Standard 521.0-B-2.

CCSDS (2018) "Cross Support Service Management — Simple Schedule Format Specification", Recommended Standard 902.1-B-1.

CCSDS (2019a) "CCSDS Cryptographic Algorithms", Recommended Standard 352.0-B-2.

CCSDS (2019b) "The Application of Security to CCSDS Protocols", Informational Report 350.0-G-3.

CCSDS (2020) "CCSDS File Discovery Protocol (CFDP)", Recommended Standard 727.0-B-5.

Cooklev, T., Normoyle, R. and Clendenen, D. (2012) "The VITA 49 Analog RF-Digital Interface", *IEEE Circuits and Systems Magazine*, vol. 12, no. 4, pp. 21–32, DOI: 10.1109/MCAS.2012.2221520.

Cutler, J.W. and Fox, A. (2006) "A framework for robust and flexible ground station networks", *Journal of Aerospace Computing, Information, and Communication*, vol. 3, no. 3, pp. 73–92, DOI: 10.2514/1.15464.

ESA (2008) "Space engineering: Ground systems and operations", Standard ECSS-E-ST-70C.

Elasticsearch (2021) "Rule types", [online], Kibana Guide, accessed 14 December 2021, https://www.elastic.co/guide/en/kibana/master/rule-types.html.

Exabeam (2021) "The essential guide to SIEM", [online], accessed 14 December 2021, https://www.exabeam.com/siem-guide/.

Fischer, M. and Scholtz, A.L. (2010) "Design of a Multi-mission Satellite Ground Station for Education and Research", *SPACOMM 2010*, pp. 58–63, DOI: 10.1109/SPACOMM.2010.13.

Fortinet (2021) "UEBA", [online], CyberGlossary, accessed 19 December 2021, https://www.fortinet.com/resources/cyberglossary/what-is-ueba.

Gartner (2021) "Security Information and Event Management", [online], accessed 12 December 2021, https://www.gartner.com/en/information-technology/glossary/security-information-event-management.

Guest, A.N. (2017) "Telemetry, Tracking, and Command (TT&C)", *Handbook of Satellite Applications*, 2nd edition, Springer International Publishing, pp. 1313–1324.

Husák, M., Komárková, J., et al. (2019) "Survey of Attack Projection, Prediction, and Forecasting in Cyber Security", *IEEE Commun. Surveys Tuts.*, vol. 21, no. 1, pp. 640–660, DOI: 10.1109/COMST.2018.2871866.

Islam, C., Babar, M.A. and Nepal, S. (2019) "A Multi-Vocal Review of Security Orchestration", *ACM Comput. Surv.*, vol. 52, no. 2, DOI: 10.1145/3305268.

Jakobson, G. and Weissman, M. (1993) "Alarm correlation", *IEEE Network*, vol. 7, no. 6, pp. 52–59, DOI: 10.1109/65.244794.

Kinman, P. (2021) "Doppler Tracking", *DSN Telecommunications Link Design Handbook*, DSN No. 810-005, 202, Rev. D, Jet Propulsion Laboratory, California Institute of Technology.

Liu, Y., Chen, Y., Jiao, Y., et al. (2020) "A Shared Satellite Ground Station Using User-Oriented Virtualization Technology", *IEEE Access*, vol. 8, pp. 63923–63934, DOI: 10.1109/ACCESS.2020.

Liu, Y., Li, Z., Pan, S., et al. (2021) "Anomaly Detection on Attributed Networks via Contrastive Self-Supervised Learning", *IEEE Trans. Neural Netw. Learn. Syst.*, early access, DOI: 10.1109/TNNLS.2021.3068344.

Lowdermilk, J. and Sethumadhavan, S. (2021) "The Gestalt: A Secure, High Performance, Low Cost Satellite Ground Station Architecture and its Implementation", *35th Annual Small Satellite Conference*, https://digitalcommons.usu.edu/smallsat/2021/all2021/108/.

Manor, Y. (2021) "Quantifying the Gap Between Perceived Security and Comprehensive MITRE ATT&CK Coverage", industry research report, https://www.cardinalops.com/siem-industry-research-report.

McDonald, G., Hacker, J., Dorame, T., et al. (2021) "Navigating space: A vision for space in defense", white paper from KPMG International and Space Foundation, https://home.kpmg/xx/en/home/insights/2021/08/navigating-space-a-vision-for-space-in-defense.html.

Microsoft (2021) "Azure Orbital: Satellite ground station and scheduling services for fast downlinking of Data", [online], accessed 12 September 2021, https://azure.microsoft.com/en-au/services/orbital/.

Miller, L. (2021) "The Gorilla Guide to Extended Detection and Response (XDR)", ActualTech Media in partnership with Fortinet, https://www.fortinet.com/resources/cyberglossary/what-is-XDR.

NIST (2018) "Framework for Improving Critical Infrastructure Cybersecurity", version 1.1, DOI: 10.6028/NIST.CSWP.04162018.

Normoyle, R. and Mesibov, P. (2008) "The VITA Radio Transport as a Framework for Software Definable Radio Architectures", *SDR 08 Technical Conference and Product Exposition*, SDR Forum.

OMG (2019) "XML Telemetric and Command Exchange Version 1.2", an OMG® XML Telemetric and Command Exchange™ Publication formal/18-10-04, Object Management Group.

Ormrod, D., Slay, J. and Ormrod, A. (2021) "Cyber-Worthiness and Cyber-Resilience to Secure Low Earth Orbit Satellites", *ICCWS 2021*, Academic Conferences International, pp. 257–266, DOI: 10.34190/IWS.21.044.

Palo Alto Networks (2022) "The Essential Guide to XDR", [online], accessed 12 January 2022, https://www.paloaltonetworks.com/content/dam/pan/en_US/assets/pdf/ebooks/cortex-ebook_the-essential-guide-to-xdr.pdf.

Pang, G., Shen, C., Cao, L. and Van Den Hengel, A. (2022) "Deep Learning for Anomaly Detection: A Review", *ACM Comput. Surv.*, vol. 54, no. 2, article 38 , 38 pages, DOI: 10.1145/3439950.

Pavur, J. and Martinovic, I. (2020) "SOK: Building a Launchpad for Impactful Satellite Cyber-Security Research", arXiv preprint arXiv:2010.10872.

Riffel, F. and Gould, R. (2016) "Satellite ground station virtualization: Secure sharing of ground stations using software defined networking", *SysCon 2016*, pp. 1–8, DOI: 10.1109/SYSCON.2016.7490612.

Robert, M., Sun, Y., Goodwin, T., et al. (2015) "Software Frameworks for SDR", *Proceedings of the IEEE*, vol. 103, no. 3, pp. 452-475, DOI: 10.1109/JPROC.2015.2391176.

Salah, S., Maciá-Fernández, G., and Díaz-Verdejo, J.E. (2013) "A model-based survey of alert correlation techniques", *Computer Networks*, vol. 57, no. 5, pp. 1289–1317, DOI: 10.1016/j.comnet.2012.10.022.

Schmitt, M., Diet, F. and Mihalache, N. (2018) "Yamcs for lean Commercial Control Centres: The ICE Cubes Control Centre", *2018 SpaceOps Conference*, DOI: 10.2514/6.2018-2682.

Sharma, B., Pokharel, P. and Joshi, B. (2020) "User Behavior Analytics for Anomaly Detection Using LSTM Autoencoder - Insider Threat Detection", *IAIT 2020*, ACM, DOI: 10.1145/3406601.3406610.

Shin, J., Choi, S.H., et al. (2019) "Unsupervised multi-stage attack detection framework without details on single-stage attacks", *Future Gener. Comput. Syst.*, vol. 100, pp. 811–825, DOI: 10.1016/j.future.2019.05.032.

Surligas, M., Papamatthaiou M., Daradimos, I., et al. (2021) "SatNOGS: Towards a Modern, Crowd Sourced and Open Network of Ground Stations", *GNU Radio Conference*, vol. 2, no. 1.

Telecommunications Industry Association (2021) "front-end processor (FEP)", [online], accessed 30 Dec 2021, http://standards.tiaonline.org/market_intelligence_/glossary/index.cfm?term=%26%23%24C%5BRR%3FN%0A.

Vasisht, D. and Chandra R. (2020) "A Distributed and Hybrid Ground Station Network for Low Earth Orbit Satellites", *HotNets '20*, ACM, pp. 190–196, DOI: 10.1145/3422604.3425926.

Vera, T. (2016) "Cyber Security Awareness for SmallSat Ground Networks", *30th Annual AIAA/USU Small Satellite Conference*, https://digitalcommons.usu.edu/smallsat/2016/TS9GroundSystems/2/.

Weston, S. (2020) "Small Spacecraft Technology State of the Art", Technical Publication NASA/TP-2020–5008734, Small Spacecraft Systems Virtual Institute, https://www.nasa.gov/smallsat-institute/sst-soa.

Physical Layer Security: About Humans, Machines and the Transmission Channel

Christoph Lipps[1] and Hans Dieter Schotten[1,2]
[1]German Research Center for Artificial Intelligence, Intelligent Network Research Department, Kaiserslautern, Germany
[2]University of Kaiserslautern, Division of Wireless Communication and Radio Positioning, Kaiserslautern, Germany
Christoph.Lipps@dfki.de
Hans_Dieter.Schotten@dfki.de

Abstract: In an increasingly interconnected and globalized world in which the volume but also the confidentiality of transmitted content is becoming ever more important, trust, confidence and trustworthiness are of fundamental importance. Particularly in human societies, this trust is established, sustained and strengthened by personal relationships and experiences. But, in a globally connected world with Cyber-Physical Production Systems (CPPS), Industrial Internet of Things (IIoT) and Digital Twins (DTs), these personal relationships do not longer exist. (Remote) access to systems is possible from anywhere on the globe. However, this implies that there have to be technical solutions to detect, identify and acknowledge entities -people and machines- in the networks and thus to establish an initial level of trust. Especially since the proliferation of appropriate use-cases, Physical Layer Security (PhySec) is becoming increasingly popular in the scientific community. Using systems' intrinsic information for security applications provides a lightweight but secure alternative to traditional computationally intensive and complex cryptography. PhySec is therefore not only suitable for the IIoT and the multitude of resource-limited devices and sensors, it also opens up alternatives in terms of scalability and efficiency. Moreover, it provides security aspects regarding the entropy H and Perfect Forward Secrecy (PFS). Therefore, this work provides insight into three major branches of PhySec: i) *Human* - Physically Unclonable Functions (PUFs) ii) *silicon/electrical - PUFs*, and iii) *Channel-PUFs*. Based on the PUF operating principle, the silicon derivatives consider the electrical properties of semiconductors. Individual and uninfluenceable deviations during the manufacturing process result in component-specific behavior, which is described in particular for Static- and Dynamic Random Access Memory (S-/DRAM). Following this PUF principle, human characteristics -biological, physiological and behavioral features-, are used to recognize and authenticate them. With respect to the wireless channel, the characteristic properties of electromagnetic wave propagation and the influences on the wireless channel -diffraction, reflection, refraction and scattering-, are used to achieve symmetric encryption of the channel. In addition to the "conventional" wireless PhySec, especially the development of the Sixth Generation (6G) Wireless Systems, opens up a wide range of possibilities in terms of PhySec, for example in relation to Visible Light Communication (VLC), Reconfigurable Intelligent Surfaces (RIS) and in general the application of frequencies in the (sub)THz range. Thus, the work provides an overview of PhySec fields of application in all areas of the IIoT: in terms of humans, machines, and the transmission channel.

Keywords: physical layer security, physically unclonable functions, Human-PUFs, Channel-PUFs, (cyber)security, trust

1. The (Industrial) Internet of Things and the matter of security

Trust is not only a fundamental requirement for social coexistence, without trustworthy and confidential relationships, economic cooperation is inconceivable. Furthermore, without trust and without doubtless identification and authentication of contacts -humans and machines-, further security goals such as confidentiality, integrity and availability (Schäfer & Rossberg, 2016) are not achievable. But, it is becoming increasingly challenging to meet the requirements: Whereas previous relationships were characterized by personal contacts, these are disappearing in a progressively interconnected and globalized communication environment. Access to systems is available via remote maintenance from anywhere around the world, a technician does not necessarily have to be physically present on the machine at the factory, she may not even have to enter the company premises, or even be in the same country.

In addition, the growing number of connected devices is increasing as well: The International Telecommunication Union (ITU) predicts that the number of Mobile Broadband (MBB) subscribers worldwide will reach more than 17 billion by 2030 and the amount of Machine-to-Machine (M2M) terminals will exceed 100 billion (this is 14 times more than in 2020) by the same timescale (International Telecommunication Union, 2015). This is accompanied by demands on the performance of the interconnected devices. Many of them are limited in their features, computing power, as well as battery lifetime and -capacity. Especially in relation of the Fifth Generation (5G) New Radio (NR) technology a new term arises: reduced capability (RedCap) devices. The

4GPP Release-17 describes this group of devices as "Mid-Speed Smart IoT application" (Valerio, 2021) addressing the Sector of smart Internet of Things (IoT).

But, especially in terms of the terminology of the IoT there are many different aspects and applications in this area. So to provide a certain structure, a distinction is made between the terms used, as illustrated in **Figure 1**: i) The **Internet of Things** encompasses the cyber networking of physical objects, including everyday (household) items such as smart light bulbs, door keys, and wearables. *Kevin Ashton* "coined" that term back in 1999 during a presentation at Procter & Gamble where he introduced the Radio-Frequency Identification (RFID) chip into the company's value chain, but also pointed out the interdependencies between computer technology and humans (Kramp, van Kranenburg & Lang, 2013); ii) **Industrial Internet of Things (IIoT),** in turn, is shifting the scope from home applications to industrial manufacturing, transportation and critical infrastructure. With the purpose of saving costs and resources, and generally minimizing risks, the IIoT primarily includes "smart sensors and actuators to enhance manufacturing and industrial processes" (Posey, et al., 2022). These "connected sensors and actuators enable companies to pick up in inefficiencies and problems sooner and save time and money" (Posey, Rosencrance & Shea, 2022); iii) Another specialization is the term **Industry 4.0 (I4.0),** which was introduced by *Wolfgang Wahlster, Henning Kagermann* and *Wolf-Dieter Lukas* in 2011 as a marketing term in "INDUSTRIE 4.0: The 4th Industrial Revolution and the Internet of Things" as part of the Hannover Fair (DFKI, 2021). I4.0 is focused on the process within production facilities with smart interconnection of devices and scenarios such as M2M and Machine-to-Service (M2S) communication.

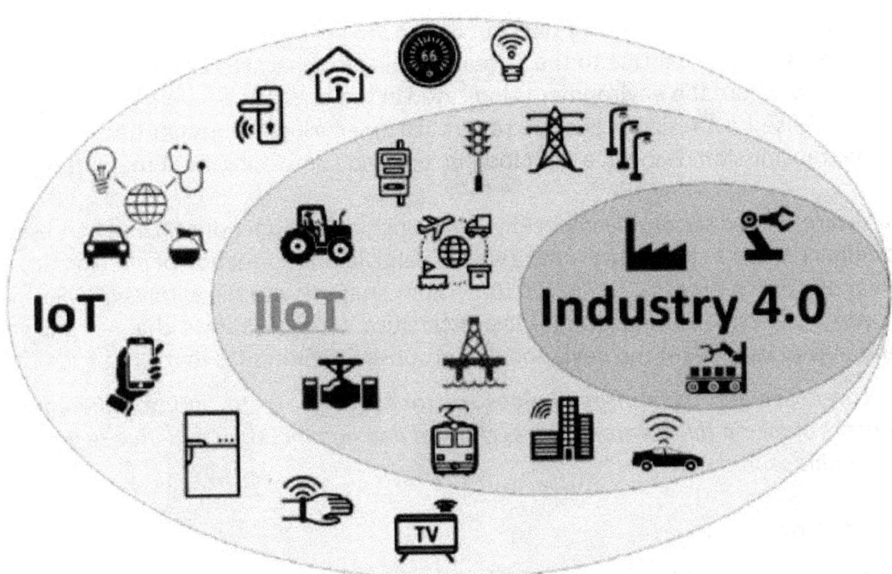

Figure 1: Linking the terms Internet of Things, Industrial Internet of Things and Industry 4.0. While the IoT primarily describes the interconnection of everyday objects, the IIoT also includes applications of critical infrastructures, means of transport (Hoffman, 2019)

The various concepts are accompanied by different threat scenarios and attack vectors, from which individual security requirements can be deduced. For instance, a smart light bulb in a home environment - switching lights on and off - does not require the same level of security as, say, controlling a robotic arm in an industrial environment - uncontrollable movements could injure/kill people. In addition to the requirements, the capabilities of the devices used are crucial. As already mentioned, they differ significantly in terms of memory, energy and computing power. A comprehensive overview of the constraints in the IIoT context is given by (Hoffman, 2019), for instance, whereas *Zou et al.* discuss security aspects in the area of radio communication (Zou, et al., 2016). *Schneier* (2015) gives a detailed insight into the field of cryptography in which he explains, for example, the differences between symmetric and asymmetric cryptography, together with the different requirements for their implementation. Especially with RedCap devices, asymmetric crypto and key exchange (Diffie & Hellman, 1976) has drawbacks that could be overcome by "modern" approaches.

A promising approach to meet these requirements is provided by Physical Layer Security (PhySec) methods, on which this work focuses. Therefore, in Section 2, a definition of the term is provided and the different aspects of the methods such as hardware related PhySec, solutions considering the wireless propagation channel and approaches which put the human body in the spotlight, are discussed. Since the authors of this work are involved in the development of the Sixth Generation (6G) Wireless Systems and consider its development as a major opportunity to incorporate PhySec methods into the corresponding standardization, Section 3 gives an insight into key enabling technologies that could lead to the breakthrough of PhySec algorithms. Section 4 discusses the approaches described and provides a classification in the security context. Finally, Section 5 concludes the work and gives an outlook for future work.

2. Physical Layer Security

In the conventional perspective on the subject of Physical Layer Security, derived from the bottommost layer of the Open System Interconnection (OSI) model, which refers to seven layers (Application-, Presentation-, Session-, Transport-, Network, Data Link-, and Physical Layer) (ITU, 1994), it describes a set of applications targeted specifically at the wireless communications channel. These applications are relevant because no security mechanisms are addressed in the OSI model, but are only added in the X.800 standard (ITU, 1991).

However, the basic principle, the utilization of inherently available information, does not only apply to the wireless channel, but can be transferred to other applications such as semiconductors or even the human body as well. That is why the authors have extended this term, as explained in Section 2.1.

2.1 Definition and classification

The topic Physical Layer Security is not a new one, but due to the proliferation of use-cases it is stepping towards the focus of researchers worldwide. Due to the "already available" characteristic and intrinsic features it offers benefits with respect to secure the environments indicated in the previous Section. As the purpose of the work, on which this paper is based, is the comprehensive approach in securing various entities –humans and machines- and in addition the transmission channel, a definition is required comprising all of the different approaches.

Based on the hardware related *silicon/electrical* Physically Unclonable Functions (PUFs), addressed in the Section below, *Halak* is defining them as a "physical entity whose behaviour is a function of its structure and the intrinsic variation of its manufacturing process" (Halak, 2018). In combination with the wireless PhySec perspective from *Hamamreh, Furqan and Arslan* to "exploit the characteristics of the wireless channel and its impairments" (Hamamreh, Furqan & Arslan, 2018) the deviation and proposed definition is that:

> *Physical Layer Security comprises various methods of how to utilize inherently present characteristics of media (hardware, wireless channel and human body) to derive secrets applicable as cryptographic primitives.*

2.2 The human factor

Although an increasing number of (I)IoT, I4.0 and Industrial Automation and Control Systems (IACS) procedures are automated - with M2S and M2M applications -, there is still a human factor involved. Starting with workers within the manufacturing sites to on-site service technicians to remote maintainers, different people are integrated into the processes and have to be identified, authenticated and authorized accordingly.

Even though biometric authentication is the simplest and oldest form of identification - even animals are able to recognize each other by smell, behaviour and voice (Schneier, 2014) -, the conditions in the (I)IoT/I4.0 context place special requirements on biometrics. On the one hand, many of the traditional authentication mechanisms such as fingerprints, iris- and facial recognition, or speech are not possible due to industrial safety equipment (helmet, goggles, gloves), and on the other hand, the general operating conditions have changed. Globalization has increased the distances for identification, although the verification is no longer carried out by humans alone; in short, the type of authentication has certainly changed.

In general, authentication is differentiated into the three areas *Knowledge, Possession* and *Inherence* factors, whereby biometrics (Jain, Ross & Prabhakar, 2004) are assigned to the *Inherence* factors. As indicated in **Figure 2,** these are in turn divided into *biological, physiological* and *behavioral* factors. In addition to the traditional methods mentioned above (fingerprinting, etc.), there are some interesting "modern" approaches, such as the Mechanomyogrm (MMG), whose mode of operation, the use of signals by muscle fibers movement is described, for instance, by *Pal, Gautam* and *Singh* (2015). *Bidgoly et al.* Examine the use of bio-electrical signals of the brain by recording Electroencephalogram (EEG) (Bidgoly, Bisgoly & Arezoumand, 2020), whereas *Sarkar, Abbott and Doerzaph* (2016) use the volumetric changes of blood in the peripheral areas via Photoplethymography (PPG).

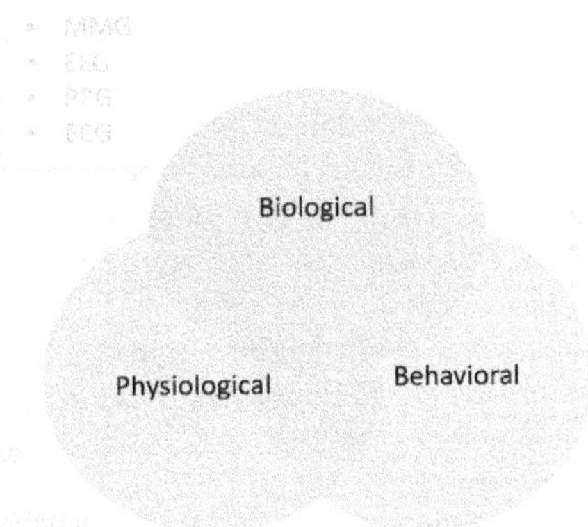

Figure 2*:* The three main categories of biometric authentication: biological, physiological and behavioral factors

Following the argumentation regarding PhySec from the previous Section, the working principle can be extended to the human body. The intrinsic, individual (bio-)physical characteristics can be used for authentication and for further cryptographic primitives. However, it is essential to consider that the human body is a living organism and is therefore continuously in a process of change: Cells are dying and new cells are being formed. Especially with regard to cryptographic applications, however, this entails certain requirements, since in such procedures a key needs to be identical every time it is required. This is where the capabilities of Artificial Intelligence (AI) in particular can contribute decisively and support the training of systems. The authors' work exemplifies the added value of such approaches and also the appropriateness of modern biometric methods.

Lipps, Herbst and *Schotten* (2021c), for instance, have developed a pressure-sensitive capable of distinguishing between people. Using an 18x6 sensor matrix, a footprint of 77 points is generated based on the form factor, which takes into account features such as pressure distribution within the foot, weight and step length. With a detection rate of about 60% after the first step, the initial setup already provides solid results, which will be further improved by i) the use of stitched conductive threads (instead of horizontal and vertical arrangement of conductive material) and ii) machine learning algorithms to train the system.

In another approach to using biometrics, *Lipps, Bergkemper* and *Schotten* (2021b) use Electrocardiogram (ECG) signals to derive conclusions about the individual. Therefore, data are derived by a 3-lead ECG, pre-processed due to noise via butterworth filtering and segmented according to the PQRST-waves. A comparison between the ML algorithms k-nearest Neighbor, Support Vector Machines (SVMs), and Gaussian Naive Bayes (GNB) indicates that the methods differ in their suitability for differentiation. For example, with a group of 20 people, SVMs can distinguish them with an accuracy of around 90%, whereas with KNN classifier only an accuracy of below 40% was achieved.

Since the individual methods already show good results with regard to the use of biometrics, it will be interesting to examine how the results are when combined and additional features are integrated. For this purpose, the authors are currently working on the use of EEG signals and the evaluation of behavior-based features.

2.3 Hardware-based Physically Unclonable Functions

In addition to humans, it is primarily the machines that are particularly relevant in the (I)IoT/I4.0 scenarios. However, the requirements for sensors, actuators and controls are different from those for human authentication, simply because machines themselves have different individual characteristics, but also have different validation mechanisms. Besides, already back in 2012, *Bonneau et al.* emphasized the importance to abolish and replace (traditional) passwords (Bonneau, et al., 2012).

This is another area where PhySec, and in particular the field of Physically Unclonable Functions (PUFs), provides a suitable solution which offers benefits compared to conventional hardware security mechanisms such as Trusted Platform Modules (TPMs) (Höller & Toegl, 2018), certificates (McLuskie & Belleken, 2018) or asymmetric cryptography (Schneier, 2015). In particular, due to the inherent availability and thus low acquisition cost - for example, Random Access Memory (RAM) is already built into almost all electronic devices -, and time until the primitives are available - a few nanoseconds -, PUFs are more than just an alternative to be considered.

Nevertheless, the overall idea is not completely new, as since the general description of the principle by *Gassend et al.* in 2002 (Gassend, et al., 2002), there are many different approaches how PUF derivatives can be used. According to *Ahr, Noushinfar* and *Lipps* (2021), a fundamental distinction of the approaches is possible into Memory-based and Timing-based PUFs. Memory-based include among others, Flip-Flop (Khan, et al., 2020), Butterfly (Kumar, et al., 2008) and SRAM-PUFs (Halak, 2018), whereas Arbiter- (Machida, et al., 2014), Ring-Oscillator- (Gao, et al., 2014) and Self-Timing PUFs are attributed to the Timing-based PUFs (Halak, 2018).

The authors' work focuses on memory-based PUFs, although here in the two different RAM types, static (see **Figure 3 a)**) and dynamic (see **Figure 3b)**). For instance, *Lipps et al.* provide an evaluation of the various external and internal influences to the SRAM-behaviour. Besides others, they examined the influences of the environmental temperature in a range between -10°C and 60°C to the Start-Up behaviour of SRAM cells. They observed fluctuations in the entropy of the stat-up values between an average of 0.87 (at -10°C) and an average of 0.96 (at +60°C). In contrast, the runtime before the restart, fluctuations in the supply voltage and the values stored in the non-volatile memory (NVM) before the restart had only little influence in their studies (Lipps, et al., 2018). In the studies of *Ahr, Lipps and Schotten* (2021), the stability of the cells could also be evaluated - in cryptographic applications, it is important that primitive cells are reproducible. In the examinations, 30 SRAMs were respectively read out 500 times and compared with regard to the stability of the start-up patterns. The results indicated that the cells are well suited for cryptographic applications with a stability of over 99.5%. In *Ahr, Noushinfar and Lipps* (2021) they highlight the differences between SRAMs and DRAMs and compare them, among others, in terms of availability, challenge-response pairs, reliability, and uniformity.

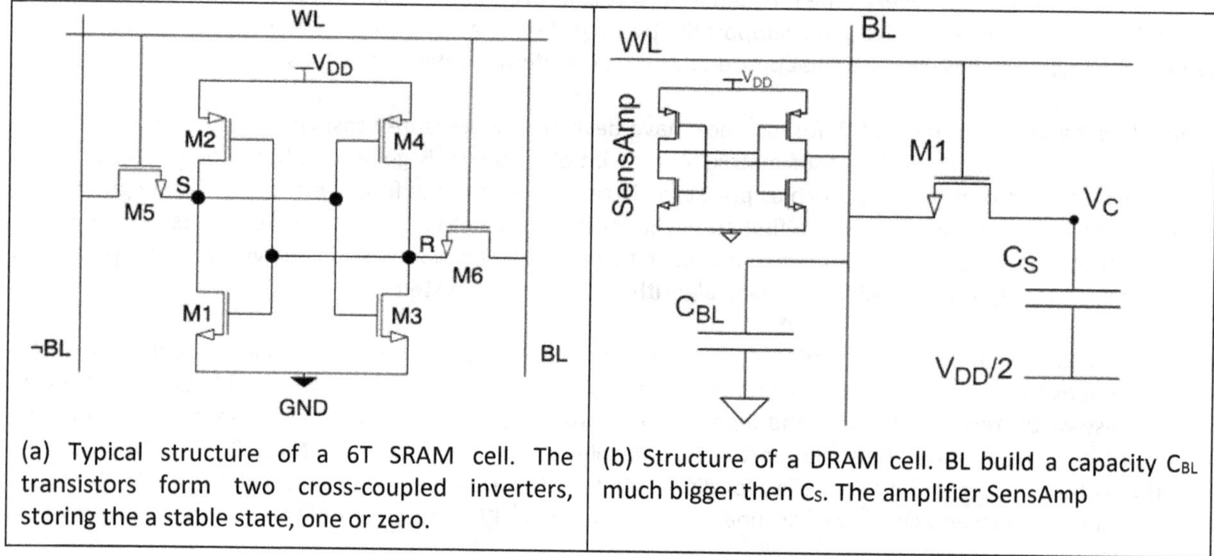

(a) Typical structure of a 6T SRAM cell. The transistors form two cross-coupled inverters, storing the a stable state, one or zero.

(b) Structure of a DRAM cell. BL build a capacity C_{BL} much bigger then C_S. The amplifier SensAmp

Figure 3: The structural design of SRAM and DRAM cells differs, among others, in the number of transistors used and their wiring

2.4 Wireless Secret Key Generation

The transmission channel does not only represent the connection between the addressed entities, the original definition of PhySec, respectively the Secret key Generation (SKG) (Ambekar & Schotten, 2014) and Channel-Reciprocity based Key Generation (CRKG) (Zenger, et al., 2015), mainly refers to wireless communication. Based on influences on the wireless channel - such as diffraction, refraction, reflection and scattering - values such as Received Signal Strength Indicator (RSSI), Reference Signal Received Power (RSRP) and Channel State Information (CSI) are measured on both sides of a channel according to the principle of channel reciprocity and used for cryptographic processes. Accordingly, this is a well-researched field, at least for the IEEE802.11 WLAN sector. For instance, *Zhang et al.,* developed a Wireless open-Access Research Plattform (WARP) based testbed to examine SKG on Wireless Local Area Network (WLAN) (Zhang, et al., 2016). Different channel types such as Multi-Antenna Channels, Multiple Access and Interference channels as well as broadcast channels are considered by (Mukherjee, et al., 2014). *Weinand, Karrenbauer and Schotten* (2018) propose a security architecture for Ultra-Reliable Low-Lattency Communication (URLLC) and a plug & trust protocol, based on PhySec algorithms.

The underlying SKG working principle, as depicted in **Figure 4**, is based on the five building blocks: Channel Measurement, Reciprocity Enhancement, Quantization, Information Reconciliation and Privacy Amplification; which are described in details for instance by (Lipps, et al., 2020). But as this is based on the propagation of electromagnetic waves, and therefore radio standard-independent, there is work transferring the "idea" to cellular systems, for instance. Here, *Lipps et al.* apply the approach in Cellular Systems as well. They were able to decrease the Bit Error Rate (BER) significantly by applying Machine Learning (ML) algorithms for the channel prediction. Considering various use-cases (static and mobile application) they evaluated the metrics RSSI and RSRP with respect to the BDR and were abel to reduce this from about 40% (RSSI static and mobile) to about 5% (RSSI Static mobile). Merely with the mobile RSSP the BDR could be reduced only to about 10% (the reasons for this are different, lie e.g. with interferences of the environment during the measurements) (Lipps, et al., 2020).

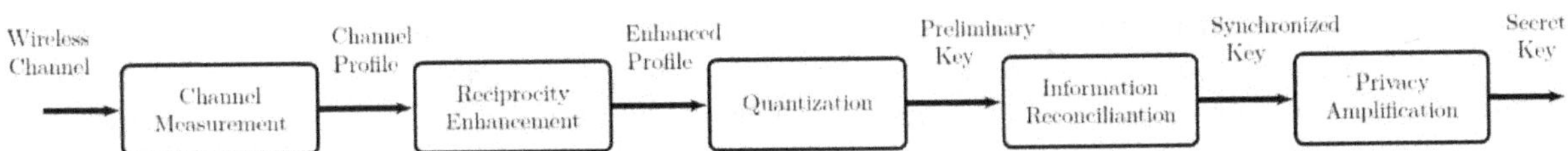

Figure 4: The Secret Key Generation building blocks

In addition to the existing radio standards, the application of PhySec will particularly be relevant during the development and subsequent standardization of the Sixth Generation Wireless Systems. As these are also wireless systems, they are discussed separately in Section 3.

3. Physical Layer Security and the Sixth Generation Wireless Systems

Despite the indicated suitability of Physical Layer Security methods and their wide potential for application, the methods have not yet become properly established as security mechanisms and unfortunately still lead a niche existence. However, the development of the Sixth Generation Wireless Systems and the (even) greater involvement of RedCab devices raises the opportunity to integrate PhySec algorithms into the standardization process. Moreover, many of 6G's claimed key enabling technologies correlate directly with PhySec methods: Reconfigurable Intelligent Surfaces (RISs), intended as passive "repeaters", open up the possibility of actively influencing the transmission channel; Wireless Optical Communication (WOC) has different properties than Radio Frequencies (RF), but still offers PhySec starting points, as known from fiber optics, for example; or the use of higher frequency bands up to the (sub)terahertz range, increases the entropy within the radio channel and thus increases the quality of the SKG key material.

3.1 Reconfigurable Intelligent Surfaces

Reconfigurable Intelligent Surfaces, Intelligent Reflecting Surfaces (IRS) or metasurfaces -depending on the scientific point of view- are "planar surfaces" (Pan, et al., 2021) composed of individually controllable passive reflecting elements. Through phase shifts or active beamforming/splitting they enable to actively influence the characteristic properties of the radio channel and thus to increase the entropy of the channel and at the same time to improve and strengthen the SKG generated keys (Lipps, et al., 2021).

3.2 Wireless Optical Communication

The visible light spectrum, in the range between 400-800 THz and wavelength range between 375-780nm, is used by the Visible Light Communication (VLC) as a subset of the WOC. One of the main advantages is the robustness against electromagnetic interference such as occurs in Radio Frequency (RF) communication. Unlike RF, however, light is hindered in its propagation by simple walls, for example, and offers advantages in terms of protection against eavesdropping. In particular, the combination with RIS offers additional opportunities to open up new use cases (Uysal, et al., 2016).

3.3 (sub)Terahertz Frequencies

Another possible way to meet the requirements of 6G is to push the communication into higher frequency ranges. The technological capabilities of (sub)terahertz communication allow signal bandwidths of more than 1GHz and data rates in the range of terabits per second. This is accompanied by advantages in terms of accurate positioning and sensing, which includes benefits in targeted communication (beamforming). Especially with regard to the entropy of the channel, on which the quality of the SKG is based, the higher frequency brings advantages up to a certain level, since the reflections further increase the entropy.

4. Discussion

Safeguarding systems, networks and end devices is becoming more and more important in an increasingly interactive world. However, with increasing complexity and heterogeneity, the requirements for the appropriate security mechanisms are growing as well. For this purpose, three different types of lightweight and inherently available and thus easy-to-implement and cost-effective security mechanisms have been described in this work. All of them with different strengths and weaknesses, even in comparison with conventional solutions. But one thing is always important to keep in mind in conjunction with the PhySec methods discussed herein: The methods can never be as strong as dedicated individual security mechanisms, but should always be considered as a complement and additional security feature.

In terms of modern biometrics, the proposed solutions such as gait-based recognition or ECG authentication offer a very simple but secure way to be used as an additional security feature, for instance, as additional contextual information. Otherwise, they are more uniquely linked to a particular person than passwords or tokens are. Passwords can be shared, tokens and mechanical keys can be stolen, which makes it easy to impersonate a fake person and gain unauthorized access. Shoes might be stolen as well, but it is almost impossible to imitate the gait profile of a person. There will always be deviations in stride length, weight and pressure distribution. Similar applies to ECG values, behavior-based features or EEG signals. In addition, biometric features (from fingerprints to gait profiles) contain more derivable information than, for example, a 128-bit password (and what user remembers a 128-bit password?).

A major advantage of the aforementioned methods is the inherent existence of the information. This applies in particular to the Physically Unclonable Functions. RAM is contained in almost all electronic devices. Certainly, there are powerful and technically sophisticated hardware modules and Trusted Platform Modules, but they all have to be introduced into an existing system from the outside. In addition, these are cost-intensive thus not economically feasible for (I)IoT/I4.0 applications that involve a large number of devices. So for "simple" security applications, PUF implementation are way more attractive. Similarly, certificates can contain sophisticated security chains, but they need to be i) inserted into a system from the outside and ii) require a non-volatile memory in the system on which they can be stored, which in turn is vulnerable to attacks. Like any of these systems, PUF security mechanisms require a validation system consisting of challenge-response pairs, for example. The prerequisite is the secure and uncompromisable storage of these pairs on the validator side (on device side the information can be generated on demand and does not have to be stored).

The advancing wireless networking and especially the just started development of the Sixth Generation Wireless Systems open up possibilities for attack vectors but also for perspectives for proposed solutions. The mentioned SKG methods provide approaches for lightweight cryptography based on inherently available information. In addition, there are advantages with regard to Perfect Forward Secrecy (PFS), as keys can be recalculated at definable intervals. However, it is also obvious that the methods require a minimum of entropy within the systems. Channel-based encryption is not possible in static and low-reflection channels that are completely free of interference and noise. However, this is where 6G technologies such as Reconfigurable Intelligent Surfaces could provide a suitable contribution.

Trustworthiness and security of the systems are becoming increasingly relevant, and the mechanisms mentioned can make a contribution to this, but they are not the solution for everything. In their respective application areas, they provide good to great results and are more than worthy of consideration as an alternative in the (I)IoT/I4.0 environments. The further development, especially of 6G, will demonstrate to what extent the methods will manage to find their way into standardization and become correspondingly applicable in everyday practice.

5. Conclusion and future work

The ongoing globalization and interconnection of all types of sensors and devices to form an (industrial) Internet of Things with Industry 4.0, Industrial Automation and Control Systems and Cyber-Physical Production Systems implies an increasing demand for secure and trustworthy communication. But, these heterogeneous landscape with its vast amount of entities -machines and humans- renders it a crucial task to identify a "one-size fits all" solution for security applications. Therefore, the respective requirements are simply too specific.

However, the methods of Physical Layer Security open up a variety of solutions, each with specific advantages tailored for the respective application: Physically Unclonable Functions as hardware-based approaches provide semiconductor-level device authentication, biometrical PhySec represent modern approaches for identifying and authenticating human entities and the Secret Key Generation, based on the propagation characteristics of electromagnetic waves, enables a symmetric encrypted communication independent of the radio standard. Nevertheless, the authors of this work are aware that the PhySec algorithms are not the answer to all problems, but they do provide a more than worthwhile and considerable complement to existing security mechanisms. They are inherently available, resource-saving and yet efficient and secure.

Here, the work provided insight into three different field of application for PhySec methods and discussed their benefits and drawbacks. The corresponding implementations, experimental results and evaluations are available in the referenced papers of the authors and demonstrate the operability and suitability of the described techniques.

Furthermore, the development of the Sixth Generation Wireless Systems with its promised key enabling technologies offers a substantial opportunity to bring PhySec out of its niche existence and to integrate them into the upcoming security standards.

Acknowledgements

This work has been supported by the Federal Ministry of Education and Research of the Federal Republic of Germany (Förderkennzeichen 16KISK003K, Open6GHub). The authors alone are responsible for the content of the paper.

References

Ahr, P., Lipps, C. and Schotten, H.D., "The PUF Commitment: Evaluating the Stability of SRAM-Cells", *European Conference on Cyber Warfare and Security,* Chester, United Kingdom, 2021.

Ahr, P., Noushinfar, M. and Lipps, C., 2021. "RAM-based PUFs: Comparing Static- and Dynamic Random Access Memory", *Workshop on Next Generation Networks and Applications*, Kaiserslautern, Germany, 2021.

Ambekar, A. & Schotten, H.D., "Enhancing Channel Reciprocity for Effective Key Management in Wireless Ad-Hoc Networks", *IEEE 79th Vehicular Technology Conference (VTC Spring).* Seoul, South Korea, DOI: 10.1109/VTCSpring.2012.7022913, 2014.

Bidgoly, A.J., Bisgoly, H.J. and Arezoumand, Z., "A survey on methods and challenges in EEG based authentication", *Computers & Security,* Band 93, DOI: 10.1016/j.cose.2020.101788, 2020.

Bonneau, J., Herley, C., van Oorschot, P. and Stajano, F., "The Quest to Replace Passwords: A Famework for Comparative Evaluation of Web Authentication Schemes", *IEEE Symposium on Security and Privacy*, DOI: 10.1109/SP.2012.44, 2012.

Diffie, W. and Hellman, M., "New directions in cryptography", *IEEE Transactions on Information Theory*, 22(6), 1976.

Djordjevic, I. B., "Physical-Layer Security and Quantum Key Distribution". 1st ed., Tucson, AZ, USA, Springer, ISBN: 978-3-030-27565-5, 2019.

Gao, M., Lai, K. and Qu, G., "A Highly Flexible Ring-Oscillator PUF", *Proceedings of the 51st Annual Automation Conference (DAC),* DOI: 10.1145/2593069.2593072, 2014.

Gassend, B., Clarke, D., can Dijk, M. and Devadas, S., "Silicon Physical random Functions", *Proceedings of the 9th ACM Conference on Computer and Communication Security*, pp. 148 -- 160, DOI:10.1145/586110.586132, 2002.

German Research Center for Artificial Intelligence, "Ten Years of Industrie 4.0 - Germany Driving Industrial AI as the means to Future Value Creation", [online] https://www.dfki.de/en/web/news/ten-years-of-industrie-4-0-interview-wolfgang-wahlster-cea-dfki, News 2021.

Halak, B., "Physically Unclonable Functions - From Basic Designn Principle to Advanced Haradware Security Apllications", 1st Edition Hrsg. Southampton: Springer, ISBN: 978-3-319-76804-5, 2018.

Hamamreh, J.M., Furqan, H.M. and Arslan, H., "Classifications and Applications of Physical Layer Security for Confidentiality: A Comprehensive Survey", *IEEE Couumnications Surveys & Tutorial,* DOI: 10.1109/COMST.2018.2878035, *2018.*

Hoffman, F., "Industrial Internet of Things Vulnerabilities and Threats: What Stakeholders Need to Consider", *Issues in Information Systems,* 20(1), pp. 119--133, 2019

Höller, A. and Toegl, R., "Trusted Platform Modules in Cyber-Physical Systems: On the Interference Between Security and Dependability", *IEEE European Symposium on Security and Privacy Workshops (EuroS&PW)*, DOI: 10.1109/EuroSPW.2018.00026, 2018.

International Telecommunication Union (ITU), "ITU-T X.200 - Data Networks and Open System Communications - Open Systems Interconnection, Model and Notation", International Telecommunication Union (ITU), 1994.

___, "X.800: Security Architecture for Open System Interconnection for CCITT Applications", *1991.*

___, "IMT Traffic estimates for the years 2020 to 2030", *M Series Mobile, radiodetermination, amateur and related satelliteservices Report ITU-R M.2370-0*, 1994b.

Jain, A., Ross, A. and Prabhakar, S., "An introduction to biometric recognition" *IEEE Transactions on Circuits and Systems for Video Technology,* 14(1), pp. 4 -- 20, DOI: 10.1109/TCSVT.2003.818349, 2004.

Khan, S., Shah, A.P., Chouhan, S.S., Gupta, N., Pandes, J.G. and Vishvakarma, S.K., "A Symmetric D Flip-Flop based PUF with improved Uniqueness", *Micoelectronic Reliability*, DOI: 10.1016/j.microel.2020.113595, 2020.

Kramp, T., van Kranenburg, R. and Lange, S., "Introduction to the Internet of Things", In: A. Bassi, et al. Hrsg. *Enabling Things to Talk: Designing IoT solutions with the IoT Architectural Reference Model.* Berlin, Heidelberg: Springer Berlin Heidelberg, pp. 1--10., 2013.

Kumar, S.S., Guajardo, J., Maes, R. Schrijen, G.-J. and Tuyls, P., "The Butterfly PUF Protecting IP on every FGPA". Anaheim, CA, USA, *IEEE International Workshop on Hardware-Oriented Security and Trust*, DOI:10.1109/HST.2008.4559053, 2008.

Lipps, C., Baradie, S., Noushinfar, M., Herbst, J., Weinand, A. and Schotten, H.D., "Towards the Sixth Generation (6G) Wireless Systems: Thoughts on Physical Layer Security", *Mobile Communication - Technologies and Applications - 25. VDE/ITG Fachtagung Mobilkommunikation*, 2021.

Lipps, C., Bergkemper, L. and Schotten, H.D., "Distinguishing Hearts: How Machine Learning identifies People based on their Heartbeat", *Sixth International Conference on Advances in Biomedical Engineering (ICABME),* Islamic University of Lebanon - Faculty of Engineering - Werdanyeh Campus, Lebanon, DOI:10.1109/ICABME53305.2021.9604855, 2021b.

Lipps, C., Herbst, J. and Schotten, H.D., "How to Dance Your Passwords: A Biometric MFA-Scheme for Identification and Authentication of Individuals in IIoT Environments", *16th International Conference on Cyber Warfare and Security (ICCWS)*, Cookeville, TN, USA, 2021c.

Lipps, C., Mallikarjun, S.B., Strufe, M., Heinz, C., Grimm, C., and Schotten, H.D., "Keep Private Networks Private: Secure Channel-PUFs, and Physical Layer Security by Linear Regression Enhanced Channel Profiles", *International Conference on Data Intelligence and Security (ICDIS),* DOI:10.1109/ICDIS50059.2020.00019, *2020.*

Lipps, C., Weinand, A., Krummacker, D., Fischer, C., and Schotten, H.D., "Proof of Concept for IoT Device Authentication Based on SRAM PUFs Using ATMEGA 2560-MCU" *1*st International Conference on Data Intelligence and Security (ICDIS), South Padre Island, TX, USA, 2018.

Machida, T., Yamamoto, D., Iwamoto, M. and Sakiyama, K., "A new Mode of Operation for Arbiter PUF to Improce Uniqueness on FPGA", *Federated Conference on Computer Science and Information Systems*, Warsaw, Poland, DOI:10.15439/2014F140, 2014.

McLuskie, D. and Belleken, X., "X509 Certificate Error Testing", *Proceedings of the 13th International Conference on Aailability, Reliability and Security,* DOI:10.1145/3230833.3232820, 2018.

Mukherjee, A., Fakoorian, A.A.A., Huang, J. and Swindlehurst, A.L., "Principles of Physical Layer Security in Multiuser Wireless Networks: A Survey", *IEEE Communications Surveys & Tutorials*, 16(3), pp. 1550 -- 1573, DOI: 10.1109/SURV.2014.012314.00178, 2014.

Pal, A., Gautam, A.K. and Singh, Y.N., "Evaluation of Bioelectric Signals for Human Recognition", *Procedia Computer Science,* Band 48, pp. 746 --752, DOI:10.1016/j.procs.2015.04.211, 2015.

Pan, C., Ren, H., Wang, K., Kolb, J.F., Elkashlan, M., Chen, M., Di Renzo, M., Hao, Y., Wang, J., Swindlehurst, A.L., You, X. and Hanzo, L., "Reconfigurable Intelligent Surfaces for 6G Systems: Principles, Applications, and Research Directions" *IEEE Communications Magazine*, 59(6), DOI: 10.1109/MCOM.001.2001076, 2021.

Posey, B., Rosencrance, L. and Shea, S., "Industrial Internet of Things (IIoT)", [Online] Available at: https://internetofthingsagenda.techtarget.com/definition/Industrial-Internet-of-Things-IIoT [Zugriff am 23 01 2022].

Sarkar, A., Abbott, A. L. and Doerzaph, Z., "Biometric authentication using photoplethysmography signals" *IEEE 8th International Conference on Biometrics Theory, Applications and Systems (BTAS),* Niagrara Falls, NY, USA, DOI: 10.1109/BTAS.2016.7791193, 2016.

Schäfer, G. and Rossberg, M., "Security in Fixed and Wireless Networks", 2nd Edition, Red Hat., ISBN: 978-1119040743, 2016.

Schneier, B., "Carry On: Sound Advice from Schneier on Security", 1st Edition, Wiley, ISBN: 978-1118790816, 2014.

___, "Applied Cryptography: Protocols, Algorithms and Source Code in C", 25th Anniversary Edition, John Wiley & Sons Inc, ISBN: 978-1119096726, 2015.

Uysal, M., Capsoni, C., Ghassemlooy, Z., Boucoucalas, and Udvary, E., "*Optical Wireless Communications: An Emerging Technology (Signals and Communication Technology)",* 1st ed., ISBN: 978-3319302003, 2016.

Valerio, P., "What is NR-REDCAP on 5G, and why is it important for IIoT?," [Online] Available at: https://iot.eetimes.com/what-is-nr-redcap-on-5g-and-why-is-it-important-for-iiot/ [Zugriff am 23 1 2022], 2021.

Weinand, A., Karrenbauer, M. and Schotten, H.D., "Security Solutions for Local Wireless Networks in Control Applications based on Physical Layer Security" *3rd IFAC Conference on Embedded Systems, Computational Intelligence and Telematics in Control CESCIT*, pp. 32 -- 39, 2018.

Zenger, C.T., Zimmer, J., Pietersz, M., Posielek, J.-F. and Paar, C., "Exploiting the Physical Environment for Securing the Internet of Things", *Proceedings of the 2015 New Security Paradigms Workshop*, pp. 44--58, DOI: 10.1145/284113.2841117, 2015.

Zhang, J., Woods, R., Duong, T.Q., Marshall, A., Ding, X., Huang, Y., Xu, Q., "Experimental Study on Key Generation for Physical Layer Security in Wireless Communications", *IEEE Access* , Issue 4, pp. 4464 -- 4477, DOI: 10.1109/ACCESS.2016.2604618, 2016

Zou, Y., Zhu, J., Wang, X. and Hanzo, L., "A Survey on Wireless Security: Technical Challenges, Recent Advances, and Future Trends", *Proceedings of the IEEE*, 104(9), pp. 1727 -- 1765, DOI: 10.1109/JPROC.2016.2558521, 2016.

On the Road to Designing Responsible AI Systems in Military Cyber Operations

Clara Maathuis
Open University of the Netherlands, Heerlen, The Netherlands
clara.maathuis@ou.nl

Abstract: Military cyber operations are increasingly integrating or relying to a specific degree on AI-based systems in one or more moments of their phases where stakeholders are involved. Although the planning and execution of such operations are complex and well-thought processes that take place in silence and with high velocity, their implications and consequences could be experienced not only by their targeted entities, but also by other collateral friendly, non-friendly, or neutral ones. This calls for a broader military-technical and socio-ethical approach when building, conducting, and assessing military Cyber Operations to make sure that the aspects and factors considered and the choices and decisions made in these phases are fair, transparent, and accountable for the stakeholders involved in these processes and the ones impacted by their actions and largely, the society. This resonates with facts currently tackled in the area of Responsible AI, an upcoming critical research area in the AI field that is scarcely present in the ongoing discourses, research, and applications in the military cyber domain. On this matter, this research aims to define and analyse Responsible AI in the context of cyber military operations with the intention of further bringing important aspects to both academic and practitioner communities involved in building and/or conducting such operations. It does that by considering a transdisciplinary approach and concrete examples captured in different phases of their life cycle. Accordingly, a definition is advanced, the components and entities involved in building responsible intelligent systems are analysed, and further challenges, solutions, and future research lines are discussed. Hence, this would allow the agents involved to understand what should be done, what they are allowed to do, and further propose and build corresponding strategies, programs, and solutions e.g., education, modelling and simulation for properly tackling, building, and applying responsible intelligent systems in the military cyber domain.

Keywords: cyber operations, cyber weapons, military operations, targeting, artificial intelligence, responsible AI.

1. Introduction

"Human technology starts with an honest appraisal of human nature. We need to do the uncomfortable thing of looking more closely at ourselves." (Tristan Harris)

Old conflicts continue in different forms and through new battles. These battles are conducted not only on conventional battlefields, but also in the information environment i.e., cyberspace. Therein, different actors i.e., state, non-state, and hybrid (Maathuis, Pieters & Van Den Berg, 2021) build skills/force for achieving goals through effective strategies. Over 100 states can launch military Cyber Operations against adversaries (Maathuis, Pieters & Van Den Berg, 2018) while having well-prepared cyber commandos and units (Smeets, 2018). In this process, intelligent technologies are used at increasing rate and scale for building intelligent systems to conducting military Cyber Operations (Brantly, 2016). Since AI is a disruptive technology containing a set of multiple technologies, one could say that is aligned with Thomas Edison's perspective on electricity: 'it is a field of fields...it holds the secrets which will reorganize the life of the world' (Schmidt et al, 2021). Thus, AI changes the world (NATO, 2021) and relationships between humans and machines, diffuses rapidly and broadly (Schmidt et al, 2021), and does these inside and through its natural environment i.e., cyberspace (Hartmann & Giles, 2020) no matter if adaptation to world's problems can be difficult since human intelligence processes are complex.

AI applications for military Cyber Operations are reconfiguring the action of an intelligent-cyber weapon if the state of an exploited vulnerability is changed by dynamically finding and exploiting another vulnerability, adapting weapon's action for limiting/avoiding unintended effects on collateral actors (Cox et al, 2019), or conducting proportionality assessment (Martellini & Trapp, 2020). However, such complex activities require vast-amounts of data, high computing power, up-to-date intelligence, advanced process knowledge, and compliance to the applicable legal-ethical frameworks. Ultimately, the ones responsible for targeting decisions are military Commanders, meaning that if they knew or should have known that the weapon used would produce massive collateral damage on civilian side, they should be responsible (Hallao et al, 2017). Additionally, AI systems are software-based, thus vulnerable to attack vectors (Reding & Eaton, 2020) through exploiting e.g., unknown software vulnerabilities, unproper communication defence, or failure of critical processes.

While the existing body of knowledge contains a rich plethora of studies relevant for grasping ethical aspects in military Cyber Operations, to the best of our knowledge, concrete definitions and assessments of challenges and corresponding solutions lack. This is the knowledge gap that this research tackles through transdisciplinary research in military Cyber Operations, military operations, AI ethics, and RAI fields where extensive literature review on scientific resources (e.g., IEEE publications/standards) and governmental resources (e.g., NATO and EU-Commission) was conducted focusing on the concepts, methods, and techniques relevant for building and conducting military Cyber Operations. Hence, following research objectives are addressed:

1. To propose a definition for RAI when building and conducting military Cyber Operations.
2. To propose an analytical model that captures the entities involved in these processes.
3. To structure and analyse challenges and recommendations for integrating RAI systems in these processes.

The remainder of this article is structured as follows. Section 2 presents related studies that consider diverse aspects for designing RAI systems when building and conducting military Cyber Operations. Section 3 proposes a definition for RAI and an analytical model in military Cyber Operations. Section 4 discusses challenges encountered when building and using RAI systems and presents recommendations applicable when using them in military Cyber Operations. Section 5 presents concluding remarks and future research ideas.

2. Related research

Research and practitioner communities formulate relevant questions and seek to build intelligent systems with a good purpose while being aware of their possible negative impact which should be prevented or eliminated when signals of its presence are detected. Hereof, Zhu et al (2022) stress that building and conducting AI-based military operations raises concerns on ethical risks associated, thus critical from a humanitarian standpoint. Additionally, the authors mention benefits like increasing accuracy and precision for decision-making, intelligence and targeting activities: facts of major importance in military Cyber Operations. Furthermore, Hartmann & Giles (2020) emphasize that due to increased data availability, computing power, and publicly available tools, cyber offenders can use successfully intelligent techniques that reach large audiences and produce significant harm e.g., deepfakes and artificial humanoid disinformation campaigns. Thusly, Hallaq et al (2017) envisage that future cyber strategies rely on AI while considering corresponding ethical issues and legal questions. These points call for diving into relevant aspects from the ongoing research and practitioner perspectives in the military ethics, AI Ethics, and RAI fields.

Dobos (2020) argues for understanding relevant aspects like power, conflict dynamics, moral conditioning and damage in war context. Furthermore, Finney & Mayfield (2018) point the importance of properly expressing self-awareness and an ethical code of behaviour e.g., the fiduciary duty of military officers when conducting military operations. Moreover, Kaurin (2016) analyse warriors' meaning in contemporary warfare which encapsulates warriors' personal identity, demands on them, and experience: aspects relevant when capturing and embedding human values when building RAI in military Cyber Operations. Petrozzino & Shapiro (2020) recommend the following actions for achieving ethical AI systems: i) creating ethical principles that drive organizational policies for supporting ethical analysis and open dialogue, ii) creating training and awareness on role-based AI ethics for leaders, policymakers, developers, and users, and iii) establishing diverse multidisciplinary AI teams to analyse ethical aspects from multi-stakeholder perspective. Canca (2020) considers that ethical principles are formed regarding autonomy, beneficence for avoiding harm and doing good, plus justice. Since responsibility has multiple meanings, Cheng, Varshney & Liu (2021) address it broader through social responsibility of AI i.e., human-value driven process where values like fairness, transparency, accountability, responsibility, safety, privacy and security, and inclusiveness are the principles, while designing socially responsible AI algorithms is the means. Peters et al (2020) propose two conceptual frameworks for integrating ethical analysis in engineering practices: the first considers integrating wellbeing support and ethical impact analysis in each engineering phase, and the second argues for wellbeing supportive design while reflecting and structuring ethical analysis. For managing AI ethical aspects through educating AI systems, Baker-Brunnbauer (2021) scrutinizes that the systems could be implicit ethical being forced preventing unethical results, explicit ethical by explicitly pointing the actions allowed/not allowed, and full ethical by benefiting free will and intention while having consciousness. Brundage et al (2020) consider institutional, software, and hardware mechanisms for building trustworthy AI systems: institutional for shaping or clarifying the incentives of people involved, software for embedding or enhancing interpretability, privacy-preserving aspects of AI systems, and hardware for securing hardware systems and processes.

IEEE developed the IEEE 7000 – 2021 standards for tackling ethical concerns during system design like the IEEE P7001 on Transparency of Autonomous Systems for developing autonomous systems able to assess own actions and understand decisions made, and IEEE P7002 on Data Privacy Process for managing privacy issues for systems collecting personal data (IEEE P7000, 2021). As Cyber Operations are software-based activities, relevant principles, guidelines, and methodologies could be proposed following such standards. Accordingly, EU aims to turn Europe into a hub for trustworthy AI as the Commissioner Thierry Breton said: "AI is a means, not an end...Today's proposals aim to strengthen Europe's position as a global hub of excellence in AI from the lab to the market" (EU Commission, 2021a). Hence, the European Commission came forward with useful programs and strategies like AI strategy, Coordinated Plan on AI, and Data Governance Act (EU Commission, 2021b). Particularly, the European Commission established seven key requirements for assuring trustworthy AI: human agency and oversight, technical robustness and safety, privacy and data governance, transparency, diversity, non-discrimination and fairness, societal and environmental well-being, and accountability (EU Commission, 2019). Moreover, NATO (2020a)'s Deputy Secretary General Mircea Geoana argues that 'there are considerable benefits of setting up a transatlantic digital community operating on AI and emerging and disruptive technologies, where NATO can play a key role as a facilitator for innovation and exchange'. NATO stresses that a dynamic adoption of new technologies like AI and their responsible governance are fundamentally important (NATO, 2020b). From the same angle, the U.S. DoD campaigns for the adoption of AI ethical principles in (non-)combat functions for upholding legal, ethical, and policy commitments in this domain. Accordingly, DoD 'will exercise appropriate levels of judgement and care, while remaining responsible' for building and using AI capabilities, plus equitable, traceable, reliable, and governable (U.S. DoD, 2020).

The above discussed studies contribute to defining RAI for military Cyber Operations and to identifying and analysing challenges and recommendations embedding academic and practitioner perspectives.

3. Definition

Since the beginning of AI, people were interested formulating questions of not only technical nature, but also ethical trying to propose answers to aspects like its capability to emulate or surpass human intelligence, design choices, and the meaning, scale, and severity of its (mis)use (Russell & Norvig, 2021). AI is changing 'the face and pace of warfare' and could be used responsibly as force multiplier to support military decision-making processes through accuracy, precision, speed, and easier integration in other battlefields (Meritalk, 2021) e.g. target localization with network/communication information and access point or even broader through a common operating picture, automatically detecting target's vulnerabilities and building corresponding exploits for efficient engagement, and collateral damage prevention on civilian infrastructure (Slayer, 2020), or using intelligent decision making support system for proportionality assessment and targeting decisions (Maathuis, Pieters & Van Den Berg, 2021). However, until now responsibility was indirectly tackled in military Cyber Operations through notions like 'attack', 'target', and 'proportionality' mainly through legal lenses. This is the underlying motivation of this article as responsibility does not only imply considering, interpreting, and integrating principles and norms, but also socio-ethical values when building military Cyber Operations while taking precautionary measures for preventing, containing, limiting, and avoiding unintended effects (Agarwal & Mishra, 2021). Correspondingly, the underlying questions would be: How to build responsible AI-based systems and solutions in respect to principles, norms, and values when developing and conducting military Cyber Operations? And, as Dignum (2019) suggests: Who or what should be responsible for AI-based systems' decisions and actions? Can an AI-based system be accountable for its actions? To find answers for such critical questions, a proper definition for RAI in this domain is required while considering specific characteristics of cyberspace e.g., being able to directly influence and impact other battlefields/domains (Brantly, 2016). Hence, Dignum (2019) calls for a human-centred approach focused on human well-being and alignment with socio-ethical values and principles.

Taking a responsible stance implies incorporating ethics in AI systems i) in design through regulatory and engineering processes that support the design and evaluation, ii) by design through established behaviour of AI systems, and iii) for designers through codes of conduct, regulatory requirements, standards, and certification processes (Dignum, 2019). The author defines RAI as 'the development of intelligent systems according to fundamental principles and values.' Similarly, Agarwal & Mishra (2021) consider that to assure the applicability, repeatability, and success of RAI systems, corresponding aspects should be integrated during their whole life cycle. Therefore, we formulate the following definition for RAI in military Cyber Operations respecting existing studies (Dignum, 2019; Agarwal & Mishra, 2021; Cheng, Varshney & Liu, 2021; Maathuis, 2022):

RAI in the military cyber domain = a sub-field of AI that deals with the integration of socio-ethical and legal principles, norms, and values when designing, developing, deploying, and using AI methods, techniques, and technologies embedded in different military cyber systems and processes.

This means that a series of agents communicate and collaborate for building military cyber tools for developing and conducting military Cyber Operations, process depicted in an analytical model in Figure 1 with its components addressed below (DARPA, 2016; Dignum, 2019; Maathuis, 2022):

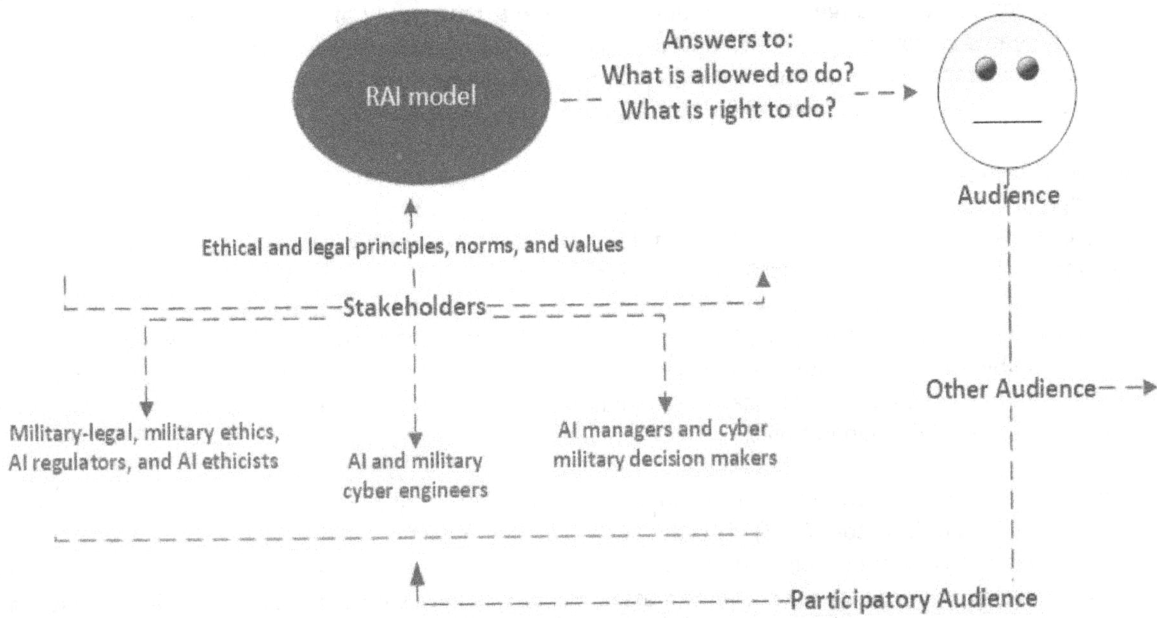

Figure 1: RAI in military cyber operations

Agents: entities participating in the design, development, deployment, and use of RAI solutions in military Cyber Operations. They are further classified considering their position:

4. *Stakeholders*: agents involved either in the process of i) design, deployment, use, standardization, and certification of the model i.e., military-legal, military-ethics, AI regulators, and AI ethicists, ii) theorizing, designing, developing, evaluating, upgrading, deploying the model, i.e. AI and military cyber engineers, or iii) design, development, deployment, and use while making sure that the model is compliant with external requirements i.e. AI managers and military cyber decision makers.
5. *Audience*: agents involved in stakeholders' processes i.e., are participatory audience or end-users, so the other audience.

The RAI model: developed AI model by corresponding agents whose life-cycle process and aim answers the following questions: What is allowed to do? and What is right to do? Important to mention is that agents have the responsibility, opportunity, and power to positively influence model's behaviour (Galliott, MacIntosh & Ohlin, 2020).

4. Challenges and recommendations

AI became an important strategic topic, and many countries are investing significant budgets in different R&D programs concerning upcoming and future trends and systems (EDA, 2021). Special attention is showed to building trustworthy, accountable, and responsible AI systems while reflecting on existing and foreseeable challenges (possibly) occurring in the lifecycle of RAI systems. This is vital when applied to military Cyber Operations since would mean e.g., mismatching a target implies spreading an intelligent cyber weapon in fractions of seconds at global scale and producing massive collateral damage on civilian infrastructure and vital processes, or directly affecting operational military processes of friendly and neutral countries. Thusly, it is necessary to understand and assess what are the challenges encountered when building RAI systems in military Cyber Operations, and from there analyse recommendations to tackle them. Hence, we consider the following categories of challenges and corresponding recommendations:

Education and Expertise: the lack of expertise, and implicitly education for properly integrating the aspects required at technical, social, and ethical levels into the design, implementation, and use of RAI systems in military Cyber Operations, could have (in)direct consequences on the built-system and other (un)related systems. It all begins with education, and more exactly, relevant and effective education. Then, the agents involved (e.g., military decision-makers) would benefit from individual and/or collective tailored training and education during mandatory training, as modular curriculum when joining/partnering with military forces, or as exchange curriculum between defence and commercial partners using gaming and simulation tools e.g., VR, AR, digital twins, or agent-based for target selection and engagement, that would allow understanding, capturing, and learning human's behaviour and values while building military Cyber Operations scenarios for effectively orienting, understanding, and acting in settings mirroring real-live contexts and environments (Dubber, Pasquale & Das, 2020; Meier et al, 2021; Reding & Eaton, 2020; Slayer, 2020).

Data: while symbolic AI systems rely on knowledge, non-symbolic AI system rely on data. Accordingly, knowledge and data are structured, represented, and further worked with as basis for understanding and tackling existing/future problems and unpredictable events that could occur considering e.g., network dynamism of cyberspace or unpredictable behaviour of AI systems as data might be errored, biased, or manipulated. Such facts could alter AI system's behaviour through unknown backdoors that allow e.g., disrupting own communication systems or improperly localizing a target (Dignum, 2019; Krasev, 2020). Then, aspects like data quality should be assured for solutions with sufficient data, and correctly balancing data e.g., oversampling with qualitative and representative technical and human-value data for solutions with scarce data like targeting decisions in military Cyber Operations.

Security: the over-reliance on AI systems conducts to an adversarial AI arms race and introduces new types of vulnerabilities (Reding & Eaton, 2020). AI-cyber vulnerabilities reflect combined and even extended cyber and AI risks to systems implemented e.g., data poisoning using open-source data possibly intentionally corrupted used for detecting advanced forms of cyber threats on military cyber systems or intelligent malware that changes its behaviour to be perceived as a legitimate behaviour and strike back into the network from where an intelligent cyber weapon was launched (Martellini & Trapp, 2020). As Norbert Wiener said: 'We had better be quite sure that the purpose put into the machine is the purpose which we really desire'. Then, intelligent systems able to both strike and defend themselves through online or hybrid learning and adaptive behaviour going through intense verification and testing processes at software, hardware, communication, and human levels represent a solution.

Cyberspace particularities: as these operations are conducted inside/through a multi-cross domain i.e., cyberspace which is dynamic, volatile, and still anonymity-friendly, then multi-domain and multi-source behavioural and value data are necessary to creating the proper picture to the agents involved in their execution along with a solid understanding on the processes involved and the effects assessed to the policy makers involved (Branthly, 2016; Slayer, 2020; Maathuis, 2021).

Trust: are issues between humans while building AI systems due to unclear, unfair, or unexpected ways of tackling the aspects and values that should be integrated, and trust issues between the humans and the AI systems implying the reliability and power of predictability of AI systems. Hence, too much trust might expose to strong unexpected behaviour and too less trust might imply using too exigent control mechanisms which would still be exposed to unexpected behaviour of AI systems (Martellini & Trapp, 2020; Bartneck et al, 2021). Then, communication and collaboration between the agents involved when building AI systems for conducting military Cyber Operations, are needed while actively integrating in a fair process all their relevant elements e.g., researchers, developers, manufacturers, technologies (Cox, 2019; Dignum, 2019; Maathuis, 2022).

The Tragedy of Metrics: statement that aims to capture and extend the classical meaning of the word 'metrics' by adding socio-ethical norms and values that AI systems should respect (Dignum, 2019). The metrics should consider not only technical and military-(ethical and legal) dimensions when building solutions for conducting military Cyber Operations, but also other social and ethical dimensions and aspects e.g., the context, aim, environment, human behaviour, rules and regulation.

Governance and Regulation: currently no specific/dedicated regulation exist for building and conducting AI-based military Cyber Operations, and this is necessary as AI is a dual-use technology that requires and impacts not only defence and industry stakeholders, but society as a whole. However, considering the tendency in the

AI domain and the upcoming awareness in the military domain towards building, using, and assessing intelligent systems on e.g., cognition, interaction, well-being, dedicated incentives in programs for analysing the suitability of current legal frameworks to intelligent systems, and their interpretation and possible adaptation to them should be considered through constructive and positive lenses while seeing intelligent systems as artefacts (Dignum, 2019). This implies collaboration between agents involved when building and using AI systems using a human-centred approach, and the further consideration of third-party RAI certification, RAI auditing, and risk management processes for implementation, testing, and approving AI systems while adopting specific principles, norms, and values in each phase of the life cycle of AI systems, plus sharing problematic incidents involving RAI systems. This further calls for diplomatic solutions for establishing international dialogue and joint of forces for developing common/compatible legal frameworks for RAI systems with defence and industry partners respecting frameworks like IHL, Human Rights Law, and societal norms and values (Hallaq et al, 2017; Petrozzino & Shapiro, 2020; Shneiderman, 2020; EU Commission, 2021a ; Schmidt et al, 2021).

Design: since existing AI systems integrated in military Cyber Operations do not consider yet a responsible approach, for the upcoming ones, to assure the effectiveness of their implementation, responsible considerations should be integrated using methods like Value Sensitive Design, Data/Design Science Research while developing and adopting a code of conduct for AI systems respecting human values and ethical considerations captured both qualitatively and quantitatively in the design, development, deployment, and use of AI systems (Dignum, 2019; Agarwal & Mishra, 2021; Zhu et al, 2022) while being protective to environment (Galliott, MacIntosh & Ohlin, 2020). This allows translating agents' values into AI development and establishing concrete features like integrating conditions or duties for limiting civilian harm, required actions like target engagement only if the conditions are satisfied, and preferences like system training for a good purpose with realistic cases. Moreover, this allows going back to a particular step if a test case (e.g., bias) fails and update the system (Anderson & Anderson, 2014; Burkhardt, Hohm & Wigley, 2019; Agarwal & Mishra, 2021).

Developments: the fact that the AI research community is somehow divided between current technologies focusing on the now and near-term AI, and future implications and technologies focusing on long-term AI based on AGI and superintelligence i.e., radical transformation of AI, creates a gap between these communities which calls for joint effort for tackling existing and emerging security problems having an eye on near and long-term future (Prunkl & Whittlestone, 2020). Hence, what would that imply and mean for targeting decisions and effects assessment in miliary Cyber Operations?

5. Conclusions

Approaching AI systems in military Cyber Operations through techno-ethical lenses allows the stakeholders involved to understand the difference between what they have right to do and what is right to do (Pottery Stewart). In this digital decade (EU Commission, 2019) and further from here since these operations are carried out at fast speeds, in silence, and embed solutions with different autonomy degrees while assessing potential risks and taking corresponding precautions (Morgan et al, 2018), it is important to accelerate education, investments, democratization, and adoption of AI systems from their design to incorporate relevant norms and values while having realistic military objectives that imply avoiding/limiting harm and embracing good purposes (Maathuis, 2022).

Hence, we present a definition and analytical model for RAI applied in military Cyber Operations, and from there analyse the challenges encountered by the agents involved and further draw recommendations that would facilitate the adoption, support, and strengthening of RAI systems in military Cyber Operations focusing on their development and execution. However, as this research focuses on the theoretical foundation of and corresponding instantiations of this topic, it further argues for involvement of academic and industry communities for properly implementing AI-based military Cyber Operations in respect to legal and ethical dimensions, and continues by assessing them for their integration in targeting decisions and controlling, limiting, and avoiding unintended effects of military Cyber Operations on military and civilian stakeholders for assuring the design, implementation, and use of trustable, accountable, and responsible intelligent systems having in mind that 'humans cannot be everywhere at once, but software can" (Schmidt et al, 2021).

References

Agarwal, S. and Mishra, S. (2021). Data and Model Privacy. In *Responsible AI*, pp. 153-170.

Anderson, M. and Anderson, S. L. (2018). GenEth: A general ethical dilemma analyzer. *Paladyn, Journal of Behavioral Robotics*, Vol. 9, No. 1, pp. 337-357.

Baker-Brunnbauer, J. (2021). Management perspective of ethics in artificial intelligence. *AI and Ethics*, Vol. *1, No.* 2, pp. 173-181.

Bartneck, C., Lütge, C., Wagner, A. and Welsh, S. (2021). Risks in the Business of AI. In *An Introduction to Ethics in Robotics and AI*, pp. 45-53.

Brantly, A. F. (2016). *The decision to attack military and intelligence cyber decision-making*. University of Georgia Press.

Brundage, M. et al. (2020). Toward trustworthy AI development: mechanisms for supporting verifiable claims. *arXiv preprint arXiv:2004.07213.*

Burkhardt, R., Hohn, N. and Wigley, C. (2019). Leading your organization to responsible AI. *McKinsey Analytics.*

Canca, C. (2020). Operationalizing AI ethics principles. *Communications of the ACM*, Vol. 63, No. 12, pp. 18-21.

Cheng, L., Varshney, K. R. and Liu, H. (2021). Socially responsible ai algorithms: Issues, purposes, and challenges. *arXiv preprint arXiv:2101.02032.*

Cox, J., Bennett, D., Lathrop, S., Walls, C., LaClair, J., Tracy, C. and Esquibel, J. (2019). The Friction Points, Operational Goals, and Research Opportunities of Electronic Warfare and Cyber Convergence. *The Cyber Defense Review*, Vol. *4*, No. 2, pp. 81-102.

DARPA. (2016). "Explainable Artificial Intelligence", [online], https://www.darpa.mil/program/explainable-artificial-intelligence.

Dignum, V. (2019). *Responsible artificial intelligence: how to develop and use AI in a responsible way*. Springer Nature.

Dobos, N. (2020). *Ethics, Security, and the War Machine: The True Cost of the Military*. Oxford University Press.

Dubber, M. D., Pasquale, F. and Das, S. (Eds.). (2020). *The Oxford handbook of ethics of AI*. Oxford Handbooks.

European Defence Agency.

EU Commision, E. (2019). Communication from the Commission of the European Parliament, the Council, the European Economic and Social Committee and the Committee of the Regions Empty.

EU Commision, E. (2021a). Europe Fit for the Digital Age: Commission Proposes New Rules and Actions for Excellence and Trust in Artificial Intelligence.

EU Commision, E. (2021b). A European Approach to Artificial Intelligence.

Finney, N. K. and Mayfield, T. O. (Eds.). (2018). *Redefining the modern military: The intersection of profession and ethics*. Naval Institute Press.

Galliott, J., MacIntosh, D. and Ohlin, J. D. (Eds.). (2020). *Lethal Autonomous Weapons: Re-Examining the Law and Ethics of Robotic Warfare*. Oxford University Press.

Hallaq, B., Somer, T., Osula, A. M., Ngo, K. and Mitchener-Nissen, T. (2017). Artificial intelligence within the military domain and cyber warfare. In *Proceedings of the 16th European Conference on Cyber Warfare and Security*, pp. 153-157.

Hartmann, K. and Giles, K. (2020). The Next Generation of Cyber-Enabled Information Warfare. In *2020 12th International Conference on Cyber Conflict,* pp. 233-250. IEEE.

IEEE (2021). IEEE Ethics in Action in Autonomous and Intelligent Systems.

Karasev, P. A. (2020). Cyber Factors of Strategic Stability. *Russia in Global Affairs*, Vol. 18, No. 3, pp. 24-52.

Kaurin, P. M. (2014). *The warrior, military ethics and contemporary warfare: Achilles goes asymmetrical*. Ashgate Publishing, Ltd..

Maathuis, C., Pieters, W. and Van Den Berg, J. (2018). A computational ontology for cyber operations. In *Proceedings of the 17th European Conference on Cyber Warfare and Security*, pp. 278-288.

Maathuis, C., Pieters, W. and Van Den Berg, J. (2021). Decision support model for effects estimation and proportionality assessment for targeting in cyber operations. *Defence Technology*, Vol. *17*, No. 2, pp. 352-374.

Maathuis, C. 2022. On Explainable AI Solutions for Targeting in Cyber Military Operations. *In Proceedings of the 17th International Conference on Cyber Warfare and Security*.

Martellini, M. and Trapp, R. (Eds.). (2020). *21st Century Prometheus: Managing CBRN Safety and Security Affected by Cutting-Edge Technologies*. Springer Nature.

Meier, R., Lavrenovs, A., Heinäaro, K., Gambazzi, L. and Lenders, V. (2021). Towards an AI-powered Player in Cyber Defence Exercises. In *2021 13th International Conference on Cyber Conflict,* pp. 309-326. IEEE.

Meritalk (2021). "Austin: DoD to Invest $1.5 Billion in DARPA AI Projects Over Five Years", [online], https://www.meritalk.com/articles/austin-dod-to-invest-1-5-billion-in-darpa-ai-projects-over-five-years/

Morgan, F. E., Boudreaux, B., Lohn, A. J., Ashby, M., Curriden, C., Klima, K. and Grossman, D. (2018). Military Applications of Artificial Intelligence. *Ethical Concerns in an Uncertain World. RAND Corporation.*

NATO (2020a). "Cooperation on Artificial Intelligence will boost security and prosperity on both sides of the Atlantic", [online], https://www.nato.int/cps/en/natohq/news_179231.htm

NATO (2020b). "Artificial Intelligence at NATO: dynamic adoption, responsible use", [online], https://www.nato.int/docu/review/articles/2020/11/24/artificial-intelligence-at-nato-dynamic-adoption-responsible-use/index.html

NATO (2020c). NATO Advisory Group on Emerging Disruptive Technologies. NATO Annual Report 2020.

NATO (2021). "Emerging and Disruptive Technologies", [online], https://www.nato.int/cps/en/natohq/topics_184303.htm

Peters, D., Vold, K., Robinson, D. and Calvo, R. A. (2020). Responsible AI—two frameworks for ethical design practice. *IEEE Transactions on Technology and Society*, Vol. *1*, No. 1, pp. 34-47.

Prunkl, C. and Whittlestone, J. (2020, February). Beyond near-and long-term: Towards a clearer account of research priorities in AI ethics and society. In *Proceedings of the AAAI/ACM Conference on AI, Ethics, and Society*, pp. 138-143.

Petrozzino, C. and Shapiro, S. (2020). *Actionable Ethics for Fairness in AI*. MITRE CORP MCLEAN VA.

Reding, D. F. and Eaton, J. (2020). *Science and Technology Trends 2020 2040: Exploring the S and T Edge*. NATO S and T Organization.

Russell, S. and Norvig, P. (2021). Artificial Intelligence: A Modern Approach, Global Edition 4th.

Schmidt, E. et al (2021). *National Security Commission on Artificial Intelligence (AI)*. National Security Commission on Artificial Intelligence.

Shneiderman, B. (2020). Bridging the gap between ethics and practice: Guidelines for reliable, safe, and trustworthy Human-Centered AI systems. *ACM Transactions on Interactive Intelligent Systems*, Vol. 10, No. 4, pp. 1-31.

Slayer, K. M. (2020). *Artificial Intelligence and National Security*. Congressional Research SVC Washington United States.

Smeets, M. (2018). Integrating offensive cyber capabilities: meaning, dilemmas, and assessment. *Defence Studies*, Vol. 18, No. 4, pp. 395-410.

U.S. Department of Defense (2020). "DoD Adopts Ethical Principles for Artificial Intelligence", [online], https://www.defense.gov/News/Releases/Release/Article/2091996/dod-adopts-ethical-principles-for-artificial-intelligence/

Zhu, L., Xu, X., Lu, Q., Governatori, G. and Whittle, J. (2022). AI and Ethics—Operationalizing Responsible AI. In *Humanity Driven AI*, pp. 15-33.

Responsible Digital Security Behaviour: Definition and Assessment Model

Clara Maathuis[1] and Sabarathinam Chockalingam[2]
[1]Open University of the Netherlands, Heerlen, Netherlands
[2]Institute for Energy Technology, Halden, Norway
clara.maathuis@ou.nl
sabarathinam.chockalingam@ife.no

Abstract: Digital landscape transforms remarkably and grows exponentially tackling important societal challenges and needs. In the modern age, futuristic digital concepts are ideated and developed. These digital developments create a diverse pallet of opportunities for organizations and their members like decision makers and financial personnel. Simultaneously, they also introduce different factors that influence users' behaviour related to digital security. However, no method exists to determine whether users' behaviour could be considered responsible or not, and in case this behaviour is irresponsible, how it could be managed effectively to avoid negative consequences. Thus far, no attempt was made to investigate this to the best of our knowledge. Then this research aims to: (i) introduce 'responsible digital security behaviour' notion, (ii) identify different factors influencing this behaviour, (iii) design a Bayesian Network model that classifies responsible/irresponsible digital security behaviour considering these factors, and (iv) draw recommendations for improving users' responsible digital security behaviour. To address these, extensive literature review is conducted through technical, ethical, and social lenses in a Design Science Research approach for defining, building, and exemplifying the model. The results contribute to increasing digital security awareness and empowering in a responsible way users' behaviours and decision-processes involved in developing and adopting new standards, methodologies, and tools in the modern digital era.

Keywords: Bayesian networks, cyber security, digital governance, digital security, responsible security, security behaviour

1. Introduction

"We are all connected by the Internet, like neurons in a giant brain." (Stephen Hawking)

During each industrial revolution, the humankind experienced a transition to a new phase of giving life to old dreams and ideas manifested as discoveries that did not seem possible to materialize and further projecting them to technological developments. In the 18th century, the First Industrial Revolution brought the use of steam engines and production mechanization. In the 19th century, the Second Industrial Revolution introduced electricity use and assembly line production. In the 20th century, the Third Revolution brought increased automatization and Information and Communication Technology (ICT) systems e.g., programmable devices like computers and robots. Nowadays, in the 21st century, the Fourth Revolution implies using ICT technologies at large scale in different industries e.g. cyber-physical production-systems and digital twins (Desoutter, 2021). Accordingly, in the last decade and during the ongoing pandemic, we see digital technologies evolve/change the way we as a society function. Digital transformation has dramatically changed the way businesses build their strategies (Puriwat, W., & Tripopsakul, 2021), and encourages further innovations through large-scale efforts while generating significant benefits and opportunities (McKinsey, 2018; OECD, 2021). Nevertheless, digital technologies could pose (un)known risks experienced by (digital)agents e.g., critical infrastructure organizations in different stages/activities implying major consequences inside/outside the digital arena. Besides the importance of organizations' strategies for building/using digital means, employees' role is vital since they could represent a direct threat to organizations unintentionally e.g., improper awareness and incorrectly/insufficiently applying security mechanisms, or intentionally e.g., insider threat (Vroom & Von Solms, 2004; Lobschat et al., 2021; OECD, 2021). It is then a shared responsibility to assure that proper mechanisms are considered and applied when needed. This relates to how agents' security behaviour could be assessed and from there we can strengthen it appropriately. While security behaviour was classified by now as malicious/non-malicious, prompted/non-prompted, informed/un/mis-informed, responsive/preventive, and actual/intended (Chowdhury, Adam & Teubner, 2020), to the best of our knowledge, no attempt to assess user's digital security behaviour as responsible/irresponsible exists. This would benefit organizations to strengthen digital security awareness, training programs, and other security mechanisms. It also aims to open a research avenue on modelling and strengthening responsible digital security behaviour for scientific and practitioner communities. Hence, the following objectives are formulated:

- To introduce and define the "responsible digital security behaviour" notion.
- To identify and classify factors that influence digital security behaviour.

- To design a BN-based model that classifies digital security behaviour as responsible/irresponsible considering these factors.
- To propose recommendations for improving/make responsible users' digital security behaviour.

To achieve these objectives, multidisciplinary research is conducted merging extensive literature review with building an intelligent model following a Design Science Research approach.

The remainder of this paper is structured as follows. Section 2 addresses relevant research regarding theoretical and practical aspects. Section 3 discusses the methodological approach taken. Section 4 introduces the proposed definition. Section 5 discusses influential factors used for classifying responsible digital security behaviour building an AI model. Section 6 presents the design and implementation of the model proposed. Section 7 provides recommendations for creating/strengthening this behaviour. Conclusively, Section 8 discusses the findings and future ideas.

2. Related work

With the increase of multi-source data availability and expansion in development and use of AI in digital security, several challenges e.g., intrusion detection and impact assessment are addressed by academic and practitioner communities. Hence, we highlight below related work positioning this research in the existing body of knowledge.

For grasping and defining responsible digital security behaviour, research on understanding security behaviour, culture, or hygiene is relevant since techno-social aspects of security behaviour are considered. Accordingly, Da Veiga et al. (2020) proposed a model for information security culture, and consider that the workforce should be aware, knowledgeable, and compliant with company's policies based on mutual trust and integrity. Aligned with this, (Vroom & Von Solms, 2004) appraise that understanding employee's security behaviour is crucial in organizations and call for assessment methods. Moreover, Lobschat et al. (2021) defined corporate digital responsibility i.e., norms and values with respect to four main processes: technology creation and data capture, operation and decision making, inspection and impact assessment, and data and technology refinement. These processes are relevant for modelling digital security behaviour. Gratian et al. (2018) correlated human characteristics with cyber security behaviour intentions e.g., password generation, proactive awareness, and updating mechanisms. Regarding employees' awareness, Li et al. (2019) found that employees are aware of their company's information security policies/procedures, and the ones aware are more effective in managing security tasks than the ones not. Conceptually, AlHogail (2015) proposed a framework for developing a comprehensive information security culture in organizations, and positions responsibility as a main dimension influencing it. Moreover, Gillam & Foster (2020) consider as avoidance factors affecting cyber security behaviours labelled as risky the following: self-efficacy, perceived susceptibility, perceived severity, perceived effectiveness, and perceived cost. Vishwanath et al. (2020) defined cyber hygiene, its dimensions e.g., use backup solutions and manage social media privacy, and advance the following approach to contain it: conceptualization, measures' development, model specification, scale evaluation and refinement, and validation. We incorporate these dimensions in this research. Conclusively, OECD (2015) and OECD (2019) addressed the issue between responsibility and capacity as some actors have responsibility but lack the necessary capacity, while other actors lack responsibility but might have the capacity to tackle digital security risks and incidents.

As the digital environment makes use at large scale of intelligent techniques, the AI research domain also includes relevant studies. Thelisson, Morin & Rochel (2020) define the digital responsibility concept to account the responsibility of actors involved and propose an index for restoring trust in a data-driven economy. Along these lines, (Dignum, 2019) argues for building and using intelligent systems for/with good intentions respecting human norms and values while not producing harm. Furthermore, the author considers that responsibility is an aspect that should be considered at individual and collective level in the processes involved. This vision is incorporated in our research. Lima et al. (2014) proposed BN-based personality prediction system considering social media text using the Big Five model which classifies personality trait as: extroversion, neuroticism, agreeableness, conscientiousness, and openness to experience. Jitwasinkul et al. (2016) developed a BN model to predict/classify work behaviour (safe/at-risk) using organizational and human factors. Peng et al. (2021) developed a BN model for predicting dangerous driving behaviour using indicators like sharp acceleration and deceleration control indicator, frequent acceleration, and deceleration control indicator. These models demonstrate BNs' potential in classification-related problems, so they are used as basis for defining model's structure.

3. Research methodology

For defining and modelling digital responsible security behaviour, this research adopts a multidisciplinary approach creating an artefact that serves social purposes objectively (Venable, Pries-Heje & Baskerville, 2017) following the Design Science Research methodology (Hevner, March & Park, 2004; Peffers et al., 2007; Peffers, Tuunanen & Niehaves, 2018). The artefact represents to the best of our knowledge the first attempt in this sense, is able to justify its design choices i.e., is a transparent and explainable model (Maathuis, 2022), and aims to support current/future digital security decision making processes (Offermann et al., 2009; Peffers, Tuunanen & Niehaves, 2018) related to assessing digital security behaviour and awareness. Accordingly, the following research-steps are considered:

- Problem identification and motivation: with the exponential increase in development and use of digital means and considering the role and impact that companies have/encounter from security incidents, is important to tackle the need for properly assessing through effective techniques the way how 'front soldiers' i.e., security employees behave and deal with corresponding digital security aspects, and from there classify their behaviour as responsible/irresponsible so that relevant mechanisms are adopted. Hence, extensive literature review was conducted in digital security, cyber security, AI, and behavioural science.
- Solution objectives: the aim is to define and assess responsible digital security behaviour through an intelligent model i.e., BN-model built from scratch given the existing body of knowledge.
- Design and development: the definition and model are developed and proposed.
- Demonstration: the context of the model is described through exemplification.
- Evaluation: the results are analysed and observing/experimenting with the model is discussed using future incident and employee data.
- Communication: the results obtained are disseminated through this article and presentations.

4. Responsible digital security behaviour definition

Bringing ideas to life and assure technological progress is what drives modern societies to innovate; in this 'solutionist' perspective, innovation is advancing in a relative future moment something improperly framed or unsolved not-yet innovated (Burgess et al. 2018). Transposing this into the digital domain implies addressing technological progress for tackling existing and future digital aspects while benefiting of the opportunities of intelligent techniques easily/directly deployed in this domain. Further scoping to addressing responsible digital security behaviour implies acknowledging that all stakeholders involved in building and using them should be accountable and responsible for their actions (OECD, 2015). Accordingly, while responsibility represents one of the fundamental issues tackled in philosophy and law, implies two aspects relevant to the digital domain: understanding the meaning of and understanding who holds responsibility, mentioning that responsibility should be defined always in a specific context (Thelisson, Morin & Rochel, 2020). The authors consider that responsibility relates to freedom performing actions respecting specific decisions, and that responsibility is: (i) negative - relates to acts/omissions that should not be done, and (ii) positive - relates to a moral relevant situation. Furthermore, Thelisson, Morin & Rochel (2020) distinguish between retrospective responsibility i.e., past action and prospective responsibility i.e., future action. Furthermore, both perspectives apply to digital responsibility, however the digital environment might create new types of (sub-)responsibilities. For security employees, the factors that cumulate to define their digital responsibility relate to context (e.g., organization, motivation), role, alignment with management and security policies (Thelisson, Morin & Rochel, 2020) as depicted in Section 5.

Moreover, Puriwat & Tripopsakul (2021) scrutinize that digital social responsibility is accepted as valuable strategic movement, and that corporate social responsibility 'encompasses the economic, legal, ethical, and discretionary (philanthropic) expectations that society has of organizations at a given point in time'. Liyanaarachchi, Deshpande & Weaven (2020) define corporate digital responsibility as "the set of values and specific norms that govern an organizations' judgements and choices in matters that relate specifically to digital issues". Moreover, Da Veiga et al. (2020) suggest that related concepts like cyber/information security culture show how people e.g., employees perceive security aspects/issues using organizational systems. Vroom & Van Solms (2004) stress that ideally employees follow established guidelines voluntarily, generally such a culture is defined on assumptions on perceptions and attitudes accepted and encouraged to incorporate security characteristics. Specifically, Da Veiga & Martins (2017) reason that information security culture contains "the attitudes, assumptions, beliefs, values, and knowledge that employees/stakeholders use to interact with the organization's systems and procedures" and this changes over time. Furthermore, in different interactions, acceptable or unacceptable behaviour occurs, fact that translates in this research as classifying digital security behaviour as responsible/irresponsible.

Aligned with the abovementioned perspectives, we define responsible digital security behaviour as:

Responsible Digital Security Behaviour = the way an agent acts in relation to a digital system in a given organizational (digital) context in respect to the agreed norms, procedures, and values of agent's organization.

The elements contained are:

- *Way:* activity/action taken by an agent.
- *Agent:* individual/group of employees of an organization.
- *Digital system:* software, hardware, or communication system belonging to an organization.
- *Organizational (digital) context:* setting in which the agent exists or deals with in an organization.
- *Agreed norms, procedures, and values:* well-established and voluntarily agreed norms, procedures, and values that represent the (digital) vision of an organization that agents should respect/embed in their daily (digital)activities as agreed.

5. Influential factors

Seeing the aim of this research, we consider the approach of (Da Veiga et al. 2020; Maathuis et al., 2018) which relates security behaviour to interaction between human, technology/infrastructure, and policy factors, as discussed below, depict in Figure 1, and express in Table 1:

- *Human:* employees make several decisions given their perspective, experience, background, and organization's vision (Donalds, C., & Osei-Bryson, 2020). These decisions (in)directly impact organization's internal/external processes, reputation, and (long-term)-existence.
- *Infrastructure:* since a significant part of the digital environment has an invisible and intangible nature (De Bruijn & Janssen, 2017), its physical projections and connections interact with employees during their design, implementation, and use (AlHogail, 2015).
- *Policy:* organizations use digital security policies by establishing procedures and guidelines that should be followed by employees for effective functioning, risk management, legal compliance, and business goals achievement (Lobschat et al., 2021).

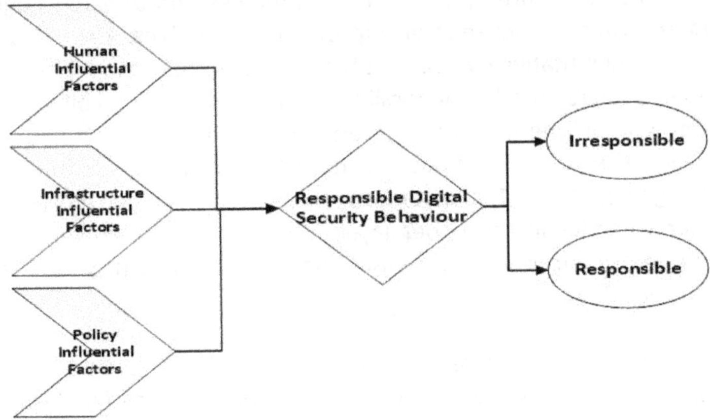

Figure 1: Influential factors

Table 1: Influential factors

Human Factors (Vroom & Von Solms, 2004; Da Veiga & Eloff, 2017; OECD, 2015; Henshel et al., 2016; FindLaw, 2018; Gratian et al., 2018; Strom, 2018; Dignum, 2019; Li et al., 2019; Trim & Lee, 2019; Chowdhury, Adam & Teubner, 2020; Da Veiga et al., 2020; Gillam & Foster, 2020; Moustafa, Bello & Maurushat, 2021)	Attitude, Motivation, Job Satisfaction, Cooperation Prior experience, Dealing with existing/similar vulnerability, Lack of knowledge/expertise Self-efficacy, Response efficacy (Un)intentional errors Time/space pressure Ethical values and conduct Decision making style: rational, avoidant, dependent, intuitive, and spontaneous Social influence, Peer behaviour, Mutual trust, Timing and manager's mood Personal traits: agreeableness, conscientiousness, neuroticism, openness, and extraversion Procrastination, Impulsivity Emotional condition, Fatigue, mood, and health Demographics, Cultural aspects (e.g., geographic, organizational, ideological expressed at individual and group levels)
Infrastructure Factors (OECD, 2015; Da Veiga & Eloff, 2007; Gratian et al., 2018; Li et al., 2019; Gillam & Foster, 2020; Vishwanath et al., 2020)	Security measures/mechanisms application Incident management Monitor software and hardware processes, Updates, Backups Device securement: locking using passwords, PINs etc., Choice of passwords, Change default passwords, Enable and use two/multi-factor authentication Proactive awareness of web content, Internet browsing using Incognito or Private mode E-mail security verification, Profile and information authenticity awareness in social media, Manage privacy settings in social media Perceived susceptibility, Perceived vulnerability, Perceived severity, Perceived effectiveness, Perceived cost
Policy Factors (Saridakis et al. 2016; Da Veiga & Eloff, 2017; Van Schaik et al. 2017; Da Veiga et al., 2020)	Security policy application/compliance Incident management Security risk management Risk perception

6. BN model structure and demonstration

This section provides a background to BNs, describes the model proposed, and demonstrates it using illustrative examples.

BNs consist of both qualitative and quantitative components (Darwiche, 2008). The qualitative component is a Directed Acyclic Graph (DAG) as in Figure 2, i.e., a set of variables and directed edges between them. The directed edges represent the cause-effect relationship between variables. Each variable has a finite set of mutually exclusive states. The quantitative component is the Conditional Probability Tables (CPTs) as in Figure 2, which includes conditional probabilities for all possible combinations of the child and parent variable states. For a variable without any parent, the CPT includes prior probabilities of the corresponding variable. This example contains two layers. The upper layer has "Pollution", "Smoking" factors/variables that increase the likelihood of a patient having lung cancer. The lower layer contains the variable which we want to query ("Lung Cancer") with evidence for variables in the upper layer. When the evidence for variables is obtained, the posterior probabilities of non-evidenced variables would be updated. This process is named belief updating, inference, or probability propagation.

BNs support four types of reasoning: predictive, diagnostic, intercausal, and combined. Predictive reasoning is from cause (e.g., smoking) to effects (e.g., lung cancer). Then, the evidence for a variable in the upper layer (cause) is provided which helps querying the posterior probabilities of non-evidenced variable in the lower layer (effect). This is relevant for our application.

BNs are used for e.g., medical decision support systems, fault diagnosis, security support. (Kahn Jr et al., 1997) built a three-layer BN (MammoNet) for breast cancer diagnosis. The upper layer contains five patient-history features, the middle layer is a target variable (breast cancer) aimed to query, and the lower layer includes two physical findings and 15 mammographic features. Once the evidence for variables in the upper/lower layer is provided, target's posterior probability is updated. (Chockalingam et al., 2021) developed a three-layer BN that helps operators in distinguishing between attacks and faults. The upper layer has contributory factors like easy physical access to sensor, lack of maintenance. The middle layer is a target variable aimed to query. And the lower layer includes observations (or test results) e.g., the redundant sensor also sends incorrect water level measurements. Once the evidence for variables in the upper/lower layer is provided, the posterior probability of the target variable is updated.

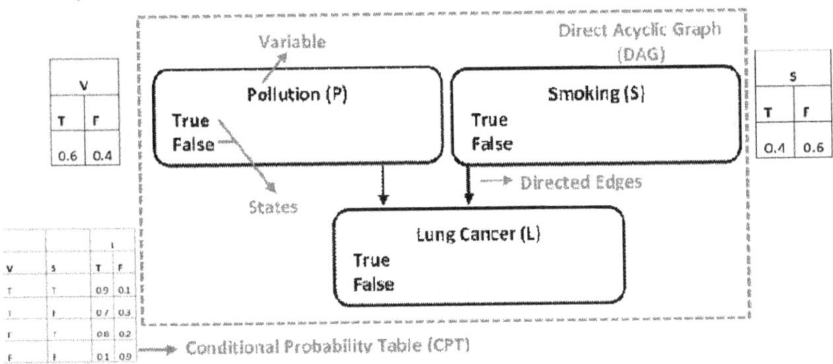

Figure 2: BN example

BNs are able to tackle our problem considering the related work plus their capability to incorporate expert knowledge and empirical data - essential in cyber security considering the well-known challenge regarding empirical data availability. Moreover, the graphical structure enhances the explainability aspect i.e., critical aspect of a model deployed. Therefore, we use BNs for modelling and consider the above-mentioned models as starting point for defining the proposed model.

Our model consists of three layers as in Figure 3. The upper layer contains factors corresponding to human, infrastructure, and policy. The middle layer has intermediate nodes (i.e., human, infrastructure, and policy) which help keeping the CPT size of the child node (i.e., the target variable) manageable using the parent divorcing technique. The lower layer embeds the target variable aimed to query with evidence for variables in the upper layer. A subset of factors from Table 1 is used for model development together with relationships between the considered variables ensuring the causality (cause-effect) aspect. The CPT values used are examples for demonstrating the BN-based approach in distinguishing digital security behaviour. Typically, CPTs are elicited from expert knowledge and empirical data.

We demonstrate the model on two illustrative scenarios to show its applicability. In the first scenario, the decision maker sets evidence for variables in the upper layer based on available information about the specific user/user-group (Lack of Job Satisfaction="True"; Time Pressure="True"; Monitor Processes="False"; Security Risk Management ="False"). Then, the posterior probability is computed by the model for other variables without any evidence. Figure 4 shows that user's digital security behaviour is irresponsible. This could be a scenario of a disgruntled employee. Additionally, this model could help selecting appropriate recommendations. Thus, putting in place appropriate measures to improve job satisfaction and monitor processes would help avoiding irresponsible security behaviour of the specific user/user-group.

In the second scenario, the decision maker sets evidence for variables in the upper layer based on available information about the specific user/user-group (Lack of Job Satisfaction="False"; Lack of Cyber Security Expertise="False"; Enforce Multi-Factor Authentication="True"; Security Policy Compliance="True"; Security Risk Management="True"). Then, the posterior probability is computed for other variables without any evidence. Figure 5 shows that user's digital security behaviour is responsible. This could be a scenario of a cyber security personnel. Moreover, this model could help selecting appropriate recommendations. Thusly, considering appropriate measures to reduce fatigue would help avoiding irresponsible security behaviour of the specific user/user-group.

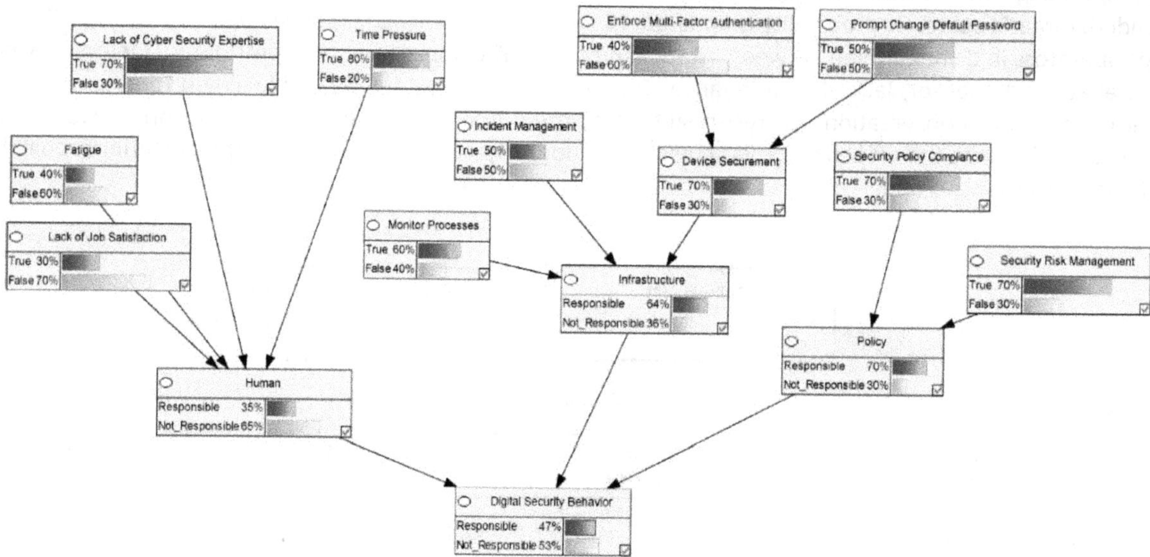

Figure 3: Developed BN model – no evidence provided

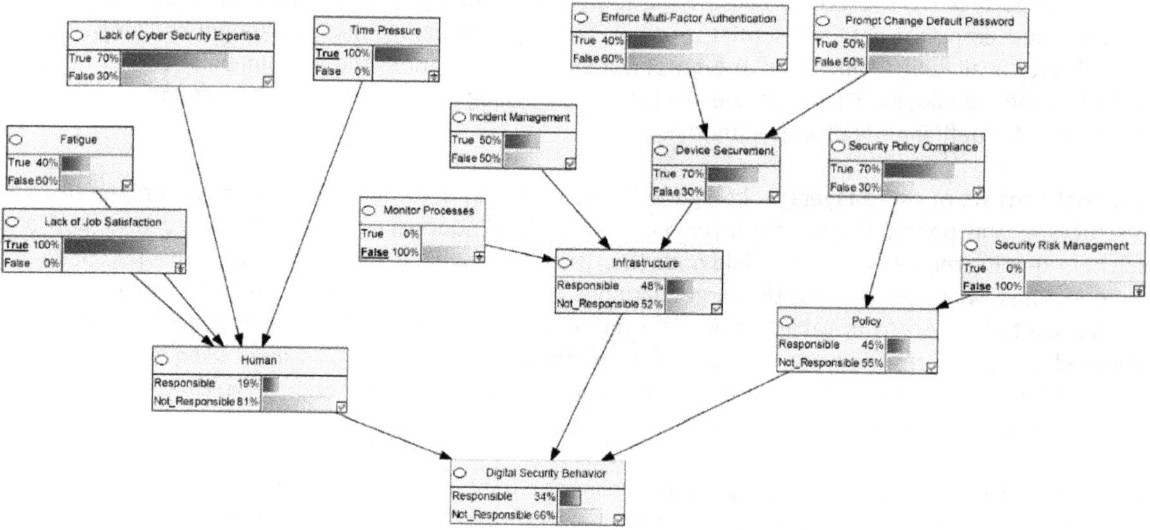

Figure 4: Developed BN model – first illustrative scenario

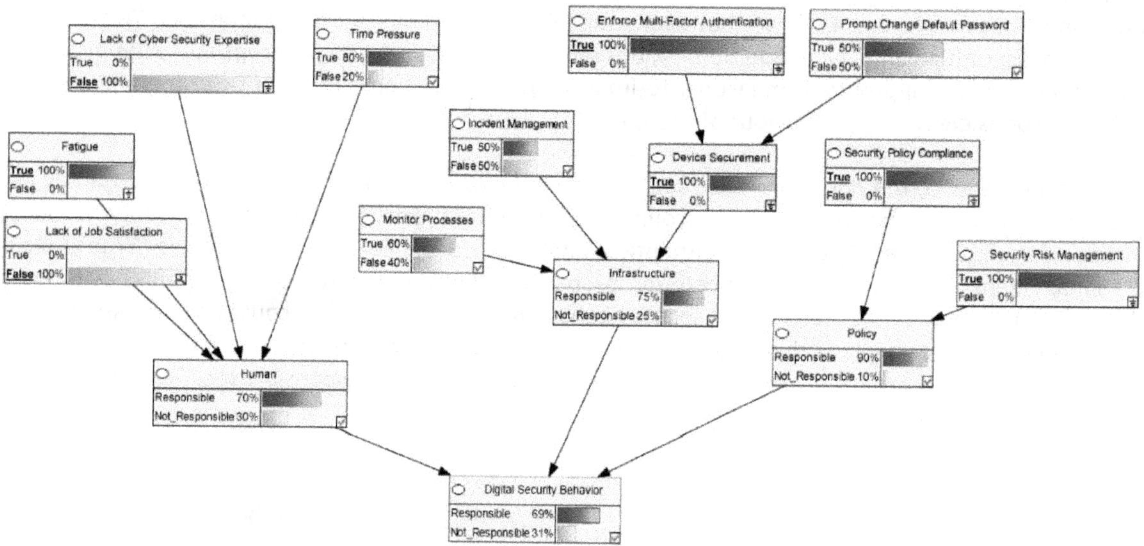

Figure 5: Developed BN model – second illustrative scenario

7. Recommendations

A model for classifying responsible digital security behaviour was introduced considering influencing factors. To facilitate model's operationalization, recommendations are considered adopting two stances: a set of recommendations that aim at controlling, limiting, or avoiding the case when irresponsible digital security behaviour is detected; and a set of recommendations that aim at supporting and strengthening when a responsible digital security behaviour is detected. The model also helps putting in place specific recommendations. To tackle both perspectives, technical-governance recommendations applicable when building and using digital technologies respecting the aspects involved in all phases of a security incident are considered. Accordingly, the following recommendations (Steen & Van de Poel, 2012; AlHogail, 2015; OECD, 2015; OECD, 2019; Alshaikh, 2020; Lobschat et al., 2021) are advised in Table 2:

Table 2: Technical and governance recommendations

Recommendation	Controlling, limiting, or avoiding irresponsible digital security behaviour	Supporting and strengthening digital security behaviour
Technical	Encourage the adoption of organization's digital security technologies, methodologies, and tools compliant with organization's policy through the use of concrete examples. Define control and advise mechanisms for tackling specific issues/factors that conduct to irresponsible behaviour.	Define challenges and competitions for tackling different digital security issues that the organization deals with.
Technical	Design and integrate modelling and simulation environments for active learning purposes.	
Technical, Governance	Encourage and support innovation through active communication and collaboration using dedicated funds and infrastructure by merging departments and teams for well-defined and scalable shared goals. Design and implement employee's assessment and monitoring digital security metrics and solutions.	
Governance	Build or adapt human capacity through relevant awareness campaigns, training, education, and R&D using international digital security standards, technologies, methodologies, and tools compliant with organization's digital security vision and policy. Define or upgrade explicit policies with concrete definitions of organization's norms, values, and applicable terms. Build a digital security network and hub at organizational level.	

8. Conclusions

While building the digital security field of science, several discourses and studies are conducted around facts like understanding the digital environment, grasping the potential and further development of diverse digital technologies during digital transformation, finding and combating (new)attack vectors, and managing risks through socio-technical lenses. While the resources studied provide a strong theoretical perspective on digital security challenges and opportunities, they do not focus on security employees which are in the first line in the battlefield against digital security incidents. Accordingly, in this research, the digital security behaviour of security employees is analysed through the eyes of responsibility by structuring factors that facilitate its classification as responsible/irresponsible. Moreover, a series of technological and governance recommendations are provided for controlling or avoiding irresponsible digital security behaviour and strengthening the responsible digital security behaviour. We do this through extensive literature review and building a novel Bayesian Belief Network model following a Design Science Research approach.

This research continues by advancing the proposed model considering additional dimensions that allow extending the variables and relationships, and conducting case studies on digital security incidents and users' behaviour in organizations with scientific and industry experts for operationalizing the model by capturing useful data and draw relevant socio-technical recommendations for controlling and strengthening digital security behaviour which contributes to a more responsible, accountable, and trustable digital environment.

References

AlHogail, A. (2015). Design and validation of information security culture framework. Computers in human behavior, 49, 567-575.

Alshaikh, M. (2020). Developing cybersecurity culture to influence employee behavior: A practice perspective. Computers & Security, 98, 102003.

Baskerville, R., Baiyere, A., Gregor, S., Hevner, A., & Rossi, M. (2018). Design science research contributions: Finding a balance between artifact and theory. Journal of the Association for Information Systems, 19(5), 3.

Burgess, J. P., Reniers, G., Ponnet, K., Hardyns, W., & Smit, W. (Eds.). (2018). Socially responsible innovation in security: Critical reflections. Routledge.

Chockalingam, S., Pieters, W., Teixeira, A., & van Gelder, P. (2021). Bayesian network model to distinguish between intentional attacks and accidental technical failures: a case study of floodgates. Cybersecurity, 4(1), 1-19.

Chowdhury, N. H., Adam, M. T., & Teubner, T. (2020). Time pressure in human cybersecurity behavior: Theoretical framework and countermeasures. Computers & Security, 97, 101931.

Da Veiga, A., & Martins, N. (2017). Defining and identifying dominant information security cultures and subcultures. Computers & Security. 70, 72-94. doi: 10.1016/j.cose.2017.05.002.

Da Veiga, A., Astakhova, L. V., Botha, A., & Herselman, M. (2020). Defining organisational information security culture—Perspectives from academia and industry. Computers & Security, 92, 101713.

De Bruijn, H., & Janssen, M. (2017). Building cybersecurity awareness: The need for evidence-based framing strategies. Government Information Quarterly, 34(1), 1-7.

Darwiche, A. (2008) Bayesian Networks, Foundations of Artificial Intelligence, vol. 3, pp. 467-509.

Desoutter (2021). Industrial Revolution – From Industry 1.0 to Industry 4.0. Available from: https://www.desouttertools.com/industry-4-0/news/503/industrial-revolution-from-industry-1-0-to-industry-4-0.

Dignum, V. (2019). Responsible artificial intelligence: how to develop and use AI in a responsible way. Springer Nature.

Donalds, C., & Osei-Bryson, K. M. (2020). Cybersecurity compliance behavior: Exploring the influences of individual decision style and other antecedents. International Journal of Information Management, 51, 102056.

FindLaw (2018). What are Legal Ethics and Professional Responsibility? Available from: https://www.findlaw.com/hirealawyer/choosing-the-right-lawyer/ethics-and-professional-responsibility.html

Gillam, A. R., & Foster, W. T. (2020). Factors affecting risky cybersecurity behaviors by US workers: An exploratory study. Computers in Human Behavior, 108, 106319.

Gratian, M., Bandi, S., Cukier, M., Dykstra, J., & Ginther, A. (2018). Correlating human traits and cyber security behavior intentions. computers & security, 73, 345-358.

Henshel, D., Sample, C., Cains, M., & Hoffman, B. (2016). Integrating cultural factors into human factors framework and ontology for cyber attackers. In Advances in human factors in cybersecurity, pp. 123-137.

Hevner, A. R., March, S. T., Park, J., & Ram, S. (2004). Design science in information systems research. MIS quarterly.

Jitwasinkul, B., Hadikusumo, B. H., & Memon, A. Q. (2016). A Bayesian Belief Network model of organizational factors for improving safe work behaviors in Thai construction industry. Safety science, 82, 264-273.

Kahn Jr, C. E., Roberts, L. M., Shaffer, K. A., & Haddawy, P. (1997). Construction of a Bayesian network for mammographic diagnosis of breast cancer. Computers in biology and medicine, 27(1), 19-29.

Lobschat, L., Mueller, B., Eggers, F., Brandimarte, L., Diefenbach, S., Kroschke, M., & Wirtz, J. (2021). Corporate digital responsibility. Journal of Business Research, 122, 875-888.

Li, L., He, W., Xu, L., Ash, I., Anwar, M., & Yuan, X. (2019). Investigating the impact of cybersecurity policy awareness on employees' cybersecurity behavior. International Journal of Information Management, 45, 13-24.

Lima, A. C. E., & De Castro, L. N. (2014). A multi-label, semi-supervised classification approach applied to personality prediction in social media. Neural Networks, 58, 122-130.

Liyanaarachchi, G., Deshpande, S., & Weaven, S. (2020). Market-oriented corporate digital responsibility to manage data vulnerability in online banking. International Journal of Bank Marketing.

Maathuis, C., Pieters, W. and Van Den Berg, J. (2018). A computational ontology for cyber operations. In Proceedings of the 17th European Conference on Cyber Warfare and Security, pp. 278-288.

Maathuis, C. 2022. On Explainable AI Solutions for Targeting in Cyber Military Operations. In Proceedings of the 17th International Conference on Cyber Warfare and Security. Academic Publishing.

McKinsey (2018). Unlocking Success in Digital Transformations. McKinsey & Company.

Moustafa, A. A., Bello, A., & Maurushat, A. (2021). The role of user behaviour in improving cyber security management. Frontiers in Psychology, 12.

OECD (2015). Digital Security Risk Management for Economic and Social Prosperity. OECD Recommendation and Companion Document.

OECD (2019). Policies for the Protection of Critical Information Infrastructure. No. 275. OECD Publishing.

OECD (2021). Recommendation of the Council on Digital Security of Critical Activities. OECD Legal Instruments.

Offermann, P., Levina, O., Schönherr, M., & Bub, U. (2009). Outline of a design science research process. In Proceedings of the 4th International Conference on Design Science Research in Information Systems and Technology.

Peffers, K., Tuunanen, T., Rothenberger, M. A., & Chatterjee, S. (2007). A design science research methodology for information systems research. Journal of management information systems, 24(3), 45-77.

Peffers, K., Tuunanen, T., & Niehaves, B. (2018). Design science research genres: introduction to the special issue on exemplars and criteria for applicable design science research.

Peng, Y., Cheng, L., Jiang, Y., & Zhu, S. (2021). Examining Bayesian network modeling in identification of dangerous driving behavior. PLoS one, 16(8), e0252484.

Puriwat, W., & Tripopsakul, S. (2021). The impact of digital social responsibility on preference and purchase intentions: The implication for open innovation. Journal of Open Innovation: Technology, Market, and Complexity, 7(1), 24.

Saridakis, G., Benson, V., Ezingeard, J. N., & Tennakoon, H. (2016). Individual information security, user behaviour and cyber victimisation: An empirical study of social networking users. *Technological Forecasting and Social Change, 102*, 320-330.

Steen, M., & Van de Poel, I. (2012). Making values explicit during the design process. *IEEE Technology and Society Magazine, 31*(4), 63-72.

Strom, B. E., Applebaum, A., Miller, D. P., Nickels, K. C., Pennington, A. G., & Thomas, C. B. (2018). Mitre att&ck: Design and philosophy. *Technical report*.

Thelisson, E., Morin, J. H., & Rochel, J. (2020). AI Governance: Digital Responsibility as a Building Block. *Science, 107*, 111.

Trim, P. R., & Lee, Y. I. (2019). The role of B2B marketers in increasing cyber security awareness and influencing behavioural change. *Industrial Marketing Management, 83*, 224-238.

Van Schaik, P., Jeske, D., Onibokun, J., Coventry, L., Jansen, J., & Kusev, P. (2017). Risk perceptions of cyber-security and precautionary behaviour. *Computers in Human Behavior, 75*, 547-559.

Venable, J. R., Pries-Heje, J., & Baskerville, R. L. (2017). Choosing a design science research methodology.

Vishwanath, A., Neo, L. S., Goh, P., Lee, S., Khader, M., Ong, G., & Chin, J. (2020). Cyber hygiene: The concept, its measure, and its initial tests. *Decision Support Systems, 128*.

Vroom, C., & Von Solms, R. (2004). Towards information security behavioural compliance. *Computers & security, 23*(3), 191-198.

A Model for State Cyber Power: Case Study of Russian Behaviour

Juha Kai Mattila
Aalto University, Helsinki, Finland
juha.mattila@aalto.fi

Abstract: The emerging cyber environment with new information channels provides a novel avenue for states to project their powers to govern their residents and fulfil their international ambitions. The recent manipulation of elections, coercing companies, blackmailing citizens, and suppressing essential infrastructure services reflects an increased activity and development both by state and non-state entities in the cyber environment. Several models for inter-state power projection are created in studies of international relationships, military strategy, and, recently, hybrid warfare. Do these models recognise the foundational transformation in international power projection? Do they explain the current national cyber strategies? Can they help foresee the possible developments of power projection in international confrontations? The paper seeks a bigger picture from other power strategies in fulfilling the state's political ambitions. Furthermore, the paper explores the evolution of the cyber environment and its possible emerging features for international power projection. A constructive research method builds a multiple domain power projection model by combining systems thinking with various models from international relationships, military strategies, business strategies to classical decision making. Finally, the feasibility of the model is tested in a case study of Russian cyber strategies and actions between 2007-2020 from a positivistic approach. As a result, the model seems to help explain the past cyber power-wielding and provide insights into current national cyber policies. Further testing is required to evaluate the model's feasibility in creating a foresight. Nevertheless, the proposed state-level cyber power projection model extends the existing models with a system dynamics viewpoint. Additionally, it adds the dimension of evolution to consider the future changes of international power projections in the information realm. Hence, the model improves the ability of national defence planners to study cyber strategies and estimate the lines of operation and impact of cyber operations.

Keywords: cyber domain, state power, international relationships, modelling, cyber strategy, and cyber operations

1. Introduction

In the early 1990s, an imagined cyberwar was perceived as a culmination of non-kinetic wars that would disarm and disable a whole society without killing masses of people. (Arquilla & Ronfeldt, 1993) That has not been the reality yet. (Rid, 2013) Nevertheless, technology and digitalisation are transforming the ways of politics (Cederberg, 2020), economy (Zuboff, 2019), social life (Dwyer & Kreier, 2015), education (McCamey, Wilson, & Shaw, 2015), industry (Schwab, 2016), and eventually also military (Fiott, 2020). For example, Russia has used cyber power as part of its operations in Estonia 2007, Georgia 2008, and Ukraine 2014. (Clark, 2020) How can one understand novel avenues of impact emerging from the seemingly volatile, uncertain, complex, and ambiguous (VUCA) (Scherrer & Grund, 2009) digital landscape?

Besides the multiple models for power in international relations, the RAND model on assessing risks of cyber terrorism Risk = Threat x Vulnerability x Consequence (Willis, 2006) may still work for the essential risk assessment. The comprehensive model for the national cyber power index, which sums and normalises the outcome of capability and intention, may provide a quantified model to compare state-level cyber powers. (Voo, Hemani, DeSombre, Cassidy, & Schwarzenbach, 2020) Estimating the relative cyber strengths of each nation by considering their cyber defence, dependence, and offence (Clarke & Knake, 2010) features may provide a strategic viewpoint to the question. Nevertheless, it is worth reviewing the model for state-level cyber power in parallel with some ongoing research projects (Tabansky, 2021), (International Institute for Strategic Studies, 2021), (Massachusetts Institute of Technology, 2020), especially when the cyber environment gains space in other realms, relationships become more complicated, the understanding of war and peace is changing, and causality between sensemaking-act-outcome becomes blurred.

The sovereign state has been the dominant global institution since the Peace of Westphalia in 1648. (Nye, 2011) Since then, the transactional relationship between states and wielding diplomatic, trade, and military power has transformed into a globalised network of interrelationships where finance, trade, transportation, manufacturing, energy, and even military cooperation is primarily driving the relationships between contemporary states. (Toffler, 1981) The last decades have seen the most improvement in global living conditions (Roser, 2020) and the least amount of militarised violence being wielded between states since 1648. (Rossling, Rossling Rönnlund, & Rossling, 2018) What kind of hard and soft powers do states wield in trying to affect the behaviour of other states in the era when military power is the least preferred mean?

On the other hand, non-state actors have been actively growing their influence at the international level. In the western hemisphere, the five most significant technology companies, Alphabet, Amazon, Facebook, Apple, and Microsoft (GAFAM) (Wikipedia, 2021), are among the most valuable public companies (Randewich & Ahmed, 2022). They play a significant role in the digital economy (Miguel de Bustos & Izquierdo-Castillo, 2019), social relationships, and so-called surveillance capitalism (Zuboff, 2019). The GAFAM and their Chinese competitors BATX (Baidu, Alibaba, Tencent, and Xiaomi) run platform ecosystems that revolutionise business-to-consumer and business-to-business trade while implementing new technologies like artificial intelligence, big data, on-line-gaming, and payment systems. (Mulrenan, 2020) They build a global cyber environment and wield powers of big data, ecosystems, R&D and end-to-end digitalisation to change the behaviour of individuals, societies, and states. How will these international organisations change the cyber realm and open or close avenues of approach at the data or information level?

Terrorists are using calculated unlawful violence or threat of violence to inculcate fear, intended to coerce or to intimidate governments or societies in the pursuit of goals that are generally political, religious, or ideological. (Theohary & Rollins, 2015) Transnational terrorist organisations, insurgents, and jihadists have used the Internet as a tool for planning attacks, radicalisation and recruitment, a method of propaganda distribution, and a means of communication, and for disruptive purposes. (Rollins & Wilson, 2007) Furthermore, Internet memes (Merriam-Webster, 2022), conspiracy theories (Oliver & Wood, 2014), and other social media-enabled avenues to social behaviour (Amedie, 2015) are opening new approaches to a variety of agents trying to manipulate crowds. What of these emerging abilities states will adopt to their box of international power-wielding tools?

Rather than diving deep into contemporary power-wielding observed in cyberspace, the research aims to approach the phenomena comprehensively. Accordingly, the hypothesis is that a state modelled as a viable system extended with the Clausewitz triangle relationship creates a better artefact to study the interrelationships and fragility of the state structure. Furthermore, the combination of international relation models and doctrines of military power explain the avenues and levels of effort. In addition, the technical and business evolution in the cyber environment needs to be included in the model. Finally, the research tests the created hypothetical model against some factual data gathered from Russian strategies and operations in the cyber environment, i.e., operations against Estonia 2007, Georgia 2008, and Ukraine 2014 (Freedman, 2017) to measure its feasibility.

The paper analyses the features of current models and provides a theoretical foundation for the hypothetical model in section 2, documents the case study of Russian behaviour in a cyber environment in section 3, explains the feasibility of the proposed model in section 4, and concludes the research in section 5.

2. Literature research

2.1 Existing models and gaps in their perception

The current knowledge base concerning the models for cyber powers, capabilities, or systems includes military, international relations, quantitative indexes, and cybercrime or security viewpoints. (van Haaster, 2016) They have all established their position over the years. However, they seem to exclude some system dynamic (Jackson, 2019, pp. 229–259) features that could explain causalities in VUCA cyberspace and its evolving role as a line of operation or domain between two states, as summarised in Table1.

Table 1: Categorising contemporary models for cyber power according to selected components of dynamic system

Contemporary models for cyber power	Source	Medium	Target	Remarks
DIMEFIL lines of operation (Armstrong, 2019)	Assumes a simple source	Primarily categorises the instruments of national power to Diplomatic, Information, Military, Economic, Finance, Intelligence, and Law enforcement.	Assumes a linear impact	A military approach to lines of operation between two states can be used as a context to the cyber realm.

Contemporary models for cyber power	Source	Medium	Target	Remarks
Betz & Stevens IR model applied to cyberspace (Betz & Stevens, 2011)	Assumes as an unvarying entity	Extends the interstate power relations with ways of Compulsory, Institutional, Structural and Productive ways of wielding power.	Assumes as an unvarying entity	The IR taxonomy of power to analyse cyber power may extend the DIMEFIL.
Nye's hard and soft power model for cyberspace (Nye, 2010)	Recognises intra cyberspace where state wields both hard and soft power. Recognises a variety of actors related to cyberspace.	Recognises escalation in relations through cyberspace: shape, create agenda, and confront.	Uses case examples of cyberspace enabling effect in other realms.	Uses both information and physical instruments to cyber power and describes the escalation model.
Belfer Center's National Cyber power index (Voo et al., 2020)	Recognises cyber-related infrastructure, public and private behaviour, and assets.	Measures 27 national cyber capabilities against 32 intents.	Recognises seven national objectives that countries pursue using cyber means.	Quantitative approach summing the product of intent and capability over 7 strategic objectives. The strategic approach may be used further.
Cyber security (Zaballos & Herranz, 2013) and crime (Mandelcorn, 2013) models	Recognises cognitive and social motivation for actors to engage in cybercrimes.	Recognises the ecosystem that operates and secures cyberspace.	Recognises preparedness and prevention, detection, and reaction processes in defending against the cyber offence.	Models for cyber security and crime prevention may be used as sub-systems.
Various cyber strategy analyses (Mattila, 2014) (Lilly & Cheravitch, 2020)	Recognises the national strategic approach related to defensive or offensive cyber activity.	Understands cyber as a part of the information domain and type of warfare.	Introduces views on how a state may perceive cyber-related threats.	Uses, e.g., Nash equilibrium or Russian military studies to make sense of some events in the real world.

It is evident that the above-reviewed models for cyber power all approach the same subject from a different viewpoint and, thus, fall short to explain the entirety of the phenomena. Therefore, the research approaches contemporary models from system science as the following gaps were recognised:

- 1. Emerging nature of cyberspace concerning time and other realms,
- 2. Dynamics of a state as a system, and
- 3. Different levels of vulnerability and maturity of abilities are available for either of the state entities.

There were other deviations from the system dynamics, but this paper's hypothetical model seeks to address the above gaps in the following subsection.

2.2 Hypothetical model

Sub-section generates a hypothetical model addressing the chosen gaps. First, cyberspace needs to be understood as an emerging, man-made realm (Scherrer & Grund, 2009) that is gradually gaining volume through digitalisation, automation, and artificial intelligence. Second, the state needs to be illustrated as a logical system that adjusts to changes in situations, environment, and relationships. Third, the state system's environment needs to be comprehended from existing and emerging threats viewpoints. Fourth, the dynamics of international relations, apparent lines of operation and chosen courses of action between two states or political entities should be considered part of the model.

Cyberspace needs to be understood as an emerging feature in the classical model of the military impact (i.e., physical, information, cognitive and social realms) (Krezer, 2021). Traditionally, militarised violence has changed social behaviour by causing material and human attrition in the physical realm. Survivors of the violence have forwarded information about horrors to other people, whose feelings and beliefs are altered based on the received information. (DoD, 2018 p. 2) When these new feelings and beliefs are confirmed within the social construction, people may change their behaviour. (Zuboff, 2019 p.93-97) That is the simple, linear approach. Whereas, in many revolutions, a force captures control over broadcasting services, starts distributing their information and changes the behaviour of society. Besides, social media has enabled terrorists to distribute videos of their physical violence for a wider audience and thus extending the impact of fear and terror. (Kaldor, 2012) Furthermore, the art of strategy (Sun, 2014, pp. 92–93) aims to conquer or suppress the adversary without fighting by indisposing the adversary's plans and preventing the junction of its forces. The physical attrition on the battlefield, especially against prepared positions, is perceived as the worst scenario. Figure 1 illustrates the cyberspace gradually extending towards the physical, information and cognitive realms and subsequently opening new avenues to create impact and change human behaviour.

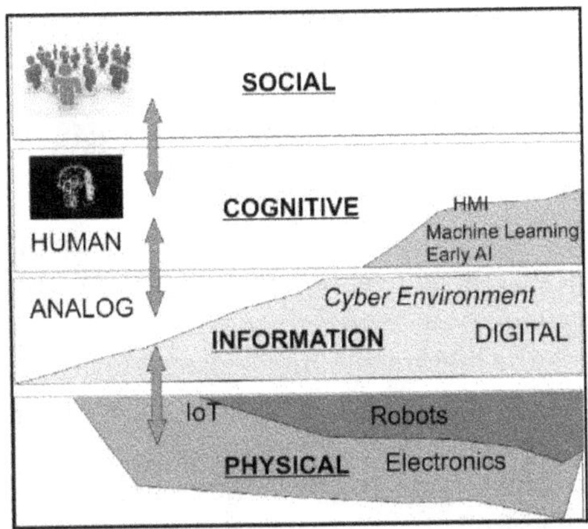

Figure 1: A model for evolution and causality in using power to change behaviour, i.e., realms of warfare

The emerging cyber environment in Figure 1 is extending, almost exponentially, both through utilisation and application in most areas of human life (e.g., electronics, robotics, digitalisation, artificial intelligence). The performance of information technology is still improving two times every 18 months (Electrical 4U, 2020). The content of WWW is increasing with over 4 million hours of content every day (Schultz, 2019) which may accumulate human knowledge base at the speed of doubling every 13 months. (Schilling, 2013) The Internet of Things and automation are foreseeing a tenfold expansion during the ongoing decade. (IOT News, 2020)

While improving the model of actor and target, the hypothesis assumes the state as a rational, hierarchical system rather than a network of autonomous nodes. Therefore, it uses Beer's Viable System Model and its improvements (Lowe, Espinosa & Yearworth, 2020) to illustrate the levels of politics, strategy, operations, and tactics as processes to manage action during the confrontation. Besides, the Clausewitz model (Clausewitz, 1984) of state elaborates the VSM with relationships between government, society, and power sources.

Systems thinking sees the open entity constantly interacting with its environment (Arnold & Wade, 2015). Hence, the state should be seen in interrelation with other political entities and threats they perceive against their interests. One of the rights of the sovereign state (Annan, 1999) includes the right to use power to prevent or deter threats to their security. Threats can be perceived differently, but one way to categorise them is existential, global/regional, and intra-state threats, as illustrated in Figure 2.

Naturally, the threat environment is dynamic, but the hypothetic model concentrates, on this occasion, on the interaction between environment and state rather than studying the evolution of threats perceived at a state level.

As a result, the state model is based on VSM (Jackson, 2019, pp. 291-343), illustrating the levels of control (tactical, operational, strategic) and interfaces between organisational bodies and the environment of their viewpoint. The VSM organisation is then elaborated with Clausewitz's state model (Clausewitz, 1984). The triangle relationship between governing entity, society and power institutes explains the lines of interaction within the state itself. It also opens the Centres of Gravity (U.S. DoD, 2015) for the adversary target analysis. The state uses its powers to impact other states at compulsory, institutional, productive, and structural levels. (Barnett & Duvall, 2005) Furthermore, the previous structure is extended with military levels of interaction in conflict: Political (Vego, 2007), Strategy (Strachan, 2013) (Clausewitz, 1984), Operational art (Strachan, 2013), Tactical (Suvorov, 2015), and Techniques (DoD, 2021, p.214).

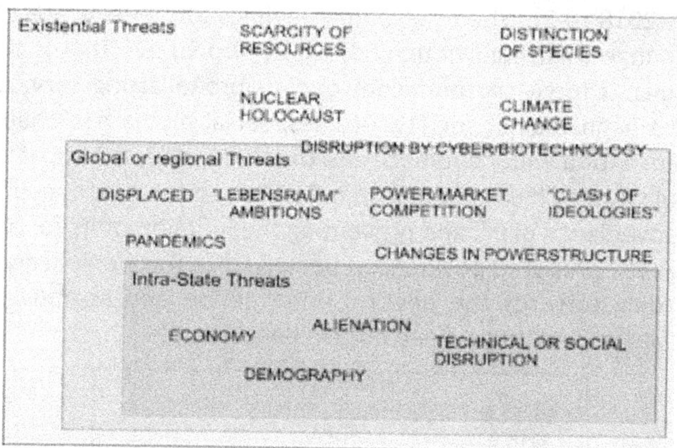

Figure 2: An example of threat environment possibly perceived by a state

In conclusion, the hypothetical model illustrates a confrontation between blue and red states using the means of DIMEFIL in the ways of compulsory, institutional, productive, or structural to create impact through evolving cyberspace that exponentially extends its range over physical, information, and cognitive realms. The means are used in ways over the medium to change the adversary's behaviour in the cognitive and social realms. The control of the applied powers follows the hierarchy of political, strategic, operational, tactical, and technics, as illustrated in Figure 3.

The model excludes grand strategy (Liddle Hart, 1991) from the control hierarchy to not open the model towards the sub-system of preparation, building, and directing of all means and ways of state powers. Subsequently, the model does not consider the escalation (Joint Chiefs of Staff, 2015) (Nye, 2010) of international relationships as it would increase the complexity of the model at this stage. Also, the current version of the model excludes individuals' perceptions and formative experiences at political decision making to allow iterative and coherent build-up. (Fuerth, 2009) The following sections will explain how the hypothetical model was tested using a case study of contemporary Russian operations in cyberspace.

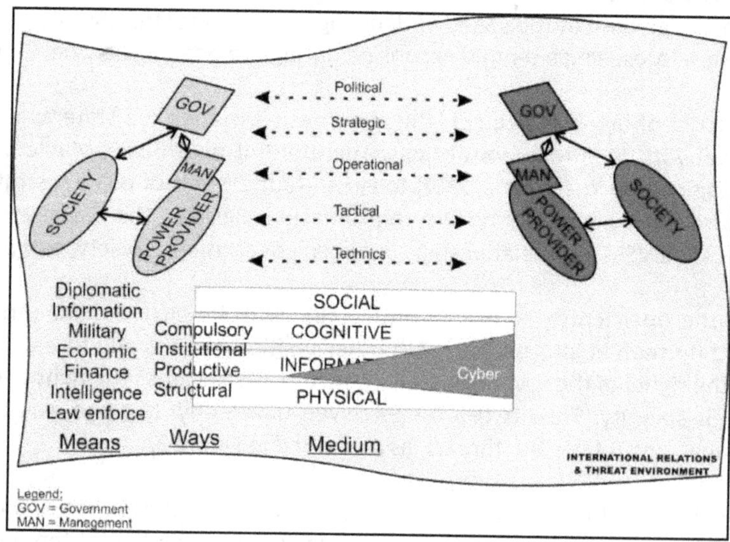

Figure 3: Hypothetical model for state-level power system using cyber domain

3. Research design

The research process follows the approach of case theory as a process of gap analysis, hypothesis, evidence collection, hypothesis testing and interpretation of results. (Gummersson, 2017, pp. 195–196) The research aims to improve the model of understanding the state-level cyber powers utilisation in a context of the broader spectrum of international relations and focus on a turbulent, man-made domain called cyberspace. The research approach is pragmatic (Creswell, 2014, pp. 10-11) to create an artefact to model the cyber domain's complex and emerging nature. The viewpoint is more from complexity than positivist posture (Gummersson, 2017 pp. 49-56). Nevertheless, the pragmatic aim requires a model that reflects meaningful simplicity amid apparent disorderly complexity (Simon, 1957).

The model of wielding state-level cyber power is assessed in a case study. Data is a sample of the latest Russian operations, especially from the action along the lines of cyber domain as part of their overall operational behaviour. Russians have used the cyber domain as one line of operation to change the behaviour of Estonian, Georgian, and Ukrainian societies. The cyber incident data should provide evidence of tactical level action. The overall conduction of hybrid operation should provide evidence on thinking at the level of operational art. Finally, the historical and contemporary strategic data should indicate the Russian approach at the strategical level of cyber policies.

Furthermore, the period from 2007 to 2014 provides a view of how cyberspace is extending and how power utilisation is evolving and exploiting the emerging features of information technology on the Internet. The span of data should ensure a sufficient longitudinal line of research to the dynamic nature of cyber operations. However, the time span also increases the complexity of the model and exposes it to ambiguities of system dynamics (Jackson, 2019, pp. 233-240).

The model attempts to meet the system dynamics' expectations by recognising the structure's four hierarchies (Forrester, 1969): boundary around the system, feedback loops within the boundary, level variables representing accumulations and rate variables representing activity within the feedback loops. Naturally, the paper is not aiming to create a holistic system model but a simplified understanding of possible real-life phenomena. The benefits of simplicity include the relations between the organisation and its environment together with sub-systems and their relations with the environment. Naturally, the thinking organisation view does not necessarily illustrate the features of society as living holacracy (Robertson, 2015). For example, it does not reflect well the confrontation between two political entities (Wittes & Blum, 2015) nor the causalities in the case of a failing nation (Acemoglu & Robinson, 2013). Moreover, data has been collected from English sources only and, therefore, possess a bias of western cyber thinking. The bias is recognised, and research tries to remedy it by sourcing from a broader selection of English publications.

4. Results and discussion

The case study of how well Russian cyber behaviour can be explained by using the features of the hypothetic model is presented in Table 2. The analysis is a sample of results focusing only on the outstanding features of the model from section 2.1 and their support in rationalising Russian behaviour. The aim is to prove the feasibility of the hypothetical power model.

Table 2: Using the hypothetical power model to explain Russian action in cyberspace

Outstanding features of the power model	Documented Russian behaviour	Explanation based on the model
A. Emerging cyberspace	1.2007 combines traditional disinformation operation with DDoS attack in Estonia.	Russia utilises emerging opportunities to divide the opponent's government from its society and create unrest. => Quick exploitation of novel avenues of attack while using criminal hackers as a power provider.
	2.2011 "Under today's conditions, means of information influence have reached a level of development such that they are capable of resolving strategic tasks." (Giles, 2016)	Indicating the strategic role, Russians see how cyberspace opens the avenues of impact to state decision making and public opinion. => Evolution of realms
	3.2011 "Disinformation is a Russian technique to manipulate perceptions and information of people." (Thomas, 2011)	Societies that use extensively social media platforms are exposed to Russian disinformation operations via troll factories. => Digitalisation of information

Outstanding features of the power model	Documented Russian behaviour	Explanation based on the model
	4.2016 Russian Information Security Doctrine defined the information sphere that includes the technical and cognitive components. (Lilly & Cheravitch, 2020)	Russia recognises a broader sphere of an effect than just the technical layer of cyberspace. => both offensive and defensive cyber power providers.
B. Dynamics of state	1. Information war aims "causing damage to information systems, processes and resources, critically important and other structures, subverting the political, economic and social systems, mass psychological work on the population to destabilise society and state, and coercing the government to take decisions in the interest of opposing side." (Giles, 2012) 2.1990 – early 2000's FSB employed illegal hackers to attack financial actors in the US and Europe. Since 2013 the GRU has been building a militarily organised information operations force. (Lilly & Cheravitch, 2020) 3.2008 Georgian operation included a reflexive control through a combination of the pressure of force, opponent's formulation of the initial situation, shaping opponent's objectives, shaping opponent's decision making and the choice of the decision-making moment. (Blandy, 2009) 4.2014 Russia established the National Defence Control Centre for central planning, coordination and command of all government agencies, state corporations and military commands. It was used to manage Russian involvement in the Syrian Civil War. (Clark, 2020) 5. As observed in the 2020s: "But its intensive focus on asymmetric measures, and in particular the utility of information warfare for exerting control without the need for overt military intervention, means that the threat from Russian expansionism is far more diverse and nuanced." (Giles, 2021)	Since the 2011 doctrine release, Russia has understood the information layer as a medium to impact the social and decision-making behaviour of the opposing side. => Medium They deem cyberspace part of the information realm (information systems) and extends to the physical realm (critical structures). =>Cyber As Internet dependability has evolved, Russia has built a professional force to run offensive operations on the cyber domain. => Initial operational capability achieved with available assets following the build-up of organised full operational capabilities. Impact through cyberspace was used as one avenue to manipulate opponent's operational level sense- and decision making. => operational course of action included actions along the cyber line of operation. Russia improves its coordination of actions through all lines of operation and shortens the feedback loop between tactical-operational-strategic levels of command. => Improving the orchestration of power provider networks in their particular lines of operation. International effect using means and ways suitable for a cyber domain are frequently applied at strategic, operational, and tactical levels. => Cyber is one domain, but the effect is created through courses of action overall domains and lines of operation.
C. Vulnerabilities and maturities of cyber-related abilities	1. "How can you successfully wage an information struggle if during Chechnya a significant part of the mass media is taking the side of the specialists? We need a law on information security." (Giles, 2012) 2. In the 2014 Ukraine operation Russia was using three different metanarratives distributed through social media, websites, and mass media in coordination with cyber suppression of opposite sources. (Pynnöniemi & Racz, 2016)	The Russian leadership is securing their control over domestic mass media while suppressing the foreign-owned social media to maintain the lines of control over Russian society. => Means & State as system Russia used open western social media to promote their narratives while suppressing Ukrainian sources through cyber means. => Exploitation of opportunities and vulnerabilities of international cyberspace

Outstanding features of the power model	Documented Russian behaviour	Explanation based on the model
	3.2015 Russian military scientists expounded that cyber weapons could endanger not only critical infrastructure but also military systems. (Lilly & Cheravitch, 2020)	Russia recognises the expansion of digitalisation both in the state's critical infrastructure and in military systems. => Digitalisation

The conclusions in Table 2 indicate that the hypothetical power model can assist in the post-analysis of cyber-related strategies and operations, at least in the case of contemporary Russian behaviour. First, understanding the emergent nature of cyberspace supported the recognition of Russian readiness to use novel ways even when it is not organised by the government but leased from the public sector. However, the agile adaptation happened within the traditional Russian doctrine in joint information and kinetic operations. Hence, the understanding of operational level action requires strategic and policy level foundation and shows that focusing only on cyber technical incidents does not provide foresight because it is not reflected in the broader picture of force projection.

Second, the long period revealed the evolution in the structure of the state concerning cyber capabilities. Russians have been building up their cyber capabilities via organising, recruiting, R&D and improving the control of joint operations. Therefore, the model needs to support the dynamics of the state system over time. Meanwhile, the political and strategical agenda has not evolved that much. On the contrary, there are some similarities in the USSR era policies excluding cyber capabilities. Hence, the model needs to understand the hierarchy of control of the emerging ways as part of the legacy means of projecting international force.

Third, the model needs to illustrate the state as an actor and a target since the case of Russia shows how they, while realising their vulnerabilities and commenced mitigation, exploited cyber vulnerabilities in other countries and the international environment. The effort Russia invested in research and development in cyber-related capabilities also indicate the need for modelling the evolution of cyber vulnerability. In conclusion, the main argument is that cyberspace should not be considered an isolated domain or line of operation. Evidently, a state-level actor like Russia uses emerging cyberspace to manipulate the opponent's behaviour as part of other means and ways.

5. Conclusion

The paper presents and validates a model to improve the understanding of state-level use of cyber power in the context of international confrontation. Since the existing models approach the topic from narrower views, the research creates a hypothetical model based on system dynamics to improve understanding in a broader context. Therefore, the model focuses on essential components of a system: state as an actor, environment, the medium that provides the lines of operation and levels of control. Once composed, the model is tested with cyber activity data of Russia spanning over the time of 2007 – 2014 to reflect the evolving nature of cyberspace.

Assessing the feasibility of the proposed model concentrates on three system dynamic features:

- 1. The emergent nature of cyberspace is essential to understand since Russia has been ready to use novel ways even when the capability is not organised but leased. However, since Russia has quickly adapted new ways as part of its traditional doctrine in joint information and kinetic operations, a narrow focus on cyber behaviour does not provide foresight because it is not reflected in the broader picture of force projection.
- 2. Dynamic state features are feasible to model and understand since Russia has been building its cyber capabilities, organising it for greater performance and improved control over joint operations, including action in the cyber domain.
- 3. Vulnerabilities and maturities related to digital capabilities are essential to comprehend as part of the model since case Russia shows how they have realised their vulnerabilities and commenced efforts to mitigate them, exploit them in an international environment and research emerging vulnerabilities.

The proposed model for cyber-related power projection at state-level confrontations is still in its early versions but is already adding value to the existing knowledge base with its system dynamics approach missing from contemporary models. Furthermore, the theoretical approach provides a base for expansion towards capturing more complex dimensions in the equation. Additionally, the benefits for cyber strategy analysts became concretely evident when analysing the Russian case study. The three system dynamic features of the model opened a more holistic foundation of understanding of each operation.

Naturally, the model and its assessment are in the early phase and have many limitations. First, the case study concerned only one actor and not a typical confrontation of two or more actors. Second, the data may be biased because of English sources. Third, the model is still far from the maturity required for being programmable. Fourth, many dimensions and effectors were left outside this model version, requiring further study. Finally, building on the selected approach and system dynamic foundation requires theoretical research, testing, and evaluation to further mature and extend the model.

References

Acemoglu, D. & Robinson, J. A., 2013. Why nations fail. 2nd ed. London: Profile Books, Ltd.

Amedie, J., 2015. The impact of social media on society. Advanced Writing: Pop Culture Intersections, Issue 2.

Annan, K., 1999. Two concepts of sovereignty. [Online] Available at: https://www.un.org/sg/en/content/sg/articles/1999-09-18/two-concepts-sovereignty [Retrieved September 2021].

Armstrong, A. H., 2019. Challenges to coordinating the instruments of national power. Baltimore: Johns Hopkins University.

Arnold, R.D., Wade, J.P., 2015. A definition of systems thinking: A systems approach. Procedia Computer Science, 44 pp. 669-678

Arquilla, J. & Ronfeldt, D., 1993. Cyberwar is Coming! Comparative Strategy, Vol 12(2), pp. 141-165.

Barnett, M. & Duvall, R., 2005. Power in International Politics. International Organization, 59(1), pp. 39-75.

Betz, D. J. & Stevens, T., 2011. Cyberspace and the state - toward a strategy for cyber-power. London: Routledge.

Blandy, C., 2009. Provocation, deception, entrapment - The Russo-Georgia five day war. Shrivenham: Defence Academy of the United Kingdom.

Cederberg, A., 2020. A comprehensive cyber security approach - The Finnish model, Helsinki: Cyberwatch Finland.

Clarke, R. A. & Knake, R. K., 2010. Cyber war. New York: HarperCollins Publishers.

Clark, M., 2020. Russian hybrid warfare. Washington DC: Institute for the Study of War.

Clausewitz, C. v., 1984. On War. New Jersey: Princeton University Press.

Creswell, J. W., 2014. Research design. London: SAGE Publication Inc.

Dwyer, D. S. & Kreier, R., 2015. Internet and Cell Phone Dependence: Too much of a good thing? Melville, NY, Stony Brook.

Electrical 4U, 2020. Moore's Law and The Exponential Growth of Technology. [Online] Available at: https://www.electrical4u.com/moores-law/[Retrieved September 2021].

DoD, 2018. Joint concept for operating in the information environment, JCOIE. Washington DC: The Joint Chiefs of Staff

DoD, 2021. DOD Dictionary of Military and Associated Terms. Washington DC: The Joint Chiefs of Staff

Fiott, D., 2020. Digitalising defence. [Online] Available at: https://www.iss.europa.eu/content/digitalising-defence

Forrester, J. W., 1969. Urban dynamics. Cambridge: MIT Press.

Freedman, L., 2017. The future of war - a history. London: Penguin Books Ltd.

Fuerth, L. 2009 "Cyberpower from the Presidential Perspective." In Cyberpower and National Security, edited by Franklin D. Kramer, Stuart H. Starr, and Larry K. Wentz, 557–62. University of Nebraska Press.

Giles, K., 2012. Russia's public stance on cyberspace issues. Tallinn, NATO CCD COE.

Giles, K., 2016. Handbook of Russian information warfare. Rome: NATO Defense College.

Giles, K., 2021. What deters Russia - Enduring principles for responding to Moscow. London: Chatham House.

Gummersson, E., 2017. Case theory in business and management. London: SAGE Publications Ltd.

International Institute for Strategic Studies, 2021. Cyber capabilities and national power: a net assessment. [Online] Available at: https://www.iiss.org/blogs/research-paper/2021/06/cyber-capabilities-national-power

IOT News, 2020. The IoT in 2030: 24 billion connected things generating $1.5 trillion. [Online] Available at: https://iotbusinessnews.com/2020/05/20/03177-the-iot-in-2030-24-billion-connected-things-generating-1-5-trillion/[Retrieved August 2021].

Jackson, M. C., 2019. Critical systems thinking and the management of complexity. Chichester: Wiley.

Joint Chiefs of Staff, 2015. The national military strategy of the United States of America 2015. Washington D.C.: U.S. DoD

Kaldor, M., 2012. New and old ward. 3rd Edition. Stanford: Stanford University Press.

Krezer, M. P., 2021. Cyberspace is an analogy, not a domain. [Online] Available at: https://thestrategybridge.org/the-bridge/2021/7/8/cyberspace-is-an-analogy-not-a-domain-rethinking-domains-and-layers-of-warfare-for-the-information-age[Accessed September 2021].

Liddle Hart, B. H., 1991. Strategy. 2nd Revised Edition. London: Penguin Books Ltd.

Lilly, B. & Cheravitch, J., 2020. The past, present, and future of Russia's cyber strategy and forces. Tallinn, NATO CCD COE.

Lowe, D., Espinosa, A., Yearworth, M., 2020. "Constitutive rules for guiding the use of the viable system model: Reflections on practice." European Journal of Operational Research, 287, pp.1014-1035.

Mandelcorn, S. M., 2013. An explanatory model of motivation for cyber-attacks drawn from criminological theories. College Park: University of Maryland.

Massachusetts Institute of Technology, 2020. Cyberpower, Cybersecurity and Cyberconflict. [Online] Available at: https://ecir.mit.edu/research/cyberpower-cybersecurity-and-cyberconflict

Mattila, J. K., 2014. Protecting National Assets against Information Operations in Post-modern World. Abu Dhabi, Proceedings of the 2nd BCS International IT Conference.

McCamey, R., Wilson, B. & Shaw, J., 2015. Internet dependency and academic performance. The Journal of Social Media in Society, pp. 126 - 150.

Merriam-Webster, 2022. Essential meaning of meme. [Online] Available at: https://www.merriam-webster.com/dictionary/meme[Retrieved January 2022].

Miguel de Bustos, J. C. & Izquierdo-Castillo, J., 2019. JC Miguel de Bustos, J Izquierdo-Castillo (2019): "Who will control the media? The impact of GAFAM on the media industries in the digital economy. Revista Latina de Comunicación Social, 74, pp. 803-821.

Mulrenan, S., 2020. China's tech giants take on the FAANGs. [Online] Available at: https://www.ibanet.org/article/D40AD0EC-8C8D-444F-8DB8-431E4F181576

Nye, J. S. J., 2010. Cyber power. Cambridge: Belfer Center for Science and International Affairs.

Nye, J. S. J., 2011. The future of power. New York: Public Affairs.

Oliver, E. J. & Wood, T. J., 2014. Conspiracy theories and the paranoid style of mass opinion. American Journal of Political Science, Vol. 58(4).

Pynnöniemi, K. & Racz, A., 2016. Fog of falsehood - Russian strategy of deception and the conflict in Ukraine. Helsinki: The Finnish Institute of International Affairs.

Randewich, N. & Ahmed, S. I., 2022. Apple's $3 trillion market value follows 5800% gain since iPhone debut. [Online] Available at: https://www.reuters.com./technology/apples-3-trillion-market-value-follows-5800-gain-since-iphone-debut-2022-01-03/[Retrieved January 2022].

Rid, T., 2013. Cyber war will not take place. Oxford: Oxford University Press.

Robertson, B. J., 2015. Holacracy. New York: Henry Holt and Company, LLC.

Rollins, J. & Wilson, C., 2007. Terrorist Capabilities for Cyberattack: Overview and policy issues, Washington DC: Congressional Research Service.

Roser, M., 2020. The short history of global living conditions and why it matters that we know it. [Online] Available at: https://ourworldindata.org/a-history-of-global-living-conditions-in-5-charts

Rossling, H., Rosling Rönnlund, A. & Rosling, O., 2018. Factfulness. Hodder & Stoughton Ltd.

Scherrer, J. H. & Grund, W. C., 2009. A cyberspace command and control model. Maxwell AFB: Air War College.

Schilling, D. R., 2013. Knowledge Doubling Every 12 Months, soon to be Every 12 Hours. [Online] Available at: https://www.industrytap.com/knowledge-doubling-every-12-months-soon-to-be-every-12-hours/3950[Retrieved September 2021].

Schultz, J., 2019. How Much Data is Created on the Internet Each Day? [Online] Available at: https://blog.microfocus.com/how-much-data-is-created-on-the-internet-each-day/[Retrieved August 2021].

Schwab, K., 2016. The fourth industrial revolution. Geneva: World Economic Forum.

Simon, H., 1957. A behavioral model of rational choice. In publication: In models of man, social and rational: Mathematical essays on rational human behaviour in a social setting. New York: Wiley.

Strachan, H., 2013. The direction of war. Cambridge: University Printing House.

Sun, T., 2014. The Art of War - Illustrated edition. New York: Fall River Press.

Suvorov, A., 2015. Voittamisen taito (Art of Winning) Jyväskylä: Docendo Oy.

Tabansky, L., 2021. Towards a theory of cyber power: Security studies, mete-governance, national innovation system. [Online] Available at: https://en-cyber.tau.ac.il/research/theoryofcyberpower

Theohary, C. A. & Rollins, J. W., 2015. Cyberwarfare and Cyberterrorism: In Brief, Washington DC: Congressional Research Service.

Thomas, T. L., 2011. Recasting the red star. Fort Leavenworth: Foreign Military Studies Office.

Toffler, A., 1981. The third wave. New York: Bantam Books.

U.S. DoD, 2015. The center of gravity - systemically understood. Middletown: U.S. Army TRADOC.

Van Haaster, J., 2016. Assessing cyber power. Tallinn, NATO CCD COE Publications.

Vego, M. N., 2007. Joint Operational Warfare. Theory and practice. Newport: Naval War College.

Voo, J. et al., 2020. National cyber power index 2020, Cambridge: Belfer Center for Science and International Affairs.

Wikipedia, 2021. List of public corporations by market capitalisation. [Online] Available at: https://en.wikipedia.org/wiki/List_of_public_corporations_by_market_capitalization

Willis, H. H., 2006. Guiding Resource Allocations Based on Terrorism Risk. [Online] Available at: https://www.rand.org/pubs/working_papers/WR371.html.

Wittes, B. & Blum, G., 2015. The future of violence. New York: Basic Books.

Zaballos, A. G. & Herranz, F. G., 2013. From cybersecurity to cybercrime - a framework for analysis and implementation. Inter-American Development Bank.

Zuboff, S., 2019. The age of surveillance capitalism. New York: Hachette Book Group, Inc.

Building Software Applications Securely With DevSecOps: A Socio-Technical Perspective

Rennie Naidoo and Nicolaas Möller
University of Pretoria, Pretoria, South Africa
rennie.naidoo@up.ac.za
nicolaas.möller@up.ac.za

Abstract: While continuous real-time software delivery practices induced by agile software development approaches create new business opportunities for organizations, these practices also present new security challenges in the DevOps environment. DevSecOps attempts to incorporate advanced automated security practices for agility in the DevOps environment. Mainstream perspectives of DevSecOps tend to overlook the collaborative role played by social actors and their relations with technologies in securing software applications in organizations. The first perspective emphasises the use of technologies such as containers, microservices, cryptographic protocols and origin authentication to secure the continuous deployment pipeline. The other dominant perspective focuses almost exclusively on the social aspects such as organizational silos, culture, and team collaboration. Such one-sided perspectives neglect the socio-technical argument that secure software applications from continuous deployment emerges when developers, quality assurers, operators and security experts combine their collective expertise together with DevSecOps technologies. The article presents a socio-technical framework of DevSecOps based on a systematic literature review. The review focused primarily, but not exclusively, on the computing and information systems literature and identified 26 peer reviewed articles from 2016 to 2020 which met the quality criteria and contributed to the analysis. The authors used a critical appraisal checklist and member checking to assess the quality of the articles. The authors then used thematic analysis to develop a comprehensive framework for DevSecOps based on the insights from these articles and a socio-technical lens. The socio-technical framework can be used by practitioners to perform a more holistic analysis of their DevSecOps practices. It highlights the key social and technical themes that underpin the effectiveness of DevSecOps and how insights about these themes can be used by practitioners to improve the instrumental and humanistic goals of DevSecOps. An interdisciplinary approach is proposed to adequately address challenging socio-technical relationships in DevSecOps. Future research can empirically test the importance of the interplay between technology and human activities to improve the overall performance of DevSecOps and other domains in cyber warfare and security.

Keywords: culture, continuous deployment, DevOps, DevSecOps, security, socio-technical

1. Introduction

Agile development practices enable organizations to continuously deliver software products and services. Since agile development teams commit many small and frequent deployments to production, failure to involve the operations team earlier in the software lifecycle tends to become a source of constraint in the software delivery process. DevOps seeks to promote cross-functional collaboration between the development and operations teams. The semi-automation and full automation of build, deployment, and testing tasks is also a critical capability in improving overall software delivery performance. However, organizations adopting DevOps practices often struggle to manage the tensions between the goals of shortening the development cycle and the faster delivery of features pursued by the development teams and the stability goals pursued by the operations teams. Of greater concern, both these teams tend to neglect security vulnerabilities that threat actors can exploit.

Neil MacDonald (2012) of Gartner initially coined the term DevOpsSec to draw attention to the need to incorporate information security within DevOps practices to balance speed, agility, and security. DevSecOps, as it is more commonly known, extends the objective of DevOps by advocating shift left security, security by design and continuous security testing. By integrating the security team with the software development and operations teams, team members can pay joint attention to information security matters throughout the software development lifecycle (Mansfield-Devine, 2018).

The distinctions between terms such as DevOpsSec, DevSecOps and SecDevOps are not clear in the academic literature. In the grey literature, the placement of "Sec" in the term appears to signify the priority given to Security (Myrbakken and Colomo-Palacios, 2017; Mohan and Othmane, 2016; Rahman and Williams, 2016). DevOpsSec is seen to prioritise development and operations at the expense of security. DevSecOps represent an improvement in the security culture but still prioritises development processes. Meanwhile SecDevOps is the ideal term for security evangelists as it prioritises security processes throughout the development lifecycle

(Mohan and Othmane, 2016). In the academic literature, these terms are often used interchangeably, and some authors have found that the term DevSecOps has become increasingly accepted by practitioners (Myrbakken and Colomo-Palacios, 2017).

In DevSecOps, information security is also emphasised early in the development lifecycle. DevSecOps also uses tools to automate the insertion of security features into software applications. Whereas the waterfall model often relied on the use of a single or few tools, Agile, DevOps and DevSecOps transformations involve an overwhelming number of diverse and specialised tools for planning, tracking, automation, and management tasks (Kersten, 2018). A Tasktop survey of 300 Enterprise IT organizations found that 70% of these organizations integrated three or more tools and that 40 percent integrated four or more tools in their toolchains (Kersten, 2018). The same survey also found that a number of software vendors have been emerging recently to provide tools to support the DevSecOps environment. While high automation has been effective in improving DevOps capabilities, some experts argue that assessing and testing security can be difficult to automate (Mansfield-Devine, 2018). For this reason, the successful transition to DevSecOps goes beyond implementing security into the DevOps toolchain by emphasising the human talent. To build an information security culture, organizations also need to address behavioural changes within the development team and the operations team (Mansfield-Devine, 2018). Integrating the security team with the development teams and operations teams to work in collaboration as an effective cross-functional team and ensuring that security is included in every stage of the software development lifecycle can be a formidable challenge.

Trends outside organizational boundaries also present a formidable challenge. According to IBM (2021), the average global cost of a data breach now exceeds $4 million. Despite increasing regulation by the EU General Data Protection Regulation (GDPR) and the European Union Agency for Network and Information Security (ENISA), the increasing trend towards developing cloud-based services and applications using agile development processes also presents major security concerns (Kumar and Goyal, 2020). The COVID-19 global pandemic that has given impetus for executing work-from-anywhere using critical software applications is adding to these security vulnerabilities (Naidoo, 2020). The same IBM report found that the average cost of breaches was $1.07 million higher in organizations supporting remote work.

Markets and Markets (2021) predicts that DevSecOps will grow at a compound annual growth rate of 31.2% reaching $5.9b in 2023. Since DevSecOps is a fairly new trend, the many challenges facing DevSecOps work practices have not been sufficiently addressed in the emerging DevSecOps literature. The review presented in this paper attempts to address these concerns by applying a sociotechnical systems (STS) approach as a framework to provide a more holistic analysis of the social and the technical challenges facing current DevSecOps practices. In addition, this review aims to provide researchers with gaps in current research.

The rest of the paper is organized as follows: first, we outline the socio-technical work system (STWS) framework as a basis for our analysis. Second, we present our systematic literature approach to review the selected DevSecOps literature in more detail. We then present and discuss the results. Finally, we draw theoretical and practical implications for the future of DevSecOps before concluding the paper.

2. Conceptual foundations

We conceive DevSecOps to be a socio-technical work system (STWS). According to Alter (2013), a work system can be defined as “a system in which human participants and/or machines perform work (processes and activities) using information, technology, and other resources to produce specific products/services for specific internal and/or external customers.” A socio-technical lens considers both the social and technical sub-systems of a work system (Sarker et al, 2019). The social sub-system is people oriented and focuses on individuals, their relationships, reward systems and authority structures (Bostrom and Heinen, 1977). The technical sub-system includes tasks, processes and technologies for achieving objectives or outcomes (Bostrom and Heinen, 1977). In a STWS, the fit between the social and the technological subsystems determines the effectiveness of the work system (Sarker et al, 2019). This requires the joint optimisation of both systems in improvement efforts. Improving the performance of STWS have instrumental and humanistic outcomes or objectives. Instrumental objectives are concerned with achieving economic objectives whereas the humanistic objectives are concerned with enhanced job satisfaction and higher quality of working life (Bostrom and Heinen, 1977). Figure 1 depicts the socio-technical work system model which will be used as an initial sensitising framework to assess how researchers are studying DevSecOps.

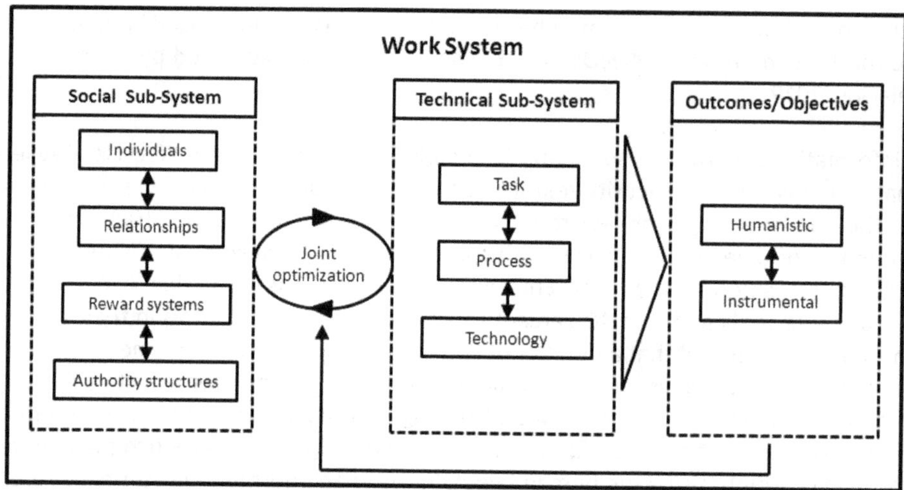

Figure 1: DevSecOps as a Socio-Technical Work System (adapted from Bostrom and Heinen, 1977; Sarker et al, 2019)

Figure 2 shows how Sarker et al. (2019) uses six types to characterise how IT researchers study the socio-technical perspective and its influence on outcomes/objectives: Type I studies are *predominantly social* and focus mainly on how human factors explain outcomes in technology-mediated work systems. Type II or *social imperative* studies consider how social aspects influence the technical component and outcomes. Type III studies consider how *social-technical factors additively deliver outcomes*. These studies assume that there is no interplay between technical and social components. Type IV studies consider how the *socio-technical interplay delivers outcomes*. Type V or *technical Imperative* studies assume that technology is a significant antecedent to social outcomes. Type VI studies are *predominantly technical* and focuses on how to develop or improve the technical component of a work system with little or no consideration of the social component. The STWS lens and these six types are appropriate for analysing the DevSecOps literature as research should strive to provide a balanced focus on both the social and the technical subsystems and the optimal interaction between these subsystems so that organizations achieve both their humanistic and instrumental objectives of DevSecOps. The purpose of this paper is to understand to what extent the literature considers the interplay of the social and technical within a DevSecOps work system in delivering outcomes or objectives.

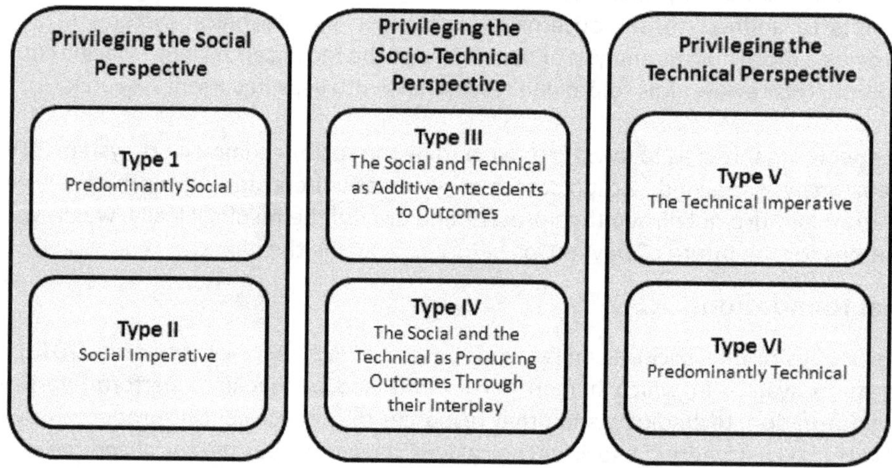

Figure 2: Types of socio-technical research (adapted from Sarker et al, 2019)

3. Research method

We conducted a systematic literature review on DevSecOps. Our inclusion criteria were as follows: The article contents should be about implementing DevOps and focus on security. Based on a preliminary analysis, our final search string was composed as follows: ("DevOps" OR "DevSecOps") AND ("security" OR "secur*" OR "cybersecur*") AND ("applications" OR "software") AND ("develop*" OR "build"). We search the following databases using the defined search strings: ScienceDirect, IEEE Xplore, and ABI/INFORM Collection. We filtered the source types to include journals only. Furthermore, we considered only English publications. 141 one articles were eligible for further analysis.

We applied the following filtering process to select the relevant literature. First, articles were identified by using our defined search string. Next, we removed duplicate articles from the source list. The remaining articles were then screened, by reading the abstract of each article. After reading the abstracts, articles that did not support the research question were excluded from the source list. Finally, the screened sources were assessed for eligibility by reading the entire article. Table 1 shows the results achieved in each step.

Table 1: Results of the filtering process

Filter process steps	Results
Identify articles	141
Remove duplicates	123
Screen abstract	60
Screen full text	28

To ensure that the article was relevant, quality assessment criteria in the form of questions were created to determine if security themes in a DevOps or similar environment was adequately discussed. One of the authors engaged in member checking to check the accuracy of the filtering process and plausibility of the thematic analysis (Yin, 2014). Table 2 depicts details of journal publications. The majority of articles are from technically oriented disciplines such as software engineering, network security, computer science and computer security.

Table 2: Journal publication details

Journal	No. of papers
IEEE Software	12
Network Security	5
IEEE Internet Computing	2
Journal of Management Information and Decision Sciences	1
Computer	1
Computer Fraud & Security	1
IET Software	1
Computers & Security	1
Journal of Systems and Software	1
Computing in Science & Engineering	1
AI & Society	1
IEEE Access	1

4. Results

We draw on the socio-technical work system framework for DevSecOps (see Figure 1) to present our results. The framework includes three practice categories and 11 practice dimensions. Figure 3 presents the six types of socio-technical perspectives and their influence on social and instrumental outcomes/objectives. Table 3 provides a condensed overview of how the framework can be used to assess the challenges facing organizations in jointly optimising their DevSecOps work system. Figure 3 shows Type VI, which refers to predominantly technical studies that lacks the consideration of humanistic outcomes, and Type III, which focuses on how social-technical factors additively deliver outcomes, were the two dominant perspectives adopted by DevSecOps researchers.

The majority of Type VI studies were focused on how to improve tasks, processes and technologies to improve the instrumental outcomes as a measure of DevSecOps performance (McGraw, 2017; Casola et al, 2020; Almuairfi and Alenezi, 2020; Kersten, 2018). Within these group of studies, references were made to security policies, threat modelling and risk assessment processes, tasks such as code reviews, application security testing, static analysis, software composition analysis and dynamic analysis, dynamic application security testing (DAST), interactive application security testing, penetration testing, and technologies such as software containers, secure cloud applications and security tools.

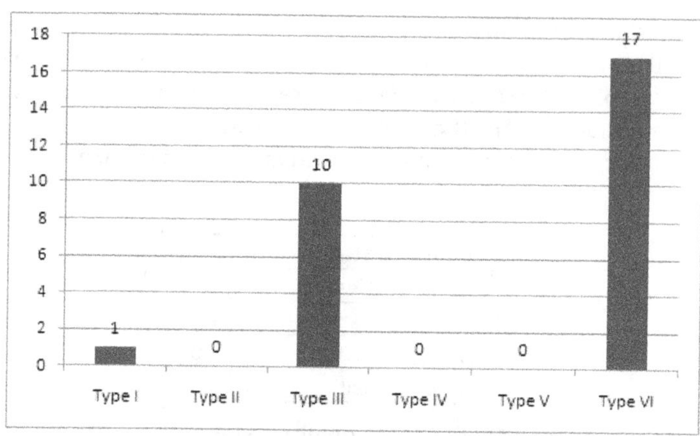

Figure 3: Number of DevSecOps studies by Types of Socio-technical Research

There was only one study for Type I where behavioural factors of developers, operators and security staff were seen to explicitly influence the outcome of the DevSecOps work system (Carter, 2017). This predominantly social study highlighted the importance of culture, inclusion, knowledge sharing, teamwork, and security training in developing a high-performance security team. A number of studies appear to belong to Type III and focused on how social and technical factors additively deliver outcomes (Bass, 2018; Mansfield-Devine, 2018; Tufin, 2020). We grouped our analysis of the social sub-system practice category by individuals, relationships, reward systems, and authority structures. The literature focuses on social actors such as individuals (Operators, Developers, Security Champion, Security Experts, End Users, Product Owners, and Project Managers) and teams (Operations teams, DevOps Teams, Security Teams). This literature also emphasises the importance of collaboration, communication and feedback between the Security team and DevOps teams in realizing continuous security. Apart from teamwork, the knowledge, skills, attitudes and behaviour of individuals and teams were highlighted (Bass, 2018; Mansfield-Devine, 2018; Tufin, 2020).

We found no study that viewed DevSecOps as explicitly consisting of two sub-systems, DevSecOps processes to achieve joint optimisation of these sub-systems, and the socio-technical interplay that generates the outcomes for the entire DevSecOps work system. Our analysis also did not reveal any Type II or social imperative studies. This is not surprising given the role of technology in achieving positive outcomes in complex DevSecOps environments. Surprisingly, there were no Type V or technical Imperative studies which focused solely on technology and its influence in achieving humanistic outcomes.

Table 3: Selected themes from the literature

Practice Category	Practice Dimensions	Themes
Social Sub-System	Individuals	The social impact of the changing work roles of developers, operators, security experts and end-users.
		The evolving security threats posed by external and internal actors.
	Relationships	The challenge of developing a collaborative cross-functional team that communicates effectively.
	Reward Systems	The challenge of aligning incentives to overall goals (safe and secure software) that conflicts with local goals (speed versus stability).
	Authority Structures	Escalating potential security threats to the product manager or business representatives.
Technical Sub-System	Tasks	Performing application security testing using code reviews, static analysis, software composition analysis, dynamic analysis and penetration testing.
	Process	The use of threat modelling to inform the risk assessment process.
	Technology	Full or Semi-Automation using Dynamic Application Security Testing (DAST) or Interactive Application Security Testing (IAST) tools.
Outcomes/Objectives	Humanistic (Positive)	The benefits of job enrichment and job enlargement
		The downside of job enlargement

Practice Category	Practice Dimensions	Themes
	Humanistic (Negative)	Teams at loggerheads (e.g. developer resistance).
	Instrumental (Positive)	Security issues are identified and fixed much earlier in the lifecycle.
		Implementing security requirements using automation to reduce delays.
		Agility and velocity in delivering time-to-market applications and services in a cost-effective manner.
	Instrumental (Negative)	Time consuming and resource intensive security activities slow down the pipeline.
		The cost of security breaches.

5. Discussion

Our results suggest that studying the interplay of the social and technical within a DevSecOps work system in delivering both instrumental and humanistic outcomes and objectives is an understudied area. Given that the majority of review articles were from technically oriented disciplines such as systems engineering, computer science and computer security, it is not surprising that these articles' focus was predominantly technical. However, many of these articles did acknowledge the importance of social factors to varying degrees. Our study suggest three important avenues for future research: First, this study complements prior work on socio-technical work systems by specifying the sub-systems, dimensions, challenges and outcomes that are more salient in the DevSecOps work system (Bostrom and Heinen, 1977; Alter, 2013; Sarker et al, 2019). Our study goes beyond prior research in DevOps and DevSecOps that emphasise people, process and technology factors by specifying a richer set of interacting elements in socio-technical work systems (Kumar and Goyal, 2020). It would be interesting to see future research studies focusing on a richer set of socio-technical concepts especially in organizational contexts where the socio-technical may interplay. Second, we infer from our analysis that DevSecOps may raise concerns about the impact of technology on the social sub-systems. For example, workers may be concerned that automation technologies will be used to replace staff. These types of social impacts have been under researched. Third, many of the studies considered DevSecOps as a vehicle to achieve instrumental objectives/outcomes without considering the social outcome, that is, how new work role transitions in DevSecOps influences job satisfaction and employee well-being. Instead, instrumental outcomes such as reducing time to market delays and cost-effectiveness were emphasized. Further studies should emphasize social outcomes at the outset for two reasons. Firstly, a socio-technical perspective strives to humanise the DevSecOps work environment which can be beneficial when teams are at loggerheads or individuals resist the change to DevSecOps. Secondly, the social dimension can improve DevSecOps performance, which has not been explicitly studied by the articles in this review.

Viewing and analysing DevSecOps using a socio-technical work system framework offers the following specific questions that could be promising for future research:

1. How do systems outside the organizational boundary influence DevSecOps work practices?
2. How do successful DevSecOps environment manage individual and/or team resistance to change?
3. What interventions are used to align the technical and social environments?
4. To what extent does organizational culture constrain and enable DevSecOps work practices?
5. To what extent does work role changes in transitioning to DevSecOps result in role-related stress?
6. How can the interplay between technology and human activities improve the performance of DevSecOps?

A sociotechnical perspective also offers new possibilities for refining prior models on cyber warfare and security that tend to be either sociocentric or technocentric (Sánchez-Gordón and Colomo-Palacios, 2020; Fletcher and Smith, 2020; Huskaj, 2019). The proposed socio-technical model offers an analytical approach that focuses on the interplay between technology and human activities, arguably providing more balanced insights about cyber challenges in domains such as cyber conflict, cyber terrorism, cyber security and information warfare (Izycki and Wallier, 2021; Huskaj, 2019). A socio-technical perspective is also salient to conceptualizing key cyber challenges. For example, by drawing attention to the interactions between the social and the technological subsystems a socio-technical perspective can offer a novel conceptualization of cyber resilience (Fletcher and Smith, 2020).

Pedagogic practices in cyber warfare and security vocational training and education can also benefit from understanding the entwined nature of social and technological relations (Avis, 2018).

Our study has four main limitations. First, our literature review might not be exhaustive due to the composition of our search string. Second, we limited our search to three databases. Therefore, the articles not found in these databases were excluded in this review. Third, in order to meet all inclusion criteria, a number of articles were filtered out manually by screening the abstract and the article. Despite member checking, our final set of articles could be prone to selection bias. Fourth, although we argue that our proposed socio-technical framework is a useful sensitising tool for further research and in practice for assessing the state of DevSecOps in organizations, our depiction and understanding of the socio-technical dimensions may lack completeness and it is also plausible that either the social or technical may be irrelevant in certain DevSecOps contexts.

6. Conclusion

Based on a systematic literature review, we found that technical studies that pay little consideration to humanistic outcomes feature prominently in the literature. Based on our coding process and thematic analysis, we developed a framework to improve our understanding of DevSecOps based on the insights from the reviewed articles using a socio-technical lens. The socio-technical framework can be used by practitioners to perform a more holistic analysis of their DevSecOps practices for realizing continuous security. It highlights the key social and technical themes that underpin the effectiveness of DevSecOps and how insights about these themes can be used by practitioners to improve the instrumental and humanistic goals of DevSecOps. An interdisciplinary approach is proposed to adequately address challenging socio-technical relationships in DevSecOps. Drawing from the socio-technical work system framework and insights from the literature, we identified avenues for future research that address social imperatives as well as humanistic objectives and outcomes. Future research can empirically test the importance of the interplay between technology and human activities to improve the overall performance of DevSecOps.

7. References

Almuairfi, S. and Alenezi, M. (2020) "Security controls in infrastructure as code", *Computer Fraud & Security*, Vol 2020 No. 10, pp 13–19.

Alter, S. (2013) "Work system theory: overview of core concepts, extensions, and challenges for the future", Journal of the Association for Information Systems, Vol 14, No. 2, pp 72–121.

Amoroso, E. (2018) "Recent Progress in Software Security", *IEEE Software*, Vol 35, No. 2, pp 11–13.

Anderson, C. (2015) "Docker [Software engineering]", *IEEE Software*, Vol 32, No. 3, pp 102–c103.

Atwood, C. A., Goebbert, R. C., Calahan, J. A., T. V. Hromadka, I., Proue, T. M., Monceaux, W. and Hirata, J. (2016) "Secure Web-Based Access for Productive Supercomputing", *Computing in Science & Engineering*, Vol 18, No. 1, pp 63–72.

Avis, James. (2018) "Socio-technical imaginary of the fourth industrial revolution and its implications for vocational education and training: A literature review." Journal of Vocational Education & Training, Vol 70, No. 3, pp 337–363.

Bass, L. (2018) "The Software Architect and DevOps", *IEEE Software*, Vol 35, No. 1, pp 8–10.

Bostrom, R. P. and Heinen, J. S. (1977) "MIS problems and failures: A socio-technical perspective. Part I: The causes", MIS quarterly, pp 17–32.

Callanan, M. and Spillane, A. (2016) "DevOps: Making It Easy to Do the Right Thing", *IEEE Software*, Vol 33, No. 3, pp 53–59.

Carter, K. (2017) "Francois Raynaud on DevSecOps", *IEEE Software*, Vol 34, No. 5, pp 93–96.

Casola, V., De Benedictis, A., Rak, M. and Villano, U. (2020) "A novel Security-by-Design methodology: Modeling and assessing security by SLAs with a quantitative approach", *Journal of Systems and Software*, Vol 163, pp 110537.

Clarke, V. and Braun, V. (2017) "Thematic analysis", *The Journal of Positive Psychology*, Vol 12, No. 3, pp 297–298.

Cope, R. (2020) "Strong security starts with software development", *Network Security*, Vol 2020, No. 7, pp 6–9.

Dyess, C. (2020) "Maintaining a balance between agility and security in the cloud", *Network Security*, Vol 12, No. 3, pp 14–17.

Ebert, C., Gallardo, G., Hernantes, J. and Serrano, N. (2016) "DevOps", *IEEE Software*, Vol. 33, No. 3, pp 94–100.

Fletcher, K. and Smith, H. A. (2020) "Cyber Resilience through Machine Learning: Data Exfiltration" In International Conference on Cyber Warfare and Security, pp. 165-XIII. Academic Conferences International Limited, 2020.

Gatrell, M. (2016) "The Value of a Single Solution for End-to-End ALM Tool Support", *IEEE Software*, Vol 33, No. 5, pp 103–105.

Huskaj, G. (2019) "The Current State of Research in Offensive Cyberspace Operations" In European Conference on Cyber Warfare and Security, pp. 660–XIV. Academic Conferences International Limited.

IBM (2021) "Cost of Data Breach Report", [online], https://www.ibm.com/security/databreach (accessed on 12 January 2021).

Izycki, E. and Vianna, E. W. "Critical Infrastructure: A Battlefield for Cyber Warfare?" In ICCWS 2021 16th International Conference on Cyber Warfare and Security, pp. 454. Academic Conferences Limited, 2021.

Jansen, C. and Jeschke, S. (2018) "Mitigating risks of digitalization through managed industrial security services", *AI & Society*, Vol 33, No. 2, pp 163–173.
Johann, S. (2017) "Kief Morris on Infrastructure as Code", *IEEE Software*, Vol 34, No. 1, pp 117–120.
Kersten, M. (2017) "Value Stream Architecture", *IEEE Software*, Vol 34, No. 5, pp 10–12.
Kersten, M. (2018) "A Cambrian Explosion of DevOps Tools", *IEEE Software*, Vol 35, No. 2, pp 14–17.
Klein, D. (2019) "Micro-segmentation: securing complex cloud environments", *Network Security*, Vol 2019, No. 3, pp 6–10.
Kumar, R. and Goyal, R. (2020) "Modeling continuous security: A conceptual model for automated DevSecOps using open-source software over cloud (ADOC)", *Computers & Security*, Vol. 97, pp 101967.
MacDonald, N., (2012) "Devops needs to become devopssec", [online], https://blogs.gartner.com/neil_macdonald/2012/01/17/devops-needs-to-become-devopssec/ (accessed on 09 November 2015).
Mackey, T. (2018) "Building open source security into agile application builds", *Network Security*, Vol 2018, No. 4, pp 5–8.
Mansfield-Devine, S. (2018) "DevOps: finding room for security", *Network Security*, Vol 2018, No. 7, pp 15–20.
Markets and Markets, (2021) "Devsecops market", [online], https://www.marketsandmarkets.com/PressReleases/devsecops.asp. (accessed on 12 January 2021).
McGraw, G. (2017) "Six Tech Trends Impacting Software Security", *Computer*, Vol 50, No. 5, pp 100–102.
Mohan, V. and Othmane, L. B. (2016) "Secdevops: Is it a marketing buzzword?-mapping research on security in devops" In 2016 11th international conference on availability, reliability and security (ARES), pp. 542-547. IEEE.
Myrbakken, H., and Colomo-Palacios, R. (2017) "DevSecOps: a multivocal literature review" In International Conference on Software Process Improvement and Capability Determination, pp. 17-29. Springer, Cham.
Naidoo, R. (2020) "A multi-level influence model of COVID-19 themed cybercrime", European Journal of Information Systems, Vol 29, No. 3, pp 306–321.
Parnin, C., Helms, E., Atlee, C., Boughton, H., Ghattas, M., Glover, A., Holman, J., Micco, J., Murphy, B., Savor, T., Stumm, M., Whitaker, S. and Williams, L. (2017) "The Top 10 Adages in Continuous Deployment", *IEEE Software*, Vol 34, No. 3, pp 86–95.
Rafi, S., Yu, W., Akbar, M. A., Alsanad, A. and Gumaei, A. (2020) "Prioritization Based Taxonomy of DevOps Security Challenges Using PROMETHEE", *IEEE Access*, Vol 8, pp 105426–105446.
Rahman, A. A. U., and Williams, L. (2016) "Software security in devops: Synthesizing practitioners' perceptions and practices" In *2016 IEEE/ACM International Workshop on Continuous Software Evolution and Delivery (CSED),* pp. 70-76. IEEE.
Rios, E., Iturbe, E., Larrucea, X., Rak, M., Mallouli, W., Dominiak, J., Muntés, V., Matthews, P. and Gonzalez, L. (2019) "Service level agreement-based GDPR compliance and security assurance in(multi)Cloud-based systems", *IET Software*, Vol 13, No. 3, pp 213–222.
Sánchez-Gordón, M. and Colomo-Palacios, R. "Security as culture: a systematic literature review of DevSecOps." In *Proceedings of the IEEE/ACM 42nd International Conference on Software Engineering Workshops*, pp. 266-269. IEEE/ACM.
Sarker, S., Chatterjee, S., Xiao, X. and Elbanna, A. (2019) "The sociotechnical axis of cohesion for the IS discipline: Its historical legacy and its continued relevance", *MIS Quarterly*, Vol 43, No. 3, pp 695–720.
Spinellis, D. (2012) "Don't Install Software by Hand", *IEEE Software*, Vol 29, No. 4, pp 86–87.
Trihinas, D., Tryfonos, A., Dikaiakos, M. D. and Pallis, G. (2018) "DevOps as a Service: Pushing the Boundaries of Microservice Adoption", *IEEE Internet Computing*, Vol 22, No. 3, pp 65–71.
van Dinter, R., Tekinerdogan, B. and Catal, C. (2021) "Automation of systematic literature reviews: A systematic literature review", *Information and Software Technology*, Vol 136, pp 106589.
Weber, I., Nepal, S. and Zhu, L. (2016) "Developing Dependable and Secure Cloud Applications", *IEEE Internet Computing*, Vol 20, No. 3, pp 74–79.
Winter, S., Berente, N., Howison, J. and Butler, B. (2014) "Beyond the organizational 'container': Conceptualizing 21st century sociotechnical work", *Information and Organization*, Vol 24, No. 4, pp 250–269.
Yin, R. K. (2014) *Case study research: Design and methods (5th ed.)*, Sage, California.
Zaydi, M. and Nassereddine, B. (2020) "DevSecOps Practices for an Agile and Secure IT Service Management", *Journal of Management Information and Decision Sciences*, Vol 23, No. 2, pp 1–16.

Effective Cyber Threat Hunting: Where and how does it fit?

Nombeko Ntingi, Petrus Duvenage, Jaco du Toit and Sebastian von Solms
Academy of Computer Science and Software Engineering, University of Johannesburg, South Africa
nntingi@gmail.com
duvenage@live.co.za
jacodt@uj.ac.za
basievs@uj.ac.za

Abstract: Traditionally threat detection in organisations is reactive through pre-defined and preconfigured rules that are embedded in automated tools such as firewalls, anti-virus software, security information and event management (SIEMs) and intrusion detection systems/intrusion prevention systems (IDS/IPS). As the fourth industrial revolution (4IR) brings with it an exponential increase in technological advances and global interconnectivity, the cyberspace presents security risks and threats the scale of which is unprecedented. These security risks and threats have the potential of exposing confidential information, damaging the reputation of credible organisations and/or inflicting harm. The regular occurrence and complexity of cyber intrusions makes the guarding enterprise and government networks a daunting task. Nation states and businesses need to be ingenious and consider innovative and proactive means of safeguarding their valuable assets. The growth of technological, physical and biological worlds necessitates the adoption of a proactive approach towards safeguarding cyber space. This paper centers on cyber threat hunting (CTH) as one such proactive and important measure that can be adopted. The paper has a central contention that effective CTH cannot be an autonomous 'plug in' or a standalone intervention. To be effective CTH has to be synergistically integrated with relevant existing fields and practices. Academic work on such conceptual integration of where CTH fits is scarce. Within the confines of the paper we do not attempt to integrate CTH with many of the various relevant fields and practices. Instead, we limit the scope to postulations on CTH's interface with two fields of central importance in cyber security, namely Cyber Counterintelligence (CCI) and Cyber Threat Monitoring and Analysis (CTMA). The paper's corresponding two primary objectives are to position CTH within the broader field of CCI and further contextualise CTH within the CTMA domain. The postulations we advanced are qualified as tentative, exploratory work to be expanded on. The paper concludes with observations on further research.

Keywords: cyber threat hunting, cyber counterintelligence, active cyber defense, cyber threat intelligence, cyber threat modelling, proactive

1. Introduction

Duvenage & von Solms (2013) rightly assert that the safeguarding of cyber space against emerging cyber threats does not only require the strengthening up of existing security measures, but also the introduction of proactive measures. Cyber threat hunting (CTH) is one such proactive and important measure that can be adopted.

The National Institute of Standards and Technology (NIST) defines CTH as a process to proactively identify and disrupt cyber threats inside the organisational ICT infrastructure and enhance security measures in order to defend against possible future threats (NIST, 2020). CTH is one of the more recently adopted methodologies implemented in the ICT sector. According to Lee & Lee (2017) the CTH field is relatively new and most security teams use unstructured methods of hunting. Lee & Lee (2017) further indicate that 45% of teams execute the threat hunting process on an *ad hoc* basis, with the main reason for this being the scarcity of published resources on structured hunting frameworks for adoption across the industry.

However, effective CTH cannot be an autonomous 'plug-in' or a standalone intervention. Its effectiveness can only be optimally leveraged when it is part of a broader cyber security approach. When it comes to finding postulations that integrate CTH with a broader cyber security approach, this is far easier said than done. In this regard, Kumar & Chacko (2020) states that while there are models, frameworks and systematic processes developed for CTH, much of the discussions published on these artefacts entail definitions and practical approaches to CTH activities *per se*. Kumar & Chacko (2020) further observes that there is a lack of propositions that integrate CTH as part of a broader approach. This observation confirms our literature review which likewise found a lack of such propositions.

This paper's first objective then is to introduce CTH and position it as an integral part of a broader CCI field. In addition to integration with a broader approach such as CCI, CTH is on a practical level dependent on close synergy with some of the elements of the Cyber Threat Monitoring and Analysis (CTMA) domain for its effectiveness. The paper's second objective is thus to contextualise CTH as a critical dimension of the CTMA domain.

The rest of the paper is structured as follows: Section 2 positions CTH as an integral part of CCI with sub-section 2.1 defining the concept of CCI and its model and sub-section 2.2 contextualising CTH within the broader field of CCI. Section 3 introduces the CTMA domain, showing interrelation amongst the various aspects that enable effective CTH processes.

2. Positioning of cyber threat hunting within cyber counterintelligence (CCI)

In order to position CTH with CCI this section commences with a conceptual explication of CCI (Sub-section 2.1). Building on this explication, sub-section 2.2 positions CTH within the CCI context. Essentially then, this section seeks to answer the question: where and how does CTH fit in as far as CCI is concerned.

2.1 CCI: Concept and model

To provide a response to the question (where and how does CTH fit in as far as CCI is concerned?), this sub-section starts by defining and explaining CCI by means of a conceptual model. The various elements and modes/postures of the CCI model are discussed. The discussion in this sub-section will thus aid in accomplishing the paper's objective of positioning CTH into a broader context of CCI. According to Duvenage & von Solms (2014) CCI is part of a multi-disciplinary counterintelligence that aims to deter, prevent, degrade, exploit and neutralise attempts by adversaries that seek to alter confidentiality, integrity and availability of sensitive and critical information through methods of cyber.

Figure 1 below depicts a model for CCI as sourced from (Duvenage & von Solms, 2015). The figure shows four distinct quadrants (indicated by the four arrows) that represent CCI's four modes/postures namely: passive-defense, active-defense, passive-offensive and active-offensive. Positioned at the centre of the arrows is "IS" which denotes Intelligence and Strategy: CCI as a subset of counterintelligence thus puts business Intelligence and Strategy at the centre of operations in order to achieve business objectives (Duvenage & von Solms, 2015). The "CI" indicated on all four quadrants represents the multi-disciplinary field of Counterintelligence that relies on passive-active and defensive-offensive measures and means. CCI methods and means are indicated as part of the broader counterintelligence field.

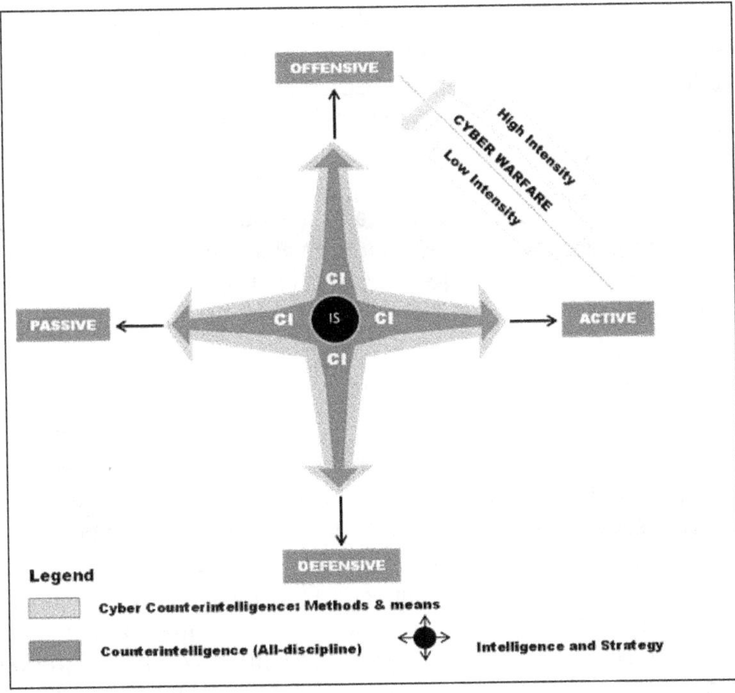

Figure 1: CCI model (Duvenage & von Solms, 2015)

According to Duvenage, Jaquire & von Solms (2016) defensive counterintelligence measures provide information and act as triggers to alert the offensive operations. Applied to this paper, the passive half of the CCI model (i.e. passive-defense and passive-offense quadrants) is deemed to incorporate the notion of 'information feeders' and 'alerts'. 'Information feeders' and 'alerts' are thus in a symbiotic relationship with, yet distinctive from, active operations.

Furthermore, it has to be emphasised that *the deployment of the CCI model is context dependent*. With the realm of nation state counterespionage, Stech & Hechkman (2018) for example, concretised the CCI model with the following application to a hypothetical NATO operation against APT 28. Table 1 describes the application of the four-sector CCI matrix. It highlights the four quadrants as depicted in Figure 1. The columns describe the passive and active modes and the rows describe the defensive and offensive modes. The columns and rows depict the CCI tools and techniques for outsmarting the adversaries. Table 1 illustrates that defensive and offensive CCI tools can be deployed both passively and actively.

Table 1: Hypothetical NATO Cyber CI Operations against cyber espionage threat (Stech & Heckman, 2018)

Modes	Passive Cyber CI	Active Cyber CI
Defensive mode	**Deny access and collect on espionage threat**	
	Passive defense:	**Active defense:**
	Harden endpoint and server Configurations	Gather intelligence on on-going intrusions
		Use honeypots to gather late-stage implants and unpatched exploits
	Share actionable indicators across NATO intelligence partners	
		Share indicators to force infrastructure and "toolkit" rotations
Offensive Mode	**Manipulate, degrade, control and neutralise espionage threat**	
	Passive offensive:	**Active offensive:**
	Use honeypots to deliver deception materials	Counter-hack hop points and control servers
		Trolling "bait victims" to lure attackers to controlled boxes
	Sinkhole APT28 hop points	
	Identify APT28 operatives	Operating controlled boxes as double agents to inject beacons, double-hacked backdoors, etc. into APT28 control environment

While Table 1 applies the CCI matrix to the international state actor arena, it nonetheless conveys the essence of, and differences between, CCI's four quadrants more generally. With certain qualifications and changes, some of the techniques noted in Table 1 are therefore also useful to CCI and CTH as employed by non-state actors. Which of these techniques are relevant to CTH will depend on its (CTH's) positioning within CCI (See Subsection 2.2).

This sub-section defined and explained CCI by means of a conceptual model. As part of this explanation specific reference was made to the interface between the CCI model's active and passive modes. This interface is at the core of our postulation on the positioning of CTH as part of CCI in the next sub-section. The sub-section further illustrated the four-sector CCI matrix as applied in the NATO case study.

2.2 CTH in the context of CCI

This sub-section expounds CTH and conceptualise it within the context of CCI. A modified CCI model that depicts how CTH fits in with CCI will be presented. NIST released a special publication on September 2020 with CTH as an official cybersecurity discipline (Secureworks Counter Threat Unit, 2020). This means that organisations can incorporate CTH in their cyber security programs and collaborate on this with other organisations (Secureworks Counter Threat Unit, 2020). Sqrrl Data Inc (2016) describes CTH as an active means of defending ICT organisational infrastructure against adversaries by proactively and iteratively seeking unidentifiable or highly-obfuscated threats that lurk in organisational systems, networks and infrastructure. This is in contrast to the traditional means such as firewalls, intrusion detection and prevention systems, quarantining malicious code in sandboxes and SIEM technologies and systems which solely rely on pre-configured rules.

As mentioned in Section 1 of this paper, for the CTH approach to be effective, it needs to co-exist and be integrated in a wider discipline such as CCI. Figure 1 above provided a background of CCI and its means and methods as a subpart of a multi-disciplinary field.

Kolthoff (2015) suggests that cyber hunt team operations should be considered as a form of offensive counterintelligence even though they fall short of hacking back. Hacking back refers to organisations tracing back the origins of the cyber-attack and taking intrusive actions against the cyber-attackers (Kassner, 2021). According to Zimski (2021), hacking back is short-sighted and may have inadvertent repercussions. Instead the focus of any cyber security operation should always be to mitigate or neutralize the immediate threat (Herping, 2021). Furthermore, a cyber-operation carried out only for the goal of retaliation or punishment may increase the danger of escalation and may be illegal under international laws (Herping, 2021). Roberts (2020), asserts that in the cyber domain the lines between offensive and defensive operations are not always clear, mainly because offensive operations are being justified as defensive, moving acceptable standards toward the offensive end of the spectrum.

In our view, CTH in some respects not only forms part of CCI's active-defense mode, but also the active-offensive mode. We based our assertion on CTH's positioning on the explanation of the CCI model in subsection 2.1 as well as the foregoing narrative explanations of CCI and CTH. As will be explained per Figure 2, CTH is thus part of active CCI. With reference to Table 1, CTH would thus include techniques such as the use of honeypots to actively gather intelligence and the luring of attackers to controlled boxes.

In the interest of clarity, we concisely recapitulate our three main key contentions, namely

- To be effective CTH needs to be part of a broader approach such as CCI
- CTH is located with CCI's active modes (i.e. active-defense and active-offensive).
- While part of active CCI, CTH relies on functions provided by CCI's passive modes (i.e. passive-defense and passive-offense).

Figure 2 represents these contentions by means of a modified CCI model. Figure 2 shows CTH as located in CCI's active-defensive and active-offensive quadrants. Intelligence and Strategy (IS) is illustrated at the centre as the main objective to be delivered by the counterintelligence. The four quadrants depicted by the four arrows are all the counterintelligence postures. CCI as the subset of counterintelligence inherits the disciplines, methods and means of CI. CTH on its part serves as a compliment of the CCI multi-disciplinary field.

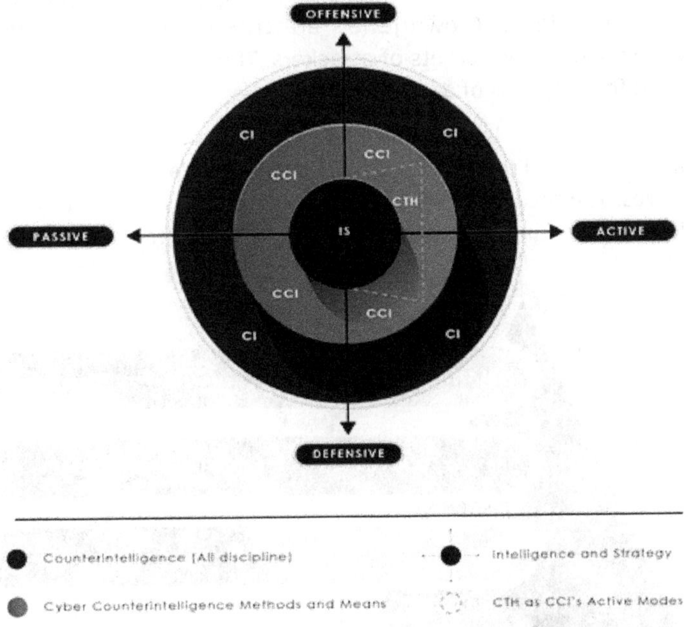

Figure 2: Cyber Threat hunting as part of cyber counterintelligence (Authors)

This section firstly focused on explaining the CCI concept and model as referenced from existing literature. CTH was then conceptualised and put in the context of CCI. A CCI model has been modified to illustrate where and how CTH fits in the four quadrants of the CCI model. Section 3 will introduce the CTMA domain and show how CTH can be contextualised within this domain. Section 3, more plainly put, will explore the question, where and how does CTH fit within the CTMA domain?

3. Introduction to the cyber threat monitoring and analysis (CTMA) domain

In this section the CTMA domain will be introduced along with some of its elements, as well as the interrelationships between these elements. This section will provide a brief background of the three CTMA domain elements in order to explain a close synergy of these elements. In order to illustrate the close synergy of the elements of the CTMA domain, this section will commence with illustrating CTMA domain in the CCI and locating the elements of the CTMA domain. The section will further explain the dependency of CTH processes to other elements of the CTMA domain. Subsections 3.2 and 3.3 will explicate CTI and CTM concepts in relation to CTH.

According to Singapore's Cyber Security Agency CSA (2021) cyber threat intelligence, cyber threat monitoring and cyber threat hunting are some of the elements of the wider CTMA domain, with CTI at the management level, CTM at the system level and CTH at an equipment or application level. Gumble (2020) states that CTM is a risk-based approach conducted to design secure systems with the purpose of identifying threats and designing mitigation strategies. According to RSI Security (2018) CTH and CTI are two of the most significant elements of cyber risk management since they enable organisations to adopt a proactive approach to hostile actors rather than reacting to issues as they arise.

As mentioned in Section 1, CTH is dependent on close synergy with some of the elements of CTMA domain for its effectiveness, Figure 3 (below) shows the location of these elements. According to CSA (2021) CTMA can be approached at three different tiers; namely management, system, equipment or application level perspectives. Prior to providing the graphic depiction (Figure 3), the elements of Figure 3 are narratively explained.

- **Management level:** The information/intelligence (cyber threat intelligence) gathered is used by Executive management for informed decision making. The intelligence gathered consist of actionable information to understand adversaries and their tactics, techniques and procedures (TTP). The purpose of data gathered is for the environmental scanning to gain understanding of the trends and cyber threats that the organisation is exposed to.
- **System level:** This level involves activities such as cyber threat modelling, it focuses on the system architecture, network and data visibility. At this level "Knowing your crown jewels" is critical, consequently data sources have to be identified. Crown jewels are critical assets to attaining the overall corporate objectives, and they are typically the targets of attackers. The crown jewels and privileged accounts need to be properly identified for purposes of data modelling.
- **Equipment or application level:** This level involves the more technical activities such as CTH activities, cyber security event logs and analytics. From Figure 3, CTH is demonstrated as a low-level activity that focusses on technical perspectives. The focus of the paper at the moment is limited to CTH. It is for this reason the "equipment or application level" in Figure 3 is highlighted with a different shade.

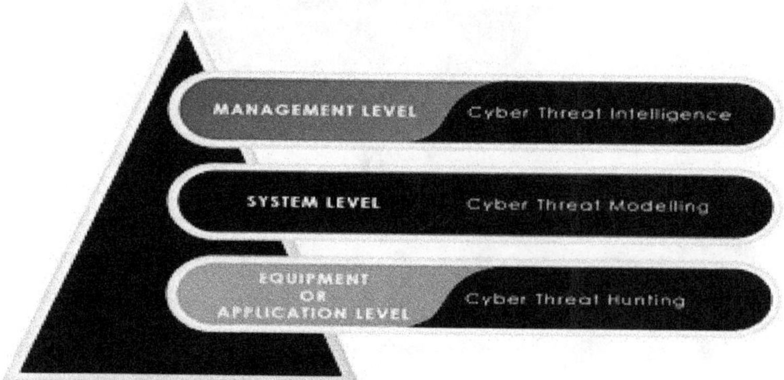

Figure 3: Elements of the CTMA domain (CSA 2021)

Figure 3 above shows the interrelationship between some of the elements of the CTMA domain. Figure 4 will now magnify Figure 3 to further illustrate how CTMA domain as a whole fit into CCI and specifically where CTH is located within the CTMA domain. Figure 4 is an adaptation of Figure 2, for simplicity the graphic depiction of Figure 4 only shows the CCI means and methods as the four quadrants represented by the arrows. The triangle displayed on its side is adopted from Figure 3 to show how the elements of the CTMA domain fit into the CCI model. The CTMA domain, which includes the CTI, CTM and CTH spans across all four quadrants to demonstrate that CTMA as a whole is executed in all the four quadrants of CCI. CTH is represented at the lower level of the CTMA domain in the active quadrants of CCI, namely (active-defensive and active-offensive). Our key contention is that CTMA is performed in all four quadrants of CCI and that CTH is located in the two active quadrants of CCI.

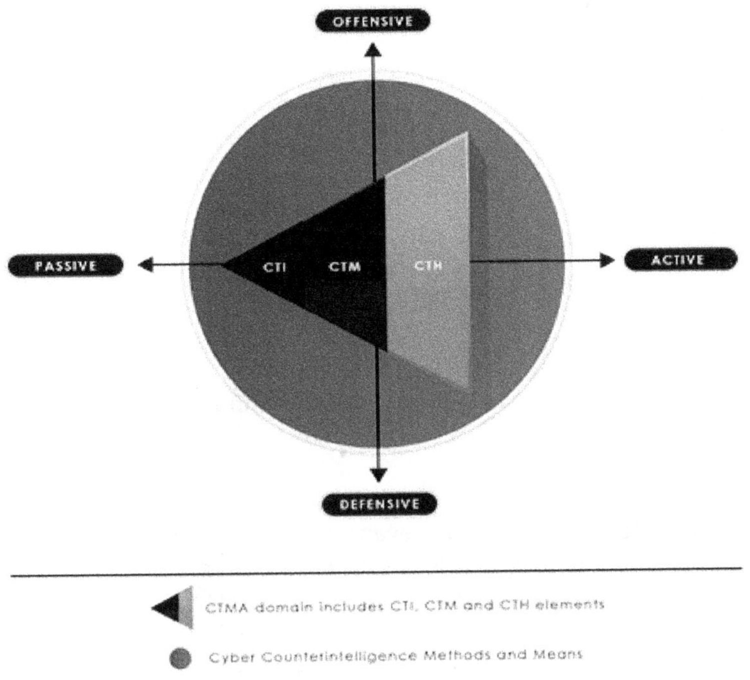

Figure 4: CTMA domain as located in the four quadrants of CCI (authors)

Figure 3 showed the interrelationship between these elements, with CTH located at the lower level of the hierarchy where the technical activities occur. According to IBM (2021) success and effectiveness of CTH depends on the accuracy and abundance of data gathered. Sapphire (2021) states that CTH leverages intelligence collected and processed during CTI process. That is to say CTI acts as an information feeder for CTH processes. Cyber threat hunters also rely on access to both host and network data sources identified during CTM to conduct hypotheses investigations, (ChaosSearch, 2021).

This section illustrated the interrelationships between the CTMA domain's elements, as well as where the CTMA domain as a whole fits within the CCI discipline's four quadrants. CTH was also shown to be in the active quadrants of the CCI model. Building on the foregoing positioning of CTMA domain in the CCI discipline and CTH as part of the CTMA domain, the next sub-sections explain in more detail CTH's relation with CTI and CTM respectively and how they contribute to the success of CTH processes.

3.1 The concept of CTI and its relation to CTH

Baker (2021) defines CTI as information that is processed, evaluated and driven by evidence about existing or emerging threats to organisational assets. This evidence-based knowledge is useful during decision making. The aim of CTI is to scan and understand the environment within which the organisations or nation states operate in. This environmental scanning helps the organisations to understand who their adversaries/competitors are and why are they being targeted. In Section 1 it was mentioned that there are dependencies for CTH processes to occur. Puzis et al (2020) states that for CTH to effectively conduct its processes such as hypotheses formulation, it relies on clear-cut evidence collected and analysed during the CTI process. Evidence based data collected during the CTI process which lies in the management level of the CTMA domain aids in making informed decisions by the executive management and serve as input to perform the CTH processes. The effectiveness of CTH processes depend on the ability of accurate collection and analysis of data.

Following explanation of CTH's relationship with CTI, we now proceed with examining the concept of CTM and how it relates to CTH.

3.2 The concept of CTM and its relation to CTH

Kost (2021) defines CTM as a proactive approach in which organisations identify potential and anticipated threats in their network security and the vulnerabilities that can be exploited by these threats. CTM can be used to prevent cyber threats from taking advantage of the system flaws by using threat modelling methods to inform defensive measures. CTM focuses on system architecture, relationships and behaviours as well as data sources such as endpoint information, logs from firewalls, and domain name services. CTH is a data driven process hence data and network visibility is of utmost importance. As mentioned in section 3, CTM which is at system level focusses on network and data visibility. According to ChaosSearch (2021), cyber threat hunters must have access to data sources that provide visibility into host and network activity, as well as telemetry data obtained by security solutions currently in use in the environment.

This section firstly introduced the CTMA domain and how the elements of CTMA relate. The interrelationships of CTI and CTM with CTH were further described in sub-sections 3.1 and 3.2, respectively. Section 3 attempted to answer the question, "Where and how does CTH fit within the CTMA domain?"

4. Conclusion

For the organisations to be more effective in defending their ICT infrastructure from an ever-expanding cyber threat base, a shift towards proactive techniques to improve their security posture should be at the forefront.

This paper explored CTH as one of the more recent such proactive methodologies. The paper highlighted the need for academic research to integrate CTH with relevant, existing fields and practices. Phrased differently, for CTH to be effective we need to be clear on where and how it fits with existing fields and practices. We thus proceeded with postulations on CTH's interface with two fields of central importance in cyber security, namely CCI and CTM. These should be viewed as tentative propositions and form part of the quest for an academic foundation for developing an active cybersecurity posture for organisations and nation-states exposed to increasingly sophisticated attacks. An integration of cyber threat hunting into other disciplines will provide a solid foundation and develop a cyber-defense active environment for organisations to expose sophisticated attack mechanisms and test their ability to detect attacks thus enabling organisations and nation states to become effective in the quest of active cyber defense activities.

References

Baker, K. 2021. What is Cyber Threat Intelligence. [Online] Available at: https://www.crowdstrike.com-/cybersecurity-101/threat-intelligence/ [Accessed 27 12 2021].

CSA Singapore 2021. A Singapore Government Agency Website. [Online] Available at: https://www.csa.gov.sg/-/media/csa/documents/legislation_supplementary_references/guide-to-cyber-threat-modelling.pdf [Accessed 08 11 2021].

ChaosSearch, 2021. The Threat Hunter's Handbook. [Online] Available at: https://www.chaossearch.io/hubfs-/ChaosSearch%20Threat%20Hunters%20Handbook.pdf [Accessed 08 02 2022].

Duvenage, P, Jaquire, V & von Solms, S. 2016. 'Conceptualising cyber counterintelligence – Two tentative building blocks' in Published Proceedings of the 15th European Conference on Cyber Warfare and Security, Munich, Germany, June.

Duvenage, P. & von Solms, S. 2013. The Case for Cyber Counterintelligence. Pretoria, IEEE.

Duvenage, P. & von Solms, S. 2014. Putting Counterintelligence in Cyber Counterintelligence. Greece, Academic Conferences and Publishing International Limited.

Duvenage, P. & von Solms, S. 2015. Cyber Counterintelligence: Back to the Future. Journal of Information Warfare, 13(4), pp. 42-56.

Gumbley, J., 2020. A Guide to Threat Modelling for Developers. [Online] Available at: https://martinfowler.com/articles/agile-threat-modelling.html [Accessed 03 02 2022].

Herping, S., 2021. Active Cyber Defense Operations. [Online] Available at: https://www.stiftung-nv.de/sites/default/files/active_cyber_defense_operations.pdf [Accessed 18 03 2022].

IBM, 2021. Cyber Threat Hunting. [Online] Available at: https://www.crowdstrike.com/cybersecurity-101/threat-hunting/ [Accessed 01 02 2022].

Kassner, M. 2021. Is hacking back affective, or does it just scratch an evolutionary itch? [Online] Available at: https://www.techrepublic.com/article/is-hacking-back-effective-or-does-it-just-scratch-an-evolutionary-itch/ [Accessed 27 12 2021].

Kolthoff, J. 2015. https://www.linkedin.com. [Online] Available at: https://www.linkedin.com/pulse/cyber-counterintelligence-hunt-team-operations-jarrett-kolthoff/ [Accessed 08 11 2021].

Kost, E. 2021. What is threat modelling. [Online] Available at: https://www.upguard.com/blog/what-is-threat-modelling [Accessed 27 12 2021].

Kumar, A. & Chacko, A. 2020. Security Intelligence. [Online] Available at: https://securityintelligence.com/posts/threat-hunting-guide/ [Accessed 12 11 2021].

Lee, R. & Lee, R. M. 2017. The Hunter Strikes Back: The SANS 2017 Threat Hunting Survey. [Online] Available at: https://www.malwarebytes.com/pdf/white-papers/sans_report-the_hunter_strikes_back_2017.pdf [Accessed 12 10 2021].

National Institute of Standards and Technology, 2020. Security and Privacy Controls for Information Systems and Organisations, Washington, D.C.: s.n. [Accessed 12 10 2021].

Puzis, R., Zilberman, P. & Elovici, Y., 2020. [Online] Available at: https://www.researchgate.net/publication-/339814086_ATHAFI_Agile_Threat_Hunting_And_Forensic_Investigation [Accessed 07 01 2022].

Roberts, M., 2020. The Cyber-Range Advantage for Defence. [Online] Available at: http://www.milsatmagazine.com/story.php?number=1777803179&msclkid=b6514886a67511ec9524dd5dd0476196 [Accessed 18 03 2022].

RSI Security, 2018. Understanding the basic components of cyber risk management. [Online] Available at: https://blog.rsisecurity.com/understanding-the-basic-components-of-cyber-risk-management/[Accessed 07 02 2022].

Secureworks Counter Threat Unit 2020. Research & Intelligence. [Online] Available at: https://www.secureworks.com/blog/threat-hunting-as-an-official-cybersecurity-discipline [Accessed 30 12 2021].

Sqrrl Data Inc 2016. A Framework for Cyber Threat Hunting. [Online] Available at: https://www.threathunting.net/files/framework-for-threat-hunting-whitepaper.pdf

Stech FJ & Heckman KE 2018 'Human nature and cyber weaponry: Use of denial and deception in cyber counterintelligence', Cyber Weaponry Issues and Implications of Digital Arms, H Prunckun (ed), Springer, Cham, CH.

Zimski, P., 2021. Why companies should never hack back. [Online] Available at: https://www.helpnetsecurity.com/2021/08/31/why-companies-should-never-hack-back/?web_view=true [Accessed 18 03 2022].

Two Novel Use-Cases for Non-Fungible Tokens (NFTs)

Alexander Pfeiffer[1 2 3], Natalie Denk[1], Thomas Wernbacher[1], Stephen Bezzina[2 3], Vince Vella[2] and Alexiei Dingli[2]
[1]Center for Applied Game Studies, Donau-Universität Krems (DUK), Krems, Austria
[2]Department of Artificial Intelligence, University of Malta (UoM), Msida, Malta
[3]B&P Emerging Technologies Consultancy Lab Ltd., St. Julian's, Malta
alexander.pfeiffer@donau-uni.ac.at
natalie.denk@donau-uni.ac.at
stephen.bezzina@um.edu.mt
Thomas.wernbacher@donau-uni.ac.at
alexiei.dingli@um.edu.mt
vince.vella@gmail.com

Abstract: Non-Fungible Tokens (NFTa) can either represent an original digital artwork, or act as a digital reference to the actual work. In both as digital references to the actual work. In both cases the record in the distributed ledger, mostly a blockchain-based database, intends to serve as a proof of ownership or transfer of rights. NFTs might also add a further purpose, which in blockchain terms is referred to as "a utility", such as access to special websites, chats or clubs in emerging metaverse platforms. This use-case paper presents a first introduction of two early stage demonstrators, set outside the common use of art images or images of historical events as NFTs. The first case shows how educational credentials can be created, in which different teachers contribute to assessment achievements. We elaborate how these partial achievements are verified separately within the actual credentials. In the second case study, we build on previous research in regard to NFTs in the music industry and show the combination of physical vinyl record special editions, in our case vinyls signed by the band, and the ownership certificate as NFT. For both demonstrators we used, in different settings, the crypto art platform NFTmagic and the blockchain-token wallet Sigbro. We developed and tested the results within the setting of a roleplay as a group and show how blockchain technologies and especially NFTs can be made useful in new ways, inspired by the ongoing process of discovering risks and opportunities in 'crypto art', thus initiating discussion on the topic and effectively bridging the cybersecurity and (digital) art communities.

Keywords: NFT, blockchain, non-fungibile-tokens, cybersecurity

1. Introduction

The white paper "Bitcoin: A Peer-to-Peer Electronic Cash System" by Satoshi Nakamoto (2008), a pseudonymous author, is regarded as one of the foundations for today's blockchain technologies. Bitcoin is an example of a public blockchain: the same digital record is replicated at many locations (nodes), and each node is managed by a different individual or corporation, at least in theory. Because the decentralized system allows for consensus, none of these people or businesses need to know or agree with each other directly. When it comes to having a tamper-proof recording of a transaction, this is accurate. But blockchain cannot verify anything other than that, such as any information provided in the attached text message or ownership of tokens besides the fact that they are stored in a legitimate blockchain wallet. A blockchain's storage mechanism is based on the constant production of fresh data blocks that are cryptographically linked to each other. The specific technique differs depending on the blockchain system in use. Some blockchain systems can do more than solely handle transactions, such as allowing users to create their own tokens and contracts (Grech and Camilleri, 2017).

In our use cases, we want to combine two worlds. The storage of data in a form where users have the possibility of data-ownership in combination with a secure digital identity. After all, what good is a secure database solution, if we cannot verify who entered the data, at least in such a way that the data was entered by a legitimate person or institution without revealing personal data?

To do so, we use the webapp NFTMagic.art[1] and the native app Sigbro.app[2], which are both facilitating the Ardor[3] blockchain, for our demonstrators. This blockchain system is a descendant of the Nxt[4] blockchain. We have two motives for this: the Ardor blockchain uses the Proof of Stake (POS) consensus algorithms, which is

[1] https://nftmagic.art/
[2] https://www.sigbro.app
[3] https://www.jelurida.com/ardor
[4] https://www.jelurida.com/nxt

considered environmentally friendly and secondly, the Ardor blockchain's Ignis[5] childchain provides the ability to execute smart transactions. With Nxt's smart transactions (which are implemented in the Ignis childchain), the code that is executed is actual software that runs in the node server. The scripts have already been incorporated into the network nodes. The user only sets those parameters that are required for the transaction or operation that he or she is attempting. As a result, Nxt may be viewed as a platform that allows developers with little or no experience with blockchain technologies to create a variety of safe decentralized services, applications, and tokens (Abctcm & APenzl, 2016). Furthermore, we have the possibility of phasing, which allows certain safe transaction to be created with conditional deferred execution based on specifically applied approval models (APenzl, 2016). With this function, we can set conditions, allowing our NFTs and utility tokens to be only sent to another account under certain conditions, thus enabling us to further greenlist all accounts in the ecosystem. This enables several possibilities to prevent fraud in our planned applications.

2. Method

The platform Nftmagic.art offers three different ways to mint the blockchain token (as NFT):

- 1. A unique token is generated for each individual case, which contains completely different asset properties and other meta data. This is the most widely accepted definition, and application, of an NFT.
- 2. However, it is also possible to generate several items, from two upwards, with the same asset properties and the same asset ID. In this case, there are only differences in the meta data (message, if attached) when the token is sent.
- 3. A third, innovative solution is to create a series of tokens (here we take the inspiration from seriality of blockchain-based art (Jutel, 2021)). This is a hybrid of both existing solutions where each token is a unique asset as in the first example, but at the moment of creation of the token it can share asset properties with other unique tokens, allowing each token to still have its own asset ID.

In all cases we have on the one hand the blockchain-based asset storing the legal data and acting as ownership certificate and on the other hand the media file, which is on an IPFS[6] server which is using a peer-to-peer hypermedia protocol provided by the platform with the possibility to clone the file with the user's very own IPFS Node. Once the asset is created (or in blockchain terms the NFT is minted), we use our Ardor Full Node to access the childchain Ignis and implement the necessary phasing conditions and approval models.

Another important feature of Nftmagic.art for our demonstrator is the Twitter verification procedure. Twitter users can apply to have a verified account on the platform which can be considers a low-level form of identity verification, following the concept of an E-ID Wallet for utility tokens from Pfeiffer and Bugeja (2020). The E-ID utility wallet's concept is based on the idea that each user can keep their utility tokens and NFTs in an easy-to-use app and combine them with digital ID in a variety of formats (if necessary). Furthermore, the user should be able to see the information about the tokens and NFTs without having any prior knowledge of blockchain technologies, and the tokens should be able to be used to securely communicate information with third parties who require read-only access to the data. To this extent, the Sigbro app acts as E-ID wallet within the demonstrated use-cases.

Figure 1 shows a schematic representation of the E-ID Wallet. You can see the wallet on the far right, in our case the Sigbro app. This app is able to hold utility tokens and display them separately in a dedicated tab, independent of the cryptocurrencies in the same wallet. At the same time, using NFTMagic and a bot technology that interacts with Twitter, we can connect the Ardor wallet address of the respective Sigbro wallet to a Twitter account. This works in such a way that you receive a private message with a QR code in Twitter. This needs to be confirmed and signed using the Sigbro app. This process takes place offline on the smartphone. Only a confirmation token is sent for verification. On NFTMagic, on the other hand, the Sigbro app is connected to the web app using a toolbox in the header. Also, via an offline signature process as described above. NFTMagic is then set up as a tool for creating and displaying NFTs.

[5] https://www.jelurida.com/ignis
[6] https://ipfs.io/

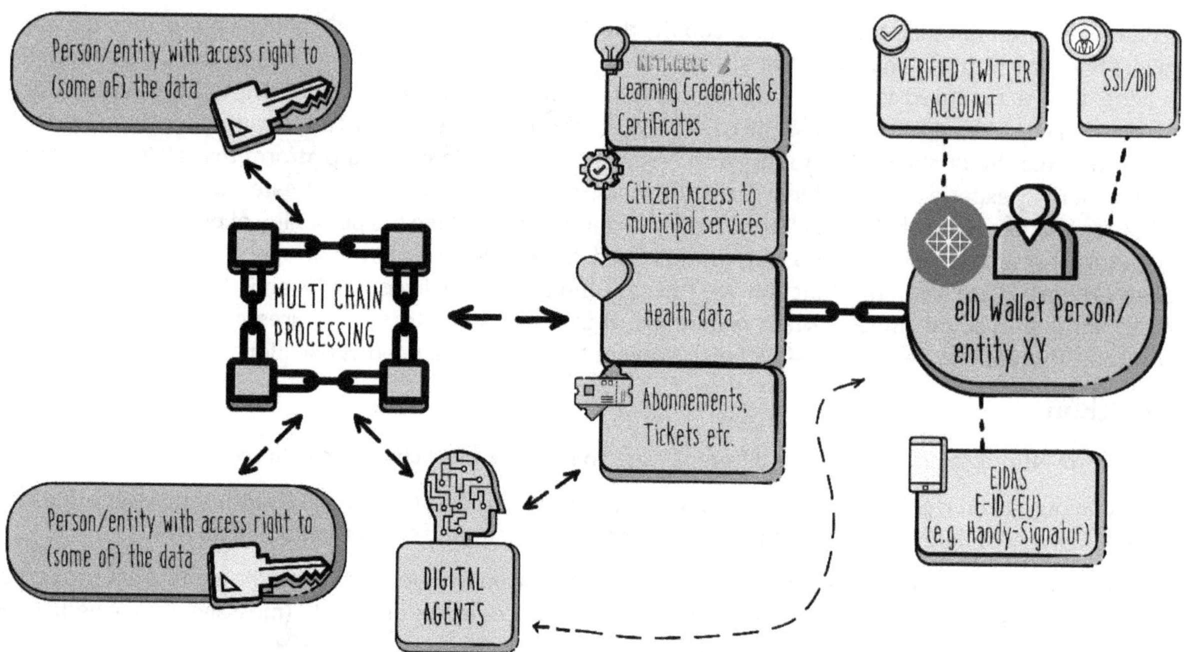

Figure 1: E-ID Wallet illustration

3. Presentation and discussion of the developed demonstrator

3.1 Use-case A: A learning certificate which provides the possibility to check the authenticity of every mark separately

The goal is to create a certificate of completion of a module which consists of two sub-modules. Each of these sub-modules is taught and graded by a different teacher. The final certificate contains a reference to these two deliverables. This means that it is now possible to see exactly how the module was completed. This is particularly interesting, when a final certificate contains achievements not only from different people but from different institutions, or for example, in the case when the final performance is based on micro credentials (Pfeiffer, 2021).

In our case, there are two partial performance certificates. Hilde Pfeiffer certifies the subject Art and Alexander Pfeiffer certifies the subject Technology. The account of the Center for Applied Game Studies then creates the final certificate based on this and sends it to the student Natalie Denk.

Sequence of the role play and transactions of the demonstrator:

ARDOR-BGWS-RQDY-KFWQ-35JST | Alexander Pfeiffer | created asset 12203487059644698664[7]

ARDOR-R7F7-R2D4-E9JE-FBCV7 | Hilde Pfeiffer | created asset 8603955761278446316

These two assets are linked to the respective PDF, which certifies the successful completion of the corresponding module. The PDF of Alexander Pfeiffer was also provided with the A-Trust citizen card signature to show how further trust can be established via using the official digital signatures following the EIDAS regulation[8].

The respective certificate can now either remain on the account of the individual teacher or be sent to the institution for which the examination was administered. In our role play, we left the certificates with the specific teacher, as read permission can be obtained via the asset IDs. In a real-life use-case, the files would have to be encrypted and the read permissions would have to be granted with shared keys or similar in addition to the asset ID.

[7] https://nftmagic.art/view?asset=12203487059644698664

[8] https://eur-lex.europa.eu/legal-content/DE/TXT/?uri=celex%3A32014R0910

ARDOR-ADH6-T5U9-L75S-4NG92 | which is the official account of the Center for Applied Game Studies created the certificate of completion including the asset-id and IPFS information (Figure 2) from both subjects completed. This certificate has been sent to the verified account of the student Natalie Denk (3). In this case a public message has been included saying "your certificate as a token". This shows the possibility that a message can be added to each transfer. With different options: encrypted or unencrypted and stored on the blockchain as long as it exists or if the message should no longer be processed after a certain block height (i.e., point in time). The message is not a fixed part of the NFT itself, it is only linked to the transaction ID, which is created uniquely for each transaction.

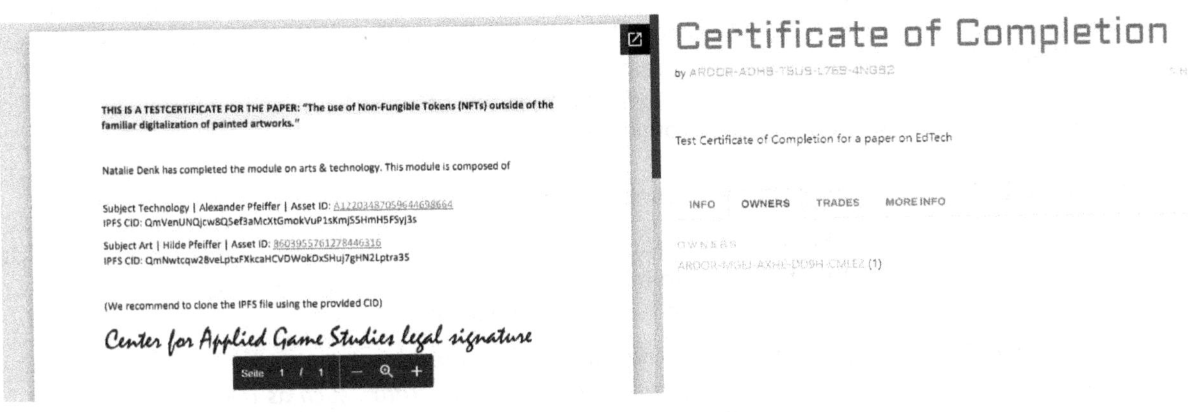

Figure 2: Screenshot certificate of completion from the NFTMagic.art webapp

Asset:	1107314914942834607
Asset Name:	Certificat
Quantity:	1
Recipient:	ARDOR-MGEJ-AXHE-DD9H-CMLEZ
Sender:	You
Type:	Asset Transfer

Public Message

Your certificate as token

Figure 3: Transfer of the NFT from the applied games wallet to the wallet of the student

ARDOR-MGEJ-AXHE-DD9H-CMLEZ | Natalie Denk | as a student has access to this certificate via the Sigbro App on her mobile phone. (see figure 4).

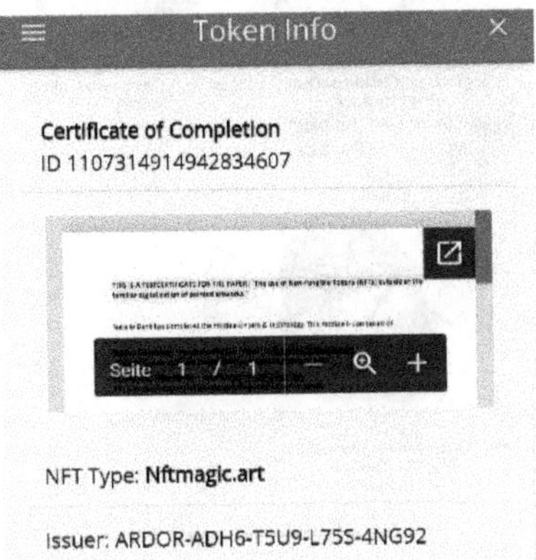

Figure 4: Screenshot of the Sigbro App from Natalie Denk, holding the certificate and the relevant meta data

As already stated for this use-case all files are readable by anyone, however in a real-world context, only the accounts involved would see the PDFs messages and the respective private data, while all other accounts can only follow transaction IDs and read the account property descriptions. Also, the certificates should not be visible on the NFTmagic.art art page, instead there would have to be an iteration of this page, possibly with secure sub-domains or directories for customers to be able to use the tool as an NFT creator and a customized branded file repository.

The Ardor Blockchain offers the possibility of integrating a chat functionality between accounts. AI-assisted automation processes could be considered at this point. These could make processes faster, for example to validate whether certain requirements for completing a grade or issuing a certificate have been met. Furthermore, work steps could be defined more transparently and fraud around the topic of grades and certificates could be minimized. Finally, the biggest advantage lies in the simple verification of the final certificate, due to the digital signatures and IDs used in the process. However, this is not a technical innovation and has already been implemented by various organisations.

3.2 Use-case B: Securing records on a blockchain basis with NFTs

This use-case is a presentation of the current status and continuation of the work presented at ECCWS21 "Use of Blockchain Technologies Within the Creative Industry to Combat Fraud in the Production and (Re)Sale of Collectibles." In the conclusion of this paper, we stated as future work:

> *"With regard to the development of the demonstrator further possibilities offered by the Ignis blockchain will be explored in further iterations of the demonstrator, such as the division into 2 token systems for the same collectible. One token will contain the private owner data (encrypted) and another token the public data around the vinyl. Both tokens can only be sent together, which is guaranteed by the corresponding approval model."*

In collaboration with the NFTMagic team, we were permitted to use the latest, not yet publicly available features to create NFTs for this second demonstrator. We described those features in the introduction of this paper. The system allows the possibility to design tokens as a series, which share an IPFS file and most of the meta data in the asset properties. This series can be then connected to each other in a collection. In addition, the two other NFT types can also be included in the same collection. This leads to completely new and interesting possibilities when it comes to mapping and managing the ownership, utilisation and usage rights of rare collectibles. What was described in the conclusion of last year's paper as two blockchain tokens, is now a blockchain token on the one hand and the public file on the IPFS server on the other. We created a collection (Figure 5) consisting of a series of three assets representing a vinyl each signed by Pfeiffer, a series of two representing a vinyl each signed by two band members, and a singleton asset (NFT of one piece) representing a single copy of a vinyl signed be all bandmembers.

Figure 5: The full collection displayed on a mobile device

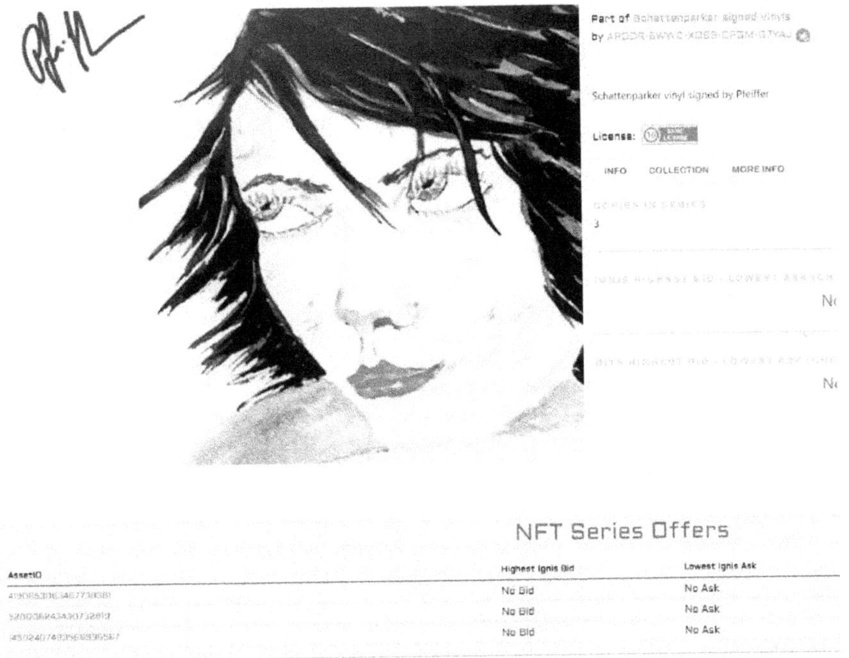

Figure 6: The Series of 3 assets representing the vinyl signed by Pfeiffer

Figure 6 shows the detailed view of the series of 3 representing the vinyls signed by Pfeiffer. On the top the account which minted (created) the NFT is visible. This is the most trustable account on the NFTMagic.art platform:

ARDOR-5WW2-XQ63-CFGM-G7YAJ, the account of the co-founder "thewiremaster". Clicking on "more info" would display the current owner, which is the verified address of Alexander Pfeiffer, member of the band and person who signed the vinyl. The license under which the NFT is issued and the other assets IDs of vinyls belonging to the same series can be seen. Now let's take a look how the asset properties that are displayed:

Asset ID: 5200062434330732819
Schattenparker signed by Pfeiffer||3|Schattenparker signed Vinyls|1|QmdgZvqkmgHCFeJVD2Yn2kPgYnLiZW46TiZC5Kcizi1rTR

Asset ID: 4190853063467738381
Schattenparker signed by Pfeiffer||3|Schattenparker signed Vinyls|2|QmdgZvqkmgHCFeJVD2Yn2kPgYnLiZW46TiZC5Kcizi1rTR

Asset ID: 14502407493566936567
Schattenparker signed by Pfeiffer||3|Schattenparker signed Vinyls|3|QmdgZvqkmgHCFeJVD2Yn2kPgYnLiZW46TiZC5Kcizi1rTR

The public information includes the respective Asset ID, the name of the series, the number within the series, the name of the collection, the total number of different assets within the collection, and the shared IPFS-CID. This demonstration with the Schattenparker vinyl in different variations is ready for a real-life application in the near future. The presented use-case, unlike the ones in education, is not exotic to the essence of the NFTMagic.art platform. In fact, it should be possible for companies and professional partners, to create a separate section for their musical works of art collector's editions. It would also be necessary to create a user-friendly possibility for the companies to choose directly in the backend from various common approval models how the digital assets may be transferred. This should also allow to deactivate the trading place with crypto assets. The goal of this development must ultimately be to secure the rights of use of real vinyls and collectibles with the help of NFTs and blockchain technology to make them trustworthy and tradable on the basis of NFT technology and to make the NFTs usable in existing online and offline shops via API interfaces. It is essential that the users themselves do not need any knowledge of blockchain technology and also do not need to hold any cryptocurrencies (in the case of this use-case the blockchain currency Ignis). Only the professional partners handle the crypto-asset, which varies depending on the blockchain system used and is necessary for the transaction fees and mint fees of the NFTs. In our specific case, a light version of the Sigbro.app would be desirable, which only shows the NFTs, separated by collection and series, but does not offer any other features of a classic wallet. Experienced crypto users, however, should be able to see and use the NFTs in their regular wallet.

4. Conclusion

In conclusion, both use-cases presented in this paper are very promising. While the use-case in the education industry is still far away from real use and is rather an inspiration, the use-case in the music industry can be put into practice much more quickly. The pervasive problem remains in choosing the right blockchain system. This must be considered individually for each use-case and the advantages and disadvantages of the different systems must be evaluated in detail and systematically.

As described in this paper, the ideal implementation would be one where end users do not need to know anything about blockchain technologies or cryptocurrency. Interested users can view transactions and can find out what benefits this offers them. This leads to many other issues such as the management of the private keys of the blockchain wallets. In the discussion among the group of authors, it was pointed out that the assets are addressed via user and password, maybe via access-tokens and that the full management is not in the hands of the end users for the time being. However, the users can apply to take over full responsibility and transfer the assets to their existing wallet. Further research is needed on both the technical process (again the correct setting of approval models) and the legal situation.

With regard to which blockchain system is chosen, one must also take into consideration whether it should be used privately, as a consortium, publicly or as a combination of these options. The consensus algorithm of the chosen blockchain is just as important as the architecture if tokens are to be created. If transaction fees are to be paid, the crypto market should be observed and calculations made as to how they can be paid optimally and in the long term. In addition, due attention must be paid to the key figures of these cryptocurrencies: number of transactions per day, amount of volume calculated in FIAT money per day and in average the last year, number of addresses that actively use and transfer assets and cryptocurrencies, number of (decentralised) apps already build on the platform and, above all, where can these cryptocurrencies be obtained and how trustworthy are the respective exchanges. Finally, it is very important whether one can operate a full node and thus not only strengthen the network, but also guarantee the respective customers that the data is accessible. This applies both to the blockchain assets and to the IPFS infrastructure used in our example.

Acknowledgements

The Center for Applied Game Studies of Donau Universität Krems has a research agreement in Q1 2022 with Jelurida SA, swiss, to explore use-cases on the Ignis Childchain in the field of education.

We like to thank the Co-Founders of Sigbro.app and NFTMagic.art, for their openness and interest in the research performed using their platforms. In particular "thewiremaster" for letting us explore the newest unreleased features on the platform.

Author contribution

Alexander Pfeiffer, Natalie Denk, Thomas Wernbacher and Stephen Bezzina developed the demonstrator and conducted the role play and discussion, while conducting the blockchain transfers.

Alexander Pfeiffer, Natalie Denk, Thomas Wernbacher and Stephen Bezzina wrote and edited the paper.

Vince Vella and Alexiei Dingli gave valuable feedback and contributions to the use-case in the educational industry in their responsibility as supervisors of Alexander Pfeiffer's PhD studies at the University of Malta.

References

Abctcm & APenzl (2016) "Introduction", In: Snapshot, NXT – unsurpassable blockchain solutions, Plaisir d'histoire, ISBN 978-2-9700947-3-9, (the authors names are pseudonyms)

APenzl (2016) "Phasing", In: Snapshot, NXT – unsurpassable blockchain solutions, Plaisir d'histoire, ISBN 978-2-9700947-3-9,)

Satoshi Nakamoto (2008) Bitcoin: A Peer-to-Peer Electronic Cash System, in Whitepaper online available https://bitcoin.org/bitcoin.pdf (Satashi Nokamoto is a pseudonym, it is not known to the general public who is behind this name.); Accessed: January, 2022

Grech, A. and Camilleri A. (2017) Blockchain in Education. https://doi.org/10.2760/60649; Accessed: January, 2020

O. Jutel (2021) Blockchain imperialism in the Pacific. Big Data & Society, 8(1), 2053951720985249. https://doi.org/10.1177/2053951720985249

A. Pfeiffer, M. Bugeja (2021) "Introducing the Concept of "Digital-Agent Signatures": How SSI Can Be Expanded for the Needs of Industry 4.0." In: Dingli A., Haddod F., Klüver C. (eds) Artificial Intelligence in Industry 4.0. Studies in Computational Intelligence, vol 928. Springer, Cham. https://doi.org/10.1007/978-3-030-61045-6_15

A. Pfeiffer, S. Bezzina, T. Wernbacher, V. Vella, A. Dingli, N. Denk (2021) The use of Blockchain Technologies to Issue and Verify Micro Credentials for Customised Educational Journeys, EDULEARN21 Proceedings, pp. 1265-1268.

A. Pfeiffer, S. Bezzina, T. Wernbacher, Thomas. (2021). Use of Blockchain Technologies Within the Creative Industry to Combat Fraud in the Production and (Re)Sale of Collectibles. ECCWS 21 proceedings. 10.34190/EWS.21.055.

Cybersecurity Risk Assessment Subjects in Information Flows

Jouni Pöyhönen, Aarne Hummelholm and Martti Lehto
Faculty of Information Technology, University of Jyväskylä, Finland
aarne.hummelholm@elisanet.fi
jouni.a.poyhonen@jyu.fi
martti.j.lehto@jyu.fi

Abstract: A modern society includes several critical infrastructures in which digitalization can have positive impacts on the levels of autonomy and efficiency in the use of infrastructure systems. Maritime transportation is an example of an infrastructure that currently needs development in the digitalization of its operations and processes. At the same time, maritime processes represent a large-scale cyber environment, thus trustable information distribution between system elements of the processes is needed. Since 2020, the Sea4Value / Fairway (S4VF) research program in Finland has been working to develop maritime digitalization which can lead to autonomy processes in the future. The first stage of the program has led to a demonstration phase of remote fairway piloting. This remote fairway piloting process, "ePilotage," is a complex system-of-systems entity. In this entity, fairway systems, ship systems and control center systems are the main processes from the operational point of view. Remote pilotage operations need support processes such as vessel traffic service (VTS) and weather forecast services. Situation awareness from other vessels and the stakeholder's processes are also essential information for the entire piloting operation. In this context, a new concept of information flows at the technical level will be based partly on cloud servers. In this paper, a cybersecurity risk assessment has been carried out at the technical level of information and communication technologies (ICT), and it concerns information transmission between a ship and a cloud server. It describes the most important topics for a comprehensive risk assessment in a specific ship-to-cloud information flow of the fairway process. The findings of the study can be considered good examples of the management of cybersecurity risks in critical information flows between all main system blocks of the fairway process. The research question is as follows: "How can the cybersecurity risks of information flows in a system-of-systems entity be described and evaluated?" The main findings are related to the risks of transmitting information from a ship to a cloud server. The methodology that has been used is based on analyzing the probabilities of cyberattacks occurring in relation to the probabilities to defend against these actions. The main risk assessment topics have been listed.

Keywords: maritime digitalization, cybersecurity, information flow, risk topics

1. Introduction

As the first stage in developing maritime autonomy in Finland, the Sea4Value (S4V) research program was started in 2020. The program is now becoming a research program of digitalization of harbor processes. At the beginning of the program, the research concentrated on creating automated remote pilotage fairway features (ePilotage). The ePilotage Act refers to the digitalization of activities related to the remote navigation of vessels on local waters. The purpose of this was to enhance the safety of vessel traffic and prevent environmental damage generated by vessel traffic (Finnpilot Pilotage Ltd, 2020).

Finnpilot Pilotage (2020) defines pilotage as follows: "As defined in the Pilotage Act, pilotage refers to activities related to the navigation of vessels in which the pilot acts as an advisor to the master of the vessel and as an expert on the local waters and their navigation. The purpose of pilotage is to enhance the safety of vessel traffic and prevent environmental damage generated by vessel traffic." The mission of the Sea4Value / Fairway (S4VF) program is to enhance towards digitalization, service innovation and information flows in maritime transport in order to prepare for advanced autonomous operations and navigation as a long-term mission. A key step towards an autonomous transport system is to ensure safe, sustainable, and efficient channel for ships to enter and leave harbors. The S4VF program improves the safe navigation for existing vessels and lays the system-of-systems foundation for autonomous vessels of the future. The ePilotage process as a remote pilotage fairway is the first step in this way.

The ePilotage environment of the S4FV project is an example of a system-of-systems in which an increased number of digital solutions are entering new environments where traditional engineering solutions are still in use. This development introduces increased risks of a malicious cyber adversaries taking deliberate actions against the system. For this reason, the threat analyses should be carried out according to the principles within the scope of the "system-of-systems threat model" (Bodeau & McCollum, 2018).

This paper describes a risk assessment approach for the remote pilotage system-of-systems environment and related threats by using an example of a subsystem and utilizing the Mitre ATT&CK framework. A remote pilotage system-of-systems configuration includes both ICT and industrial control systems (ICS) networks and components.

In this paper, cybersecurity risk assessment has been carried out on an organization's technical level as an example of the importance of trustworthy information chains in the system-of-systems architecture. It concerns information flow from a ship to a cloud server. One of the ways to use this information is to control the ePilotage process. This paper follows our previous papers on the S4V research program and describes the most important topics for comprehensive risk assessment in specific cloud information flows of the fairway process. The findings of the study can be considered good examples of cybersecurity risk assessment work in critical information flows between the main system blocks of the fairway process. The research question is as follows: "How can the cybersecurity risks of the main information flows in a system-of-systems entity be described and evaluated?"

The main findings are related to the risks of transmitting essential process information from a ship to a cloud server. The methodology used is based on analyzing the probabilities of a cyberattack in relation to the probabilities to defend against such actions in the use of ICT. The main risks assessment topics have been listed.

2. System-of-systems ICT network architecture and general attack vectors

Figure 1 shows the general communication network architecture and the main attack vectors related to the architecture. In this context, ship systems are like a home, factory, and so on. Ships and ship systems are important parts of the ePilotage process because they are remote pilotage attributes. At the same time, ships are connected to land-based communication and control systems and satellite systems. In this case, the ship is connected to land-based control systems via the cloud. The ship systems are also vulnerable to cyber-attacks in a similar way as other ICT and ICS systems, applications and devices are. Attacks can come from either inside the ship or outside the ship. Cyber-attackers can harm the control information that is needed for operations. The main attack vectors at the system level are shown in Figure 1.

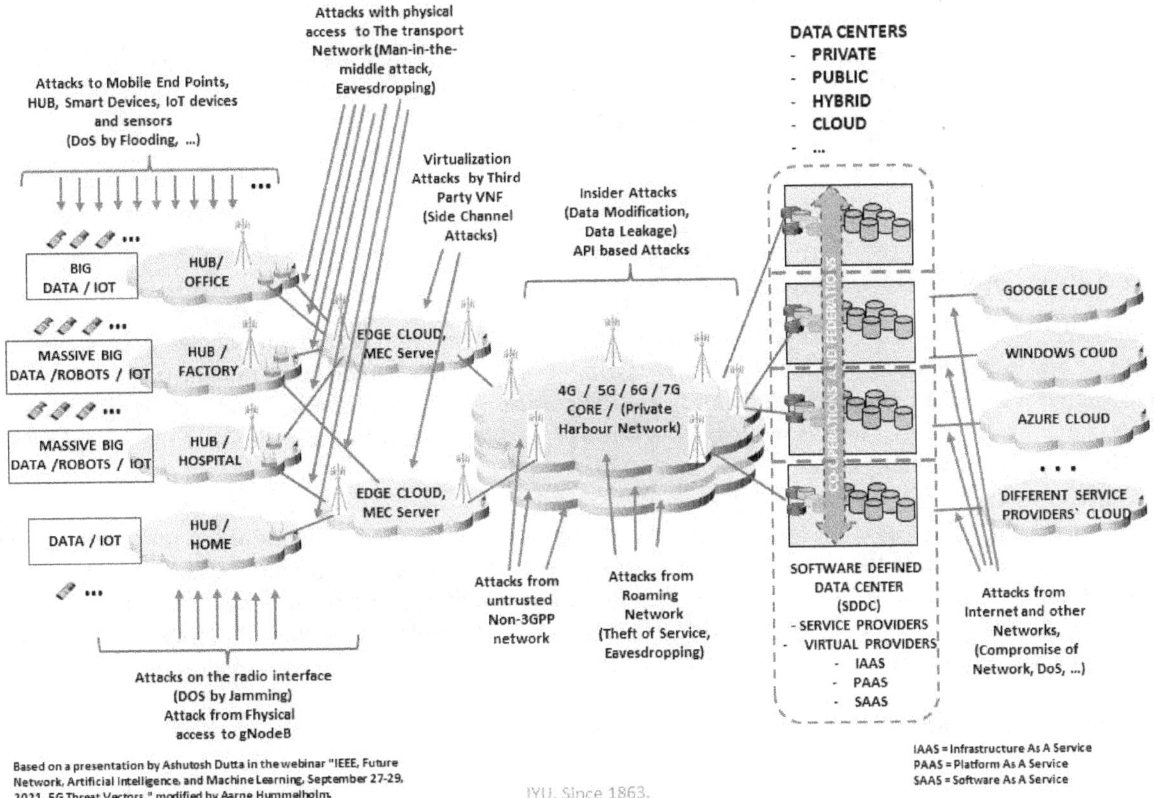

Figure 1: System-of-systems communication network architecture and the main attack vectors (Dutta, 2021, modified)

Understanding the motivation aspect enables situations that heighten the risk of a cyber-attack to be predicted (Casey, 2015). By combining the motivational factors for each attack archetype, it is possible to discover different events being triggered by attacks. Many cyber-attacks are associated with social, political, economic, and cultural backgrounds. It is crucial to identify comprehensively different kinds of circumstances that might trigger an attacker archetype. This can be valuable for risk assessments related to various situations. The motivation affects the attacker's targeting and methods. A vandal seeks visibility by defacing a website, but a spy wishes to stay unnoticed to gain information. The varying level of capability restricts some of the attackers from achieving their goals (Bodeau, McCollum & Fox, 2018). Therefore, being motivated does not mean that an attack is possible for the attacker. Understanding the motivations and capabilities of different archetypes limits the number of scenarios and thus makes evaluation feasible for the defender.

In the case of cybervandalism, the arrival of a controversial ship in a fairway might trigger actions mainly from ideological motivations. The controversy might be with the cargo, the ship's operations, or the owner. For cybercrime, valuable cargo is more tempting because financial gains act as the motivation. Cyberterrorism or sabotage can include business or political motivations. Political factors may arise from national or international issues. In the worst case, international tensions in the region could escalate to military cyber operations against maritime traffic. The parameters of the attacker archetypes for this case are presented in Table 1. (Kovanen, Pöyhönen & Lehto, 2021a)

Table 1: Attributes of the attacker archetypes (Kovanen, Pöyhönen & Lehto, 2021 a)

	Vandalism	Crime	Terrorism	Sabotage
Motivation and goal	Trying to make political change based on personal political or ideological motives. Egoism gain	Making money through fraud or from the sale of valuable information. Financial gain	Gaining social instability and influencing political decision-making. Anarchy gain	Causing instability, chaos, political change, and infrastructure paralysis. Paralysis gain
Target	Digital services of governments and companies, individuals' information systems	Digital services of governments and companies, individuals' information systems	Data and information about governments and companies. Critical infrastructure	Nation's critical infrastructure
Impacts	ICT: Defacement ICT: Network Denial of Service ICS: Loss of Productivity and Revenue	ICT: Data Encrypted for Impact (ransom) ICT: Resource Hijacking (mining cryptocurrencies) ICS: Manipulation of Control	ICS: Loss of Safety	ICS: Damage to Property (shipwreck)

3. The study of ship information

In many ways, the cyberworld challenges organizations, processes, and the use of technologies. An organization can use its own capabilities to develop security in its cyber domain. They can do so by enhancing its capabilities that are applicable to the operational domain. These can consist of people, processes, and technology meant to achieve outcomes or effects (Jacobs, von Solms & Grobler, 2016). It important to note that these capabilities can also include cyber vulnerabilities.

The remote pilotage process, ePilotage, is a special environment with a large network of separate systems and stakeholders in the cyber domain. By examining the impacts of cyberthreat actions and thus risks assessment in this connected environment, it is obvious that the threat impacts affecting one subsystem are propagated to affect other systems. For that reason, people, processes, and technologies should all be considered in risk assessment work, even if we have just one organization's technical level under risk evaluation.

3.1 People: Stakeholders

Management must recognize that clear, rational, and risk-based decisions are necessary from the point of view of process continuity. Understanding and dealing with risks are part of an organization's strategic capabilities and key tasks when organizing the operations. This requires, for example, the continuous recognition and understanding of security risks at the different levels of management. The security risks may be targeted not only at the organization's own operation but also at individuals, other organizations, and society as a whole—and in this case the entire ePilotage process. (Joint Task Force Transformation Initiative, 2011) The Joint Task Force Transformation Initiative (2011) recommends implementing an organization's cyber risk management as a comprehensive operation, in which the risks are dealt with from the strategic to the tactical level. In this way, risk-based decision-making is integrated into all parts of an organization. In research by the Joint Task Force Transformation Initiative, the follow-up operations of the risks are emphasized on every decision-making level. For example, on the tactical level, the follow-up operations may include constant threat evaluations about how the changes in an area can affect the strategic and operational levels. The operational level's follow-up operations, in turn, may contain, for example, the analysis of new or current technologies in order to recognize the risks to business continuity. The follow-up operations on the strategic level can often concentrate on an organization's information system entities, the standardization of the operation and, for example, on the continuous monitoring of the security operation. (Joint Task Force Transformation Initiative, 2011)

3.2 Process: ePilotage, (subprocess: Ship information flow)

ENISA emphasizes maritime transport as a crucial activity for the European Union economy. The global digitalization trend has led authorities to set recent policies and regulations to maritime processes to face new cybersecurity challenges with regards to IT and ICT (ENISA, 2019). Development, implementation, and maintenance of a cyber risk management program is essential part of organizations processes. The management of the process by the organizations senior experts should stay engaged to it throughout the process to ensure that the protection and contingency planning are balanced to manage risks within an acceptable limit (BIMCO et al., 2021). Processes are key to the implementation of an effective cybersecurity strategy. Processes are crucial in defining how the organization's activities, roles, and documentation are used to mitigate the risks to the organization's information. Processes also need to be continually reviewed: cyberthreats change quickly, and processes need to adapt with them. Processes, however, are nothing if people fail to follow them correctly (Dutton, 2017).

3.3 Technology: Ship systems, ICT-systems, Cloud Service

Technology is obviously crucial when it comes to cybersecurity at the organization's tactical level. By identifying the cyber risks that an organization faces, it can then start to look at what controls to put in place, and what technologies will be needed to do this. Technology can be deployed to prevent or reduce the impact of cyber risks, depending on your risk assessment and what you deem an acceptable level of risk (Dutton, 2017). Figure 2 presents the current technology at the subsystem level of information and information flow from ship to cloud and after that in use by the remote control center (RCC). Data storage in cloud and cloud computing services are used also for many other purposes of the remote pilotage process. The detailed configuration on the technical level is described Figure 3. Cloud computing services are platforms of two-sided markets connecting users with developers of complementary products or services. The resultant user-side transactions allow providers of Infrastructure as a Service (IaaS), Platform as a Service (PaaS), and Software as a Service (SaaS). (Arce, 2020)

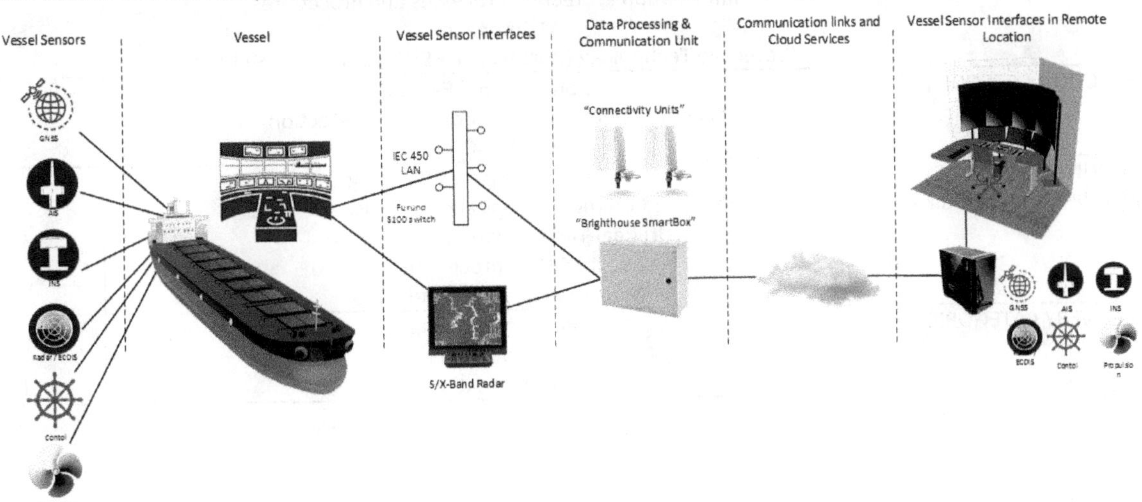

Figure 2: Ship sensors' information flow to the control center (Brighthouse Intelligence, 2021)

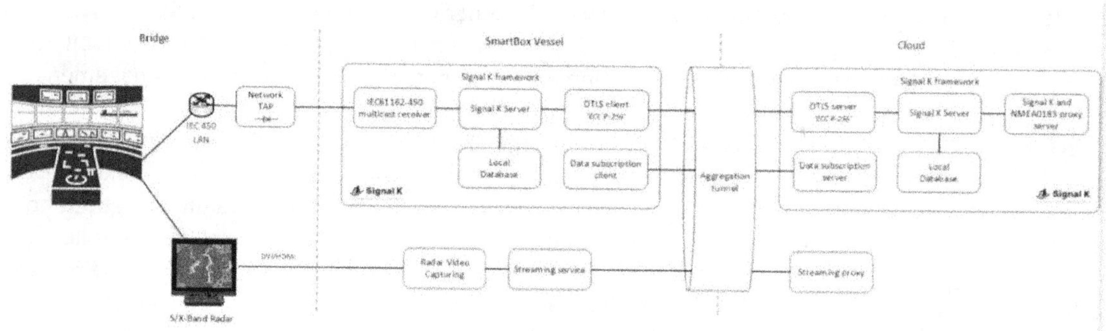

Figure 3: Ship sensors' connection to cloud, communication technology (Brighthouse Intelligence, 2021)

These two figures present the subsystem components of information flow for risk assessment work. These are ship network, ship LAN/WLAN, ship data process, transmitting tunnel, transmitting 4G or 5G, and cloud service. The information flow consists of data from ship status and live camera pictures.

4. Risk assessment method

In this paper, cybersecurity risk assessment has been applied to ship sensor information flow by investigating probabilities and using the elements in Table 2. This method is described in our paper "Assessment of cybersecurity risks: Maritime automated piloting process" (Pöyhönen & Lehto, 2022). The table has been used as a risk assessment tool by investigating the probabilities of each element of it. Probability tree principles have been applied as well. In Figure 4, the probability tree is described as using Defense probability P_D\` against Attack probability P_A in the evaluation process. Cyberattacks (A) in the Sea4Value ePilotage process are the same as the "Attack Identification" and located on all levels of the stakeholder's responsibilities (Strategy, Operational, Tactical/Technical) concerning the ship sensor information transmission to the cloud. The P_A attack probability (P_{SOT}) to defend against attack probability P_D\` (P_P, P_D, P_M, P_R) is related to the combination of cybersecurity capabilities (people, processes, and technologies), using "Protection" (P), "Detection" (D), "Countermeasure" (M) and "Recovery" (R) activities according to Table 2. The entire risk assessment process has been done by experienced cybersecurity professionals related to the case.

Table 2: Ships risk probabilities (NIST, 2018; Hummelholm, Pöyhönen, Kovanen & Lehto, 2021)

ACTION	EXAMPLES	NOTATION
ATTACK IDENTIFICATION	Attacks at strategy level (S) Attacks at operational level (O) Attacks at technical/tactical level (T)	A
PROTECTION CATEGORIES	Identity Management and Access Control Awareness and Training Data Security Information Protection Processes and Procedures Maintenance Protective Technology (Port scan, FIREWALL, IDS, IPS, SIEM...)	P
DETECTION CATEGORIES	Anomalies and Events Security Continuous Monitoring and Detection a) SOC	D
COUNTERMEASURE (RESPOND) CATEGORIES	Conducting Response Planning Communications and Analysis: a) Real-Time Situation Awareness b) OODA procedure Mitigation and Improvements	M
RECOVERY CATEGORIES	Recovery Planning Improvements Communications	R

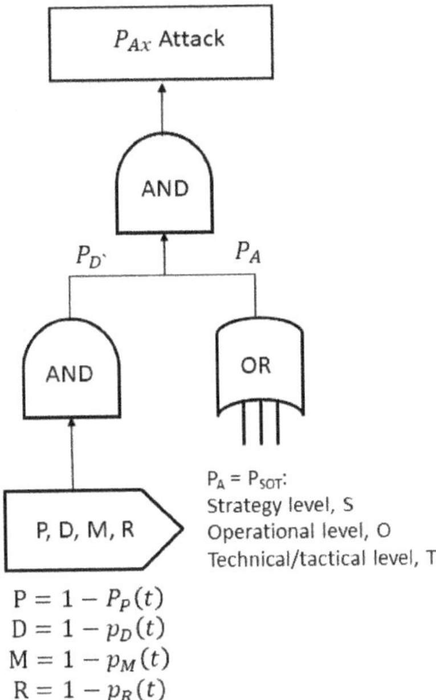

Figure 4: Probability tree; Defense probability $P_{D'}$ against Attack probability P_A in the ePilotage process (Wang & Liu, 2014, Pöyhönen & Lehto, 2022)

The probabilistic success of attacks, P(t), against the defense of system x can now be evaluated and calculated as follows, adapting the principle in "Threat Analysis of Cyber-Attacks with Attack Tree+" (Wang & Liu, 2014, mod.)

$$P_{Ax}(t) = P_A P_{D'} = (P_{SOT})(1 - P_P(t))(1 - p_D(t))(1 - p_M(t))(1 - p_R(t)) \quad (1)$$

5. Making threat analyses and risk levels estimation

In this case we have used Delphi method principle in order to make relevant threat analysis and risk-level estimations from the ship camera system. The members that have been involved in this analysis process are researchers and research methods. Delphi is advocated by cybersecurity experts from the S4V program: "The Delphi method is an iterative process to increase consensus-building and at the end to have consensus among an experts from an examine case. The Delphi method is part of quantitative as a means to achieve an optimally reliable expert consensus." It could have on one of three objectives (Garson, 2012):

- 1. forecasting future events
- 2. achieving policy consensus on goals and objectives within organizations or groups
- 3. identifying diversity in and obtaining feedback from stakeholders in some policy outcome.

Table 3 includes the results of Delphi method research on the ship-to-cloud subsystem. It has been done in order to forecast future events conducted as part of the risk evaluation process. Cybersecurity researchers and expert's contributions are related to the main threats/attacks, the impacts of them, the main defense categories and risk level columns. The probability estimation has been done by the cybersecurity researchers and experts according to the formula (1) principal. In this evaluation it is expected that stakeholders have normal cybersecurity solutions and procedures in use. We have used the OWASP Risk Rating Methodology to identify a security risk. The evaluation has collected information about the threat agent involved, the attack that will be used, the vulnerability involved, and the impact of a successful exploit on the operation of the system. The risk levels are divided into three categories (LOW, MEDIUM, HIGH) depending on the estimated severity of the attack impacts and occurrence of harm after estimated defense capabilities. The determination of the risk level is based on elements within each factor, such as the motive and ability of the attacker, the ease of finding vulnerabilities, the loss of the CIA, and damages to the system. Each factor has a set of options, and each option has a likelihood rating from 0 to 0,9 associated with it. The 0-to-0,9 scale is split into three parts: 0 to <0,3 = LOW, 0,3 to <0,6 = MEDIUM, and 0,6 to 0,9 = HIGH (OWASP, 2022).

Table 3: Ship systems to cloud service; main threats/attacks, related impacts (Mitre, 2019, 2020; Kovanen, Pöyhönen & Lehto, 2021b), main defense categories (NIST, 2018) and risk levels

Subsystem/ Ship-to loud	Main threats/attacks	Impacts	Defense Categories	Risk level
Ship Network	Brute Force Credential Theft	Manipulation of View Denial of Service System Shutdown	Identity Management and Access Control Security Continuous Monitoring and Detection	LOW
Ship LAN or WLAN	Man in the Middle Jamming	Denial of Service	Data Security	MEDIUM
Ship, Data process	Physical Access	Service Stop	Access Control	LOW
Transmitting Tunnel	Credential Theft	Loss of Safety	Identity Management and Access Control	LOW
Transmitting 4G or 5G, 3rd party	Insider Attacks Attacks from Rooming Network API based Attack	Data Destruction Denial of Service	Identity Management and Access Control Communications and Analysis	MEDIUM
Cloud service, 3rd party	Attacks from Internet Insider Attacks Credential Theft DoS Attacks API-based Attack Ransomware Attacks	Data Manipulation Data Destruction Denial of Service Data Encrypted for Impact	Data Security Communications and Analysis Recovery Planning Awareness and Training Security Continuous Monitoring and Detection	HIGH

In an ICS environment, the Mitre framework uses the terms *denial*, *loss*, and *manipulation*. Denial is a condition which occurs only while the attack is active. Loss refers to sustained loss of an asset that continues after the active malicious interaction has ceased. Manipulation alters the asset and can be either loud and easy to detect or subtle and longer sustained. According to the paper "Cyber-Threat Analysis in the Remote Pilotage System" (Kovanen, Pöyhönen & Lehto, 2021b), we have described the impacts as follows:

- **Manipulation of view** is a more subtle attack type than denial or loss of view. Slightly falsified data are harder to detect than missing data. Therefore, the attack can continue for longer periods of time undetected. Consequently, the operator of the affected system loses correct situational awareness and makes decision based on false data. The effect spreads to all connected systems and operators using the manipulated view.
- **Denial of Service** attacks can be carried out by affecting the endpoint or the network that leads there. In either case, the service is unavailable for use. All other systems that depend on the affected system experience difficulties. If an alternative system is available and the deployment is designed and implemented, the effects of this attack type decrease.
- **Data destruction**, as well as **data encrypted for impact**, disk wipe and **service stop**, all prevent the use of the data and services. These can also prevent the use of the whole system in case the action is targeted at, for example, disk structure rather than the data itself. **System shutdown**/reboot can be used to make systems inaccessible faster by, for example, rebooting after wiping the master boot record. The severity of this type of an event depends on the system and time the restoration from backups takes. If similar data or service is served from alternative systems, the overall resilience increases.
- **Loss of Safety** is dangerous especially with cyber physic systems as the result may cause injuries or death when the safety mechanism of a system is disabled. Even a threat of this type of circumstance can delay reaction to other impact types if a human operator is not able to initiate countermeasures due to a fear of unsafe working conditions.
- **Data manipulation** is harder to detect than data destruction if the manipulation is subtle. Systems and operators can continue to act but they base their decisions on false data. For example, location information could be manipulated to lead a ship off course. Depending on the magnitude of the manipulation and the availability of location information from unaffected systems (and the correct comparison checks), the time until detection varies.

Table 4 illustrates the risk levels after the evaluation process in three categories (LOW, MEDIUM, HIGH), selected defense categories, recommendations for relevant risk-level mitigations, and recommendations for action priority. Recommendation priorities are (1) direct, (2) as soon as possible, and (3) when the action is convenient to perform. The action to be taken can be reasonably divided into development circles. The residual risks should be evaluated after the first round and beyond that after every round.

Table 4: Ship system to cloud Service; risk-level mitigation, action priority

Subsystem/ Ship to Cloud systems	Risk level	Defense Categories	Risk level mitigation	Action priority 1–3
Ship Network	LOW	Identity Management and Access Control Security Continuous Monitoring and Detection	Network Segmentation Role Based Access Control Ports Hardening	3
Ship LAN or WLAN	MEDIUM	Data Security	Authentication Policy Access Control WPA2+PSK Security	2
Ship, Data process	LOW	Access Control	Physical Security	1
Transmitting Tunnel	LOW	Identity Management and Access Control	Multi-Factor Authentication VPN update procedure	3
Transmitting 4G or 5G, 3rd party	MEDIUM	Identity Management and Access Control Communications and Analysis	Service Agreement Audit of Service	2
Cloud service, 3rd party	HIGH	Data Security Communications and Analysis Recovery Planning Awareness and Training Security Continuous Monitoring and Detection	Zero Trust Network Access Service Agreement Audit of Agreement	1

Protective technology (Port scan, FIREWALL, IDS, IPS, SIEM) and update procedures of these solutions as well as other relevant resiliency actions, such as cybersecurity policies, are needed in daily life as well as in the use of the ICT environment. In addition to these actions, however, mitigation of the risks identified in Table 4 is highly recommended. These actions may require special attention from every stakeholder of ship information transmission. The priority 1 actions should be performed as soon as possible in the first round of cybersecurity development of ship information flow. This category includes either critical actions or actions that are very easy to perform. The other priorities can be addressed after the first round of development actions, depending on the resources for the development. Residual risks should be determined after every development round.

6. Conclusion

In the first stage of Finnish maritime digitalization, the Sea4Value / Fairway (S4VF) research program has been launched to create automated remote fairway pilotage features. It is called the ePilotage research process. This process is an essential part of the critical maritime traffic and transportation supply chain. The fairway and its stakeholders' systems are together a complex system-of-systems entity, characterized by a conglomeration of interconnected networks and operational dependencies. The research program increases the level of various digital solutions, stakeholders, and processes in maritime fairways. However, there will also be a continuing need for traditional engineering solutions for a long time to come. This environment increases the risks of all levels of people, processes, and technology.

A system-of-systems technical environment is a comprehensive cybersecurity entity, and it should be considered as a common structure for all operationally related stakeholders of the pilotage process. Therefore, in this maritime research case concerning the information flow of ship sensors the systems communication way between process elements is used as example. In this risks assessment evaluation work, we have viewed all risks in such a way that they can been seen as well as the same way between other information flows as they relate to secure communication. In that sense, the paper exploits the risk assessment method where cyberthreats are considered in relation to defense capabilities.

This paper has established a research framework for the cybersecurity risk assessment of maritime automated remote ePilotage fairway systems and processes. The case of the framework is an example and uses risk probability evaluation in one of the most important information flows between the main fairway systems. The risk assessment methodology that has been used is based on attack probabilities against the probabilities to defend against adversarial actions in the use of communication technologies. Risk assessment factors have been identified and the risk assessment tool has been described. It is a way of thinking about risks and risk prioritization. These are needed to answer the research question: "How can the cybersecurity risks of automated remote piloting fairway operations be evaluated?"

Protecting the system-of-systems environment against its cyberthreats implies measures taken based on risk assessment of the system-of-systems, and eventually all critical information flows between those elements. It ensures confidentiality, integrity, and the availability of primarily digital information in the operating processes, achieving operational continuity and the reliability of activities being examined.

References

Arce, G. D. (2020). Cybersecurity and platform competition in the cloud. Computers & Security 93 (2020) 101774.

BIMCO, INTERCARGO, INTERTANKO, ICS, IUMI, OCIMF, OCIMF, Sybass and WORLD SHIPPING COUNCIL (2021). The Guidelines on Cyber Security Onboard Ships. Version 4. https://www.bimco.org/about-us-and-our-members/publications/the-guidelines-on-cyber-security-onboard-ships, retrieved 1.3.2022

Brighthouse Intelligence, (2021). Future Fairway Flash Event, Sea4Value Fairway project report 19.11.2021.

Bodeau, D. J. and McCollum, C. D., (2018). System-of-systems threat model. The Homeland Security Systems Engineering and Development Institute (HSSEDI) MITRE: Bedford, MA, USA.

Bodeau, D.J., McCollum, C.D. and Fox, D.B. (2018) "Cyber Threat Modeling: Survey, Assessment, and Representative Framework", Mitre Corp, Mclean.

Casey, T. (2015) "Understanding Cyber Threat Motivations to Improve Defense", Intel White Paper.

Dutta, A. (2021) Future Network, Artificial Intelligence, and Machine Learning. IEEE webinar. September 27-29. 2021.

Dutton, J. (2017) Three pillars of cyber security. Available from: https://www.itgovernance.co.uk/blog/three-pillars-of-cyber-security, retrieved 25.1.2022

ENISA. (2019). *PORT CYBERSECURITY*. Good practices for cybersecurity in the maritime sector. November 2019.

Finnpilot Pilotage Ltd, (2020). Available from: https://finnpilot.fi/en/pilotage/what-is-pilotage/, retrieved 25.1.2022

Garson, G. D. (2012). The Delphi method in quantitative research. Asheboro, NC: Statistical Associates Publishers. Available from: https://faculty.chass.ncsu.edu/garson/PA765/delphi.htm, retrieved 25.1.2022

Hummelholm, A., Pöyhönen, J., Kovanen, T. & Lehto, M. (2021). Cyber Security Analysis for Ships in Remote Pilotage Environment. Presented of ECCWS 2021 - 20th European Conference on Cyber Warfare and Security. 24th - 25th June 2021, Chester, UK.

Jacobs, P. C., von Solms, S. H. & Grobler, M. M., (2016). Towards a framework for the development of business cybersecurity capabilities. International Conference on Business and Cyber Security (ICBCS), London, UK. The Business and Management Review, Volume 7 Number 4, 51–61.

Joint Task Force Transformation Initiative, (2011). NIST Special Publication 800-39: Managing Information Security Risk - Organization, Mission, and Information System View, Gaithersburg: National Institute of Standards and Technology.

Kovanen, T., Pöyhönen, J. & Lehto, M. (2021 a). Cyber Threat Analysis in the Remote Pilotage System. Presented in ECCWS 2021 - 20th European Conference on Cyber Warfare and Security. 24th - 25th June 2021, Chester, UK.

Kovanen, T., Pöyhönen, J. & Lehto, M. (2021 B). ePilotage System of Systems' Cyber Threat Impact Evaluation. Proceedings of the 16th International Conference on Cyber Warfare and Security ICCWS p. 144-153.

Mitre. (2019). "Impact", Available from: https://attack.mitre.org/tactics/TA0040/, retrieved 25.1.2022

Mitre (2020). Impact. Available from: https://collaborate.mitre.org/attackics/index.php/Impact, retrieved 25.1.2022

National Institute of Standards and Technology, NIST. (2018). Framework for Improving Critical Infrastructure Cybersecurity, April 16, 2018

OWASP Risk Rating Methodology, Available from: https://owasp.org/www-community/OWASP_Risk_Rating_Methodology, retrieved 25.1.2022

Pöyhönen, J. & Lehto, M., (2022). Assessment of cybersecurity risks - Maritime automated piloting process. Submitted to be published in ICCWS 2021 - 17th International Conference on Cyber Warfare and Security. 17th - 18th March 2022, Albany, New York, USA.

Wang, P. & Liu, J. C. (2014). Threat analysis of cyber-attacks with attack tree+. Journal of Information Hiding and Multimedia Signal Processing, 5(4).

Exploring Care Robots' Cybersecurity Threats From Care Robotics Specialists' Point of View

Jyri Rajamäk and Marina Järvinen
Laurea University of Applied Sciences, Espoo, Finland
jyri.rajamaki@laurea.fi
marina.jarvinen@student.laurea.fi

Abstract: Care robots can perform tasks related to physical or mental care; assisting in daily tasks or rehabilitation, independently or semi-automatically. Care robots are exploitable in home-care, nursing homes, or other care facilities. Care robots have the potential to solve several challenges related to aging people. However, care robots suffer have similar cybersecurity problems as other information and communication technology (ICT) devices. In addition, the cybersecurity threats of care robots have been studied less than those of industrial robots. This study's purpose is to map cybersecurity threats related to care robots from the perspective of care robotics specialists. The study consists of thematic interviews of six purposive-selected specialists in care robotics. A semi-structured thematic interview guide based on the literature view of previous studies, facilitates the conversations at the interviews. All interviews were transcribed verbatim, analyzed by deductive content analysis, and the remaining material was analyzed by inductive content analysis. According to the interviewed specialists, care robots' cybersecurity threats are associated with the same risks and threats as the use of other ICT devices or robots. Most potential threats are considered to be remote access of care robots, spying, and eavesdropping. Network connectivity is seen as the main interface to the realization of cybersecurity threats in care robotics. New features such as artificial intelligence and machine learning are considered to create more opportunities for new threats. Experts also highlight the underlying human factors behind cybercrime. According to the results, more studies exploring the motives for cybercrime against care robots and the potential benefits derived from it are needed to determine the likelihood of the realization of threats to care robots are needed. Cybersecurity is a race against cybercrime and finding a balance between significant and acceptable risks. In the future, a service ecosystem should be developed which guarantees the safety of care robots throughout their life-cycle: during the design and development phase, deployment and user guidance, maintenance, and reuse of the robot. Additionally, it is important to take into account how new robust operating models can withstand failures and how critical services can be secured in the event of a cybersecurity threat.

Keywords: care robot, cybersecurity of robots, cybercrime, thematic interviews, hijacking, rehabilitation systems

1. Introduction

As the population ages, care for the elderly will face new challenges as the demand for care increases. Care robots are cyber-physical systems that can perform tasks related to physical or mental care. They can be used to care for a person at home, in a care home, or in a care facility. They can also facilitate the work and co-operation of carers between home and care, for example through remote connections (Van Aerschot & Parviainen, 2020; Särkikoski, Turja & Parviainen, 2020).

Robots can make everyday life easier, create a sense of security, and perform a variety of tasks, but as a misfortune, they suffer from similar cybersecurity issues that computers have suffered from for a decade (Lera, Llamas, Guerrero & Olivera, 2017). Cybersecurity threats related to service and care robots have been studied much less than, for example, the cybersecurity of robots intended for industrial environments (Lera et al. 2017; Fosch-Villaronga & Mahler, 2021).

This qualitative study highlights the current view of care robotics experts on the cybersecurity of care robots and the real cybersecurity threats associated with their use. The research question is, "What are the cybersecurity threats to care robots?" The goal of threat mapping is to increase awareness, making it easier for both service providers and end-users to critically assess the threats associated with the use of devices and the risks they are prepared to take.

2. Literature review

A 'robot' refers to a reprogrammable mechatronic device that influences its environment by means of sensors and actuators (Särkikoski et al., 2020). Typically, robots are divided into industrial and service robots depending on whether they are used for the benefit of industry or perform tasks useful to humans (Fosch-Villaronga & Mahler, 2021). Robots are directly involved in human life and raise crucial ethical problems for our society (Tzafestas, 2018). Table 1 presents examples of healthcare robot applications according to the Policy Department for Economic, Scientific and Quality of Life Policies of the European Parliament.

Table 1: Examples of healthcare robot (Dolic, Castro & Moarcas, 2019).

Healthcare robot applications	
Robotic surgery	Allowing more accurate, less invasive and remote interventions relying on the availability and assessment of vast amounts of data
Care and socially assistive robots	Allowing to meet the expanding demands for long-term care from an aging population affected by multi-morbidities
Rehabilitation systems	Supporting the recovery of patients as well as their long-term treatment at home rather than at a healthcare facility
Training for health and care workers	Offering support for continuous training and life-long learning initiatives

Several articles talk about service robots, social robots, and care robots in the same context. A 'service robot' is a robot that is able to perform partially or completely independently services that are beneficial to human well-being or the environment. A 'social robot' complements, increases, or replaces human social interaction (Särkikoski et al. 2020). 'Care robots' can perform tasks related to physical or mental care independently or semi-automatically, such as assisting in daily tasks, rehabilitation, or mental care (Van Aerschot & Parviainen, 2020; Särkikoski et al. 2020).

Robots can often be programmed to be modified for different uses, so it may not be completely unambiguous to determine whether a robot is a care robot based on a purpose other than the use or environment (Fosch-Villaronga & Mahler, 2021). Thus, if the robot is used, for example, to care for children, the elderly and the disabled, it is a care robot, even if the same robot could be programmed elsewhere to deliver, for example, the work of a lobby clerk. A social robot can also be a care robot if its purpose is related to mental care.

Care robots can help with daily activities, provide companionship and a sense of security. Efforts are being made to develop and bring robots to the market, and care robots have been tested in Finland, e.g. in the care of the elderly (Schönberg, 2017) and the care of people with memory problems (Forum Virium, 2020). So far, however, it has not been possible to develop a care robot that could completely replace human work in helping the elderly in their daily activities. Monitoring devices, automated drug dispensers, robotic pets, cell phone attendance devices, and hospital logistics are already in use, but they are only capable of simple spoken language interactions or modest repetitive tasks, not situationally demanding day-to-day operations (Van Aerschot & Parviainen, 2020).

Robots can identify, process, and store the world around them, and they are constantly collecting data. Robots suffer from similar cybersecurity issues that computers have suffered from for a decade (Lera et al., 2017). Robot operation (such as navigation, speech, object recognition, etc.) requires heavy computing enabled by cloud services. As the number of interconnected systems and devices increases, so does the potential for vulnerabilities in the systems and the risk of malicious attacks (Fosch-Villaronga & Mahler, 2021).

3. Research method

This qualitative study explores the views of care robot experts (researcher, developer, or service provider) on what cybersecurity threats pose to care robots by thematic interviews lasting 35 to 75 minutes, carried out in the fall of 2021. The body of the interview questions was created based on literature. The themes were:

- cybersecurity of care robots at the present,
- potential cybersecurity threats to care robots, and
- the biggest risks to the cybersecurity of care robots.

With six interviews, saturation was observed. The accumulated material was analyzed with a theory-based analysis, in which the themes of the interviews also served as categories to which the material was related. The results of the interviews were also compared with the literature. The remaining material after the theory-based analysis was also analyzed by material-based content analysis to reveal the main aspects of the material that did not fit into the analysis framework described above.

4. Results

Figure 1 summarizes the results of the interviews. Experts' views on cybersecurity in care robotics fall into the three theoretical categories outlined above. In addition, the 'cybercrime' perspective emerged in the evidence-based analysis.

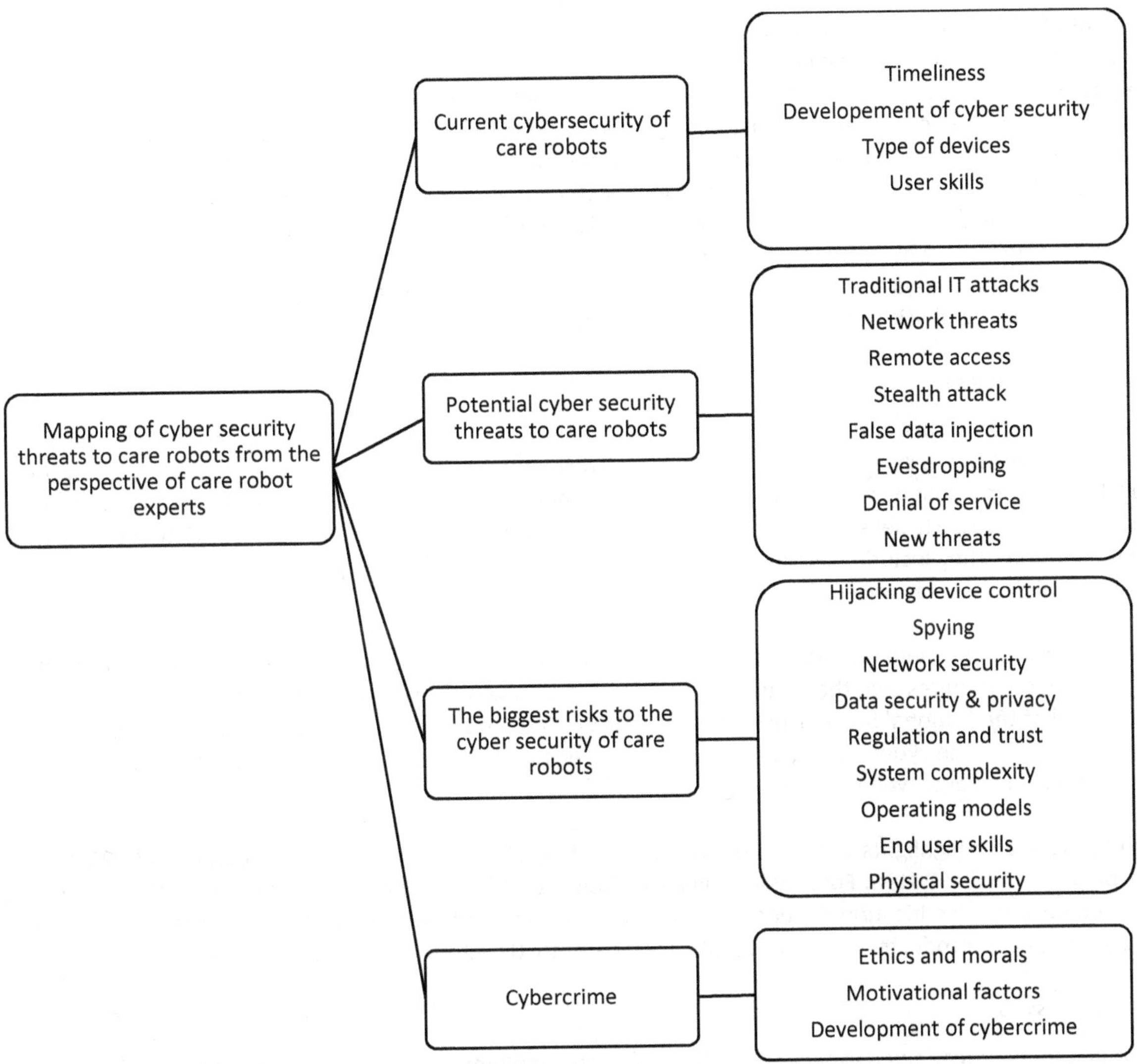

Figure 1: Care robot experts' views on the cybersecurity of care robots

4.1 Current cybersecurity of care robots

4.1.1 Timeliness

The interviewed experts are not worried about the cybersecurity of care robots at the present because there are few care robots in use or available on the market. The experts have not heard of the threats to care robots and feel that there is little talk about the subject. On the other hand, the experts view that the cybersecurity of care robots is a topical issue, as the use of care robots, like all other devices, involves risks, and as the number of care robots increases, so does the interest of cybercriminals in care robots. At present, however, it was felt more likely that cybercrime would target other devices that are already in wide use.

The views of interviewed experts on the use of the devices currently support previous studies. For example, Van Aerschot and Parviainen (2020) have found that care robots are hardly in use yet, although care robots are a possible solution to the future shortage of nurses and home carers as the population ages.

4.1.2 Development of cybersecurity

According to the interviewed experts, the development of functionality is currently a priority in the design and development of care robots, and thus the cybersecurity aspects of care robots are receiving less attention. The experts point out that cybersecurity focuses on data-related threats and security and does not take into account the threats that the care robot may pose to the physical environment.

Care robots are cyber-physical systems that combine hardware and software components, network and communication processes, mechanical actuators, controllers, operating systems, and sensors to interact with the physical world (Quarta et al., 2017). The view that robot's cybersecurity issues focus more on data-related threats than on physical-related threats has not been presented in previous studies. In general, however, the view of the experts is in line with Fosch-Villaronga and Mahler (2021) that the development of care robotics focuses more on the development of functionality than safety. Investing in data security is understandable because a lot of personal data is processed in social and health care, as well as sensitive data related to customers, which is why confidentiality, privacy, data integrity, and accessibility are key (Vuorinen, 2019). However, threats to physical security need to be increasingly addressed so that they do not become a problem in the future as the physical characteristics of care robots develop.

4.1.3 Type of devices

Currently, the term 'care and socially assistive robotics' includes certain types of medical, wellness technology, and other devices. This is problematic for equipment development. Telepresence and remote access robots are available for homes and home environments, but the experts have no information on where or how much these robots are used. Experts believe the security of remote access robots is at the same level as that of typical remote connections, i.e. they include base-level protection, and are vulnerable to cybersecurity threats. The safety of non-medical devices was uncertain, according to experts.

Experts consider the implementation of cybersecurity for medical devices to be mandatory, as the regulation of medical devices requires that the security of the devices be taken care of. In Finland, medical devices are under the control of the National Supervisory Authority for Welfare and Health (Valvira). Experts believe that there is no guarantee that the cybersecurity of a device will be ensured if it is not classified as a medical device because there is no control over welfare technology.

The literature also highlights the problematic nature of legislation on care robots (Holder et al., 2016; Fosch-Villaronga and Golia, 2019a; Fosch-Villaronga and Golia, 2019b). It is unclear how care robots can be legally classified because classification depends on their intended use, so these robots could be seen as either medical devices or general products, which are regulated differently (Fosch-Villaronga and Mahler 2021).

4.1.4 User skills

According to the interviewed experts, user skills affect the implementation of cybersecurity in care robots in two ways; when purchasing the device and when using the device. If a device does not have a rating for a medical device, the end-user is practically unable to deduce the level of security of the device to be acquired. User skills also affect the safety of equipment during use: it is difficult to make robots safe if they are not used following the principles designed by the manufacturer.

Clear communication about the level of cybersecurity of care robots is most important to the implementation of cybersecurity, as it is unclear whether robot vendors can assess the level of robot security (Fosch-Villarongan & Mahler, 2021; Lera et al. 2017). The experts emphasize the importance of clear communication from the perspectives of the service provider and end-user. The knowledge and understanding of these parties may not be sufficient to enable them to make informed decisions and risk assessments regarding the use of the equipment.

4.2 Potential cybersecurity threats to care robots

Care robots differ technically very little from other information and communication technology ICT equipment already in use making the attacks and the methods used on them technically the same as on other devices. So, care robots are subject to the same risks as other ICT devices. According to experts, traditional, purely computer-based information security attacks are also possible on care robots; the methods for implementing the attacks are the same as for other ICT devices. Interviews show that care robots are exposed to virtually all traditional security threats as well as threats related to the physical nature of robots. The possible threats perceived by experts are therefore in line with the literature (c.f. Lera et al.,2017; Rousku, 2014). The following is a more detailed discussion of potential cybersecurity threats that come up from interviews.

4.2.1 Remote access and network threats

Controlling the robot remotely brings with it threats. For example, hijacking device control with remote access is one of the potential cyber threats to care robots. Although all experts interviewed say hijacking is a potential threat, opinions are divided on whether the threat is likely or not.

Network connections expose care robots to a variety of threats, and experts feel that networks play a key role in how secure different devices are. The need for a network connection depends on the purpose and characteristics of the care robots, and the more complex and interactive devices are developed, the more important the network connection will be. Experts see that all of the above threats are more likely to occur if the device is connected to the Internet. The view of the interviewed experts is in line with the literature (c.f. Lera et al., 2017; Rousku, 2014) that all robots connected to the network can be a risk to their users because an outsider can access the devices via the network.

4.2.2 Stealth attack

A stealth attack, i.e., an attack in which an attacker gets to manipulate the operation of the robot's sensors and thus, for example, cause a robot to collide, is also a potential threat. Some of the interviewees think that since similar attacks are possible on other mobile robots, they can also be targeted at care robots, especially if the robot is poorly protected. Other experts find it difficult to find the motivation to carry out such attacks.

4.2.3 False data injection

Attacks in which the data processed by the robot can be modified are possible and, in certain situations, even easy to implement. Experts agree that the consequences of such a threat may be critical, but opinions differ on whether the threat is likely.

4.2.4 Eavesdropping

Eavesdropping and watching in secret are the most prominent issues that arise from the interviews. They are potential and probable threats because similar attacks have been carried out on other robots with microphones and cameras. According to experts, it is likely that particularly vulnerable remote access robots could be used for espionage, as their level of security may not be enough at present.

4.2.5 Denial-of-service

Two perspectives on denial-of-service attacks exist: care robots can be used to implement denial-of-service attacks, and denial-of-service attacks can be used to block the service provided by the robot. Experts see denial-of-service attacks as potential threats from both perspectives. On the other hand, devices that are not connected to the Internet are safe from the point of view of the denial of service. Experts believe that if the device is not connected to the Internet, for example, blocking remote connections will have little effect on the operation of the device. Also, offline devices cannot be used to carry out denial-of-service attacks.

4.2.6 New threats

The constant development of care robotics also brings new threats that have not been identified in previous studies. In this study, experts highlighted that the integration of new technologies, such as the use of cloud services, artificial intelligence, and machine learning in care robotics, creates opportunities for new attacks. In addition, unlike previous studies, experts stressed that in the future, physical risks will become more central to cybersecurity as the physical properties of care robots are developed.

4.3 The biggest cybersecurity risks of care robots

4.3.1 Hijacking device control and spying

Experts see that one of the biggest threats to cybersecurity for care robots is the hijacking of device control. Through the hijacking of a care robot, a cybercriminal can do practically all the same things that a user could do with the device, for example, access device information and spy on device users. In the case of a physically mobile device, an attacker can also cause physical damage to the robot environment, or even to the end-user of the robot, through the hijacking of the device.

Experts say another vast threat is spying. Using a care robot to eavesdrop on and/or watch in secret is easy to implement and possible to implement for a wide range of people. According to experts, the benefit-to-input ratio obtained by criminals is greatest through spying.

4.3.2 *Network*

Experts think network security, especially remote connections, is one of the most significant risks in the cybersecurity of care robots. The network connection is often the interface that allows the safety and security of care robots to be compromised. Isolated local area networks are slightly more secure than remote internet access, but experts see that it is also possible to access them on the spot. According to experts, the security of current remote applications is at a basic level, which makes them possible targets of hacking.

4.3.3 *Data security and privacy*

Experts are concerned about the security of the data collected by care robots. In particular, data protection issues arise when personal data is stored in cloud services. The concern is how to be able to secure and be sure that the data is recovered, stored, and disposed of correctly and that the data is not accessed by the wrong parties. Experts also raise concerns about maintaining data integrity.

4.3.4 *Regulation and trust*

In previous studies, the cybersecurity of care robots has often been addressed from a fairly technical perspective. However, all the risks associated with care robotics are not limited to the care robots themselves and their use. The current lack of legislation on cybersecurity underlines the trustfulness of the manufacturer and supplier of the care robot. Experts representing the service provider's point of view think that at the moment, you have to trust that the manufacturer of the care robot has taken safety and security considerations into account and has made the devices as safe and secure as possible. A risk factor for non-medical devices is that the only guarantee for the safety and security of the devices is the supplier's promise.

4.3.5 *System complexity & operating models*

Care robots may be part of a large system with a complex operating model. In this case, the interference may be to some other part of the system, which will cause the care robots to malfunction. According to experts, changes in operating models are risk factors. When the implementation of operations is designed with the help of robots, the damage can be vast if, for some reason, the robots do not work. Returning to the old operating model can be difficult and time-consuming. From the end-user's point of view, the malfunctions of the care robot can be very critical, depending on the criticality of the tasks performed by the care robots. Experts also see the risk that systems and service robots are subject to vulnerabilities in both the upgrade and operating systems throughout their lifecycle.

This is in line with Fosch-Villaronga & Mahler (2021); as the number of interconnected systems and devices increases, so do the cyber risks.

4.3.6 *End-user*

According to experts, end-users of care robots are associated with risk factors from a safety and security point of view. Experts are not convinced that end-users will necessarily take care of the security of their own homes and networks. Experts are also not convinced that end-users will act following data security principles.

Experts representing the development and manufacturing of care robots point out that it is difficult to make the equipment safe if it is not used following the principles designed by the manufacturer. The view is in line with previous studies (Lera et al., 2017; Fosch-Villarongan & Mahler, 2021). Vuorinen (2019) also states that e.g. rigid software can entice users of devices to circumvent information security mechanisms, which may result in users not changing their default passwords or using the same user ID. In other words, the user's competence is also central to the realization of the cybersecurity of the devices in use. When the end-users of care robotics are the elderly, concerns arise about their potentially deficient digital skills. Also, human memory and other characteristics do not usually improve with aging. According to experts, it is likely that as a person ages, he or she may no longer remember or know how to use the device and take care of data security.

4.3.7 Physical security

Experts see that the risks associated with the physique of robots will increase as the physical properties of the robots, such as the ability to manipulate objects, improve. In addition, experts think that if a cybercriminal has, for example, a state-level incentive to cause physical harm, it is possible through a care robot.

4.4 Cybercrime

The development of cybersecurity is a competition against cybercrime, where the analysis of various risk analyzes and motivational factors would lead to a better understanding of the factors behind the crime and potentially focus resources on the development of cybersecurity to address the most likely threats. The following are issues related to cybercrime that emerge from the interviews.

4.4.1 Ethics and morals

According to experts, cybercrime is viewed with blue eyes in Finland. People's attitudes towards cybercrime are affected by uncertainty about what is already possible in the field of cybercrime. The general belief is that people act ethically and thus do not want to cause harm to other people, especially the vulnerable.

4.4.2 Motivational factors

The motivating factors behind cybercrime emerge from all the interviews on several occurrences. Experts often consider different threats based on which motivational factors affect a criminal's actions and what benefits can be achieved from a criminal's perspective when a particular threat materializes. The general view of experts is that if a criminal has sufficient motivation and access to resources, the likelihood of various threats materializing will increase. At present, however, it is seen that care robotics is not yet such a tempting opportunity as to attract cybercriminals, as the benefits of crime are currently small. Experts have a hard time imagining why anyone at all would want to attack devices designed to help the vulnerable. However, the care robots of very significant persons can be exclusions.

4.4.3 Development of cybercrime

Experts are aware that cybercrime currently occurs in a variety of contexts. Although care robotics is not currently considered to be a particularly topical target for cybercrime, experts believe that where care robotics is evolving, cybercrime is also evolving.

5. Conclusions

The purpose of the paper is to map the experts' opinions on the cybersecurity threats and risks associated with care robotics. As the interviews progressed, saturation was observed in the research data, meaning that six interviews were sufficient for this study.

The use of care robots has the same risks and threats as the use of other ICT devices or robots, which supports the results of previous studies on the cybersecurity threats of care robots. The biggest threats are related to the hijacking of control of care robots, and the fact that they can be used for spying and eavesdropping, but other threats are also possible. In addition, network connectivity, and new features such as artificial intelligence and machine learning create more opportunities for cybercrime. In the future when the physical characteristics of robots, such as the ability to manipulate objects improve, threats to the physical environment will increase.

Although care robots have the same threats as other ICT devices already in use, there is a risk that care robots will be statistically less exposed to cybersecurity threats because there are few care robots. From this perspective, it seems that cybercrime is currently targeting other devices. However, the situation will change in the future as the use of care robots increases.

Care robotics as a robot class is so far very vague. At present, there are no rules or regulations that specifically stipulate that the safety and security of care robots must meet any requirements. If new devices that do not have a medical device classification are introduced, special attention should be paid to their cybersecurity. For example, if telepresence devices and other remote access robots are used for care, special care must be taken concerning their safety and security, as the experts interviewed in this study typically do not aim for a high level of data security. If such devices are used for care, care must be taken to ensure that connections cannot be broken because sensitive conversations or the like may occur between the caregiver and the client.

Compared to previous studies, this study highlighted more aspects related to the human factors underlying cybersecurity, and that cybersecurity is a race against cybercrime. When deploying care robots and assessing their cybersecurity, it should be borne in mind that there is usually some motive behind criminal activity. If we assess risks narrowly (e.g., what worst can occur) and do not consider the motives behind the actions, we can conclude that there should be no care robots because they can cause things that are almost impossible to prevent. Using this logic, we could also conclude that there should be no cars because someone can sabotage the car so that it doesn't work as it should and the driver can crash. Therefore, the mapping of motives and the benefits of crime is key to striking a balance between acceptable and unacceptable risks in care robotics as well.

Further research is needed on the distribution of responsibilities related to the cybersecurity of care robots as more care robots are deployed; what kind of concept or entity ensures the safety and maintenance of the equipment throughout its life cycle, and how the responsibility should be divided between the different parties. This perspective should be further explored and mapped out what kind of service ecosystem would ensure the safety of care robots during development, deployment, and maintenance, and enable end-users to receive the guidance and support they need on the use of care robots.

Acknowledgements

This work was supported by the SHAPES project, which has received funding from the European Union's Horizon 2020 research and innovation programme under grant agreement no. 857159.

References

Dolic, Z., Castro, R. and Moarcas, A. (2019) "Robots in healthcare: a solution or a problem?, Study for the committee on environment, public health, and food safety", *Policy Department for Economic, Scientific and Quality of Life Policies*, European Parliament, Luxembourg.

Forum Virium (2020) "Hoivarobotti viihdytti muistisairaita ja helpotti hoitajien työtaakkaa Kustaankartanossa", [online], City of Helsinki 13.1.2020, https://www.hel.fi/uutiset/fi/helsinki/hoivarobotti-helpotti.

Fosch-Villaronga, E. and Golia, A. Jr. (2019a) "Robots, standards and the law: rivalries between private standards and public policymaking for robot governance", *Comput Law Secur Rev*, 35 (2) (2019), pp. 129-144

Fosch-Villaronga, E. and Golia, A. Jr. (2019a) "The intricate relationships between private standards and public policymaking in the case of personal care robots. Who cares more", in P. Barattini, F. Vicentini, G.S. Virk, T. Haidegger (Eds.), *Human-robot interaction: safety, standardization, and benchmarking*, CRC Press.

Fosch-Villaronga, E. and Mahler, T. (2021) "Cybersecurity, safety and robots: Strengthening the link between cybersecurity and safety in the context of care robots", *Computer law & security review* (41). DOI: 10.1016/j.clsr.2021.105528

Holder, C., Khurana, V., Harrison, F., Jacobs, L. (2016) "Robotics and law: key legal and regulatory implications of the robotics age (Part I of II)", *Comput Law Secur Rev,* 32 (3), pp. 383-402.

Lera, F., Llamas, C., Guerrero, Á., and Olivera V. (2017) "Cybersecurity of Robotics and Autonomous Systems: Privacy and Safety", *InTech Open*. DOI: 10.5772/intechopen.69796

Quarta, D., Pogliani, M., Polino, M., Maggi, F., Zanchettin, A. and Zanero, S. (2017) "An experimental security analysis of an industrial robot controller", *Proceedings of the 2017 IEEE symposium on security and privacy*, pp. 268-286.

Rousku, K. (2014) *Kyberturvaopas - Tietoturvaa kotona ja työpaikalla*, Talentum, Helsinki.

Schönberg, K. (2017) "Vanhukset ottavat robotin ilolla vastaan – hoitajat epäillen", [online], YLE, https://yle.fi/uutiset/3-9720927.

Särkikoski, T., Turja, T. and Parviainen, J. (2020) *Robotin hoiviin? Yhteiskuntatieteen ja filosofian näkökulmia palvelurobotiikkaan*, Vastapaino, Tampere.

Tzafestas, S. (2018) "Roboethics: Fundamental Concepts and Future Prospects", *Information,* 9 (148). DOI:10.3390/info9060148

Van Aerschot, L. and Parviainen, J. (2020) "Robots responding to care needs? A multitasking care robot pursued for 25 years, available products offer simple entertainment and instrumental assistance", *Ethics and Information Technology,* (22), pp. 247–256.

Vuorinen, S. (ed.) (2019) *Cyber security: Guidance for operators in the healthcare and social welfare sectors*. Publications of the Ministry of Social Affairs and Health, Helsinki.

Cyberterritory: An Exploration of the Concept

Jori-Pekka Rautava[1] and Mari Ristolainen[2]
[1]University of Oulu, Finland
[2]Finnish Defence Research Agency, Riihimäki, Finland
jrautava@student.oulu.fi
mari.ristolainen@mil.fi

Abstract: What does the future of cyberspace look like? The idealistic notion of cyberspace as a 'free' and 'open' global infrastructure is progressively challenged by projecting territoriality and conveying traditional nation-state models of governance into cyberspace. The aim of this interdisciplinary paper is to examine the process of cyberspace territorialisation and to present a conceptual definition of a theoretical 'cyberterritory' as a bounded sovereign entity that operates under the jurisdiction of a certain nation-state. Firstly, we explain the different views of the cyberspace governance and summarize the latest developments in the UN's efforts to bring order over cyberspace. Secondly, we analyse the different views on 'digital sovereignty' and show how several nations have felt the need to express publicly their views on sovereignty in cyberspace. Thirdly, we discuss the possibility of new techno-economic alliances, because only few (if any) nation-states could have sufficient resources to be 'sovereign' in cyberspace. Finally, we present a conceptual definition of a theoretical 'cyberterritory' that encompasses political, legal and technical aspects. The significance of this paper is in its contribution to the discussion of future cyberspace governance by presenting a definition of a theoretical 'cyberterritory' as an entity of its own - a new nation-state 'digital terrain' of the future.

Keywords: territorialisation of cyberspace, 'cyberterritory', cyberspace governance, digital sovereignty

1. Introduction

Originally, the purpose of the global Internet was to combine geographically isolated intranets into a single network of networks. Ideally, the aim of a global Internet was to erase geographical areas, borders, and state control, and to create a 'global commons' in which data could move and be stored freely across national (geographical) territories (Kahin & Nesson 1997; Goldsmith & Wu 2006). In this so-called 'deterritorialization process' the network of networks developed to be a 'space' outside of the territorial (geographical) space of nation-states. Hence, cyberspace is on one hand an independent and integrated, but also a complex, partly overlapping and confusing combination of public and private services and critical infrastructure that is used by everyone from security authorities to individual citizens.

Despite its idealistic goals, the discussion about cyberspace governance and control has been ongoing since the 1980's (Radu 2019). The main objective has been to find ways to ensure the stability and security of cyberspace. In general, the discussions of the cyberspace governance can be divided into supporters of a 'multistakeholder' and a 'multilateral' governance system. Discussions within the UN (United Nations) have sometimes taken one-step forward and then two steps backwards. Moreover, concepts such as 'fragmentation' and 'balkanization' have been often used when evaluating the future development of the cyberspace (Mueller 2017). Depending on the perspective, the fragmentation of the global cyberspace has been seen as rapid or slower, yet almost inevitable development. Generally, cyberspace fragmentation is divided into three different forms (Drake, Vinton & Kleinwächter 2016). 'Technical fragmentation' is related to the development of the Internet infrastructure that affect the interoperability of devices and the data mobility. 'Governmental fragmentation' encompasses all state actions that restrict or prevent access to the Internet and control the data mobility. 'Commercial fragmentation' involves measures that prevent or impede the use of the Internet and data mobility by various commercial operators (ibid). All the different fragmentation developments serve the interests of different actors. However, it has also been suggested the real aim of the whole fragmentation debate is to subjugate the governance of the Internet under nation-state jurisdiction and to create nation-state power structures in the cyberspace and to supress the global information flows and data movement (Mueller 2017). This development can be called as a 'process of cyberspace territorialisation' (cf. Ristolainen 2021) and the outcome as 'cybered Westphalian age' (Demchak & Dombrowski 2011; Demchak & Dombrowski 2013).

The aim of this interdisciplinary paper is to examine the process of cyberspace territorialisation and to present a conceptual definition of a theoretical 'cyberterritory' as a bounded sovereign entity that operates under the jurisdiction of a certain nation-state. Firstly, we explain the different views of the cyberspace governance and summarize the latest developments in the UN's efforts to bring order over cyberspace. Secondly, we analyse the different views on 'digital sovereignty' and show how several nations have felt the need to express publicly their views on sovereignty in cyberspace. Thirdly, we discuss the possibility of new techno-economic alliances, because only few (if any) nation-states could have sufficient resources to be 'sovereign' in cyberspace. Finally, we present a conceptual definition of a theoretical 'cyberterritory' that encompasses political, legal and technical aspects.

2. Transformation of the cyberspace governance models

The main objective of the discussion about cyberspace governance and control has been to find ways to ensure the stability and security of cyberspace. In general, the discussions of the cyberspace governance can be divided into supporters of a 'multistakeholder' and a 'multilateral' governance model (Glen 2014). However, it is important to notice that the debate is also indirectly influenced by multinational companies (including Google, Alibaba, Yandex) and global civil society networks (including criminal and extremist groups).

Proponents of the state-led governance model aim for regional or national Internet segments of countries or groups of countries and strive for state sovereignty in cyberspace. Sovereign nation-states and their geographical boundaries are to be given priority in the national and global regulation of cyberspace. In this 'multilateral' governance model, driven by Russia and China in particular, cyber-related decision-making would be done by the so-called 'multilateral community', i.e. by the ITU (International Telecommunication Union) (Singh 2009; Glen 2014).

The 'multistakeholder' model, which emphasizes the cyberspace's global, open and interoperable systems, is opposing cyberspace territorialisation. According to the supporters of the 'multistakeholder' model, cyberspace governance should not be dependent on the control of individual state governments. In the 'multistakeholder' model, led predominantly by the US, the cyberspace governance is organized under the 'global community' or the 'stakeholder community', such as ICANN (Internet Corporation for Assigned Names and Numbers). Supporters of the 'multistakeholder' model oppose regional or national Internet segments and seek to preserve the 'freedom and openness' of cyberspace (Strickling & Hill 2017).

Essentially, cyberspace governance models are issues of international law that are being resolved within the UN. Since 1998, within the UNODA (United Nations Office for Disarmament Affairs), Russia has been pushing for a resolution that instead of global transparency a national sovereignty applies in cyberspace (Korzak 2021). No such resolution has been adopted so far, but since 2004, every two years, a Group of Governmental Experts (GGE) has been writing 'consensus reports' on cyberspace governance for the UN General Assembly to approve. These reports have analysed how the international law affects states' actions in cyberspace and tried to find ways to promote compliance with existing cyber standards that are acceptable to all (see, e.g. Ruhl et al 2020).

In 2010, 2013 and 2015, the UN General Assembly confirmed, on the recommendation of the GGE, that international law applies and regulates the activities of nation-states in the cyberspace. At the same time the need to discuss specific features of the cyberspace, such as speed, interdependence, complexity and anonymity, was recognized. The 2015 resolution established voluntary, non-binding standards for responsible state behaviour in cyberspace. However, after the 2015 resolution, the debates stalled. In 2017, the GGE group did not reach consensus and their work was suspended. The application of international law in the cyberspace became a particular problem. In 2018, the UN General Assembly accepted a resolution led by Russia to form a new OEGW (Open-Ended Working Group) to analyse the previous GGE reports in order to identify new standards and to explore to form a new dialogue between different UN institutions (Ruhl et al 2020). However, in 2018, the UN General Assembly also passed a US-led resolution that was in part inconsistent with the Russian-led resolution. The US-led resolution decided to re-establish a new GGE-group to write unanimous reports on state action in cyberspace. Nevertheless, there were significand differences in the composition of the newly established groups. All the UN Member States (193) were invited to join the OEGW-group, while only 20-25 countries have been involved in the GGE-group. The groups were also designed to operate in a different way. The OEGW group will operate until an agreement is reached; the GGE process has a two-year time limit at the time (ibd.).

When the OEGW process started, many individual states felt the need to express their national views on the international law in cyberspace. At this stage, various national openings and new initiatives began to emerge among the proponents of the 'multistakeholder' model. These new openings and initiatives considered in particular state sovereignty in cyberspace. Although the supporters of the original 'multistakeholder' cyberspace governance model still relied on the security deriving from the global system, some, mostly European countries, saw cyber threats more and more in the national framework than at the global level. This 'new thinking' resembled in a way the views of the 'multilateral' model of governance proponents. It seems that the transformation of the cyberspace governance models began when some kind of combination of 'partial multistakeholder' and 'partial multilateral' started to emerge. This development is affected by the fact that cyber threats are targeted at national critical infrastructure, economic competitiveness, national security and citizens, both directly and indirectly. Nowadays, many nation-states are interested in safekeeping of their own national cyberspace and own national systems. The national controllability of cyberspace and the independent defence of various systems add to the sense of cybersecurity. This in turn reflects the original ideas of the proponents of the 'multilateral', i.e. state-led governance model of cyberspace. The efforts of nation-states to resists the influence of multinational corporations, and civil society networks, as well as criminals and terrorists also bring superpowers together and make the state-led governance model an attractive option for many nation-states.

In the summer of 2021, both the OEGW and GGE groups came to some conclusions (United Nations 2021a). It was decided to continue the OEGW for another five years, and the group reached a final statement which was accepted both the US and Russia (United Nations 2021b). Similarly, the GGE experts approved their final report that was also confirmed by the UN General Assembly in the summer of 2021. However, the GGE final report is, in fact, a reduplicate of the 2015 report (United Nations 2021c).

In both OEGW and GGE final statements sovereignty in cyberspace is considered only as a general principle that could not lead to legal consequences in the cyberspace. Nor was the role of UN extended in the cyberspace governance. Nevertheless, the supporters of the state-led cyber governance model extended the process to the entire UN level (e.g. OEGW process), created a certain dissimilarity among the original supporters of the 'multistakeholder' model and gained the opportunity to influence more countries' views on the future of the cyberspace governance. This can be seen as an effort by the proponents of the 'multilateral' model to increase the legitimacy of their model of governing the cyberspace, which seems to be a planned normative line of effort (Kukkola, Ristolainen & Nikkarila 2017).

3. Sovereignty in cyberspace

The recent change in the governance models of cyberspace reflects a change in how different nations see sovereignty in cyberspace and how they desire to use similar international laws that direct the relations between states in the physical geographical environment also in cyberspace. However, this poses certain problems because the so-called 'national cyberspace' does not necessarily follow the physical borders of a nation-state and there is no internationally unified understanding of what sovereignty means in cyberspace.

Cyberspace is associated with very different conceptions of sovereignty - often referred as 'digital sovereignty' (Pohle & Thiel, 2020). However, by 'digital sovereignty' is referred frequently to national regulation of data mobility, i.e. 'data sovereignty' (Braud et al 2021). In one context, 'digital sovereignty' refers to much broader 'information sovereignty' (see, e.g. Efremov 2017), and in another context, 'digital sovereignty' denotes a 'national segment of the Internet' that can be disconnected from the global network (see, e.g. Kukkola 2020). Thus, there is no single conception of what sovereignty in cyberspace means, although the same concept 'digital sovereignty' is used.

As noted earlier, several nations have felt the need to express publicly their views on sovereignty in cyberspace after the two conflicting UN resolutions in 2018. Many European countries have expressed the view that the principle of State sovereignty applies in cyberspace. For instance, Tallinn Manual (2017) states that "a State enjoys sovereign authority with regard to the cyber infrastructure, persons, and cyber activities located within its territory, subject to its international legal obligations."[1] Cyber infrastructure is defined as "Cyber infrastructure: The communications, storage, and computing devices upon which information systems are built and operate" (ibid.). According to this definition, the state would have a jurisdiction of all information and communication technology (ICT) and the persons operating it located in its territory, which sounds rather totalitarian.

[1] This statement refers to the 2015 GGE report; see United Nations 2021c.

According to Tallinn Manual (2017), sovereign authority includes 'cyber infrastructure' within the physical geographical state territory, where, however, the ownership of 'cyber infrastructure' is divided between public administration, business, organizations and individuals both nationally and internationally. Moreover, it is not clear whether the sovereign authority includes strategic data and services in global cloud services that are not physically located within the physical territory of the state. On the other hand, a physical territory tied definition is problematic in a situation where the physical territory of the state contains 'cyber infrastructure' owned by a foreign state and 'persons involved in cyber activities' that are foreign citizens. Likewise, satellites, electromagnetic spectrum or marine cables not located on the physical territory of the state that are critical to functioning of the 'cyber infrastructure' of different countries. It could be possible that one country defines these as its sovereign authority, which can lead to uncomfortable situations and expose the physical territory owner to hybrid influencing. Sovereignty in cyberspace requires a definition of 'cyber borders' that has not been done by so far.

The definition of a cyberterritory should be as broad as possible. If the definition is too strict, it is easy to argue that the only way to implement a cyberterritory is totalitarian ownership of the infrastructure. A cyberterritory should be implemented in a way that technical solutions allow having control over the infrastructure without taking over the infrastructure from companies owning it. The owner should be able to achieve three goals to define borders of a cyberterritory: 1. Delimitation and demarcation; 2. Protection; 3. Control (Kukkola & Ristolainen 2018).

First, delimitation means that the place of borders is decided between nation-states through negotiations. With demarcation is meant that the cyberterritory owner must be able to set strict boundaries for when the network traffic enters a cyberterritory. Demarcation is based on delimitation agreements that set the agreed boundaries (Kukkola & Ristolainen 2018). We argue that for the definition of a cyberterritory, it is not necessary to define how the demarcation is done. Second, protection means that the owner must be able to protect the cyberterritory for which it has set boundaries (ibid.). There is some entity that has the responsibility to arrange the measures for protection. The entity responsible of the protection is also responsible of the control of traffic across the border. Finally, control means that the owner must have control over the cyberterritory (ibid.). We argue that the control and protection are strictly tied together. Without control, it is not possible to efficiently protect the area and without protection, it is naive to say that one would for sure have control over the area.

4. Techno-Economic alliances in future cyberspace

When assessing potential future developments, it seems possible that the state-led model of cyberspace governance could gain more supporters. This can lead to a formation of new techno-economic alliances, because only few (if any) nation-states could have sufficient resources to be 'sovereign' in cyberspace. Techno-economic alliances could be based on competing technical platforms and national solutions (MGIMO 2019). Techno-economic alliances could develop new technologies at national (or alliance) level and build services based on national (alliance) technology. Many countries are striving for self-sufficiency and 'digital sovereignty'. However, in order to remain competitive, they must either join or strengthen emerging techno-economic alliances.

Potential techno-economic alliances could be formed around the 'Anglosphere' led by the US, around China, around Russia and around EU (ibid.) US, Canada, Great Britain, Australia and New Zealand are economically tightly integrated and their alliance could also be attractive to countries such as Mexico. This alliance would use its privileged position in the world to create the best conditions for itself. Similarly, China is expanding its alliance with neighbouring countries tying them to China's economy and infrastructure. The Chinese model is based on the absolute self-sufficiency and it would have access to enormous markets that are largely closed to competitors' technology and user data. Russia wishes to remain an independent global actor and this would be possible only as part of some kind of techno-economic alliance. Russia's alliance would be dependent on the domestic market and public investments. Russia's rational partners would be the members of the Eurasian Union, i.e. the individual states of the former Soviet Union (Belarus, Azerbaijan, Kazakhstan, Uzbekistan, Tajikistan, and Moldova) (ibid.). The EU's Digital Strategy of 2020 states that the EU must strengthen its digital sovereignty and set standards instead of lagging behind others. EU's alliance would focus on data protection, technology and infrastructure (Shaping Europe’s digital future 2020). The aim would be to strengthen Europe's technological capacity, independence and confidence, and to improve Europe's position in the global competition. However, the future of an EU's techno-economic alliance seems rather uncertain. Furthermore, the role of South-America and Africa in techno-economic alliances remains ambiguous. Both continents will be under heavy influence of several alliances in the future.

The border between techno-economic alliances is technological. However, the delimitation process is still between nation-states and ratified by the UN. It can be argued that cyber borders run in the competition between competing critical infrastructure technology platforms controlled by governments at the national or at the alliance level. Due to national security, it will be impossible to allow third parties access to independent critical infrastructure. Alliances would require a desire to use the classical principles of international law in the cyberspace, i.e. all the similar law that govern state relations in the physical geographical environment (cf. state-led 'multilateral' cyberspace governance model). Therefore, techno-economic alliances are also military and political alliances and have strategic importance (Kukkola 2021).

5. Definition of a theoretical 'cyberterritory'

Based on the above presented developments and discussions, it can be estimated that territoriality and traditional nation-state models of governance are conveyed more and more into future cyberspace. Nevertheless, territorialisation of cyberspace is relatively new phenomenon, i.e. it requires a comprehensive conceptualization. In the spring of 2021, a research workshop for military and civilian experts was held at the Finnish Defence Research Agency (FDRA), where the aim was to find a conceptual definition of a theoretical 'cyberterritory' that encompasses political, legal and technical aspects (FDRA 2021).

In the research workshop was formed a preliminary definition of a cyberterritory as *"an entity of networks and technical infrastructure containing services which is controlled by a sovereign nation-state"*. Next, we elaborate the definition by first defining what entity or who can own a cyberterritory. Big multinational corporations might have bigger and more complex networks than those of small nation-states. The corporations also have independent control over their networks. For example, Google has its datacentres all around the globe spanning its networks to almost every part of the globe (Safenames Ltd, 2017). Large corporation networks could be dealt as cyberterritories in this sense but there is one major piece missing. Big corporations do have control only over the infrastructure in their own premises, but when the connections leave from the premise, they move to infrastructure which is controlled by someone else, usually in many different nation-states. This infrastructure cannot be controlled by the companies, which introduces a critical difference between nation-states and corporations. Moreover, as the Russian and Chinese examples show, companies can be forced to open their networks and to control their networks in the way nation-states want. Therefore, nation-states can have at least some control over the infrastructure through legislation in their territory even when the infrastructure belongs to some company.

Cyberterritory must be tied to an actual nation-state and its jurisdiction (FDRA 2021). Nation-states have their borders and embassies are part of the nation-state for which they belong. Similarly, all the infrastructure should belong to a cyberterritory which is controlled by a sovereign nation-state. The nation-state controlling cyberterritory has undisputed and sovereign control over the infrastructure in the region of their cyberterritory. Yet, the 'international cyberspace' outside of national cyberterritories evokes questions: Does all infrastructure belong to some nation-state or is there parts of international cyberspace which do not belong to anyone particularly i.e., nobody has sovereign control over the part of the cyberspace? This is strictly related to cyberterritory spreading outside of the borders of the nation-state controlling a cyberterritory. Embassy can have direct connection to the original cyberterritory for example via VPN (Virtual Private Network), but the infrastructure which the VPN uses to create the connection is not (completely) controlled by the owner of the cyberterritory (FDRA 2021).

Moreover, in the FDRA research workshop, a series of questions, concerning data in a cyberterritory was raised: Is it possible to have cyberterritory without data? Alternatively, is the data what makes some entity of infrastructure a cyberterritory? Thus, we must consider the importance of data in the context of a cyberterritory. We produce more data every day than ever before. There are huge amounts of data moving around in networks and that data is used to sell people things, to gather information of behaviour of individuals, to make business decisions and many more. Our whole world is run by data. Data is something that is saved and moved in a cyberterritory and between cyberterritories (FDRA 2021). Data is also something that can be stolen from a cyberterritory. If data can be stolen from a cyberterritory, then is a part of a cyberterritory stolen also? A cyberterritory inevitably contains massive amounts of data. Data in a cyberterritory always belongs to someone.

For a cyberterritory to be a place where citizens want to belong the nation-state cannot own all data in a cyberterritory. In a cyberterritory, the data belongs to entities who have produced it or have legal right to own it. Because the owner of a cyberterritory cannot own all the data in a cyberterritory and the data in a cyberterritory can be stolen, we argue the definition of data is not necessary for the definition of a cyberterritory. This does not mean that data would not be a crucial part of a cyberterritory. Without data a cyberterritory is not very useful and thus data is integral part of the actual implementation of a cyberterritory even if it is not a part of the definition.

6. Discussion: 'Cyberterritory' - a new nation-state 'digital terrain' of the future?

The purpose of a cyberterritory is to provide more control for the owner of the cyberterritory. Owner of the cyberterritory would be able to control the data flowing to and from the cyberterritory. This of course raises a problem if the controller of the cyberterritory is so-called totalitarian or authoritarian nation. A totalitarian nation in this context is a nation that prohibits the existence of opposition parties and treats dissidents as criminals. It is safe to assume that those totalitarian nation-states will find ways to oppress the people if they want to even without a cyberterritory. The purpose of controlling data flow is strictly related to security. If a conflict escalates, the variety of tools used to influence the enemy is wide. The cyberspace is inevitably one medium for different tools for impacting the enemy. Thus, controlling the cyberspace would benefit the defender since they would be able to create asymmetric situation towards the attacker (Kukkola 2020; Kukkola 2021).

A cyberterritory should allow citizens to maintain their privacy. A cyberterritory should also provide a possibility to block harmful content. It is up to the owner of a cyberterritory to decide what is considered harmful. A cyberterritory should provide tools to redirect traffic to 'border crossing points'. The great firewall of China (GFW) should not be considered as a 'border fence' of a cyberterritory because firewall does not provide sufficiently fine-grained control. The existing GFW can either block or allow traffic. A cyberterritory should be able to simultaneously block and allow content from the same source, something that firewall is not able to do. E.g., Russia used BGP (Border Gateway Protocol) and geoblocking when defending its 'cyberterritory' in the war against Ukraine in spring 2022 that allows more effective control over network flow than a firewall such as GFW (Goodin 2022).

It is noted that the idea of a cyberterritory is, in some extent, first proposed through the supporters of 'multilateral' governance model. Some of those initial supporters are rather authoritarian nation-states. Therefore, it is obvious that cyberterritory could be used for terror, surveillance, and oppression of citizens. Nevertheless, some of the original supporters of the 'multistakeholder' governance model have also started to move towards the idea of controlling nation-state networks. Sovereignty in cyberspace has had negative connotation because -states like China, Iran and Russia have been supporting the nation-state-controlled Internet (Litvinenko 2021).

It is clear that controlling the cyberspace in a cyberterritory enables authoritarian control. However, the control in cyberspace also enables securing the network from external and internal threats. Companies already have security operation centres (SOC), which monitor what happens in the internal networks of a company. Similarly, SOCs could be established in the context of a cyberterritory to monitor traffic in nation-state wide networks to find anomalies and malicious activities. For example, some telecommunication operators already have SOCs to monitor their networks (Elisa 2021; Telia 2021; AT&T 2022), these could be chained together in nation-state SOC that would get information from different telecommunication providers and then gather the information together in one place to monitor the nation-state status in networks. Sovereignty comes from the ability to decide what happens on the area controlled by sovereign nation-state. Thus, digital sovereignty demands the ability to have control in the networks of a nation-state.

The benefits of a cyberterritory when implemented by non-totalitarian nations are greater than the possibility of misuse; the nation-states who want to control their citizens will find the ways to do it anyway. Through a cyberterritory it would be possible to better protect citizens and companies in a cyberterritory from cybercrimes and other malicious actors. In case of crisis, it is easier to manage the internal affairs when the amount of data from outside can be reduced. For example, misinformation campaigns can be shut down before they even start to reduce the damage produced by misinformed people. In normal situation the non-totalitarian cyberterritory has not any difference from the current state of networks. However, in case of conflict or escalation the nation-state could better control their digital territory by controlling the traffic in a cyberterritory (cf. Kukkola 2021). Thus, in the non-totalitarian cyberterritory the benefit of control and surveillance would fully emerge only when the political situation is something else than normal. The legislation should be updated according to technical aspects of a cyberterritory to reduce the risk of unwanted surveillance towards citizens.

The core responsibilities of a nation-state are derived from the universal declaration of human rights that serves as cornerstone for law making in many nation-states. The core responsibilities for nation-state towards its citizens are to protect the peace of a society, to protect right to life and liberty for everyone and to provide equal justice for everyone regardless of their socio-economic status (United Nations 1948). To achieve these responsibilities in cyberspace the nation-state should have visibility to nation-state-level networks in its cyberterritory.

The idea of a cyberterritory has evolved through 2010s and the evolution will continue through 2020s (United Nations 2021a). It makes sense to define and produce a technical framework that can be agreed upon by many countries. The framework should be something that approved by both supporters of 'multistakeholder' and 'multilateral' governance models. When creating the framework together, it is possible to preserve the open nature of Internet and give the possibility of territorialisation. Creating the framework together also could prevent the birth of techno-economic alliances, which could prevent conflicts and competition in cyberspace. Cyberterritory can be implemented in such manner that regular user knows no difference in user experience when compared to the current situation. Through mutual interests of nation-states, it is possible to create safer implementation for a cyberterritory since there are more observers during the development. In this process, a theoretical 'cyberterritory' as an entity of its own can turn into a new nation-state 'digital terrain' of the future.

References

AT&T (2022) Security operations Center, [online], https://cybersecurity.att.com/solutions/security-operations-center, [Accessed January 13 2022].

Braud, A., Fromentoux, G., Radier, B., & Le Grand, O. (2021) "The Road to European Digital Sovereignity with Gaia-X and IDSA", *IEEE Network*, Vol 35, No. 2, pp 4-5.

Demchak, C., & Dombrowski, P. (2011) "Rise of a Cybered Westphalian Age", *Strategic Studies Quarterly*, Vol 5, No. 1, pp 32-61.

Demchak, C., & Dombrowski, P. (2013) "Cyber Westphalia: Asserting State Prerogatives in Cyberspace", *Georgetown Journal of International Affairs,* International Engagement on Cyber III: State Building on a New Frontier, pp 29-38.

Drake, W. J., Vinton, C. G., & Kleinwächter, W. (2016) "Internet Fragmentation: An Overview", *World Economic Forum*, Davos, [online] https://www3.weforum.org/docs/WEF_FII_Internet_Fragmentation_An_Over-view_2016.pdf, [Accessed October 12 2021].

Elisa (2021) "Elisa's Cyber Security Services", [online], https://yrityksille.elisa.fi/en/cyber-security, [Accessed January 12 2022].

FDRA (2021) Finnish Defence Research Agency: *Research workshop for military and civilian experts on cyberterritory*, 29.-31.3.2021.

Glen, C. M. (2014) "Internet Governance: Territorializing Cyberspace?", *Politics & Policy*, Vol 42, No. 5, pp 635-657.

Goldsmith, J., & Wu, T. (2007) *Who Controls the Internet? Illusions of a Borderless World*, Oxford University Press, New York.

Goodin D., "After Ukraine recruits and 'IT Army', dozens of Russian sites go dark", [online], https://arstechnica.com/information-technology/2022/02/after-ukraine-recruits-an-it-army-dozens-of-russian-sites-go-dark/, [Accessed March 18, 2022].

Kahin, B., & Nesson, C. (1997) *Borders in Cyberspace: Information Policy and the Global Information Infrastructure*, The MIT Press, Cambridge, Massachusetts and London, England.

Korzak, E. (2021) "Russia's Cyber Policy Efforts in the United Nations", *Tallinn Paper No. 11*, CCDCOE.

Kukkola, J. (2020) *Digital Soviet Union: The Russian national segment of the Internet as a closed national network shaped by strategic cultural ideas*, National Defence University, Helsinki.

Kukkola, J. (2021) *Rakenteellisen kyberasymmetrian strategiset vaikutukset: Venäjän kansallinen internetsegmentti sotilasstrategisena ilmiönä*, Finnish Defence Research Agency, Riihimäki.

Kukkola, J., & Ristolainen, M. (2018) "Projected Territoriality: A Case Study of the Infrastructure of Russian Digital Borders", *Journal of Infromation Warfare*, Vol 17, No. 2, pp 83-100.

Kukkola, J., Ristolainen, M., & Nikkarila, J.-P. (2017) *Game Changer: Structural transformation of cyberspace*, Finnish Defence Research Agency, Riihimäki.

Litvinenko, A. (2021) "Re-Defining Borders Online: Russia's Strategic Narrative on Internet Sovereignty", *Media and Communication*, Vol 9, No. 4, pp 5-15.

MGIMO (2019) "Mezhdunarodnye ugrozy 2020: Kazhdyi za sebia", *Laboratoriia analiza mezhdunarodnykh protsessov MGIMO MID Rossii*, [online], https://mgimo.ru/upload/iblock/2ac/int-threats-2020.pdf, [Accessed October 11 2021].

Mueller, M. (2017) *Will The Internet Fragment? Sovereignty, Globalization and Cyberspace*, Polity Press, Cambridge.
Pohle, J., & Thiel, T. (2020) Digital Soverignty, *Internet Policy Review*, Vol 9, No. 4, pp 1-19.
Radu, R. (2019) *Negotiating Internet Governance*, Oxford University Press, Oxford.
Ristolainen, M. (2021) "Softaa kyberrajalle! Katsaus kybertilan valtioalueellistamisprosessiin meillä ja maailmalla", *Tutkimuskatsaus 1/2021*, Puolustusvoimien tutkimuslaitos, [online], https://puolustusvoimat.fi/web/tutkimus/tutkimuslaitoksen-julkaisut#tutkimuskatsaukset, [Accessed December 17 2021].
Ruhl, C., Hollis, D., Hoffman, W., & Maurer, T. (2020) "Cyberspace and Geopolitics: Assessing Global Cybersecurity Norm Processes at a Crossroads", *Working Paper.* Carnegie Endowment for International Peace, [online], https://carnegieendowment.org/files/Cyberspace_and_Geopolitics.pdf, [Accessed December 15 2021].
Safenames Ltd. (2017) "Google Data Center FAQ", [online], https://www.datacenterknowledge.com/archives-/2017/03/16/google-data-center-faq-part-2, [Accessed January 12 2022].
Scott, K. (2021) "Connected, Continual Conflict: Towards a Cybernetic Model of Warfare", *ECCWS 2021 20th European Conference on Cyber Warfare and Security*, pp. 375-381.
Shaping Europe's digital future (2020) "Shaping Europe's digital future", *Publications Office of the European Union*, [online], https://ec.europa.eu/info/sites/default/files/communication-shaping-europes-digital-future-feb2020_en_4.pdf, [Accessed December 15 2021].
Singh, J. (2009) "Multilateral Approaches to Deliberating Internet Governance", *Policy & Internet*, Vol 1, No. 1, pp 91-111.
Strickling, L. E., & Hill, J. (2017) "Multi-stakeholder internet governance: success and opportunities", *Journal of Cyber Policy*, Vol 2, No. 3, pp 296-317.
Tallinn Manual 2.0. (2017) *Tallinn Manual 2.0 on the International Law Applicable to Cyber Operations.* Cambridge University Press, Cambridge.
Telia (2021) Cyber Security, [online], https://www.telia.fi/business/one-large/security?intcmp=b2b-en-large-grid-cyber-security, [Accessed January 12 2022].
United Nations (1948) "The Universal Declaration of Human Rights", [online], https://www.un.org/en/about-us/universal-declaration-of-human-rights, [Accessed January 12 2022]
United Nations (2021a) "Developments in the field of information and telecommunications in the context of international security", [online], https://www.un.org/disarmament/ict-security/, [Accessed December 17 2021]
United Nations (2021b) "Open-ended Working Group", [online], https://www.un.org/disarmament/open-ended-working-group/, [Accessed December 17 2021]
United Nations (2021c) "Group of Governmental Experts", [online], https://www.un.org/disarmament/group-of-governmental-experts/, [Accessed December 17 2021].
Yefremov, A. (2017) "Formirovanie kontseptsii informatsionnogo suvereniteta gosudarstva", *Zhurnal Vysshei ekonomiki*, No. 1, pp 201-215.

Researching Graduated Cyber Security Students: Reflecting Employment and job Responsibilities Through NICE Framework

Karo Saharinen, Jarmo Viinikanoja and Jouni Huotari
JAMK University of Applied Sciences, Jyväskylä, Finland
karo.saharinen@jamk.fi
jarmo.viinikanoja@jamk.fi
jouni.huotari@jamk.fi

Abstract: Most research and development on Cyber Security education is currently focusing on what should be taught, how much, and where within the degree programmes. Different Cyber Security frameworks are currently evolving to include Cyber Security education parallel to older paradigms of Computing Education, existing alongside with such as *"Information Technology"* and *"Software Engineering"*. Different Cyber Security specialisations or even whole degree programmes have started within universities before the frameworks have been defined into standardised degree structures. This is mainly the result of a dire industry need of well-educated cyber security personnel, a phenomenon affecting the industry globally. Our research concentrates on Finnish alumni students who have already graduated from a bachelor's degree programme in Information Technology with a specialisation in Cyber Security in Finland. Within our gathered research data, we analysed what is the industry sector where their current job resides, and what are the cyber security responsibilities in their current work. The questionnaire also contained an after-reflection section where the graduated students could choose what they would study were they about to start and plan their studies again. The results verify that Cyber Security is still the most favoured specialisation within the former Cyber Security alumni students. Slight variation is evident from the data, which in the authors' perspective, verifies the multifaceted nature of Cyber Security. When analysing alumni students' job responsibilities, the main category of work resides in the *"Protect and Defend"* category of the NICE Framework, which in the terms of the conference, relates to Critical Infrastructure Protection being the main subject of employment for fresh graduates. These results give insight to other education organisations on how to develop their curricula to further emphasise the employment of students or to offer modules which are of interest for newly employed Cyber Security professionals. In addition, it gives an insight of industry demand for freshly graduated students within the target group.

Keywords: cyber security, degree programme, cybersecurity skills

1. Introduction

Cyber Security capability building is a world-wide phenomenon where different nations are either gathering or developing tools, training people (Catota et al, 2019) and perfecting their processes to an extent that some might even call a cyber arms race (Limnéll, 2016). This paper concentrates on researching the training of cyber security professionals through the education systems of a country. An undertaking which is simultaneously answering to an evident workforce need of a functioning industry (Jaurimaa et al, 2020) and the national cyber resilience levels of a country (Whyte, 2020). Both of which are targeted by threats coming from the cyber domain affecting e.g. the critical infrastructure of a country or the information security of a nation.

To answer this need of cyber security professionals, degree programmes fully dedicated to Cyber Security are being established in the Higher Education institutions of different countries. European Union Agency for Cybersecurity (ENISA) established Cybersecurity Higher Education Database (CyberHEAD) to map these degree programmes (Zan De & Di Franco, 2019). Criteria for degree programme approval were:

- 25 percent of *cyber security topics* for bachelor's degrees
- 40 percent of *cyber security topics* for master's degrees
- and research on *cyber security topics* for PhD students

At the time of writing this paper, there are 139 programmes in 25 countries that are approved in CyberHEAD (ENISA, 2021). Within these 139 programmes the word cyber security appeared in the title of 13 out of 23 bachelor's degrees, 56 out of 105 master's degrees and in none of the three PhD programmes (and 0 out of 8 specialiation postgraduate courses). This emphasises that almost half of the degrees are titled and focused on other areas of Computing. However, they contain the percentage required in *cyber security topics* to be a part of CyberHEAD.

2. Literature review

As described by the introduction chapter, the *cyber security topics* in use at CyberHEAD were defined by ENISA (Zan De & Di Franco, 2019) to be aligned with Joint Task Force on Cybersecurity Education called CSEC2017 (Associate for Computing Machinery, 2017), which is published by the Association for Computing Machinery (ACM) in their collection of curricula recommendations (Associate for Computing Machinery, n.d.). These recommendations were published to emphasise Cyber Security as a paradigm of Global Computing Education. An aspect which was lacking in the ACM Curricula Recommendations of 2005 (Shackelford et al, 2005). ACM recently published their Curricula Recommendations 2020 (CC2020 Task Force, 2020), which stabilised the presence of Cybersecurity as a full paradigm of computing next to older topics such as *"Information Technology"* and *"Software Engineering"* to name a few.

Alongside these developments the National Institute of Science and Technology (NIST) released National Initiative for Cybersecurity Education (NICE) Cybersecurity Workforce Framework in 2017 (Newhouse et al, 2017), which described the knowledge, skills and tasks in different categories and workroles within Cyber Security. The framework aspired to partner academia, private and public sector to provide comprehensive material to improve workforce development in education and training within the United States of America. In the UK, as put forth by the UK Cyber Security Strategy (United Kingdom, 2016), The Cyber Security Body of Knowledge was released in 2019 by version 1.0 (Rashid et al, 2019) and 2021 with version 1.1.0 (Rashid et al, 2021) respectively. It is a simplification if stated that both are quite similar in their agendas and goals.

Similar projects were conducted in the European Union in two different research and development projects; SPARTA and Cyber Security for Europe (CS4E). SPARTA released their *"Cybersecurity skills framework"* in 2020 (Piesarskas et al, 2020), but based a part of their work on the NICE framework. A very similar undertaking was developed under the Cyber Security for Europe (CS4E) project in Work Package 6 with a topic of Cybersecurity Skills & Capability Building. The work package released a Deliverable on *"Design of Education and Professional Framework"* in 2021 (Karinsalo et al, 2021).

Many skills framework documents were motivated by the worldwide need of Cybersecurity Workforce. This topic was declared as follows: *"The cybersecurity skills shortage and gap are well-documented issues that are currently having an impact on national labour markets worldwide"*, a direct quote from a publication of ENISA released on 24th of November 2021 titled "Addresing Skills Shortage and Gap Through Higher Education" (Nurse et al, 2021), released just prior to writing this research paper. This shortage was referenced by seven different sources, divided regionally here to be from European Union, UK, North America, Central and South America, Asia and Australia. This emphasises the fact that there is a world wide need of Cyber Security Professionals. Even the newly published cyber security strategy of the European Union (European Comission, 2020) states this lack of professionals, but with fewer references. These parallelly generated frameworks, curriculum guides, and different publications prove an evident background and need of establishing cyber security focused education.

Finland published its first Cyber Security Strategy on 24 January 2013 as a government resolution (The Security Committee of Finland, 2013). The strategy declared different goals and operation models to meet the challenges of the cyber domain and ensure the fuctionality of the cyber domain. The first version of the strategy contained a sentence declaring *"The study of basic cyber security skills must be included at all levels of education"*. This was further enforced in the updated strategy (The Security Committee of Finland, 2019) that all cyber and information and communications technology (ICT) related training/degree programmes will be strengthened.

The first Finnish strategy can be seen as a clear point in time when Higher Education institutions in Finland began to start degree programmes purely dedicated to cyber security. JAMK University of Applied Sciences (JAMK) and University of Jyväskylä (JYU) both launched a master's degree programme on purely cyber security in 2013 (JAMK University of Applied Sciences, 2013) (University of Jyväskylä, 2013). JAMK also started a bachelor's degree with a cyber security specialisation in 2015 (JAMK University of Applied Sciences, 2015). Afterwards many other Universities of Applied Sciences and Universities in Finland followed with their own offering of Cyber Security, be it degree-oriented curricula (South-Eastern Finland University of Applied Sciences, 2021) or just specialisation studies for life-long learning with no official degree completion (Metropolia University of Applied Sciences, 2021). This timeline of frameworks and degree oriented higher education is further visualised in the Figure 1.

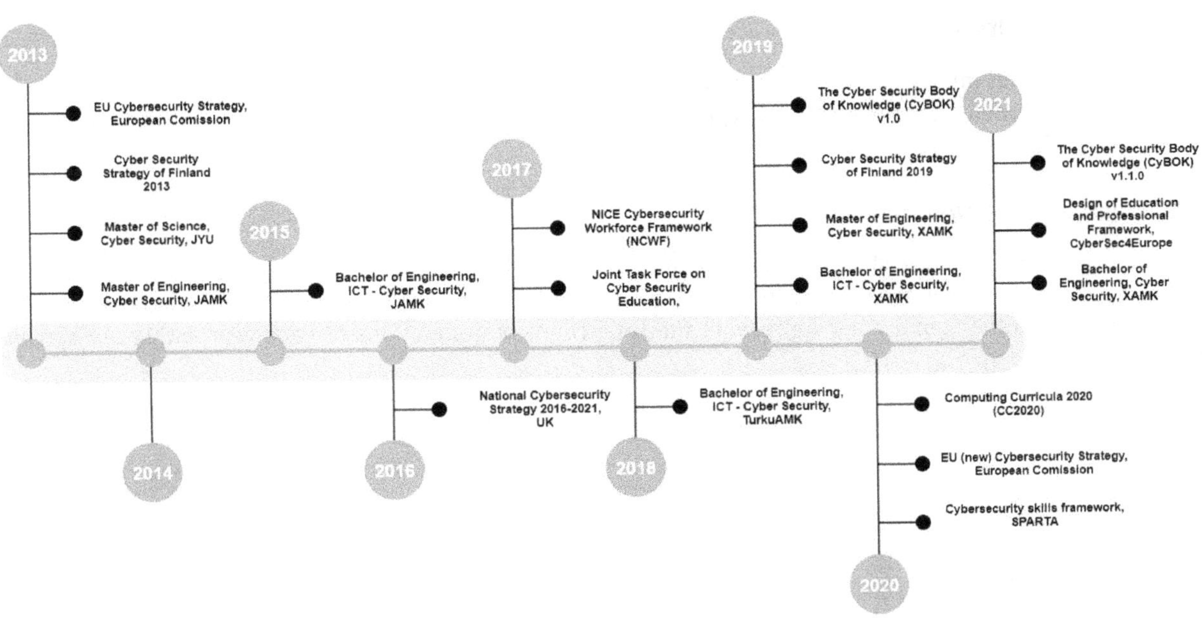

Figure 1: Timeline of cyber security frameworks, Finnish cyber security strategies and degree programmes

3. Survey research on graduated students

This research was scoped to concern graduated bachelor's degree students of JAMK University of Applied Sciences with a cyber security specialisation in their degree. The degree programme was started in 2015 and has a recommended length of 4 years (240 ECTS). Thus, the first students to graduate according to the recommended timetable should have been around 2019. Noteworthy is that these students were the first graduates within Finland to have a cyber security focus in their bachelor's degree.

The research was designed to directly involve the university in contacting the students, however it proved to be a troublesome task. Cyber Security students, by the nature of their studies presumably, had marked that their contact information should not be used for research purposes, nor should they be contacted later by the university. Thus, the research permission process of the university granted no results for student contact information.

This result of the permission process forced the researchers of this paper to contact the students through different social media platforms; asking the students publicly to inform of their willingness for the research by contacting the researchers personally. Luckily few active students could be found which then forwarded the request to attend the research to more specific and limited messaging groups of the students. To increase the reliability of the research, one aspect was that the questionnaire was handed only to graduated students who had directly contacted the researchers and been identified as former students. This resulted in 19 respondents out of 68 graduated, thus sample size from the total possible participants was 27.94%.

The research method used was a survey containing mostly quantitative measurements of the participants. Research ethics were used design the questionnaire in a way that would give the researchers the necessary information, and then the replies were generalised (e.g., specific company to be "private/public company") so that no singular student could be identifiable from the data.

As the literature review stated, there are multiple frameworks possible of data categorisation/analyzation. In this research the NICE Framework was chosen to categorise and analyse the work roles of researched participants. The Framework has descriptive terminology on each work responsibility assigned to each work role. Because of this, the NICE framework was something that the students were requested to examine, if they had doubt in selecting what work role described their profession the best.

4. Survey results and analysis

This chapter divides into two different sections; place of employment answers the research question "Where are graduated students employed?" and type of work answering, "What kind of work responsibilities do the students have?".

4.1 Place of employment

First question concerned the student's starting year and graduation year to get a glimpse of the length of their studies. This is visualised in the Figure 2 in which the darker color shows the total count of started degree studies of each year, and the lighter color represents the total count of graduations of each year within our sample group.

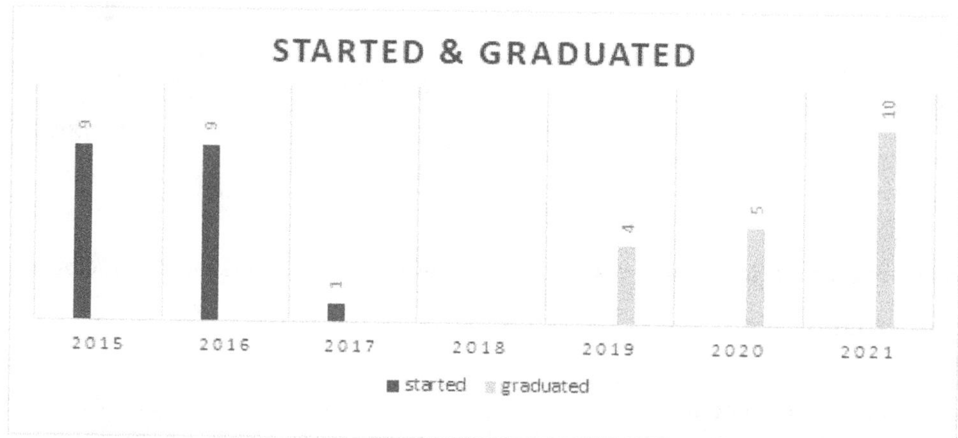

Figure 2: Starting year compared to graduation year

It is evident that even though almost half of the respondents started their studies in 2015, still many graduated behind schedule in 2020 and 2021. The authors interpret this could be the result of e.g. fast employment during studies as ICT degrees do not need to be finished to start working in the industry which results to delaying the graduation of a student. Although other reasons might be as plausible as proved by other research (Willoughby et al, 2021). Unfortunately, within our research this reason of delay was not a separate question.

One of the main research objectives was to find out where the bachelor's degree students get employed, which industry sector and company size. These are apparent in the data gathered and visualised in the Figure 3.

Figure 3: Organisation size and sector

Within these results, neither the employment sector nor company size surprised the authors. In Finland, the growing cyber security sector seems to follow the same footsteps as the global phenomenon. The European Cyber Security Strategy (European Comission, 2020) states that *"Over two-thirds of companies, in particular Small to Medium Enterprises (SMEs) are considered 'novices' in cyber security..."*. This stated need for protection can be witnessed from the service offering of private cyber security companies in Finland. These provided services need workforce behind them and thus, graduated bachelor's degree students get employed.

Given the employment, these companies can be dissected further based on their industry sector. A proposal for a European Cybersecurity Taxonomy (Nai Fovino et al, 2019) declared industry sectors which were utilised in the data categorisation of this research. Students' employment information was translated into these sectors as represented by figure 4.

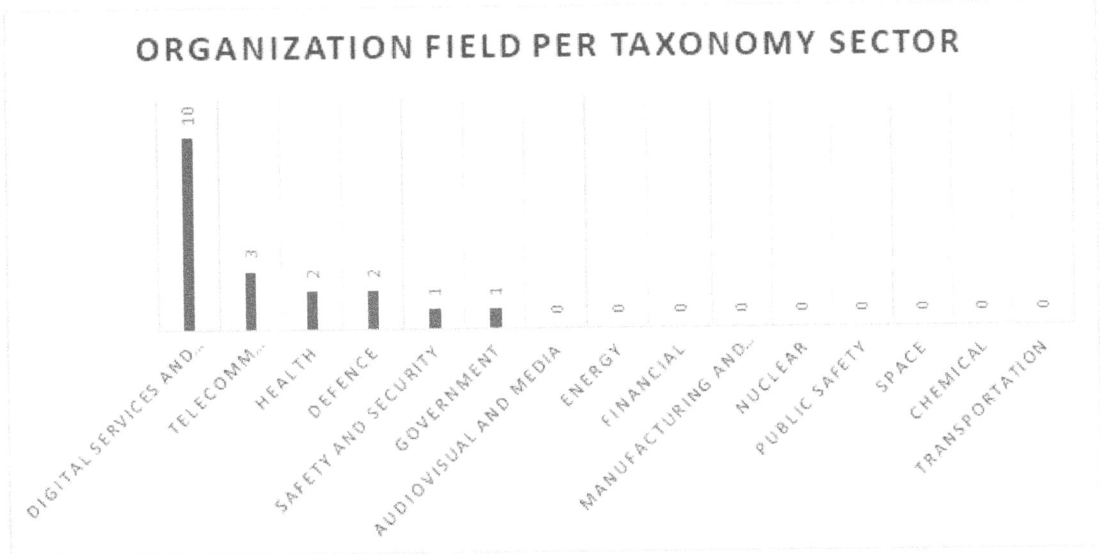

Figure 4: Employment sector of students based on a proposal for a European cybersecurity taxonomy

Out of these sectors, the business to business (B2B) companies were most apparent. Most of them categorising under the *"Digital services and platforms"* sector of the taxonomy. *"Telecomm infrastructure"* had significant Internet Service Providers (ISPs) of Finland recruiting some students, but for reliability sake it is worth metioning that some of the *"digital services and platforms"* were subsidiary organisations of the previously mentioned ISPs. Thus, based on analysis interpretation of the organisations, these two were the largest employers. *"Health"* and *"Defence"* sectors have employed two students both with *"Defence"* being the majority of public organisation employers. *"Government"* and *"Safety and Security"* sectors followed, but there are no results of other sectors within this survey scope.

4.2 Type of work

As for the following results, we asked the students to place emphasis on the question of "what work role of the NICE framework describe their work the most?". As the quantitative grading scheme, we asked them to place the work roles in 1st to 5th order where the 1st being the most descriptive work role for their current work and 2nd being the second most descriptive work role etc. Figure 5 shows a graph of the whole data.

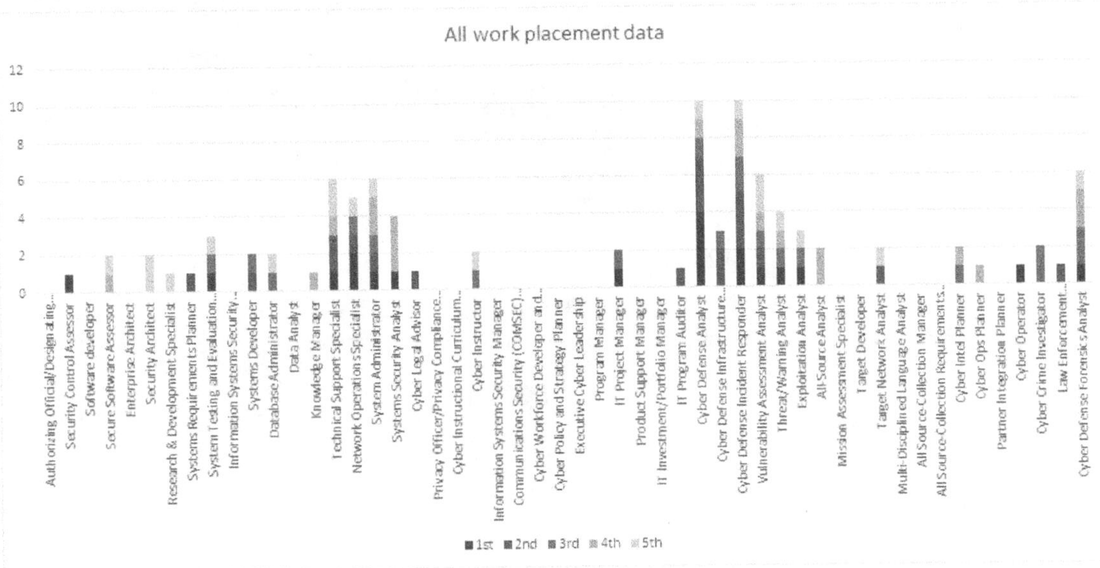

Figure 5: Top NICE categories based on what best fit their work

As there are 52 work roles described by the NICE Framework, Figure 5 is quite extensive or even hard to differentiate, but clear emphasises can already be observed from the visualization. To illustrate the work responsibilities more informatively, we used the frameworks categories to further delve into the data and order it from the most hit category (up top) and the least hit category (on bottom) as visualised in the Figure 6.

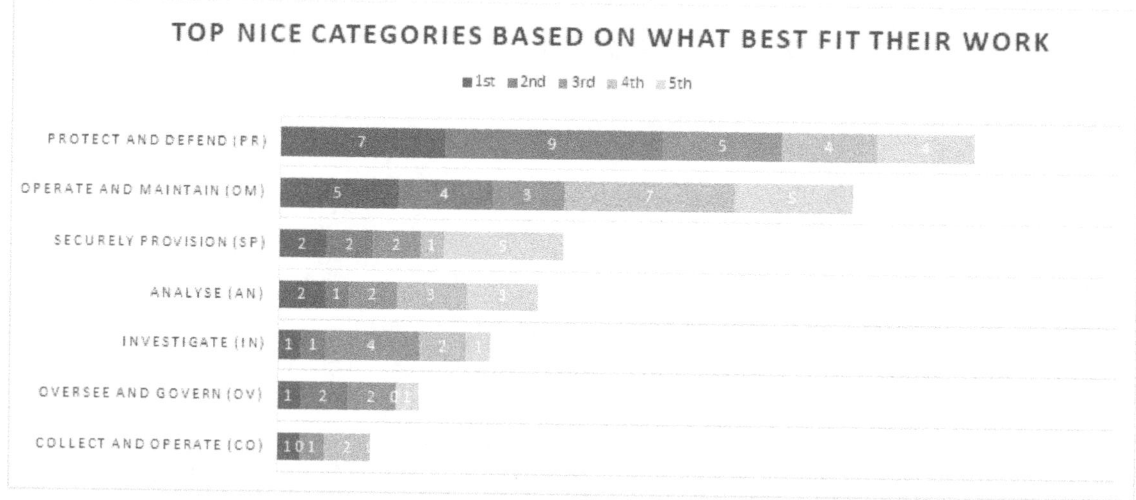

Figure 6: Top NICE categories based on what best fit their work

With this analysis of the data one can see that the bachelor's degree students are clearly employed in the *"Protect and Defend"* and *"Operate and Maintain"* categories with *"Securely Provision"* category closely behind them. *"Investigate"* and *"Analyse"* are categories that closely tie in with one another, thus they are quite similarly represented in the data. *"Oversee and Govern"* is quite administrative or managerial category with executive work roles; thus, the authors are not surprised that the bachelor's degree students do not work in that category immediately at the end of their studies. *"Collect and Operate"* category has described as intelligence gathering and offensive operations performed within the cyber domain, and as such it was the lowest category to receive answers.

To get a better view of the most frequent work roles we filtered 3rd to 5th selections from the data (still visible in figure 5) to get an understanding of what are the primary work roles of the respondents. This visualization can be seen in figure 7.

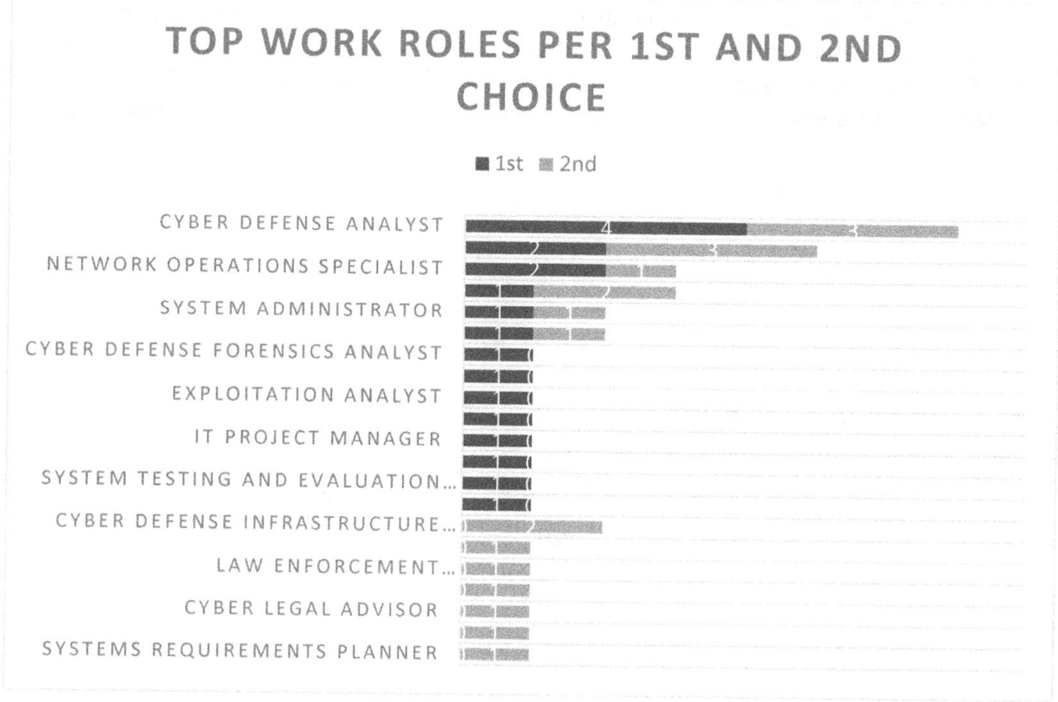

Figure 7: Top NICE categories based on what best fit their work

Almost all the data was either in *"Cyber Defense Analyst"* or *"Cyber Defense Incident Responder"* work roles with both belonging to *"Protect and Defend"* category. The authors would assume these two work roles of NICE Framework to be essential parts of the current establishment of Security Operations Centres (SOCs) within Finland (Carson, 2014), a growing private and public sector functionality within the field of Cyber Security (Jauhiainen, 2021). The newly graduated would most probably be workforce to create, upkeep or provide this service.

4.3 Cyber security specialisation in retrospective

At the end of the survey, the hindsight of the students is asked; "How would you choose your specialisation modules nowadays, with all the knowledge of your current work occupation and your hindsight of the studies". The student could choose two 30 ECTS modules but not the same module twice.

Noteworthy is that the module selection is available at the University of Applied Sciences they graduated from, but from a newly updated curriculum (JAMK University of Applied Sciences, 2021). The students were asked to familiarise themselves with the updated curriculum and then make their module selections. One central theme of the curriculum is to divide the modules into the *"DevSecOps"* ideology (Sánchez-Gordón & Colomo-Palacios, 2020) within ICT; the acronym standing for *"Dev"* being developers, *"Sec"* meaning (cyber) security and *"Ops"* as Operations. Results from the student answers are visualised in the Figure 8.

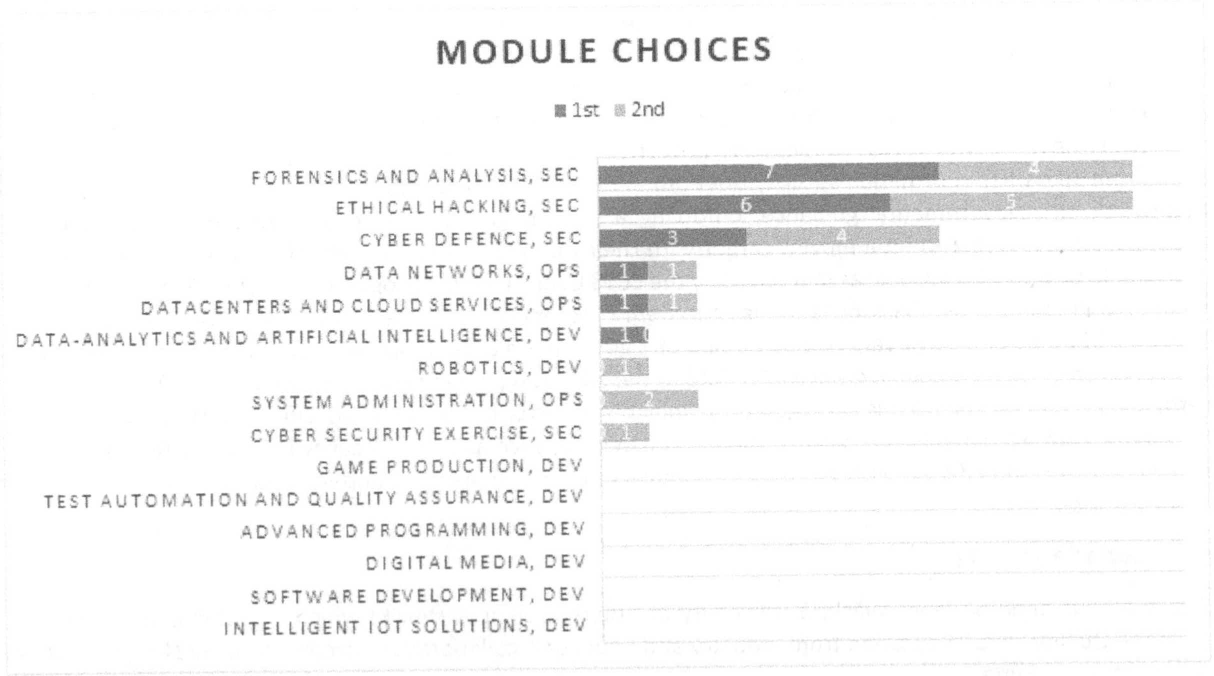

Figure 8: Specialisation modules of the bachelor's degree

An encouraging result from the survey data is that the module choices would still focus on Cyber Security. These top three modules are considered by the ICT degree programme coordinators to be a part of Cyber Security specialisation. The following two choices *"Data Networks"* and *"Datacentres and Cloud Services"* are more in the Operations section of the curriculum but noticeably related to some students' current work. This ties well with the results in Figure 6 as *"Operate and Maintain"* category was the second most descriptive category of their work.

One misstep of the authors was that we did not ask the alumni students about what modules they already had studied. Without this information it is impossible trace back Figure 8 data as being a new selection of the participants or did they just confirm that they would choose the same module they did once again, supposing they were freshly started students. Still, it would indicate the trend, that they wish to further emphasise their studies in Cyber Security. And as an education organisation it would give a confirmation to the university that these specialisation studies should be offered to the industry as a part of life-long learning.

5. Conclusion

Given the results and analysis, one can conclude that the cyber security students graduate and get employed to *"Protect and Defend"* the Critical Infrastructure through ISPs and work with the safety of *"Digital Services and Platforms"* in Finland. Our research data can be interpreted to prove that students are employed to be ensuring the functionality of the Finnish Cyber Domain as the Cyber Security Strategy of Finland stated. Through our research the education organizers (at JAMK) can now have a better understanding of the work placements of their former students in the field of Cyber Security. Adjustments of the curriculum can be based on researched data. By researching the education landscape (Saharinen et al, 2020), industry need (Jaurimaa et al, 2020) and student employment and satisfaction data the degree programme coordinators can verify their curricula to be up to date, have an ongoing discussion with the industry and provide current students of the degree programme information about their module choices. The timeline of the different, parallel cyber security frameworks gives a view of the evolving atmosphere around cyber security education. Different frameworks have varying amount of scientific research behind them, and this is typically stated in publication of the framework. The authors of this paper would like to conclude that all additions are of course an enrichment of the field, but for an education organisation; it would be preferrable to establish a basis of education on one of the frameworks (Saharinen et al, 2019) and proceed with the chosen framework consistently throughout the curriculum.

6. Discussion

The lifespan of a bachelor's degrees varies from three to four years in the Finnish education system (Ministery of Education and Culture - Finland, 2021). Given the degree completion length of the participating students in the research, there are six different cyber security frameworks published as visualized in Figure 1. The authors would assume that many of these Finnish cyber security degree programmes were started purely to respond to an industry need, however, also to meet this governmental resolution in Finland. Their formation might have come from an earlier information security orientation degree background, rather than a guiding cyber security framework or a governmental guidance, enforcing a clear degree structure and content. Thus, education organizations are trying to hit a moving target with their module and course structures within their curricula, that should be publicly available as mandated by the ECTS Users' Guide (European Comission, 2015). Finland has a national graduands feedback questionnaire system (Rectors' Conference of Finnish Universities of Applied Science, 2021) in place; however, the questionnaire is generalised to cover all education fields in the Universities of Applied Sciences. Although it gives useful data to the educating organisations, it rarely has relevant data on a certain degree field. The data is aggregated to Finland's Ministry of Education specific "Fields of Steering" and thus it does not even mention a specialisation of the degree, such as Cyber Security in ICT. Our research in this paper could and should be replicated to various universities to gain a better understanding of the graduands of cyber security.

Acknowledgements

This work has been done in Jyväskylä University of Applied Sciences (JAMK) which is participating in LIPPA - project – Quality to ICT Education from Industry and Education collaboration (project code S22466) funded by European Social Fund.

The authors would like to thank Tuula Kotikoski for her contribution in proofreading the English language on the paper.

References

Associate for Computing Machinery. (2017) *Cybersecurity Curricula 2017 - Curriculum Guidelines for Post-Secondary Degree Programs in Cybersecurity.* New York: Associate for Computing Machinery.

Associate for Computing Machinery, n.d. *Curricula Recommendations.* [online] https://www.acm.org/education/curricula-recommendations

Carson, Z. (2014) *Ten Strategies of a World-Class Cybersecurity Operations Center.* s.l.:The MITRE Corporation.

CC2020 Task Force, 2020. *Computing Curricula 2020: Paradigms for Global Computing Education.* s.l.:Association for Computing Machinery.

Catota, F.E., Morgan, M.G. and Sicker, D.C. (2019) Cybersecurity education in a developing nation: the Ecuadorian environment, *Journal of Cybersecurity*, 5(1), p. tyz001. doi:10.1093/cybsec/tyz001.

ENISA (2021) *CYBERHEAD - Cybersecurity Higher Education Database.* [online] https://www.enisa.europa.eu/topics/cybersecurity-education/education-map

European Comission, 2015. *ECTS Users' Guide.* [online] https://ec.europa.eu/assets/eac/education/ects/users-guide/docs/ects-users-guide_en.pdf

European Comission, (2020) *The EU's Cybersecurity Strategy for the Digital Decade.* [online] Available at: https://digital-strategy.ec.europa.eu/en/library/eus-cybersecurity-strategy-digital-decade-0

JAMK University of Applied Sciences. (2013) *Cyber Security, Master of Engineering.* [online] https://www.jamk.fi/en/Education/Technology-and-Transport/Cyber-Security-Masters-Degree/

JAMK University of Applied Sciences. (2015) *Bachelor of Engineering, Information and Communications Technology.* [online] https://www.jamk.fi/en/Education/Technology-and-Transport/information-and-communication-technology-bachelor-of-engineering/

JAMK University of Applied Sciences. (2021) *Bachelor's Degree Programme in Information and Communications Technology.* [online] https://opetussuunnitelmat.peppi.jamk.fi/en/48/en/5290/TTV2021SS/year/2021

Jauhiainen, J. (2021) *List of SOC service providers.* [online] https://csoc.fi/

Jaurimaa, J., Saharinen, K. and Kotikoski, S. (2021) Critical Infrastructure Protection: Employer Expectations for Cyber Security Education in Finland, in *Proceedings of the 2021 20th European Conference on Cyber Warfare and Security. European Conference on Cyber Warfare and Security*, United Kingdoms: Academic Conferences International Limited. doi:10.34190/EWS.21.015.

Karinsalo, A. et al. (2021) [online] Available at: https://cybersec4europe.eu/wp-content/uploads/2021/06/D6_3_Design-of-Education-and-Professional-Frame-work_Final.pdf

Limnéll, J. (2016) The cyber arms race is accelerating – what are the consequences?, *Journal of Cyber Policy*, 1(1), pp. 50–60. doi:10.1080/23738871.2016.1158304.

Metropolia University of Applied Sciences (2021) *Cyber Security specialization studies.* [online] https://www.metropolia.fi/fi/opiskelu-metropoliassa/osaamisen-taydentaminen/erikoistumiskoulutukset/kyberturvallisuus

Ministery of Education and Culture - Finland (2021) *Finnish Education System.* [online] https://okm.fi/en/education-system

Nai Fovino, I. et al. (2019) *A Proposal for a European Cybersecurity Taxonomy.* [online] https://publications.jrc.ec.europa.eu/repository/bitstream/JRC118089/taxonomy-v2.pdf

Newhouse et al. (2017) *National.* s.l.:National Institute of Science and Technology.

Nurse, J. R. et al. (2021) [online] https://www.enisa.europa.eu/publications/addressing-skills-shortage-and-gap-through-higher-education/@@download/fullReport

Piesarskas, E. et al. (2020) [online] https://sparta.eu/assets/deliverables/SPARTA-D9.1-Cybersecurity-skills-framework-PU-M12.pdf

Rashid, A. et al. (2019) *The Cyber Security Body of Knowledge.* [online] https://www.cybok.org/media/downloads/CyBOK-version-1.0.pdf

Rashid, A. et al. (2021) *The Cyber Security Body of Knowledge.* [online] https://www.cybok.org/media/downloads/CyBOK_v1.1.0.pdf

Rectors' Conference of Finnish Universities of Applied Science, 2021. *University of Applied Sciences Graduand Feedback Questionnaire.* [online] https://avop.fi/en

Saharinen, K., Backlund, J. & Nevala, J. (2020) *Assessing Cyber Security Education through NICE Cybersecurity Workforce Framework.* New York, NY, USA, Association for Computing Machinery, p. 172–176.

Saharinen, K., Karjalainen, M. & Kokkonen, T. (2019) *A Design Model for a Degree Programme in Cyber Security.* New York, NY, USA, Association for Computing Machinery, p. 3–7.

Sánchez-Gordón, M. & Colomo-Palacios, R. (2020) *Security as Culture: A Systematic Literature Review of DevSecOps.* New York, NY, USA, Association for Computing Machinery, p. 266–269.

Shackelford, R. et al. (2005) In: *Computing Curricula 2015.* s.l.:The Association for Computing Machinery (ACM); The Association for Information Systems (AIS); The Computer Society (IEEE-CS).

South-Eastern Finland University of Applied Sciences (2021) *Bachelor of Engineering, cyber security.* [online] https://www.xamk.fi/koulutukset/insinoori-amk-kyberturvallisuus/

The Security Committee of Finland (2013) *Finland's Cyber security Strategy.* [online] https://www.defmin.fi/files/2378/Finland_s_Cyber_Security_Strategy.pdf

The Security Committee of Finland, 2019. *Finland's Cyber security Strategy 2019.* [online] https://turvallisuuskomitea.fi/wp-content/uploads/2019/10/Kyberturvallisuusstrategia_A4_ENG_WEB_031019.pdf

United Kingdom (2016) *National Cyber Security Strategy 2016-2021.* [online] https://assets.publishing.service.gov.uk/government/uploads/system/uploads/attachment_data/file/567242/national_cyber_security_strategy_2016.pdf

University of Jyväskylä (2013) *Cyber Security, Master of Philosophy.* [online] https://www.jyu.fi/it/fi/opiskelu/maisteriohjelmat/kyberturvallisuus

Whyte, C. (2020) Cyber conflict or democracy "hacked"? How cyber operations enhance information warfare, *Journal of Cybersecurity*, 6(1), p. tyaa013. doi:10.1093/cybsec/tyaa013.

Willoughby, T. et al. (2021) A Long-Term Study of What Best Predicts Graduating From University Versus Leaving Prior to Graduation, Journal of College Student Retention: Research, Theory & Practice. doi: 10.1177/1521025120987993.

Zan De, T. & Di Franco, F. (2019) *Cybersecurity Skills Development in the EU.* [online] https://www.enisa.europa.eu/publications/the-status-of-cyber-security-education-in-the-european-union/@@download/fullReport

ZTA: Never Trust, Always Verify

Char Sample[1,2], Cragin Shelton[2], Sin Ming Loo[1], Connie Justice[3], Lynette Hornung[2] and Ian Poynter[2]
[1]Boise State University, USA
[2]Independent, USA
[3]Indiana University at Purdue, USA
charsample50@gmail.com
drcragin@icloud.com
smloo@boisestate.edu
cjustice@iupui.edu
lhornung@gmail.com
ianpoynter@gmail.com

Abstract: Zero Trust Architecture (ZTA) deployments are growing in popularity, widely viewed as a solution to historical enterprise security monitoring that typically finds attackers months after they have gained system access. ZTA design incorporates multiple industry security advisories, including assuming network compromise, using robust identity management, encrypting all traffic, thwarting lateral movement, and other security best practices. Collectively, these features are designed to detect and prevent attackers from successfully persisting in the environment. These features each offer solutions to various ongoing security problems but individually are not comprehensive solutions. When designed for cloud services ZTA holds the promise of outsourcing security monitoring. However, some observations about ZTA suggest that the component solutions themselves have flaws potentially exposing systems to additional undetected vulnerabilities, providing a false sense of security. This paper addresses vulnerable paths using a bottom-to-top approach, listing problem areas and mapping them to attacker goals of *deny, deceive, disrupt, deter,* and *destroy*. The paper then addresses residual risk in the architecture. Based on the findings the paper suggests realistic countermeasures, offering insights into additional detection and mitigation techniques.

Keywords: zero trust architecture, vulnerabilities, attack, component, system

1. Introduction

Cybersecurity has a history of *magic bullet* solutions for protecting information from myriad intentional and inadvertent damaging situations (Simmonds, 2019). Highlights include bandwagons formed for encryption, discretionary access control (DAC), firewalls, virtual private networks (VPN), secure socket layer / transport layer security (SSL/TLS), public key infrastructure (PKI), blockchain, artificial intelligence (AI), et cetera. A theme in this history is recognition at each phase of the level of trust assumed for internal and external entities (humans and systems) interacting with the system of interest. Not surprisingly, a current magic bullet, zero trust architecture (ZTA) has reached the pinnacle of proposing system design and operation based on trusting no entity at any time in any situation.

Each of the technologies named above has been both a legitimate contribution to the field and a popular marketing term. Today ZTA is in the same boat, both as a design principle and a marketing term. ZTA is not a new design philosophy; instead ZTA strengthens existing trust technologies by adding additional decision points and enforcing temporal limits. The previous *defense-in-depth* approach did the same, only in smaller domains, creating stovepipes that made sharing more difficult while preserving physical security. We argue that the binary trust/distrust (or perhaps more appropriately distrust/trust) model does not reflect the complexity of work environment relationships (Campbell, 2020).

In spite of ZTA offering no new technologies, the rapid adoption of ZTA proceeds possibly because of Executive Order 14028 (Biden, 2021) which directed each U.S. federal agency leader to create plans to implement ZTA in their respective agencies. This order was part of an initiative to facilitate inter-agency information sharing while maintaining a superior security posture (Ibid). The balancing act between sharing and security is a long-standing security challenge where assumptions and implementation details create exploitable vulnerabilities. Previous design philosophies of perimeter defense and defense-in-depth emphasized need-to-know (Bell & LaPadula, 1976); ZTA prioritizes the need to share (Biden, 2021). So, the agencies may attempt to share, but individual program managers may be reluctant to do so since they are responsible for the program's security.

Adding confusion, products and services advertise ZTA as an offering, when products can at best support only component aspects of ZTA. ZTA is a set of design principles, not something that can be implemented with a single product. (Kindervag, Balaouras, & Colt, 2010). Services can also fail at ZTA by not addressing the ramifications of design decisions. Butcher (2021) noted that business requirements over time weakened perimeter defenses, giving rise to other solutions where trust is too restricted. The likelihood of business requirements undermining ZTA remains. Business goals and cybersecurity objectives require a balancing act extending beyond technical solutions into organizational behaviors, ultimately shaping the security policy that the ZTA supports.

If a site currently has a weak security posture, ZTA principles may offer an improvement but cannot assure complete security. In short, ZTA won't guarantee freedom from cybersecurity problems and in certain case ZTA could introduce additional problems. Enterprise architects need a baseline of competence in understanding technologies, underlying assumptions, and inherent security gaps, residual risk, and the operating environment before applying the ZTA.

2. Background

Kindervag, Balaouras, & Colt (2010) first defined ZTA for Forrester in 2010. Butcher (2021) presented ZTA as a *design philosophy* for security architects, a philosophy flexible enough to allow various instantiations based on site requirements. However, cybersecurity has a history of implementations varying from envisioned designs; ZTA is no exception. A quick overview of strengths and concerns associated with ZTA follows to assist in framing the discussion.

Figure 1 (Rose, et al., 2020, Figure 2) depicts the top-level overview of the NIST ZTA. This graphic shows the overall security system with input and output elements, processing functions, and the interactions between elements and functions. Examination of the system provides attack targets from the system supply chain, through hardware, software, and ultimately humans. Trust is complicated and thus a vulnerability (Campbell, 2020). Traditional mechanisms to grant trust are multi-faceted, but the trust decision is binary; this is still true in ZTA implementations. Trust is granted based on the specific transaction, the confirmed identities of participating entities (human and system), the circumstances surrounding the transaction, and time.

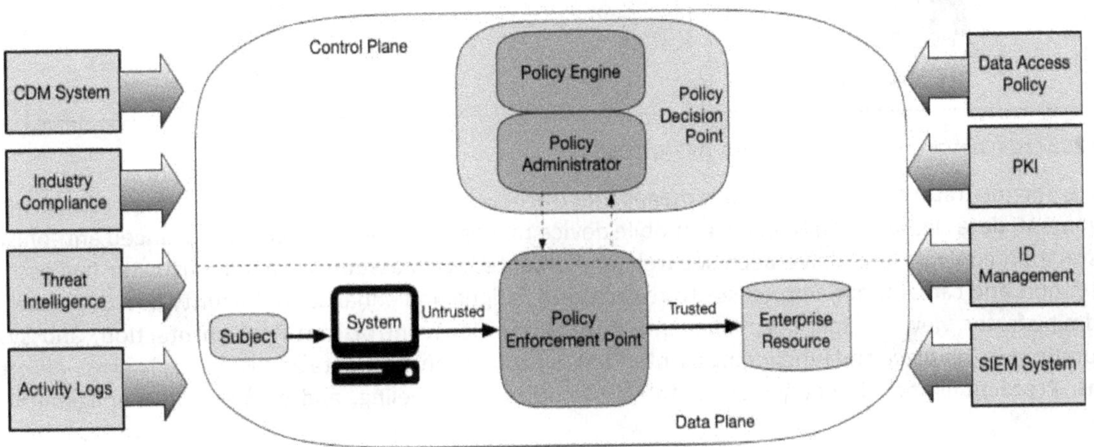

Figure 1: Zero trust architecture overview (Rose, et al., 2020, Figure 2)

2.1 Strengths

The NIST Zero Trust Architecture (Rose et al., 2020) recognized that ZTA will be an evolving approach, requiring resiliency and ongoing evaluation as technology, security threats, and protection tools change. The NIST ZTA supports security with identification, authentication, and authorization for users, assets, and resources. Also important are data and service protection to manage risks, protecting credentials and endpoints with encryption, Multi-Factor Authentication (MFA) and other mechanisms. Continuous Diagnostics and Mitigation

(CDM) is used to ensure patches and fixes are applied. The NIST Risk Management Framework (RMF) (Ross, et al., 2018), and Privacy Framework (Lefkovitz & Boeckl, 2020) are used with ZTA as needed to ensure sensitive data and certificates have appropriate encryption and other critical data controls are implemented. The Trusted Internet Connection (TIC) (Weichert, 2019) has expanded to include cloud and mobile environments, also critical elements in U.S. Federal government systems ZTA.

The use case shown in Figure 2 (NSA, 2021) illustrates ZTA with security safeguards in place, following the Cyber Kill Chain (CKC) steps (Lockheed Martin, 2015). Thus, once an adversary gains access and initiates lateral movement inside the now-compromised system, the policy enforcement point (PEP) can block that movement. When an attacker tries to impersonate a legitimate user, seamless MFA prompts are at play and automatic blocking operates as intended.

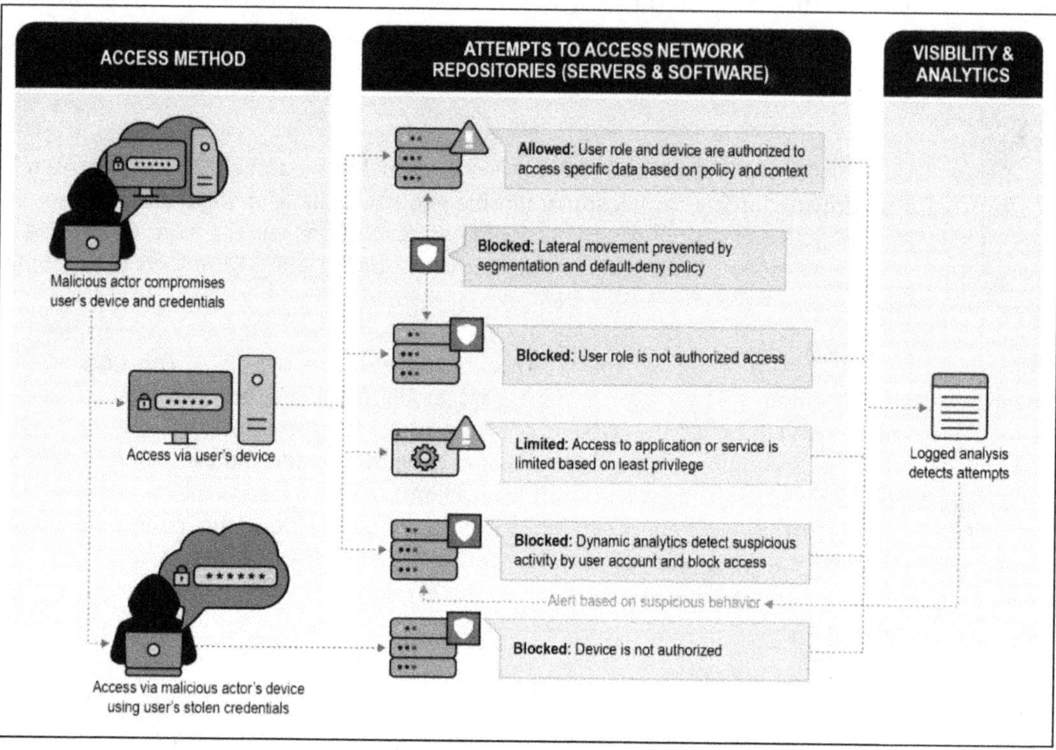

Figure 2: Zero trust architecture use case (NSA, 2021)

From the administrator's point of view, this means that the user uses MFA and the various configurations (device management, data classification & labeling, mobile device management, email security, advanced anti-phishing, impersonation controls, etc.) have been securely implemented. This assumes the administrator has access to security tools and can automate processes to reduce alert fatigue and enhance the security posture under ZTA. The administrator views of identity management, authorization control, certificate protection, and system configurations are in place and functioning as intended to protect sensitive data as it is shared in cross-enterprise systems. Protections include encryption, data classification and labeling, and ex-filtration management and control.

2.2 Concerns

Good ideas, especially high-level ideas, can break down during design and implementation. While ZTA strengths are noteworthy, ZTA is similarly vulnerable to failures during both design and implementation phases. The gaps between vision and implementation can be thought of as *touchpoints* (McGraw, 2006b, p.83). McGraw applied the term touchpoint(s) to software gaps, one of the three pillars of software security; the other two are applied risk management and knowledge (McGraw, 2006a). The same three-pillar philosophy can be applied analogously to ZTA. This philosophy requires an understanding of risk, knowledge of the system & environment, and recognition of the touchpoints. Designing for security, risk analysis/vulnerability testing (Ibid) applies to ZTA as well as software. The security architect must design a ZTA solution that goes beyond controlling unauthorized access to countering the attacker's goals of *deny, deceive, disrupt, destroy,* and *deter*. Some of these attacker goals can be achieved even in ZTA-compliant implementations.

Figure 1 allows the reader to focus on the touchpoints where vulnerabilities could be invoked. A quick view of Figure 1 shows the numerous touchpoints that serve as attack targets. Section 4 of this paper discusses these components in greater detail.

Best practice-defined defense-in-depth solutions offer many of the same advantages of ZTA solutions, such as strong access control, separate enclaves, and proper use of encryption. However, existing solutions often rely on physical separation of systems. The growth of cloud services, software-defined networks (SDN), and software defined containers (SDC) has made such physical and hardware solutions less available. The dynamic nature of SDCs is attractive and frequently mentioned in ZTA solutions. This trade-off is an example of a touchpoint for further examination of layer 2 attacks. The physically separated defense-in-depth solution may suffer layer 2 vulnerabilities but is better positioned to isolate the exploit.

Security architectures are instantiations of overarching security policies designed to reach across the organization (Sherwood, Clark & Lynas, 2005). Such broad policy statements, when applied in the organizational environment where requirements become situational, can be adapted to the single session accesses associated with ZTA. Focusing on specific session details introduces additional touchpoints where interfaces and interactions are not easily standardized. Real value is achieved when the relationship between policy and architecture is carefully considered. Thought leadership must reside amongst architects and policy makers who truly understand real world implications of ZTA. For example, policy makers should have real world experience and be politically neutral, so they may understand how ZTA can be used not just by government agencies, but also by private industry, from small businesses to multinational enterprises. The policy makers should be aware as well of hostile entities who will seek to undermine the security afforded by ZTA.

A remaining factor not depicted, and typically not included, in ZTA discussions is the varying nature of attacker behavioral profiles. The assumption that attackers always seek to gain access, move laterally, then persist in-system, is long-standing (NIST CSRC), reflecting defender biases that do not apply universally. This decades-old characterization of attackers only explains certain hacker behaviors (Sample, et al., 2016). This view contributes to planners failing to imagine alternate attacker goals and behaviors, especially those goals that may be less ambitious but equally effective. A well-timed disruption or denial of service through locking out an important user at a critical moment can be as effective as traditional access breaches.

3. Method

The method used in this study reflects a hypothetical case study. Using the ZTA reference architecture (Department of Defense, 2021) as the test case, this study examines the components and their processing from the adversarial view as a means of identifying potential vulnerabilities. This study presents vulnerability data on ZTA components followed by an exemplar case to step through several examples of activities using multiple architectural views: general user, administrator, and owner. Each case is a hypothetical representation of real-world activities, along with examples of effective intrusions and how they would work in the ZTA example. The selected attacks reflect the attacker goals of deny, deceive, disrupt, destroy, and deter, and are evaluated for efficacy in perimeter, defense-in-depth, and ZTA architectures. Attacks at various layers form the sample.

The hypotheses are:

H_0: ZTA is unbreakable

H_1: ZTA is susceptible to vulnerabilities

4. Findings

Before addressing the components consider the list of security concerns known to impact ZTA and often not sufficiently addressed.

- 1. Signatures vs anomaly detection. Many security products are signature-based, and anomaly detection products have not yet found an effective way to deal with false positives. Byzantine fault tolerance (Veronese et al., 2011) has been used in various domains to create a baseline of acceptable use applicable to ZTA systems.
- 2. Supply chains

- a) Hardware supply chain. When detected, software vulnerabilities can, in many cases, be rebuilt in hours or days. Hardware vulnerabilities, however, may require weeks to months to replace or repair the affected components (Dixon, 2021). Furthermore, nominally identical hardware may be assembled in multiple locations, either in sequential steps or in parallel full production. The absence of public hardware baseline data makes it impossible to verify the purity of a particular piece of hardware. This lack of baselined values makes it possible for attackers to remain undetected as they compromise ZTA component systems (Bhunia et al. 2014).
- Internal network design and firmware of chips used in hardware are generally outside the scope of enterprise security architects designing systems using ZTA principles. Dixon (2021) described a proposed solution being developed by the industry consortium DMTF, the Security Protocol and Data Model Architecture (DMTF, 2022). The driver interface to the hardware must enforce security adherence. As Dixon (2021) suggested, ZTA principles must be designed in before hardware is manufactured.
- b) Software supply chain. Software supply chain vulnerabilities are well documented. A recent identified vulnerability, Log4j (Bing, Satter, & Menn, 2021), is an example showing how ZTA could detect or miss the vulnerability depending on the behaviors involved. Should the vulnerability be invoked, and lateral movement were to follow, the ZTA solution would operate as advertised. However, if the vulnerability were invoked and logging of a specific activity on the server were erased this would make detection more difficult.
- c)Infrastructure. Infrastructure now goes beyond hardware, wires, radios, and traditional infrastructure services such as routing and DNS, to include hypervisors, virtual machines, SDNs and SDCs. A recent compromise of VMWare (Larkshamanan, 2022) listed privilege escalation as one of the effects. When trusted platforms can provision new containers or networks the ability for intruders to collect information indicating which users or hosts are critical becomes easier and less detectable since most security products work above layer 2.
- 3. Zero Day Attacks (0day). 0day attacks are typically undiscovered for 10 months (Greenberg, 2012, Halpern, 2021). In some cases, this time interval is longer. This makes possible intruders hiding themselves by simply implanting without moving and embedding exfiltrated data into “good” payloads.
- 4. Data Centric Attacks. Methods that include data poisoning and other techniques, these attacks take advantage of insufficient data checking against known good or baselined responses.
- 5. Artificial Intelligence/Machine Learning attacks. Data poisoning and model manipulation are well-known methods to subvert AI. Poisoned data can cause AI to generate false negatives while model manipulation can create false or misleading positives.
- 6. Human attacks. Digital deception is becoming increasingly more difficult to detect (Willingham 2022). This suggests that should automated processes fail, the human override can also fail.

5. Vulnerability roadmap

Cyber vulnerability exploitation rarely occurs as an individual event; rather vulnerabilities are exploited in concert as part of a cyber campaign. Campaigns can have kill chains that differ from the cyber kill chain. For this reason, Table 1 lists the ZTA components, their vulnerabilities, and the effects of exploitation of those vulnerabilities.

Table 1: ZTA vulnerability roadmap

ZTA Component and Function	Vulnerabilities	Ramifications
CDM – Detect and mitigate problems. Feeds PE.	Signature based leaves site open to 0-day attacks. Layer 1 & layer 2 attacks	Deny, destroy, deceive.
Industry Compliance – Best practices and standards adherence. Feeds PE	Compliance or best practices not-equal secure. Attackers design campaign around known practices and standards	Disrupt.
Threat Intelligence – Internal and external threat actor feeds including TTPs, indicators of compromise, malware, ransomware. Feeds PE	Signature based leaves site open to 0-day attacks Emergence of new stealth actors False flag operations	Deceive.
Activity Logs – system logs, messages, alarms, notification. Real Time (RT) security posture. Feeds PE.	False flag entries. Attackers remove evidence. Attackers overwhelm logs with entries	Deceive.

ZTA Component and Function	Vulnerabilities	Ramifications
PEP – RT executor of policy Policy Engine – permit or deny access based on CDM, IDAM. Policy Admin - session token creator	PE – Trick via inaccurate feeds. PE - Standard software attacks to breach. PA – gain knowledge of the token creation process to subvert the process.	Deny and deceive.
Data Access Policy – read, write, execute, and delete are granted as least privilege. Users are only included in groups that are needed. Roles are carefully considered.	Privilege escalation via 0-day operating system or application feeds, instead of lateral movement lies in wait. Data centricity results in many groups overlapping, malicious users exploiting transitive trust relationships.	Deny, destroy, and deceive.
PKI - encryption key management	Unknown vulnerabilities in implementation. Intruders gain access to all communications	Disrupt and deceive.
Identity Management Entity credentials, certificates, attributes, roles, etc.; integrates with PKI.	Stolen credentials Lock out key users (e.g., admin) In machine-to-machine communications information gathering (reconnaissance)	Disrupt and deceive.
Security Information & Event Management (SIEM) - Collects security information for analysis and warning.	Misses attack due to obfuscation Ignores attack	Deceive.
Security Orchestration Automated Response (SOAR) - Updates security posture based on SIEM outputs.	Confused response when conflicting data encountered. Poisoned data results in bad decisions Algorithms are manipulated with good data in deceptive weights.	Deny and deceive.

6. Conclusions

ZTA may or may not improve existing security architectures. Traditional physically separated sites may paradoxically increase their risk profile when transitioning to ZTA, while other sites may see an improvement. Architects should have a deep understanding of technology subversion with trade-offs for each instantiation. Hardware and hypervisor vulnerabilities can undermine the carefully crafted separation between data and control planes. Typically undetected for long periods, these attacks are usually missed by security products such as intrusion detection systems and SIEMs, which can only detect indicators of compromise. Similarly, the domains or sandboxes created by SDCs and networks suffer the same fate.

0day attacks undermine the integrity of the components CDM, compliance, activity logs, SIEM, and SOAR, all of which feed the PEP. New threat actors and different behaviours bring unanticipated attacker goals, and previously unseen tactics are missed by of threat intelligence feeds. Encryption may assure privacy and data integrity but once the systems are compromised the same encryption that makes private the communications may also cloak malicious activity.

Incorporating accurate, well-defined baselines into ZTA strengthens the security guidance through an understanding of known good attributes and behaviours. One example is the incorporation of Byzantine fault tolerance methods as mentioned above. Other possible improvements include extending the ZTA principals beyond the traditional software and networking realm into hardware and firmware, where those principles are not regularly addressed, thereby, offering a potentially lasting ZTA benefit.

References

Bell, D. E, and La Padula, L. J. (1976) "Secure computer system: Unified exposition and Multics interpretation." Technical Report ESD-TR-75-306, MITRE Corporation, Bedford, MA. https://csrc.nist.gov/csrc/media/publications/conference-paper/1998/10/08/proceedings-of-the-21st-nissc-1998/documents/early-cs-papers/bell76.pdf [Accessed 14th February 2022]

Bhunia, S., Hsiao, M.S., Banga, M. and Narasimhan, S., 2014. Hardware Trojan attacks: Threat analysis and countermeasures. Proceedings of the IEEE, 102(8), pp.1229-1247

Biden J, (2021) Improving the Nation's Cybersecurity. Executive Order 14028, The White House, https://www.federalregister.gov/documents/2021/05/17/2021-10460/improving-the-nations-cybersecurity [Accessed 14th February 2022]

Bing, C., Satter, R., and Menn, J. (2021) Widely used software with key vulnerability sends cyber defenders scrambling. Reuters. https://www.reuters.com/technology/widely-used-software-with-key-vulnerability-sends-cyber-defenders-scrambling-2021-12-13/ [Accessed 24th February 2022]

Butcher, Z. Zero Trust Architecture, https://www.tetrate.io/white-paper-zero-trust-architecture/ [Accessed 14th February 2022]

Campbell, M. (2020) "Beyond zero trust: trust is a vulnerability. *Computer*, 53(10), pp.110-113.the Department of Defense (DOD) Zero Trust Reference Architecture, V. 1.0 (2021) https://dodcio.defense.gov/Portals/0/Documents/Library/(U)ZT_RA_v1.1(U)_Mar21.pdf [Accessed 14th February 2022]

Dixon, M.G., (July 21, 2021). "A zero trust approach to architecting silicon". https://www.intel.com/content/www/us/en/newsroom/opinion/zero-trust-approach-architecting-silicon.html#gs.q7qp87 [Accessed 24th February 2022]

DMTF (2022) DMTF Releases Security Protocol and Data Model (SPDM) Architecture as Work in Progress. https://www.dmtf.org/content/dmtf-releases-security-protocol-and-data-model-spdm-architecture-work-progress [Accessed 24th February 2022]

Greenberg, A., (June 16, 2012). "Hackers exploit software bugs for 10 months on average before they are fixed", *Forbes*. https://www.forbes.com/sites/andygreenberg/2012/10/16/hackers-exploit-software-bugs-for-10-months-on-average-before-theyre-fixed/?sh=72a6d2daee1a [Accessed 22d February 2022]

Halpern, S., (January 25, 2021). "After the solarwinds hack, we have no idea what cyber dangers we face", The New Yorker. Website https://www.newyorker.com/news/daily-comment/after-the-solarwinds-hack-we-have-no-idea-what-cyber-dangers-we-face

Kindervag, J., Balaouras, S., and Colt, L. (2010) "No more chewy centers: Introducing the zero trust model of information security", Forrester Research, https://media.paloaltonetworks.com/documents/Forrester-No-More-Chewy-Centers.pdf [Accessed 14th February 2022]

Larkshamanan, R. (February 16, 2022). "VMWare issues security patches for high-severity flaws affecting multiple products", The Hacker News. Website: https://thehackernews.com/2022/02/vmware-issues-security-patches-for-high.html

Lefkovitz, N. and Boeckl, K. (2020) "NIST Privacy Framework: An Overview" NIST ITL Bulletin https://csrc.nist.gov/CSRC/media/Publications/Shared/documents/itl-bulletin/itlbul2020-06.pdf [Accessed 14th February 2022]

Lockheed Martin (2015), Gaining the Advantage: Applying Cyber Kill Chain® Methodology to Network Defense, https://www.lockheedmartin.com/content/dam/lockheed-martin/rms/documents/cyber/Gaining_the_Advantage_Cyber_Kill_Chain.pdf [Accessed 14th February 2022]

McGraw, G. (2006a) Three Pillars of Software Security. http://www.swsec.com/resources/pillars/ [Accessed 14th February 2022]

McGraw, G. (2006b) Software Security: Building Security In. Addison Wesley.

NSA – National Security Agency (2021), Embracing a Zero Trust Security Model, v. 1.0, https://media.defense.gov/2021/Feb/25/2002588479/-1/-1/0/CSI_EMBRACING_ZT_SECURITY_MODEL_UOO115131-21.PDF [Accessed 14th February 2022]

Rose, S., Borchert, O. , Mitchell, S. and Connelly, S. (2020) Zero Trust Architecture, Special Publication (NIST SP) 800-207, National Institute of Standards and Technology, Gaithersburg, MD, [online], https://doi.org/10.6028/NIST.SP.800-207 , [Accessed January 3d, 2022]

Rose, S. (2019) Zero Trust 101, https://csrc.nist.gov/CSRC/media/Presentations/zero-trust-architecture-101/images-media/Zero%20Trust%20Architecture%20101%20-%20Scott.pdf [Accessed 14 February 2022]

Ross, R., et al. (2018) Risk Management Framework for Information Systems and Organizations: A System Life Cycle Approach for Security and Privacy, NIST SP 800-27 Rev. 2. https://doi.org/10.6028/NIST.SP.800-37r2 [Accessed 14th February 2022]

Sample, C., Cowley, J., Watson, T., and Maple, C., (2016) "Re-thinking Threat Intelligence". In 2016 International Conference on Cyber Conflict (CyCon US) (pp. 1-9). IEEE.

Sherwood, J., Clark, A., and Lynas, D. (2005) Enterprise Security Architecture: A Business-Driven Approach, CMP Books, San Francisco

Simmonds, P. (2019). The Fallacy of the "Zero Trust Network. RSA Conference 2019. https://youtu.be/tFrbt9s4Fns [Accessed Feb 8, 2021]

Veronese, G.S., Correia, M., Bessani, A.N., Lung, L.C. and Verissimo, P., 2011. Efficient Byzantine fault-tolerance. IEEE Transactions on Computers, 62(1), pp.16-30.

Weichert, M. (2019) Update to the Trusted Internet Connections (TIC) Initiative, OMB Memorandum M-19-26, https://www.whitehouse.gov/wp-content/uploads/2019/09/M-19-26.pdf [Accessed 14th February 2022]

A Collaborative Design Method for Safety and Security Engineers

Taito Sasaki, Takashi Hamaguchi and Yoshihiro Hashimoto
Nagoya Institute of Technology, Japan
t.sasaki.176@nitech.jp
hamaguchi.takashi@nitech.ac.jp
hashimoto.yoshihiro@nitech.ac.jp

Abstract: The number of cyberattacks has been increasing not only on information systems but also on physical systems. Safety must be considered as an influence of cyberattacks. Vulnerabilities exploited in cyberattacks continue to occur day by day even if systems were developed securely. Security engineers must eliminate vulnerabilities even if the vulnerabilities occur after the developed systems are released. Vulnerabilities must be managed throughout system life cycle. But it takes time to apply its security patch. Safety engineers are required to ensure safety even when vulnerabilities exist. Therefore, collaboration between safety and security (S&S) engineers is necessary to manage corresponding S&S in operation process. S&S should be considered simultaneously in early stage of development process. Collaborative discussion is useful to mitigating risk of reworks. It is an example of reworks by inadequate S&S discussion that the braking system might be redesigned to promote the response in order to compensate for the delay caused by encryption. Therefore, this paper proposes common models effective for the collaboration throughout system life cycle. A management approach using the models is also proposed. Common model is represented by data flow diagram (DFD) because a module under cyberattacks can adversely affect other modules only through data flows. In the proposed method, the three improvements contribute to supporting management throughout system life cycle. Firstly, the models are applied to safety analysis and security analysis. Secondly, vulnerability occurrence is managed at the level of modules. System structures are designed based on modules. Module abnormalities caused by cyberattacks on the vulnerabilities are managed as causes of safety corruption. To indicate critical points for system to be considered, the points from a safety perspective must be identified. Processes and information are traced from the points in DFD. Finally, a module, which performs sets of functions, is outsourced. For each module, it must be considered who will manage vulnerabilities. The proposed method is illustrated using a development of a self-driving wheelchair as an example. In this paper, the collaborative design method for S&S engineers of products and their management based on modules are described to ensure safety even when unexpected vulnerabilities exist.

Keywords: cybersecurity, safety, collaborative design, system life cycle, vulnerability, management

1. Introduction

Cyber threats are regarded as significant issues for not only information systems but also physical systems. Cyberattacks can cause invalidation of safety functions, service outages, deterioration of product quality, or life-threatening incidents. Conventionally, in development phase of physical systems, safety is focused on, but not security. In information systems, the products are developed with care not to include vulnerabilities that can be exploited in cyberattacks.

However, in both developments, new vulnerabilities continue to occur day by day even if systems were developed securely. "The number of vulnerabilities registered in Common Vulnerabilities and Exposures (CVE) exceeds 18,000 a year and continues to increase" (Omo, 2021). Throughout system life cycle, it is necessary to consider that vulnerabilities will occur and continue to be managed. When a vulnerability is detected by security engineers, it should be eliminated. But it takes time to apply its security patch. Safety engineers are required to ensure safety even when vulnerabilities exist. An example of the vulnerability responses is as follows: When a vulnerability is detected in a module, it must be decided whether the operation is stopped or continued without the module. To make such decision, collaboration between safety and security (S&S) engineers is necessary.

If conventional development is applied, security is considered after development based on safety. Security requirements might lead to reworks. For example, encrypting communications causes a risk of delaying brake response. The braking system might be redesigned to improve the risk. To avoid reworks, S&S should be considered simultaneously. In addition, some measures are effective for both S&S. In terms of cost, it is also effective to consider S&S simultaneously.

"In the actual field, few engineers are familiar with both S&S" (IPA,2018). Therefore, collaboration between S&S engineers is required. However, engineers in S&S domains have different perspectives, cultures, and backgrounds. There are hurdles to achieve effective collaboration.

This paper proposes a management approach using common models. The communication using the models is expected to promote collaboration.

In Section 2, a systematic approach for system life cycle safety and security management is proposed. In Section 3, the proposed approach is illustrated using a development example of a self-driving wheelchair.

2. A life cycle management approach for safety and security

In system development, "V-model represents the phase of implementing while disassembling from the whole plan to detailed parts, the phase of proceeding with verification and validation while integrating partial elements into the overall structure" (JSA, 2020). Although V-model usually represents only system development process, some V-models to deal with system life cycle were proposed (Graessler, 2018). A new V-model with deep discussion of safety against cyberattacks is needed. To keep consistent management throughout life cycle, common models which could be utilized in whole process must be designed. In this paper, a method to design such models is proposed. Figure 1 shows the process flow over the lifecycle in order design, implementation, verification, operation, considering security and safety. In implementation process, modules are developed to satisfy specifications by applying secure coding. The first step in verification process is module validation. It verifies not only that the module meets specifications, but also that the supplier is capable of continuing to manage vulnerabilities that may occur in the module in the future. The following steps are verification using integration test and system test. In operation process, maintenance and upgrades are continued in order to ensure safety even when unexpected vulnerabilities exist. In the term of design process, all conditions to succeed the management of implementation, verification and operation must also be designed.

2.1 Overview of design process

System behaviour should be represented by inter-object communication. A module under cyberattacks can adversely affect other modules only through data flows. Hence, data flow diagram (DFD) is suitable as one of the representations of system. The same measure can be effective regardless of whether the failure is caused by a breakdown in safety or security. If such measures are taken, S&S can be considered simultaneously. Some measures can continue service even when malfunctions occur. Others can stop service safely. The discussion about such measures is the discussion about the structure of multiplexing and so on. Although the protection measures for safety or security are also discussed, these measures are not discussed in this paper. For example, in the case of safety, there is the selection of material, strength, thickness, and the margin of the motor and brake. For security, there is the introduction of anti-virus, firewalls, whitelists, and so on.

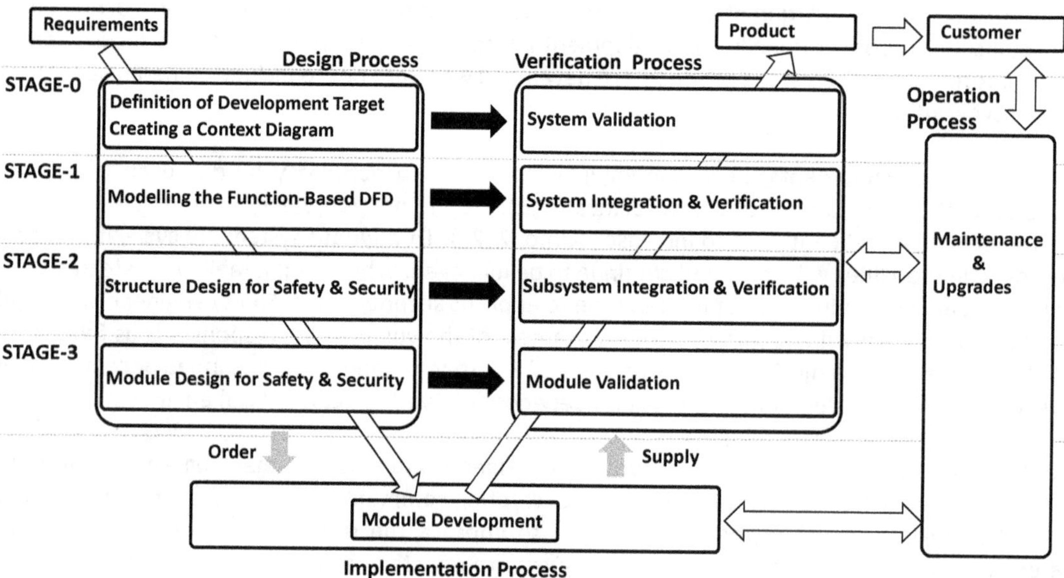

Figure 1: Life cycle process for safety and security

According to the flow shown in Figure 1, the design method of the model structure in the design process is described below.

2.1.1 STAGE-0

In STAGE-0, it is necessary to gather information related to system development. STAGE-0 defines the requirements for system development, such as the scope of the development system, environment, users, and service scenarios. It is important to identify and clarify the scope of development. If the requirements for system development are unclear, the development is delayed, or the quality of the deliverable doesn't meet the customer's expectations. As a result, the risk of the collision may become a reality. To avoid such a situation, it is necessary to identify and clarify the scope of development depending on the situation.

In addition, a context diagram based on definitions to visualize the situation is created. From this development phase, a common model is required for safety engineers and security engineers to be involved in the development and consider each risk. The context diagram is created as a common model at the context level. The context diagram describes the prerequisites of things assumed as a system. Because of a simple model, they can understand the status of development targets and possible external entities. This common model can contribute to lowering the hurdles for communication between development members. Engineers also table descriptions of the context to keep records and reduce cognitive discrepancies. Using this common model and descriptions, the safety engineers and security engineers can reach common ground from the initial phase and reduce the discrepancy in recognition.

The created model is a visual representation model that doesn't require a technical point of view or knowledge. This model does not represent the entire system. Therefore, it is desirable to add what you noticed from each viewpoint of development members to the model as necessary and to clear the entire system.

2.1.2 STAGE-1

In STAGE-1, engineers model the function-based DFD up to the granularity of information needed, to discuss the necessary level of S&S. In STAGE-0, they define the requirements for system development and table descriptions of the model. Based on such information, they organize the functions to be developed necessary to meet customer requirements. Then they evaluate the validity of the deployed functionalities to the ambiguous customers' demands and model the deployed functionalities as a DFD model.

2.1.3 STAGE-2

In STAGE-2, structure discussion such as multiplexing is conducted to realize the required levels of safety and security in each function discussed in STAGE-1. The first step is to consider countermeasures that can be taken even if a function fails. Such measures should be effective regardless of the cause, whether it is a safety failure or a security failure. For multiplexing, both the function and the communication between functions are subject to consideration. With regard to security, not only multiplexing but also diversity is necessary. In the next, safety analysis and security analysis on the designed structure are executed simultaneously.

2.1.4 STAGE-3

In STAGE-3, module construction to be outsourced is focused on. Modules, which perform sets of functions, are designed. Generally, the supplier should manage vulnerabilities in the outsourced module during implementation process. During operation process, unexpected vulnerabilities might occur in the module. If the module was a black box for system developer, that outsource module implementation to supplier, it is difficult for system developer to manage the vulnerabilities. The contracts of outsourcing specify whether the supplier continue to manage vulnerabilities after implementation process. If not continued, the supplier should provide not only outsourced module but also enough information including all design documents to manage unexpected vulnerabilities. If continued, the supplier should provide not only outsourced module but also proof of ability to continue the management. In addition, if system developer considers that the module is critical for safety, the system developer might require the supplier to provide the necessary information to ensure the safety of the module. During operation process, vulnerabilities within all modules, regardless of contract type, must be monitored.

When vulnerabilities are detected, the response to them should be considered in the design process. Regardless of the type of unexpected vulnerabilities, the consideration of the response is simplified by assuming that the module is unavailable while the vulnerabilities exist. Even when some modules are unavailable, operations might be able to continue with the other modules. Or it might be necessary to shut down the entire operation. Thus, the response is designed on the basis of modules.

3. Explanation of the proposed life cycle management approach using an example

Section 3 shows an example of the application of the proposed approach to the development of a self-driving wheelchair for shopping mall service.

3.1 STAGE-0 in design process

STAGE-0 defines the development target and creates a context diagram to identify the scope of the development system.

Figure 2: The context diagram of the situation between the self-driving wheelchair and external entities

Table 1: Description of specific elements in Figure2

Element	Description	
Shopping Mall Environment	Database(DB)	System History Log
		Data such as MAP and route
		User's personal data
		Facility usage status
	Facilities, Equipment	Shopping mall structure
		Shopping mall facilities
	Fluid Things	People (adults, children, the elderly)
		Obstacles (carts, trash cans, poles, signboards, dropped objects, etc. that are not on the map)
Users	Users are handicapped or elderly people who can come to shop but have difficulty walking for a long time. Those who already live in wheelchairs are likely to have their own wheelchairs. Therefore, these people can be excluded from consideration.	
Supervisory Control System	People and Systems that execute online maintenance or emergency intervention	

Some information about order side requirements, users, service scenarios, development targets, and shopping mall environments are described.

<Order Side Requirements>

Some ordering side's requirements are as follows.

1. The wheelchair can automatically move to a destination specified by the user.
2. The wheelchair can be used without installing lines on the floor.
3. The wheelchair can automatically return to the parking lot after service.
4. Safety and Security of the wheelchair and its users must be ensured.
5. Maintenance for Safety and Security must be ensured during operation process.

<Wheelchair User>

6. Users are handicapped or elderly people who can come to shop but have difficulty walking for a long time. Those who live in wheelchairs are likely to have their own wheelchairs. Therefore, these people can be excluded from consideration.

<Overview of Services Scenarios>

7. A user who gets on the self-driving wheelchair at the specified location in the shopping mall selects the desired destination in the shopping mall from the map on the tablet terminal. Then, the wheelchair transfers the user to the user's destination. After arriving at the destination, a clerk supports the user's shopping. The user can also manually move around to shop or eat. If the user wants the wheelchair to transfer automatically, the user selects the destination in the shopping mall from the map on the tablet terminal again when you leave the store. When the user finishes using the wheelchair and gets off, the wheelchair will return to its designated location.

<The Shopping Mall Environment>

8. The shopping mall environment can include databases, facilities, equipment, fluid things, and so on. In this example, the control of the self-driving wheelchair is assumed in a normal state, not in an abnormal situation such as a disaster. In addition, for example, the risk of falling on steep slopes can be excluded from consideration, in the shopping mall.

<Development Target>

9. The Product is a self-driving wheelchair. The electric wheelchair body and sensors required for self-driving are regarded as outsourced modules. Although the specifications of test items are designed by the system developer, the design and implementation of the modules are consigned to suppliers. The main target to develop is the self-driving control system.

<Out of Scope>

10. In STAGE-0, it is important to identify scope of the development. In some cases, it might be effective to mention out of scope. In this development, it is exempted from development that the self-driving wheelchair goes up and down the floor by elevator. Since the self-driving wheelchair needs to link with the elevator system for calling or designating a destination, it is given up to install an elevator boarding system.

Figure 2 shows the context diagram created to understand the situation between the self-driving wheelchair and external entities. In terms of safety, risks of collisions with shoppers, fall on stairs, and so on can be discussed. In terms of security, risks such as wheelchairs being tampered with, stolen, etc. can be discussed. In Table 1, the specific elements in Figure 2 are explained.

3.2 STAGE-1 in design process

In STAGE-1 the function-based DFD up to the granularity of information needed is generated. The context diagram is deployed until the main functions of the self-driving wheelchair are identified. By using a method such as quality function deployment (QFD) (Akao, 1997) as a method of developing functions from order side requirements, the functions have been deployed to the following sets of functions. Figure 3 shows the DFD model based on the functions.

11. Interface for specifying the target position
12. Capability to detect the position of the wheelchair itself
13. Capability to plan a movement route to the target position
14. Motion control to move the wheelchair along the specified route
15. Function to correct deviation from the route
16. Capabilities to detect unplanned events or the excessive tilt of the wheelchair and stop or decelerate
17. Function to make an emergency stop by people in the vicinity
18. Map information to determine proximity to walls, stairs, and so on
19. Capabilities to scan for obstacles or falling areas in the vicinity
20. Function to identify the user without receiving personal information

The function of identifying the Users without receiving personal information (j) is in the issue of information security. In this paper, this function is excluded from discussion because safety is focused on.

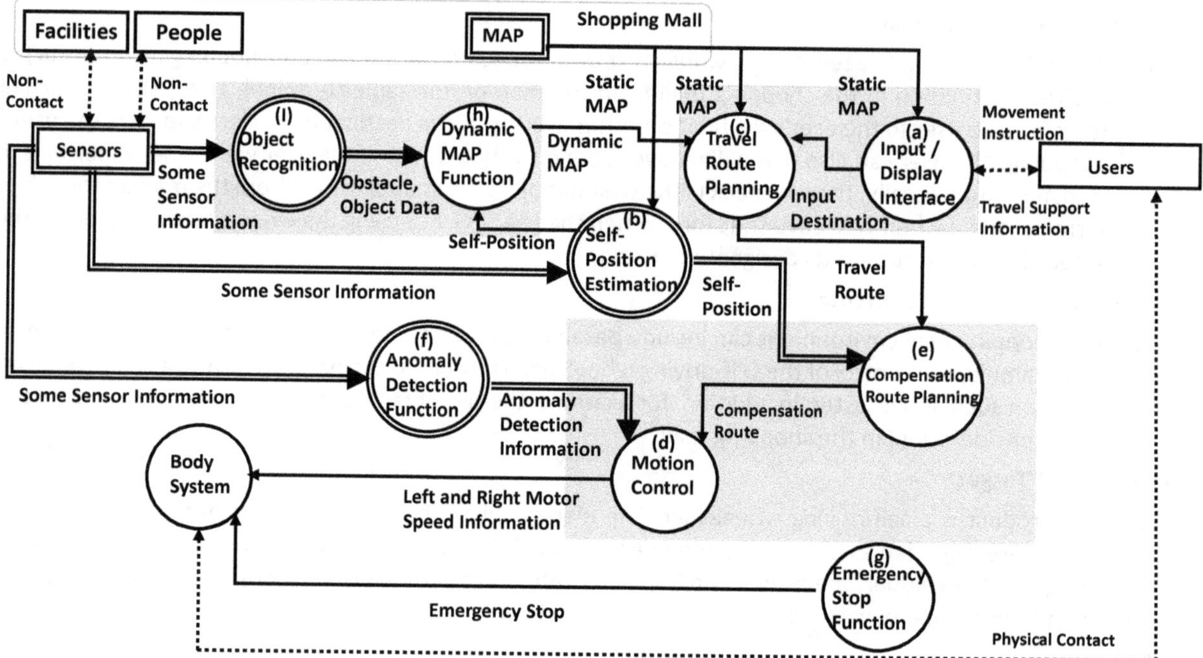

Figure 3: Function-Based DFD & structure with module

3.3 STAGE-2 in design process

In STAGE-2, it is determined how to structure the functions in the model from the viewpoints of S&S domains. This stage focuses on the functions that are important for ensuring safety. Here, the "risk that normal self-driving cannot be performed due to an abnormality in the self-driving control system" is taken as an example. In terms of safety, the causes of failure are erosion, corrosion, environmental stress, malfunction, maloperation, and so on. In terms of security, the causes of failure are malware targeting vulnerabilities, remote operation from the internet, denial of service (DoS) attacks, physical destruction, and so on. Since the system involves human lives, safety is the top priority. From Figure 3, S&S engineers can agree with the idea that the communication of "Left and Right Motor Speed Information" from "Motion Control" (d) to "Body System" of the wheelchair is the critical point that affects the safety of the whole system. A cover might be provided as an anti-breakage (Safety) and anti-disconnection (Security) measure for communication cables. From this example, it can be understood that the same measures can be effective whatever the type of causes of the failure.

Even self-driving systems should incorporate elements that humans can contribute to safety. "Emergency Stop Function" (g) in Figure 3 allows the system to stop and ensure safety in the event of a system malfunction.

From here, the procedure for simultaneously analyzing safety and security is explained according to Figure 3. The reason why the outputs of a function become abnormal is either the existence of abnormality in the inputs, a failure in the function, or a cyberattack. The analysis goes back from the critical point to the input side. Since the critical point is "Left and Right Motor Speed Information", its cause can be traced back to "Motion Control" (d). The causes of abnormal output are breakage or disconnection of the cable and falsification of communication, etc. The next function is "Motion Control" (d). The causes in it are computer module failure and program bugs, etc. in terms of safety, and deletion and falsification of data or program, etc. in terms of security. The next target of analysis is either "Anomaly Detection Information" or "Compensation Route". The self-driving wheelchair cannot move without routes. Therefore, "Compensation Route" is important for "Motion Control" (d). A measure is taken to prevent cyberattacks on this communication by making it a linkage within the program, rather than using communication lines. This measure is expressed with a grey square in Figure 3. There are other measures such as installing a firewall on the communication line or applying encryption to the communication. The next target is "Anomaly Detection Information". Anomaly detection is achieved by using multiple functions such as cameras and radar. In this way, even if one function is lost, anomaly detection can be achieved with other functions. This multiplexing measure is represented by double lines in Figure3. The above discussion will be deployed thereafter. Figure 3 shows the structure determined in the above discussion.

3.4 STAGE-3 in design process

In STAGE-3, each circle and grey square shading is assigned to a module. The suppliers of "Anomaly Detection Function" (f) modules did not accept to pass design documents or sources. And they contracted to continue vulnerability management in operation process. From the supplier of "Emergency Stop Function" (g) module, all design documents and sources were supplied. Because the module of "Motion Control" (d) and "Compensation Route Planning" (e) is important, the system developer implemented it by itself.

3.5 Implementation process

This process is executed by suppliers. Supplier should implement similar V-shaped development.

3.6 Verification process

In verification process, the test items determined in design process are checked. In STAGE-3, the test items for the vulnerability management in operation process are included.

3.7 Operation process

In operation process, all modules should be monitored whether unexpected vulnerabilities occur in them. When it is detected, the system developer informs the customer when the patch will be provided and suggest how to respond until applying the patch. The response based on modules was already designed on STAGE-3 in design process.

4. Conclusions

This paper's topic is S&S management throughout system life cycle. Especially, the response to vulnerability in operation process was focused on. This paper doesn't deal with secure development, such as secure coding, which aims for the product to be secure at the point of release. The purpose of the proposed security management is to ensure safety even when unexpected vulnerabilities exist in operation process. To achieve the purpose, from design process, it is necessary to collaborate between S&S engineers. This paper sought to manage continuously safety throughout system life cycle by presenting a management approach using common models based on DFD. A module under cyberattacks can adversely affect other modules only through data flows. Therefore, DFD is applied. In the proposed method, the three improvements contribute to supporting management throughout the system life cycle. Firstly, systems are represented by DFD. The models are applied to both safety analysis and security analysis. Communication using common models is expected to promote collaboration. Secondly, vulnerabilities are regarded as causes of safety corruption. Safety engineers must consider how abnormalities of the module which contains a vulnerability affect system safety. They must evaluate its risk and respond it. Security engineers must consider detection of vulnerabilities in modules, development of their security patches and their application. Both engineers collaborate to design system structure based on modules. To indicate the points to be considered, critical points from a safety perspective are identified. Processes and information are traced from the points in DFD. Finally, a module, which performs sets of functions, is outsourced. For each module, it must be considered who will manage vulnerabilities throughout system life cycle. If module developers don't want to continue vulnerability management after release, they are required to provide enough information to manage unexpected vulnerabilities in their module. In this case, the provided information is verified. Product developer should decide whether to manage the modules by itself or to outsource the management. If module developers continue vulnerability management, their management system is verified. In operation process, all modules are monitored whether new vulnerabilities occur in them. When vulnerabilities are detected, to ensure safety, the response will be implemented on a module-by-module basis.

The proposed method with collaboration between safety and security from design process is desired to contribute to the advancement of safety management against cyberattacks throughout system life cycle.

References

Akao, Y1997) "QFD: Past, Present, and Future", International Symposium on QFD '97, Linköping

Graessler, I., J. Hentze and T. Bruckmann (2018) "V-models for Interdisciplinary Systems Engineering", International Design Conference-Design 2018, Design Process, pp747-756

Information-technology Promotion Agency Japan (IPA) (2018) Control System Safety and Security Requirements Study Guide (Basic) https://www.ipa.go.jp/files/000064728.pdf

Japanese Standards Association (JSA) (2020) JIS X 0170:2020 System Life Cycle Processes

Omo, K (2021) Trends and Considerations in the Number of CVEs (2021 Edition) Part1 https://security.sios.com/security/cve-total-info-20211203.html

Siamese Neural Network and Machine Learning for DGA Classification

Lander Segurola-Gil, Telmo Egues, Francesco Zola and Raúl Orduna-Urrutia
Digital Security Department, Vicomtech Foundation, Basque Research and Technology Alliance (BRTA), Donostia/San Sebastian, Spain
lsegurola@vicomtech.org
tegues@vicomtech.org
fzola@vicomtech.org
rorduna@vicomtech.org

Abstract: Domain Generation Algorithms (DGA) are systems used to create immediate multiple and varying domain names. Such "artificial" domains can be then used for siting command and control servers which in turn oversee recruiting/infecting devices, and finally turning them into new resources to be exploited. In this sense, identifying DGA domain names can be crucial, to avoid cyberattacks like Phishing, Spam sending, Bitcoin mining, and many other. Usually, domain names generated by DGAs, are comprised by illegible character strings, but new "intelligent" DGAs tend to generate names using combination of words in dictionaries making its detection a challenging task. For this reason, in this work, we propose to address this problem using a combination of Machine Learning algorithms for improving the classification of DGAs domains. In particular, we propose to combine Siamese Neural Networks and traditional supervised Machine Learning algorithms in order to expand the input domain into separable n-dimensional data points and then achieve the domain classification. The proposed approach can be separated into 3 phases. In a first phase, domain names are encoded, by a one-hot encoder and a variation of this, named probabilistic one-hot encoder, which are implemented separately. Then, in the second phase, Long Short-Term Memory and Convolutional Siamese embedders are tested and compared. In particular, the first one is combined with the one-hot, while the Convolution algorithm is applied with the probabilistic one-hot encoded data. In the final step, five Machine Learning algorithms are tested using the two ways embedded data. Both embedder approaches reach very high results in terms of F1-score and Accuracy (about 91%) depending on the implemented classifier. The promising results obtained by the application of the proposed method shows that it is possible to perform DGA domain classification uniquely over the domain names, without considering external information such as DNS packets features.

Keywords: Siamese Neural Network, DGA classification, cybersecurity

1. Introduction

In the actual world, paradigms like 5G, IoT, Industry 4.0 and so on have increased the interconnectivity among millions of devices, facilitating many operations and increasing the productivity (Rao, 2018). However, being connected everywhere and at every time has triggered increasingly disruptive cyberattacks, which can generate huge impact in the daily life of many people (Bada, 2020). Hacker groups or individuals have the power to create massive botnets and zombie farms that wait until their leader gives the order to attack a website, take down a server farm or DDOS an opposing faction (Mirkovic, 2004). To create such zombie farms, hackers need to infect, recruit and take control of thousands and thousands of devices (Perrone, 2017). This goal can be reached by deploying several cyberattacks like Phishing and Spam sending (Aaron, 2010), which are based on fooling the trustworthiness of the user by using a seemingly legitimate-looking message from a trusted-looking sender (Baykara, 2018).

DGAs (Sood, 2016) comprise a family of algorithms with the aim of immediately creating vast, varying amount of domain names. Created domains often are used for locating command and control servers, targeting devices to be recruited/infected, turning them into new resources to be exploited. Usually, even if DGAs generate domain names comprised by illegible character strings, later trends base these algorithms in word dictionaries, creating human readable domain names. For this reason, in this work, we propose a novel method for DGAs classification, based on combining Siamese Neural Network (SNN) and Machine Learning (ML) algorithms.

The proposed approach is divided into three phases: in the first one, the input data (domain names) is encoded using the one-hot. In the second phase, a Siamese Neural Network is trained for converting the encoded data in embedded features, and finally, in the third phase, these embedded features are used for training and testing different machine learning classifiers.

To the best of our knowledge, this is the first work that addresses the DGA classification combining one-hot encoder and Siamese technologies, for firstly embedding and the classifying data. This study not only shows promising results that can be used for improving the state of the art of phishing detection, but also implement and evaluate different architectures to highlight which is more tied for this specific use case.

The rest of this paper is structured as follows: in Section 2, basic concepts for the compression of the work and related works are introduced. In Section 3, the followed methodology is presented, detailing every component in the process, starting from the encoding phase, passing through the embedding phase, to end with the classification phase. In Section 4, the dataset, experiments and the validation process are introduced. In Section 5, the results of the experiments are illustrated, and finally, in section 6, the conclusions and the future work guidelines are drawn.

2. Background

In this section, several concepts are introduced, which conform the basis of this work. In Section 2.1, a general description of one-hot encoding and Siamese Neural Networks structure are presented and in Section 2.2 the state of the art related to this research is shown.

2.1 Preliminaries

In this paper, two innovative technologies are exploited during the experiments, *one-hot encoder* used for extracting numerical information from the initial domain name dataset and *Siamese Neural Network (SNN)* used for converting such numerical information into embedded features.

One-hot Encoding: The one-hot encoder encodes data by generating a canonical basis for a set of terms, i.e., it is a function mapping every term in the set of terms, to a vector with dimension equal to the set size, where every value of the vector is 0 except for a value, which is equal to 1, and no image of the set of terms is equal. This can be extended for every sentence generated by the combination of these terms, by concatenating terms encodings, generating a matrix of dimensions length of the sentence times the size of the terms set.

When working on categorical data, there is a need of encoding it, due to how Machine Learning works. Several encoding methods can be found, but in the scenario posed in this work, an order preserving encoding might be differential. Due to the nature of the one hot encoding (concatenation of term encodings), the order of terms in each sentence is preserved in the encodings, making this choice especially suitable for this task.

Siamese Neural Networks: The Siamese Neural Networks (SNN) (Chicco, 2021) are specially structured neural networks, usually consisting of two parallel neural networks connected to an output layer. The last one calculates the distance between the outputs of each neural network, trying to minimize distance between points belonging to a same set and maximizing the distance between points belonging to different sets.

Siamese Neural Networks learn how to separate point sets in the training. Each of the networks learns to embed input belonging to a same family in different spatial places. This makes this kind of structure interesting, which can help to enhance classification or clustering tasks.

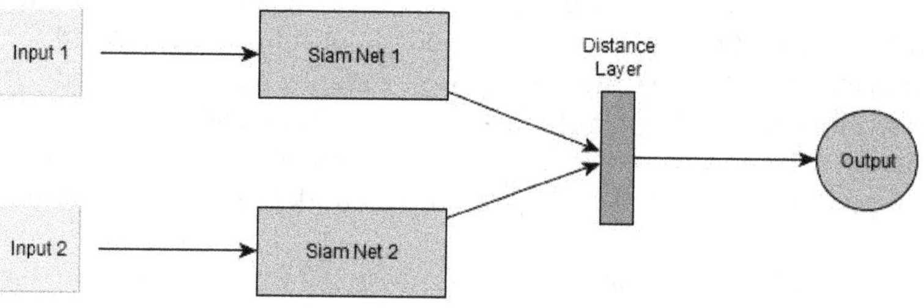

Figure 1: Siamese Neural Network

They belong to the set of supervised learners, receiving data as pairs of points, labelled with the aim of representing equality or inequality between both points. Binary labels are used usually for training these types of structures.

2.2 Related works

In (Ravi, 2021), botnets using DGA domains and DNS homographs are detected using deep learning. They use this technology to detect the DGA and verify the model's viability by testing them against common adversarial attacks. (Vinayakumar, 2019) shows an approach to DGA detection also using deep learning. However, they create a whole detection chain by not only detecting the domains through semantic similarity, but also embedding the domains in case the semantic similarity did not trigger the alarms. On the same line, (Hwang, 2020) presents a method to detect and classify DGA by extracting features and passing them to a CNN-based model that labels the domains as DGA or legit. More deep learning based DGA detection research works can be found, such as (Tuan, 2022) where LSTM based techniques are used, or (Aravamudu, 2022) where various ML classifiers are tested against this task. Techniques involving more conventional Machine Learning techniques can be found too. For example, in (Anand, 2020), lexical and statistical features are extracted for a total of 44 features from domain names, and then conventional algorithms such as the RF are applied. In (Chin, 2018), it is proceed in a similar way. The authors extract some lexical features from the domain names, together with some DNS features, for DGA ML based classification and clustering.

The technology underlying these detectors are commonly related to deep learning (Yu, 2018). In particular, Siamese Networks perform exceptionally well in word embedding projects. (Neculoiu, 2016) is a clear example of this, where a bidirectional LSTM is used to embed the words into a fixed space. (Amin, 2019) uses a Manhattan LSTM Siamese model to extract features and information from domains while in (Benajiba, 2019), the authors tackle the semantic pattern similarity problem by implementing a Siamese model to detect SQL pattern alikeness. (Zhu, 2018) goes even beyond by making a comparison between a D-LSTM model that extracts primary sentence information and generates basic sentence components through a standard LSTM to determine which works better in compositional knowledge. In (Pontes, 2018), a combination of Siamese CNN and LSTM are used in order to detect textual similarity; while CNN notices the local context of words, the LSTM Siamese takes into account the global context of the sentence. (Vinayakumar R. A., 2021) implemented a two-level deep-learning framework involving a Siamese network to detect botnets in Internet of Things networks. Close to these approaches, in (Ling, 2019), the authors implemented a multi-stage banking trojan botnet detection system using deep learning to identify DGA domains and (Kwon, 2016) presents a botnet detection implementation that can be used in large-scale DNS traffic environments.

3. Methodology

In this section, a novel methodology is presented, which encodes and Siamesely learns to embed domain names, with the aim of classifying these into DGA based name or into a legit one. Specifically, in Section 3.1 the motivation of carrying this work and the contributions made are presented. In Section 3.2, the encoding phase is explained, followed by the Section 3.3, which deepens the Siamese training phase and finishes in the Section 3.4, which details the ML algorithms used for the classification.

3.1 Motivation and contributions

When trying to detect DGA created domain names, several techniques have been studied. Common old techniques include blacklisting or Regular Expression (regex) matching. Latest techniques include ML based techniques for filtering anomalous domain names. For example, in (Lin, 2019), they combine both techniques to enhace detection. However, DGA techniques keep evolving, creating more sophisticated DGAs, such as the dictionary-based ones. For this reason, in this work, we propose a classification enhancing approach, based on Siamese embeddings.

3.2 Encoding phase

In a first phase, domain names need to be encoded in order to get numerical features, so the ML algorithms can potentially learn the patterns from data. For this purpose, two approaches have been taken, one-hot encoder and a probabilistic one-hot encoder.

One-hot encoder: As mentioned in Section 2.1, the one hot encoder has been the one chosen for vectorizing natural language data for the training of the LSTM based Siamese. This will encode data by considering lowercase letters, numbers and hyphens as the set of terms to be encoded, for a dictionary of 39 terms, precisely comprising the characters permitted for the domain names.

Probabilistic one-hot encoder: Let D be a set of documents d, and t terms in d. Then the amount of documents d in D containing the term t is defined as the document frequency:

$$\mathrm{df}(t, D) = \frac{|\{d \in D : t \in d\}|}{N}$$

Equation 1: probability of a term appearing in an arbitrarily chosen document

where N = |D|. I.e. the probability of a term appearing in a given random document d in D. For this encoding, instead of encoding each of the terms of the dictionary as the canonical basis of 39 dimensioned Euclidean space, the 1 appearing with a normal one-hot is substituted by df(t,D). This encoding is the one selected for the Convolutional Siamese training.

In both cases matrixes will have 39 columns corresponding to the vocabulary size. Rows amount will vary, matching with domain names length. Then, truncating and padding operations will be performed over the encoded data.

3.3 Siamese training phase

As introduced in Section 2.1, the common SNN are trained by comparing the results of two dense NN. However, in this work, due to the complexity of the input data, as well as the dependency that each letter has with its previous and next character, we explore two different Siamese architectures: one based on Long-Short Term Memory (LSTM) and another based on Convolutional Neural Network (CNN). More specifically, the first one is used together with the one-hot encoder, while the second is used with the probabilistic one-hot encoder. The normal one-hot fits well with the LSTM Siamese due to its structure. However, the Convolutional Siamese can potentially increase its performance when a value is given to each letter, so the max pooling layer does not trivialize data. This is the reason why this election is made for this work.

LSTM based Siamese model: For the LSTM based model, two parallel LSTM models are implemented composed of two LSTM layers where the first one comprises a number of neurons equal to the vocabulary size (39) and a second layer of a hundred neurons. The scalar product between the output of both models is computed in a connecting layer, which is connected to a final neuron with a sigmoid activation, for the minimization of the loss mean absolute error.

Convolutional Siamese Model: For the convolutional model, two parallel convolutional models are developed, both having two blocks of a convolutional layer (with 32 filters for the first layer and 64 for the second, two sized in both cases) and a max pooling layer, connecting both to a flatten layer and finishing in a hundred neuroned layer. The output of both layers is connected to a layer composed of a hundred neurons, which connects with an output neuron with a sigmoid activation function, and a mean squared logarithmic error loss, which replaces the usual distance layer.

Both models will have as input the matrixes generated by the encoding phase. When embedders are extracted, those will embed those matrixes into a 100 dimensional Euclidean space, with the aim of providing this data to the machine learning classifiers.

3.4 Machine learning classifiers

For the classification task, 5 common machine learning classifiers are trained with the embedded data. In this way, it is possible not only to validate the information generated by the two SNNs, but also to evaluate it through different classifiers and choose which is the best in this specific use case. The selected learners comprise Random Forest (RF), Multilayer Perceptron (MLP), Decision Tree (DT), Support Vector Classifier (SVC) and Linear

Regression (LR). More concretely, each model is configured with particular parameters: RF has 50 trees with a max depth of 25 nodes, the MLP is composed by a hidden layer of 100 neuron, with a sigmoid activation function, a batch size of 250, is trained within 100 epochs, and the RMSProp optimizer and mean squared error loss, the DT has max depth of 25 nodes, the SVC is chosen with a radial basis function is used as the kernel, and finally the LR, where the intercept is chosen to be fitted.

4. Experimental study

In this Section, the dataset and the pre-processing operations are introduced (Section 4.1), as well as the used evaluation metrics (Section 4.2). Finally, in the Section 4.3, the two experiments are drawn.

4.1 Data overview and pre-processing

The dataset used in this research was gathered by investigators of the Polytechnic University of Marche in order to develop a DGA detector and DGA-family classifier based on n-gram features (Cucchiarelli, 2021). The dataset is a collection of DGA-based and benign domains, where the DGA-based domains are divided into 25 different DGA families and the benign domains have been provided by Alexa's top site. Both mischievous and legit domains are equally distributed and have been shuffled so none of the DGA families, nor the valid domains are grouped together in the list. Malicious domains were extracted from the 360 Netlab Opendata Project repository, a repository maintained by the 360 Netlab Research Lab (https://netlab.360.com/). The whole dataset consists of 674,898 domain names, where 337,398 are domains labelled as benign domains by Alexa, and the remaining 337,500 comprise the malicious domains set, uniformly distributed for each DGA family, with 13,500 samples.

For the experiments, the initial dataset is split into a Siamese training dataset, a ML train dataset and a test dataset with a proportion of 10%, 30% and 60%, respectively. More specifically, the Siamese train dataset is used for training the SNN, then, the ML train dataset is embedded by the one of the Siamese models, and the output is used for training the classifiers. This process is repeated on the ML test set, which will be used to evaluate the whole chain (SNN and ML classifier). This configuration allows us to obtain validation results over a complete unseen dataset increasing the generalization of the results and decreasing the probability of overfitting results. Furthermore, a 5-fold cross validation process is performed, which means that each evaluation is repeated 5 times changing the used dataset composition but leaving unchanged their overall representation (keeping the indicated proportions). Finally, it is important to highlight that, for the SNN training, pairs of samples are generated from the Siamese training set considering all possible combinations. However, due to the huge amount of possible combinations (the square of the size), just a 0.005% of all combinations are used, randomly selecting samples.

4.2 Evaluation metrics

For the evaluation of the performance of the models, 4 metrics are used, comprising the Accuracy, Precision, Recall and F1-score, which are derived from the confusion matrix (Figure 2) related to binary classification.

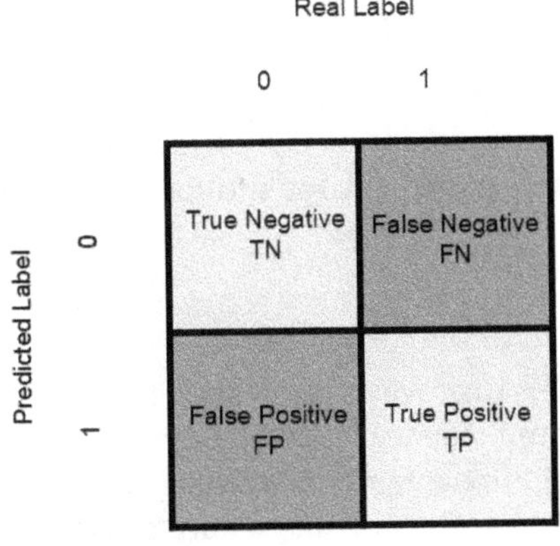

Figure 2: Confusion matrix

The values observed in Figure 2 comprise

- **True Negative (tn)**: The predicted and the real value equals to 0 (negative class)
- **False Negative (fn)**: The predicted value equals to 0 but the real value is 1 (positive class)
- **False Positive (fp)**: The predicted value equals to 1 but the real value equals to 0.
- **True Positive (tp)**: Both, the predicted and real values equal to 1.

These four metrics are computed from these values, over the ML models and both Siamese models.

Accuracy: It measures the hit rate. It is computed by summing *tp* and *tn* divided by the total amount of samples (the sum of all four values), as it can be observed in Equation 2.

$$\frac{tp + tn}{tp + fp + tn + fn}.$$

Equation 2: Accuracy

Precision: It measures the positive hit rate in relation of true positives. It is computed by dividing the amount of *tp* by the total amount of positives (the sum of *tp* and *fn*) as in Equation 3.

$$\frac{tp}{tp + fn}.$$

Equation 3: Precision

Recall: It measures the positive hit rate in relation of predicted positives. It is computed by dividing the amount of *tp* by the total amount of predicted positives (the sum of *tp* and *fp*) as it can be seen in Equation 4.

$$\frac{tp}{tp + fp}.$$

Equation 4: Recall

F1-score: The harmonic mean of the precision and recall. It is calculated by the formula shown in Equation 5.

$$2 \cdot \frac{precision \cdot recall}{precision + recall}.$$

Equation 5: F1-score

Those are the four metrics used for the measurement of the performance of the models. In particular, the measure for the ML classifiers is performed over the test set and for the Siamese model, the validation set is used for the computation of these metrics.

4.3 Experiments

First experiment: The aim of this experiment is to prove that a convolutional embedder can potentially work for DGA classification. For this, domain names are firstly vectorized by the probabilistic one hot encoding. For this, the df(t,D) (Equation 1) is calculated over all the domains comprising the training set of the Siamese. In other words, every domain forming the pairs for the Siamese training set are considered for computing the probability of a term of the previously defined characters appearing in a domain is computed. Then probability one-hot is applied to these, which will generate the matrixes for feeding the convolutional Siamese model.

Once the model is trained, one of both Siamese models is arbitrarily extracted to embed the remaining data. Once the embedder is extracted, the remaining data (the training and testing sets for the Machine Learning Classifiers) is one-hot encoded and Convolutionally embedded.

Finally, the embedded training set is provided to the selected five ML classifiers and the performance of these is measured over the test set, considering the four metrics presented in Section 4.2.

Second experiment: This experiment is like the previous one. This time, domain names are encoded using the one hot encoder for the generation of the matrixes. Once again, even in this case they are binary matrixes, those are used to feed the LSTM based Siamese models.

AS in the first experiment, when the Siamese Network has finished its training, an LSTM based embedding is obtained by the subtraction of one of the Siamese models. Again, the splits of the remaining data, the training and testing datasets for the classifiers, is one-hot encoded and LSTM based embedded.

Finally, and keeping the same structure, the embedded training set is used to feed the five ML classifiers, and the four metrics introduced in Section 4.2 are used for evaluating the classifiers over the embedded test set.

5. Results

In this section, the results of the evaluation of the proposed method are presented, in terms of the presented metrics. In the Section 5.1, the results for each experiment will be divided into two tables, showing the values for the respective Siamese model in a table, and the scores of the classifiers in another. The values are computed by calculating a mean and standard deviation among the 5 folds. In Section 5.2 a discussion of the obtained results is written.

5.1 Experiments' Results

- **First experiment**

For the Convolutional Siamese model, the results gotten are the ones observed in Table 1.

Table 1: Convolutional Siamese results

	Accuracy	Precision	Recall	F1-score
Convolutional Siamese	0.946±0.004	0.95±0.004	0.941±0.006	0.945±0.004

As it can be noticed, the Convolutional Siamese discerns with a high rate (95% of accuracy) the DGA domains from the legit domains. In the second phase of the proposed method, the Table 2 shows the performance indicators for all five machine learning classifiers.

Table 2: Performance of the classifiers (after Convolutional embedding)

	Accuracy	Precision	Recall	F1-score
RF	0.893±0.002	0.918±0	0.863±0.003	0.889±0.002
MLP	0.897±0.002	0.918±0.006	0.873±0.008	0.895±0.002
DT	0.839±0.002	0.835±0.002	0.843±0.003	0.83±0.002
SVC	0.888±0.002	0.919±0.002	0.851±0.003	0.883±0.002
LR	0.88±0.001	0.909±0.001	0.854±0.003	0.884±0.001

Results show that the one getting the best overall results has been the Multilayer perceptron, closely followed by the Random Forest, where the first one has an 89.7% of accuracy, and the second a 89.3%, both have 91.8 of precision, 87.3% and 86.3% of recall respectively, and 89.5% in the F1-score for the MLP and 88.9% for the RF. The Supported Vector Classifier and the Linear Regression achieve similar results compared to the best two classifiers, whereas the Decision Three falls behind, getting the scores diminished by a 5%-7%.

- **Second experiment**

For the LSTM based Siamese model, the results can be observed in Table 3: LSTM based Siamese results:

Table 3: LSTM based Siamese results

	Accuracy	Precision	Recall	F1-score
LSTM Siamese	0.822±0.176	0.823±0.177	0.722±0.367	0.855±0.117

The results gotten by the classifiers when LSTM based embedded data is provided, can be seen in Table 4 .

Table 4: Performance of the classifiers (after LSTM embedding)

	Accuracy	Precision	Recall	F1-score
RF	0.915±0.039	0.929±0.037	0.90±0.043	0.912±0.04
MLP	0.885±0.092	0.894±0.097	0.875±0.088	0.884±0.092
DT	0.873±0.061	0.874±0.061	0.871±0.061	0.873±0.061
SVC	0.868±0.116	0.879±0.116	0.85±0.125	0.864±0.121
LR	0.902±0.055	0.917±0.044	0.881±0.072	0.899±0.059

The results are less stable than in the first case, getting high standard deviations specially for the Siamese part (±0.367 in Recall and ±0.177 in the rest). The best classification model in this experiment has been the RF (even if the MLP achieved greater results when the Siamese model learnt well). Scores about 90%-92% can be observed in the four metrics for the RF

5.2 Discussion

As it can be perceived in the first experiment, the Convolutional Siamese discerns quite good the legit domains and the DGA based ones (95%) in the validation set. When classifying with the ML models, the results are still of about a 90% in the best classifiers, even if performance has been diminished. This could be since the distance layer in the Convolutional Siamese is computed by another dense layer. This fact may convert both Convolutional models "lazier" in a sense that this layer may not only fit the distance function, but help in the classification of the pairs, making both models suboptimal. This could potentially explain, even if it is low, the reduction of performance of the classifiers when compared to the Convolutional Siamese.

For the second experiment, higher values can be perceived in the standard deviation and lower in the mean. In fact, the highest values for the LSTM based Siamese model were higher than for the convolutional (about a 97% in the four metrics) and a higher enhancement was observed in the classifiers (about a 95% in all the four metrics for the Multilayer Perceptron). However, the learning phase was not steady in the Siamese part. It learnt well in three of the five folds but behaved strangely in the remaining two. Optimal parameter tunning and structural modifications could be potential solutions for this issue.

6. Conclusion and future work

The idea of this work has been to observe the behaviour of some of the most common and state of the art classifiers when applying Siamese embedders to data for DGA domain classification. The results show that the method should be strongly considered. In fact, and in particular, the good results gotten by the Linear Regression (even if in the first case are a bit worse) classification with both Siamese embedders, indicate good separability of both (DGA and legit) sets. This leads to interesting future work. First, an optimization of the Convolutional Siamese networks could be achieved, by building a better structured network. Optimal parameter tunning could be found for both cases too, in particular for the LSTM case, in which data folds changing produced irregular results. Then, a multiclassification task could be interesting. Differentiating between DGA families may not be such a critical matter, but the separation could be done into legit, random DGAs and dictionary based DGAs. These three subfamilies may be embedded into separable data sets by Siamese embedders.

Acknowledgements

This work has been partially supported by the Basque Country Government under the ELKARTEK program, project TRUSTIND (KK-2020/00054).

References

Aaron, G. (2010). The state of phishing. *Computer Fraud & Security*, 5--8.

Amin, K. L. (2019). Advanced similarity measures using word embeddings and siamese networks in CBR. *Proceedings of SAI Intelligent Systems Conference* (pp. 449-462). Springer.

Anand, P. M. (2020). An ensemble approach for algorithmically generated domain name detection using statistical and lexical analysis. *Procedia Computer Science*, 1129--1136.

Aravamudu, P. a. (2022). Exploring and Comparing Various Machine Deep Learning Technique algorithms to Detect Domain Generation Algorithms of Malicious Variants. *Computer Science and Information Technologies*.

Bada, M. a. (2020). The social and psychological impact of cyberattacks. *Emerging cyber threats and cognitive vulnerabilitie*, 73--92.

Baykara, M. a. (2018). Detection of phishing attacks. *2018 6th International Symposium on Digital Forensic and Security (ISDFS)* (pp. 1--5). IEEE.

Benajiba, Y. S. (2019). Siamese networks for semantic pattern similarity. *2019 IEEE 13th International Conference on Semantic Computing (ICSC)* (pp. 191-194). IEEE.

Chicco, D. (2021). Siamese neural networks: An overview. *Artificial Neural Networks*, 73--94.

Chin, T. a. (2018). A machine learning framework for studying domain generation algorithm (DGA)-based malware. *International Conference on Security and Privacy in Communication Systems*, {433--448.

Cucchiarelli, A. a. (2021). Algorithmically generated malicious domain names detection based on n-grams features. *Expert Systems with Applications*, 114551.

Hwang, C. K. (2020). Effective DGA-Domain Detection and Classification with TextCNN and Additional Features. *Electronics*, 1070.

Kwon, J. L. (2016). A scalable botnet detection method for large-scale DNS traffic. *Computer Networks*, 48-73.

Lin, H. a. (2019). Detection of application-layer tunnels with rules and machine learning. *International Conference on Security, Privacy and Anonymity in Computation, Communication and Storage* (pp. 441--455). Springer.

Ling, L. G. (2019). *An AI-based, Multi-stage detection system of banking botnets.* arXiv:preprint arXiv:1:1907.08276.

Mirkovic, J. a. (2004). A taxonomy of DDoS attack and DDoS defense mechanisms. *ACM SIGCOMM Computer Communication Review*, 39--53.

Neculoiu, P. V. (2016, August). Learning text similarity with siamese recurrent networks. *Proceedings of the 1st Workshop on Representation Learning for NLP.*, (pp. 148-157).

Perrone, G. a. (2017). The Day After Mirai: A Survey on MQTT Security Solutions After the Largest Cyber-attack Carried Out through an Army of IoT Devices. *IoTBDS*, 246--253.

Pontes, E. H.-M. (2018). *Predicting the semantic textual similarity with siamese CNN and LSTM.* arXiv:preprint arXiv:1810.10641.

Rao, S. K. (2018). Impact of 5G technologies on industry 4.0. *Wireless personal communications*, 145--159.

Ravi, V. A. (2021). Adversarial Defense: DGA-Based Botnets and DNS Homographs Detection Through Integrated Deep Learning. *IEEE Transactions on Engineering Management.*

Sood, A. K. (2016). A taxonomy of domain-generation algorithms. *IEEE Security & Privacy*, 46--53.

Tuan, T. A. (2022). On Detecting and Classifying DGA Botnets and their Families. *Computers & Security*, 102549.

Vinayakumar, R. A. (2020). A visualized botnet detection system based deep learning for the Internet of Things networks of smart cities. *IEEE Transactions on Industry Applications*, 4436-4456.

Vinayakumar, R. S. (2019). Improved DGA domain names detection and categorization using deep learning architectures with classical machine learning algorithms. *Cybersecurity and Secure Information Systems* , 161-192.

Yu, B. a. (2018). Character level based detection of DGA domain names. *2018 International Joint Conference on Neural Networks (IJCNN)* (pp. 1--8). IEEE.

Zhu, W. Y. (2018). Dependency-based Siamese long short-term memory network for learning sentence representations. *PloS one*, e0193919.

Probability of Data Leakage and its Impacts on Confidentiality

Paul Simon and Scott Graham
Air Force Institute of Technology, Wright-Patterson AFB, Ohio, USA
paul.simon.ctr@afit.edu
scott.graham@afit.edu

Abstract: A multi-channel communication architecture featuring distributed fragments of data is presented as a method for improving security available in a communication architecture. However, measuring security remains challenging. The Quality of Secure Service (QoSS) model defines a manner by which the probability of data leakage and the probability of data corruption may be used to estimate security properties for a given communication network. These two probabilities reflect two of the three aspects of the IT security triad, specifically confidentiality and integrity. The probability of data leakage is directly related to the probability of confidentiality and may be estimated based on the probabilities of data interception, decryption, and decoding. The number of listeners who have access to the communication channels influences these probabilities, and unique to the QoSS model, the ability to fragment and distribute data messages across multiple channels between sender and receiver. To simulate the behaviors of various communication architectures and the possibility of malicious interference, the probability of data leakage and its constituent metrics require a thorough analysis. Even if a listener is aware that multiple channels exist, each intermediate node (if any) simply appears to have one input and one output. There may be one or more listeners, and they may or may not be working cooperatively. Even if the listener(s) gains access to more than one channel, there is still the challenge of decrypting, decoding, or reassembling the fragmented data. The analysis presented herein will explore the probability of confidentiality from both the authorized user's and the adversary's perspective.

Keywords: confidentiality, communications modeling, probability, security, metrics, data leakage

1. Introduction

Accurate and repeatable metrics describing confidentiality and integrity of communications through contested environments would be useful. The Quality of Secure Service (QoSS) model, defined by Simon et al. (2021) and Simon et al. (2022), provides a quantitative method to describe security of communications available to authorized users. Security, in this context, refers to confidentiality, integrity, and availability of the path between transmitter and receiver. While availability metrics are common, confidentiality and integrity are difficult to quantify, therefore surrogate metrics of probability of data leakage and probability of data corruption are used. The QoSS model further considers existence of adversarial listeners and malicious disruptors. While it may be relatively easy to recognize when a disruptor is injecting malicious data, it is challenging to detect adversarial actors eavesdropping on communications. It is also difficult to know an adversary's capability or intention. Tools exist to limit what eavesdroppers are able to receive. However, the QoSS model adds the ability to quantify how much data the eavesdropper may be able to receive. Due to the difficulty in quantifying security, probabilistic models are useful in conjunction with simulations. Exercising simulations many times provides insight into unexpected emergent behaviors.

The primary contribution of this work is to explore the impact data fragmentation has on the probability of data leakage from both authorized user's perspective and adversarial listener's perspective. Since the probability of leakage is a surrogate for the probability of confidentiality, this simulation analysis provides insight into available communication system security, despite adversarial interaction. Section 2 of this paper presents an overview of the QoSS model. Section 3 relates simulation results of multiple communication architectures based on the QoSS model. Section 4 provides analysis on how those simulation results may be perceived by an authorized user or an adversary. Section 5 highlights future research and provides a conclusion.

2. Background

The QoSS model details a manner by which the probability of data leakage and the probability of data corruption may be calculated for a given communication network. These two probabilities reflect two of the three aspects of the IT security triad, specifically confidentiality and integrity. Other authors, such as Almerha et al. (2010), attempt to frame the IT security triad based on arbitrary routing metrics or, like Hughes et al. (2013), by developing security requirements for confidentiality and integrity. Leon et al. (2010) attempt to develop an all-encompassing organizational matrix to measure security, whereas Wang et al. (2008) use the Common Vulnerability Scoring System (CVSS) to quantify vulnerabilities as a surrogate. Unlike most traditional models, availability in the QoSS model reflects strictly physical capabilities of an architecture. This, in turn, leaves all malicious interactions with the architecture, notably eavesdropping, jamming, or spoofing, considered under confidentiality and integrity.

According to the QoSS model, the probabilities of leakage and corruption are based on six characteristics of a communication architecture, possible interactions with malicious listeners and disruptors, and the number of communication channels available. The characteristics are probability of interception, probability of decryption, probability of decoding, probability of injection, probability of suppression, and probability of noise. The probability of leakage, specifically, is estimated based on the probabilities that a transmitted message may be intercepted, decrypted, and decoded. A receiver, authorized or not, is not guaranteed to successfully receive, decrypt, or decode messages. This approach is inspired by the analysis developed by Sweet et al. (2018).

Some communication systems allow for multiple parallel heterogeneous channels to exist between transmitter and receiver. One example is Signaling System 7 (SS7), detailed by Modarressi et al. (1990) and Russell (2002), used in analog telephone networks. Another example developed by Khisti et al. (2012) uses two independent parallel channels to limit the amount of information leakage if either are intercepted. Redundant Array of Inexpensive Disks (RAID) systems, detailed by Hennessy et al. (2002), allow the ability to fragment and distribute data across multiple disks for security purposes. Even TCP/IP networks allow for fragmentation of data into packets that comply with the Maximum Transmission Unit (MTU) of the network, although those strategies, highlighted by Creedon et al. (2009), are strictly to optimize the performance of a single TCP/IP connection. Modern networking also employs the use of intermediate nodes to provide flexible relaying and routing across broad networks. In a complex architecture that features multiple channels and multiple intermediate hops similar to the one shown in Figure 1, even if a listener is aware that multiple channels exist between endpoints **A** and **D**, each intermediate node (if any) simply appears to have one input and one output. There may be one or more listeners, and they may or may not be working cooperatively. Further, even if the listener(s) gains access to more than one channel, there is still the challenge of decrypting and decoding or reassembling the fragmented data. As such, the overall calculation for the probability of leakage, *P(l)*, is defined as

$$P(l) = \frac{P(int) \cdot P(dcr) \cdot P(dco) \cdot \sigma}{n}$$

where *P(int)* is the probability of interception, *P(dcr)* is the probability of decryption, *P(dco)* is the probability of decoding, σ is the number of adversarial listeners who are able to access the channel(s), and *n* is the number of channels between transmitter and receiver. These metrics comprise the probability of leakage, and

$$P(C) = 1 - P(l)$$

where *P(C)* is the probability of confidentiality. We assume that the maximum number of listeners, whether they are cooperating or not, cannot exceed the number of channels.

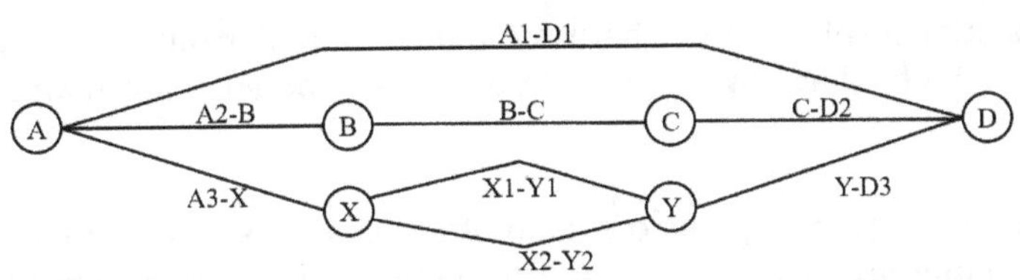

Figure 1: Hybrid communication network featuring three heterogeneous channels between **A** and **D**

This research focuses on security of communications as defined by the QoSS model. We are developing a representative simulation environment of communication architectures with varying complexity and featuring multiple heterogeneous channels and intermediary nodes. This architecture is in contrast to separately routing critical information over secure channels or maintaining multiple other levels of security, as proposed by Winjum et al. (2008), which is similar to the multi-level security model presented by Jin et al. (2012). Typically, communication systems are simulated using Markov chains because the mathematical models are tractable. However, due to the difficulty in defining security, understanding security within an architecture similar to the one shown in Figure 1 for both authorized and unauthorized users becomes intractable. The goal of the simulation environment is to quantify confidentiality with an eye toward developing more secure communication architectures through the creation of a comprehensive protocol that splits data across multiple channels with varying amounts of cross-channel duplication, checksum, or CRC overhead. Messages must also be received correctly and discourages eavesdropping despite the possible malicious injections and recognizing which channel is being targeted. This simulation represents one of many possible approaches to implementing and quantifying the security available within a communication network.

By simulating various architectures ranging from single point-to-point connections to multi-channel, multi-hop architectures, and by varying the data transmitted across those channels, it becomes clear that the probability of leakage and its constituent metrics have a complex relationship to system confidentiality. The simulations provide anecdotal evidence that a multi-channel architecture may limit the useful information an eavesdropper receives. The probabilities of interception and decoding, and the amount of fragmentation and duplication across channels demonstrate useful emergent performance characteristics of confidentiality from both the authorized user's and the adversarial listener's perspective.

3. Analysis of simulation

Test cases in the simulation environment observe the security aspects of various communication architectures, ranging from a single point-to-point channel to a single channel with five intermediary relay nodes, to five parallel heterogeneous channels with five intermediary relay nodes. Additional nodes and channels do not appear to add insight. Each communication link, for example **A1-D1**, **C-D2**, or **A3-X** in Figure 1, is programmed to have specific probabilities of interception and decoding. To simplify the simulation process, probability of decryption is set to 1, implying that no encryption is used in these systems. For simplicity and brevity, test cases presented herein are limited to a two-channel and a three-channel system using ASCII-encoded data to observe the simple readability of the alphabet as transmitted through the system. These rudimentary tests reflect the data confidentiality.

3.1 Messages and fragments

The same 8-bit ASCII-coded message is used for all simulations. It is transmitted across the architecture, and, based on the probability of interception, an adversary intercepts the message or it does not. Based on the probability of decoding, the adversary does or does not decode the message. It is irrelevant if the adversary is able to decode a message if they do not receive it. It is possible to have duplicative or cooperative listeners. Multiple listeners may work cooperatively and share intercepted data, but that may affect the probability of decoding since properly merging the intercepted data may prove difficult. For this study, we assume one adversarial listener, but that listener may have access to one, several, or all of the channels within an architecture. The listener accesses the communications between relay nodes, for example **A2-B**, **B-C**, or **C-D2** from Figure 1, simulating on-the-wire or over-the-air connections. If the listener were located in nodes, that would allude to compromised hardware, which is a different challenge outside the scope of this research. If the listener is on any one link, then the entire channel is compromised because all information from that channel is available. In that case, we only consider the worst probability of interception across any given channel.

These aspects of the simulation point directly to harnessing the channels in different ways. One method is simply to duplicate data over multiple channels. This method is ideal for ensuring tamper-free messages during transmission.

Another method involves splitting messages across the multiple channels to prevent unauthorized access to the information. There are numerous methods of fragmenting and distributing messages across the available channels. Poly et al. (2016) developed a detailed analysis of multi-channel communications with respect to maintaining privacy. Castro-Medina et al. (2019) perform a review of fragmentation and duplication for cloud-based systems. Kapusta et al. (2015) describe fragmentation for distributed storage systems. Feng et al. (2015) describe self-adaptive fragmentation for large files to reduce overhead. Another approach uses the Trivium cypher with TLS to distribute data across channels, detailed by Hayden et al. (2020). For simplicity, the following examples use a round-robin technique of distributing data across channels. These techniques are intended to further increase security by shuffling what data appears on which channels.

For example, message M is eight 8-bit ASCII-encoded characters, for a total of 64 bits of data. For this example, there are n=2 channels and the duplication factor, DF=1, meaning that no data is duplicated on the two channels. Assuming that M is split into two, four, or eight equally sized fragments, k, then the total number of bits per fragment is 32, 16, or 8 bits, respectively. These fragments maintain the byte-wise alignment of the ASCII characters. As there are two channels, each channel transmits one, two, or four fragments. With more fragments, there are more opportunities for rudimentary scrambling. A mechanism to reassemble the original message correctly is assumed contained in the receiver. Therefore, in the examples shown in Figure 2, if an eavesdropper has access to only one of the channels, then the eavesdropper only intercepts at most half of the original message, equal to the average loading (*AL*) of the channel.

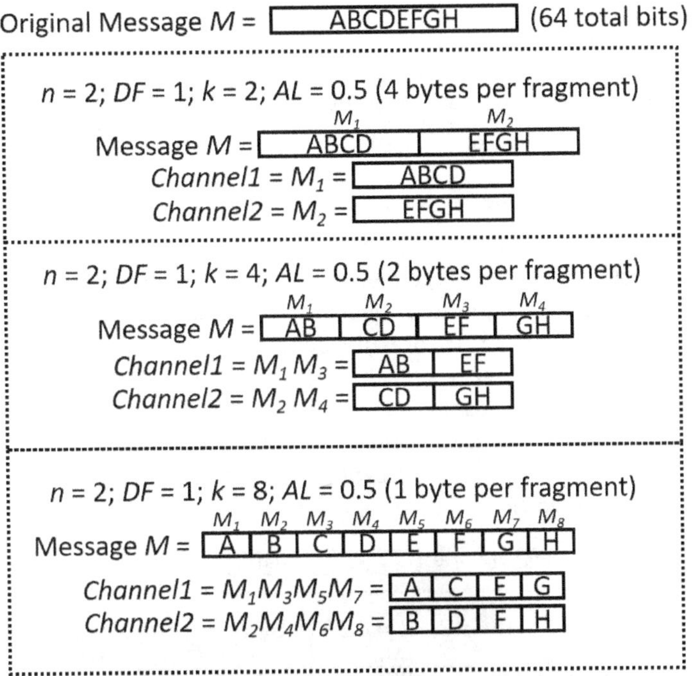

Figure 2: Byte-wise fragmentation of original message across two channels

Figures 3 and 4 show the 64-bit message *M* split into smaller fragments, such that each fragment is 4 bits or 1 bit, respectively. If the sixteen 4-bit fragments were distributed across two channels, an eavesdropper would again only have access to half of the total message, although the bits appear random if ASCII-encoding is assumed. It is interesting to note that in Figure 2, the lightly scrambled data on channel 1 appears to be a series of ASCII letter "D", while meaningful information is transmitted on channel 2.

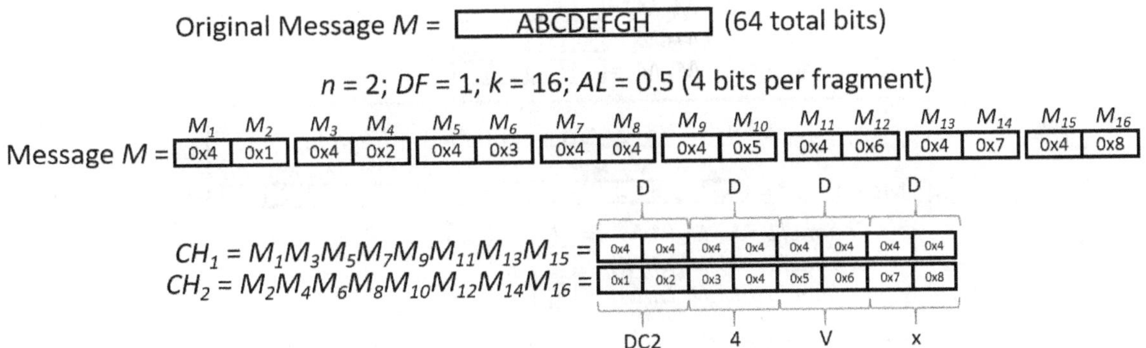

Figure 3: 4-Bit fragmentation of original message across two channels

If the 64 one-bit fragments shown in Figure 4 are distributed across two channels, the amount of data on each channel remains half of the original message. In this case, the reassembly protocol must interleave the two channels as they are received, increasing the complexity of the receiver. The scrambled bits on each channel do not appear to be in any pattern, especially if they are assumed to be ASCII-encoded text.

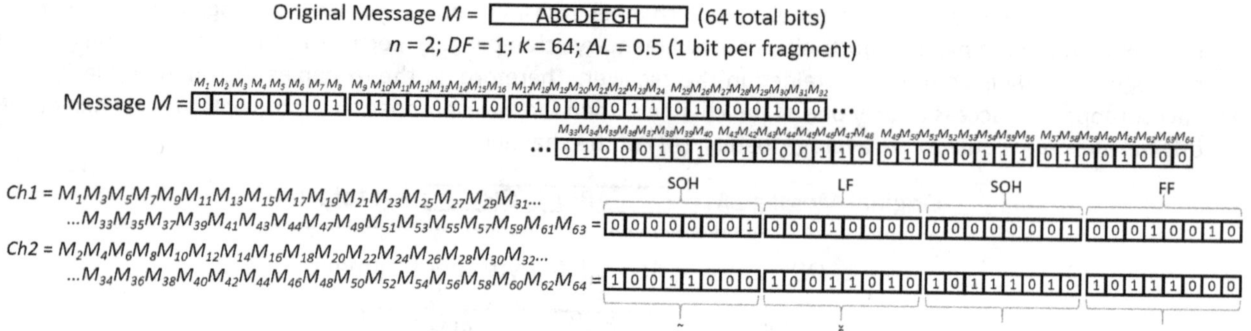

Figure 4: 1-Bit fragmentation of original message across two channels

Figures 5 through 7 show another example system utilizing three parallel channels. To avoid uneven division, message M is twelve 8-bit ASCII-encoded characters, for a total of 96 bits. For this example, there are n=3 channels and DF=1, meaning no data is duplicated. Figure 5 shows the examples with M split into two, four, or eight fragments, k, for a total number of 32, 16, or 8 bits per fragment, respectively. These fragments maintain the byte-wise alignment of the ASCII characters. Each of the channels transmits one, two, or four fragments. Because there are three channels, if an eavesdropper accesses one channel, they intercept at most one third of the original message, equal to the average loading (AL) of the channel. However, a key difference between this example and that of Figure 2 is if the eavesdropper captures two channels, they only receive two thirds of the original message, and there may be no clear manner of reassembly.

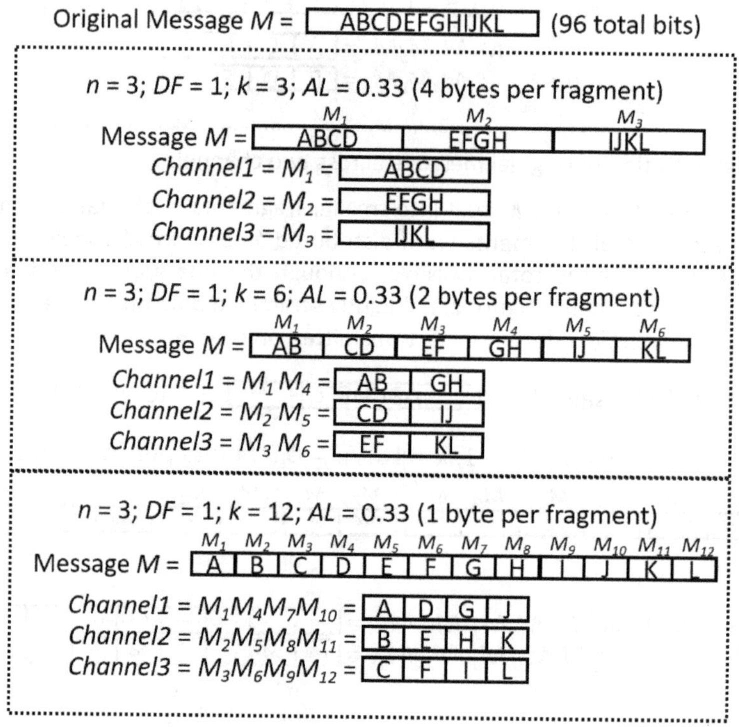

Figure 5: Byte-wise fragmentation of original message across three channels

The 96-bit message M is split into smaller fragments, as shown in Figures 6 and 7, such that each fragment is four bits or one bit, respectively. An eavesdropper would again have access to one third of the total message bits if a single channel were intercepted. An unexpected alignment in the ASCII-encoded text occurs between channels 2 and 3 of the 1-byte fragment example and channels 1 and 3 of the 4-bit fragment example. In both examples, channel 3 provides the same information, although context for correct reassembly does not exist. Additional research is needed to verify if this is a coincidental anomaly or indications of a broader pattern.

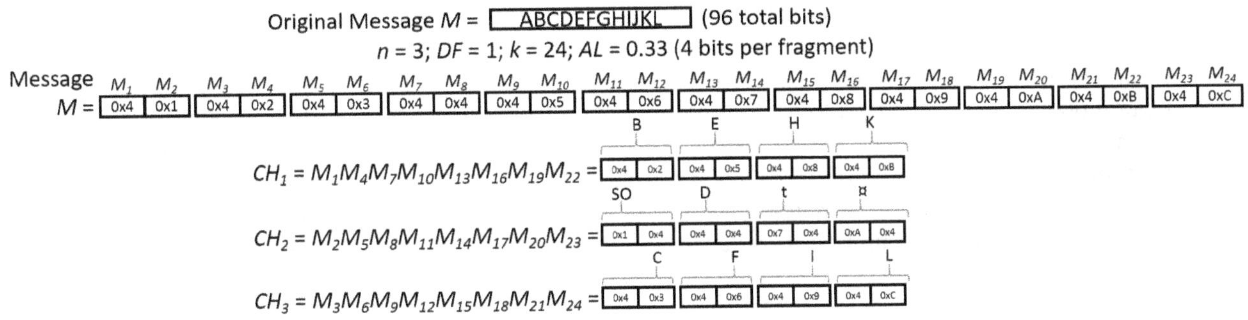

Figure 6: 4-Bit fragmentation of original message across three channels

If the 96 one-bit fragments shown in Figure 7 are distributed across three channels, the amount of data on each channel remains one third of the original message. The scrambled bits on each channel do not appear to be in any pattern, especially when compared to the examples shown in Figures 5 and 6. If an adversary intercepts two of three channels in this example, the information received jumps to two-thirds of the original message. As every third bit of the original message is missing, reconstructing and decoding the message remains difficult.

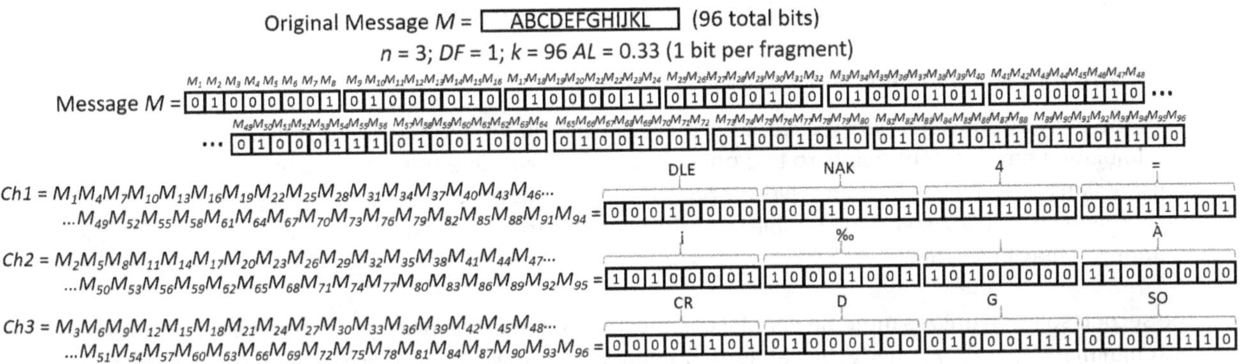

Figure 7: 1-Bit fragmentation of original message across three channels

3.2 Probability of interception and probability of decoding

To estimate data leakage, the probabilities of interception and decoding and amount of fragmentation are applied to the multiple channels. This is a surrogate estimation for the overall security available on a communication architecture.

As an example, we assume the probability of interception, *P(int)*=0.5, for all channels, the probability of decoding, *P(dco)*=0.5, and the probability of decryption, *P(dcr)*=1, which means there is no encryption on any of the channels. The calculation for *P(l)* becomes

$$P(l) = \frac{0.5 \cdot 1 \cdot 0.5 \cdot \sigma}{n} = \frac{0.25 \cdot \sigma}{n}$$

which indicates that *P(l)*=0.25 for any channel, yielding a probability of confidentiality, *P(C)*=0.75. If σ=0 listeners, there is no leakage. In the example network where *n*=2 channels, *P(l)* becomes

$$P(l) = \frac{0.5 \cdot 1 \cdot 0.5 \cdot \sigma}{2} = \frac{0.25 \cdot \sigma}{2} = 0.125 \cdot \sigma.$$

Therefore, the maximum *P(l)*=0.25. If σ=1 listener, *P(l)*=0.125. For every additional channel used, *P(l)* is further reduced. This assumes data is not duplicated across channels; duplicated data requires modifying the equations.

By comparison, in the example network where *n*=3 channels, *P(l)* becomes

$$P(l) = \frac{0.5 \cdot 1 \cdot 0.5 \cdot \sigma}{3} = \frac{0.25 \cdot \sigma}{3} = 0.083 \cdot \sigma.$$

Therefore, when σ=1 listener, *P(l)*=0.083; when σ=2 listeners, *P(l)*=0.166. The maximum *P(l)*=0.25.

The simulation environment confirms these results for messages transmitted across various network configurations. Based on network metrics, the observation is that leakage is approximately the percentage of bits of a message that an adversary successfully receives and decodes. Therefore, as suggested by simulation and despite the possible presence of adversarial listeners and the added complexity of the transmission/reception protocol, increasing the number of channels and maintaining a minimal amount of duplication across channels significantly reduces the probability of leakage, in turn increasing the probability of confidentiality.

4. Implications

The results may be viewed from two perspectives, that of the authorized user and that of the eavesdropper. The authorized user, in most cases, wants as little information to be intercepted by the eavesdropper as possible. The ideal case is exactly zero bits being intercepted, although a limited amount of intercepted data may be acceptable as long as it is not useful. Conversely, the eavesdropper wants to collect as much information as possible. The ideal case is to collect, decrypt, and decode all information. However, this suggests a specific minimum amount of data that must be intercepted to be useful. Because of these opposing viewpoints, the implications must be assessed from both the authorized user's perspective and from the adversarial perspective.

4.1 From the user's perspective

From the authorized user's perspective, to achieve secure communications and reduce possible data leakage, the simulation environment points to two possible solution sets. One solution is to use strong encryption for all transmissions. It is irrelevant if an eavesdropper is able to intercept messages if they are not able to decrypt them. However, many systems are unable to perform necessary encryption processes. In these cases, the second solution set may be useful.

The second solution utilizes multiple parallel channels and distributes message fragments across those channels with minimal duplication. The technical hurdles contained in this solution include: the transmitter must set up and maintain numerous separate channels; the receiver must accurately reconstruct the data; the system must maintain minimal latency; and the system must appear as a single point-to-point connection. Modern communication networks have mastered the ability to minimize latency, whereas this architecture would harness that requirement across multiple connections, while also managing possible timing disparities, jitter, or potentially lost packets or channels. Pitkanen et al. (2008) addresses some of these challenges.

The probability of leakage presents a trade space between security and network complexity. As demonstrated in the simulation environment, with data split evenly (*DF*=1 and *AL*=0.5 across two channels), an adversary with access to only one channel has access to 50% of the data. Across three channels with the same amount of fragmentation, the adversary's challenges become harder. Furthermore, due to the size of the fragments and the manner those fragments are assembled into the transmission packets, the probability of decoding the data also decreases. Since the original message is split evenly across two or three channels, the overall time the connection must be maintained is reduced, thus possibly reducing the temporal opportunity for exploitation. These benefits continue to increase with four or more channels.

Clearly, the more channels that are instantiated, the more complex the system becomes. If each channel is routed differently, there is no guarantee that data packets from each channel will be received at the same time. This obviates the need for a reassembly protocol that enumerates the received packets and recombines the fragments correctly. These are technical trade-offs that exist within most technologies. However, in situations where using encryption is not an option, at least there is some opportunity to thwart eavesdroppers through the use of rudimentary scrambling and multiple channels.

A third solution set does exist and may be applied to situations where driving data leakage to zero is imperative. In those cases, to ensure near-absolute security, the original message may be encrypted and then fragments of the encrypted message may be distributed across multiple channels, similar to the techniques that Ciriani et al. (2010) propose. Each fragment may also be encrypted before transmission for added confidentiality. This would have the ultimate effect of layering protection mechanisms, possibly to a degree beyond the capabilities of the eavesdropper and driving the probability of leakage to near zero. This solution adds additional technical challenges to achieve exceptionally high levels of security.

4.2 From the adversary's perspective

Estimating capabilities and intentions of an adversarial listener is always challenging. Hu et al. (2019) develop a model to analyze the possibility, goals, and capabilities of an eavesdropper. Without any additional information about the adversary, a logical assumption is that an adversary has equal or greater knowledge or capabilities than an authorized user. In this research, a foundational assumption is that the adversary is aware of other channels, and of the potential to fragment data across multiple channels. A second assumption is that they are aware of the protocols used in the transmitter and receiver. However, the number of utilized channels, the channel characteristics, and the size and distribution of the data fragments remain unknown to the adversary.

The challenge for the eavesdropper is to align the probabilities of interception, decryption, and decoding. If strong encryption is used, then the data the eavesdropper intercepts, regardless of number of channels or fragments, will appear to be randomized data. However, in the cases where encryption is not used, the adversary has a reasonable opportunity to intercept data, and systems that do not utilize encryption will be targeted.

Ignoring the challenges of receiving specific coherent transmissions on different physical media, exploiting a communication system begins with intercepting a message transmission. If the architecture utilizes a single wired or wireless channel between transmitter and receiver, the adversary must tap into the physical media without revealing themselves. If, for example, data is transmitted across a single channel that has multiple intermediate relay nodes, any of those connections will provide the same data, thus the whole channel is compromised. If the architecture utilizes two channels and splits the data equally between the two channels, the eavesdropper with access to one of the two channels has access to at most 50% of the data. If the two channels have multiple intermediary nodes, those provide additional points for infiltration. If the eavesdropper gains access to both channels simultaneously, then the eavesdropper has access to 100% of the data. As more channels are introduced, it becomes more challenging to intercept all those channels simultaneously, especially if they feature path diversity. The eavesdropper may be able to discern the existence of additional channels based on the message and packet formation, but that does not guarantee successfully finding and intercepting them.

The probability of decoding is the other key factor in the adversary's calculations, specifically reassembling and decoding the intercepted message. Much like an eavesdropper listening into a conversation, they need to figure out the content and context of the communications. They cannot assume that data is ASCII-based text, or even English language based. The data may appear scrambled, but additional information may be available about the messages elsewhere. Based on partial reassembly of the message or based on educated guess, the adversary may determine the existence of other channels, or they may determine what portions of the message are missing. Depending on the value of the target, the eavesdropper may take additional measures to discover techniques used in transmitting the messages. It must be assumed that the eavesdropper is relentless and, with a sufficient portion of data intercepted, will eventually discover any secret.

5. Conclusion

These simulation results point to improved data security. Fragmenting data across multiple communication channels demonstrates potential improvements to confidentiality through a method of scrambling across a distributed attack surface. The challenge of reconstructing data from multiple channels may push an adversary to reach a point of diminished returns. Future research will focus on incorporating data fragmentation across multiple channels into a protocol suite that will reside between the application layer and the TCP/IP stack. These initial simulation results will guide future protocol development to further improve communication security with various feedback mechanisms or encryption. In the end, no security mechanism is completely secure, but better mechanisms require more time and effort to defeat. Understanding what security is available in a communication network will allow designers to focus on developing techniques and technologies to keep adversaries one step behind or with a small fragment of their ultimate goal: useful information.

Acknowledgements

Funded in part by the Air Force Institute of Technology, Center for Cyberspace Research. The views expressed in this paper are those of the authors, and do not reflect the official policy or position of the United States Air

Force, Department of Defense, or the U.S. Government. This document has been approved for public release, case number 88ABW-2022-0037

References

Almerhag, I. A.; Almarimi, A. A.; Goweder, A. M. and Elbekai, A. A. (2010). Network security for QoS routing metrics. *International Conference on Computer and Communication Engineering, ICCCE'10*, (May), 11–13. https://doi.org/10.1109/ICCCE.2010.5556868

Castro-Medina, F.; Rodríguez-Mazahua, L.; Abud-Figueroa, M. A.; Romero-Torres, C.; Reyes-Hernández; L. Á., and Alor-Hernández, G. (2019). Application of data fragmentation and replication methods in the cloud: a review. In *2019 international conference on electronics, communications and computers (CONIELECOMP)* (pp. 47-54). IEEE.

Ciriani, V.; Vimercati, S.D.C.D.; Foresti, S.; Jajodia, S.; Paraboschi, S. and Samarati, P. (2010) Combining fragmentation and encryption to protect privacy in data storage. ACM Transactions on Information System Security (TISSEC), Vol. 13, 1–33.

Creedon, E. and Manzke, M. (2009). Impact of fragmentation strategy on ethernet performance. In 2009 Sixth IFIP International Conference on Network and Parallel Computing (pp. 30-37). IEEE.

Feng, L.; Zhang, Y. and Li, H. (2015) Large file transmission using self-adaptive data fragmentation in opportunistic networks. In Proceedings of the 2015 Fifth International Conference on Communication Systems and Network Technologies, Gwalior, India, 4–6 April.

Hayden, M.; Graham, S.; Betances, A. and Mills, R. (2020) Multi-Channel Security through Data Fragmentation. In Proceedings of the IFIP International Conference on Critical Infrastructure Protection, Arlington, VA, USA, 16 March; pp. 137–155.

Hennessy, L.J. and Patterson, D.A. (2011) Co.mputer Architecture: A Quantitative Approach; Elsevier: Amsterdam, The Netherlands.

Hu, Z., Vasiliu, Y., Smirnov, O., Sydorenko, V., & Polishchuk, Y. (2019). Abstract Model of Eavesdropper and Overview on Attacks in Quantum Cryptography Systems. In *2019 10th IEEE International Conference on Intelligent Data Acquisition and Advanced Computing Systems: Technology and Applications (IDAACS)* (Vol. 1, pp. 399-405). IEEE.

Hughes, J. and Cybenko, G. (2013) Quantitative metrics and risk assessment: The three tenets model of cybersecurity. Technology Innovation Management Review. Vol. 2, 3, 15–24.

Jin, J. and Shen, M. (2012). Analysis of Security Models Based on Multilevel Security Policy. *Management of E-Commerce and E-Government (ICMeCG), 2012 International Conference on*, 95–97. https://doi.org/10.1109/ICMeCG.2012.72

Kapusta, K. and Memmi, G. (2015). Data protection by means of fragmentation in distributed storage systems. In *2015 International Conference on Protocol Engineering (ICPE) and International Conference on New Technologies of Distributed Systems (NTDS)* (pp. 1-8). IEEE.

Khisti, A. and Liu, T. (2014). Private broadcasting over independent parallel channels. *IEEE Transactions on Information Theory*, *60*(9), 5173–5187. https://doi.org/10.1109/TIT.2014.2332336

Leon, P.G. and Saxena, A. (2010) An approach to quantitatively measure information security. In Proceedings of the 3rd India Software Engineering Conference, Mysore, India, 25–27 February.

Modarressi, A.R. and Ronald, A.S. (1990) Signaling system no. 7: A tutorial. IEEE Communication Magazine, Vol. 28, 19–20.

Pitkanen, M.; Keranen, A. and Ott, J. (2008) Message fragmentation in opportunistic DTNs. In Proceedings of the 2008 International Symposium on a World of Wireless, Mobile and Multimedia Networks, Newport Beach, CA, USA, 23–26 June. (pp. 1-7). IEEE.

Pohly, D.J. and Patrick, M. (2016) Modeling Privacy and Tradeoffs in Multichannel Secret Sharing Protocols. In Proceedings of the 2016 46th Annual IEEE/IFIP International Conference on Dependable Systems and Networks (DSN), Toulouse, France, 28 June–1 July.

Russell, T. (2002) Signaling System # 7; McGraw-Hill: New York, NY, USA; Vol. 2.

Simon, P.M.; Graham, S.; Talbot, C. and Hayden, M. (2021) Model for Quantifying the Quality of Secure Service. Journal of Cybersecurity and Privacy, Vol. 1, 289–301.

Simon, P.M. and Graham, S. (2022) Extending the Quality of Secure Service Model to Multi-Hop Networks. . Journal of Cybersecurity and Privacy, Vol. 1-4, 793-803.

Sweet, I.; Trilla, J.M.C.; Scherrer, C.; Hicks, M. and Magill, S. (2018) What's the Over/Under? Probabilistic Bounds on Information Leakage. In International Conference on Principles of Security and Trust; Springer: Cham, Switzerland.

Wang, J.A.; Xia, M. and Zhang, F. (2008) Metrics for information security vulnerabilities. Journal of Applied Global Research. Vol. 1, 48–58.

Winjum, E. and Berg, T. J. (2008). Multilevel security for IP routing. *Proceedings - IEEE Military Communications Conference MILCOM*, 1–8. https://doi.org/10.1109/MILCOM.2008.4753318.

An Analysis of the Prevalence of Game Consoles in Criminal Investigations in the United Kingdom.

Iain Sutherland [1], Huw Read[1,2] and Konstantinos Xynos[3]
[1]Noroff University College, 4612 Kristiansand S, Agder, Norway
[2]Norwich University, Northfield, Vermont, USA
[3]MycenX Consultancy Services, Stuttgart, Germany
iain.sutherland@noroff.no
hread@norwich.edu
kxynos@mycenx.com

Abstract: There is a body of current research on the technical analysis of computer games consoles to determine if information present might be of value in a criminal investigation. This research has highlighted the potential forensic value of the various consoles depending on the type of crime and the capabilities of the console. There is also anecdotal information, presented in the media, on various crimes that have been prosecuted using evidence obtained from games consoles. However, there appears to be no recent study examining the degree of involvement of games consoles in actual criminal activity, cases being investigated or their use in court cases. This paper presents the results of a Freedom of Information request using the UK Freedom of Information Act (2000) and the Freedom of Information (Scotland) Act 2002. The Freedom of Information Act request was aimed at obtaining an overview of the criminal misuse of game consoles during 2020. This request was sent to the 49 Police forces that cover England, Scotland, Wales and Northern Ireland, seeking details on games consoles included in cases that they have investigated. Current results provide limited information on the involvement of game consoles in cybercrime in the United Kingdom. In examining the prevalence of different types of games consoles in police investigations, the potential need for further work on game console forensics is discussed along with possible factors affecting both the data collection and the patterns observed in the study.

Keywords: games consoles, digital forensics, FOIA request

1. Introduction

There has been considerable work to date (Conrad 2010, Davies 2015, Moore 2014, Read 2016, Pessolano 2019, Barr-Smith 2021) highlighting the possible evidence sources that can be found in computer games consoles and best practices for extracting, analysing and presenting this information in a forensically sound manner. This previous research has shown that for a number of these systems the functionality of these game consoles makes them a potentially useful source of evidence, one that may be overlooked. There is also a body of anecdotal evidence (Urquhart 2012, Telegraph 2012, Holmes 2020, BBC News 2021) where specific cases reported in the news media, have highlighted prosecutions supported by evidence sourced from game consoles. However there appears to be no information on actual numbers of cases and the types of information and devices encountered by police forces. This presents several potential challenges:

- No clear indication of the significance of game consoles as a useful field of study, or whether further research efforts should be continued in this field.
- No clear indication of which consoles should be the focus of investigative research, i.e. where is the greatest need for additional research.

This paper outlines the results of a Freedom of Information Act 2000 (FOIA) request for the 49 police forces in England, Wales and Northern Ireland and Freedom of information (Scotland) Act 2002 (FOISA) request for Police Scotland. The objective is to obtain a snapshot of game consoles' involvement in cases that the forces have investigated within the UK.

2. Capability and misuse of games consoles

Modern games consoles include capabilities beyond simple game play. These types of devices now incorporate network communication between various parties, as well as sharing and downloading of information. Unfortunately, this has the potential for misuse and so maybe a potential evidence source. There are a number of media examples: A 14-year-old boy from Austria used a PlayStation to download bomb-making plans (Nasralla, 2015). The FBI applied for a search warrant to compel Sony for data concerning a PlayStation 4 user, who was suspected of using the messaging application in the PlayStation to communicate and arrange cocaine transactions (Brousil, 2019). Emails released by the Anonymous hacking group in 2012 identified how US police were using Xbox and PlayStation consoles in investigations (Urquhart, 2012). There are various examples in the media where even less capable consoles have provided useful evidence (BBC News 2021, Holmes 2020). In 2012, a paedophile was caught after the 10-year-old victim took pictures of the abuse using her Nintendo DSi (Telegraph, 2012). Nintendo deactivated their popular "Swapnote" email-like service on the Nintendo 3DS globally as child predators were allegedly using this feature to send offensive material (Ashcraft, 2013).

3. UK Freedom of Information Act

The Freedom of Information Act (2000) is United Kingdom legislation that enables members of the public to request information from public authorities. The Act covers information held in England, Wales and Northern Ireland. A comparable piece of legislation, the Freedom of Information (Scotland) Act 2002, provides the same legal right of access in Scotland. Recorded information includes computer records, emails and printed documents (Information Commissioner 2021a). Detailed criteria defining the operation of the Act can be found in the Information Commissioner Office Code of practice (Information Commissioner's Office, 2021b). The definition of public authorities also includes the territorial police forces. There are several organisations in the UK that are exempt from responding to the Freedom of Information Act requests, including some national law enforcement bodies (Freedom of Information Act, 2000). The National Crime Agency (NCA) (NCA, 2021) and various commands of the NCA, for example the Child Exploitation Online Protection Command (CEOP) are exempt from the FOIA. Due to the role and function of these agencies, it is likely that they may encounter evidence of criminal activities on computer game consoles, although it would be impossible to know since they would not respond positively to information requests.

4. Methodology and information sought

An FOIA request was formulated based on advice provided on the UK information commissioner's website on making FOIA requests (Information Commissioner's Office, 2021c) and with a list of Police FOIA contacts provided by Burgess (2021). This included the forces responsible for territorial regions in England, Scotland, Wales and Northern Ireland (45) and the (4) non-territorial forces; the Port of Dover Police, The British Transport Police, the Civil Nuclear Constabulary and the Ministry of Defence Police. The FOIA request sought the information outlined in Figure 1.

1. The total number of cases handled by your cybercrime unit for each of the following years: 2018, 2019, and 2020.
2. The number of cases where evidence was sought from game consoles for each of the following years: 2018, 2019, and 2020.
3. For each of the three years the following details on cases:

2018 (example)

Console type seized in case	Relevant Legislation	What type of evidence was sought on the Games console
Playstation	Computer Misuse Act 1990, S1	Email
Nintendo DS	Serious Crime Act 2015 c. 9, s. 67	Photos

Figure 1: Initial FOIA request made to one UK force

The request was sent to a single UK police force initially to determine the response as there are several possibilities within the FOIA that allow for the responding organisations to decline to provide information. This is termed a refusal of request (section 17 of the Act). There are various grounds for refusal, based on exemptions outlined in part II of the Act. These exemptions include; where the cost of compliance exceeds the appropriate limit, (section 12 of the Act); vexatious or repeated requests (section 14 of the Act) or where another exemption applies such as those for law enforcement (section 31 of the Act). The result of the preliminary request contained in Figure 1 was a refusal with section 12 cited, that is the cost to retrieve the requested data exceeded the reasonable limit, a nationally specified threshold currently set at £450 which can be equated to 18 hours of work. (The Freedom of Information and Data Protection (Appropriate Limit and Fees) Regulations 2004).

The FOIA request was then revised to reduce the amount of information requested to a single year, 2020, as outlined in Figure 2. This revised request was then sent out to the remaining 48 police forces listed in Burgess (2021). The revised request was not sent to the force that had received version 1, in Figure 1.

1. The total number of cases handled by your Force in 2020.
2. The number of cases (cyber enabled and cyber dependant) where evidence was sought from game consoles.
3. The types of game console (playstation, Xbox, Wii, Switch, DS) involved in these cases

Figure 2: FOIA Request made to the 48 Police Forces in England, Scotland and Wales

5. Results

Under the terms of the Freedom of Information Act, all UK public bodies are required to respond within 20 days of receiving the request. However, this period does not include any time spent if the force needs to seek clarification on the wording of the request. Therefore, in practice this can take more than 20 days. Some forces responded immediately with automatic acknowledgements and FOIA reference numbers. The FOIA request had also included a reference to section 16 of the Act, which is the legal obligation to provide guidance and assistance to those requesting information. In all, 20 of the 48 forces that were sent the second modified request, in Figure 2, responded seeking clarification of the word 'cases'. Others provided responses without any request for clarification. Those forces that requested clarification were provided with a reworded request where the term 'case' was clarified as 'offence'. During the process some forces advised they would be slower in replying due to issues connected with the COVID-19 pandemic. At the time of writing there are 13 requests outstanding and 36 responses have been received, (all from forces in England and Wales). A summary of the FOIA responses is shown in table 1.

Table 1: Responses to FOIA Requests

Type of Response	Reason / Response	Number of Responses
Refusal to provide information	Declined under section 12, exceeding the cost to retrieve and provide the data.	14
	Declined under section 31, Law Enforcement	1
Information not held	This non-territorial force did not hold the type of information requested.	1
Full information provided	Full information on all three questions was provided	13
Partial information provided	Summary figures for question 1 was provided	7
Total number of responses		36

A total of 15 police forces declined to provide the data, of these 14 applying the exemption under section 12 of the Act; excessive cost involved in retrieving and collating the data. Most of the refusals provided a rough calculation on the amount of effort required to demonstrate the point:

> *"The cost of providing you with the information requested in respect of your request is above the amount to which we are legally required to respond i.e. the cost of locating and retrieving the information exceeds the "appropriate level" as stated in the Freedom of Information (Fees and Appropriate Limit) Regulations 2004. It is estimated that it would exceed 18 hours (i.e. minimum of **761.8 hours**) to comply with your request. "*

One force cited section 31(3) as a reason for a refusal. Section 31 relates to Law Enforcement and a public interest test is applied to determine if the data can be released. (ICO Law Enforcement (Section 31), 2022). In this case the information was refused as:

> *"...To confirm or deny information is held would compromise law enforcement tactics in a specific area of policing which includes elements of organised criminality, which would hinder the prevention and detection of crime..."*

One of the non-territorial forces replied outlining the nature of their work meant that they did not investigate this type of crime, that was rather the jurisdiction of the different territorial forces.

At the point of writing, 12 forces provided data for all three questions. A further 7 forces provided partial information. The majority of forces that provided partial information provided some information on question 1 and then refused to provide more information on questions 2 and 3 citing the section 12 cost exemption. Although one force responded with answers to questions 2 and 3 and no data relating to the total number of crimes committed. One force provided some additional data on the types of crime they encountered relating to games consoles in 2020, shown in Table 2. These are offence groups, the types of offence that this might include can be determined by comparing these against the Offence Classification Index (UK Home Office, 2013). It should be stressed that this is from the perspective of one police force, for one year. It is useful to see such detailed information, in comparison to the majority of other forces that have refused to provide information due to the estimated cost and time taken. This could be due to the open-ended nature of the questions put forward.

Table 2: Type of crimes relating to games consoles observed by one UK force

Type of crimes relating to Games Consoles in 2020
Sexual Offences
Violence Against The Person
Misc. Crimes Against Society
Theft Offences

There are some interesting conclusions that can be made based on the data returned by the various forces.

- 1. Question 1, in Figure 2, has clearly been interpreted in several different ways. We have seen figures in excess of 100,000 records spanning a whole force, to specific actions taken by the cyber crime unit (e.g., 287 for one force).
- 2. In question 2, cases/offences where information was sought from games consoles; for the 13 forces that provided data this totalled to 1379 cases/offences for 2020. It should be noted that for two of the 13 forces responding, 2 were non-territorial forces and had no cases involving game consoles. Therefore, the figures above actually reflect 1379 occasions where evidence was sought from games consoles from 11 forces.
- 3. The response to question 3 included some forces that provided figures according to console types. Other forces provided a more detailed view including versions of the consoles (e.g., more specific models). In figure 3, these have been combined to provide a total for the different types of game consoles.

Some forces that had calculated that the request would exceed the section 12 limitation advised on other additional data sources such as the Office for National Statistics Annual Crime report (Office for National Statistics, 2022a). These crime statistics do provide outline figures on fraud and computer misuse (Office for National Statistics, 2022b). However, these official statistics do not provide the required degree of granularity into specific devices such as game consoles.

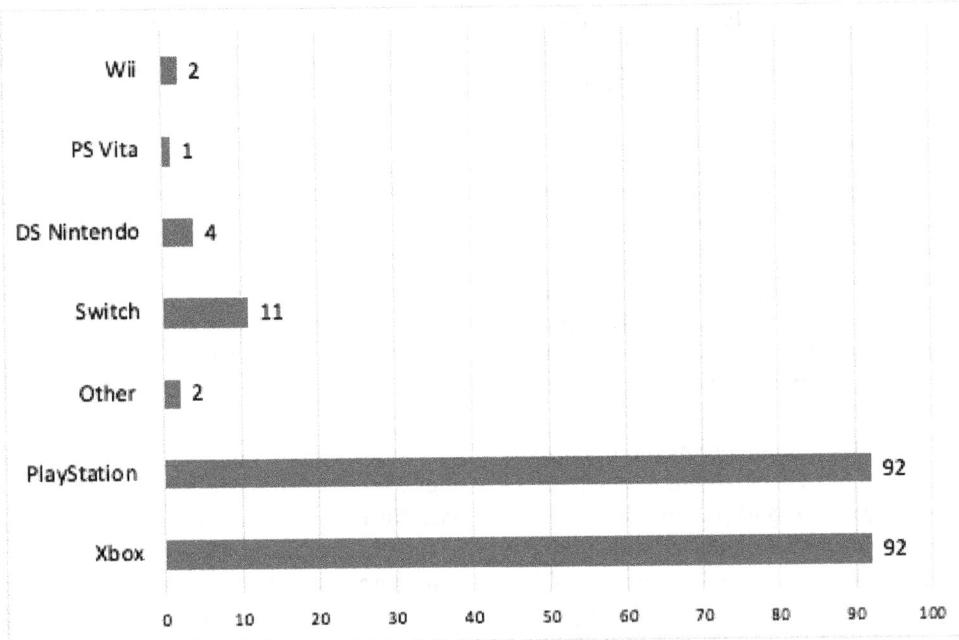

Figure 3: Cases/offences involving games consoles 2020 by console type (from 13 responses)

6. Limitations on the data collected

There are several challenges in collecting data in this manner, meaning the following limitations have been identified:

Despite the UK government guidance on the cataloguing and recording of offences the FOIA responses imply that the different police forces in the UK have different local methods for cataloguing the data. Indeed, a number of forces highlighted that information is stored differently with different police forces and information disclosed should not be compared between forces. One force made this very clear in their response.

> *"However, the systems used for recording these figures are not generic, nor are the procedures used locally in capturing the data. It should be noted that for these reasons this forces response to your questions should not be used for the comparison purposes with any other response you may receive. "*

Therefore, at best these figures can provide a rough guide to what is being analysed and should be considered as such. Though the authors endeavoured to obtain comparable results, it is important to highlight to the academic community the challenges in drawing direct comparisons between the data provided by the different forces.

The FOIA request has clearly been interpreted differently by the regional FOIA officers; some have requested clarification on the phrasing of the question (the issue being the word 'case' which was clarified to the word 'offence') while other forces did not seek any clarification. Some forces provided information on how the data was retrieved and provided additional information extending to several pages of text. Other forces limited their response to just the requested statistics, making it difficult to compare the method of data extraction. There is a suggestion that one force included theft of devices in the figures. These were considered for removal, but it could be argued that the device may still need to be examined to determine the rightful owner. This would also be a subjective decision by the authors. Therefore, from a digital forensics perspective these numbers are useful and have been included.

This study was limited to territorial (45) non-territorial (4) forces that fall under the FOIA. The FOIA does not apply to the National Crime Agency (NCA, 2021) or those under its command, which could be argued that due to the technical nature of the cases these organisations may handle many of these types of investigations/cases. The NCA was contacted as part of this study, but the organisation declined to provide any information citing exemption from the Freedom of Information Act.

One force pointed out that they had reviewed the records to eliminate false positives as searches for 'DS' would also match 'Detective Sergeant' and 'Switch' would match 'light switch', indicating keyword searching was used to retrieve the data. It is not clear if every force examined and checked the information at this level of detail.

7. Discussion and conclusions

The challenges relating to the quality of the data mean that only limited and broad conclusions can be drawn. This is in part to the wording used in the FOIA request, which this work clearly highlights, needs to be unambiguous, tightly focussed with closed ended and specific/targeted questions. We recommend that a review of public court cases involving games consoles would also assist in the formulation of focused questions. Future requests would need to provide targeted questions and example end results/keywords, therefore limiting the amount of data records that need to be reviewed by the officers. This, in turn, would narrow the search or time frame with the aim of avoiding refusals (using section 12).

The picture for the UK is incomplete as the majority of forces refused the request under section 12 of the Act. However, the number of 1379 occasions where games consoles were examined illustrate the situation that these devices are clearly recognized as possible sources of evidence for certain types of cases. The frequency of the different consoles reflects the capability and connectivity of the PlayStation and Xbox consoles. It is interesting however to note that the hand-held devices; the Nintendo Switch and Nintendo DS are also of interest, although the latter is no longer being manufactured. Despite the potential challenges of data quality this study provides a perspective on the relative importance and interest in specific games consoles.

8. Future work

In addition to the response to the FOIA request, forces made suggestions on revising the FOIA scope to provide a more focussed and manageable FOIA request. The aim is to repeat the request in a revised form in future years, to be more specific with the intention of avoiding refusal due to the excessive cost (section 12) and where possible gather more comparable data between forces. Court records could be examined along with FOI legislation in other countries to provide an international perspective. However, in other countries the law enforcement exclusion that applies to the NCA in the UK may be broader and apply to all police agencies, hindering data gathering from that country's police forces. One example is the Freedom of Information Act in Norway (Lovdata 2021) which appears to include law enforcement as an entity subject to the Norwegian Freedom of Information Act which would limit the amount of available information.

Acknowledgements

The authors would like to thank the freedom of information teams of the 49 United Kingdom police forces for the time and effort expended in answering the requests, for the additional helpful FOIA advice and highlighting other possible data sources.

References

Ashcraft, B. (2013) Accused Child Predator Allegedly Used Nintendo's Swapnote Service, Kutaku, November, 2013. Available online: https://kotaku.com/child-predators-were-using-nintendos-swapnote-service-1459304126

Barr-Smith, F. Farrant, T. Leonard-Lagarde, B. Rigby, D. Rigby, S. Sibley-Calder, F. (2021) Dead Man's Switch: Forensic Autopsy of the Nintendo Switch, Forensic Science International: Digital Investigation, Volume 36, Supplement, 2021.

BBC News (2021) Special Olympics winner caught with abuse images for second time, BBC News, 3 August 2021. Available online: https://www.bbc.com/news/uk-scotland-tayside-central-58383718

Brousil, B. (2019) Application for a Search warrant to the United States DIstrict Court for the Western District of Missouri (Case No. 19-SW-00364-JTM) Available Online: https://www.documentcloud.org/documents/6565970-PlayStation-Seach-Warrant-Application.html

Burgess M., (2021) FOI Directory,. Available online: https://www.foi.directory/

Conrad, S. Dorn G. and Craiger P., (2010) Forensic Analysis of a Playstation 3 Console, Advanced in Digital Forensics VI, K. Choi and S. Shenoi (Eds.), Springer, NY, ch. 5, 2010.

Davies, M., Read, H. Xynos K., Sutherland I., (2015) Forensic analysis of a Sony PlayStation 4: A first look, Digital Investigation, vol. 12, no. 1, pp. 81-89, 2015.

Freedom of information Act 2000. Available online: https://www.legislation.gov.uk/ukpga/2000/36/contents , last accessed: 03/02/2022

Freedom of information Act (Scotland) 2002. Available online: https://www.legislation.gov.uk/asp/2002/13/contents , last accessed: 03/02/2022

Holmes W., (2020) Blackpool child rapist tried to use secret Nintendo DS camera to record people on toilet two months after release, Blackpool Gazette, 20 February 2020. Available online:

https://www.blackpoolgazette.co.uk/news/crime/blackpool-child-rapist-tried-use-secret-nintendo-ds-camera-record-people-toilet-two-months-after-release-1884849

Information Commissioner's Office (2021a). Available online: https://ico.org.uk/ , last accessed: 29/12/2021

Information Commissioner's Office (2021b) Code of Practice, Data Sharing. Available online: (https://www.gov.uk/government/publications/code-of-practice-on-the-discharge-of-public-authorities-functions-under-part-1-of-the-freedom-of-information-act-2000)

Information Commissioner's Office (2021c) How to access information from a public body. Available online: https://ico.org.uk/your-data-matters/official-information/ , last accessed: 29/12/2021

The Freedom of Information and Data Protection (Appropriate Limit and Fees) Regulations 2004. Available online: *https://www.legislation.gov.uk/uksi/2004/3244/contents/made*

Information Commissioner's Office (2022) Law Enforcement (Section 31), Version 1. Available online: https://ico.org.uk/media/for-organisations/documents/1207/law-enforcement-foi-section-31, last accessed: 03/02/2022

Lovdata (2021) Act relating to the right of access to documents held by public authorities and public undertakings (Freedom of Information Act). Available online: https://lovdata.no/dokument/NLE/lov/2006-05-19-16#KAPITTEL_2 (English Version)

Moore, J. Baggili, I., Marrington A., Rodrigues A., (2014) Preliminary forensic analysis of the Xbox One, Digital Investigation, vol. 11, pp. 5765, 2014.

Nasralla, S. Stonestreet, J. (2015) Teenager in Austrian 'Playstation' terrorism case gets two years, Reuters, May, 2015. Available Online (archived): https://web.archive.org/web/20150526195103/http://www.reuters.com/article/2015/05/26/us-mideast-crisis-austria-idUSKBN0OB0LK20150526

National Crime Agency (2021) Freedom of Information. Available online: https://www.nationalcrimeagency.gov.uk/contact-us/3-freedom-of-information

UK Home Office (2013) Offence Classification Index. Available online: https://assets.publishing.service.gov.uk/government/uploads/system/uploads/attachment_data/file/977202/count-offence-classification-index-apr-2021.pdf

Office for National Statistics (2022) Crime in England and Wales: year ending September 2021. Available online: https://www.ons.gov.uk/peoplepopulationandcommunity/crimeandjustice/bulletins/crimeinenglandandwales/yearendingseptember2021.

Office for National Statistics (2022b) Nature of crime: fraud and computer misuse. Available online: https://www.ons.gov.uk/peoplepopulationandcommunity/crimeandjustice/datasets/natureofcrimefraudandcomputermisuse .

Pessolano, G. Read, H. Sutherland, I. Xynos, K. (2019) Forensic Analysis of the Nintendo 3DS NAND, Digital Investigation, Volume 29, Supplement, 2019, Pages S61-S70.

Read, H. Thomas, E. Sutherland, I. Xynos K. Burgess, M (2016) *A Forensic Methodology for the Analysis of a Nintendo 3DS,* Twelfth Annual IFIP WG 11.9 International Conference on Digital Forensics, New Delhi, India January 4-6, 2016, www.ifip119.org

Telegraph (2012) Paedophile caught after victim takes picture using Nintendo game, The Telegraph, March 28, 2012. Available Online: https://www.telegraph.co.uk/news/uknews/crime/9171452/Paedophile-caught-after-victim-takes-picture-using-Nintendo-game.html

Urquhart, L. (2012) US Police use games consoles in crime investigations, naked security by Sophos, January 2012. Available Online: https://nakedsecurity.sophos.com/2012/01/26/us-police-use-games-consoles-in-crime-investigations/

How are Hybrid Terms Discussed in the Recent Scholarly Literature?

Ilkka Tikanmäki[1, 21] and Harri Ruoslahti[12]
[1]Security and Risk Management, Laurea University of Applied Sciences, Espoo, Finland
[2]Department of Warfare, National Defence University, Helsinki, Finland
Ilkka.Tikanmaki@laurea.fi
Harri.Ruoslahti@laurea.fi

Abstract: Hybrid threats range from cyber-attacks on critical systems to disruption of critical services (such as energy and financial services), influencing public confidence, and polarization within society. Awareness, resilience, and response to threats are central to countering hybrid threats. Hybrid warfare is not a new phenomenon, it has existed throughout the history of warfare, however, hybrid threat and hybrid warfare were re-defined as the western concept, as discussed in this paper, in 2014. Securing vital functions of society, i.e., managing overall security includes preparing for threats, and managing and recovering from disruptions and emergencies. Energy policy, which relies on cross-border energy transmission infrastructures (e.g., Russian gas line imports to Europe), can be a tool to influence foreign policy (Geo-economics). Trolls and cyber weapons can be used to impact information and elections, and their activity are based on supranational Information Technology (IT) infrastructure. The vital functions of society are prime targets for political, economic, and military pressure from external actors. Hybrid warfare deliberately blurs the boundaries between peacetime and wartime, which makes it difficult for targeted organizations and countries to plan appropriate and timely countermeasures. The threat of hybrid disruptions can be addressed with resilience. Multifaceted hybrid threats require planning and testing one's defensive possibilities, so that the various actors of society will be able to respond to possible hybrid attacks and commit all areas of society for an effective defence. Identifying and understanding hybrid warfare is challenging. Situation awareness is a prerequisite, so societies and their organization can meet these challenges

Keywords: hybrid threats, hybrid influence, hybrid interference, hybrid warfare, hybrid operations

1. Introduction

Hybrid threat and hybrid warfare were re-defined as western concepts in 2014 when Russia occupied the Crimea peninsula and started military actions in Eastern Ukraine by using so-called "green men", deputy warriors (rebels), cyber-attacks and information warfare. Russia achieved its political goals without using its military power (Raitasalo 2019). Raitasalo (2019) notes that hybrid warfare contains elements of cyber warfare and information warfare. He envisions it as a necessary conceptual tool for Western nations, who have been lulled in believing that international relations had entered a new era of co-operation after the end of the Cold War, to plan and prepare to defend against hybrid threats. Finland, for example, published the first Strategy for securing the vital functions of society to guide Finnish authorities, businesses, and organizations design, prepare, and practice long-term responses to a wide range of possible security threats (Finnish Government 2003). Securing vital functions of society (i.e., managing overall security) includes preparing for threats, and managing and recovering from possible disruptions and emergencies (Terminology Centre 2017). Critical functions in society include leadership, international and European Union (EU) action, defence capabilities, internal security, economy, security of infrastructure and supply, capabilities of and services for the population, and mental resilience (Terminology Centre 2017).

This study is organized as follows. Section 1 presents the content of the study and outlines the case study method used. Section 2 summarizes hybrid threats, influence, interference, warfare, and operations. The findings are discussed and concluded in Section 3. The study is a literature study, and the research question of this study is "How are hybrid terms discussed in recent scholarly literature?"

1.1 Methods

The study was conducted as a qualitative study and the research method is descriptive. The results of qualitative research are based on the researcher's reasoning (Huttunen & Metteri 2008). The research problem is discussed in the constructive research approach. According to Yin (2009), there are six sources for case studies: documentation, archival records, interviews, direct observation, and participation in observing and physical

[1] https://orcid.org/0000- 0001-8950-5221
[2] https://orcid.org/ 0000-0001-9726-7956

objects. It is recommended that several sources of evidence be used in the case study. The research data was collected from scientific reports, collected articles and literary reviews.

2. Hybrid terms in recent scholarly literature

This chapter explains hybrid terms, which in the context of this study are hybrid threats, hybrid influencing, hybrid interference, hybrid warfare and hybrid operations.

2.1 Hybrid threats

A hybrid threat is the simultaneous and adaptive use of an embedded combination of (1) political, military, economic, social, and media, and (2) traditional, irregular, terrorism, and disruptive/criminal conflict methods (GAO 2010). Hoffman (2007, p. 8) argues that hybrid threats are "Threats that incorporate a full range of different modes of warfare, including conventional capabilities, irregular tactics and formations, terrorist acts including indiscriminate violence and coercion, and criminal disorder, conducted by both states and a variety of non-state actors".

Hybrid threats range from cyber-attacks on critical systems to disruptions of critical services (e.g. energy and financial services), which can influence public confidence, and even create polarization within society. Threats to critical vulnerabilities seek to hamper effective decision-making (European Union 2018). One defining characteristic is the continuous utilization of identifiable asymmetries, both during a non-violent phase and in actual war. Asymmetric warfare refers to the usage of Special Forces, precision bombs and missiles, electronic warfare, guerrilla warfare, terrorist attacks, and the utilization of biological and chemical weapons (Mack 1975; Arreguín-Toft 2001). Asymmetries are utilized as a combination of surprise, abuse, and deception (Cederberg & Eronen 2015).

According to key efforts to combat against hybrid threats on the EU level are awareness, resilience, and response. Combined, these efforts can improve abilities to detect and understand adverse actions at early stages and help improve the sustainability of critical infrastructure, societies, and institutions. These above actions are essential in improving the abilities of the EU to withstand and recover from attacks. Action by EU Member States and closer cooperation between them, and with partner countries and NATO are needed to combat hybrid threats (European Union 2018).

2.2 Hybrid influencing

The Finnish Institute of International Affairs (FIIA) divides hybrid influence into geo-economics, information, and electoral impact; Energy policy, which relies on cross-border energy transmission infrastructure, can be used as an instrument to influence foreign policy (geo-economics); for example Russian gas imports to Ukraine and Europe. Trolls and cyber weapons can be used for information and electoral impacts that are based on a supranational IT infrastructure. (FIIA 2018.) The Security Strategy for Society defines hybrid engagement as an activity that pursues its own goals through a variety of complementary means and by exploiting the weaknesses of the target. Means of hybrid influence can be economic, political, or military, and can be used simultaneously or sequentially with technology and social media (Security Committee 2017).

Puistola (2018) presents the operation line of the hybrid influencing (Figure 1) on Diplomatic, Informational, Military and Economic (DIME) levels in three phases of pressure.

As shown in Figure 1, the operation line in the "New normal" -stage may start with pressure (1), which is followed by reconnaissance and recruitment (2), denigration (3), troop deployment, exercises, acquisition of strategic areas (4), pressure (5), and finally disruptions in the energy market and cyber-attacks (6). In the "Growing tensions" phase, activities become followed by import and export bans (7), territorial violations, cyber-attacks, acts of terrorism (8), false reporting, suppression of international aid, agitation (9), and invitations to negotiate (10). The "Exceptional conditions" stage could continue with the use of special forces (11), use of conventional armed forces (12), crushing of defender morale (13), and compelling peace (14). Sub-goals 1-4 aim at the final goal, where the target country makes military, political and economic national and international decisions in the interests of the influencer. (Puistola 2018.)

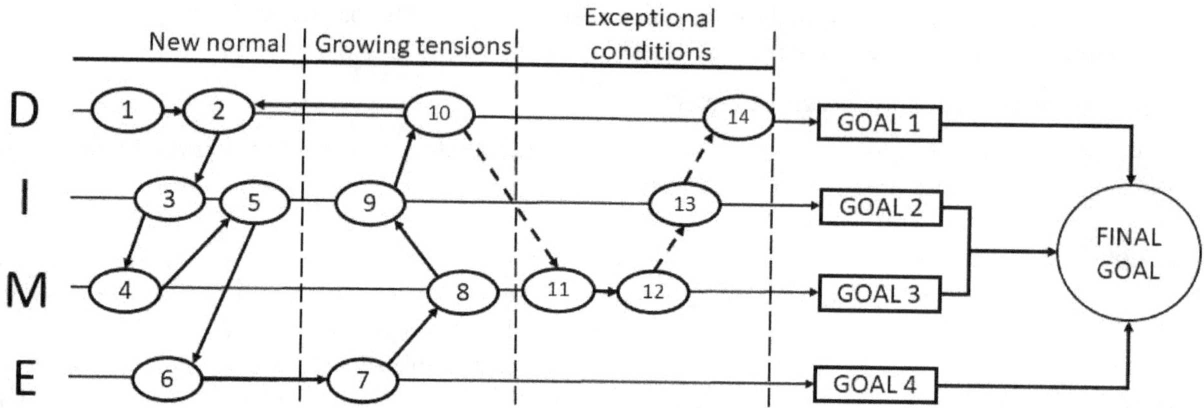

Figure 1: Operation line of the hybrid influencing (Adapted from Puistola, 2018).

Mäkelä (2018, p. 13) describes hybrid action as "a systematic activity in which a state or non-state actor can simultaneously utilize various military means or, for example, economic or technological pressure, as well as information operations and social media." In conclusion, Mäkelä concludes that often the aim is to keep the hybrid effect at a level where escalation into open conflict does not occur; to apply effects in ways that are disputable. Puistola (2018) sees information influence as a strategic deception in which the subject's perception of reality is obscured to achieve the attacker's strategic goals.

Social media can be a powerful contributor in shaping civic opinion. As reported by Puistola (2018) the means used for strategic deception are 1) deception and prohibition; 2) hiding of final goals; 3) maintaining the legality facade; 4) military defiance and threats, and 5) global input of one's narrative through various mediums. Tiilikainen (2020) states that hybrid influence is characterized by the use of shadow regions: a threat can come from just somewhere you cannot expect. Therefore, the Finnish model, combining different administrative sectors, is a policy that could be used in the wider world.

Hybrid influence, as reported by the Institute for International Affairs (2018), means synchronized use of multiple ways of harassment, aimed at generating deep dividing lines, within the societies of targeted countries, by leveraging non-military techniques. An actor seeking to influence will want to disrupt the political process in the target country so that the target country paralyses its decision-making ability itself. Influencing can be sabotaged, seeking to disintegrate communities, and emphasizing its aggressive nature. (FIIA 2018.) Hybrid interference may include the elements presented in Table 1.

Table 1: Elements of hybrid interference (FIIA 2018)

Concept	Description
Hybrid influence	Synchronized harassment aimed at generating deep dividing lines within the societies of the target countries by leveraging non-military techniques, the aspirant wants to disrupt the political process in the target country so that the target country itself paralyses its decision-making ability.
Elements of hybrid interference	cyber operations spreading false news dissemination of propaganda financing political extremism influencing key institutions of democracy (e.g., elections and governance) control of critical infrastructure providing financial incentives
Key efforts to counter hybrid disruptions	The threat of hybrid disruption can be addressed with resilience.

Multifaceted hybrid threats require planning of defenses to respond to hybrid threats by committing all areas of society to be alert and defend against all threats. Comprehensive (hybrid) defense requires patiently building national capabilities. In the short term, deficiencies in hybrid defense can be reduced by leveraging allied capabilities and performance. (Cederberg & Eronen 2015; Lalu & Puistola 2015) Cederberg and Eronen see hybrid warfare as a concept that the West is trying to classify based on Ukrainian events. Hybrid warfare can be long lasting because quick victory over an opponent is not necessary, also the active and passive stages of the conflict can vary. (Cederberg & Eronen 2015.)

The vital functions of society are the prime targets of political, economic, and military pressure from an external actor. The threat of hybrid disruption can be addressed with resilience. Resilience refers to the ability of society to resist, withstand and recover quickly from malfunctions (FIIA 2018). According to de Bruijne et al. (2010, p. 9), "resilience refers to the ability of a social system (such as an organization, city, or society) to proactively adapt to and recover from disorders that it considers to be beyond normal and expected disorders."

2.3 Hybrid warfare

Gärdström (2018, p. 2) states that "hybrid warfare is the combination of instruments and means of influence to subvert states, institutions, and societies". An article by Hyytiäinen (2018) describes a model for hybrid action and hybrid warfare that can be used in other potential security situations and can be used as a tool for preparedness. Table 2 shows the components of hybrid warfare as presented by Hyytiäinen and the Munich Security Conference (2015), which for the most part are very similar. The main difference is that Hyytiäinen also includes infrastructure and energy as concepts of hybrid warfare.

Table 2: Hybrid warfare concepts comparison. (Munich Security Conference 2015; Hyytiäinen 2018)

Munich Security Conference (2015)	Hyytiäinen (2018)
Special forces	Special Forces
Irregular forces	Non-State Forces (Rebels)
Support of local unrest	Social harmony
Information warfare / propaganda	Information influence and propaganda
Diplomacy	Diplomacy
Cyberattacks	Cyberattack
Economic warfare	Economic Sanctions
Regular military forces	Military Action
	Infrastructure and energy

The model of Hyytiäinen is based on the hybrid warfare model presented by NATO, with infrastructure and energy added. The hybrid warfare model of Hyytiäinen does not include extensive military influence: it describes the non-military components of hybrid warfare and the supporting military activities. (Hyytiäinen 2018.) The Munich Security Conference (2015) hybrid warfare model corresponds to the NATO model. Infrastructure and energy are integral functions of society, and as such justified as additional components. Hybrid warfare blends conventional and irregular warfare throughout the conflict. Figure 2 shows an example of approaches that could be included in hybrid warfare, which combines elements of irregular pressure and conventional warfare to create pressure without conventional warfare.

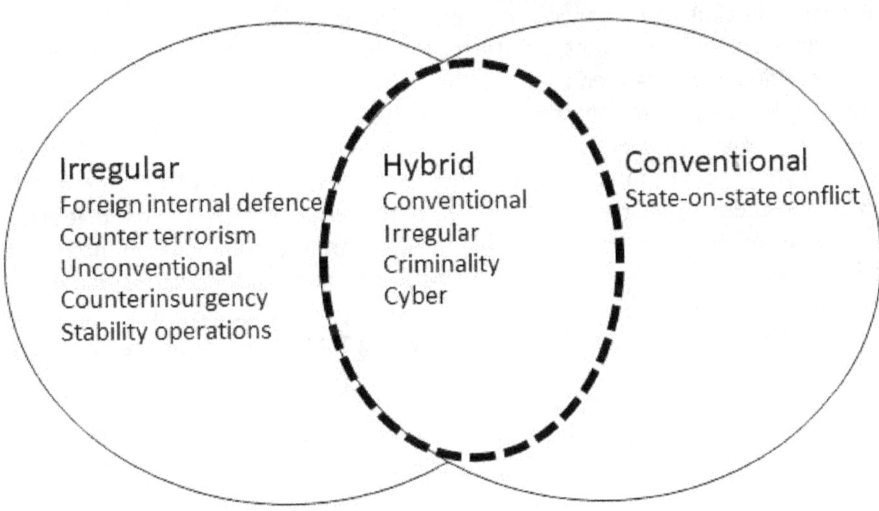

Figure 2: The hybrid warfare concept (Source: GAO 2010)

As seen in Figure 2, conventional warfare is a form of warfare between states that uses direct military confrontation to defeat an opponent's armed forces, destroy an opponent's warfare capacity, or conquer or maintain an area to force a change in an opponent's government or policy. "Conventional warfare can also be called "traditional" warfare". Irregular warfare, in turn, is a violent struggle between state and non-state actors to gain legitimacy and influence over the populations concerned. As stated by the US Air Force "Hybrid warfare is more powerful and complex than irregular warfare due to increased pace, complexity, diversity, and broader orchestration across national borders". Hybrid warfare can be described as conflicts that carry either state and/or interstate threats that use multiple forms of warfare, including conventional abilities, irregular tactics, and criminal disruption. (GAO 2010, pp. 16-18.)

Hybrid warfare is primarily strategic, but it also has an impact on the operational and tactical levels. Hybrid warfare can begin long before armed actions begin and can even offer the opportunity to win a war despite its defeat. Hybrid warfare has existed throughout the history of war and thus, is not a new phenomenon. The invasion of disguised Russian soldiers ("green men") in the Crimea, with Russia denying its involvement in the invasion, introduced hybrid as a concept to the broad public. (Hybrid CoE 2019)

The ultimate purpose of hybrid warfare is to achieve a set of goals without fighting or with little use of force. Thus, in hybrid warfare, it is impossible to say when battles or violence takes place, as can be identified in the classic form of war. Hybrid warfare blurs the line between traditional Western thinking of peace and civilian activities, and military operations. The blurring of boundaries is achieved by combining both violent military actions and non-violent means, without crossing the threshold of war (Cederberg & Eronen 2015). Open democratic societies are particularly prone to hybrid warfare. Hybrid warfare uses strategic domains and sources of power: politics, diplomacy, intelligence, information, defense forces (including military actions), economics, financial elements, technology, culture, legal, psychology, morality, and other means of influence. Hybrid warfare also involves the use of force (Hybrid CoE 2019).

After 2014, the term hybrid threat was intended to refer to hidden vulnerabilities in Western countries that could be exploited by potential adversaries. While projecting your own vulnerabilities into an opponent's means of selection, the Western world today speaks of hybrid warfare. Raitasalo's (2019) conclusion is that hybrid warfare is a Western concept that seeks to conceptualize and understand the surprise of traditional Russian superpower policies in the West.

2.4 Hybrid operations

In the first phase of typical hybrid operations, the weaknesses of the target country (political, economic, social, infrastructure) are explored. The second phase is to attack the target country's administration and seize critical military and civilian targets. In the third phase, the stabilization phase, an apparent government is formed in the conquered region (Järvenpää 2017). According to Chekinov and Bogdanov (2013) new generation (e.g. hybrid or information) operations begin with a months-long non-military campaign against the target country. There are several ways to pressure the target, such as information, moral, psychological, ideological, diplomatic, economic, etc. Propaganda aims to influence the population, armed forces, and administration of the target country. There are provocations, insecurities, and terrorist acts in the target country. Prior to an armed conflict, critical objects are identified and paralyzed by armed force. Following the operation, military force invades the target country, isolating key targets and stopping potential resistance (Chekinov & Bogdanov 2013).

The structure of the hybrid operation consists of three main phases of pressure, where the aggressor uses multiple means of hybrid action against various functions of the targeted state and its society to achieve desired strategic goals. Sub-goals for hybrid operations are 1) Creating a threatening diplomatic climate; 2) Increasing distrust of government leadership; 3) Increasing the ambiguity of the situation picture, and 4) Increasing distrust of services and the functioning of critical infrastructure. The final goal is that the policies of Finland and the Baltic Sea states favour the policies of the influencing state. Following Table 3 describes the phases and events of the hybrid operation.

Table 3: The events and their descriptions of the hybrid operation. (Puistola, 2018)

Phase	Description / Elements
I	Fake news at a meeting of heads of state Agitating controversies between Churches Child Abduction Event Banking disruption, card payment interruptions Cyber-disruption in news communications, alternative media
II	Military exercises in the Baltic Sea - quick implementation Land acquisition attempts near power grid hubs - data for 3 years Information leakage in government - revealed 5 years later Airspace violations - multiple within 3 weeks Restricting the movement of merchant ships in the Baltic Sea
III	Terrorist's strike False news from ministers, spoofing campaigns Unidentified persons near power network nodes Disruption of gas supply Fostering citizen insecurity between botnets and critical communities

Hybrid operations aim to outperform the opponent's strengths and attack specific weaknesses with varying tactics and tools and intensity (FIIA 2018). A hybrid operation involves the means of achieving the desired political outcomes of two or more states. The range of means may include political and economic instruments, cyber warfare, the use or threatening of military force, cyber operations, and the use of Special Forces. The focus of hybrid operations will be identified weaknesses or weaknesses in the target country. (Cederberg & Eronen 2015.)

Tikanmäki and Ruoslahti (2021) note that internal and external security are becoming increasingly difficult to separate, as the operational environment is constantly changing in today's globalized world; there are situations, where external security can be influenced by internal security and vice versa.

The Ministry of the Interior of Finland (2016) recommends that actors and authorities exchange staff to better develop operating models towards improved situational awareness, and lessening cross-sector barriers that prevent information exchange between administrative sectors. Tikanmäki and Ruoslahti (2021) present a model, where situation understanding (aided by e.g., common information sharing systems) is recognised as a building block for deeper collaboration between authorities to focus on societal preparedness and combine both internal and external perspectives of security (Figure 3).

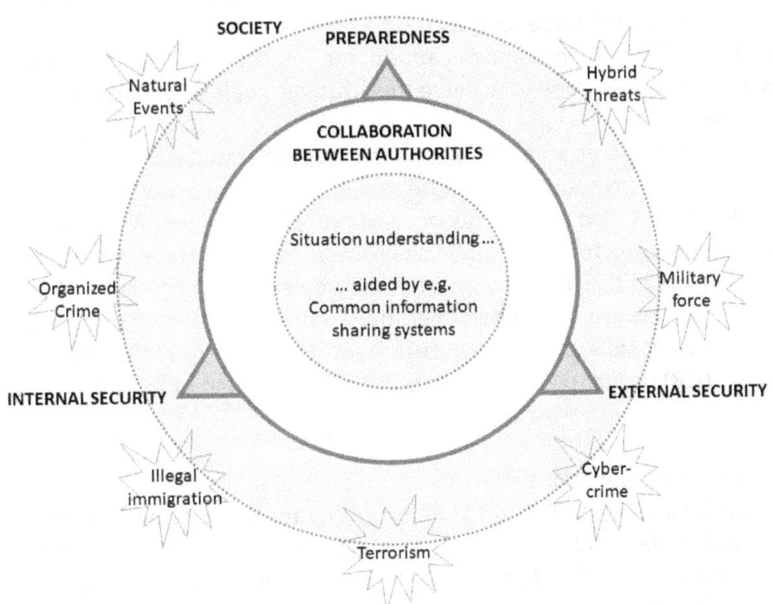

Figure 3: Authority collaboration and societal preparedness in internal and external security (Source: Tikanmäki & Ruoslahti 2021).

As seen in Figure 3, enhanced cooperation between authorities, through e.g. national and EU-wide security databases, is important in strengthening overall security against hybrid influence and threats. "Rapid and up-to-date exchange of information between security authorities is needed to maintain situational awareness." (Tikanmäki & Ruoslahti 2021, p. 340). The Ministry of the Interior (2016) aims at strengthening the exchange of information between authorities and with relevant actors that may need information across organizational boundaries (Ministry of Interior 2016).

3. Discussion and conclusions

The key efforts to counter hybrid threats – awareness, resilience, and response – are addressed on state and multi-state levels. In the EU, this calls for close cooperation first between the Member States, but also on wider international platforms of collaboration.

Each Member State has the responsibility to build systems that detect and alert of possible hybrid threats. This and building resilience and effective responses to these threats call for statewide coordination and cooperation. Especially those companies and actors that provide critical infrastructure and services (e.g. communications, energy and food supply) must find ways to share information on threats and build resilience together as a system-of-systems, rather that each actor by themselves.

Some main ways of hybrid influence can be geo-economics, information, or electoral impact, with the means of economic, political, or military pressure. These can be used all at once simultaneously or sequentially in phases. Today's high reliance on technology and open usage of social media, have increased possibilities to take advantage of possible weaknesses in resilience to create societal unrest. For example, cyber-attacks can be used to create continuity problems for critical infrastructure and services, or social media used to create division between different groups of society, or to steer unrest and demonstrations.

Hybrid influencing can occur on many levels. Puistola (2018) for example, lists four levels, which are diplomatic, informational, military, and economic (DIME), and it can be argued that technology could be added to this list to make it DIMET.

Hybrid influencing requires hybrid engagement or hybrid action, which can be defined as activities that pursue certain set goals and use a variety of means that complement each other to exploit the weaknesses of the target. The target begins to see signs of and experience systematic activity by a state or non-state actor where various military, economic or technological pressure, as well as information operations and social media campaigns are applied simultaneously. As discussed, the main ways to counter hybrid threats are resilience and response, and these require awareness, it is vital that the actors within and the state have an open culture and practice of information exchange to spot hybrid influencing as soon as possible. This can be easier said than done, as when targeted by strategic information influence and deception, the subject's perception of reality becomes obscured, and society becomes divided. This is done to achieve the strategic goals of the attacker.

Seeing signs of hybrid interference at an early stage can help the state and society counter with the needed resilience and effective defensive response. The state and its main actors can be targeted by synchronized use of multiple ways of harassment that aim to generate deep dividing lines within the society, possibly also leveraging techniques that threaten to disrupt political processes that paralyze its decision-making ability. These are listed in Table 2, as is a list of the main means of strategic deception by Puistola (2018). Thus, some of the warning signs of hybrid interference and cyber operations are false news, propaganda, political extremism, influence on elections or governance control over critical infrastructure, and financial incentives. The Institute for International Affairs (2018) is clear on its recommendation: the threat of hybrid disruption can be addressed with resilience.

Hybrid warfare can be a combination of instruments and means of influence to subvert states, institutions, and societies. Hybrid warfare is used in conflicts on state or interstate levels. It poses threats of multiple forms of warfare that exploit hybrid threats and apply means of hybrid influencing. Some components of hybrid warfare can be military action, special forces, rebel non-state forces, breaking of social harmony, information influence and propaganda, diplomacy, cyber-attacks against e.g. infrastructure and energy, and economic sanctions.

Defense against hybrid threats requires planning defenses and the commitment of a wide spectrum of actors in the society. The first line of defense is being alert and ready to defend against possible threats. Building national capabilities is a co-creative effort across all wakes of society.

Hybrid operations may proceed in phases. In the first phase, possible weaknesses of the target country (political, economic, social, infrastructure) are explored. The second phase will produce attacks against administration and infrastructure and will aim to seize critical military and civilian targets. The third and final phase is stabilization, where an apparent government is formed in the conquered region. The process will begin with a months-long non-military campaign, where several ways of applying pressure can be identified: information, moral, psychological, ideological, diplomatic, and economic. Table 4 lists possible forms of hybrid operations.

Table 4: Forms of hybrid operations

Hybrid operations		
Concept	Description / Elements	Source
Hybrid operations	1st phase: weaknesses of the target country (political, economic, social, infrastructure) are explored. 2nd phase: attack target country's administration and seize critical military and civilian targets. 3rd phase: stabilization: an apparent government is formed in the conquered region.	Järvenpää (2017)
	Begin with a months-long non-military campaign against the target country. There are several ways to pressure the target, such as information, moral, psychological, ideological, diplomatic, economic, etc.	Chekinov and Bogdanov (2013)
Hybrid operation events	1st goal: Creating a threatening diplomatic climate 2nd goal: Increasing distrust of government leadership 3rd goal: Increasing the ambiguity of the situation picture 4th goal: Increasing distrust of services and the functioning of critical infrastructure.	Puistola (2018)
Hybrid operations aim	At outperforming opponent's strengths and attacking specific weaknesses with varying tactics and tools and intensity	The Finnish Institute of International Affairs (2018)

Hybrid operations aim to outperform the strengths of the target state and attack its specific weaknesses. Varying tactics, tools, and intensity are used to achieve this aim and modern European societies should be aware of possible hybrid threats, build resilience against hybrid influencing, and have capabilities, societal unity, and international collaboration to defend against possible hybrid warfare.

This research contributes to the academic body of knowledge onhybrid threats and warfare by examining hybrid concepts and providing a basis of building some practical measures to build awareness, resilience, and defenses against hybrid influencing and activities. Future research could address societal resilience from the perspective of intentional hybrid disturbances, and co-creation of state and multi-state resilience. One interesting area of study is the study of situation awareness and situation understanding, and especially with the increasing importance of IT solutions, in the field of cyber security, where e.g. Early Warning Systems, and other methods of identifying cyber threats can increase cyber security and build resilience against hybrid threats and activity.

References

Arreguín-Toft, I. (2001)" How the weak win wars: A theory of asymmetric conflict." *International Security* 26, no. 1 (2001): 93-128. doi:10.1162/016228801753212868.

de Bruijne, M., Boin, A., and van Eeten, M. (2020)" *Resilience: Exploring the concept and its meanings*." In Designing Resilience: Preparing for Extreme Events, eds. Louise K Comfort, Arjen Boin, and Chris C Demchak, 13-32. Pittsburgh: University of Pittsburgh Press.

Cederberg, A. and Eronen, P. (2015)" How can Societies be Defended against Hybrid Threats?" *Strategic Security Analysis*, no. 9 (September 2015). (Geneva: Centre for Security Policy, GCSP).

Chekinov, S.G. and Bogdanov, S.A. (2013)" The Nature and Content of a New-Generation War." *Military Thought* 4 (2013): 12-23.

European Union. (2018)" A *Europe that Protects: Countering Hybrid Threats.*" Available at: https://eeas.europa.eu/topics/economic-relations-connectivity-innovation/46393/europe-protects-countering-hybrid-threats_en. (Accessed 13 November 2019)

GAO. (2010) *Hybrid Warfare*, GAO-10-1036R, United States Government Accountability Office. Washington, DC September 10, 2010, 16-18.

Gärdström, J. (2018)*" Hybridisodankäynti - uutta vai vanhaa*?" (Master's Thesis, National Defence University, 2018), 2.

FIIA. The Finnish Institute of International Affairs. (2018)" Hybridivaikuttaminen ja demokratian resilienssi - ulkoisen häirinnän mahdollisuudet ja torjuntakyky liberaaleissa demokratioissa". *FIIA Report* 55/2018. ISBN 978-951-769-567-1.

Finnish Government. (2003)" *Strategy for Securing the vital functions of society*". (In Finnish: Yhteiskunnan elintärkeiden toimintojen turvaamisen strategia), Valtioneuvoston periaatepäätös 27.11.2003. Available at: https://www.defmin.fi/files/248/2515_1687_Yhteiskunnan_elintArkeiden_toimintojen_turvaamisen_strategia_1_.pdf. (Accessed 9 January 2020)

Hoffman, F.G. (2007) *Conflict in the 21st Century: The Rise of Hybrid Wars*. Arlington: Potomac Institute for Policy Studies.

Huttunen, M. and Metteri, J. (eds.*).* (2008) *Ajatuksia operaatiotaidon ja taktiikan laadullisesta tutkimuksesta. Maanpuolustuskorkeakoulu*. Taktiikanlaitos. Julkaisusarja 2 no 1/2008. Helsinki: Edita Prima Oy.

Hybrid CoE. (2019)" *Hybrid Warfare – a very short introduction*." The European Centre of Excellence for Countering Hybrid Threats. COI Strategy & Defence Conception Paper, May 2019. ISBN 978-952-7282-20-5.

Hyytiäinen, M. (2018)*" Hybridivaikuttaminen*", in Turvallinen Suomi 2018 - Tietoja Suomen kokonaisturvallisuudesta. Helsinki: Lönnberg Print & Promo.

Järvenpää, M. (2017) Viranomaisten toimivaltuudet kohteiden suojaamisessa hybridiuhkia vastaan. *Tiede ja Ase* 74 (February 2017). Available at: https://journal.fi/ta/article/view/60630. Accessed 9 January 2020)

Lalu, P. and Puistola, J-A. (2015)" On the concept of hybrid warfare.*" Finnish Defence Research Agency Research Bulletin* 01-2015. Helsinki: Finnish Defence Research Agency.

Mack, A. (1975)" Why big nations lose small wars: The politics of asymmetric conflict.*" World Politics* 27, no. 2 (1975): 175-200. doi:10.2307/2009880.

Ministry of Interior. (2016) Interdependence of Internal and External Security. Will the operational culture change with the operational environment? Available at: http://julkaisut.valtioneuvosto.fi/bitstream/handle/10024/79230/37_2017_Interdependence%20of_nettiin.pdf. (Accessed 4 March 2020)

Munich Security Conference. (2015)" *Munich Security Report 2015: Collapsing order, reluctant guardians*?" Published on the occasion of the MSC 2015. Available at: https://securityconference.org/en/publications/munich-security-report-2015. (Accessed 19 July 2020)

Mäkelä, J. (2018)*" Merelliset hybridiuhat*." CMD Juha Mäkelä's presentation at Sotatieteenpäivät 23.5.2018.

Puistola, J-A. (2018)" *Kokonaisturvallisuus ja hybridivaikuttaminen*." CAPT (N) Juha-Antero Puistola's presentation at Sotatieteenpäivät 23.5.2018.

Raitasalo, J. (2019)" Hybridisota ja hybridiuhat – paljon vanhaa, onko mitään uutta?" in *Tiede ja Ase* 76 (January 2019). Available at: https://journal.fi/ta/article/view/7754. (Accessed 31 January 2020)

Security Committee. (2017)" *The Security Strategy for Society*. Yhteiskunnan turvallisuusstrategia". Valtioneuvoston periaatepäätös 2.11.2017. Available at: https://turvallisuuskomitea.fi/wp-content/uploads/2018/02/YTS_2017_suomi.pdf. (Accessed 1 February 2020)

Terminology Centre. (2017) *Vocabulary of Comprehensive Security*. ISBN 978-952-9794-36-2 (PDF). Available at: http://www.tsk.fi/tiedostot/pdf/Kokonaisturvallisuuden_sanasto_2.pdf. (Accessed 1 February 2020)

Tiilikainen, T. (2020)" *Hybridivaikuttaminen on kylmäävän laajaa*". In ERVE Uutiset 21.1.2020. Available at: https://erveuutiset.erillisverkot.fi/teija-tiilikainen-hybridivaikuttaminen-on-kylmaavan-laajaa/?utm_source=creamailer&utm_medium=email&utm_campaign=Erve+Uutiset+Maaliskuu+14+2019&utm_content=%5Bemail%5D. (Accessed 7 February 2020)

Tikanmäki, I. and Ruoslahti, H. (2021) Interdependence of Internal and External Security. Proceedings of the 20th European Conference on Cyber Warfare and Security (ECCWS 2021), University of Chester 24th - 25th June 2021, pp. 425-432.

Yin, R.K. (2009) *Case Study Research. Design and Methods*. London: SAGE Publications.

Application of Geospatial Data in Cyber Security

Namosha Veerasamy, Yaseen Moolla and Zubeida Dawood
CSIR, Pretoria, South Africa
nveerasamy@csir.co.za
ymoolla@csir.co.za
zdawood@csir.co.za

Abstract: Geospatial data is often perceived as only being related to maps, compasses and locations. However, the application areas of geospatial data are far wider and even extend to the field of cybersecurity. Not only is there an ability to show points of interest and emerging network traffic conditions, geospatial data also has the ability to model cyber crime growth patterns and indicate affected areas as well as the emergence of certain type of cyber threats. Geospatial data can feed into intelligence systems, help with analysis, information sharing, and help create situational awareness. This is particularly useful in the area of cyber security. Geospatial data is very powerful and can help to prioritise cyber threats and identify critical areas of concern. Previously, geospatial data was primarily used by militaries, intelligence agencies, weather services or traffic control. Currently, the application of geospatial data has multiplied, and it spans many more industries and sectors. So too for cyber security, geospatial data has a wide number of uses. It may be difficult to find patterns or trends in large data sets. However, the graphic capabilities of geo mapping help present data in more digestible manner. This may help analysts identify emerging issues, threats and target areas. In this paper, the usefulness of geospatial data for cyber security is explored. The paper will cover a framework of the key application areas that geospatial data can serve in the field of cyber security. The ten application areas covered in the paper are: tracking, data analysis, visualisation, situational awareness, cyber intelligence, collaboration, improved response to cyber threats, decision-making, cyber threat prioritisation and protect cyber infrastructure It is aimed that through the paper, the application areas of geospatial data can be more widely adopted.

Keywords: geospatial data, cyber security, geoinformatics, GIS

1. Introduction

The advancement and integration of information and communication technology (ICT) in peoples' daily lives has led to a growing cyberthreat landscape. This pushes the need for better solutions and techniques to combat threats. Integrating geospatial data into existing tools and techniques could strengthen software systems.

Information in a digital format has become more valuable. Organisations can gain tremendous insight and awareness from information that is presented and visualised in a useful representation. Information that has been analysed and communicated into a relevant format with dataflow pipelines which are designed for rapid continuous updates can vastly improve decision making and establish priorities.

Decision making is driving organisations and core to this is accurate information. Strategically organisation recognise the value of information as an asset. This brings forth the continuous need for novel data sources and solutions. There is a persistent pursuit to find new ways to use data, find relationships and identify trends. Taking geospatial data into consideration, this field provides vast areas for note-worthy data visualisation. Previously, geospatial data was primarily used by militaries, intelligence agencies, weather services or traffic control. Currently, the application of geospatial data has multiplied, and it spans many more industries and sectors. So too for cyber security, geospatial data has a wide number of uses. For instance, the location of cyber-attacks can be identified, and patterns can be revealed towards predicting future attacks.

The contributions of the paper are summarised as follows:

- Insight into how GIS data can be applied to the cyber security field
- Generation of ideas on new techniques and technologies that can be used to represent cyber security (incidents, attacks and crime)
- Show how geo-spatial mapping can help with the monitoring of cyber security incidents
- Indicate how cyber security threats can be studied to create awareness on risks and communicate pivotal information about cyber threats frequency and impact

This paper looks at these application areas to the domain of cyber security. The remainder of the paper is structured as follows: the next section provides some background on geospatial data. Thereafter, the framework of application areas is discussed. The researchers then conclude the paper and propose future work.

2. Geospatial data

Geospatial data typically combines location information (usually coordinates on the earth) and attribute information (the characteristics of the object, event or phenomena concerned) with temporal information (the time or life span at which the location and attributes exist) (IBM 2022).

With geospatial analytics, timing and location can be added to common data types to produce useful visualisations. Visualisations can be in the form of maps, graphs, statistics and cartograms that can show changes over a period of time, as well as shifts in development. These visualisations offer more insight as many aspects can be missed in a long list of data. Patterns can be identified, and trends detected. This can lead to faster and more reliable predictions and influence decision-making.

Examples of geospatial data include (IBM 2022):

- Vectors and attributes: Descriptive information about a location such as points, lines and polygons
- Point clouds: A collection of co-located charted points that can be re-contextured as 3D models
- Raster and satellite imagery: High-resolution images of our world, taken from above
- Census data: Released census data tied to specific geographic areas, for the study of community trends
- Cell phone data: Calls routed by satellite, based on GPS location coordinates
- Drawn images: CAD images of buildings or other structures, delivering geographic information as well as architectural data
- Social media data: Social media posts that data scientists can study to identify emerging trends

Maps are a common practice for presenting spatial data as they can easily communicate complex topics. They can help validate or provide evidence for decision making, teach others about historical events in an area, or help provide an understanding of natural and human-made phenomena (Safe.com 2022).

A key capability of geospatial mapping is the use of choropleth maps. Choropleth maps are able to show differences, consistencies and patterns. Classified areas in a choropleth map will have distinct boundaries whereas heat maps, which demonstrate the concentration or density of a phenomenon, have indistinct boundaries. Different colour schemes and classes can be used to represent issues (Figure 1).

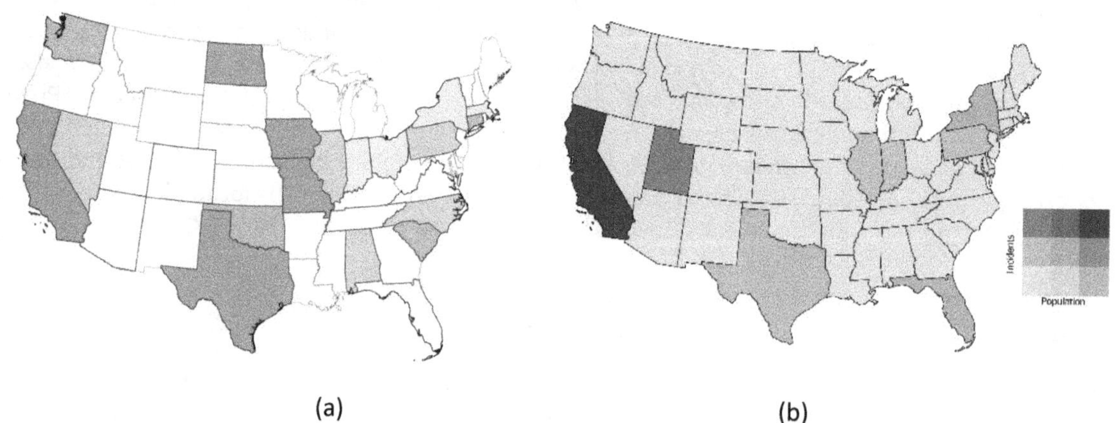

Figure 1: Example Choropleth maps. (a) shows a monovariate choropleth map; and (b) shows a bivariate choropleth map, which visualises the correlation between two variables through two overlayed colour scales

The world is advancing at a rapid rate and there are more developments in the technological fields. With the rapid growth of ICT, comes the associated risks of cyber threats. Cyber threats also evolve as attackers adapt their methods of operation. Geospatial data can help better understand the changes that are taking place and support the defence against threats through data analysis and powerful visualisations. Some questions that can be considered are:

- Where are cyber-attacks taking place?
- What are the locations of targets?
- What are the locations of the sources of attacks?
- What are the locations of intermediary infrastructure in cyber-attacks, e.g., proxy servers, digital post-boxes, and command and control servers?
- What type of cyber-attacks are occurring?
- What are the prime targets?
- What are the main methods used?
- How is the cyber-attack pattern changing over time?

Geospatial data and analytics can provide insight into these important questions by identifying the locations, types, targets, methods, as well performing comparisons.

Previously spatial data was cumbersome to use and required advanced software. However, with advances in processing and ICT, more domains can turn to spatial data to provide better insight and find solutions for problem areas. Smartphones, vehicle tracking, and satellites are helping to unlock the world of spatial and location data for organizations big and small (Safe.com).

Geospatial data now performs the functions that maps, compasses and smoke signals once fulfilled. Tracking people, populations, topography changes, events, storms and traffic can all be performed using geospatial data.

A key application area of geospatial data in the field of cyber security is tracking. Geospatial data can play a very strategic role for data analysis. It can help visualise the impact of a hacking group or see the plot of propaganda in social media feeds. Visualisation of the spread of cyber-attacks can create awareness of the scope and impact of these events. With geospatial data, cyber-attack data can be represented in an impactful manner. Location intelligence can be key to developing cyber intelligence. Connections can be shown, and links found, through the use of maps, descriptions and data fields. Graphical symbology is very useful in conveying critical pieces of information that makes it easier for the viewer to process.

Cybersecurity specialists can now also rely on geospatial data to improve security and build defensive mechanisms. Geospatial data is undervalued. Most organisations embrace the latest technology, but the vital capabilities of geospatial data can be missed. There is scant literature on using GIS for cyber security. Within the Florida University, researchers have mapped cyber-attacks to geospatial data (Zhiyong Baynard, Hongda & Fazio 2015). The geospatial analysis of this data had revealed spatial patterns, and identified countries that were more prone to cyber-attacks. Furthermore, hotspots within the United States were identified. Xui and Li performed some preliminary analysis on mapping IP addresses to a spatial database for geolocating cybercrimes (Xui W 2014). Bhargava et al. developed a framework for defining various types of cybercrimes based on laws in India, and analyse the spatial distribution of cybercrimes across India (Bhargava 2015). In the military domain, German cybersecurity experts integrated GIS with cybersecurity tools to discover patterns from cyber attacks (Conklin B 2022). In the commercial space, Kaspersky (Kaspersky 2021), Bitdefender (Bitdefender 2022), and Fortinet (Fortinet 2022) provide maps to visualise threats that are detected by their respective software suits. It is unclear what further processing and analysis, beyond processing, is performed on their geospatial data.

Geospatial data can the ability to strengthen cyber security by providing a wealth of information. This paper summarises various benefits and application areas for the use of geospatial data in the field of cybersecurity. The proposed framework that discusses the applications are as follows.

3. Framework

The authors were able to gauge the usefulness of geospatial data during an introductory course on a geo-mapping. However, the literature review revealed that scientific literature for demonstrating the use of geospatial data within cyber security is lacking. To solve this, the authors prescribe using a framework for achieving seamless integration between geospatial data and cyber security.

Geospatial data can be used to represent information from various domains such as: (Trajectory Magazine 2022):

- Retail: income, housing/rent prices, population, age
- Weather patterns: hurricanes, tornadoes, extreme winter weather
- Site identification: traffic patterns, foot traffic, number of residents, competitor information
- Healthcare: water location, drug users, environmental hazards, vaccines,
- Financial services: visualise real estate, track construction over time, analyse investments without travel
- Logistics/ Transportation: vehicle tracking, expedite schedule, route analysis

Exposure to these functional areas showed tremendous capability to further apply geospatial data to cyber threats and cyber security.

Unequivocally, by looking at the capabilities and features of geospatial data, it can be further extended to strengthen an organisation's line of cyber defence and security. Many national security agencies may already utilise geospatial data. Defensive organisations, like the military or national intelligence services, emergency services and infrastructure safety control groups may already include geospatial data in their systems. The field of cyber security can also reap the benefits of utilising geospatial data. A framework is proposed in Figure 2 that encapsulates the core application areas of geospatial data to cybersecurity. A discussion follows on these application areas.

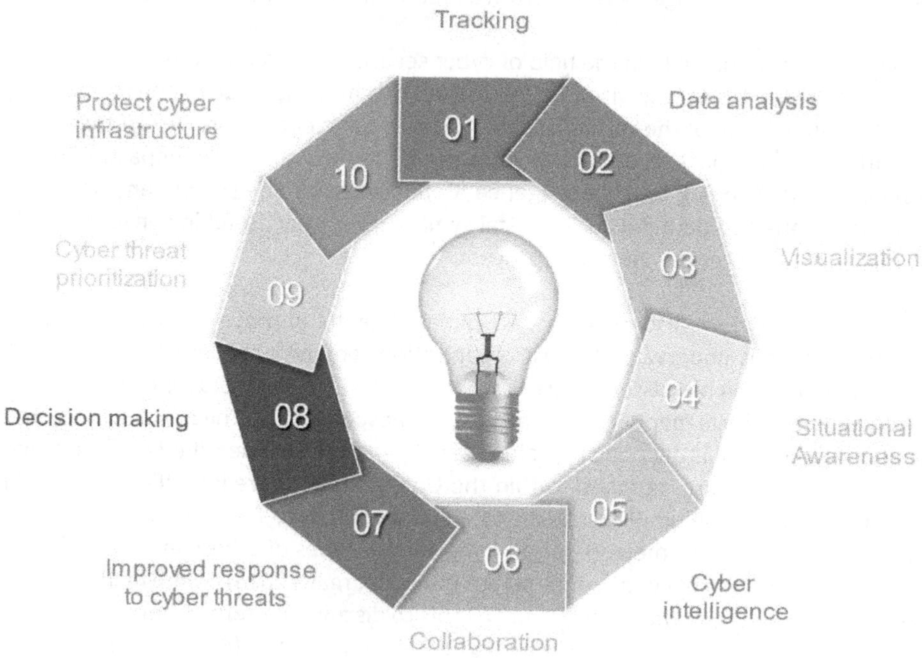

Figure 2: Application of geospatial data to cyber security

The framework summarises 10 principal application areas which are:

- Tracking
- Data analysis
- Visualisation
- Situational awareness

- Cyber intelligence
- Collaboration
- Improved response to cyber threats
- Decision-making
- Cyber threat prioritisation
- Protect cyber infrastructure

These are elaborated on in the next few sections. The framework provides a snapshot of how geospatial data can be integrated into stronger cyber defence capabilities. It encapsulates the main benefits of using geospatial data to visualise, analyse and grow cyber intelligence. The framework is not rigid in that it can include many more application areas. The main aim of the framework is to show the value of geospatial data so that it can be further utilised and contribute to the development of stronger cyber defence capabilities, as well as create awareness of threats.

3.1 Tracking

A key application area of geospatial data in the field of cyber security is tracking. "By implementing defence systems that include mappable and traceable physical locations in the digital sphere, security experts can more effectively follow and track possible threats (Brode 2021)."

Whitelisted zones can serve as "geo-fences" to only permits access within specific ranges. Access attempts from blacklisted areas can be restricted and flagged.

Furthermore, when threats are detected from certain locations, perpetrators can be tracked to identify patterns or core targets. With the use of geospatial data, trends can be identified, and pre-emptive action taken when a potential attack is identified.

Certain areas can be identified to be hotspots for specific threats. Closer monitoring and resources can be deployed to hotspot locations or those locations more susceptible to certain vulnerabilities.

3.2 Data analysis

Geospatial solutions provide the capability to integrate information from multiple sources, channels and also use existing data towards establishing correlations between objects and events. These various sources help provide more clarity and provides for insightful analysis.

For example, data can be presented in tables showing various locations of a cyber-attack. However, once this data is plotted into a map or grouped, aggregrated and processed by geographic region, then the prevalence of a specific cyber-attack in certain locations can be seen. Geospatial analysis provides the ability to combine data to produce intelligence that can be used for information sharing and the creation of new knowledge.

3.3 Visualisation

Large files with numbers are usually hard to read and make it difficult to spot patterns easily (Datumize 2020). The graphic representations in geospatial data are able to present extensive sets of data in a clear and cohesive manner. This further provides for comprehension of the data which can be used to draw conclusions and grasp different perspectives. Visual representation of data also provides the ability to detect anomalies. For instance, if a certain town shows 2 million attacks but it is relatively small town with only 1000 residents, this will clearly raise a red flag that a data capturing error has occurred. This is more readily detectable in a visual representation of the data than a long table or text list. Thus, graphical representations can be used to avoid cognitive overload with cyber security data.

3.4 Situational awareness

Endsley has proposed one of the most widely used views of situational awareness (1995). It is shown as a three-step model in Fig 3.

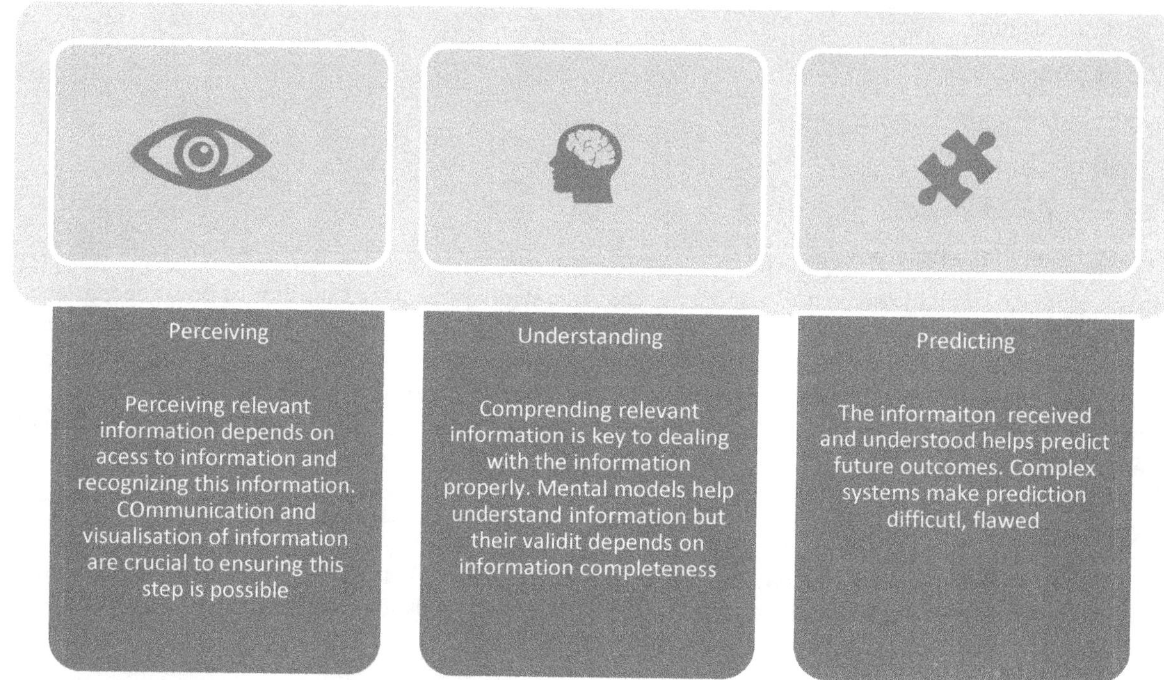

Perceiving
Perceiving relevant information depends on acess to information and recognizing this information. COmmunication and visualisation of information are crucial to ensuring this step is possible
Understanding
Comprending relevant information is key to dealing with the information properly. Mental models help understand information but their validit depends on information completeness
Predicting
The informaiton received and understood helps predict future outcomes. Complex systems make prediction difficutl, flawed

Figure 3: Steps in situational awareness (Endsley 1995)

The first steps of situational awareness is to perceive the relevant information by being able to access it and being able to recognize it. A critical requirement is communication and proper visualisation (CQ Net 2022). With geospatial data, cyber security threat information can be communicated and visualised more effectively

According to Visual Teaching Alliance (Shiftelearning, 2021):

- The brain can see images that last for just 13 milliseconds.
- Our eyes can register 36,000 visual messages per hour.
- We can get the sense of a visual scene in less than 1/10 of a second.
- 90% of information transmitted to the brain is visual.
- Visuals are processed 60,000X faster in the brain than text.
- 40 percent of nerve fibers are linked to the retina

The second step of situational awareness entails the comprehension of the information. This is largely dependent on the knowledge to handle the incoming information as well how the information is presented. A person can create a mental model or update an existing one depending on how the information is presented (CQ Net 2022). Visuals have been found to improve learning by up to 400% (Shiftelearning, 2021). By representing the information graphically and visually, intelligence can be created by taking in more information and synthesising it into new knowledge.

The final step for situational awareness is the prediction of possible outcomes depending on the information received and comprehended. There may be complex dependencies but, with useful information, insightful forecasts can be made to assist with decision- making.

Situational awareness is critical for cyber security to gain an understanding of the environment to support decision-making. For instance, if geospatial models demonstrate the number of incidents in the USA, using a map such as the one shown in Figure 4, one can clearly distinguish the states that are more at risk.

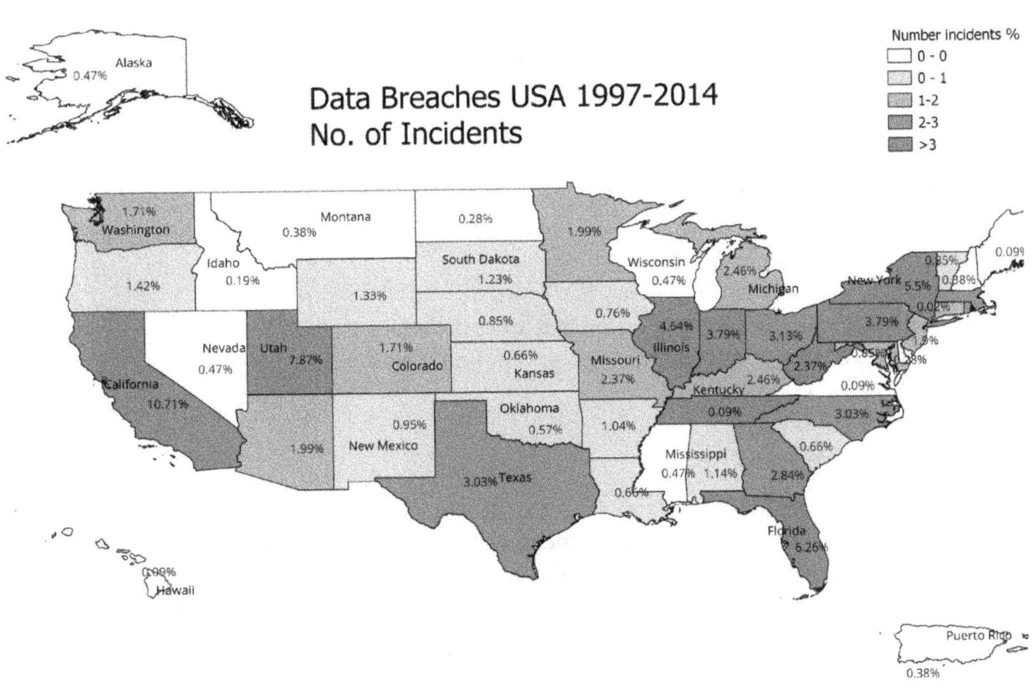

Figure 4: Map to indicate data breaches in states of the USA. Map developed using QGIS

3.5 Cyber intelligence

Raw data is collected and processed to form information. Using GIS, this information can be processed and analysed in conjunction with other data sources in order to create intelligence. With proficient visualisation, strong insight can be gained.

Some conclusions can become more evident when shown in a geospatial manner. Other issues may need to be investigated to gain more clarity. For example, a time lapse of attacks may show a decline or increase in a threat over time. This may correlate to a certain event or cyber vulnerability.

3.6 Collaboration

Analysis of certain attacks can reveal a pattern. For example, a particular switch or phone could be targeted. Equipped with the data analysis results, cyber specialists can reach out to the affected organisations/ service providers and aim to work together in tracking, tracing and monitoring the threat.

If a specific type of threat is occurring, organisations around the world can join forces to work together and implement controls to try and deter or stop the more rampant rise. With the use of geospatial data, security response centres, intelligence bureaus and security organisations would be able to see which specific sectors are vulnerable and how to carry out an operations plan to reduce the spread of a threat.

3.7 Improved response to cyber threats

The benefits of visualisation in geospatial data are that the visualisation capabilities allow us to identify emerging trends and react more quickly based on what has been identified. These patterns are easier to consume in a visual format as we are able to more closely correlate parameters.

3.8 Decision-making

The insight gained from geospatial data analysis can contribute to more effective decision-making. Instead of using intuition, decision makers can rely on more discerning findings. Geospatial data can give a good sense of the findings and increase visibility and understanding of core issues.

For example, geospatial data can reveal certain sites or locations that are being targeted by a certain attack type. More resources and defensive mechanisms can be deployed. Awareness campaigns can be targeted to warn users about the threat of a specific type of cyber-attack.

3.9 Cyber threat prioritisation

Since better insight can be gained from geospatial data, it can be used to prioritise threats. For example, a rise in a specific threat can be detected. Also, the location of prevalent threats can be found. By analysing the data, specialists can classify which threats need more attention.

With the pattern detection and trend observation in geospatial analytics, cyber security specialists can monitor a threat to see if it is increasing or slowing down over a period. This can help identify critical threats and those that need urgent attention.

3.10 Protect cyber infrastructure

By providing insight into trending attacks and targets, key decision makers can assign resources and deploy defensive techniques. Geospatial data thus has the capability to help improve security and protect cyber infrastructure by providing key information about focus areas and methods of attack. This information can feed into defensive strategies and protection mechanisms of key locations.

4. Conclusion

With the rise of cyber-attacks and cybercrime, it is important to find new knowledge areas that can be used in the field of cyber security. With the identification of new application areas, stronger awareness of cyber security threats can be created. Using GIS within cyber security systems is an emerging trend.

There are many benefits of using geospatial data for cyber security. With geospatial data, cyber threat data can be captured, for example, the location of attacks, the number of attacks, the proportion, dates, types and various other factors. Encapsulated into visual representation, the information can be better interpreted and processed. With geospatial data information can be organised better and ideas can be communicated more effectively. This helps enhance comprehension, interpretation and processing. It also increases the ability to find patterns and relationships. For example, when attacks are plotted into a map, frequent attacks on more urbanised locations can be identified or a pattern of a certain attack in specific locations that are susceptible to a vulnerability. The literature about using GIS within cyber security is limited. A strong foundation is required to enable the use of geospatial data for cyber security. To assist with this, the researchers provide a multi-dimensional framework that can be used as a starting point for using for integrating geospatial data for cyber security. Future work includes applying the framework to various use-cases.

5. Future work

Future work includes applying the framework to various use-cases. For the South African context, future work can entail linking geo-mapping to another aspect with the development of a system whereby users can report when they become a victim of cybercrime. The two systems can then plot incident reports with location, impact and frequencies. This can also be extended to show different scales of losses from cyber incidents. Other future work would include, first collecting more recent cybercrime incident data, preferably for South Africa. Further statistical analysis will be done to determine dependencies and correlations between various demographic factors and cybercrime incidents. Models can then be developed for predictions across time and space domains.

References

Bitdefender, 2022 Cyberthreat Real-time Map, [Online] Available at https://threatmap.bitdefender.com/, Accessed 26 January 2022.

Bhargava, N., Bhargava, R., & Tanwar, P. S. ,2015. Analysing and Implementing Spatial Distribution of Cyber Crime Trends in India. *International Journal of Advanced Research in Computer Science*, *6*(4).

Brode B, 2021, Why cybersecurity experts want more geospatial data, Geospatial World, [Online] Available at https://www.geospatialworld.net/blogs/why-cybersecurity-experts-want-more-geospatial-data/, [Accessed 19 January 2022].

Conklin B, 2022 Cybersecurity: The Geospatial Edge, [Online] Available at https://www.esri.com/about/newsroom/blog/german-cybersecurity-experts-use-gis/ Cybersecurity: The Geospatial Edge, [Accessed 25 January 2022].

CQ Net, 2022, Situational awareness: What it is and why it matters as a management tool, [Online] Available at https://www.ckju.net/en/dossier/situational-awareness-what-it-and-why-it-matters-management-tool, [Accessed 20 January 2022].

Datumize, 2020, The top five advantages of data visualisation, [Online] Available at https://blog.datumize.com/top-five-advantages-of-data-visualization, [Accessed 20 January 2022].

Fortinet, Inc. ,2022, Fortiguard Map, [Online] Available at https://threatmap.fortiguard.com/, [Accessed 26 January 2022].

Endsley, MR , 1995, Toward a Theory of Situation Awareness in Dynamic Systems. Human Factors Journal 37(1), 32-64. Human Factors: The Journal of the Human Factors and Ergonomics Society. 37. 32-64. 10.1518/001872095779049543.

IBM, 2022, What is geospatial data?, [Online] Available https://www.ibm.com/topics/geospatial-data, [Accessed 21 January 2022].

Kaspersky Labs ,2021, Cyberthreat Real-time Map, [Online] Available at https://cybermap.kaspersky.com, Accessed 26 January 2022.

Safe.com, 2022, What is spatial data?, [Online] Available https://www.safe.com/what-is/spatial-data/, [Accessed 21 January 2022].

Shiftelearning, 2021, Studies Confirm the Power of Visuals to Engage Your Audience in eLearning, [Online] Available at https://www.shiftelearning.com/blog/bid/350326/studies-confirm-the-power-of-visuals-in-elearning, [Accessed 20 January 2022].

Trajectory Magazine, 2022 The Past, Present, and Future of Geospatial Data Use, [Online] Available at https://trajectorymagazine.com/past-present-future-geospatial-data-use/, [Accessed 21 January 2022].

Xui W 2014 Xiu, W., & Li, X. (2014, April). The design of cybercrime spatial analysis system. In *2014 4th IEEE International Conference on Information Science and Technology* (pp. 132-135). IEEE.

Zhiyong H, Baynard, CW, Hongda H, Fazio, M , 2015. GIS mapping and spatial analysis of cybersecurity attacks on a Florida university, *IEEE 2015 23rd International Conference on Geoinformatics* , Wuhan, China (2015.6.19-2015.6.21).

Layer 8 Tarpits: Overwhelming Malicious Actors With Distracting Information

Toni Virtanen and Petteri Simola
Finnish Defence Research Institute, Tuusula, Finland
toni.virtanen@mil.fi
petteri.simola@mil.fi

Abstract: This paper presents a concept for utilising falsified documents and disinformation as a security measure by diminishing the utility of the stolen information for the attacker. Classical definition of tarpitting honeypots is to create virtual servers attractive to worms and other malware that answer their connection attempts in such a way that the machine on the other end becomes stuck. A common extension to the OSI model is to refer the user as the layer 8 on top of the application layer. By generating attractive looking but falsified documents and datasets within our secured network along with the real information, we could be able to force the malicious user on the other end similarly to be 'stuck' as they need to dig through and verify all the information they have managed to steal. This in effect slows down the opponents' decision making speed, can make their activity in the network more visible and possibly even mislead them. The concept has similarities to the Canary trap or Barium Meal type of tests, and using Honey tokens to help identify who might be the leaker or from which database the data was stolen. However, the amount of falsified data or fake entries in databases in our concept is significantly larger and the main purpose is to diminish the utility of the stolen data or otherwise leaked information. The requirement to verify the information and scan through piles of documents trying to found the real information among them can give more time to the defender to react if the attack was noticed. It will also reduce the value of the information if it is just dumped in the open, as its contents and authenticity can be more easily questioned. AI powered methods such as the GPT-3 that can generate massive amounts very realistic looking text which is hard to differentiate from human generated texts could make this type of concept more feasible to the defender to utilise. The shortcoming of this concept is the risk that legitimate end-users could also confuse the real and falsified information together if that is not prevented somehow.

Keywords: tarpit, honeypot, information security, disinformation, decision making

1. Introduction

Securing critical information and assets is paramount for operational security in the military and other defence critical organizations. According to a recent Ponemon Institute study, 52% of all data breaches are caused by malicious or criminal attacks, 23% happen due to human error, and 25% are caused by both IT and business process failures and system glitches at an average total cost of $4.27 million (Ponemon Institute, 2020). Defenders have long been using monitoring and intrusion detection systems to alert against unauthorized access (Carol & Raouf, 2014). These can simply be sensors that are spread across the network to alert the Cyber Security Operations Centres (CSOC) if they detect traffic that are flagged as suspicious by the detection scripts monitoring the traffic. In addition to detecting anomalies within the network, various kinds of honeypots and honey tokens can be further used to bolster the detection capability of the defenders. These could be files, network locations and services, which look interesting to the hacker and their value lies in the fact that they are probed and attacked by the hackers. Honeypots can therefore be considered as a deceptive cyber security systems and are valuable in distracting attackers away from more valuable targets, provide early warnings on new exploitation methods and allow in-depth examination of adversaries tactics during and after an attack (Qassrawi & Hongli, 2010). Honeypots can be classified according to the level interaction they simulate for the attackers, from low-interaction honeypots that simulate a limited number of network services and vulnerabilities to medium- and high-interaction honeypots that provide increasing levels of emulation of a real operating system for the attacker to interact with. (Mokube & Adams, 2007)

The OSI model (Open Systems Interconnection model) is a seven-layer abstract model that describes an architecture of data communications for networked computers (ISO, 1994). It consists of physical layer, data link layer, network layer, transport layer, session layer, presentation layer and application layer. Although the seven-layer OSI model is not as common as the more widely used five-layer TCP/IP model (Braden, 1989), quite often the application layer is still referred as the seventh layer. The OSI model have also been extended by referring the layer 8 as the "user" layer. (Bauer & Patrick, 2004) This user layer represents the Human-Computer Interaction elements of the network communication architecture and some have even added political aspects to the layer 8. As the key factor for any honeypot is to deceive the attacker into believing the intrusion is real (Odemis, et al., 2018). Therefore, especially with the high interaction honeypots, it is crucial that in addition to looking technically realistic, they need to target the emotional and cognitive processes of the hackers through deception. One such deceptive measure is to lure the attacker towards a honeypot by presenting that as a more easy or valuable target instead of the real system. This is often done by leaving easy, but not suspiciously easy, vulnerabilities in the honeypot. It is quite fitting therefore that the term honeypot comes from human intelligence (HUMINT) terminology, referring to a strategy where an attractive male or female agent is used to seduce individuals and exploit the created sexual relationship to coerce them to cooperate (Odemis et al 2018).

Tarpit honeypots on the other hand are generally virtual servers attractive to worms and other malware that answer their connection attempts in such a way that the machine on the other end becomes stuck in a connection loop. Although especially effective against automated scripts that might be stuck for a very long time, tarpits can also delay human hackers significantly before they realize there is something wrong (Zobal, et al., 2019). A seminal example of a classical tarpit presented by Tom Liston was the LaBrea tarpit that tricked the attacker with a TCP three-way handshake by answering the connection establishing SYN-packets with a SYN/ACK packet while not actually opening any connection and ignoring the final ACK packet of the handshake. It was developed especially to delay and hinder worms and similar malware wasting the resources of those malicious actors using them. (Liston, 2002) For a more interactive tarpit honeypot, this type of delaying and hindering the attacker's reconnaissance of the system would be to use fake error messages and exaggerated delays (Rowe & Goh 2007).

As the title suggests, this paper discusses whether similar tactics used in honeypots and tarpits can be used against the actual malicious person or organization behind an attack, e.g. the 8-layer. By generating attractive looking but falsified documents and datasets within a secured network along with the real information, it could be possible to force the malicious user on the other end similarly to be 'stuck' as they need to dig through and verify all the information they might have managed to exfiltrate out of the secured network.

Although this concept is not technically a preventative measure, it could act as a deterrent by reducing the significance and utility of the exfiltrated data. It could also help to mitigate potential repercussions from a possible data leak, as it would be easier to question the authenticity of the leaked information. An example of this type of cyber-blurring has already been documented with the Macron's e-mail leak just 24 hours before the 2017 French presidential election (AFP, 2017). Macron's campaign team knew they could not protect themselves 100 % from hacking and sophisticated spear-phishing attacks, so they created false accounts, documents, and e-mails that would at least slow the attackers, as they would need to verify all of them. (Nossiter & Sanger, 2017) They were also proactive by often stating in the media that they have been under hacking attempts during the whole 2-year campaign period. Effectively making a narrative that some leaks are eventually going to happen, and when it did, they manage to take initiative by making statement just hours after the leak. Admitting that although there was a leak, it was mixed with falsified information "in order to sow doubt and disinformation" (Vilmer & others, 2019). With their fast reaction to the leak, they managed to change the discussion on the content of the leak to the trustworthiness of the leak. If the data were already mixed with falsified information, taking this type of initiative would be even easier to the defender.

2. Theory

2.1 Increasing opponent's workload

From a theoretical perspective, honeypots and tarpits can be consider as a form of deception. In military deception, it is all about gaining and maintaining initiative. Usually attackers have that initiative, and thus has an advantage over the defenders, who often are forced into a reactive position. The Macron´s campaign case presented earlier is a good example of such attempt to keep the initiative, even if a breach or leak does occur. There are several different ways in defining deception and military deception. One way is to categorise deception in two main categories: A-type and M-type deception. A-type, which is also called ambiguity increasing deception is designed to create general confusion and to distraction. M-type, which is also called misleading deception is designed to mislead an enemy into believing a specific deception plan (Daniel, et al., 1980). The Layer 8 tarpits can be considered to work mainly as A-type deception. Attackers are unable to distinguish what is real and relevant and what is fake or irrelevant and thus are forced to work through all documents, which is time and resource consuming. If the material is properly prepared, the selection of correct documents are always also prone to so-called human errors. In specific cases, the layer 8 tarpits can also be prepared to act as a type-M deception. However, such cases would require good threat intelligence on the malicious actors so that misleading information could be created specifically to them. The type-M layer 8 tarpits would need more active curating and would be time limited, as they would need to be tightly linked to upcoming events.

The concept for layer 8 tarpits is on generating information overflow and on our humane inability to handle such overflow to the malicious actors. American psychological association defines information overflow as *"the state that occurs when the amount or intensity of information exceeds the individual's processing capacity, leading to anxiety, poor decision making, and other undesirable consequences"* (APA). When dealing with vast amount of information we are prone to use certain cognitive rules of thumb and shortcuts that ease our decision-making (Gigerenzer & Gaissmaier, 2011). However, these shortcuts also makes our decision making biased and vulnerable to errors. Knowing these biases can help defender create content that is most appealing to attackers.

There are literally hundreds of cognitive biases and going through them all would require a whole book. Instead we will present three of the most well-known and studied biases that are relevant in the scope of this paper. One of the most studied bias is the so-called confirmation bias, which causes us to seek and accept information that we expect to find and that supports our previous assumptions (Nickerson, 1998; Tversky & Kahneman, 1973). Thus, if we know what attacker is hoping to find we can create fake data that fits in attackers assumptions. Second well-studied bias is the negativity bias. Negative information gains our attention more often than positive (Baumeister, et al., 2001; Rozin & Royzman, 2001). Therefore, the valence of information may be crucial when the malicious actors goes through the material and any negative information from their viewpoint gains more attention than other information. The third bias presented here is related to the social nature of our thinking. We prioritize social information over any other, so piece of information contains something social, for example information about relationships or scandals are more likely to gain attention thus to become selected as significant information (Raichle, et al., 2001; Schilbach, et al., 2008).

Our decision-making is not as rationale as we wish to think. In cognitive demanding cases such as information overflow, we are prone to use certain heuristics of which above-mentioned are just few examples. Knowing human decision-making can help us create tarpits that targets to the "layer 8", e.g. the malicious users and organizations behind the attacks. Obviously, one can argument that well educated attacker also knows these heuristics or biases. Fortunately, even though we are somewhat able to recognize erroneous thinking in others, we are often blind to our own vulnerabilities. (West, et al., 2012).

2.2 Adding more workload with encryption

Encrypting both the falsified and real documents in a similar manner would even further increase the time and resources needed to utilize the intercepted information by malicious actors. Even if they would be able to intercept and decrypt the information with massive computational resources, they still would need to use even more resources to try to verify which of the intercepted documents contain the genuine information and which are decoys. This type of encryption by obfuscation would add another layer of security especially for data transfers and could be used in a similar manner as with the canary trap and barium meal tests. Where sending varying versions to different receivers to reveal an insider leak. In those cases, the documents are purposefully varied by using synonyms and other small differences, without changing the actual meaning of the content. However, if information about the contents would leak out, one could see whose version matches the leaked document. One could also just send multiple distorted copies along with the real document and use shared symmetric-keys to identify which document is the genuine one and which are decoys, while also encrypting the whole message using asymmetric encryption methods. The drawback on this would be the increased requirements for bandwidth in data transfers and the need to have robust version control for the real documents and decoys for both parties.

2.3 Decreasing defenders workload

Machine Learning and Natural Language Processing systems could ease the workload of generating massive amounts very realistic looking text. Especially by utilizing Generative Pre-training, by first training the system with massive amounts of diverse corpus of unlabelled text, followed by a more context depended corpuses and fine tuning the model for a specific task. (Radford, et al., 2018) As an example, the current generation GPT-3 that has been claimed by its developers to generate so realistically looking text that it can be difficult to determine whether it was written by a human (Brown, et al., 2020). Using insights from deception and human decision making along with the typical document style and language within the organization, we could develop a process that creates tempting and realistically looking false documents that tap on the cognitive biases and vulnerabilities of the potential malicious actors. To make the distinction even more difficult, it could be possible to use the real users' documents and file structures as a baseline and generate similar but falsified duplicates that mimic their style and behaviour. To further ensure that an opponent could not train their own algorithm to recognize the fake documents from real ones, it would be best to use more than one method of generate the fake documents.

3. Concept model for layer 8 tarpit

A simplified network topology layout for using deceptive tarpit networks is presented below. Real user data is fed to a pre-trained machine-learning algorithm as a context and situation specific training data (Distorted mirroring). The ML generates realistic but distorted images of the user data and virtual machine images, which are then mashed-up together with the original virtual workstations. Hardware or other means of authentication are then used to filter out the falsified entries from the authorized user's view. The layer 8 tarpit virtual desktops would also need to have some level of authentication, so that they would not be easy to distinct from the genuine virtual desktops. (Figure 1)

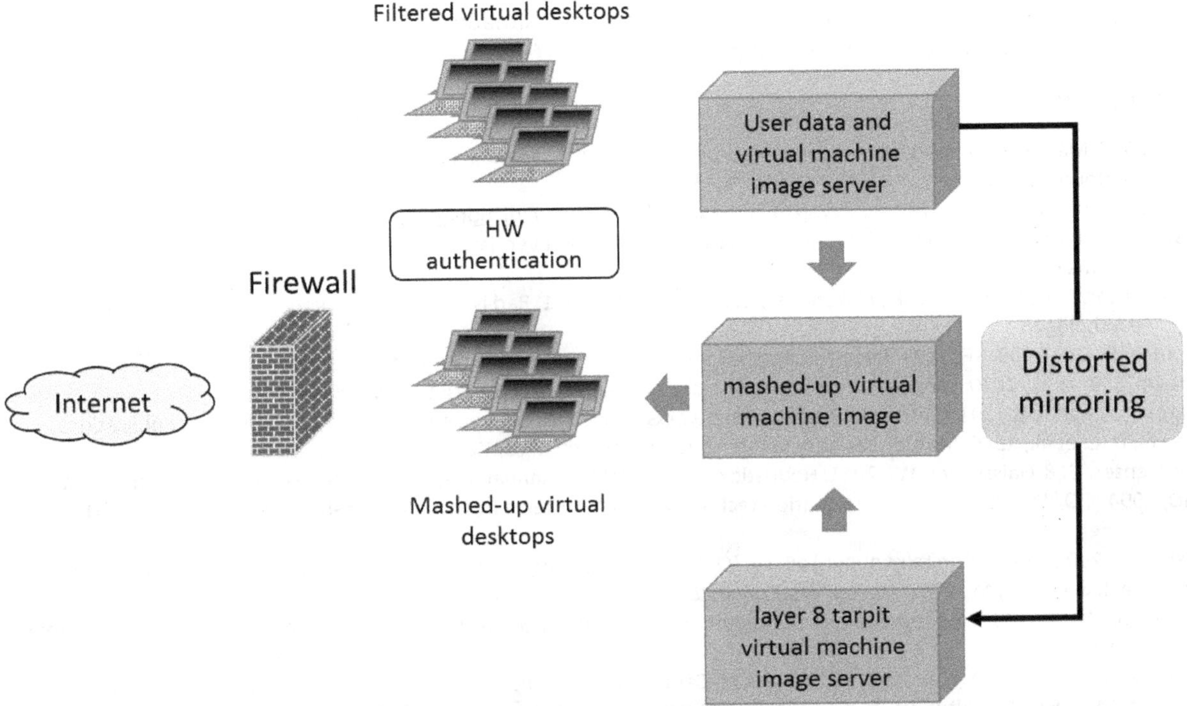

Figure 1: Simplified virtualized network topology model for layer 8 tarpits. User data is mirrored through a ML system that distorts most of the information to a realistically looking layer 8 tarpit counterpart. The real virtual machine image is then mashed-up with the layer 8 tarpit virtual machine. Hardware authentication is then used to give filter out the falsified entries from authorized users view.

4. Discussion

This paper presented a concept for using deception and falsified information as a way to delay and hinder the malicious actors behind and attack. In other words, targeting the human from the extended OSI network communications layer model e.g. the 8-layer. It discussed on using theories and understanding from psychology to enhance the falsified information so that it would create an information overflow to the opponent depleting their resources and reducing the utility of the information. AI powered methods such as the GPT-3 that can generate massive amounts very realistic looking text which is hard to differentiate from human generated texts could make this type of concept more feasible to the defender to utilise in the future. The presented concept for a network topology model illustrates a hypothetical application for using disinformation and obfuscation as a security measure. Overwhelming the attacker with similar looking targets and introducing ambiguity for not knowing whether they have managed to exfiltrate genuine information or just information from a decoy tarpit.

The obvious shortcoming of this type of defence by obfuscation is the added costs of space and bandwidth that the deceptive information would take within the defender's own network. Our estimate is that this type of system would require as much resources as running a state-of-the-art honeynet system. This type of defensive measure should also never be used as the only line of defence, but as an additional layer security that mainly would mitigate the damage after a successful attack. Another problem comes with the risk of authorized users confusing the real and falsified information if there is not good processes to keep them separated or identify which is real and which is not. The final issue with using deception as a defensive measure comes from potential legal ramifications that this type of approach might have on corporate, national or international level. As an example, there might be national legislations on misconduct where a public servant should not mislead citizens or give a false statement when in office.

This paper did not focus on how deception could be used as a pre-emptive defensive measure, by introducing false error messages, demands and stalling in high-interaction honeypot systems. This type of active honeypots could be effective in giving more time to the defenders to react once an attack has been detected. (Rowe & Goh, 2007). However, although our concept of targeting the human factor of the malicious actors and using deception to overwhelm the opponent with falsified information is mostly reactive. It can level the cyber battlefield by hindering the opponents' decision-making speed, alter their situational awareness and give an opportunity for the defender to claim the initiative to mitigate the damages from a leak in the information space.

References

AFP, 2017. Macron says hacked documents have been mixed with false ones to 'sow doubt and disinformation'. TheJournal.ie, 6 5.

APA, n.d. APA Dictionary of Psychology. [Online] Available at: https://dictionary.apa.org/

Bauer, B. & Patrick, A. S., 2004. A human factors extension to the seven-layer OSI reference model. Retrieved January, Volume 6.

Baumeister, R. F., Bratslavsky, E., Finkenauer, C. & Vohs, K. D., 2001. Bad is stronger than good. review of general psychology.

Braden, R., ed., 1989. RFC1122: Requirements for Internet hosts-communication layers. s.l.:IETF.

Brown, T. B. et al., 2020. Language models are few-shot learners. arXiv preprint arXiv:2005.14165.

Carol, F. & Raouf, B., 2014. Intrusion Detection Networks : A Key to Collaborative Security.. s.l.:Auerbach Publications.

Daniel, D. C. et al., 1980. Multidisciplinary perspectives on military deception, s.l.: Calhoun.

Gigerenzer, G. & Gaissmaier, W., 2011. Heuristic decision making. Annual review of psychology, Volume 62, p. 451–482.

ISO, 1994. ISO/IEC 7498-1: 1994. Information technology-Open systems interconnection-Basic reference model: The basic model.

Liston, T., 2002. Tom Liston talks about LaBrea. [Online] Available at: https://labrea.sourceforge.io/Intro-History.html

Mokube, I. & Adams, M., 2007. Honeypots: concepts, approaches, and challenges. s.l., s.n., p. 321–326.

Nickerson, R. S., 1998. Confirmation bias: A ubiquitous phenomenon in many guises. Review of general psychology, Volume 2, p. 175–220.

Nossiter, A. & Sanger, D. E. &. P. N., 2017. Hackers Came, but the French were prepared. The New York Times, 9 5.

Odemis, M., Yucel, C., Koltuksuz, A. & Ozbilgin, G., 2018. Suggesting a Honeypot Design to Capture Hacker Psychology, Personality and Sophistication. s.l., s.n., p. 432–438.

Ponemon Institute, 2020. 2020 cost of data breach study: Global analysis. [Online] Available at: www.ponemon.org

Qassrawi, M. T. & Hongli, Z., 2010. Deception Methodology in Virtual Honeypots. s.l., s.n., pp. 462-467.

Radford, A., Narasimhan, K., Salimans, T. & Sutskever, I., 2018. Improving language understanding by generative pre-training.

Raichle, M. E. et al., 2001. A default mode of brain function. Proceedings of the National Academy of Sciences, Volume 98, p. 676–682.

Rowe, N. C. & Goh, H. C., 2007. Thwarting cyber-attack reconnaissance with inconsistency and deception. s.l., s.n., p. 151–158.

Rozin, P. & Royzman, E. B., 2001. Negativity bias, negativity dominance, and contagion. Personality and social psychology review, Volume 5, p. 296–320.

Schilbach, L. et al., 2008. Minds at rest? Social cognition as the default mode of cognizing and its putative relationship to the "default system" of the brain. Consciousness and cognition, Volume 17, p. 457–467.

Tversky, A. & Kahneman, D., 1973. Availability: A heuristic for judging frequency and probability. Cognitive psychology, Volume 5, p. 207–232.

West, R. F., Meserve, R. J. & Stanovich, K. E., 2012. Cognitive sophistication does not attenuate the bias blind spot.. Journal of personality and social psychology, Volume 103, p. 506.

Vilmer, J.-B. J. & others, 2019. "Macron Leaks" Operation: A Post-Mortem. Atlantic Council.

Zobal, L., Kolář, D. & Fujdiak, R., 2019. Current State of Honeypots and Deception Strategies in Cybersecurity. s.l., s.n., pp. 1-9.

Cybersecurity Threats to and Cyberattacks on Critical Infrastructure: A Legal Perspective

Murdoch Watney
University of Johannesburg, South Africa
mwatney@uj.ac.za

Abstract: Over the years cybersecurity threats to and cyberattacks on the critical infrastructure by state and non-state actors have escalated in intensity and sophistication. Cyberattacks, such as the 2017 NotPetya ransomware attack, the 2020 SolarWinds software supply chain attack and the 2021 Colonial Pipeline ransomware attack, illustrate the vulnerability of critical infrastructure to cyberattacks. Most cyberattacks are committed across borders involving criminal hackers or state supported hackers. Furthermore, critical infrastructure is increasingly interconnected and interdependent. Connectivity brings about the risk of a cyberattack, demonstrated by the 2021 Colonial Pipeline ransomware attack. Interconnectedness also means that the compromise of one critical infrastructure asset can have a domino effect that degrades or disrupts others and results in cascading consequences across the economy and national security. Operational continuity is essential and this may have been one of the reasons why Colonial Pipeline paid a ransom to cyber-attackers. A cyberattack on the critical infrastructure of a state cannot be seen in isolation as the consequences of the attack may impact other states, this was illustrated by the 2017 WannaCry and NotPetya ransomware attacks. The level of sophistication of cyberattacks has increased over the years as shown by the 2020 SolarWinds software supply chain attack. The escalation of attacks has served as a catalyst for governments to address the risk to critical infrastructure. Countries need to have strong government bodies which supervise cybersecurity in their country and work together with their counterparts in other countries by sharing information regarding threats and attacks against critical infrastructure. The discussion focuses on the challenges that threats to and attacks on critical infrastructure present, the possible solutions a government may implement in addressing cyberattacks on critical infrastructure and the accountability of state and non-state actors of cyberattacks on critical infrastructure. The issues are discussed from a legal perspective.

Keywords: critical infrastructure, cybersecurity threats, cyberattacks, ransomware attacks, software supply chain attack, state and non-state cyber-attackers

1. Introduction

Private companies and governments are concerned about the vulnerability of critical infrastructure to the threat of cyberattacks by nation and non-nation-states. An attack on critical infrastructure can have a devasting impact on society's social well-being, health, security, and safety to name but a few.

The first cyberattack against a state took place in 2007 when Estonia became the victim of a Distributed Denial of Service (DDoS) attack committed across borders involving compromised computers from 178 countries (Haatja, 2009). Although the DDoS attack did not cause physical damage or destruction to critical infrastructure as the 2010 Stuxnet, it impacted on the Estonian critical infrastructure where daily operations of various organisations, including banks, government departments and small businesses were seriously impaired (Herzog, 2011; Haatja, 2009).

The discovery of the Stuxnet malware in 2010 — which resulted in a nuclear facility in Iran having its centrifuges damaged via compromised programmable logic controllers (PLCs) — demonstrated that critical infrastructure could be targeted by a cyberattack and cause physical damage or destruction (Cox, 2021). Cox (2021) opines that at the time of the 2010 Stuxnet, critical infrastructure industries used computers designed to ensure operational continuity with little regard for cyber security, because at that stage the threat of a cyberattack on the critical infrastructure may have seemed either low or non-existent. Since then, a number of attacks targeting industrial environments have emerged on the global threat landscape (Cox, 2021).

With the severe and continuing threat that cyberattacks present to critical infrastructure, and the increasing calls to address this threat, the following issues will be discussed from a legal perspective, such as

- The challenges cybersecurity threats to and attacks on critical infrastructure present;
- The possible solutions a government may implement to address cyberattacks on critical infrastructure; and
- The legal position concerning the accountability of state and non-state actors that commit cyberattacks on critical infrastructure.

2. Conceptualising terminology

It is important that terminology relevant to the discussion is conceptualised as it will serve as a point of reference with regards to the discussion of cybersecurity threats to and cyberattacks on critical infrastructure from a legal perspective.

Critical infrastructure is a term used by governments to describe assets, systems and networks - such as communications, data storage or processing, financial services and markets, water and sewerage, energy, healthcare and medical, higher education and research, food and grocery, transport, space technology; and the defence industry sector - whether physical or virtual, which are considered so vital that their incapacitation or destruction would have a debilitating effect on security, national economic security, national public health or safety, or any combination thereof (see Rege and Bleiman, 2020). Depending on the country, the definition of what constitutes critical infrastructure varies slightly.

Critical infrastructure components are to a large extent dependant on one another. Communication, information technology, financial, commercial and public services are closely linked, for example, agriculture requires the supply of dam water and water purification and pumps require electricity. Interference with transportation systems may cut off supplies for critical manufacturing and medical supplies. A disruption of a single critical infrastructure can trigger a series of effects, that together can have far worse consequences.

Critical infrastructure is vulnerable to cyberattacks. Haatja (2019) defines cyberattacks as "deliberate computer-enabled actions to alter, disrupt, deceive, degrade or destroy adversary computer systems of networks or the information and/or programs resident in or transiting these systems or networks". Cyberattacks therefore seek to compromise computer security in one or more ways by undermining the integrity of information, undermining the operation of computer systems and/or disrupting the flow of information within a network.

The cyberattacks on critical infrastructure may be committed by criminal (non-state) hackers and state or state-sponsored hackers. Shakarian (2020) draws a distinction between criminal hackers and state or state-sponsored hackers which is relevant when evaluating examples of cyberattacks on critical infrastructure (see par. 3 hereafter). Criminal hackers focus on near-term financial gain as was the case in the Colonial Pipeline ransomware attack (see par. 3 hereafter). Criminal hackers use techniques, such as ransomware, to extort money from their victims, steal financial information and harvest computing resources for activities, including sending spam emails or mining for cryptocurrency. Criminal hackers exploit well-known security vulnerabilities that, had the victims been more thorough in their security, could have been prevented, for example when the hackers used a compromised username and password to breach Colonial Pipeline's network (Turton and Mehrotra, 2021). Criminal hackers typically target organisations with weaker security, such as health care systems, universities and municipal governments. Medical systems tend to use specialty medical devices that run older and vulnerable software that is difficult to upgrade (Shakarian, 2020). On the other hand, hackers associated with national governments have entirely different motives. They look for long-term access to critical infrastructure, gather intelligence and develop the means to disable certain industries. They also steal intellectual property – especially intellectual property that is expensive to develop in fields such as high technology, medicine, defence and agriculture. The amount of effort required to infiltrate one of the SolarWinds victim firms is a telling sign that this was not a mere criminal hack (see par. 3 hereafter).

3. Examples of cyberattacks on critical infrastructure

The following three examples of cyberattacks highlight the changing landscape and growing threats to critical infrastructures.

Example 1: NotPetya ransomware attack

The reason why reference is specifically made to the 2017 NotPetya ransomware attack is that it was described at that time as the most financially damaging cyberattack in history (Haatja, 2019; Shakarian, 2020). It caused billions of US dollars' worth of damage and major disruptions to global shipping and trade.

The NotPetya malware infected computers in a range of government and private organisations in Ukraine and spread to companies and organisations around the world. NotPetya was unique as it disguised itself as a form of ransomware, such as the 2017 WannaCry, but was capable of simultaneously deleting user data (Shakarian,

2020). In 2019, the US disclosed that Russia was responsible for the attack, but Russia denied attribution (see par. 6).

Example 2: 2020 SolarWinds software supply chain attack

SolarWinds, a major software company, was the subject of a cyberattack that spread to its clients, which may have started in 2019 but was only discovered in December 2020. Hackers, believed to be tied to the Russian government, gained access to SolarWinds systems and added a malicious code into the company's software system (Thompson, 2021). The system, called Orion, is widely used by companies to manage IT resources. Orion updates, which included the hacked code, were received by as many as 18000 SolarWinds customers (Mehrotra, 2021). The code created a backdoor to customer's information technology systems, which hackers then used to install even more malware that helped them spy on companies and organisations. The hacking campaign that infected numerous government agencies and tech companies with malicious SolarWinds software had also infected more than a dozen critical infrastructure companies in the electric, oil and manufacturing industries running the same software (Zetter, 2021.) The consequence was the penetration of multiple networks.

SolarWinds is used as an example as it was considered as one of the most devastating cyberattacks in history (Thompson, 2021). It exposed vulnerabilities in global software supply chains that affected government and private sector computer systems and constituted a major breach of national security (Shakarian, 2020). The hack revealed gaps in the US cyber defences and could be the catalyst for rapid, broad change in the cybersecurity industry (Oladimeji and Kerner, 2021; see par. 5; and https://www.npr.org/2021/04/29/991333036/biden-order-to-require-new-cybersecurity-standards-in-response-to-solarwinds-att).

Example 3: 2021 Colonial Pipeline ransomware attack

In May 2021, a ransomware attack was launched on Colonial Pipeline, a private company that controls a significant component of the US energy infrastructure and supplies nearly half of the East Coast's liquid fuels. The FBI attributed the attack to a Russian cybercrime gang (Thompson, 2021; see par. 6).

The Colonial Pipeline ransomware attack illustrates the vulnerability of OT and connectivity. Although the ransomware attack did not directly target the OT – the devices that drive gas flows - but the IT, the consequence of the attack was the shutting down of the OT to prevent the risk of the attack and threatening the safety of the OT. This highlights that threats to both human and environmental safety, along with the uncertainty as to the scope of infection, present as risk factors for sensitive industrial environments (Cox, 2021).

4. Challenges facing the protection of critical infrastructure

The following interlinked challenges are identified:

1) Cybersecurity risk mitigation

The starting point of government and a private company that owns critical infrastructure should be to conduct cybersecurity risk mitigation which involves the use of security policies and processes to reduce the overall risk or impact of a cybersecurity threat. As indicated, NotPetya resulted in billions of US dollars' worth of damage and major disruptions to global shipping and trade which may be the consequence of the risk of a cyberattack not being adequately assessed. In the Colonial Pipeline attack, the OT was shut down to prevent the risk of the attack spreading to the OT (see par. 3). Although the SolarWinds attack appeared to be aimed at the theft of emails and other data, the nature of the intrusions created "back doors" which presents the risk of attacks on physical infrastructure.

Cybersecurity risk mitigation may be separated into four elements: prevention, detection, response and recovery.

With respect to the prevention element: It may be difficult to completely prevent the threat of an attack as protecting all systems from any attacker may not be possible (Hemsley and Fisher, 2018; Thompson, 2021). Hemsley and Fisher (2018) opines that a key lesson learnt from Stuxnet is that a well-financed, sophisticated threat actor can likely attack any system that it desires. Thompson (2021) opines that preventing ransomware attacks, such as the Colonial Pipeline attack, would require US intelligence and law enforcement to infiltrate every organized cyber-criminal group in Eastern Europe.

Regarding the detection, response and recovery elements: An important take away from the Stuxnet attack (discussed at par. 1) and the subsequent cyberattacks (discussed at par. 3) is the ability to detect, respond and recover from a cyberattack (Hemsley and Fisher, 2018). Where a critical infrastructure has been infiltrated, it should be detected as soon as possible. In this regard, SolarWinds was only detected after some time which impacted negatively on recovery.

2) Interconnectivity and interdependence

Connectivity without proper or insufficient safeguards creates significant vulnerabilities. Tidy (2021) opines that the simplest way to protect OT is to keep it offline with no link to the internet at all. OT networks were traditionally segregated from the Internet in what is known as an 'air gap.' Malware may be installed manually via external media, such as a USB which was used in Stuxnet. Malware external media installation doubled in 2021 with 79% of these holding the potential to disrupt OT. Cox (2021) indicates that the threat of a cyberattack is not completely eliminated in instances where the OT network is segregated.

Over the years operational demands have made critical infrastructure more vulnerable. These include the convergence of IT and OT, the adoption of devices in the Industrial Internet of Things (IIoT), and the deprecation of manual back-up systems (OT). This means that OT can be disrupted by cyberattacks that first target IT systems, rather than having to be installed manually via external media (Cox, 2021).

3) Skill level of threat actors against critical infrastructure and defenders of critical infrastructure

The skill level of sophisticated threat actors is also increasing, as are the frequency of attacks targeting critical infrastructures and the systems that control them. The defenders of these systems need to have equally advanced skills and knowledge to protect essential resources.

4) Outsourcing software

Thompson (2021) opines that SolarWinds did not consider the risk associated with outsourcing software development to Eastern Europe, including a company in Belarus. Russian operatives have been known to use companies in former Soviet satellite countries to insert malware into software supply chains. Russia used this technique in the 2017 NotPetya attack.

5) Investing in cybersecurity of critical infrastructure

Cyber threats are very real, and appropriate investments in cybersecurity should be made by the companies and government.

6) Paying ransom

The issue of paying ransomware is contentious. There are calls that government must prevent ransoms being paid in secret (Tidy, 2021).

The Colonial Pipeline's CEO acknowledged that his company paid a $4.4 million ransom to hackers who were an affiliate of a Russia-linked cybercrime group known as DarkSide, as executives were unsure how badly its systems were breached or how long it would take to restore the pipeline. The hackers also stole nearly 100 gigabytes of data from Colonial Pipeline and threatened to leak it if the ransom was not paid (Turton and Mehrotra, 2021). The latter is an example of extortion ransomware which may be the manner in which these attacks progress in future.

Cox (2021) opines that critical infrastructure— ranging from power grids and pipelines to transportation and health care — must maintain continuous activity. The 2021 ransomware attack against Colonial Pipeline demonstrates why the company paid the ransom as the closure of the pipeline resulted in dangerous panic buys, long lines at the pump and gas shortages. There was also the issue of extortion being linked to the payment of the ransom.

7) Fragmented approach to cybersecurity of critical infrastructure

Thompson (2021) points out that the fragmentation of the US authorities for national cyber defence evident in the SolarWinds hack is a strategic weakness that complicates cybersecurity for the government and private sector and invites more attacks on the software supply chain.

The official strategy at the time of the 2020 SolarWinds attack is to split cybersecurity responsibilities between the Pentagon for defence and intelligence systems and the Department of Homeland Security (DHS) for civil agencies. Execution of the strategy relies on the Department of Defence's US Cyber Command and DHS's Cyber and Infrastructure Security Agency (CISA). DOD's strategy is to "defend forward", that is, to disrupt malicious cyber activity at its source. CISA, established in 2018, is responsible for providing information about threats to critical infrastructure sectors. Neither agency appears to have sounded a warning or attempted to mitigate the attack on SolarWinds. A private cybersecurity firm called FireEye was the first to notice the breach when it noticed that its own systems were hacked (Jibilian and Canales, 2021). The government's response came only after the attack (see par. 5).

8) Attribution of cybersecurity threats and accountability for cyberattacks

Cybersecurity threats are mostly cross-border and the hacker, either a criminal hacker or state sponsored hacker, will be outside the victim's country borders. In cyberspace, there are no borders at all; hackers in a certain country may well use servers and other digital infrastructure in other countries for their operations (van der Meer, 2020). The SolarWinds attack was launched from inside the US on servers Russian actors had rented from places such as Amazon and GoDaddy, but the actors executed the attack outside the US border. By renting servers in the US, the hackers were able to slip past the National Security Agency (NSA) early warning systems as the NSA is not allowed to conduct surveillance inside the US (see https://www.npr.org/2021/04/29/991333036/biden-order-to-require-new-cybersecurity-standards-in-response-to-solarwinds-att).

Countries such as Russia, Iran and North Korea, are regularly accused of harbouring ransomware groups. For example, federal investigators and cybersecurity agents believe a Russian espionage operation was responsible for the SolarWinds hack (Oladimeji and Kerner, 2021). The Russian government denied any involvement in the attack, releasing a statement that said, "Malicious activities in the information space contradicts the principles of the Russian foreign policy, national interests and understanding of interstate relations." They also added that "Russia does not conduct offensive operations in the cyber domain" (Oladimeji and Kerner, 2021, see par. 6).

There have been calls that state and non-state cyber-attackers on critical infrastructure are held accountable (see par. 6; van der Meer, 2020).

5. Private companies and government response to cybersecurity threats to and attacks on critical infrastructure

The response of the EU, Australia and US are briefly discussed hereafter.

5.1 European Union (EU)

The EU are taking cybersecurity seriously by implementing various initiatives (see https://digital-strategy.ec.europa.eu/en/policies/cybersecurity-policies).

ENISA ('European Union Agency for Network and Information Security') is the EU agency that deals with cybersecurity. It provide support to member states, EU institutions and businesses in key areas, including the implementation of the NIS Directive. The Cybersecurity Strategy strengthens the role of ENISA. The agency now has a permanent mandate, and is empowered to contribute to stepping up both operational cooperation and crisis management across the EU.

In 2016, the European Commission as part of the EU Cybersecurity Strategy proposed the EU Network and Information Security (NIS) directive. The NIS Directive (see EU 2016/1148) was the first piece of EU-wide cybersecurity legislation. The goal was to enhance cybersecurity across the EU. The NIS directive was adopted in 2016 and the national transposition by the EU member states happened in 2018. In 2020, an updated NIS (referred to as NIS2) was introduced (see https://digital-strategy.ec.europa.eu/en/policies/cybersecurity-policies; https://www.consilium.europa.eu/en/press/press-releases/2021/12/03/strengthening-eu-wide-cybersecurity-and-resilience-council-agrees-its-position/) as well as a Directive on the resilience of critical entities. The proposed NIS2 will set the baseline for cybersecurity risk management measures and reporting obligations across all sectors that are covered by the directive, such as energy, transport, health and digital infrastructure. The revised directive aims to remove divergences in cybersecurity requirements and the implementation of cybersecurity measures in different member states. To achieve this, the directive sets out minimum rules for a regulatory framework and lays down mechanisms for effective cooperation among relevant authorities in each member state. It updates the list of sectors and activities subject to cybersecurity obligations and provides for remedies and sanctions to ensure enforcement. The directive will formally establish the European Cyber Crises Liaison Organisation Network, (EU CyCLONe) which will support the coordinated management of large scale cybersecurity incidents.

The Commission with ENISA is also working on an EU wide certification framework. The Cybersecurity Act outlines the process for achieving this framework.

5.2 Australia

In 2021, the *Security of Critical Infrastructure Act* (SOCIA) *2018* was amended to ensure better protection of critical infrastructure and systems of national significance. The SOCIA provides for a Register of Critical Infrastructure Assets – the register builds a clearer picture of critical infrastructure ownership and control in high-risk sectors, and supports more proactive management of the risks these assets face. Furthermore it imposes mandatory cyber incident reporting – following recent amendments to the SOCI Act, responsible entities for critical infrastructure assets may be required to report critical and other cyber security incidents to the Australian Cyber Security Centre's online cyber incident reporting portal (https://www.homeaffairs.gov.au/about-us/our-portfolios/national-security/security-coordination/security-of-critical-infrastructure-act-2018

5.3 US government solutions to address critical infrastructure protection

The Department of Homeland Security, Cybersecurity and Infrastructure Security Agency (CISA) leads the coordinated national effort with public and private sector critical infrastructure partners to enhance the security and resilience of the nation's critical infrastructure. Following the SolarWinds and Colonial Pipeline attacks, the US government has started various initiatives aimed at improving cybersecurity and addressing the risk to critical infrastructure.

In 2021, the government issued an executive order (available at https://www.whitehouse.gov/briefing-room/presidential-actions/2021/05/12/executive-order-on-improving-the-nations-cybersecurity/) which is aimed at modernising the nation's cybersecurity:

- Where a company does business with federal government, the software must comply with certain standards which improves security of software sold to the government.
- IT service providers are required to inform the government about cybersecurity breaches that could impact US networks, and removes certain contractual barriers that might stop providers from flagging breaches.
- Federal government must upgrade to secure cloud services and other cyber infrastructure, and mandates deployment of multifactor authentication and encryption with a specific time period.
- The establishment of a "Cybersecurity Safety Review Board" which comprises public- and private-sector officials, which can convene after cyberattacks to analyse the situation and make recommendations.
- Information-sharing within the federal government by enacting a government-wide endpoint detection and response system.

In 2021, the Cyber and Infrastructure Security Agency (CISA) published the "Rising Ransomware Threat to OT Assets" fact sheet in response to the recent increase in ransomware attacks targeting operational technology

(OT) assets and control systems. The guidance (available at https://www.cisa.gov/publication/ransomware-threat-to-ot):

- provides steps to prepare for, mitigate against, and respond to attacks;
- details how the dependencies between an entity's IT and OT systems can provide a path for attackers; and
- explains how to reduce the risk of severe business degradation if affected by ransomware.

In 2021, the US Department of Energy (DOE) launched an initiative to enhance the cybersecurity of electric utilities' industrial control systems (ICS) and secure the energy sector supply chain. (see https://www.energy.gov/articles/biden-administration-takes-bold-action-protect-electricity-operations-increasing-cyber-0).

6. Accountability of state and non-state cyber-attackers

Where a cyberattack has been attributed to a cyber-attacker, the state and non-state actors should be held accountable. Holding cyber-attackers accountable shows that a victim state will not tolerate such activities and involvement, and it may also deter other states from engaging in such cyberattacks (van der Meer, 2020).

Although this is a topic that warrants a discussion on its own, the most important issues relevant to the discussion are highlighted.

6.1 State actors

The US government levelled sanctions against Russia for the SolarWinds attack. The White House indicated that there would be more "seen" and "unseen" responses to the breach. The unseen responses — for example, whether the government is preparing a reprisal attack against Moscow in cyberspace — was not disclosed (see https://www.whitehouse.gov/briefing-room/statements-releases/2021/04/15/fact-sheet-imposing-costs-for-harmful-foreign-activities-by-the-russian-government/). Whether a reprisal attack for a non-material cyberattack is permissible within the ambit of the international law, is debatable.

Article 2(4) of the United Nations Charter prohibits the use of force in international relations. Whether a cyberattack constitutes as the use of force and amount to warfare within the ambit of the present international law, has been debated (Haatja, 2019). In instances where the cyberattack does not result in material effects such as damage or destruction to a physical object, death or injury to human beings, it would not constitute the use of force within the present international law and is not specifically prohibited by the international law. Haatja (2019) is of the opinion that the present legal position should be reformed. Today society relies on information and communication technology (ICT) and a non-material cyberattack on critical infrastructure has the potential to impact negatively on society. Haatja (2019) opines that the harm caused by non-material cyberattacks should be considered within the scope of Article 2(4) and recognised as a new form of violence that the international law should limit states from engaging in.

6.2 Non-state actors

Van der Meer (2020) indicates that the majority of cyber-attacks in the world are launched by non-state actors, especially criminals looking for money (see par. 2). However, van der Meer (2020) opines that state actors increasingly hire non-state actors to launch more severe cyber-attacks with potentially damaging effects for societies abroad. He indicates that effectively responding to state-launched cyber-attacks is already a complicated task which becomes even more difficult when states hide behind non-state actors.

Van der Meer (2020) discusses a policy brief which explores the problems in dealing with non-state cyber-attackers. He provides the following 7 policy options available to states responding to cyber-attacks which are convincingly attributed to non-state actors:

- Requesting the host state to take action against the attacker;
- If the 'host' state is willing, but is not able to take effective measures against the non-state cyber-attacker, then the requesting state may assist the state in doing so;
- If the 'host' state is doing little or nothing after the request for assistance, while it should be able to do so, a diplomatic protest may be a viable option;

- Employing legal measures, such as an indictment. Indictments will generally occur at a national level, for example under national criminal law;
- Sanctions could also be used to target the non-state actor and/or the state that is not taking effective action against this actor;
- A victim state could also respond to a large-scale cyber-attack by retaliating with a counter- attack; and
- A state could use conventional military retaliation, for example through a proportional strike against a specific location related to the non-state actor behind the cyber- attack or the state from which it operates.

The victim state must consider the risks and benefits of each option before engaging with one of the 7 options.

7. Conclusion

The question today remains how to best achieve robust cyber defence. The different cyberattacks referred to in the discussion demonstrate that the scope of protection and security must include systems that process data (information technology (IT)) and those that run the vital machinery that ensures our safety (operational technology (OT)).

Although a cyberattack cannot be completely eliminated, prevention, detection, response and recovery must be prioritised. An attack must be reported and therefore a mandatory reporting obligation must be considered. There must be a central body assisting with attacks on critical infrastructure. A distinction should not be drawn between a private company owning critical infrastructure or it being operated by government since an attack on critical infrastructure affects national security. The risk of a cyberattack on critical infrastructure owned by a private company cannot be only that company's responsibility, but government must provide support.

Over the years the threat of ransomware attacks has escalated (Rege and Bleiman, 2020). Ransomware attacks have become lucrative because a government body or private company cannot afford to lose data that is locked up and will therefore pay the ransom, especially in instances where it is linked to extortion. The manner in which a country should deal with a non-state or state-sponsored actor who resides in a foreign country presents challenges. An attack on a critical infrastructure should not be held without accountability and here the national and global response should be considered (see par. 6).

Countries need to have strong government bodies that supervise cybersecurity and work together with their counterparts in other countries by sharing information because the protection of critical infrastructure is not only a national security issue, but a global one.

References

Cox, O. (2021) "How cyberattacks take down critical infrastructure", [online] https://www.darktrace.com/en/blog/how-cyber-attacks-take-down-critical-infrastructure/.

Haatja, S. (2019) Cyberattacks and International Law on the Use of Force. Routledge: New York. p. 1 – 10, 52 - 111 – 127, 194, 198.

Hemsley, K.E., and Fisher, R.E. (2018) "History of Industrial Control System Cyber Incidents", [online], https://www.osti.gov/servlets/purl/1505628.

Hertzog, S. (2011) "Revisiting Estonian cyberattacks: digital attacks and multinational response", *Journal of Strategic Security* Vol. 4, No. 2, Strategic Security in the Cyber Age, pp. 49-60, [online], https://www.jstor.org/stable/26463926?seq=10#metadata_info_tab_contents.

Jibilian, I. and Canales, K. (2021) "Here's a simple explanation of how the massive SolarWinds hack happened and why it's such a big deal", [online], https://businessinsider.mx/heres-a-simple-explanation-of-how-the-massive-solarwinds-hack-happened-and-why-its-such-a-big-deal/.

Mehrotra, K. (2021) "SolarWinds hack leaves critical infrastructure in the dark on risks", [online], https://www.bloomberg.com/news/newsletters/2021-01-05/solarwinds-hack-leaves-critical-infrastructure-in-the-dark-on-risks.

Oladmeji, S. and Kerner, SM. (2021) "SolarWinds hack explained: everything you need to know", [online], https://whatis.techtarget.com/feature/SolarWinds-hack-explained-Everything-you-need-to-know.

Rege,A., and Bleiman, R. (2020) "Ransomware attacks against critical infrastructure", Proceedings of the 19th European Conference on Cyberwarfare and Security, pp. 324 – 333.

Shakarian, P. (2020) "The Sunburst hack was massive and devastating – 5 observations from a cybersecurity expert", [online], https://theconversation.com/the-sunburst-hack-was-massive-and-devastating-5-observations-from-a-cybersecurity-expert-152444.

Thompson, J. (2021) "The Colonial Pipeline cyberattack and the SolarWinds hack were all but inevitable. Why national cyber defence is a 'wicked' problem", [online], https://www.virginiamercury.com/2021/05/13/the-colonial-pipeline-cyber-attack-and-the-solarwinds-hack-were-all-but-inevitable-why-national-cyber-defense-is-a-wicked-problem/.
Turton, W. and Mehrotra, K. (2021) "Hackers breached Colonial pipeline using compromised password", [online], https://www.bloomberg.com/news/articles/2021-06-04/hackers-breached-colonial-pipeline-using-compromised-password.
Tidy, J. (2021) "Colonial Hack: How did cyber-attackers shut off pipeline", [online], https://www.bbc.com/news/technology-57063636.
Van der Meer, S. (2020) "How states could respond to non-state cyber-attackers", [online], https://www.clingendael.org/sites/default/files/2020-06/Policy_Brief_Cyber_non-state_June_2020.pdf.
Zetter, K. (2020) "Infected critical infrastructure, including power industry", [online], https://theintercept.com/2020/12/24/solarwinds-hack-power-infrastructure.

Cyber Security Norms: Trust and Cooperation

Allison Wylde
Cardiff University, UK
wyldea@cardiff.ac.uk

Abstract: As cyber crime becomes ever more sophisticated and a significant asymmetric threat, the need for effective cyber security is of vital importance. One important cyber security response is through cyber norms. At the same time, calls for multi-sector and multi-domain trust and cooperation are widespread. Yet research on the nature of trust and cooperation in cyber security norms appears to be underdeveloped. Key questions remain concerning the emergence and nature of trust and cooperation in norms. In addressing this gap, the article first considers how we can understand trust and cooperation in cyber norms through leveraging well-established theory from management research on trust building. Next, the paper examines the SolarWinds breach, as an example, to evaluate norms, trust and cooperation. The paper then applies principles from prominent trust-building theory to examine the antecedents, processes of outputs involved in building trust and cooperation. The contribution of this work presents a foundational conceptual framework, to allow the dynamics of norms, trust, and cooperation in managing cyber crime incidents to be studied. In doing so, the literature on examining trust and cooperation in norms is extended. Other researchers' interest is encouraged as is an agenda for further research on norms, trust, and cooperation to support cyber security management. Implications may help the cyber security community as they construct and manage norms, trust, and cooperation.

Keywords: cyber security, norms, trust, cooperation, dynamics

1. Introduction

The SolarWinds breach of 2021 highlights an instance of an ongoing and large-scale extraction of sensitive material from carefully targeted organizations across government departments, financial institutions and public health and education institutions. The full impact may never be known due to the sensitivities involved. Government responses were rapid. Although the implementation of norms in cyber security is assumed, in practice, explicit understanding of the central and underpinning elements of trust and cooperation is limited. The puzzle at the heart of this paper concerns an important question; if norms for cyber security rely on implicit trust and cooperation, how is this understood by the various international actors involved? Indeed, is there agreement on a shared understanding? The SolarWinds breach is dawn on to provide an example through which to explore trust and cooperation. The author explored the SolarWinds breach during as part of the UN IGF workstream 2. Several other breaches were also examined along with policy (UN IGF BPF, 2021a; b).

The development of norms has progressed. As a start, the United Nations Government experts 2015 (UN, 2015) and later the Global Commission on Cyberstability (GCSC, 2019) set out important norms for cybersecurity. However, arguably, the continued proliferation of norms without an underpinning framework for evaluation poses a challenge for a common understanding.

To answer the questions raised, this paper next revisits the central literature on norms associated with cyber security to explore the roles of trust and cooperation. This is followed by a review of trust building theory drawn from the management literature. The Integrative Trust Model (IT) (Mayer et al. 1995), together with foundational research in conflict management Deutsch (1958; 2006) are selected due their ability to disentangle the various elements and processes involved in trust building. Given that trust is central in cooperation, these models are reexamined with a focus on extracting the key elements that can shed light on processes of cooperation. As the central terms trust and cooperation are multiplex, definitions in the context of this paper are next presented. The study is operationalised through an exploration of the SolarWinds breach with a focus on the norms as implemented together with the underlying trust and cooperation. In the final section, the conclusion, the work as executed is discussed together with limitations, future directions, and implications.

2. Key literature

2.1 Norms for cyber security

Although norms are widespread and generally understood to relate to shared beliefs and actions founded on what is correct or proper, some confusion remains. To add clarity, what follows is a brief overview of norms theory and a focus on the development of norms for cyber security.

Social scientists view social norms as constructs of three main types, descriptive norms, simply representing what other people do and injunctive norms, what people should do (Smith and Lewis, 2008). A third category is subjective norms. These norms concern an individual's perception of another actor's approval (or not) of their own actions and the motivating factors that are involved (Cialdini et al, 1991). In subjective norms, an individual's perception of approval (from a referent) may prompt that individual to behave in a way that may be encouraged by the referent (Cialdini et al, 1991). This thinking, extended in the Technology Acceptance Model (TAM) (Ajzen, 1991), suggests that individuals may be motivated to copy or mimic the behaviour of a referent group out of a belief that in doing so their own status will be enhanced (Venkatesh and Davis, 2000). This is often seen in the release of new technology - people want the latest gadget, 'everyone has it' and individuals may feel better, and indeed enhanced, once they possess the item. Indeed, a similar way, some individuals may act as enforcers and implement sanctions if a particular norm is not followed (Ellickson, 1999) while other studies have found that the negative action may be performed if an individual believes a referent group approves of those actions, for example, illegally downloading streaming services (Wired, 2021).

Summing up, norms are viewed as beliefs and motivators - in some cases the right thing. Norms are dynamic. Change may arise through change agents, from self-motivated leaders to norm entrepreneurs and finally, opinion makers (Ellickson, 1999). In provide a working definition, this paper views norms as based on the key considerations; a norm itself is agreed, norms act as informal rules (Smith and Lewis, 2008), norms act as drivers and can prompt referents to behave in a particular way (Ajzen, 1991; Cialdini et al, 1991), and importantly, that norms are subject to change by change agents (Ellickson, 1999).

For cyber security, norms are seen as agreed methods and shared beliefs of how to behave and operate in the cyber domain and referents are motivated to follow the guidelines (Smith and Lewis, 2008; Ajzen, 1991; Cialdini et al, 1991). Active norm creation has occurred through the establishment of agencies such as UN the Internet Governance Forum (UN IGF) (UN, 2015) which aimed to help support the development and practices of norms for cyber security. Numerous individual states and organizations developed their own norms along with, in 2015 the UN's Framework for Responsible State Behaviour (UN, 2015; GCSC, 2019). The foundational cyber security norms include (1) the non-interference of the public core of the internet (2) the protection of electoral services (3) the avoidance of tampering (4) agreement not to commandeer ICT devices into botnets and (6) a reduction and mitigation of significant vulnerabilities (GCSC, 2019). In 2021 the UN IGF BPF undertook research to identify and map the frequency of policy actions for each norm, the highest frequency was cooperation, adherence to human rights, reporting vulnerabilities and providing remedies (UN IGF BPF, 2021a; 2021b).

As highlighted norms are dynamic, norms also have arisen organically as best practice among practitioners, notably the adoption of zero trust approaches among cyber security practitioners (Wylde, 2021). A driver to establish a norm may arise internally through practitioners' actions as norms entrepreneurs (Ellickson, 1999). As a gap is identified practitioners cooperate to provide a solution, as in the example of zero trust. Alternatively, policy makers may implement a new norm (NCSC, 2021a). This gives rise to questions concerning underpinning assumptions in cyber security norms. Are all norms trusted? Is trust and cooperation necessary for the formation and subsequent operation of norms, are all norms based on trust and cooperation, which would in turn promote trust, so forming a positive feedback loop? Figure 1, below, presents a first conception of norms formation and the role of trust and cooperation. The question here, concerns how to understand the processes trust and cooperation involved in norms.

2.2 Norms: Trust and cooperation

Although recent findings point to the underpinning and implicit role of trust and cooperation in norms (UN IGF BPF, 2021a) there remains important gaps concerning what this may mean in practice. In the extensive literatures on both trust and cooperation, questions concerning definitions and dimensions remain. Important studies are considered next with a focus on teasing out and drawing together the key strands that will form the conceptual basis for this paper.

2.2.1 Trust

In an extensive literature, trust is viewed as related to uncertainty and risk (Mayer et al, 1995) in an interactive process (Dietz, 2011), as an enabler of cooperation and an alternative to formal governance (Vanneste, 2016) and indeed control (Mayer et al, 1995). The definition, context and conceptual model for this paper, cyber security, is discussed next.

A definition for trust as viewed in this paper, is provided by the integrative trust building model of Mayer Davis and Schoorman (1995) and the foundational research by Deutsch (1958;2006). These authors define trust as "the willingness of a party to be vulnerable to the actions of another party based on the expectation that the other will perform a particular action important to the trustor, irrespective of the ability to monitor or control the other party (Mayer et al, 1995, p.712). The element of willingness to be vulnerable is further clarified as "a psychological state comprising the intention to accept vulnerability based on positive expectations of the intentions or behaviour of another" (Rousseau et al, 1998, p. 395). Drawing these themes together, a psychological state and intention (Rousseau et al, 1998) and willingness to accept vulnerability; irrespective of the ability to monitor or control (Mayer et al, 1995). From the conflict resolution literature from which norms have arguably evolved (Wylde, 2021) trust is viewed as founded on an individuals' ability to trust along with their experience of trust (Deutsch, 1958), cooperation or a lack of cooperation (Deutsch, 2006). Important also are considerations of norms in society and alignment with the individual trustor's beliefs (Deutsch, 1958; 2006). This approach allows researchers to evaluate key antecedents, processes, and outcomes in these multi dimensional and understudied concepts.

Important research has also argued for clarity regarding the scale and referent in trust relations. Fulmer and Gelfand's (2012) work set out the different levels of trust relations, trust in individuals or teams or organizations and at the level of institutions. Non-person-based trust has also been studied, including the examining nature of trust in technology (Mckight, 2011). For this paper, the scope of the trust relations examined are confined to trust in institutions and in policy. Importantly as Vanneste (2016) clarifies, it is the people in the organizations who trust and prompt others to trust.

For the context of cyber security, important beliefs about trust are seen as based on confident positive expectations of the other's trustworthiness based on an assessment of ability, benevolence and integrity moderated by an antecedent actors' propensity to trust (Mayer et al, 1995) and psychological intention and willingness to accept vulnerability; irrespective of the ability to monitor or control (Mayer et al, 1995; Rousseau et al, 1998). Together with a trustor's experiences and beliefs (Deutsch, 1958; 2006).

Yet, this foundational view of trust may be at odds with current practices in cyber security, which rely on zero trust, in other words, non-presumptive trust (Wylde, 2021). In a similar sense as trust, in zero trust, the central thinking is founded on a psychological state based on experience and societal norms. Importantly in implementing zero trust vulnerability or risk are not accepted, rather, there is continuous monitoring, assessment, and authentication (NCSC, 2021a). In Table 1, below, the dynamics of trust and cooperation building: antecedents, processes, and goals (DT&CB) model, sets out key elements from the trust and collaboration models. Cooperation discussed next, together with trust form the basis of the conceptual frame in this paper, presented in Table 1 below.

2.2.2 Cooperation

Cooperation has been studied across different disciplines, in particular management and public administration. Like trust it is subject to numerous competing definitions. An attempt to add clarity is provided next.

Clear conceptual understanding is essential to allow the distinct elements involved in cooperation and collaboration and coordination to be discerned. Currently the terms are entangled (Castañer and Oliveira, 2020). Indeed, cooperation and collaboration are used interchangeably in many studies of policy documents on norms, (UN IGF BPF, 2021a). As Dietz (2011) highlights, even trust and cooperation may be conflated. Common conceptualisations on relationships based on cooperation include perceived risk, risk taking and that at least two-parties are involved (Dietz, 2011). This position is further confused as some researchers see cooperation as an umbrella term, encompassing both collaboration and coordination (Gazley, 2017). This is based on a relationship involving mutual goals, and the management of activity to achieve jointly agreed outcomes (Gazley, 2017) while others suggest that coordination is a deliberate process among partners based on order to achieve goals (Gulati et al, 2012) This thinking is expanded by public administration researchers, who see collaboration as involving cooperation and comprising discrete dimensions, an event horizon comprising, antecedents, processes, and outcomes (Thompson, 2006).

Coordination and coordination are also considered in strategic alliances, with the authors' arguing that cooperation suggests the pursuit of private goals at the expense of the collective (Kretschmer and Vanneste,

2017). Given space constraints, this discussion is not taken further here, what is important to note is the importance of disentangling these terms, as indicated, for this paper, the emphasis is on cooperation.

For clarity, in this paper, further discussion is limited to cooperation as defined through the meta study undertaken by (Castañer and Oliveira, 2020) which suggests that individuals' voluntarily helping others (Ring and Van de Van, 1994). The authors point to the goal as the unit of analysis, in the context of interorganizational relations (IOR). This is whether a common goal or a private goal. Since the focus of this paper concerns common and/ or collective goals, for parsimony, private goals are considered out of scope.

Table 1. Dynamics of trust and cooperation building: antecedents, processes and goals, DT&CB model (extending Deutsch, 1958; 2006[1]; Mayer et al, 1995[2], and, for zero trust, Wylde, 2021[3])

Antecedents	Processes	Goals
Beliefs and willingness to act[1]	Trust 'fit' with personality[1] (zero trust)[3]	Trust and cooperation (zero trust)[3]
Ability to trust[1]	Trust mirrors prevailing societal rules and norms[1] (zero trust)[3]	Trust and cooperation (zero trust)[3]
Experience of trust[1] (zero trust)[3]	Ability[2]	Trust (zero trust)[3]
Confident positive expectations of trust[2,] (negative expectations)[3]	Benevolence[2]	Trust (zero trust)[3]
Propensity to trust[2] (zero propensity to trust)[3]	Integrity[2]	Trust (zero trust)[3]
	Acceptance of vulnerability[2] (non acceptance)[3]	Trust and cooperation (zero trust)[3]
	Risk taking behaviour[2] (no risk taking)[3]	Trust and cooperation (zero trust)[3]

Examples, the pursuit of common and or collective goals can be seen in the actions of state and inter-state organizations. As an example, the Council of Europe was set up in 2001 as a formal initiative to encourage international cooperation (EU, 2001). This was followed by the formation of the NATO cooperative defense center in Estonia (Schmitt, 2013). Subsequent bodies have followed and created policy for cooperation (UN IGF BPF, 2021a). In sum, cooperation is seen as reliant on common and collective goals.

2.2.3 *Trust and cooperation*

Taken from the discussions above and drawing from Levine's (2019) view of the integrative social contract theory, together with Donaldson and Dunfee (1994) who suggest that individuals' follow norms; a conceptual frame is proposed to elaborate on our thinking on norms-building in the context of cyber security.

The underlying assumptions are as follows: organizations follow norms, in this context, norms for cyber security, and in doing so promote trust, trustworthiness and cooperation (Levine, 2019). The mechanisms involved here are based on understanding the processes and elements involved in the formation of trust and, as is suggested here, by implication, in the formation of cooperation. As highlighted, other studies on norms in cyber security are founded on assumed trust and cooperation.

In addressing this gap, Figure 1 below, presents norms viewed as a continuous process. The start point arises from a driver for formation such as an incident or event, this is followed by the emergence of trust and cooperation as a necessary condition to push forward the development of the norm and then its implementation. As the norm is implemented trust and cooperation form, in doing so these processes drive the formation of additional trust and cooperation, seen here as a positive feedback loop. The figure illustrates the role of positive feedback in driving the whole process forward such that new norms are formed. The NTAC+ model, Figure 1, below, forms part of the contribution of this paper through extending our thinking on how we can examine the dynamics of norms, trust, and cooperation (notably, Vanneste, 2016).

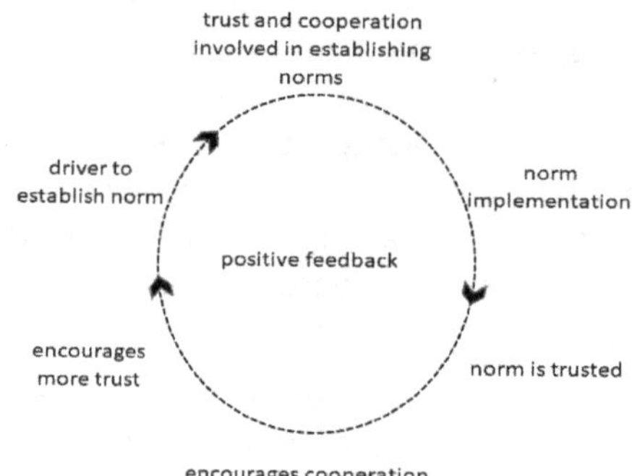

Figure 1: Simplified framework: The dynamics of norms, trust and cooperation, and positive feedback (NTAC⁺)

As highlighted earlier, in this conception, NTAC⁺ (Figure 1, above), control and monitoring are absent (Mayer et al, 1995). Vanneste supports this view, arguing that trust enables cooperation as an alternative to formal governance (Vanneste, 2016).

3. SolarWinds

During 2020 a breach took place on the cyber security firm SolarWinds with more than 1,800 of their client organizations, healthcare, education, the military and governments and prominent listed companies worldwide, affected. The breach occurred as routine updates were released. A second wave followed, targeting organizations selected in the first wave for further exfiltration of sensitive material (NCSC, 2021b).

Responses by the US Government followed, measures implemented included attribution, financial sanctions, and the expulsion of diplomats (NCSC, 2021b). Some argued the punitive response meant that US President Biden broke the norms of US foreign policy (NCSC, 2021b). As one official explained the hack was 'beyond the boundaries' due to the level and severity of the breach (Volz, 2021). Subsequent reactions included the creation of the US agency responsible for cybersecurity and infrastructure security (CISA, 2021).

The breach of SolarWinds is considered to have acted as a driver for the implementation of norms (BPF 2). In this paper the subsequent responses by governments and agencies are examined to evaluate the presence of trust and cooperation in cyber norms (UN IGF BPF, 2021b).

4. Evaluation of norms, trust, and cooperation

In the immediate outcomes from the SolarWinds breach, several governments and agencies implemented punitive measures, created policy and later, established agencies to drive implementation of the norm, zero trust (NCSN, 2021b). The central question of this paper concerns understanding the roles of trust and cooperation in norms. The conceptual model as set out in Table 1 (DT&CB model), above, is drawn on next, to disentangle and explore the separate, conditions, processes and outcomes concerning the lead-up to, and aftermath of the SolarWinds breach.

If we start by examining the antecedent conditions, the first elements in the model concern the ability to trust, and experiences of trust and cooperation (drawn from Deutsch, 1958; 2006). Looking back, prior to the SolarWinds breach, we can see a history of norms for cyber security under development, reliant on trust and cooperation by participants (EU, 2001; Schmitt, 2013; UN IGF 2015; GCSC, 2019). At the same time, norms were in development organically, notably through the actions of individual practitioners, later followed by cyber security organizations, such as FireEye, Microsoft and the NCSC (NCSC, 2021b). These examples are taken to demonstrate the presence of the ability to trust. Agencies, organizations, and practitioners were indeed working together and cooperating (NCSC, 2021b); providing evidence of experience of trust and cooperation.

In engaging in norms development, cooperation can be considered to represent an expectation, and indeed what can be termed, as a positive expectation of trust. Or indeed if we consider zero trust, no presumptive expectation of trust (NCSC, 20211). If we dig deeper, the element propensity to trust, may be viewed as tempered by need and desire (Dietz, 2011). In the case of zero trust, which emerged organically from practitioners, we see the approach being adopted and refined arguably as an indication of trusted practices as assessed through the next stage the processes (Table 1, the DT&CB model).

In the next phase, the processes implemented in the aftermath are examined. The processes include and are demonstrated by beliefs and a willingness to act. In the SolarWinds breach, beliefs were voiced, attributions made, and reactions were quick (Voltz, 2021). The processes include assessing if the measures or responses fit with the actor and the prevailing norms of society (Deutsch, 1998; 2006).

Trust itself is assessed through the three dimensions of ability, benevolence, and integrity (Mayer et al, 1995). If we consider a norm, say zero trust, the ability dimension asks, is zero trust able to do the job? In addressing a breach, a change to zero trust, though arguably not so simple (NCSC, 2021a). This is followed by the question of benevolence, will applying zero trust bring benefits that fulfill the responsibilities of an agency? Lastly the question of integrity, can the referent be sure that the provider is working with the best interests of the referent in mind? What example can demonstrate this?

Next, what evidence can be identified regarding accepting vulnerability and taking a risk? Actions in the response to an incident occur in a domain of the unknown. Respondents look for solutions that may involve accepting vulnerability and or risk taking (NCSC, 2021a). Arguably issuing a public consultation or encouraging referents to review material published by private organizations demonstrating such an example.

Turning to the final outcomes, these include trust, and cooperation, both are evident throughout the timeframe, from antecedent through processes and beyond into the future. Summing up, the model as enabled an evaluation of the complex and separate elements of the dynamics, and in doing so, extended the literature (notably, Vanneste, 2016).

5. Conclusion

The principle of norms for cyber security are to provide a trusted guide and an agreed set of rules (GCSC, 2019). Yet, as highlighted, the underpinning foundations rely on the understudied role of trust and cooperation. Through leveraging well-established theory from management and conflict resolution the contribution of this paper has addressed important gaps in our understanding to date and identified and a solution.

Returning to the early research in conflict resolution in the aftermath of World War II and the use of atomic bombs in Japan literature is timely. This work highlighted the key role of mutual trust and cooperation in overcoming suspicion and creating norms of cooperation (notably, Deutsch, 1958; 2006).

In addressing our limited understanding in this domain, this paper has sought to provide a first foundation to evaluate the role of the central elements of trust and cooperation in norms. The contribution of this paper extends the literature (Vanneste, 2016) through providing a conceptual model that makes explicit the dynamics involved in trust and cooperation building, the DT&CB model (Table 1). The NTAC^{+} framework (Figure 1) allows an understanding of the dynamics of norms, trust, and cooperation. Application of the DT&CB model together with the underpinning conceptual framework (NTAC^{+}) has proved capable in disentangling the multifaceted elements involved in the SolarWinds breach. The evaluation has highlighted the various actions by organizations both public and private in developing norms and in driving their implementation. The development of zero trust as a norm for cyber security was also discussed (NCSC, 2021a), in the context of its emergence as an organic norm (Wylde, 2021).

In this paper discussions are inevitably limited due to space constraints. It is important to add that further work is necessary to allow elaboration and to address any shortcomings. Several promising avenues for further research arise, notably the opportunity for empirical studies to examine in detail, the specific contexts of trust and cooperation activities as norms are practiced. For example, organizational structure, strategy or culture. Further work could usefully explore the application of machine learning to assist analysis through automation. Although these conditions could be incorporated here, their development is beyond the scope of this paper

The significance of the contribution of this paper is reflected in the recent Vienna talks (Joint Comprehensive Plan of Action) (TASS, 22). Speaking about how "joint statement would enhance mutual trust", one foreign minister stated that "replacing competition among the great powers would help with cooperation...and the building of major-country relations" (TASS, 2022). Finally, it is hoped that this work will help provide colleagues, policy makers and practitioners with a starting point to help disentangle the important constructs of trust and cooperation in norms for cyber security

References

Ajzen, I. (1991) "The theory of planned behavior", *Organizational Behavior and Human Decision Processes*, Vol 50, No 2, pp. 179-211.

Castañer, X. and Oliveira, N. (2020) "Collaboration, coordination and cooperation among organizations: establishing the distinctive meanings of these terms through a systematic literature review", *Journal of Management*, Vol 46, No 6, pp. 965-1001.

Cialdini, R., Kallgren, C.A. and Reno, R. (1991) *A focus on normative conduct: Theoretical refinement and reevaluation of the role of norms in human behaviour*, in M. P. Zanna (Ed.) *Advances in Experimental Social Psychology*, pp. 201-234.

CISA (2021) Cybersecurity and infrastructure agency, [online], CISA, https://www.cisa.gov/

Deutsch, M. (1958) "Trust and suspicion", *The Journal of Conflict Resolution*, Vol 2 No 4, pp. 265–79.

Deutsch, M. (2006) *Cooperation and competition*, in, M. Deutsch, P.T. Coleman and E.C. Marcus (Eds.), *The handbook of conflict resolution: Theory and practice*, pp. 23-42, San Francisco, Jossey-Bass.

Dietz, G. (2011) "Going back to the source: Why do people trust each other?", *Journal of Trust Research*, Vol 1, No 1, pp. 215-222.

Donaldson, T. and Dunfee, T.W. (1994) "Towards a unified conception of business ethics: Integrative social contracts theory", *Academy of Management Review*, Vol 19 No 2, pp. 252-284.

Ellickson, R.R. (1999) *The evolution of social norms: A perspective from the legal academy*, in M. Hechter and K. Dieter (Eds.) *Social Norms*, pp. 35-75

EU (2001) "Convention on cybercrime, Budapest", [online], EU, https://www.europarl.europa.eu/meetdocs/2014_2019/documents/libe/dv/7_conv_budapest_/7_conv_budapest_en.pdf

Fulmer, A. and Gelfand, M. (2012) "At what level (and in whom) we trust: trust across multiple organizational levels", *Journal of Management*, Vol 38, No4, pp. 1167-1230 (2012).

Gazley, B. (2017) "The current state of interorganizational collaboration: lessons for human service research and management", *Human Service Organizations: Management Leadership and Governance*, Vol 41, pp. 1-5.

GCSC (2019) "Advancing Cyberstability", [online], Global Commission on the Stability of Cyberspace https://cyberstability.org/norms/#toggle-id-6

Gulati, R., Wohlgezogen, F and Zhelazkov, P. (2012) "The two facets of collaboration: Cooperation and coordination in strategic alliances", *Academy of Management Annals*, Vol 6, No 1, pp.531-583.

Kretschmer, T. and Vanneste B.S. (2017) *Collaboration in strategic alliances: Cooperation and coordination,* In Mesquita, L.F. and Ragozzino, R. and Ruer, J.J. (Eds.) *Collaborative strategy: Critical issues for alliances and networks*, pp. 53-62. Edward Elgar, Cheltenham, UK

Levine, L. (2019) "Digital Trust and Cooperation with an Integrative Digital Social Contract, *Journal of Business Ethics*, Vol 160 No 2, pp. 393-407.

Mcknight, D.H., Carter, M., Thatcher, J.B. and Clay, P. (2011) "Trust in a specific technology: an investigation of its components and measures", *ACM Transactions on management information systems*, Vol 2, No 2, pp. 1-25.

Mayer R., Davis, J. and Schoorman, F. (1995) "An integrative model of organizational trust", *Academy of Management Review*, Vol 20, No 3, pp. 709-734.

NCSC). (2021a) "Zero trust architecture design principles", [online], NCSC, https://www.ncsc.gov.uk/collection/zero-trust-architecture

NCSC (2021b) "NCSC statement on the SolarWinds compromise", [online], NCSC, https://www.ncsc.gov.uk/news/ncsc-statement-on-solarwinds-compromise

Ring, P.S. and Van de Ven, A.H. (1994) "The developmental processes of cooperative interorganizational relationships", *Academy of Management Review*, Vol 19, No 1, pp. 90-118.

Rousseau, D.M., Sitkin, S.B., Curt, R.S. and Camerer, C. (1998) "Not so different at all: a cross discipline view of trust", *Academy of Management Review, Vol* 23, No 3, pp 393-404.

Schmitt, M.N. (Ed.) (2013) *Tallinn Manual on the international law applicable to cyber warfare*, Cambridge University Press, Cambridge.

Smith, J. and Louis, W.R. (2008) "Do as we say and as we do: the interplay of descriptive and injunctive group norms in the attitude-behaviour relationship", *British Journal of Social Psychology*, Vol 47, pp. 647-666.

TASS (2022) "Nuclear power's arms control efforts to boos mutual trust - China's foreign ministry", [online], TASS, https://tass.com/world/1383711

Thompson, A.M. and Perry, J.L. (2006) "Collaboration processes: inside the black box", *Public Administration Review*, Vol 66, Issue S1, pp. 2032.

UN General Assembly (2015) "Group of governmental experts on developments in the field of information and telecommunications in the context of national security", [online], General Assembly, UN, https://www.ilsa.org/Jessup/Jessup16/Batch%202/UNGGEReport.pdf

UN IGF BPF (2021a) "Mapping and analysis of international cybersecurity norms agreements", [online], UN. IGF BPF, Workstream 1, https://www.intgovforum.org/en/filedepot_download/235/19830

UN IGF BPF. (2021b) "Testing norms concepts against cybersecurity events", [online], UN. IGF BPF, Work stream 2, https://www.intgovforum.org/en/filedepot_download/235/20025

Venkatesh, V. and Davis, F.D.A. (2000) Theoretical extension of the technology acceptance model: four longitudinal field studies, *Management Science*, Vol 46, No 2, pp. 186-204 (2000).

Vanneste, B.S. (2016) "From interpersonal to interorganizational trust: the role of reciprocity", *Journal of Trust Research*, Vol 6, No 1, pp. 7-36.

Voltz, A. (2021) "In punishing Russia for SolarWinds, Biden upends US convention on cyber espionage", [online], Wall Steet Journal, https://www.wsj.com/articles/in-punishing-russia-for-solarwinds-biden-upends-u-s-convention-on-cyber-espionage-11618651800

Wired. (2021) "Netflix's password sharing crackdown has a silver lining", [online],
WIRED, https://www.wired.com/story/netflix-password-sharing-crackdown

Wylde, A. (2021) *Zero trust: never trust always verify*, in C. Onwubiko T. Lynn P. Rosati A. Erola X. Bellekens P. Endo G. Fox and M. G. Jaatun (Eds.) 2021, *Cyber Science 2021, CyberSA for Trustworthy and Transparent Artificial Intelligence (AI)*, C-MRiC.ORG 2021:UK.

A Managerial Review and Guidelines for Industry 4.0 Factories on Cybersecurity

Najam Ul Zia, Ladislav Burita, Aydan Huseynova and Victor Kwarteng Owusu
Tomas Bata University in Zlin, Czech Republic
zia@utb.cz
burita@utb.cz
husenova@utb.cz
owusu@utb.cz

Abstract: The Fourth Industrial Revolution (Industry 4.0) has created a rebellion in traditional factories by introducing the Internet of Things (IoT) and Cyber-Physical Systems (CPS). This revolution has caused increased automation and customized production, which has occurred through a synergy between customer demands, stocks, and supply chains. This synergy has also exposed factories to potential cyber-attack threats. Although there is extensive literature available on the topic of cyber security, however, business owners still assume cyber security as business preservation. This study sheds light on a step-by-step cyber security aspect of manufacturing factories with Industry 4.0. The study presented possible vulnerabilities and threats to the networks and devices used in a factory by dividing them into various common parameters. We reviewed the proposed literature and provided solutions to Industry 4.0 factories regarding cybersecurity challenges. The reviewed articles are divided into four segments, starting from the purpose of the proposal, the adopted methodology, the proposed cyber security solution, and finally the author's evaluation. The study reports on a state-of-the-art cyber security solution for Industry 4.0 factories. The characterization of cybersecurity is also proposed concerning management aspects, by showing that every level of organization has its role. The study also highlighted that cybersecurity could play a crucial role in the creation of value for businesses. It is suggested that despite adding an expert system paradigm for cyber security solutions, factories should also adopt new innovative ways, such as machine learning, digital twins, and honeypots. This review highlights that cyber security is not only a technical concern, but it also needs support from multiple actors of the organization to add it to the comprehensive strategy of an Industry 4.0 factory, and every user must be trained and aware of the cybersecurity risks.

Keywords: industry 4.0, cyber security, cyber solution, internet of things

1. Introduction

In an industry 4.0 environment, the cyber-physical system plays a crucial role in performing decentralized decisions to maximize the customized production capacity of smart factories (Kannengiesser and Müller, 2018). To achieve this important task, the logical systems in an internet of things (IoT) interact and collaborate in real-time to apply all kinds of operational processes, organizational services, and intelligent production solutions(Banafa, 2018). IoTs interconnects sensor, devices and instruments which combine with industrial applications like energy and production management to automate the process at a higher level (Banafa, 2018). This IoTs connection moves on to collect data, exchange it and analyse to facilitate the production performance in the production chain of a factory. It also facilitates the manufacturing section to innovate and produce those parts that looked impossible previously. To fully transform the supply chain to a fully IoT based supply chain, there should be an uninterrupted exchange of information from every step of the production scale. Therefore, for a fully automated system, IoT systems are combined with a multilevel architect, hardware level, network level and upper layers. The hardware-level comprises physical systems like sensors, control systems, actuators, and security mechanisms etc. The network-level consists of physical networking like a combination of wired and wireless networking. Finally, the upper layers collect and transmit data and information from this communication network (Tsiknas *et al.*, 2021). This continuous boost of communication in an Industry 4.0 factory creates a strong need for industrial systems protection from cyber-attacks (Juárez, 2019). All the industrial systems that control the process of production, have continuous access to the internet, and these devices are known as industrial control systems (ICS)(Kargl *et al.*, 2014). SCADA (supervisory control and data acquisition) is known to be the most common type of ICS which are used to collect measurement and support process information (Falco, Caldera and Shrobe, 2018). All these systems are interconnected to IoTs that facilitate the remote monitoring and management of processes. Due to this network and connectivity, the operational efficiency of the production system improves, but at the same time, it poses major challenges to secure this infrastructure regarding integrity, confidentiality, and availability (Falco, Caldera and Shrobe, 2018). Another important point is that all the machines and devices are prepared with an objective to enhance smooth production, but not in a mind to secure the devices, which further deteriorates the integrity of system networks

(Tsiknas *et al.*, 2021). This exploitation of machines and devices to external cyber threats means a compromise that may result in malfunctioning or destruction of the whole production system (Panchal, Khadse and Mahalle, 2018). The current literature focuses on security risks in IoT based factories, however, there is a paucity of literature on the knowledge and clear understanding of threats associated with IoT systems. In this aspect, our study highlights the ways of industrial application attacks and the available solutions in the literature. The paper contributes by providing literature for researchers and for organizations dealing with IoTs technologies on cyber threat issues and also the solutions for protecting these industrial applications and instruments.

The study organizes as follows, section 2 provides a detailed explanation of main industrial IoT environment tasks and the possible effective solutions that are taken out from the available literature. Section 3 shows the results of the study, and then the last section comes with a conclusion and future research possibilities.

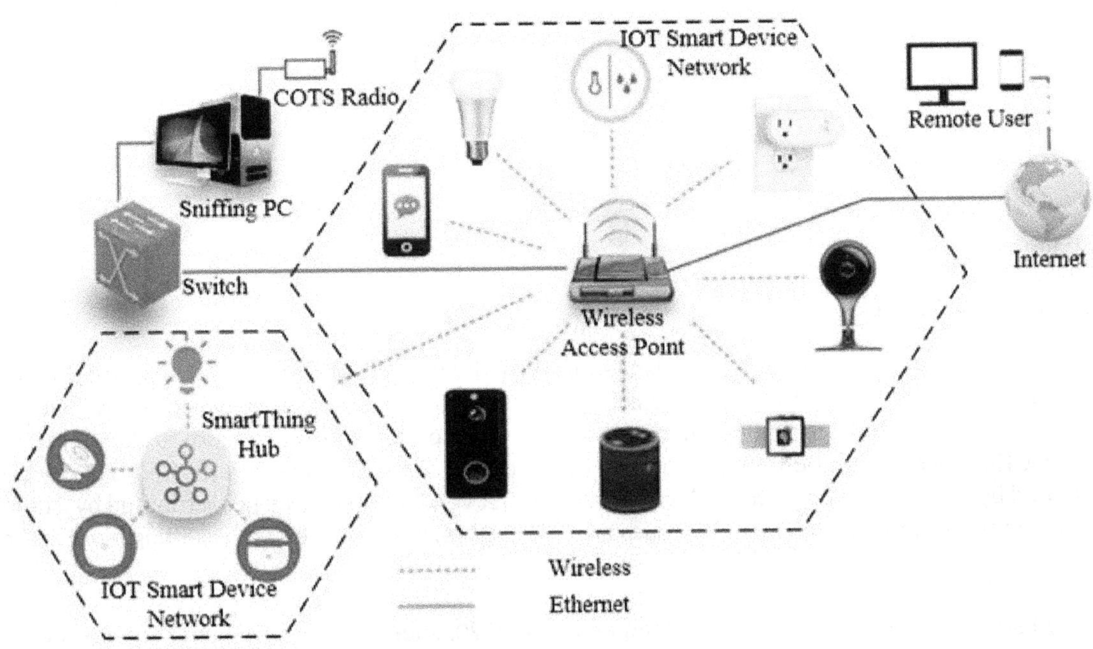

Figure 1: Internet of Things layers (Calix *et al.*, 2020)

2. Cyber threats and possible solutions

To achieve customized production and quality milestones, automation and remote control are considered as most crucial methods in an Industry 4.0 factory (Mikhalevich and Trapeznikov, 2019). This system requires efficient management of IoT systems consisting of maximum accuracy, security, and reliability. The digital infrastructure that is part of these IoT systems improves the critical infrastructure efficiency but meanwhile requires securing the infrastructure against possible cyber-attacks. Not only this brings a need to protect the local digital infrastructure, but it also directs to protect the general crucial digital infrastructure of the country. In the below sections, we have categorized the IoT threats into phishing attacks and supply chain attacks. This categorization presents a clear comprehensive and clear information about cyber risks and the solutions of protection in an Industry 4.0 environment.

2.1 Phishing attacks

This is a typical method of stealing sensitive information from consumers. It occurs when a hacker impersonates a trustworthy entity (Roman *et al.*, 2009) and dupes individuals into entering personal information on a fake website or downloading an attachment, resulting in the installation of malware or the disclosure of sensitive information. Advanced social engineering tactics known as compromised attacks are used by specialized phishers to target important infrastructures. They target both the absence of specific active security measures by systems and the lack of information or attention of users. Generally, a cyber attacker tries to approach the IoT systems through the front end level. Several papers have highlighted the website crawling based techniques. A new technique called PHONEY is proposed by Chandrasekaran, Chinchani and Upadhyaya, (2006) that can automatically detect and highlight phishing attacks. This technique keeps the main idea of a web browser extension, which gives information on website security certificates, quality of websites or a misleading URL.

Another technique is introduced by McRae and Vaughn (McRae and Vaughn, 2007) detects phishing contents sites by using honey tokens. A more promising technique referred to as URL embedding was proposed by Yan *et al.* (2020). They used an algorithm to investigate the correlation between various domain names for a calculation of the correlation coefficient between various URLs.

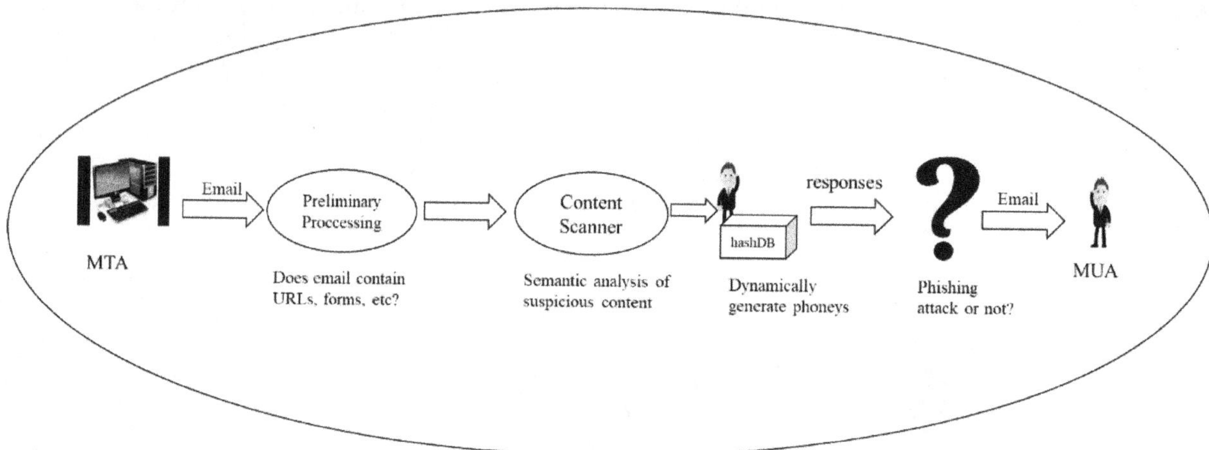

Figure 2: PHONEY architecture

2.2 Supply chain attacks

Supply chain attacks are considered as most dangerous. Security is the major challenge in the supply chain of Industry 4.0. It is difficult to find the hardware chips with implanted malicious code because this code can be executed for a long time without being noticed. Another cause of security weaknesses is the stakeholders' involvement. The device acquisition system is not unique and centralized. Because different kinds of devices are manufactured by different vendors, then assembled by another vendor and at the end distributed by a different vendor. Due to this situation, many security issues arise. Therefore, risk management is getting more and more attention day by day.

A study by Farooq and Zhu, 2019; Kieras, Farooq and Zhu (2020, sheds the light on supply chain threats and suggest various approaches concerned with risk management methods. The study highlighted the risks involved to IoT supply and define it as extremely diverse. Though the study has a general understanding of the risks of the supply chain, however, it did not provide the possible solutions or countermeasures to address these kinds of attacks in an Industry 4.0 environment. In a study by Radanliev *et al.* (2020), a self-adapting and dynamic supply chain system is introduced which is supported by real-time intelligence, machine learning (ML), and artificial intelligence (AI). This approach is castoff for small and medium enterprises (SMEs) to grow a transformational roadmap for the Industry 4.0 Industrial Internet of Things, as small companies lack resources to mitigate the high cyber-attack risks. The cyber risk measurement is due to the weakness of understanding Industry 4.0 supply chain operations. Kieras, Farooq and Zhu (2020) stated in their study about the risk analysis of IoT supply chain threats. They introduced an adoption of attack free technique associated with vendors and suppliers. They intend to highlight and uncover the threats that are associated with potential supplier collusion

A vendor can incorporate backdoor routes in their equipment, implant viruses, or supply defective chips. The hazards in the supply chain are difficult to detect and control. As the IoT ecosystem becomes more complicated, the risk spreads from one device to the next. Another challenge is dissecting the supply chain linkages in IoT, which means that determining the relationships between devices, suppliers, and among them is always challenging. They also underline the implications and repercussions of IoT hazards, and as a protective measure, they recommend seeing the ecosystem from a supply chain perspective and then taking appropriate risk-control measures. They distinguish between two approaches: the top-down method, which is more centralized, and the bottom-up approach, which emphasizes decentralization. This study provides a broad knowledge of supply chain risks, but it does not include technological remedies for dealing with these sorts of attacks in a context that is already facing this danger and cannot adapt 's entire risk management system. In this paper, Radanliev *et al.* (2020), propose a dynamic and self-adapting supply chain system powered by artificial intelligence (AI), machine learning (ML), and basic intelligence for predicted cyber risk analytics. This method is used to create a transformative roadmap for the Industrial Internet of Things in Industry 4.0 supply chains of small and medium businesses (SMEs) because these organizations typically lack the resources required to successfully combat the significant risks posed by cyber-attacks. The inability of existing cyber risk impact assessment methods to measure the impact of supply chain infrastructure is an intriguing topic of debate from the major findings. Furthermore, due to a lack of understanding of supply chain activities in Industry 4.0, there is inconsistency in quantifying supply chain cyber threats. Kieras, Farooq and Zhu (2020) in this study introduced the RIoTS (risk analysis of IoT supply chain threats) technique in their paper, which is risk analysis methodology in network infrastructures such as the IoT that arise from single components providers. They believe that risk analysis should move away from a vulnerability-centred strategy and toward modelling suppliers and elements as a system. They suggest modifying attack tree methodologies to account for the risk associated with suppliers and supplier groupings. Their goal is to expose and uncover hidden dangers to the IoT ecosystem caused by potential supplier collaboration. As we've seen, the majority of research concentrate on risk management measures for supply chain assaults.

3. Discussion

The general security of the infrastructure and the dependability of the intended solutions stated should not be taken for granted, since the cyber security of the IoT ecosystem is a multifactorial dilemma (Nakamura and Ribeiro, 2018). Particularly, due to the landscape of the IoT and the extensive series of exposures that can happen from the intricacy of the systems tangled in it, significant structures related to multifaceted patterns, systems, or procedures are recognized and preserved, which do not develop in parallel with the overtime and which are possible weaknesses of the entire network (Sengupta, Ruj and Bit, 2020). More commonly, the problem deceits in the fact that in the specific high intricacy atmosphere under inspection, while adjustment systems are multivariate, high assortment happens and is preserved, as this can be accredited to the age of systems that have not been promoted, to the composite connection that defines them, and the delicate alterations that differentiate them (Lee and Chen, 2019).

Among the threats stated, the supply chain incidents are turning out to be a severe concern, because substantial issues like complexity and stealth do not offer simple solutions. To diminish these types of attacks, typically risk management methods are utilized. Another main disadvantage is the fact that older industrial systems, which in most cases do not have security as a precondition in their structure stipulations, are spinning points of the complete security of the system, suggestively growing the overall hazard of attacks, even if access control or encoding methods are added in them (McLaughlin *et al.*, 2016). In addition, the adjustment and organizing events with the existing established standards raise thoughtful alarms, as most of the current IoT systems have an extraordinary degree of dependency on their development company, which generates difficulties of reorganization or revision of their mechanisms, such as functions that they contain or can support (Lee and Chen, 2019). Moreover, due to the real-time process and progress of the IoT, the supervision of data with the time difference, taking into account correlations from other instruments or devices that may be incorporated in the data flow categorization, creates additional requirements in the ways of confirming accuracy and integrity of information. While offering tight requirements, the encrypt (Nakamura and Ribeiro, 2018) and key management approaches suggested and utilized in the IIoT environment fall behind in the development of mechanisms that will be implemented fast and without much complexity, allowing them to be employed by low-resource devices. Subsequently, a further important finding through the use of most of the machine learning techniques presented in the literature is that only statistics on the system or network traffic are used (Zhou and Guo, 2018), resulting in ineffective smoothing because the metrics trained do not include a variety of aspects from different uses or behavioural parameters of the system overall. The error originates from the mistaken assumption that the initial version and all of its updated duplicates had identical features distributions, and so the current statistics could be shared with all of the intelligent learning inner current adjustments. Another better approach, which was used in the suggested method, is to save statistics throughout stages and read the optimizing parameters methodically to every inner loop iteration.

4. Conclusion

Given the increasing complexity of threats in the ever-changing environment of the Industrial IoT, as well as the parallel vulnerability of existing security systems to detect significant threats of increasing magnitude and duration, it is necessary to recognize the risks that threaten particular infrastructure and services and provide industrial data confidentiality (McLaughlin *et al.*, 2016). Likewise, while there is a chance that hackers may get access to the manufacturing process, perhaps with disastrous, if not incalculable, effects, most industrial organizations seek security know-how to defend their infrastructure. It should have been highlighted that IoT designs and industrial systems in general (Kargl *et al.*, 2014; Falco, Caldera and Shrobe, 2018) require a separate type of protection than regular networks, because traditional security solutions, such as virus scanners and traditional firewalls, do not match industry norms and criteria. A thorough description of attacks against Industrial IoT systems was carried out in this study, considering the most important features and vulnerabilities that they incorporate, as well as a thorough analysis of indicative solutions against these vulnerabilities, as proposed in the most recent literature. It is a proven reference framework and an indicative scientific presupposition in this context for the identification and evaluation of hazards associated with the ever-changing industrial environment. One factor that might be addressed in the future extension of this research is the analysis of unconventional ways of attack or innovative techniques of combined approach of unknown assaults such as zero-day attacks. Lastly, the research might be broadened by looking into unique protective strategies against IoT, the physical security of IoT devices, from malicious setup of mechatronic subsystems that are part of this network, with the goal of exploitation by a third – party for example vendors.

References

Banafa, A. (2018) '2 the industrial internet of things (IIoT): challenges, requirements and benefits'.

Calix, R. A. *et al.* (2020) 'Cyber Security Tool Kit (CyberSecTK): A Python Library for Machine Learning and Cyber Security', *Information*, 11(2), p. 100.

Chandrasekaran, M., Chinchani, R. and Upadhyaya, S. (2006) 'Phoney: Mimicking user response to detect phishing attacks', in *2006 International Symposium on a World of Wireless, Mobile and Multimedia Networks (WoWMoM'06)*. IEEE, pp. 5-pp.

Falco, G., Caldera, C. and Shrobe, H. (2018) 'IIoT cybersecurity risk modeling for SCADA systems', *IEEE Internet of Things Journal*, 5(6), pp. 4486–4495.

Farooq, M. J. and Zhu, Q. (2019) 'IoT supply chain security: Overview, challenges, and the road ahead', *arXiv preprint arXiv:1908.07828*.

Juárez, F. A. B. (2019) 'Cybersecurity in an Industrial Internet of Things Environment (IIoT) challenges for standards systems and evaluation models', in *2019 8th International Conference On Software Process Improvement (CIMPS)*. IEEE, pp. 1–6.

Kannengiesser, U. and Müller, H. (2018) 'Towards viewpoint-oriented engineering for Industry 4.0: A standards-based approach', in *2018 IEEE Industrial Cyber-Physical Systems (ICPS)*. IEEE, pp. 51–56.

Kargl, F. *et al.* (2014) 'Insights on the security and dependability of industrial control systems', *IEEE security & privacy*, 12(6), pp. 75–78.

Kieras, T., Farooq, M. J. and Zhu, Q. (2020) 'RIoTS: Risk analysis of IoT supply chain threats', in *2020 IEEE 6th World Forum on Internet of Things (WF-IoT)*. IEEE, pp. 1–6.

Lee, J.-C. and Chen, C.-Y. (2019) 'Exploring the determinants of software process improvement success: A dynamic capability view', *Information Development*, 35(1), pp. 6–20.

McLaughlin, S. *et al.* (2016) 'The cybersecurity landscape in industrial control systems', *Proceedings of the IEEE*, 104(5), pp. 1039–1057.

McRae, C. M. and Vaughn, R. B. (2007) 'Phighting the phisher: Using web bugs and honeytokens to investigate the source of phishing attacks', in *2007 40th Annual Hawaii International Conference on System Sciences (HICSS'07)*. IEEE, pp. 270c-270c.

Mikhalevich, I. F. and Trapeznikov, V. A. (2019) 'Critical infrastructure security: Alignment of views', in *2019 Systems of Signals Generating and Processing in the Field of on Board Communications*. IEEE, pp. 1–5.

Nakamura, E. T. and Ribeiro, S. L. (2018) 'A privacy, security, safety, resilience and reliability focused risk assessment methodology for IIoT systems steps to build and use secure IIoT systems', in *2018 Global Internet of Things Summit (GIoTS)*. IEEE, pp. 1–6.

Panchal, A. C., Khadse, V. M. and Mahalle, P. N. (2018) 'Security issues in IIoT: A comprehensive survey of attacks on IIoT and its countermeasures', in *2018 IEEE Global Conference on Wireless Computing and Networking (GCWCN)*. IEEE, pp. 124–130.

Radanliev, P. *et al.* (2020) 'Cyber risk at the edge: current and future trends on cyber risk analytics and artificial intelligence in the industrial internet of things and industry 4.0 supply chains', *Cybersecurity*, 3, pp. 1–21.

Roman, R. *et al.* (2009) 'Trust and reputation systems for wireless sensor networks', *Security and Privacy in Mobile and Wireless Networking*.

Sengupta, J., Ruj, S. and Bit, S. Das (2020) 'A comprehensive survey on attacks, security issues and blockchain solutions for IoT and IIoT', *Journal of Network and Computer Applications*, 149, p. 102481.

Tsiknas, K. *et al.* (2021) 'Cyber Threats to Industrial IoT: A Survey on Attacks and Countermeasures', *IoT*, 2(1), pp. 163–186. doi: 10.3390/iot2010009.

Yan, X. *et al.* (2020) 'Learning URL embedding for malicious website detection', *IEEE Transactions on Industrial Informatics*, 16(10), pp. 6673–6681.

Zhou, L. and Guo, H. (2018) 'Anomaly detection methods for IIoT networks', in *2018 IEEE International Conference on Service Operations and Logistics, and Informatics (SOLI)*. IEEE, pp. 214–219.

A Cyber-Diplomacy and Cybersecurity Awareness Framework (CDAF) for Developing Countries

Hendrik Zwarts, Jaco Du Toit and Basie Von Solms
University of Johannesburg, South Africa
hzwarts@yahoo.com
jacodt@uj.ac.za
basievs@uj.ac.za

Abstract: Cybersecurity is high on the agenda of national and international security policy discussions – mostly lead by diplomats. The practise of diplomacy has evolved since the Internet has become the backbone of society as we know it. Technological evolution has resulted in a significantly bigger and more accessible cyberspace, but the ability of governments and institutions to respond to and function in an expanding cyberspace seems to be lagging behind. The practice of diplomacy has similarly changed fundamentally and created a cyber-diplomacy environment where there is an increased utilization of inter alia social media platforms to achieve foreign policy goals. There is not enough attention given to practical processes to guide the new breed of diplomats in the evolving world of cyber-diplomacy and there is a need to improve the cybersecurity awareness of diplomats in all countries, but this article will focus primarily on developing countries. To mitigate potential cyber threats to diplomacy, diplomats need to be subjected to cyber-diplomacy orientation as well as functional cyber awareness training. Preliminary research conducted suggests that there is a gap between the existing and required cyber-diplomacy and cybersecurity awareness levels of diplomats from developing countries. The purpose of the article is to present a cyber-diplomacy and cybersecurity awareness framework (CDAF) that can be used by developing countries to equip their diplomats to play a more constructive role within the international cyber-diplomacy domain. The CDAF comprises of two distinct components, namely cyber-diplomacy and cybersecurity awareness, but this article will focus primarily on the cyber-diplomacy capacity building aspect of the CDAF. The CDAF was developed by following a design science research approach where a real-world problem was identified followed by an in-depth literature review to identify objectives and possible solutions to the problem. The subsequent outcomes were used to design and development of the CDAF. The article concludes with a critical evaluation of the proposed framework as well as how it can be incorporated into the developing cybersecurity knowledge modules of the Global Forum on Cyber Expertise (GFCE).

Keywords: cyber-diplomacy, cybersecurity, awareness, framework, cyber conflict, CDAF

1. Introduction

Countries should first and foremost be responsible for their citizens' safety and security; in an ever-changing world that is becoming more and more digitalised this includes protecting their citizens in cyberspace (Buzatu, 2021). While the world's economies continue to develop an increasing dependence on technology the development of cybersecurity capacity across the whole of cyberspace is critical to avoid what the Global Cyber Security Capacity Centre (GCSCC) refer to as *"cyber-ghettos"* – environments where cyber-harm may become prevalent and from where cyberattacks can easily be launched (Global Cyber Security Capacity Centre (GCSCC), 2021).

However, many countries lack the capacity to protect their own information and communication technologies (ICT) networks, to learn about cybersecurity threats and respond effectively to them through bilateral, regional and global engagements at both a technical and/or diplomatic level. The absence of such capacity leave state institutions and critical sectors (Buzatu, 2021). This lack of cybersecurity expertise is not isolated to technical issues, but also relates to a lack of cyber-diplomacy capacity – especially in developing countries.

This article will present a cyber-diplomacy and cybersecurity awareness framework (CDAF) that can be used by developing countries to equip their diplomats to play a more constructive role within the international cyber-diplomacy domain. The CDAF comprises of two components, namely cyber-diplomacy and cybersecurity awareness, but this article will focus primarily on the cyber-diplomacy capacity building aspect of the CDAF. The opening segment of the article reflects on the evolution of cyber-diplomacy followed by a discussion on different cybersecurity capacity building initiatives. The focal part of the article is the discussion and presentation of the CDAF.

2. Cyber-diplomacy – the new frontier

This section deals with evolvement of traditional diplomacy into cyber-diplomacy and the critical role that it plays within the modern, digitalised world. Cyber-diplomacy can be defined as *"diplomacy in the cyber domain and the use of diplomatic resources and the performance of diplomatic functions to secure national interests with regard to the cyberspace"* (Barrinha & Renard, 2017). It is closely interrelated to cybersecurity which is defined by the EU Cybersecurity Act (Regulation 2019/881, Article 2(1)) as *"the activities necessary to protect network and information systems, the users of such systems, and other persons affected by cyber threats"*. The rational of diplomacy being practiced in cyberspace is indisputable and yet it is a relatively new field. A brief recap of how diplomacy changed over time will contextualise the importance for countries to participate and contribute towards diplomacy in cyberspace i.e., cyber-diplomacy.

Wight (1979) defined diplomacy as *"the attempt to adjust conflicting interests by negotiation and compromise"* (Wight, 1979, p. 89). The objectives of diplomatic practice are:

- to enable communication in international politics,
- to negotiate treaties,
- to collect intelligence and information on countries,
- to circumvent and reduce conflict in international relations and,
- to represent the actuality of a civilization of states (Barrinha & Renard, 2017).

This view on diplomacy evolved over time and in 2004 Jönsson and Langhorne stated that diplomacy was no longer an activity solely undertaken by a select group of people as it has changed to encompass wider relationships and dialogues involving a broader array of entities such as regional and international intergovernmental (IGOs) or non-governmental (NGOs) organizations, multinational firms, pressure groups, advocacy networks and influential individuals (Jönsson and Langhorne, 2004). From this it can be deduced that (cyber) diplomats interact with any combination of the following role players on cyber related issues:

- Diplomats from other countries through bilateral or multilateral forums such as the UN.
- Various non-state actors such as leaders of internet companies (Facebook, Google)
- Technology entrepreneurs.
- Civil society organizations.
- Individuals

Over the last 20 years the Internet has become the backbone of society as we know it; it impacts on everyday economic-, social and political life, global mobility, communication, the Internet of things (IoT) and the storage of big data. This technological evolution has also resulted in a significantly bigger and more accessible cyberspace (Nye, 2014). Governments and institutions were initially slow to respond to and manage the swift advances in technology related to cyberspace (Nye, 2014) which resulted in a cyber-governance void in terms of the practice of diplomacy in cyberspace.

The practice of diplomacy created a cyber-diplomacy environment where there is a *"growing use of social media platforms by countries to achieve its foreign policy goals and proactively manage its image and reputation"* (Adesina, 2017). The revolution in ICTs is one of the significant factors currently impacting on diplomatic processes. It revolutionized the way people communicate and share information, thereby changing local and international political, social and economic playing fields. With the increased digitalization of diplomacy, the cyber threat to diplomacy also increased significantly (Adesina, 2017).

Nowhere as the Internet was initially unregulated and its governance mostly informally managed by software engineers over time governments became involved and cyberspace became more structured. As the occurrence of international meetings increased discussions by government IT experts on cyber issues also increased. The exponentiation and institutionalization of these meetings together with the expansion of cyber topics resulted in an increase in "online" political tussles which paved the way to cyber diplomacy (Deibert, 2015). Cyber issues were initially treated as technical issues then as peripheral aspects of national policies, before they were acknowledged as key foreign policy focus areas; diplomats stepped in because cyberspace became a diplomatic sphere (Barrinha & Renard, 2017). The development and growth of various social media platforms have also created a mutual jurisdiction for diplomacy between states and within states (Goundar, Chandra, Bhardwaj, Saber, Appana, 2020).

Unfortunately, the ability of governments and institutions to respond to and manage these rapid advances in cyberspace has lagged behind (Nye, 2014). This cyber governance void also transpires to the practice of diplomacy. The practice of diplomacy has likewise changed fundamentally as a result of digitalisation and created a cyber-diplomacy environment where there is a *"growing use of social media platforms by countries to achieve its foreign policy goals and proactively manage its image and reputation"* (Adesina, 2017). This revolution in information and communication technologies (ICTs) was one of the significant factors impacting on diplomatic processes. It revolutionised the way people communicate and share information, thereby changing local and international political, social and economic playing fields. With the increased digitalisation of diplomacy, the cyber threat to diplomacy also increased significantly (Adesina, 2017). Diplomats are busy changing traditional policy environments, methods, and processes to deal with the impact of digitalisation. This also necessitates a reform in the modus operandi of diplomatic services (Leira, 2019). The development and growth of various social media platforms have also created a jurisdictional communality for diplomacy between states as well as within a state (Goundar, Chandra, Bhardwaj, Saber, Appana, 2020). The "business as usual" days of conducting diplomatic functions have made way for a new frontier i.e. cyber diplomacy and there is a need to improve the cyber awareness of diplomats. To mitigate potential cyber threats to diplomats, they need to be subjected to cyber-diplomacy orientation as well as functional cyber awareness training. This will augment their level of cyber diplomacy as well as the security and integrity of dialogue, negotiations and other diplomatic processes (Al-Muftah, Weerakkody, Rana, Sivarajah, Irani , 2018).

From the discussion on the evolvement of cyber-diplomacy it is clear that all countries should participate and contribute towards diplomacy in cyberspace. Existing literature suggest that there are limited cyber awareness tools, training or support available for diplomats from developing countries to improve the overall quality of the dialogue and the benefits drawn from international discussions in support of better cybersecurity in the international arena. The only way to fill this void is through the development and utilisation of a structured capacity building framework focused on cyber-diplomacy and cybersecurity awareness. The next section will look at various capacity building components that could be used to formulate and develop such a CDAF.

3. Building cyber-diplomacy and cybersecurity capacity

Cybersecurity capacity building initiatives are fairly new, starting in 2013 with the formation of the Oxford GCSCC. Since then developing cybersecurity capacity is high on the agenda of most governments, international organizations (IOs) and companies. Numerous capacity-building initiatives, such as the International Telecommunications Union (ITU), the Potomac Institute, the Diplo Foundation, the Australia Strategic Policy Institute and the GFCE attest to the progress made towards capacitating countries in terms of cybersecurity. The main components encompassed in these capacity enabling efforts include aspects such as policy and strategy, sociocultural outlooks, knowledge and skills, protocols, law enforcement and technical criteria and capabilities (Dutton, Creese, Shillair & Bada, 2019). Dutton et al (2019) suggest that cybersecurity capacity building can be approached by differentiating between the dimensions defining cybersecurity, the broader policy environment that is influencing cybersecurity and the diversity of actors relevant to each of these dimensions. By adapting this approach critical attributes can be identified for incorporation into a CDAF. The following three subcategories will be discussed briefly:

- Dimensions of cybersecurity capacity building
- Capacity building role-players
- Cyber-diplomacy in developing countries

3.1 Dimensions of cybersecurity capacity building

Cybersecurity capacity building has evolved to the point where emergent frameworks include much more than technical cybersecurity aspects. Digitalisation and social media uptake in almost all aspects of modern-day living necessitated a paradigm shift w.r.t. cybersecurity capacity building to include aspects such as cybersecurity strategy and policy, awareness, legal and regulatory structures, threat response etc. (Dutton et al., 2019). The digitalisation of diplomatic processes has also questioned the capacity and preparedness of cyber-diplomats to effectively function in cyberspace; including the effectiveness of traditional diplomatic training methods. In an effort to identify capacity building dimensions and themes for the CDAF the following models and programs were dismembered:

- The Cybersecurity Capacity Maturity Model for Nations (CMM)

- The GFCE's agenda for cyber capacity building
- The Association of Southeast Asian Nations (ASEAN) Cyber Capacity Programme (ACCP)
- The twenty-one knowledge areas (KAs) of the CyBOK

The CMM differentiates between five capacity building dimensions, namely "*developing cybersecurity policy and strategy, boosting responsible cybersecurity culture, creating cybersecurity knowledge and capabilities, creating operational legal and regulatory frameworks and risk mitigation through standards and technologies*" (Global Cyber Security Capacity Centre (GCSCC), 2021, p5).

The GFCE's agenda for cyber capacity building revolves around inter-connected themes that can be applied to national, regional and/or global cyber security developments i.e.:

- Cybersecurity policy and strategy
- Incident management and infrastructure protection
- Cybercrime
- Cybersecurity culture and skills, and
- Cybersecurity standards.

The focus areas of the ACCP include aspects such as cyber policy, cyber legislation, strategy development and cybersecurity incident response (Interpol, 2021)

Perhaps one of the most frequently used models to elucidate cybersecurity awareness training components is the twenty-one KAs of the CyBOK as it outlines a range of topics within the general scope of cybersecurity (Martin et al., 2021). The components of the CyBOK that cross cuts with the CMM, GFCE agenda and ACCP include standards, best practices, risk assessment and mitigation, regulatory requirements, social and behavioural factors impacting security, security culture and awareness, protecting if databases and data, malware and attack technologies etcetera. The following section will briefly look at the potential role-players that can contribute towards cybersecurity capacity building.

3.2 Capacity building role-players

Although the CDAF does not specifically address who is responsible for cybersecurity capacity building it is important to keep it in mind when developing specific KAs or learning outcomes. The Annual Conference of the GCSCC that was held at the University of Oxford in 2018 elaborates on the composition and interactions between various actors that should contribute towards build cybersecurity capacity (Global Cyber Security Capacity Centre, 2018). In the end the collection of actors can be simplified into three distinct groups, namely donors, implementers and recipients. The individuals responsible for cyber-diplomacy and cybersecurity capacity building will vary from country to country as well as from organisation to organisation. Table 1 presents the potential links between some of the more prominent cyber-diplomacy and cybersecurity awareness dimensions/themes and the potential role-players involved with each dimension/theme.

Table 1: Cyber-Diplomacy and cybersecurity awareness role-players

Dimensions/Themes	Role-Players						
	Experts	Researchers	Trainers	Networkers	End Users	Sponsors	Policymakers & Diplomats
Cybersecurity Policy & Strategy	x	x		x			x
Cybersecurity Culture	x	x		x	x		x
Legal & Regulatory Framework	x	x					x
Cybersecurity Standards	x	x	x	x			x
Cybersecurity Capacity Building	x	x	x	x		x	x
Incident/Risk Management	x	x					x
Cybercrime & Cyberattacks	x	x			x		x

Note: Derived from Dutton et al. (2019), Interpol (2021), GFCE (2017) and GCSCC (2021)

From Table 1 it is clear that policymakers and diplomats play a vital role within in cyberspace and cybersecurity. Whether all diplomats are suitably equipped to engage on all the intricacies of these dimensions is debateable. To contextualise the target audience of the CDAF the next section will take a concise look at cyber-diplomacy and cybersecurity in developing countries.

3.3 Cyber-diplomacy in developing countries

Every country in the world is exposed to cyberspace and is reliant on it to function in an interconnected world. The significance of cybersecurity awareness has increased over the last decade as governments, businesses and individuals' day-to-day activities around the world have moved more and more online. The challenge is that in most emerging economies these entities lack organizational, technological and human resources to secure their online activities and systems (Veale & Brown, 2020). The developing world has joined first world countries in relying on ICTs for the improvement of their populations' quality of life and economic growth. To access cyberspace developing countries predominantly uses services made by the developed world making them dependent on western-designed systems and protocols to regulate their actions in cyberspace. Cyber-diplomacy is needed to maintaining a constant dialogue between countries to develop norms of accountable government behaviour in cyberspace and addressing disagreements between role-players (Barrinha & Renard, 2020). Diplomatic interaction in global affairs is therefore an international security priority. Within this inter-connected world all countries strive towards advancing their own political and economic agendas – a function that is primarily steered by diplomats - in cyberspace. Cyberspace provides digital apparatuses to facilitate the effective execution of diplomatic strategies (Attatfa et al., 2020). Diplomats are busy changing traditional policy environments, methods, and processes to deal with the impact of digitalisation (Leira, 2019). Cyber-diplomacy is therefore an emerging field that gained momentum through its online application. Each country's uptake and implementation of cyber-diplomacy follows a different approach due to differences in foreign policies and perspectives on technology, but developed countries have progressed at a more rapid rate than most developing countries (Al-Muftah, Weerakkody, Rana, Sivarajah, Irani, 2018). It appears that there is no specific and/or not enough attention given to edify new diplomats in the evolving movement towards cyber-diplomacy. In the next section a CDAF will be presented that can be utilised to improve the cyber-diplomacy and cybersecurity awareness capacities of diplomats in developing countries.

4. A cyber-diplomacy and awareness framework for developing countries

Literature research conducted on the skills and capacity of diplomats in developing countries suggests that there is a gap between the existing and required cyber security awareness levels of these countries (Attatfa et al., 2020), (Zhang et al., 2021), (Sabillon et al., 2019). The integration of developments in cyber diplomacy and cybersecurity, as well as the available capacity building models resulted in the conception of the following CDAF that can promote the practice of cyber-diplomacy within developing countries:

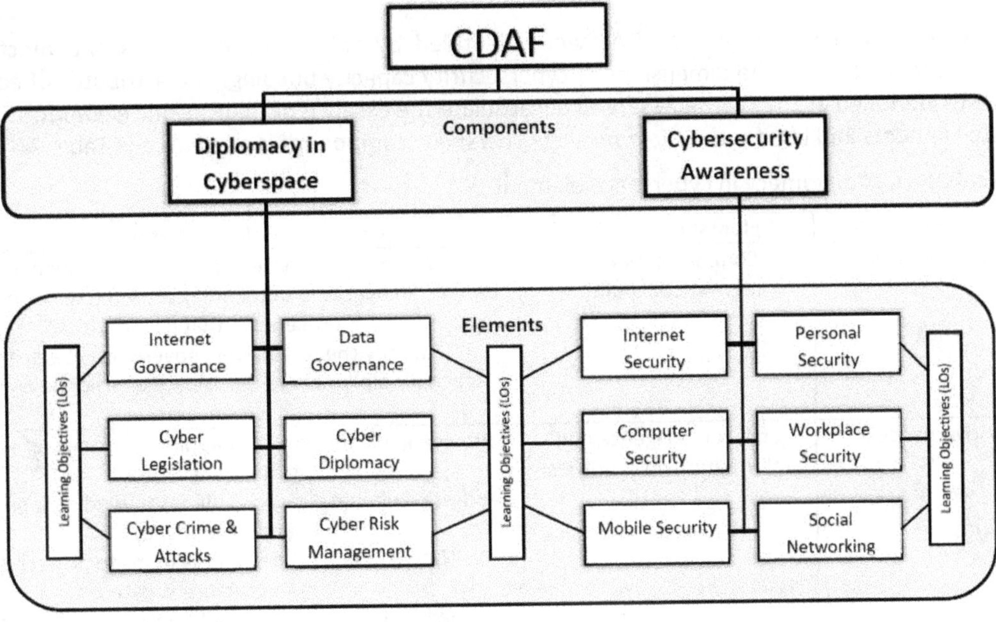

Figure 1: High-level view of cyber diplomacy and awareness framework (CDAF)

The CDAF can be broken down into components, elements and learning objectives:

- Components depict the combined subcategories of the core elements that encompass a country's cyber-diplomacy and cybersecurity awareness focus. In this case two components were identified, namely *Diplomacy in Cyberspace* and *Cybersecurity Awareness.*
- Elements provides a breakdown of the components into concentrated entities that are easier to delineate and understand. Certain elements can be more important or less important for a specific country depending on their cyber and diplomacy maturity level. The CDAF lists the following six elements under each of the two components:

Diplomacy in Cyberspace:

- 1. Internet Governance
- 2. Data Governance
- 3. Cyber Legislation
- 4. Cyber Diplomacy
- 5. Cyber Crime and Attacks
- 6. Cyber Risk Management

Cybersecurity Awareness:

- 1. Internet Security
- 2. Personal Security
- 3. Computer Security
- 4. Workplace Security
- 5. Mobile Security
- 6. Social Networking

Only the elements identified under *Diplomacy in Cyberspace* will be discussed further in the next section in an effort to explain how the framework can be tailor-made and/or adapted to meet specific requirements.

- Learning Objectives (LO) denotes the most basic parts of the CDAF and describes the attributes that could be included under an element. LO are not rigid nor finite and as such could be changed depending on the target groups' skills level on the particular element.

4.1 Constructing the CDAF

The components and elements of the CDAF were identified by linking the objectives and functioning of diplomacy in cyberspace with the dimensions of cybersecurity capacity building. The attributes of each of the elements were identified as the LOs necessary to encapsulate the essence of that specific element. The process to formulate elements and LOs for the *Diplomacy in Cyberspace* component is presented in Table 2.

Table 2: Breakdown of diplomacy in cyberspace element

Elements	Diplomatic Objective(s)	Learning Objectives
Internet Governance	Communication on international politics	Define cyberspace and its major components Provide an overview of the Internet & the major policies & procedures that governs it Identify the role players governing cyberspace Capacitate diplomats to participate in dialogue on internet governance issues
Data Governance	Collect intelligence and information on countries	Provide guidelines for the handling & storage of sensitive state-owned data Address diplomats' responsibilities toward data protection and data privacy Develop procedures and systems to protect ICT infrastructure & data Identify threats to information that is posted online

Elements	Diplomatic Objective(s)	Learning Objectives
		Describe how open-source information can be used for diplomatic purposes
Cyber Legislation	Circumvent and reduce conflict in international relations	Provide an overview of all the legislation/ policies w.r.t. cyber diplomacy Contextualise cyber legislation in relation to the regulatory framework of the diplomats' own country Formulize and institute cybersecurity legislation Establish regional & international jurisdictional cybersecurity cooperation Derive implementable regulations/SOPs in support of legislation Capacitate diplomats to engage their counterparts on legal issues w.r.t. cyberspace
Cyber Diplomacy	Represent the actuality of a civilization of states Negotiate treaties Communication on international politics	Differentiate between cyber diplomacy and digital diplomacy Contextualise diplomatic functions within cyberspace Explain how the interconnectivity of systems and interdependence of actors across cyberspace influences cyber diplomacy Describe cyberspace operations as a standard tool of diplomacy Explain how online tools, software, applications and systems can be applied to diplomatic activities Include best practices on digital negotiation Identify opportunities within cyberspace to augment diplomatic functions
Cyber Crime & Attacks	Circumvent and reduce conflict in international relations	Identify cyber security threats for online diplomatic actions as well as best practices on how to counter them Identify the major types of cybercrimes and cyber attacks Contextualise the role of diplomacy in cyber attacks List and discuss the main role players involved in cybercrime and cyberattacks Describe best practices to counter cybercrimes and cyber attacks Establish procedures and systems to deal with cyber incidents
Cyber Risk Management	Circumvent and reduce conflict in international relations	List cybersecurity threats for online diplomatic actions as well as best practices on how to counter them Provide an overview of the best practices regarding cyber risk management Provide guidelines for the development of a Cyber Risk Management Plan for diplomats Raise awareness on the benefits and dangers of AI & the IoT

Only the elements and LOs identified under *Diplomacy in Cyberspace* will be discussed further to explain how the framework can be applied and adapted to meet specific requirements of diplomats in a particular country.

4.2 Application of the CDAF

To demonstrate how the CDAF can be used the Cyber Diplomacy element will be applied to a South African (SA) perspective.

In SA the Department of International Relations and Cooperation's (DIRCO) is responsible for the recruitment, training and deployment of diplomats. DIRCO's mission is to *"formulate, coordinate, implement and manage South Africa's foreign policy and international relations programmes, and promote South Africa's national interest and values and the African Renaissance"* (DIRCO, 2020, p20)They have a Diplomatic Academy and

International School that presents training programmes aimed at capacitating SAs diplomats to contribute to the country's domestic priorities. Their 2020/2021 Annual Report elaborates on challenges concerning adapting to online activities. A number of references to digital-, economic-, public and virtual diplomacy are made, but the word "cyber" is mentioned only once in the 358 page report (DIRCO, 2020). The fact that cyber-diplomacy is not very high on DIRCOs agenda should be a concern. It does not suggest that SA diplomats are not involved in discussions on cybersecurity and/or cyberspace; diplomats from SA has been involved in various forums and committees related to cybersecurity and on on 11 June 2021 a virtual discussion with a SA delegate that attended the UN OEWG on Cyber Diplomacy indicated that a CDAF could add value to the efforts of DIRCO to capacitate their diplomats in terms of a broader view of cyber diplomacy as well as the technical and cyber awareness issues related to practising diplomacy in cyberspace. This poses the question whether there is a focused approach to equip SA diplomats to engage their international counterparts on issues related to cyberspace? The CDAF can provide a framework to either verify that the most important aspects of Cyber Diplomacy for instance are covered by the Diplomatic Academy or it can be used to create content-specific workshops or for altering existing programs.

Some of the following LOs that are listed in the CDAF under the Cyber Diplomacy element can be adopted to a SA context:

- 1. Differentiate between cyber diplomacy and digital diplomacy. Available literature submits that SA diplomats are familiar with the digital aspect of diplomacy – they know how to conduct online meetings, post information on different social media platforms, utilise digital data etcetera, but their understanding of cyber diplomacy as a concept seems to be lacking. The CDAF makes a clear distinction between cyber diplomacy and digital diplomacy.
- 2. Contextualise diplomatic functions within cyberspace. SA diplomats are trained in certain diplomatic programmes where they focus on South Africa's foreign policy and engagements with other international role-players. In this regard they are part of a number of forums such as the Southern African Development Community (SADC), African Union (AU), United Nations Security Council (UNSC) and BRICS. Cyber diplomacy in some of the countries that SA have diplomatic relations with has adapted swiftly and cyber issues are now decisively on their diplomatic agendas. SA and most of the developing countries face several challenges in this regard hampering their contribution and participation in the cyber diplomacy space (Borg Psaila, 2021). The CDAF can provide diplomats with a broader view of cyberspace and identify opportunities to augment their diplomatic functions.
- 3. Explain how the interconnectivity of systems and interdependence of actors across cyberspace influences cyber diplomacy. There needs to be a clear understanding amongst diplomats on cybersecurity aspects countries face; this includes legislative issues, cross-border investigations (because cyberspace transcends all borders), cyber threats to critical infrastructures, cyber countering measures etcetera (Borg Psaila, 2021). The CDAF can lay a solid foundation and create cybersecurity awareness in the digital environment where diplomats perform their functions.
- 4. Explain how online tools, software, applications and systems can be applied to diplomatic activities. Technology creates opportunities to interact with anyone - leading to cyberspace as a ground for international diplomacy. Social media networks are driving the growth of cyber diplomacy and in 2018 it was estimated that more than 70% of the world's head of states were using Twitter (Norwich University, 2018). The CDAF can provide diplomats with a holistic view of the latest and most secure online tools, software, applications and systems can be applied to diplomatic.
- 5. Include best practices on digital negotiation. The way nations relate diplomatically is changing. This is because cyber diplomacy is changing the way in which diplomats connect with their counterparts and the way governments communicate to their citizens. Statements by DIRCO that "*it is important that every effort is made to ensure the continuation of traditional diplomacy*" (DIRCO, 2020, p45) suggests that there is a void in terms of the value and application of cyber diplomacy. The CDAF can capacitate diplomats to effectively utilise digital negotiation to reach their diplomatic objectives.

As the CDAF is still under development, it can only be critically evaluated after a full draft version is available. At that stage a contact within DIRCO will be used to evaluate the CDAF in the live environment.

5. Conclusion

This article presented a CDAF that can be used by developing countries to equip their diplomats to play a more constructive role within the international cyber-diplomacy domain. Throughout the article it was echoed that different countries might have progressed further or are lacking behind on the terrain of cyber diplomacy, but the fact remains that every country in the world is present in cyberspace – whether a country represents himself in cyberspace or is a spectator depends on the cyber diplomacy capacity of that country.

References

Adesina, O. S. (2017). Foreign policy in an era of digital diplomacy. *Cogent Social Sciences*, *3*(1). https://doi.org/10.1080/23311886.2017.1297175

Al-Muftah, H., Weerakkody, V., Rana, N. P., Sivarajah, U., & Irani, Z. (2018). Factors influencing e-diplomacy implementation: Exploring causal relationships using interpretive structural modelling. *Government Information Quarterly*, *35*(3), 502–514. https://doi.org/10.1016/j.giq.2018.03.002

Attatfa, A., Renaud, K., & de Paoli, S. (2020). Cyber diplomacy: A systematic literature review. *Procedia Computer Science*, *176*, 60–69. https://doi.org/10.1016/j.procs.2020.08.007

Barrinha, A., & Renard, T. (2017). Cyber-diplomacy: the making of an international society in the digital age. *Global Affairs*, *3*(4–5), 353–364. https://doi.org/10.1080/23340460.2017.1414924

Barrinha, A., & Renard, T. (2020). The Emergence of Cyber Diplomacy in an Increasingly Post-Liberal Cyberspace. *Council on Foreign Relations*, 1–6. https://www.cfr.org/blog/emergence-cyber-diplomacy-increasingly-post-liberal-cyberspace

Borg Psaila, S. (2021). *Improving the practice of cyber diplomacy: Training, tools, and other resources PHASE I*. www.diplomacy.edu

Buzatu, A.-M. (2021). Promoting Openness , Prosperity , Trust and Peace and Security in Cyberspace. In *International Cyber Security Policy and Diplomacy Capacity BuildingProgram since 2014* (Issue April). http://www.ict4peace.org/

DIRCO. (2020). *ANNUAL REPORT 2020 / 21 DEPARTMENT OF INTERNATIONAL RELATIONS AND COOPERATION (DIRCO)*. http://www.dirco.gov.za/department/report_2020-2021/annual_report2020_2021.pdf

Dutton, W. H., Creese, S., Shillair, R., & Bada, M. (2019). Cybersecurity Capacity: Does It Matter? *Journal of Information Policy*, *9*(May 2021), 280–306. https://www.jstor.org/stable/10.5325/jinfopoli.9.2019.0280

Global Cyber Security Capacity Centre (GCSCC). (2021). *Cybersecurity Capacity Maturity Model for the Nations (CMM) 2021 Edition* (Issue CMM 2021 Edition). https://gcscc.web.ox.ac.uk/cmm-2021-edition

Global Cyber Security Capacity Centre, G. (2018). Collaborative Approaches to a Wicked Problem: Global Responses to Cybersecurity Capacity Building. *Annual Conference of the Global Cyber Security Capacity CentreSecurity Capacity Centre, February*, 1–35. https://doi.org/10.2139/ssrn.3660000

Interpol. (2021). *Fostering regional cooperation against cybercrime in Southeast Asia*. https://www.interpol.int/en/Crimes/Cybercrime/Cyber-capabilities-development/ASEAN-Cyber-Capacity-Development-Project

Leira, H. (2019). The Emergence of Foreign Policy. *International Studies Quarterly*, *63*(1), 187–198. https://doi.org/10.1093/isq/sqy049

Martin, A., Rashid, A., Chivers, H., Schneider, S., Lupu, E., & Danezis, G. (2021). *Introduction to CyBOK Knowledge Areas* (p. 22). The National Cyber Security Centre 2021. www.cybok.org

Norwich University. (2018). *Norwich University online*. https://online.norwich.edu/academic-programs/masters/information-security-assurance/resources/articles/role-of-computer-forensics-in-crime

Nye, J. S. (2014). The Regime Complex for Managing Global Cyber Activities. *CIGI Publications*, *1*, 1–15.

Sabillon, R., Serra-Ruiz, J., Cavaller, V., & Cano, J. J. M. (2019). An effective cybersecurity training model to support an organizational awareness program: The Cybersecurity Awareness Training Model (CATRAM). A case study in Canada. *Journal of Cases on Information Technology*, *21*(3), 26–39. https://doi.org/10.4018/JCIT.2019070102

Veale, M., & Brown, I. (2020). Cybersecurity, Internet Policy Review. *Alexander von Humboldt Institute for Internet and Society*, *9*, 0–22. https://doi.org/https://doi.org/10.14763/2020.4.1533

PhD Research Papers

Assessing Information Security Continuous Monitoring in the Federal Government

Tina AlSadhan and Joon Park
Syracuse University, Information School (iSchool), USA
talsadha@syr.edu
jspark@syr.edu

Abstract: To confront the relentless and increasingly sophisticated cyber assaults from cybercriminals, nation-state actors, and other adversaries, the U.S. Federal Government must have mechanisms to reduce or eliminate compromise and debilitating consequences. Information Security Continuous Monitoring (ISCM) leverages technology to rapidly detect, analyze, and prioritize vulnerabilities and threats and deliver a data-driven, risk-based approach to cybersecurity. Although monitoring information system security became a requirement for government agencies over 20 years ago and billions of dollars are being spent annually for cybersecurity, ISCM remains at a low maturity level across the Federal Government. This research framework presented is part of ongoing doctoral research. The research seeks to identify the challenges achieving an effective ISCM program and inform measures needed to optimize ISCM. The research involves conducting an ISCM Program Assessment in a Department of Defense (DoD) organization using the recently published National Institute of Standards and Technology (NIST) ISCM Assessment (ISCMA) methodology and the companion assessment tool ISCMAx. An ISCM doctrine placement is presented, derived from the NIST ISCM assessment elements, to more clearly articulate and visualize the doctrine of a well-designed and well-implemented ISCM program. This research will also contribute to the knowledge base for assessing ISCM in the Federal government and the functionality of the ISCMAx tool.

Keywords: cybersecurity, continuous monitoring, security assessment

1. Introduction

In an age of frequent and persistent attacks riddled with determined adversaries, cyber defenses must be agile and responsive to explosive rates of new vulnerabilities, persistent threats, determined enemies, as well as the dynamic state of technologies and mission/business functions (Joint Task Force Transformation Initiative, 2011). Defenders must have timely information available to address the cyber situation before attackers do. An incomplete, inaccurate, or delayed representation of the current security state may contribute to a compromise, increased debilitating effects, and delayed responses (Center for Internet Security (CIS), 2021). As the government's attack surface continues to profoundly expand, it is becoming even more apparent that timely, relevant, and accurate cybersecurity data is vital to deliver effective and efficient cyber defenses.

The Federal Information Security Act (FISMA) of 2002 required continuous monitoring and assessing security controls at a frequency commensurate with risk to implement leaner, more responsive, and more effective information security practices (FISMA, 2002). Federal agencies are mandated to incorporate automation to enable risk-based decisions, transitioning from compliance-based to real-time situational awareness and response (Dempsey et al, 2011; Office of Management and Budget (OMB), 2011). FISMA of 2014 emphasized continuous monitoring and securing information systems commensurate with risk (FISMA, 2014). As required by the FISMA of 2014, Circular A-130 was updated emphasizing the use of risk-based cost-effective security controls and a shift from compliance-based security to continuous-risk-based programs (OMB, 2016).

Although security controls and their assessment and monitoring can be manual, semi-automated, or automated, ISCM leverages technology for greater efficiencies and enhanced cyber operations. Technology can enable continuous monitoring by collecting, aggregating, correlating, analyzing, and reporting of volumes of disparate security data in ways not manually feasible (Dempsey et al, 2011). Technology can augment security processes conducted by cyber professionals, reduce the efforts on redundant tasks, recognize patterns and relationships that may not be apparent, and free up time to work on efforts requiring the human analyst (Dempsey et al, 2011).

An effective ISCM program maintains a picture of an organization's security posture, presents security-related data, and incorporates results to portray greater situational awareness (Mell et al, 2012; Dempsey et al, 2020). Through constant observation and analysis, ISCM provides visibility into assets, awareness of vulnerabilities and threats, and improves overall cyber situational awareness. These capabilities enable organizations to transition from compliance-driven security to data-driven risk management, enabling the prioritization of efforts to focus on what can most improve the security posture.

There are notable efforts in the Federal Government to implement ISCM per directives and NIST guidance. Cybersecurity and Infrastructure Security Agency's (CISA's) flagship ISCM program *Continuous Diagnostics and Mitigation* (CDM) is a comprehensive program providing access to a set of cybersecurity continuous monitoring tools, integration services, and dashboards to Federal civilian agencies (the *.gov domain), as well as state, local, and tribal governments. The program, initially launched in 2013, is estimated to be a $10.9 billion program by 2031 (Government Accountability Office (GAO), 2020a; CISA, 2021). CDM is intended to help government agencies reduce security threats, increase cybersecurity posture, improve response capabilities and streamline FISMA reporting requirements (OMB, 2021a).

Although there are new requirements to influence adoption of CDM program (OMB, 2020), it is not essential to participate in CDM to have an ISCM program. The Department of Defense (DoD), for instance, does not participate in the CDM program, however, deploys a multitude of tools and technologies that can be leveraged for ISCM data-driven risk-based decision making. Any organization, whether public or private, can implement ISCM mechanisms per the NIST or other Continuous Monitoring framework.

2. Research problem

Although monitoring information system security became a requirement for government agencies over 20 years ago and billions of dollars are being spent annually for cybersecurity, ISCM remains at a low maturity level across the Federal Government (OMB, 2020). In 2019 and 2020, the average score of 23 of the largest Federal agencies for the "Detect" capability, corresponding to ISCM, was 2.56 and 2.78 out of 5, respectively (OMB, 2020; OMB, 2021b). In Fiscal Year 2020, only 5 of these agencies met or exceeded Level 4: Managed and Measurable, what is considered an effective level of security (CISA, 2020a; OMB, 2021b). Many government agencies still lack the capabilities to effectively collect, aggregate, correlate, and analyze security-related information to enhance real-time threat detection and response and data-driven risk-based decision making (GAO, 2020a; GAO, 2020b). Multiple agencies have not yet implemented the vital practices for cybersecurity risk management, including establishing a cyber risk management strategy and policies for assessing, responding to, and monitoring cyber risks (GAO, 2019).

According to a GAO study, although federal agencies were using the CDM vulnerability scanning tools to identify vulnerabilities, there was insufficient visibility of the remediation of identified vulnerabilities (GAO, 2020a). The agency dashboards were not utilized as intended to provide agency level awareness of hardware and software configurations and vulnerabilities for security-related decisions. The hardware inventory capability, an essential component, was not fully implemented. The feature to compare configuration settings against both an agency-approved baseline and the federal baseline was not functional. Due to the shortfalls in achieving fully operational CDM capabilities, the CDM agency dashboard data was inaccurate leading to an inaccurate federal dashboard. The study concluded that the billion-dollar CDM program would likely not fully achieve expected benefits without CISA further assisting agencies with implementation (GAO, 2020a).

There is concern that the government cybersecurity capabilities are insufficient to respond to the increasingly sophisticated and egregious attacks. Russian nation-state hackers compromised the supply chain of a widely used network traffic monitoring and management platform *SolarWinds* to exploit networks in at least nine government agencies (CISA, 2020b; The White House, 2021; GAO, 2021). Despite the extensive cyber measures available to these agencies, to include participation in CISA's CDM program, neither the vulnerability nor the anomalous activity was detected by the government. "...the fact remains that despite significant investments in cyber defenses, the federal government did not initially detect this cyberattack," (U.S. Senate Committee on Homeland Security & Governmental Affairs, 2021). More robust ISCM capabilities are needed to understand what the security posture is, reduce the attack surface, and get in front of the threat and adversary.

3. Research design

This research is designed to identify the ISCM elements that need to be improved and recommendations to optimize ISCM. Utilizing the recently published NIST ISCM Assessment (ISCMA) methodology and the accompanying ISCMAx tool, the researcher will conduct an ISCM program assessment within a DoD organization, hereafter referred to as ORG. The research will also provide insight into using the NIST ISCMA methodology, the assessment elements, and the ISCMAx tool.

3.1 ISCM program assessment instrument

The NIST ISCMA methodology provides an operational approach to assess "ISCM strategies, policies, procedures, implementations, operational procedures, analytical processes, specific reporting, presentation results, risk assessment and risk scoring, risk response, and the program improvement process" (Dempsey et al, 2020, p. iv). NIST has identified 128 ISCM program assessment elements which are foundational statements reflective of a well-designed and well-implemented ISCM program. The NIST assessment elements are drawn from authoritative documents such as FISMA, those from OMB and NIST, and also incorporate the knowledge of ISCM experts. Each assessment element includes the evaluation criteria supporting the determination statement, i.e., "Determine if..." or "There is...", which is an ISCM principle or element that an ISCM program should meet. The element's assessment procedure includes a stated objective, potential assessment methods, and objects (evidence) to facilitate the judgment of the element. The assessment elements will be judged as either *Satisfied* or *Other Than Satisfied*. Each element references its authoritative source(s). An assessment element example is shown in Figure 1.

ID	Assessment Element Text	Level	Source	Assessment Procedure	Discussion	Rationale for Level	Parent	Critical Element in NISTIR 8212 / ISCMAx Tool	Chain Label	Chain Sort
1-002	There is an ISCM program derived from the organization-wide ISCM strategy.	Level 1	NIST SP 800-137	**ASSESSMENT OBJECTIVE** Determine if there is an ISCM program derived from the organization-wide ISCM strategy. **POTENTIAL ASSESSMENT METHODS AND OBJECTS** **Examine:** Organization-wide ISCM strategy, ISCM policy and procedure documentation, ISCM design documents, ISCM concept of operations (CONOPS) **Interview:** Level 1: SAISO, ISCM point of contact (POC)	The ISCM program is comprised of the ISCM policies and procedures derived from the organization-wide ISCM strategy and includes the ISCM documents that guide ISCM implementation (e.g., ISCM technical architecture and ISCM CONOPS).	Level 1 is responsible for defining the ISCM program.	*The Define Step has no parent element*	N	ISCM Program Management	03.01-002

Figure 1: Assessment element example (Dempsey et al, 2020)

NIST links together related assessment elements into traceability chains to depict the parent/child relationship between elements. Figure 2 is the traceability chain related to Control Assessment Rigor, with the parent element being 1-003 and 3 different chains, the first linking elements 2-003 and 3-002, the second linking element 2-003a, and the third linking 2-004, 2-004a, and 3-035.

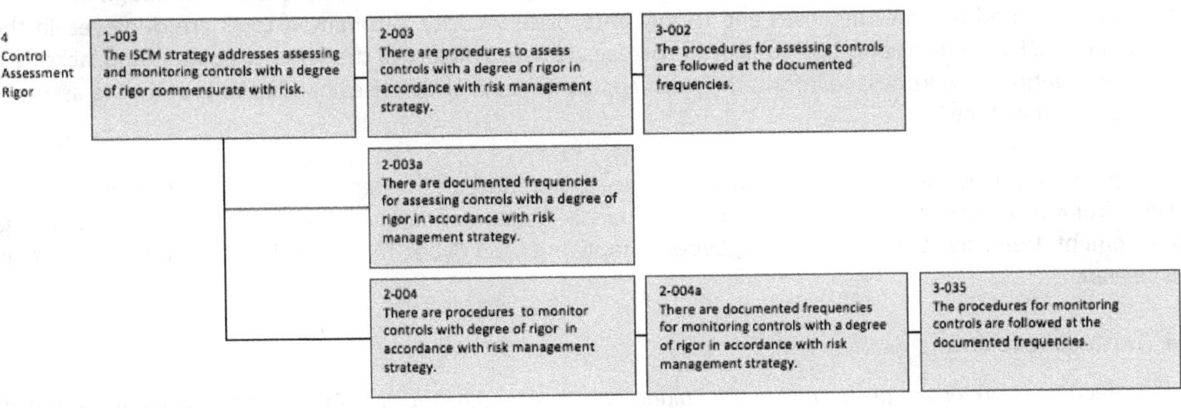

Figure 2: Control Assessment Rigor traceability chain (Dempsey et al, 2020)

The ISCMAx tool will be leveraged to record element judgments, capture annotations, aggregate results from assessment participants, and facilitate data analysis using its scorecard and graphical output (Dempsey et al, 2021). The ISCMAx tool is a ready-to-use Microsoft Excel application containing the 128 ISCMA program assessment elements. It is a companion tool to NISTIR 8212 *ISCMA: An Information Security Continuous Monitoring Program Assessment*. Within the tool, the user can view the element and the guidance, record judgments, and enter notes and recommendations (Dempsey et al, 2021). In addition, the ISCMAx tool combines judgments, calculates scores, and generates a workbook with the complete assessment results (Dempsey et al, 2021).

3.2 Action research

The ISCM program assessment will be conducted as an action research project, a collaboration between the researcher and ORG. The purest form of action research is one that "participants of a social system are involved in a data collection process about themselves and utilize the data that they have generated to review the facts about themselves in order to take some form of remediation or developmental action" (Lippet, 1979, as cited in Coghlan and Brannick, 2014, p. 47). The organization members have the knowledge of the problems and will live with changes made; therefore, it is important that they be part of the effort (Shani and Pasmore, 1985).

Coghlan and Brannick's (2014) action research cycle: Initial pre-steps and the steps of a) *constructing*, b) *planning action*, c) *taking action*, and d) *evaluating action*, and the corresponding ISCM program assessment actions that will be conducted as part of this research are identified in Table 1.

Table 1: ISCM assessment action research cycle

Cycle	**Steps**		
Pre-Step Constructing	ISCM assessment schedule, identify SMEs to participate, concurrence to conduct ISCM assessment in ORG in conjunction with academic research		
Cycle	*Planning Action*	*Taking Action*	*Evaluating Action*
ISCM Assessment	Design research and ISCM program assessment plan, information/kick off meetings with SMEs and leadership	Artifact review (strategy, policies, processes, plans, reports, former assessment results, etc.), discussions with SMEs, analyze information, element judgement, clarifying notes	Report the findings and recommendations, identify ISCM elements needing improvement and what is needed to optimize ISCM

3.3 Subject organization and participants

ORG is one of several Federal Government organizations facing challenges to achieve the objectives of ISCM and transition to ongoing/continuous assessment and authorization of its information systems. ORG serves as a Cyber Security Services Provider and is responsible for executing the full range of cybersecurity services for its Enterprise information systems. ORG leverages multiple technology solutions to deliver cybersecurity services. ORG has made significant efforts to ensure ORG's information systems are always prepared to pass cybersecurity inspections by increasing compliance audits followed by the remediation of deficiencies. Although extensive resources dedicated to cyber defenses and robust enterprise security automation tools are deployed in the environment, ORG is reliant on labor-intensive processes to gain visibility of the security posture and lacks near real-time situational awareness to inform the most appropriate cyber defenses needed to protect the assets at any given point of time.

ORG's government civilian cybersecurity and IT professionals, hereafter referred to as Subject Matter Experts (SMEs), from focal areas such as Cybersecurity Governance, Assessment and Authorization, Cyber Audit, Change Management, Defensive Cyber Operations, Cyber Compliance, and Cyber Threat Intelligence, will participate in this research.

3.4 Data collection

Data collection activities are performed in the *Taking Action* research step, see Table 1. ORG's artifacts relevant to ISCM program assessment elements will be reviewed, to include cybersecurity policies, procedures, and inspection findings. Artifacts which NIST identifies element "potential assessment objects", such as the ISCM strategy, ISCM policy, system security plans, security-related data from monitoring systems, reports, and briefings will support element judgments (Dempsey et al, 2020, p. 37-38). Based on the review of supporting artifacts, or absence thereof, initial judgments, recommendations, and notes will be recorded by the researcher in the ISCMAx tool.

The ISCM program assessment elements, inclusive of the preliminary judgments, recommendations, and notes compiled from the artifact review will be provided to the SMEs by focal areas via email. The SMEs will be requested to review the data in advance of an online meeting. Updates and/or corrections can be submitted to the researcher via email. The researcher will facilitate discussions with the SME(s) to reach a consensus as to whether the element is *Satisfied* or *Other Than Satisfied*. The discussions will provide knowledge into the lived experiences and perspectives in actual practice that may not be apparent through a review of artifacts alone. If the assessment element includes discussion points, they may be used to clarify the wording or the intent of the assessment element. The discussions will be grouped into element reviews by traceability chains or sections to facilitate the flow of the assessment, the focus of the SMEs, and the linkage between the elements when discussing and preparing evidence. The judgment and relevant insight from discussions with SMEs will be captured as annotations or recommendations in the ISCMAx tool. These notes may identify shortfalls, indicate what further actions would satisfy the determination statement, and contain other information that may be helpful to the development of ISCM program recommendations.

3.5 Data analysis

The ISCMAx tool facilitates data analysis through a detailed scorecard and other visualizations to summarize the assessment (Dempsey et al, 2020, p. 28). The ISCMAx tool scores the elements and aggregates scores for all assessed risk levels and elements. The ISCMAx tool can display the *Other Than Satisfied* judgments in a single worksheet, identifying the weakest areas of the program. The annotations and recommendations captured during the data collection will be particularly helpful to provide additional clarification of what the challenges are faced in regards to the element and/or what is necessary to satisfy the element and to improve ISCM in ORG.

The ISCM Process Steps Scores for ORG are depicted in Figure 3. ORG's scores for the Define, Establish, and Implement steps are all below 15%. These three process steps comprise 77% of the assessment elements. These scores reflect ORG's current state of being in the initial stages of its program. ORG does not yet have a Risk Management strategy or an ISCM strategy. Key foundational governance for security controls, security monitoring tools, security status-monitoring, ISCM data, ISCM technical architecture, and others have yet to be established. ORG scored 37.5% and 45.5%, respectively in the *Analyze/Report* and *Respond* ISCM process steps. These scores reflect ORG's robust security investment and acquisitions processes and cyber threat program. The scores reflect shortfalls in analyzing effectiveness risk responses and security controls, reviewing ISCM metrics and findings, and having ISCM training and resource plans that align to ISCM objectives.

	Define	Establish	Implement	Analyze / Report	Respond	Review / Update	Totals
Elements	24	43	32	10	9	10	128
Raw Score	4.0	9.0	5.0	6.0	5.0	0.0	29.0
Max Score	42.0	65.0	48.0	16.0	11.0	14.0	196.0
Percentage Score	9.5%	13.8%	10.4%	37.5%	45.5%	0.0%	14.8%

Figure 3: ISCM process step scores

The scored Traceability Chains will provide a graphical representation of the related elements, such as Security Status Monitoring, Security Monitoring Tools, ISCM metrics, etc., further identifying the challenge areas, see Figure 4. The Scored Traceability Chain provides the score for each Level assessed in the square; Level 1 – the Organization, Level 2 – the Mission/Business and Level 3 – the System, and then an overall score for the Element. If an assessment element is not applicable to a level, it will be appear as ⦸. *Satisfied* elements will appear as a "1" and *Other than Satisfied* elements will appear as "2" for each level. If the aggregate score for an element is *Satisfied*, it will appear in Green and if *Other Than Satisfied*, it will appear in Red.

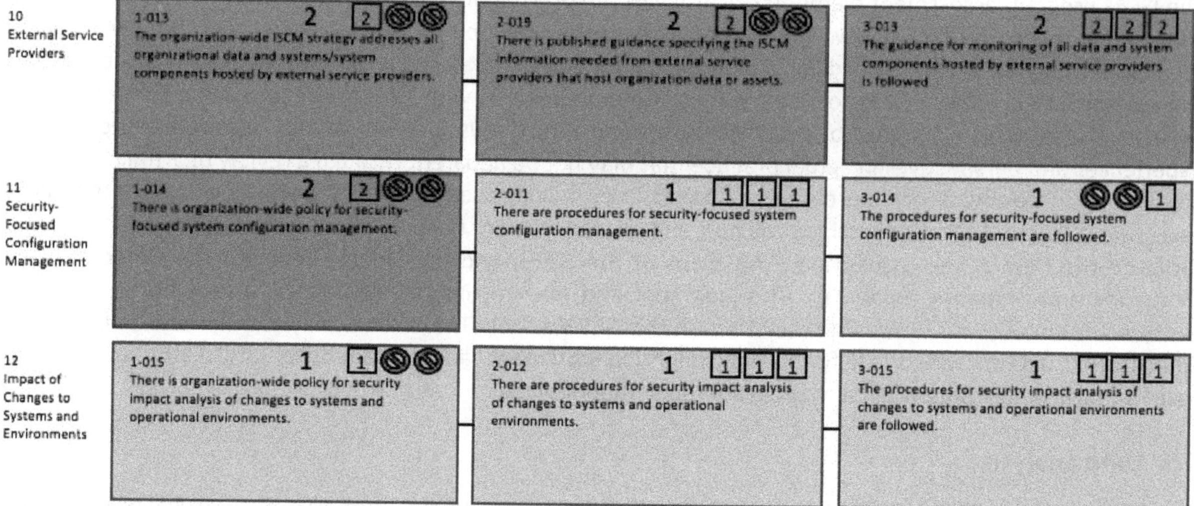

Figure 4: Scored traceability chain example

For the Security-Focused Configuration Management chain, ORG lacks a Configuration Management policy. ORG does, however, have documented procedures to establish organizational security-configuration baselines, identify the risk of the deviations from the DoD baseline, and an organizational process to accept the risk of deviations. ORG's Cyber Readiness initiative includes auditing Technology implementations against the DoD security-configuration baselines prescribed in the Defense Information Systems Agency (DISA) Standard Technical Implementation Guides (STIGs) and the approved organizational baselines. ORG is in the process of defining and documenting the frequency to conduct security configuration audits and defining the sample size for each review. Although Security Content Automation Protocol (SCAP) enabled tools do facilitate automated checks of some security configurations, several technologies require manual checks that are time consuming to conduct. ORG utilizes findings from security-configuration baseline checks as input to project its score based on the Command Cyber Readiness Inspection (CCRI) criteria, which is an indicator of its security posture.

3.6 Scope and limitations

The research focus is on ISCM concepts, governance, structure, and requirements levied by FISMA and OMB rather than the evaluation of ISCM technologies or security controls. The research does not delve into specific vendor technologies or solution sets, assessing security data, assessing actual security controls, or examining control assessments. The research is not an audit of compliance with policy or information security controls, or a risk assessment.

The ISCM program assessment will be conducted in a single DoD organization as a self-assessment facilitated by the researcher. Assessment results are dependent on element scoring of the organization, its mission/business, and systems and are unique to the organization. The ISCM program assessment process, rather that the scoring of the subject organization, will most contribute to the community.

4. Expected outcomes

It is expected that ORG's ISCM program assessment score will be low. Although ORG is performing the full spectrum of cybersecurity services, it is only in the initial stages of implementing as ISCM program. Furthermore, nearly 20% of the ISCM assessment elements are dependent on an organization-wide ISCM strategy and Risk Management strategy. Organizations assessed under the ISCM assessment elements will not satisfy 20% of the ISCM elements without these foundational ISCM governance documents in place. Even if the organization has developed other areas of their ISCM program, scoring will be impacted without this governance.

It is expected that SME's participating in the research will have gained knowledge of ISCM upon completion of review of the assessment elements. With this new knowledge, it is anticipated that they may address the governance requirements in subsequent updates of their policies, procedures, and other documentation.

5. Contributions

An ISCM program assessment conducted in a federal organization will contribute to understanding the elements that need to be improved and how to optimize ISCM. It is anticipated that these findings will not only be helpful to government agencies but will also apply to other public and private organizations to improve their cybersecurity operations. This research may be utilized to assist agencies in the performance of their assessments and ultimately mature programs to meet federal compliance requirements for ISCM implementation and deliver cybersecurity in a more responsively and effectively.

The research is expected to provide a deeper understanding of the useability and limitations of the ISCM Program Assessment methodology and the ISCMAx tool. The ISCMAx tool was released by NIST in 2021. There are no other assessments conducted within the DoD known to the research using this tool. As a result of this research, an ISCM Doctrine Placemat, described in 5.1 will facilitate a better understanding of the scope of doctrine involved in a well-developed ISCM program.

5.1 ISCM doctrine placemat

Several of the NIST ISCM assessment elements involve strategies, policies, procedures, and other documentation. An ISCM Doctrine Placemat based on key doctrine identified within the NIST ISCM assessment elements has been developed as part of this research, see Figure 5. The ISCM Doctrine Placement facilitates a single-page visual representation of the key foundational ISCM program doctrine and components without having to delve into the individual assessment elements. The placemat could visually represent the state of an organization's ISCM doctrine by color coding the doctrine that satisfies the requirements in Green and the items that are deficient in Red. The ISCM doctrine is grouped into categories: *Organizational Strategy, Operationalize, Resource, Report, Respond,* and *Related Functions. Organizational Strategy* contains the Risk Management Strategy and the ISCM Strategy, the most foundational doctrine of an ISCM program. *Operationalize* doctrine contains the documents which provide structure to the ISCM mechanisms. *Resource* includes funding and budgeting the ISCM program as well as resourcing it with human capital. *Report* covers ISCM reporting mechanisms and *Response* addresses the doctrine for risk-based response activities.

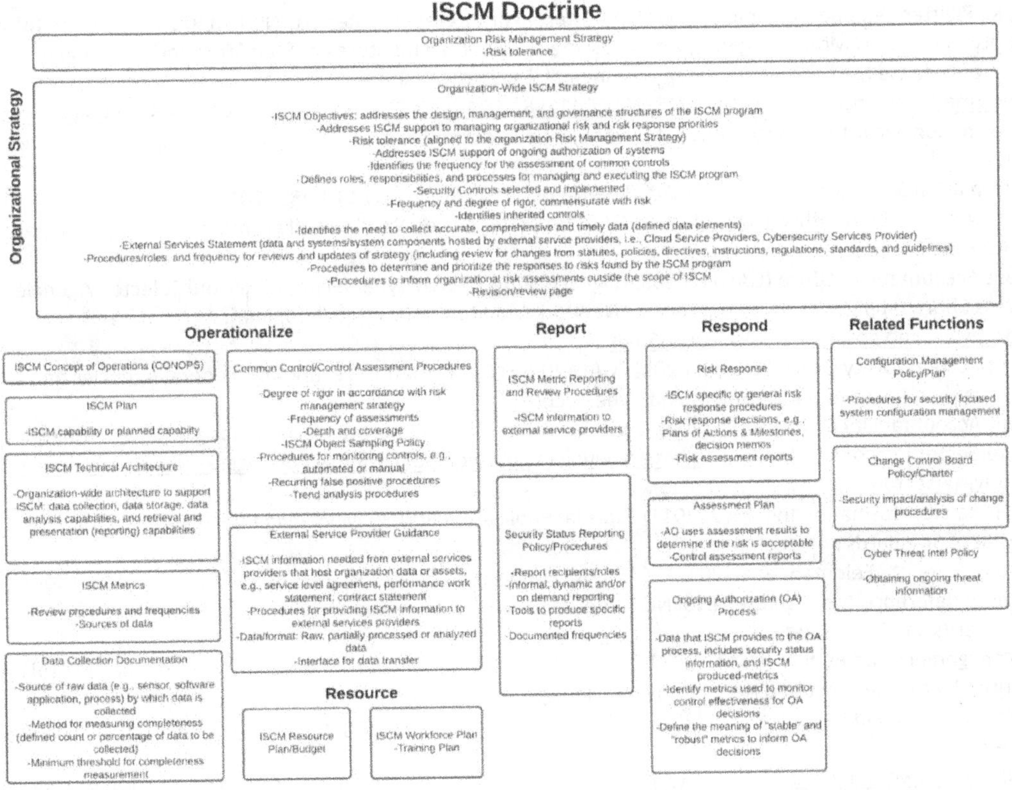

Figure 5: ISCM doctrine placemat

6. Stage of the research

The research design is presented as part of the research conducted for a doctoral thesis. The research proposal has been approved by a doctoral committee. The data collection for all 128 of the ISCM assessment elements has been completed and data analysis is ongoing. Preliminary research findings have been presented at the IEEE 2021 International Conference on Computational Science and Computational Intelligence, Symposium on Cyber Warfare, Cyber Defense & Cybersecurity (AlSadhan and Park, 2021). The final thesis is expected to be defended in 2022. By participation in the research Consortium, the researcher is looking to gain insight for narrative analysis and formulating the research findings so that they are insightful and connect back to the research objectives.

References

AlSadhan, T. and Park, J.S. (2021, December) "Leveraging Information Security Continuous Monitoring to Enhance Cybersecurity", In Proceedings of the 8th Annual Conference on Computational Science and Computational Intelligence (CSCI) Symposium on Cyber Warfare, Cyber Defense & Cyber Security (CSCI-ISCW).

AlSadhan, T. and Park, J.S. (2015, March) "Leveraging Information Security Continuous Monitoring for Cyber Defense", In Proceedings of the 10th International Conference on Cyber Warfare and Security, p 401.

Center for Internet Security (CIS) (2021, May) "CIS Controls Version 8.0", [online], http://www.cisecurity.org/controls/

Coghlan, D. (2001) "Insider Action Research Projects: Implications for Practicing Managers", *Management Learning, 32*(1), pp 49-60.

Cybersecurity and Infrastructure Security Agency (CISA) (2020a, April 17) "FY 2020 Inspector General Federal Information Security Modernization Act of 2014 (FISMA) Reporting Metrics Version 4.0", [online], https://www.cisa.gov/sites/default/files/publications/FY_2020_IG_FISMA_Metrics.pdf

Cybersecurity and Infrastructure Security Agency (CISA) (2020b, December 13) "Mitigate SolarWinds Orion code compromise (Emergency directive 21-01)", [online], https://cyber.dhs.gov/ed/21-01/

Cybersecurity and Infrastructure Security Agency (CISA) (2021a) "CDM Program Overview", [online], https://www.cisa.gov/sites/default/files/publications/2020%2009%2003_CDM%20Program%20Overview_Fact%20Sheet_1.pdf

Dempsey, K., Chawla, N., Johnson, A., et al (2011) "Special Publication 800-137 ISCM for Federal Information Systems and Organizations", National Institutes of Standards and Technology.

Dempsey, K., Pillitteri, V. Y., Baer, C., et al (2020) "Special Publication 800-137A Assessing Information Security Continuous Monitoring (ISCM) Programs", National Institutes of Standards and Technology.

Dempsey, K., Pillitteri, V., Baer, C., et al (2021) "NIST Internal or Interagency Report (NISTIR) 8212 ISCMA: An Information Security Continuous Monitoring Program Assessment", National Institutes of Standards and Technology.

Department of Defense (2018) "Cyber strategy summary", [online], https://media.defense.gov/2018/Sep/18/2002041658/-1/-1/1/CYBER_STRATEGY_SUMMARY_FINAL.PDF

Federal Information Security Management Act (FISMA) Title III of the e-Government Act of 2002., Public Law 107-347 (2002)

Federal Information Security Modernization Act (FISMA) of 2014, Public Law 113-283 (2014)

Government Accountability Office (GAO) (2019) "GAO-19-384 Agencies Need to Fully Establish Risk Management Programs and Address Challenges"

Government Accountability Office (GAO) (2020a, August) "GAO-20-598 Cybersecurity DHS and Selected Agencies Need to Address Shortcomings in Implementation of Network Monitoring Program", [online], https://www.gao.gov/assets/710/708885.pdf

Government Accountability Office (GAO) (2020b) "GAO-20-691T Information Technology Federal Agencies and OMB Need to Continue to Improve Management and Cybersecurity"

Government Accountability Office (GAO) (2021, April 22) "SolarWinds Cyberattack Demands Significant Federal and Private-Sector Response", [online], https://www.gao.gov/blog/solarwinds-cyberattack-demands-significant-federal-and-private-sector-response-infographic

Joint Task Force Transformation Initiative (2011) "Special Publication 800-39 Managing Information Security Risk", National Institute of Standards and Technology.

Mell, P., Waltermire, D., Feldman, et al (2012) "CAESARS Framework Extension: An Enterprise Continuous Monitoring Technical Reference Architecture" (No. NIST Internal or Interagency Report (NISTIR) 7756 (Draft)), National Institute of Standards and Technology.

Office of Management and Budget (OMB) (2011) "GAO M-11-33 FY 2011 Reporting Instructions for the Federal Information Security Management Act and Agency Privacy Management".

Office of Management and Budget (OMB) (2016) "Circular No. A-130 Managing Information as a Strategic Resource", [online], https://obamawhitehouse.archives.gov/sites/default/files/omb/assets/OMB/circulars/a130/a130revised.pdf

Office of Management and Budget (OMB) (2020) "Fiscal Year 2020-2021 Guidance on Federal Information Security and Privacy Management Requirements", [online], https://www.whitehouse.gov/wp-content/uploads/2020/11/M-21-02.pdf

Office of Management and Budget (OMB) (2021a) "Annual Report to Congress: Federal Information Security Act of 2014 Fiscal Year 2020", [online], https://www.whitehouse.gov/wp-content/uploads/2021/05/FY-2020-FISMA-Report-to-Congress.pdf

Office of Management and Budget (OMB) (2021b) "FY 2020 Annual Cybersecurity Performance Summary", [online], https://www.whitehouse.gov/wp-content/uploads/2021/05/FY2020FISMAAnnualCybersecurtiyPerformanceSummaries.pdf

Shani, A. B. and Pasmore, W. A. (1985) *Organization inquiry: Towards a New Model of the Action Research Process. Contemporary Organization development: Current Thinking and Applications, Scott, Foresman, Glenview, IL,* pp 438-448.

United States Senate Committee on Homeland Security & Governmental Affairs (2021, April 6) "Portman, Peters Request Information on Federal Government's Response to SolarWinds Orion and Microsoft Exchange Cyberattacks," [online], https://www.hsgac.senate.gov/media/minority-media/portman-peters-request-information-on-federal-governments-response-to-solarwinds-orion-and-microsoft-exchange-cyberattacks

The White House (2021, May 12) "Executive Order on Improving the Nation's Cybersecurity", [online], https://www.whitehouse.gov/briefing-room/presidential-actions/2021/05/12/executive-order-on-improving-the-nations-cybersecurity/

Expectations and Mindsets Related To GDPR

Pauliina Hirvonen
University of Jyväskylä, Finland
pauliina.a.hirvonen@student.jyu.fi

Abstract: The aim of this qualitative case study is to examine the initial expectations and assumptions related to General Data Protection Regulation (GDPR) of the European Union from the perspectives of selected Finnish organizations: what were the initial expectations of GDPR, how were they adapted/refined over time, and what was the impact on organizational planning and resourcing. There are no precise earlier studies on the subject. The research question was: What were the organizations' initial expectations of GDPR - and how have they affected the efforts made? GDPR can be described as an input that forms images, preconceptions and views among other things, through various active and passive communication flows. As the empirical results indicate GDPR has been a legal issue, mainly due to the inadequate and unspecific active, official, communication flows. As a result, organizations have experienced difficulties to scale the necessary GDPR efforts. The results of this research can benefit both privacy and information security managers and personnel responsible for aligning policies and practices, and to evaluate organization-specific actions on GDPR compliance. The results can support regulators and authorities in the future GDPR and other policy work and provide ideas for service providers.

Keywords: GDPR early stage issues, Legal Issues, privacy development, privacy regulation, GDPR expectations, GDPR and organizations

1. Introduction

General Data Protection Regulation (GDPR) early stage issues were examined in this article. The aim of this case study was to examine the initial expectations and assumptions related to GDPR from the perspectives of selected Finnish organizations. The research question was: What were the organizations' initial expectations of GDPR - and how have they affected the efforts made?

The keywords of the research are shortly defined here. GDPR early stage issues refer to the time when first official guidance on GDPR requirements appeared available. The study will then focus on the challenges that rose in connection with GDPR during the so-called preparation period, 2016 - 5/2018. Legal issues refer to the general nature of GDPR and the study seeks to explain why GDPR is generally perceived as a challenging legal problem. Referring to its role in privacy regulation and development, GDPR was expected to become globally as one of the most significant and comprehensive privacy tools of today. GDPR expectations include, in particular, organizations' perceptions and initial experiences of the forthcoming regulation. GDPR and organizations include the focus on the organizational approach and awareness.

Understanding the expectations and mindsets is essential, as they affect the quantity and quality of organizations' later GDPR activities. The gap of early work reveals that existing academic research does not include similar approaches to the GDPR issue. The early stage of GDPR can be seen as a kind of igniter in this research which, through various active and passive communication flows, forms images, expectations and views.

As the empirical results indicate, the experiences of Finnish organizations confirm that GDPR has been a real legal issue, as the quality of activity regarding formal communication flows in particular has been inadequate and vague. As an outcome, organizations have had challenges sizing GDPR resources. The article is structured as follows: theoretical background is shortly discussed next. The research implementation is presented in the third section. The fourth section includes the analysis. The discussion and conclusions are in the fifth section.

2. Theoretical Background

The aim of the literature review was to create a base and reasoning for further empirical research. Research problems related to integrating the existing literature to GDPR expectations for that conclusions could be made. The research question was: What earlier studies have shown from organizational initial GDPR expectations?

The review was a descriptive (traditional) review, without strict and precise rules and the materials used were extensive. The method was suitable as the phenomenon studied can be described extensively and the properties of the phenomenon can be classified, if necessary (Salminen, 2010). Compared to a systematic review an integrative review provides a broader scope of the existing literature (Evans 2008: 137). According to Whittemore (2008: 149), the perspectives of research material can be more varied and broader than in a systematic review and summarising the main research material as a basis for the review is possible, due critical appraisal (Birmingham 2000: 33–34).

The review process was implemented paraphrasing by Cooper´s model (1989: 15), including the following phases: research problem setting, data acquisition, evaluation, analysis, and interpretation and presentation of the results. Academic peer-reviewed articles in information security discipline that were published between the years 2016 – 2020 were included in the review. Research material was collected from academic databases and journals, such as Computers & Security, Information and Computer Security, The Journal of Information System Security, Information Security Journal (A Global Perspective), Association for Computing Machinery Digital Library (ACM) and Institute of Electrical and Electronics Engineers and Institution of Engineering and Technology (IEEE) Xplore Digital Library. After the assessment, the articles that had no connection to GDPR were eliminated.

Results showed that GDPR had generally inspired writers and researchers both in academia and practice. Hundreds of scientific articles had been published under different disciplines, through several different academic publishing channels with varying themes, qualities and methods. The material reflected the approaches of law and legislation, health care/(bio-) medicine, banking, security and privacy, business and economy, information systems/technology and socio, showing that many academic disciplines were taken into concern.

Very few articles had even indirectly examined the research topics from an information system or information security perspective. Based on the analysis, the existing research could generally be divided for two by its chronological order: time before GDPR was enforced (before 25.5.2018); and time after GDPR officially was enforced (25.5.2018 and after). Several studies were designed to interpret either GDPR as a phenomenon, specific GDPR requirements, or offer solutions to specific challenges. Existing studies were mainly implemented so that the research dilemma was narrowly observed, highlighting that the early work did not enlighten GDPR theme in its wide meaning. Also practical relevance and quality were typically limited. As an example, some of these articles did not have comprehensive methodological groundings, missed the analysing method or did not explain the composition of the research material. Overall, the early work included lots of academic shortcomings.

It can be concluded that there is a gap of precisely understanding GDPR expectations and those impacts on organizations. Even though the focus of these studies was not in the expectations of GDPR, the side notes revealed the most important finding: GDPR guidance was felt as inappropriate, consuming and unclear. Regulation was seen just as unclear for researchers and practitioners both before and after GDPR turned into the force. This message occurred in most of the works. The amount of the articles aiming to increase transparency of the unclear content of the regulation remained high despite the timing.

The research question was: What does early work know of organizational early stage GDPR expectations? Existing literature knew a little of the organizational initial expectations. Early work lacked information system and security discipline approaches, academic quality, content and reasoning, wide perspectives, proper definitions and most of all, research on precisely on this topic, was missing. Based on the findings of the next best articles, the following hypothesis could be built: Organizations' expectations and perceptions of GDPR's requirements were unclear and varied due to confusing and inadequate official guidance and communication. Empirical and concrete experiences are required for future work to supplement the hypothesis so that a sufficiently comprehensive image can be created of what the effects these expectations have had.

3. Principles of implementing empirical research

The principles and rationale for implementation are discussed in this section, including presenting shortly a qualitative research process and methods, quality aspects and the framework of the study. Case study method was selected, as the focus of this research was on human interpretations and attributed meanings (Walsham, 1995). The aim was to examine the initial expectations and assumptions related to GDPR from the perspectives of selected Finnish organizations: what were their initial expectations regarding GDPR, how were they adapted/refined over time and what were the impacts on organizational planning and resourcing. The answer for the research question and the relevancy of the hypothesis were constructed through ten interview questions, presented in the section 4, and exploring the data during the whole process.

Case study followed the process presented below in the figure 1, and the model has been modified from Yin´s (2009), including six interconnected phases. The differences to Yin (2009, p. 1), were this process combined the planning and designing phases, because of the need for decreasing the delay between these phases. Reporting was added as a new and separate phase to underline its active role in the process ensuring that experiences gained there will be considered in planning and designing. The third difference was that all phases have a direct connection to the design and planning phase, so interferences and experiences of each phase were documented and considered in subsequent rounds.

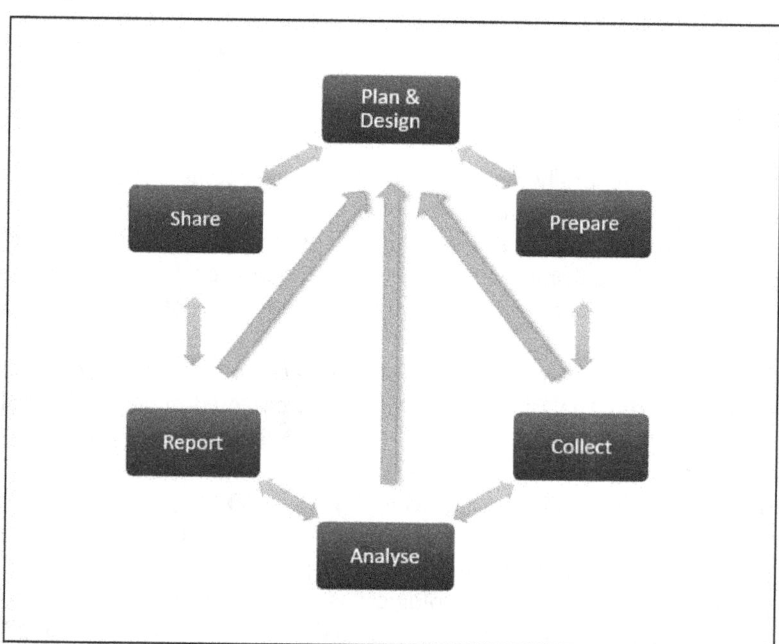

Figure 1: The case study process modified from (Yin, 2009, p. 1)

Target group included eight large or medium-sized Finnish organizations either originating from or located in Jyväskylä. Organizations varied by their background, industries, branches and culture. The titles of the participants varied from coordinator, experts, manager, officer, chef to director. All participants represented the excellence of GDPR matters of their own organization and were the people in practical charge of GDPR. The roles covered e.g. the following functions: law, HR, quality, privacy/data protection/info security, logistics or environment. The names and details of the organizations and participants were anonymized for ethical and privacy reasons. The results were then pooled for analysis. Research data from participants was collected by semi-structured thematic interviews regarding ten research areas. The same open-ended questions were asked from all interviewees, but the order varied if the interviewees themselves naturally moved on to the following questions. This was done to allow them to share experiences as freely as possible. The interviews were recorded. The participants reviewed transcribed interviews and, if necessary, supplemented them.

Eriksson and Koistinen (2014) argued that any analysing methods, or different methods at the same time, can be used in case study research. Research data was analysed using an explanation building technique supplemented with an iterative data enrichment concept and cross-case comparisons. Data from several cases were cross-compared in analysis and analysis is made while collecting data. Explanation building, according to Yin (2009), is a pattern matching aiming to analyse the research data by building an explanation of the case. The General Accounting Office (1990) emphasised the observe, think, test, and revise (OTTR) concept for case study data collection and analysis. Researchers should assess the meanings of information continuously, so that data enrichment needs, in order to confirm existing or create alternative interpretations, could be considered as the process progresses (GAO, 1990). Causality would be achieved through the internal consistency and plausibility of explanation (GAO, 1990), and Miles and Huberman (1984) defined the word ´explaining´ as a process of building a set of causal reasonings for how or why things progressed as they did. The research framework is presented in figure 2. It consists of the research question, the hypothesis built in the literature review, the target group and the 10 interview questions.

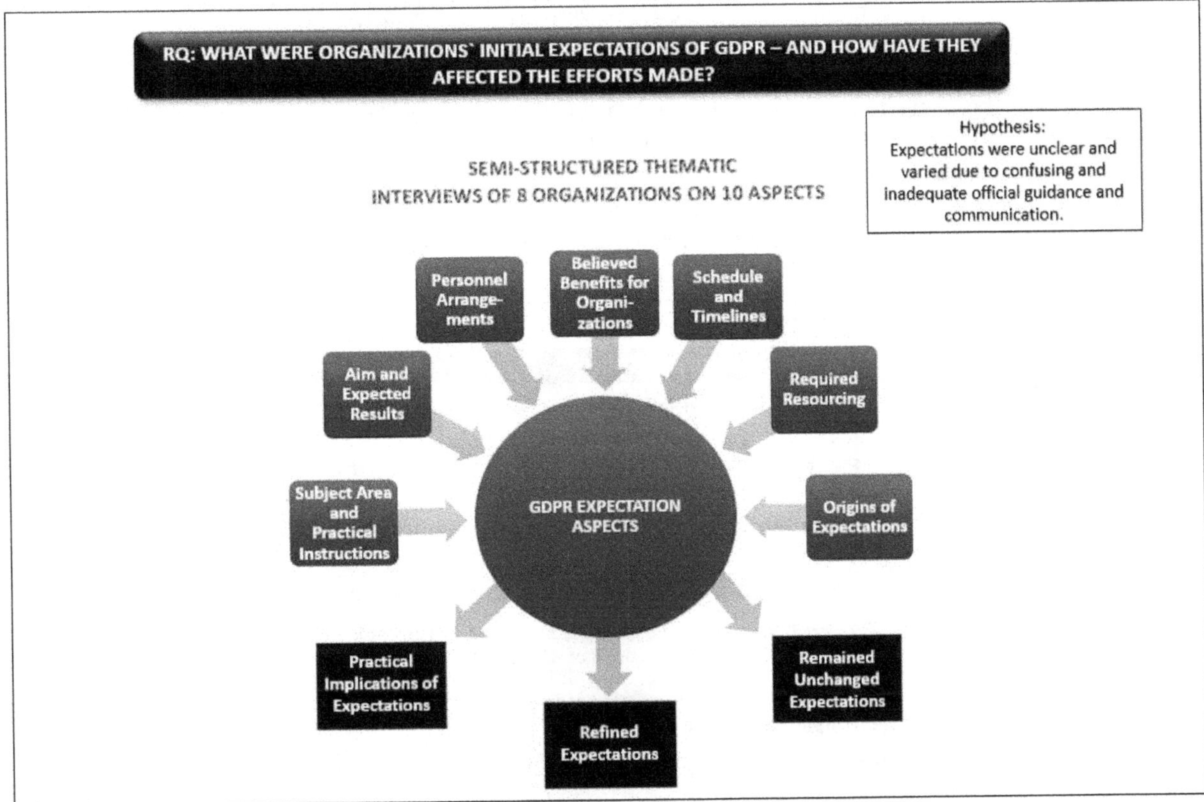

Figure 2: Research framework

The quality of the empirical work consists of (internal, construct & external) validity and reliability (Edmonds & Kennedy, 2012). Internal validity of this work was confirmed by using multiple data sources, which, according to Yin (1984) is an important way to implement a triangulation of case study. Denzin argued (1989: 236–37) that triangulation can occur with data, investigators, theories, and even methodologies. For ensuring construct validity, among multiple source usage Yin (2009) proposed reviewing key informants in the report and documenting clearly and fairly evidence of research, so that reviewers would be able to track initial research questions from conclusions (Sarker & Lee, 1998). GAO´s (1990) guidance of using triangulation in methodology and data source (including cross-case comparisons) was applied to increase internal validity (GAO, 1990). External validity embraces generalisability of the findings to other cases (Tellis, 1997), but unambiguous methodological guidance of achieving it is commonly missing (Halkier, 2011). The external validity is addressed by selecting a sufficiently large research population to allow comparability. Reliability was achieved by documenting and implementing a case study protocol carefully, including e.g. an overview of the project, case study questions, field procedures and applying guidance for addressing descriptive or explanatory questions as its narrative format (Yin, 1994, p. 64).

4. Analysis

The research question was faced by looking at 10 research aspects (Table 1) empirically with certain questions. The analysis of the results and the findings to these aspects are discussed here question by question and the research implementation analysis follows that. Through aspects 1 - 7, the initial expectations were examined. Looking at aspects 8 - 10, the effects of expectations were identified.

The first question addressed the organization's expectations of GDPR and its guidance. In practice, the preparations could be divided into first (2016 - 5/2018) and second (6/2018 - 2019) period. Finding reliable and usable information in the information management process, as well as using GDPR as a sign of responsibility, were seen as essential. Among other things, the general lack of guidance, the lack of interpretation of the text of the regulation, the open issues of liability, the emphasis on sanctions and the oversupply of consultants made the regulation initially very vague, difficult to understand, complex and confusing. Formal communication was a negative threat, as well as too impractical and abstract. Therefore, even before the end of the transition period, GPDR was seen as a mandatory evil with risk factors. The invisibility of the Finnish Office of the Data Protection Ombudsman (ODPO) opened the door for consultants, and the number was large, and knowledge of the content of GDPR was variable and in many respects´ poor. Service providers also lacked the required organization-specific knowledge. The selection criteria for useful partners were e.g. reliability and competence and were wanted to be involved throughout the process. It can be concluded that the formal guidelines should have sought to be more concrete than on levels 1 and 2, as well as to clarify the benefits and possibilities of the Regulation. Certification or training of consultants would also have facilitated their use. Expectations are likely to affect how organizations prepare and attitude, and how realistically the role and amount of activity is assessed. It is possible that preparatory efforts and resources were wasted if these expectations differ significantly from those of the legislators. If the will of the authorities is vaguely expressed, it is also possible that the original requirement is not met.

Table 1: Research aspects and questions

RESEARCH ASPECTS	RELATING RESEARCH QUESTIONS
1. Subject Area and Practical Instructions	What were the initial expectations of GDPR and its official guidance?
2. Aim and Expected Results	Describe the initial rendering of GDPR´s expected goal and outcome.
3. Personnel Arrangements	How many people in your organization work for GDPR at some level? How were the persons in charge of GDPR identified and defined?
4. Believed Benefits for Organizations	Describe the initial expectations of the usefulness of GDPR for your organization.
5. Schedule and Timelines	Describe the initial expectations of GDPR-related schedules.
6. Required Resourcing	Describe the initial estimations of the GDPR-related investment/ resources required from your organization.
7. Origins of the Expectations	Where the expectations came from and what they were based on?
8. Practical Implications of Expectations	What role did the initial expectations have for concrete design and implementation?
9. Refined Expectations	Describe how the expectations have become more precise over time. What have been the changes, in particular, that were not in line with the initial expectations?
10. Remained Unchanged Expectations	What expectations have been met as expected?

The second issue addressed GDPR's aim and expected results. The results revealed three perspectives: the strengthening of individuals´ rights, the business potential of GDPR, and the factors that employ or limit. The goals and outcomes were not clear to everyone, which was connected to inadequate official communication. The communication was initially unsuccessful, as it was not able to demonstrate clearly the benefits and added value of GDPR, compared to the old regulation. Second, the negative tone was emphasised in the initial communication with emphasis on sanctions, which caused fear and frustration. Clear expectations affect how an organization invests in an issue, how it prioritises doing so, and how individuals engage in it. It is estimated that the perceptions will guide the mental and concrete preparation of organizations for ongoing GDPR activities.

The third question addressed the internal personnel and organizational arrangements that have met GDPR's challenges. Actors outside the organization were excluded from the review. The results suggest that the interpretation of GDPR expectations have had a significant impact on the allocation of GDPR responsibilities by organizations. In addition, staffing practices varied significantly between organizations, pending on the industry, business, size, stage of operation and financial success, organizational arrangements, management models, human resources and expertise. In addition, due to the nature of the organizations' activities, the amount and sensitivity of the personal data processed varied significantly. It can be concluded that existing data protection and security practices were affected. Consequently, the solution required varied significantly. A few of the answers revealed that in addition to the actual, there were challenges in performing the data protection task, e.g. adequacy of time, workload and status (whether authority is sufficient). Organizations also felt that data protection affected the external layers of the organization, for example as service providers or partners. The participants did not reveal the practice of involving these third parties, sometimes very significantly involved in the organisation's activities, as members of the data protection organization, if necessary. GDPR matters were still dealt with mainly by third parties. From the responses, it can be concluded that the industry and the nature of the activities of the organization play a role in whether GDPR is a more one-off or continuous process.

The fourth question explored the potential benefits of GDPR for organizations. The benefits expected were as difficult to identify as the overall objectives of GDPR and the expected results. Believed benefits for organizations were threefold: uselessness, minimum requirement & necessity, and opportunity. GDPR is not automatically perceived as useful to organizations. With regard to usability, the existing legislation, the lack of a formal justification for information and the poor financial measurability of data protection were highlighted. Critical or negative attitudes can also potentially slow down the take-up of activities and reduce staff commitment in organizations. GDPR as a minimum requirement could be related to the risk management process, or the avoidance of sanctions. While maintaining GDPR minimum requirement, it is possible that the implementation of privacy investment proposals become challenging. In terms of opportunities, the values of data and responsibility were mentioned as intangible assets and the development of information management with the help of law.

The fifth question concerned the adequacy timetables. It was seen that there was time to prepare, but in practice many left the action close to the deadline and in the end there was a rush. The lack of formal ready-made practices and solutions also made it difficult to implement the requirements in time, as it was not possible to interpret the workload of the activities, or to measure the time required for implementation. It was mentioned that in December, none of the interviewees indicated that there was too much time or that all was done. It emerged that the operating environment outside the organizations was not yet ready for the regulation at the end of 2017.

The sixth issue anticipated the resources required by GDPR. The main required resourcing concerned the personnel and others were money and time. Although organizations had expectations of GDPR, it was felt challenging to evaluate and plan for accurate resources, and views have only strengthened and refined over time. The quality of resources was easier to assess than the quantity. A project was set up for GDPR, which in many organizations has since been transformed into a continuous process, making GDPR a permanent part of the organisation's operations. Some found the start-up clearly easier which had to do with human resources as well as adequate legal expertise. Those who did not have enough of the former relied on consultants. Consultants were used in particular to assess the workload and time required, to support the identification of responsibilities and tasks of those in charge, and to tailor implementation to the needs of the organization. The large number and varying quality of the consultants posed challenges in assessing the qualifications of the consultants.

The seventh question identified the sources from which GDPR expectations arose. The answers created a picture of the information channels, different actors involved in the process, the nature of the information and the impact on the organizations. Origins of the Expectations were primary sources (ODPO and EU-level GDPR authorities) as well as secondary sources (including consultants and other stakeholders) that communicated, edited, and interpreted information from the primary source to organizations. The different data protection and security maturities of the organizations as well as human resources influenced the choice of the source of the information and whether the information was interpreted by themselves or with the help of consultants. The invisibility of the EU and the national authority created a widespread need for the use of consultants.

The eighth question examined the practical GDPR implications for organizations, as they have made significant efforts to meet GDPR requirements. Practical implications of expectations were that organizations were able to initiate GDPR activities but could not accurately quantify the amount of resources required, could make interpretations incorrectly, could arise unnecessary input, or waiting for guidance caused a delay in implementation. This caused frustration and could internally question GDPR accountants within organizations because they lacked the capacity to put things into practice effectively and independently. These factors also had a potential impact on other operations and businesses, as GDPR affected the organization as a whole.

The ninth question mapped how GDPR expectations changed and evolved during the initiation process. GDPR was perceived as more demanding than estimated. The changed expectations were e.g. monitoring the implementation of the regulation and imposing sanctions, organizations being left alone in the transition phase, the organisation's own interpretation activity, the ongoing monitoring of legal issues and the importance of internal policies in handling the regulation, practical implementation challenges (personal data mapping, Refined expectations stemmed from delayed and inadequate official guidance and communication, which many felt was invisible for the first 18 months. Clarifications and lawsuits from the authorities had been long awaited and the solutions obtained were perceived as complex (e.g. if personal data were processed abroad). It was also mentioned that the EU lacked an official that would provide clear guidance on how individual organizations should operate.

The tenth question mirrored expectations in retrospect, and the answers show what facts and images were retained in organizations throughout the GDPR process. The question clarified views on the relevance and implications of interpretations related to GDPR requirements. The remaining expectations that had not changed were mainly related to organizational needs to seize initiative and actively interpret the regulations, to bring about a constant change in the nature of GDPR, to emphasise the importance of documentation and to guide GDPR through the process.

Overall, the observations from the 10 areas provided a comprehensive overview of the topic. GDPR meant challenges, additional work, minimum obligations and also opportunities. Especially in the beginning, it put pressure on organizations and they felt left alone. The initial communication was unsuccessful due to lack of clarity on the benefits and added value of GDPR and the concrete meaning of the content. The negative tone highlighted in the communication caused fear and frustration. In addition to the lack of clarity on formal guidance on the purpose, expected impact and outcomes of GDPR, it was not clear how much resources were expected from organizations. The responses provided an insight on how the guidance and communication supported organizations in identifying the necessary preparations and resources. The interpretation of GDPR expectations has had a significant impact on GDPR activities implemented by organizations in all areas. In addition to identifying expectations, the responses provided an overview of the organization-specific factors that influence the structure of organizations 'expectations. The responses also revealed how the expectations impacted the organizations.

Due to the descriptive literature review, the broad level understanding of findings and gaps of it was achieved. The modified iterative case study process, sample and framework were functional, as the desired scope and thoroughness were achieved. Through the interviews, diverse and detailed information on the research topic was obtained. Based on the feedback and responses, the questions and semi-structured thematic interviews were functional. Interviewees had the opportunity to check and correct the transcripts, thus ensuring the integrity of the information and the correct understanding between the researcher and the interviewer. The quality of a case study is generally challenging to ensure, but here the quality factors were considered as reported in section 3. During data collection and analysis, interpretations were made by progressively deepening and retrieving triangulation of actual similar research results through deliberately overlapping research questions. Iteratively performed analysis was reasonably comprehensive. However, the analysis was demanding and slow to implement in this way.

5. Discussion and conclusion

The aim of this work was to examine empirically the initial expectations of GDPR from the perspectives of selected Finnish organizations. The research question was: What were the organizations' initial expectations of GDPR - and how have they affected the efforts made? The answer is that GDPR has been a legal issue, mainly due to the inadequate and unspecific active, formal communication flows. It can be concluded from the results that the interpretation of GDPR expectations has had a significant impact on GDPR activities of organizations. Organizations have experienced difficulties in scaling the necessary GDPR efforts. The hypothesis was confirmed in the empirical study. The research question was examined using ten aspects and sub-research questions. An analysis provided a comprehensive overview of the phenomenon, as well as answers as to why and how causality occurred. The key expectations of the organizations, the factors that influenced the expectations and the effects of the expectations were comprehensively listed in the analysis of the results.

The key finding is the importance of interpretation of legal requirements in detail at the earliest possible stage of the process to influence the success of subsequent decisions and investments taking into account the specific organizational traits and requirements. Poor, untimely and incomplete communication can lead to wasted investment, hinder initiating activities and reduce personnel's willingness to commit. Empirical findings suggest that the initial phase of GDPR was easier, if data protection and security, an understanding of legal issues, adequate human resources, and a strong knowledge of the organization were already in place. These factors were likely to support the planning and implementation of GDPR activities, reduce delays, friction and pressure.

Detailed analysis can benefit both privacy and information security managers and responsible persons to align policies and practices, and to evaluate organization-specific actions on GDPR compliance. The results can support regulators and authorities in the future GDPR and other policy work and provide ideas for service providers. The organizations themselves, as well as the governing bodies, should be able to consider the factors identified in this study, which varies from organization to organization, and regulates the ability to receive guidance. In addition, it would likely be beneficial to create concrete instructions for different levels, step-by-step example plans and e.g. quality certification of consultants. Implementing procedures like this in the future can be challenging based on organizations 'negative experiences with GDPR if urgent fixes are implemented. The results contradict official GDPR expectations at an early stage, although over time, officials will likely seek to develop policies based on feedback and experience.

5.1 Restrictions and future study

The limitations of the study are related to the generalizability of the results. Objectivity and quality could also have been refined more precisely, which is typical when using a case study method. The data collection for the literature review ended in 2020, and several qualified studies have been published after that, the results of which had to be excluded from this work due to time resources. Because the interviews yielded significant amounts of research findings, it is clear that the findings in the results needed to be significantly condensed to fit into the article. In this case, there is a risk that something essential was excluded.

In later studies, the topic could be addressed by a broader sampling and by means other than the case study method. Since 2020, implementation studies have been published in various countries, they could be compiled to gain a broader understanding of the subject. Comparative studies could be carried out abroad, including in non-EU countries. In addition, the issues of beneficial use of consults when implementing actions such as GDPR, considering organizational needs when drafting guidelines for regulations and engaging organization's external parts as part of the organization's data protection team could also be examined.

References

Birmingham, P. (2000). Reviewing the Literature. In: *"Researcher's Toolkit: The Complete Guide to Practitioner Research"*, 25–40. Edit. David Wilkinson.

Cooper, H. (1998). *Synthesizing Research: a Guide for Literature Reviews*. Thousand Oaks: Sage Publications, Inc.

Denzin, N. K. (1989). *The Research Act.* Third edition. Englewood Cliffs, N.J. Prentice-Hall.

Edmonds, W. A., & Kennedy, T. D. (2012). *An applied reference guide to research designs: Quantitative, qualitative, and mixed methods*. Thousand Oaks, CA: Sage

Eriksson, P. & Koistinen, K. (2014). *Monenlainen tapaustutkimus*. Kuluttajatutkimuskeskus.

Evans, D. (2008). Overview of Methods. In: *"Reviewing Research Evidence for Nursing Practice: Systematic Reviews"*, 137–148. Edit. Christine Webb & Brenda Ross. Oxford: Blackwell Publishing.

General Accounting Office (GAO). (1990). *Case study evaluations.* Washington, DC.

Halkier, B. (2011). *Methodological practicalities in analytical generalization*. Qualitative Inquiry, 17(9), 787-797.

Miles, M. B., & Huberman, A. M. (1994). *Qualitative data analysis: An expanded sourcebook*. Thousand Oaks, CA: Sage.

Salminen, A. (2010). *Mikä on kirjallisuuskatsaus.* Johdatus kirjallisuuskatsauksen tyyppeihin ja hallintotieteellisiin sovelluksiin. Vaasan yliopiston julkaisuja. Opetusjulkaisuja 62. Julkisjohtaminen 4.

Sarker, S., & Lee, A. S. (1998*). Using a positivist case research methodology to test a theory about IT-enabled business process redesign*. Paper presented at the International Conference on Information Systems, Helsinki, Finland

Tellis, W. (1997). *Introduction to Case Study*. The Qualitative Report. 3. 10.46743/2160-3715/1997.2024.

Yin, R. (1984). *Case study research: Design and methods* (1st ed.). Beverly Hills, CA: Sage Publishing.

Yin, R. (1994). *Case study research: Design and* methods (2nd ed.). Beverly Hills, CA: Sage Publishing.

Yin, R. K. (2009). *Case study research: Design and methods* (4th ed.). Los Angeles, CA: Sage.

Walsham, G. (1995). *Interpretive case studies in IS research: Nature and method.* European Journal of Information Systems, 4(2), 74-81.

Whittemore, R. (2008). Rigour in Integrative Reviews. In: "*Reviewing Research Evidence for Nursing Practice: Systematic Reviews*", 149–156. Edit. Christine Webb & Brenda Ross. Oxford: Blackwell Publishing.

A Cyber Counterintelligence Competence Framework

Thenjiwe Sithole and Jaco Du Toit
Academy of Computer Science and Software Engineering, University of Johannesburg, South Africa
thenjiwes@icloud.com
jacodt@uj.ac.za

Abstract: The increased use of cyberspace and technological advancement are fundamentally changing the cyber threat landscape. Cyberattacks are becoming more sophisticated, frequent, and destructive. Internationally, there is a growing acceptance that Cyber Counterintelligence (CCI) is essential to counter cyber-attacks optimally. Therefore, in addition to government intelligence and security agencies, more companies are incorporating a CCI approach as a critical element of their posture for engaging cyber threats. However, the successful adoption of a CCI approach depends on the availability of skilled CCI professionals equipped with the requisite competences. The creation of such CCI professionals, in turn, requires a framework for developing the necessary CCI competences. At least in as far as reviewed academic literature is concerned, there is no existing postulation on a framework to develop the CCI competences, specifically for developing countries. Given the complexity and multi-disciplinary nature of the emerging CCI field, such a framework needs to provide two distinctive skillsets linked to CCI's two distinct areas of expertise, namely cyber (security) and counterintelligence. The paper presents a high-level Cyber Counterintelligence Competence Framework (CCIC Framework) that outlines dimensions of CCI, functional areas, job roles and requisite competences (knowledge, skills, and abilities), and tasks for each CCI job role. The CCI framework also outlines five levels of proficiency expected for each job role. The identification of competences and levels of proficiency are integral to the successful implementation of the framework and workforce development. The CCIC Framework is intended to be used as a tool to retain, assess, and monitor knowledge, skills, and abilities for CCI workforce development. In addition, the CCIC Framework can be used to assist in providing the basis for individual performance management, education, training, and development pathway, as well as career progression. Therefore, this paper presents a CCIC Framework which is an overarching, integrative construct that synergistically combines different components required to develop a competent workforce for the emerging field of CCI.

Keywords: cyber counterintelligence, cyber counterintelligence job roles, competence framework, proficiency levels, cybersecurity

1. Introduction

Hyperconnectivity, sharing of big data and information at high speed and in real-time, amongst other things, is the order of the day in today's world shaped by cyberspace and inevitable advanced technologies. Therefore, this implies cyberspace and advanced technologies are fundamentally and increasingly integrating into the global population, positively changing the way people live and communicate with one another and improving the quality of services in organisations (Li & Liu, 2021; Petrillo, et al., 2018). In 2020 the world experienced an unprecedented acceleration of internet connectivity due to the novel Covid-19 pandemic (ITU, 2021). Public and private sector organisations and educational institutions had to rethink, adapt, and accelerate the use of technology and cyberspace, in magnitude, for remote working and teaching and learning.

Certainly, cyberspace and technology advancement bring opportunities, but at the same time, they come with cyber risks, fundamentally changing the cyber threat landscape. There has been an unprecedented increase in cyberattacks, which are becoming frequent, complex, on a large scale, destructive and stealthy, with cyber attackers becoming more efficient (Meier, et al., 2021; Check Point, 2018). State and non-state actors continuously leverage disruptive technologies to develop advanced and sophisticated tools for expediting their efforts in conducting cyber espionage, cyber warfare, cyber-surveillance, or cybercrime (Public-Private Analytic Exchange Program, 2019). These actors may launch attacks on any private and public organisation globally.

Conventional cybersecurity alone is no longer sufficient and effective. The recent events of unprecedented cyberattacks should highlight the significance of new thinking on countering these cyberattacks and cyber risks. Therefore, in this increasingly hyperconnected cyber world, it is imperative for organisations to adopt a proactive approach in responding to cyber risks, preventing cyberattacks before they even happen and protecting their critical information systems. More particularly, the threat landscape, in which threat actors of various types have expanding intelligence capabilities, is underlining the need for organisations to incorporate counterintelligence (CI) – and thus cyber counterintelligence (CCI) - as part of their security approaches (Duvenage et al., 2020a; Duvenage & von Solms, 2015; Jelen, 2020). Duvenage (2019) defines CCI as "the subset of multi-disciplinary CI aimed at deterring, preventing, degrading, exploiting and neutralising adversarial attempts to collect, alter or in any other way to beach the C-I-A of valued information assets through cyber means."

This paper introduces a high-level outline of the cyber counterintelligence competence (CICC) Framework and its constituents. The CCIC Framework is an overarching, integrative construct that combines different constituents required to develop a competent workforce for the emerging field of CCI.

This paper is structured as follows: Section 2 motivates for development of the CCIC Framework. Section 3 introduces the CCIC Framework structured into four elements: Dimensions, Functional Areas, Job Roles, and Competences (knowledge, skills, abilities, and attitude). The section also presents five levels of proficiency required for each job role. Section 4: provides a conclusion on the proposed CICC framework.

2. Motivation for developing the CCIC framework

CCI is a developing field gaining ground internationally (Duvenage, et al., 2020a). Therefore, more and more companies are implementing a CCI approach as part of their endeavour to engage and neutralise proliferating threats. This escalating trend is increasingly pushing the availability of a skilled CCI workforce equipped with the necessary capabilities to the centre. However, the CCI field lacks the fundamental body of knowledge relating to a professional and capable workforce, such as describing and understanding the requisite CCI competences (Duvenage, et al., 2020b; Duvenage, et al. 2019; Black, 2014). A review of the CCI industry and peer-reviewed academic research - conducted for the purposes of this paper - found no existing CCI framework explicitly referring to the requisite competences (knowledge, skills, and attitude) for CCI workforce development in the consulted literature. As substantiated in further paragraphs of this section, identifying such competences is a critical requisite for developing a CCI capable workforce and is thus a defining feature of the CCIC Framework.

Note should be taken that literature shows the two similar terms – "competency" or "competence" – being used interchangeably or to define two different concepts. This paper adopts "competence" for consistency and to avoid confusion.

Competence can be defined as a demonstrated ability to apply relevant characteristics for achieving recognisable performance to the levels of a set appropriate standard and can be improved through continuum education, training and development (Apollo Education Group, 2015; CWA, 2014). Within the context of this definition, 'characteristics' refer to applicable knowledge, skills, and attitudes. In further elucidating the definition, 'standard' denotes the degree of proficiency required for different competences or job roles. Proficiencies, in turn, can be a valuable tool for career planning and career pathing (Griffiths & Washington, 2015).

The identification of CCI competences will assist organisations in determining the kind and level of capabilities required for the CCI workforce to execute their jobs effectively. Therefore, it is essential to correctly identify such CCI competences as they are paramount to elevating performance in countering escalating cyber threats and incidents. CCI competences will differ according to assigned job roles and different levels of employment (i.e., CCI Execution levels: strategic, operational, and tactical) (Kansal & Singhal, 2018). CCI competences are based on the CI and CCI dimensions, CI categories as well as CI functions (Prunckun, 2019; Stech & Heckman, 2018; Duvenage & von Solms, 2014; Kuloğlu, et al., 2014; US Army Publishing Directorate, 2009) and blend with other relevant cybersecurity competences. The next session will provide an overview of the CCIC Framework.

3. Overview of the CCIC framework

CCI is becoming a significant part of combating cyber risks and cyber-attacks. Therefore, A CCIC Framework is essential for the development CCI workforce. The notion of CCI is multi-disciplinary and incorporates two distinct fields of expertise – Cyber security and CI. Therefore, effective CCI depends on the two corresponding distinctive skillsets of cyber security and CI (Black, 2014). The CCIC Framework is aligned to the underlying four dimensions/categories of CI and CCI Framework: defensive, offensive, active, and passive. This section will expound these as dimensions of the underlying structure that underpins the CCIC Framework's other elements.

Governments and the private sector have developed several cybersecurity skills frameworks to improve cybersecurity practices, providing a comprehensive range of cybersecurity roles, tasks, and competences. However, none of the frameworks explicitly make mention of the CCI and CCI competence or skills or workforce development. The concept for CCIC Framework takes inspiration from international frameworks such as the NIST

NICE Cybersecurity Workforce Development Framework, CyBOK, Skills Framework for the Information Age (SFIA) and Chartered Institute of Information Security (CIISec) Skills Framework, but it is modified to highlight CCIC (SFIA, 2021; CyBOK, 2019; NIST, 2017; CIISEC (IISP), 2010).

The CCIC Framework is a construct that comprises four elements: dimensions, functional areas, job roles, competences (knowledge, skills, abilities, and attitude) and tasks. The CCIC Framework is organised into four dimensions, which is an overarching structure of the CCIC Framework. Each CCI dimension comprises one or more CCI Functional Areas. Each CCI functional area has various associate job roles. Each job role has associated competences and numerous tasks. The relationship between the elements of the CCIC Framework is depicted throughout this section. The mapping of the framework *per* four elements considers, but further evolves and adds to, notions advanced in the international frameworks cited in the previous paragraph. Figure 1 illustrates an overview of the CCIC Framework showing these four elements. Each of these elements is discussed in sections 3.1 to 3.4.

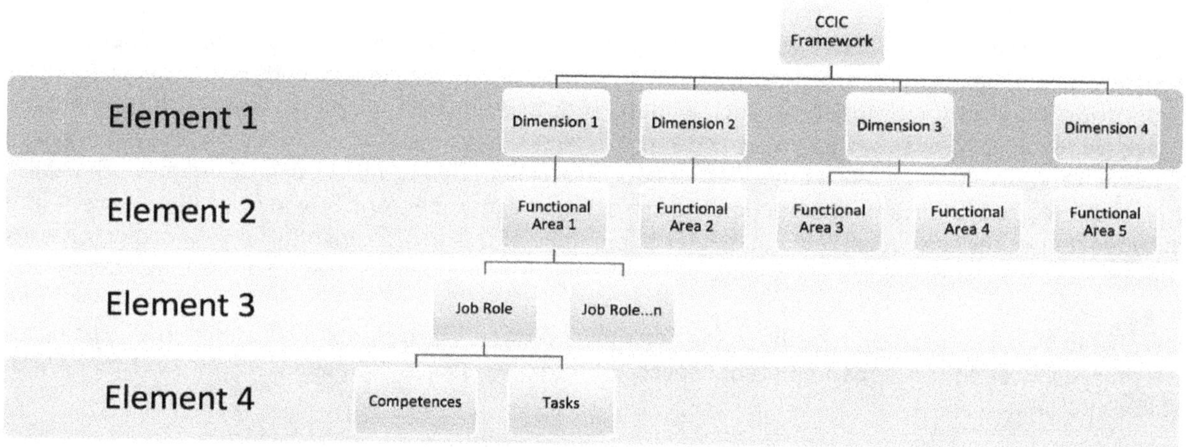

Figure 1: An overview of the CCIC Framework

3.1 Element 1: Dimensions

Element 1 provides an underlying structure of the CCIC Framework, referred to as Dimensions. As mentioned earlier in this section, the CCI incorporates two distinct fields, CI and cybersecurity. Basically, CCI uses CI principles, concepts, and functions within the cyberspace. Therefore, the CCI dimensions are derived from the two fundamental categories of CI, defensive CI and offensive CI (Prunckun, 2019; Sims, 2009).

Defensive CI/CCI applies the detection and deterrence techniques to deny access and collect information on espionage threats (Prunckun, 2019; Duvenage & von Solms, 2014). Defensive CCI is concerned about preventing the adversaries' endeavour from penetrating the organisation's information systems and, at the same time, collecting intelligence against the adversaries and minimising the threat landscape. An example of defensive CCI is performing vulnerability assessments and threat analysis (Lee, 2015). The offensive CI/CCI applies techniques to detect, deceive, and neutralise espionage threats and covert threats (Prunckun, 2019; Duvenage & von Solms, 2014). Offensive CCI is concerned about detecting and directly gathering intelligence about adversaries' covert, espionage, or cyber operations or deceiving and manipulating them. This can be done inter alia by creating honeypots containing files with misinformation (Lee, 2015).

The defensive CI and offensive CI can be implemented in both passive and active measures. Thus, formulating the four CI dimensions - active defensive, passive defensive, active offensive, and passive offensive (Prunckun, 2019; Stech & Heckman, 2018; Duvenage & von Solms, 2014; Sims, 2009; US Army Publishing Directorate, 2009). The dimensions are also derived from various conventional defensive and offensive cybersecurity processes and procedures (Jaquire, et al., 2018). The four CI/CCI dimensions are illustrated as quadrants in Figure 2.

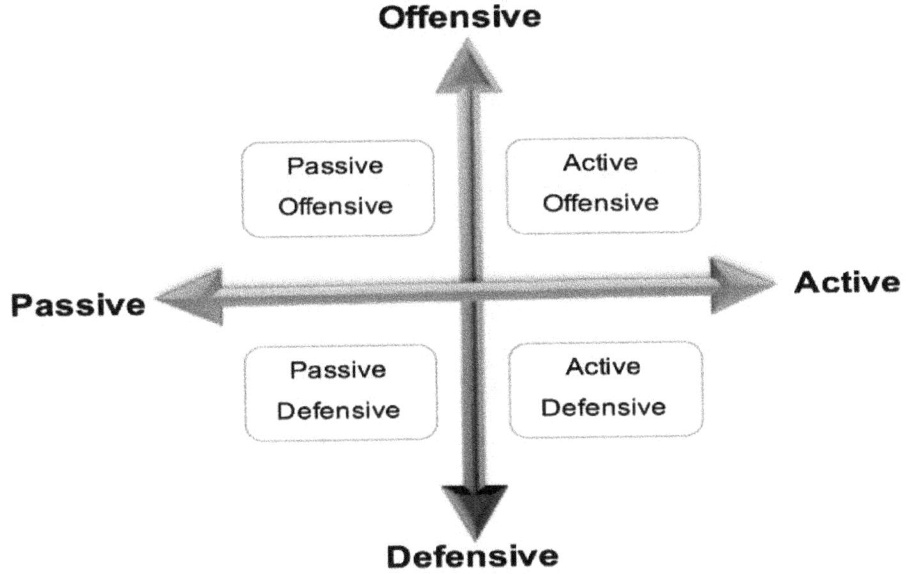

Figure 2: The four dimensions of CCI (Duvenage & von Solms, 2014; Sims, 2009)

As depicted in Figure 2 that both defensive and offensive CI/CCI have passive and active modes. These four dimensions can be explained in tabulated format (Table 1) as follows:

Table 1: Four-sector counterintelligence matrix (Duvenage, et al., 2020a)

Passive Defensive	**Active Defensive**
Denies the adversary access to information through physical security measures and other security systems and procedures.	The active collection of information on the adversary to determine its sponsor, modus operandi, network, and targets. Methods include physical and electronic surveillance, dangles, double agents, moles, and electronic tapping
Passive Offensive	**Active Offensive**
Reveals selected information to the adversary. This could range from selective exposure of actual information to decoys and dummies. The adversary is thus left to draw its own inferences and interpretations.	The adversary is fed with disinformation and its interpretation thereof manipulated. Disinformation can be channelled through, for example, double agents and moles. Active offensive CI could include some forms of covert action.

The identification of the four dimensions forms the first element of the CCIC Framework – passive defensive, active defensive, passive offensive, active offensive and is graphically depicted in Figure 3.

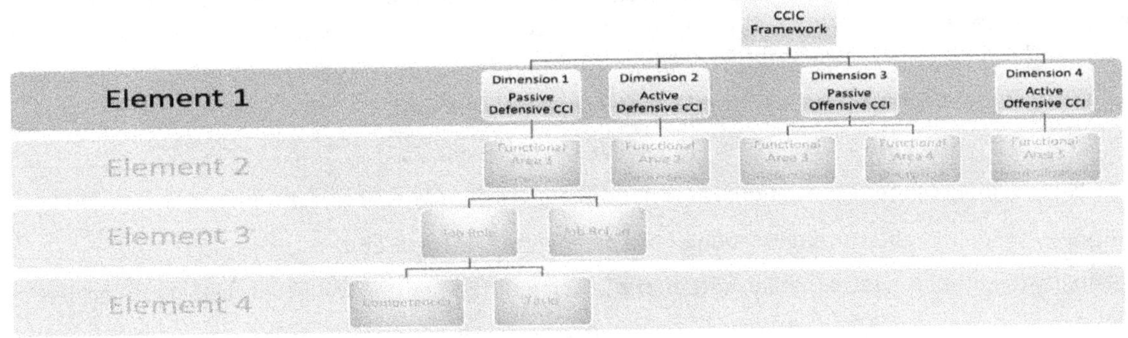

Figure 3: Element 1 of the CCIC framework - the four dimensions

3.2 Element 2: Functional areas

The four dimensions mentioned above have one or more associated functional areas. The functional areas are underpinned by the principles of CI to accomplish CI missions. The principles are detection, deterrence, deception, and neutralisation (Stech & Heckman, 2018). Mapping the dimensions with the principles, defensive CCI is concerned with detection and deterrence, whereas offensive CCI is concerned with detection, deception, and neutralisation (Prunckun, 2019; Stech & Heckman, 2018). The four CCI dimensions and the identified corresponding functional areas form the second element of the CCIC Framework illustrated in Figure 4.

- Dimension 1: Passive-defensive CCI
- *Functional Area1: detection*
- Dimension 2: Active-defensive CCI
- *Functional Area 2: deterrence*
- Dimension 3: Passive-offensive CCI
- *Functional Area 3: detection*
- *Functional Area 4: deception*
- Dimension 4: Active-offensive CCI
- *Functional Area 4: neutralisation*

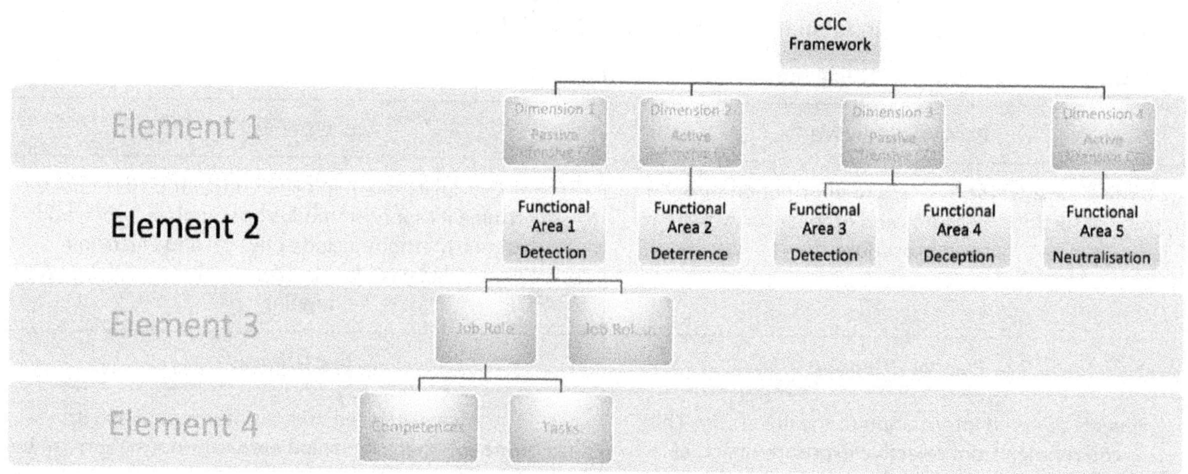

Figure 4: Element 2 of the CCIC framework - the five functional areas

*3.2.1 Functional Areas 1 and 3: Detection**

To discover the existence of any cyber activities, anomalies and threats targeting the information systems and associated compromise or possible compromise of the confidentiality, integrity, and availability of the information systems. Prunckun (2014) offers five premises that include the detection principle, and for the objective of this paper, identified as the following abilities:

- 1.Ability to identify the cyber event, activity or incident of concern;
- 2.Ability to identify the person(s) who are involved in the event;
- 3.Ability to identify the organisational association of the person(s) of interest;
- 4.Ability to identify the current location of the person(s) of interest; and
- 5.Ability to gather the facts that indicate that the person(s) committed the event, activity or incident.

*Cyber detection constitutes a functional area within Passive Defensive (Functional Area 1) and Passive Offensive CCI (Functional Area 3). In both these areas, 'cyber detection' has the same meaning and requires the five premises stated above. For this reason, Functional Area 3 is not discussed in a separate subsection. Detection under passive defensive is reactive, the process of identifying or discovering occurrences of CCI events has no human interaction, and it is not in real-time. In contrast, detection under passive offensive is proactive and there is real-time human interaction and information gathering or actively hunting for threats to learn from and respond to the adversary.

3.2.2 Functional Area 2: Deterrence

Deterrence is the ability to dissuade an adversary from attempting intrusive cyber operations on information systems or by preventing an adversary from conducting cyber intelligence by ensuring that the adversary perceives the risks and costs of their action outweighing the benefits or that the advantages they expect (Soesanto & Smeets, 2020; Jensen, 2012). Deterrence is active defensive because it proactively stops adversaries before they can achieve their objectives. Deterrence can be achieved through (Prunckun, 2019; Jensen, 2012):

- 1. Deterrence by punishment – that is, threatening to retaliate, an organisation must be able to conduct an attack back on its adversary.
- 2. Deterrence by denial – that is, discouraging the adversary and denying the benefits of an attack. This approach must be perceived by an adversary and must look credible to succeed.

3.2.3 Functional Area 4: Deception

Deception is used to mislead and confuse the adversary about operations, capabilities, intentions, plans or vulnerabilities through manipulation, distortion, or falsification to make them believe what is fabricated, is accurate and make them either take action or not so that these actions prove ineffective (Prunckun, 2019; Heckman, et al., 2011). Deception is a passive offensive measure as it can deliberately mislead an advisory and conceal certain information from the adversary.

3.2.4 Functional Area 5: Neutralisation

Neutralisation renders the adversary's cyber activities and capabilities inactive, failure, collapse. Stech & Heckman (2018) assert that neutralisation of adversary cyber activities can be achieved by "destruction, paralysis, loss of interest or loss of confidence that collection will be able to achieve its objective". Neutralisation is active offensive because it proactively and in real-time counteract or destruct an adversary's activities.

3.3 Element 3: Job roles

Sections 3.1 and 3.2 gave an overview of the first two elements of the CCIC Framework. These two elements, the Dimensions, and the Functional Areas, are the foundation for identifying Job Roles aligned to the CCI approach. That is, each Functional Area has corresponding Job Roles. Furthermore, each Job Role has a job title, job purpose, levels of proficiency and associate competences (knowledge, skills, abilities, attitudes (KSA), and tasks required for optimal performance of the role. The third element of the CCIC Framework is illustrated in Figure 5.

3.3.1 Job role

The term 'job role' denotes a detailed grouping of related jobs consisting of competences (knowledge, skills, and abilities), a group of defined tasks and levels of proficiency required to achieve the job role. The job role has the following components:

- Job title – means the name used to refer to a particular job.
- Job code – a unique code identifier used to reference the job
- Job purpose – comprises an overview of what the job entails: the primary purpose and objectives of the job.
- Tasks – a detailed, specific list of primary responsibilities needed to be performed for the job role.
- Level of proficiency – description of the levels of proficiency required for each job role for tasks and competences.
- Competences – a detailed list of all competences that will be required to carry out the tasks.

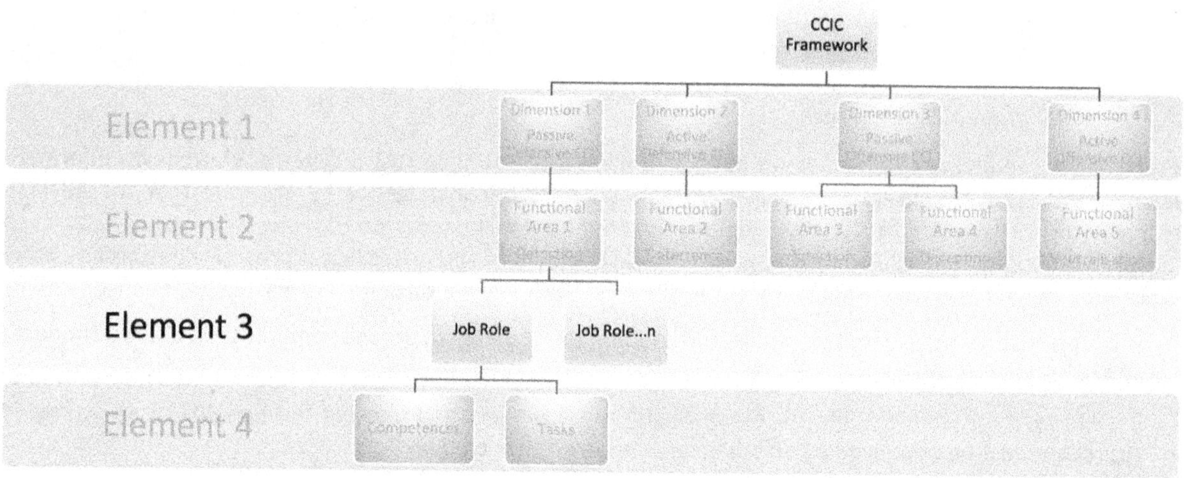

Figure 5: Element 3 of the CCIC framework - the job roles

3.3.2 Levels of proficiency

Levels of proficiency describe an individual's level of expertise for a particular job role. The levels of proficiency are essential for the successful implementation of the CCIC Levels of proficiency. Furthermore, proficiency levels represent the degree of the development of competences and the complexity of the tasks (Carretero, et al., 2017). The levels of proficiency are defined per competence per each job role. The CCIC Framework proposes five levels of proficiency described in Table 2. The levels of proficiency presented in Table 2 draw from those advanced in some other frameworks (SFIA, 2021; Bashir & Miyamoto, 2020; Department of Higher Education and Training, 2016; United Nations Population Fund, 2003)

Table 2: Five levels of proficiency

Proficiency Level	Description
Level 1: Awareness - All employees, including management	The learning point involves being cognizant or knowledgeable of the field and the required skills without practical experience. Follow instructions or others.
Level 2: Foundation - Junior Practitioner	Acquired basic knowledge and understanding. Ability to assist, demonstrate or apply basic knowledge and skills for comprehensible tasks. Work under the direct supervision and focus on enhancing knowledge and skills.
Level 3: Intermediate - Practitioner	Ability to apply knowledge and skills to basic tasks without supervision and complex tasks with limited supervision. Ability to work independently.
Level 4: Advanced - Senior Practitioner	Acquired a deep understanding of the knowledge associated with the skill. Ability to apply advanced knowledge and skills in a range of complex tasks with no supervision. Capable of leading, advising, initiating, and influencing any task. A seasoned practitioner with a track record and the ability to make decisions and take responsibility.
Level 5: Specialist /expert	Acquired broad and deep knowledge, skills, and experience to apply in extensive and diversified circumstances. Deep understanding of the implications associated with the field and industry. Has authority and is accountable for all functions and decisions. Lead, advise, initiate, and influence implementations, innovations, transformation, and developments. Has a sustained track record that leads to extensive recognition in the field and industry.

3.4 Element 4: Competence and tasks

The last element of the CCIC Framework provides an overview of the competences (knowledge, skills, abilities, and attitudes - KSA) and the tasks required to perform a specific job role. Numerous specific tasks need to be completed in a particular job role. The element is also about identifying competences (KSA) required to execute tasks within a particular CCI job role and deliver superior performance. The competences are acquired through learning and development. Competences for each job role can be used as a guideline to design and develop competence-based education, training and development curriculum required to be competent in the job role (Parsona, et al., 2018). To achieve effective performance: cognitive competence, functional competence, social competence are the guide in identifying or developing a set of competences (KSA) for each job role. This relation can be expounded as follows:

- Knowledge: relates to the set of cognitive competences required for a specific job role in conjunction with the ability to apply the knowledge effectively. Knowledge (theoretical, facts, concepts, and information) must be known and understood to accomplish job roles and tasks effectively.
- Skills: pertains to the set of functional competences required to execute the tasks of the specific job role effectively, and the tasks can be effortlessly demonstrated. Skills are about how to use the knowledge gained.
- Attitude: denotes an individual's mindset to execute and yield outstanding performance. Attitudes refer to an individual's way of thinking, belief, or feeling that they can use knowledge and skills gained to perform tasks effectively. Attitude relates to social competences and influences ability, motivation, commitment, confidence, responsibility, performance, and adaptability.

To have outstanding performance and a competent workforce requires complete continuous integration of these three competences. The cognitive and functional competences correspond to the professional aspect, as they are about the functional expertise in the job role. Social competence corresponds to the personal aspect as they have to do with personal attitude, behavioural traits, and motives (Arifin, et al., 2017).

The addition of competence and tasks, is the fourth element that completes the CCIC framework, is graphically depicted in Figure 6.

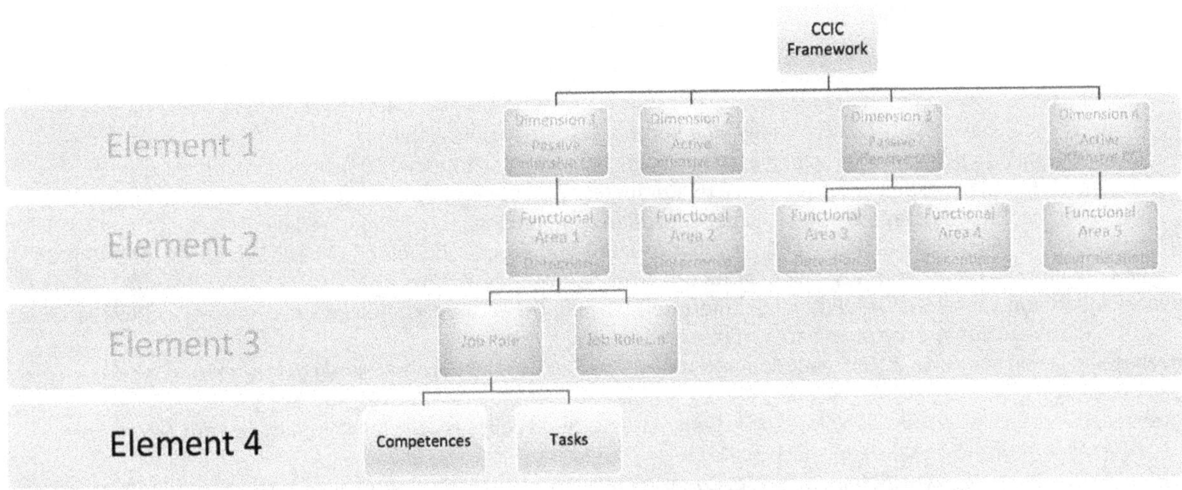

Figure 6: Element 4 of the CCIC Framework - competences and tasks

This section presented an overview of the CCIC Framework, a construct of four elements. These elements have a symbiotic relationship to achieve effective performance, and each element is integral to the successful implementation of the CCIC Framework. The four elements are Dimensions, Functional Areas, Job Roles, and Competences (KSA) and Tasks.

4. Conclusion

Cyber counterintelligence is gaining momentum in both the public and private sectors. A skilled workforce is required for the effective execution of CCI to counter advanced and unprecedented cyber threats. A comprehensive competence framework is essential for CCI workforce development at all CCI employment or execution levels to create such a skilled workforce.

This paper gave an essential introduction to the CCIC Framework and outlined four elements. The first and second elements have been identified using the four principles of CI. Subsequently, the CCIC Framework was progressed and expanded with the addition of job roles (element three) as well as competences and tasks (element four). Ongoing research is focused on detailing all of the CCIC frameworks elements. The CCIC Framework can be implemented by any type and size of a public or private sector organisation. The CCIC Framework will assist in identifying competence gaps, designing and developing competence-based education, training, and development curricula, providing career progression guidance, and developing CCI practitioners with the requisite knowledge, skills, and attitude for optimising CCI performance. Future research will explore the following job roles, task analysis and competence, and legal ramifications of attacking back and neutralisation.

Acknowledgements

This research benefitted, in part, from support from the Faculty of Science at the University of Johannesburg.

References

Apollo Education Group, 2015. Competency Models for Enterprise Security and Cybersecurity Research-Based Frameworks for Talent Solutions, University of Phoenix: Apollo Education Group, Inc..

Arifin, M. A., Rasdi, R. M., Anuar, M. A. M. & Omar, M. K., 2017. Addressing Competency Gaps for Vocational Instructor through Competency Modelling. International Journal of Academic Research in Business and Social Sciences, 7(4), pp. 1201-1216.

Bashir, S. & Miyamoto, K., 2020. Digital Skills: Frameworks and Programs, Washington: International Bank for Reconstruction and Development / The World Bank.

Black, J. M., 2014. The Complexity of Cyber Counterintelligence Training, Unpublished Thesis Submitted for Master of Science in Cybersecurity, Utica, New York: Utica College.

Carretero, S., Vuorikari, R. & Punie, Y., 2017. DigComp 2.1 The Digital Competence Framework for Citizens With eight proficiency levels and examples of use, s.l.: Luxembourg: Publications Office of the European Union.

Check Point, 2018. the Generation Cyber Attacks are Here and Most Businesses are Behind: A New Model For Assessing and Planning Security. [Online] Available at: https://www.checkpoint.com/downloads/product-related/whitepapers/preventing-the-nextmega- cyber-attack.pdf [Accessed 3 December 2021].

CIISEC (IISP), 2010. *IISP INFORMATION SECURITY SKILLS FRAMEWORK.* [Online] Available at: https://apmg-international.com/sites/default/files/documents/products/iisp_skills_framework_v1_0.pdf [Accessed 19 March 2021].

CWA, 2014. European e-Competence Framework 3.0: A common European framework for ICT Professionals in all industry sectors, s.l.: European Committee for Standardization (CEN) Workshop Agreement (commonly abbreviated CWA).

CyBOK, 2019. *The Cyber Security Body of Knowledge Version 1.0.* [Online] Available at: https://www.cybok.org/media/downloads/CyBOK-version-1.0.pdf [Accessed 19 February 2020].

Department of Higher Education and Training, 2016. Competency Framework for Career Development Practitioners in South Africa, Pretoria: Department of Higher Education and Training.

Duvenage, P. 2019. A conceptual framework for cyber counterintelligence, Unpublished Thesis Submitted for PhD (DCom) in Computer Science, University of Johannesburg, South Africa.

Duvenage, P. & von Solms, S., 2014. Putting Counterintelligence in Cyber Counterintelligence. University of Piraeus, Greece, 14th European Conference on Cyber Warfare and Security.

Duvenage, P. & von Solms, S., 2015. Cyber Counterintelligence: Back to the Future. Journal of Information Warfare, 13(4), pp. 42-56.

Duvenage, P., Jaquire, V. & von Solms, S., 2020a. Cyber Counterintelligence Matrix for Outsmarting Your Adversaries. Journal of Information Warfare, 19(1), pp. 1-11.

Duvenage, P., Jaquire, V. & von Solms, S., 2020b. Counterintelligence: Some Contours towards the Academic Research Agenda. Chester, UK, 19th European Conference on Cyber Warfare and Security ECCWS 2020.

Duvenage, P. Jaquire, V.& von Solms, S., 2019. Towards a literature review on cyber counterintelligence. Journal of Information Warfare, 17(2), pp. 1-12.

Duvenage, P., von Solms, S. & Corregedor, M., 2015. The Cyber Counterintelligence Process - a Conceptual Overview and Theoretical Proposition. University of Hertfordshire, Hatfield, Proceedings of the 14th European Conference on Cyber Warfare & Security, 2-3 July 2015.

Griffiths, B. & Washington, E., 2015. Competencies at Work Providing a Common Language for Talent Management. New York, NY: Business Expert Press, LLC.

Heckman, K. E. et al., 2011. Cyber Denial, Deception and Counter Deception A Framework for Supporting Active Cyber Defense. Advances in Information Security, Volume 64, p. 2011.

ITU, 2021. easuring digital development Facts and figures 2021. [Online] Available at: https://www.itu.int/en/ITU-D/Statistics/Documents/facts/FactsFigures2021.pdf [Accessed 2 December 2020].

Jaquire, V., Duvenage, P. & Solms, S. v., 2018. Building the Ideal Cyber Counterintelligence Dream Team. University of Oslo Norway , 17th European Conference on Cyber Warfare and Security ECCWS 2018. 28 29 June 2018 .

Jelen, S., 2020. Cyber Counterintelligence: When Defense Alone is No Longer Sufficient. [Online] Available at: https://securitytrails.com/blog/cyber-counterintelligence [Accessed 4 July 2020].

Jensen, E. T., 2012. Cyber Deterrence. Emory International Law Review, 26(2), pp. 773-824.

Kansal, J. & Singhal, S., 2018. Development of a competency model for enhancing the organisational effectiveness in a knowledge-based organisation. International Journal of Indian Culture and Business Management, 16(3), pp. 287-301.

Kuloğlu, G., Gül, Z. & Erçetin, Ş. Ş., 2014. Counterintelligence as a Chaotic Phenomenon and Its Importance in National Security. In: Chaos Theory in Politics. Dordrecht: Springer, pp. 178-188.

Lee, R. M., 2015. Cyber Intelligence Part 4: Cyber Counterintelligence From Theory to Practices, s.l.: Tripwire.

Li, Y. & Liu, Q., 2021. a comprehensive review study of cyber-attacks and cyber security; Emerging trends and recent developments. Energy Reports, Volume 7, pp. 8176-8186.

Meier, R. et al., 2021. Towards an AI-powered Player in Cyber Defence Exercises. Tallinn, Estonia, 2021 13th International Conference on Cyber Conflict Going Viral.

NIST, 2017. *National Initiative for Cybersecurity Education (NICE) Cybersecurity Workforce Framework.* [Online] Available at: https://www.nist.gov/file/359276 [Accessed 2 July 2018].

Parsona, L., Childs, B. & Elzie, P., 2018. Using Competency-Based Curriculum Design to Create a Health Professions Education Certificate Program the Meets the Needs of Students, Administrators, Faculty, and Patients. Health Professions Education, Volume 4, pp. 207-218.

Petrillo, A., De Felice, F., Cioffi, R. & Zomparelli, F., 2018. Fourth Industrial Revolution: Current Practices, Challenges, and Opportunities. In: Digital Transformation in Smart Manufacturing. s.l.:BoD – Books on Demand, pp. 1-20.

Prunckun, H., 2019. Counterintelligence Theory and Practice. Lanham, Maryland: Rowman & Littlefield Publishing Group, Inc.

Public-Private Analytic Exchange Program, 2019. Geopolitical Impact on Cyber Threats from Nation-State Actors. Commodification of Cyber Capabilities A Grand Cyber Arms Bazaar. [Online] Available at: https://www.dhs.gov/sites/default/files/publications/ia/ia_geopolitical-impact-cyber-threatsnation- state-actors.pdf [Accessed 2 July 2020].

SFIA, 2021. *Skills Framework for the Information Age SFIA 8 The framework reference.* [Online] Available at: https://sfia-online.org/en/sfia-8/documentation/sfia-8-the-framework-reference-v8-0-sfiaref-en-210928.pdf/@@download/file/SFIA%208%20The%20frame [Accessed 11 January 2022].

Sims, J. E., 2009. Twenty-first-Century Counterintelligence The Theoretical Basis for Reform. In: J. E. Sims & B. Gerber, eds. Vaults, Mirrors, and Masks: Rediscovering US Counterintelligence. Washington, D. C.: Georgetown University Press., pp. 19-50.

Soesanto, S. & Smeets, M., 2020. Cyber Deterrence: The Past, Present, and Future. In: F. Osinga & T. Sweijs, eds. NL ARMS Netherlands Annual Review of Military Studies 2020 Deterrence in the 21st Century—Insights from Theory and Practice. The Hague: T.M.C. ASSER PRESS, pp. 385-400.

Stech, F. J. & Heckman, K., 2018. Human Nature and Cyber Weaponry: Use of Denial and Deception in Cyber Counterintelligence. In: Cyber Weaponry, Advanced Sciences and Technologies for Security Applications. Cham: Springer, pp. 13-27.

US Army Publishing Directorate, 2009. Army Publishing Directorate. [Online] Available at: https://armypubs.army.mil/ProductMaps/Pubform/Details.aspx?PUB_ID=85844 [Accessed 2 July 2018].

United Nations Population Fund, 2003. Competency Framework, s.l.: Office of Human Resources United Nations Population Fund.

The Cyber Era`s Character of War

Maija Turunen
Finnish National Defence University, Helsinki, Finland
maijaturunen@yahoo.com

Abstract: The nature of war is often considered unchanged, although in the cyber era the concept of war, weapon, and fighter have become blurred. Instead, the character of war is constantly changing and is always unique. The character of war is not similar at different battle domains or levels of warfare, which complicates the course of war. A serious deviation from a strategic-level perception of war character in relation to an operational or tactical level perception character of war can result in defeat. The fog of war has intensified, although the situational awareness of conventional battlefields has clarified due to advances in technology. Technology is a key factor in shaping the war character of the cyber era, depending on the point of view, in 4th or 5th generation warfare. The nature of the next generation warfare and the formation of the character of war may be determined by the Artificial Intelligence or other Emerging and Disruptive Technologies, which itself develops and uses technology or some other technology, not yet known to us. This paper seeks to find factors that influence the formation of the cyber era`s war character and its transformation in Western and Russian military thinking. The aim is to describe the opportunities and challenges associated with the use of advanced technology in the military purpose. This review is based on the NATO`s and the Russia`s strategy papers. Theoreticallly, this paper draws on the theory of the character of war, which is applied to the question under study through the theory of strategic culture. An integrative literature analysis has been used as the research method. The key findings of the paper are that Russia and the West share the view that a warlike battle is already under way in the cyberspace. That requires an faster and better capacity to utilize advanced technologies as part of or in support of weapons systems. Russia and the West are struggling with the moral, legal, and technical problems associated with the use of advanced technology, but are aware of its necessity in the cyber warfare.

Keywords: character of war, cyber warfare, Russia, NATO, artificial intelligence

1. Introduction

The motives behind to prepare for war and the warfare: honour, fear, and interests (Chance, 2012, 13), and some features, like as dominant role of policy and strategy, psychological factors, irrationality, violence and uncertainty (Vego, 2011, 64) have remained the same and universal throughout the history of warfare. The nature of war is often considered unchanged, although in the cyber era the concept of war, the weapon, and the fighter have become blurred.

The character of war is not stable but in constant change. The character of war is changing with the development of an international and interactive political process and technology driven and conditioned by social and economical changes. The character of war is tied to its creator: knowledge of what is available, the ability to interpret and utilize this knowledge. The character of war is a position and culture bound concept that can be shaped for example through strategic communication, reflexive control or information operations. Because the war is always a unique, circumstantial, and dynamic activity between two or more parties, involving the unpredictable changes, each party to the war has its own character of war. (Gray, 2010, pp. 12-13; Vego, 2011, 61, 64; Gerasimov, 2013) The character of war is not similar at different battle domains or levels of warfare, which complicates the course of war.

The history of modern warfare can be divided into eras or generations, at the beginning of which there was a significant change in the way a war was waged and in the character of war. Modern warfare can be considered to have begun in the mid-17th century, when the Treaty of Westphalia sought to create a straightforward and systematic means of war based on nation states. The first generation of warfare was characterized by mass armies, mobile footsteps, and static fire stations. The second-generation warfare can be considered to have begun in the 19th century, when the motorization of the troops, the development of weapons and communication systems allowed for faster movement and fire, and smaller troop divisions. Something about the pace of change in the development of warfare is that the third-generation warfare can be considered to have begun as early as the 20th century, when the advanced aeronautical technology made it possible to strike at enemy targets faster and over long distances. The fourth-generation warfare was characterized by an expanding range of means of asymmetric strategies and tactics designed to challenge the values and social system of the adversary using information, psychological, and lawfare methods. (Paronen, 2016; Lind & all, 1989, 23; Ahvenainen, 1994, 96-97)

The fifth-generation warfare is marked by its invisibility, which blurs and creates uncertainty about the difference between war and peace. The focus is on non-kinetic interference, such as information and cyber operations, which seek to undermine the non-military resilience of the adversary and obscure the adversary's situational awareness, as well as deny the adversary right to use military force. (Paronen, 2016; Abbott, 2010, 20). The strong and fast development of information, cyber, autonomous and hypersonic weapons, as well as artificial intelligence (AI), quantum computing, and robotics, can also be considered in the fifth-generation warfare, which current phase could also be called the cyber era`s warfare. Even the term "sixth-generation warfare" (Slipchenko, 2013) has been used in the Russian scientific debate. However, the sixth-generation warfare is counted as the New Generation Warfare. (Bērziņš, 2019, 167, 170, 176).

There is no broad consensus on the division of the strategic, operational, tactical and technical levels of warfare into generations or what stage they are at. However, it can be seen that generational change in warfare involves key ideas common to all changes, such as changes in perception on the battlefield, changes in the mobility and the speed of hostilities, and the goals (enemy forces, political system, critical infrastructure), and the objectives (destroy vs. collapse). (Lind & all, 1989, 23; Paronen, 2016). Gray (2010, 11-12) has listed five significant changes in the contexts that shape contemporary war and strategy. They are: 1) The development of cyber power (all future wars will harbor integral cyber warfare); 2) Space warfare; 3) The rise of a global electronic media with real-time access to events; 4) An information-led revolution in the military affairs (RMA); and 5) Belligerents and irregular warfare.

The era of cyber warfare can be considered to have begun in the 1960s with the proliferation of computers and the invention of the Internet. The cyber warfare is affected by the general functional dimensions of warfare, such as human, technology, organizational skills, logistics, knowledge, doctrines, time, space, and energy. In the cyber domain, there is an ongoing struggle between attackers and defenders. The durability and extensibility of the "Red Lines" set by the states are tested all the time. The success of this struggle in the militarization of information and new technologies, especially artificial intelligence, acts as a game changer (Edmonds, 2021, 79-83).

1.1 Theoretical background

This research is based on the application of the theory of character of war. The character of war can be defined as follows: The character of war means the common perceptions in the international system of the nature, needs and possibilities of the use of armed forces, as well as the effective principles and operating models of the armed forces. In the theory of character of war, war was viewed as a pragmatic and changing phenomenon. The war character is connected in the international system and security environment, as well as in operational logic, strategic communication, rules, and influence of the new technological advances. It also constructs association with identities of the actors. Through the character of war, the creator of the character of war seeks to outline the prevailing military threats against which one must be able to wage war; the situation in which war may be waged; the methods by which war is to be fought; the factors that can be used to increase credible military power; and the objectives of warfare. (Raitasalo – Sipilä, 2008, 9; Vego, 2011, 64; JDN 1-8, 2018, I-4).

The theory of strategic culture plays a central role in understanding Russian art of warfare, but is also suitable for explaining the factors behind the Western perception of the character of war. The strategic culture can be explained as a set of persistent and consistent historical patterns of how the state`s leadership thinks about the use of force to achieve political goals. The preferences originate in the historical experiences related to the threat and use of force by the state and are influenced by the philosophical, political, cultural, and cognitive experiences and characteristics of the state (Kari, 2019, 71; Johnston, 1995).

Integrative literature analysis were used as the research method. By integrating and analysing the strategic documents and literature on the current perception to future cyberwarfare, the aim is to consider what kind of factors influence the formation of the war character of the cyber era and its transformation in Western and Russian military thinking. The purpose is to position the examination of the research question as part of the scientific debate on the topic.

2. Russian perception of the character of cyberwar

The National Security Strategy (2021), the Russian Military Doctrine (2014), the Information Security Doctrine (2016) and National Strategy for the Development of Artificial Intelligence for the period until 2030 (2019) are the main official public documents outlining the formation of the Russian perception about the character of cyber war and warfare. They highlight in particular the various threats as well as identify and provide to the authorities, including army, and obligation to respond vigorously to these threats in all domains. They also include clear measures for managing the cyber environment and consider the entire world as a theatre of operations and information space as the battlefield, not just the information space under Russian control. In other words, these documents are used by Russia to legitimize its own actions (Bērziņš, 2014, 3; Thomas, 2016, 22). Strong legalism is emphasized as part of Russian society in its power structures and their supporting activities. Unlike most Western countries where the powers of the authorities are limited by the law, in Russia the law is confirmed obligations to the public authorities from doctrines and strategies, while giving them powers to carry out these tasks.

Russia's new National Security Strategy (NSSRF) can be described as a manifesto, a defiant declamation to the rest of the world and a narrative for citizens emphasizing to victimizing and to sacrifice for Russian sovereignty and traditional values. The strategy explicitly states that: "Space and information space are being actively explored as new spheres of warfare." Development of a safe information space, protection of Russian society from destructive information and psychological impact are mentioned as one of Russia's national interests and strategic national priorities. (NSSRF, 2021, 4-6, 8). Particular attention is paid to the timely consideration of trends in the changing nature of modern wars and armed conflicts, the creation of conditions for the fullest realization of the combat capabilities of troops (forces), the development of requirements for prospective formations and new means of armed combat and ensuring the technological independence of the defense-industrial complex of the Russia. (NSSRF, 2021, 11-14)

Russia accuse foreign states for computer attacks, resistance to their initiatives in the field of international information security and activities of special services to conduct reconnaissance and other operations in the Russian information space. Also Russia complains, that armed forces of foreign states are practicing actions to disable critical information infrastructure facilities of Russia. Russia intends to fight against such activities, e.g. by the development of forces and means of information confrontation and improvement of means and methods of information security based on the use of advanced technologies, including artificial intelligence and quantum computing technologies. And finally, the Strategy states ominously: "The Russian Federation considers it legitimate to take symmetric and asymmetric measures necessary to suppress such unfriendly actions and to prevent their recurrence in the future." (NSSRF, 2021, 19-22, 39)

The General Staff Commander of the Russian Army, General Gerasimov, has stressed the importance of researching the nature of modern warfare, military and non-military means of waging war, and the problems of strategic deterrence, which means finding ways to prevent hybrid pressure and the ability to maintain a strategic initiative for the possibility of admission (Thomas, 2016, 18-19; Krasnaya Zvezda, 2019). Sukhankin (2019, 332) considers with reference Gerasimov (2016), the militarization of information as Russia's information strategy the most important pillars. Worth noting is that Russia considers cyber and electronic warfare capabilities to be part of information security and Russia'swde21 Anti-Access/Area-Denial (A2/AD) strategy. The difference between technological and psychological information control is clear, but both are crucial to achieving the goals already in the initial period of war by taking control of the adversary's information space. Where Western countries emphasize the freedom of information, the right to information, the protection of privacy, and information as the key to truth, in Russian thinking, the information weapon is a weapon like others are and the moral-based and self-restraining attitude of the Western states towards information opens up vulnerabilities that can be exploited through information operations (Kucharsky, 2018, 2).

Russian military thinking also involves creating an alternative reality or realities. The idea is that in a state under threat of war, society's support for the state's strategic goals - in other words, the legitimacy of war - is essential to achieve a victory. The Russians have placed the idea of influence on the center of its operative planning. Examples of such influence are both internal and external communication, deception operations, and psychological operations. (Bērziņš 2019, 166-167, 170-171). Also Kasapoglu (2015, 5-6) estimates that Russia's hybrid warfare is aimed at creating a fog of hallucinatory warfare and it consists of consistent delusions not intended to paralyze Western intelligence and proactive capabilities, but to changes Western analysis results and perceptions of Russia's strategic intentions.

The Russian Military Doctrine (2014) sees that there are military dangers and threats are increasingly moving into the information space and information acts as a justification for military action. The use of cyber methods is suitable for the implementation of the basic principles of Russian operative thinking, like surprise and confusion. These methods have subtle nature and they left space for speculation. The use of cyber methods and the need to protect against them are most evident in the Information Security Doctrine (2016), which defines that Russia's national interests in the information space contain e.g. maintaining continuous and smooth information operations in information infrastructure.

Unlike Russia's Information Security Doctrine, their National Strategy for the Development of Artificial Intelligence for the period until 2030 (2019), despite its ambitious goals, is quite modest when it comes to utilizing AI for the military use. Robotics and unmanned vehicle control are mentioned in related areas of the use of AI (NSDAI, 2018, 4). The Russian Artificial Intelligence Strategy also includes the principles which are obligatory during the implementation of that strategy: a) the protection of human rights and liberties; b) security; c) transparency; d) technological sovereignty; e) innovation cycle integrity; e) reasonable thrift; and g) support for competition (NSDAI, 2018, 7-8). Russia's military thinking recognizes the risks associated with the development and use of AI, but considers it inevitable that some military systems will become completely autonomous. The Syrian war provided a good opportunity for Russia to test its autonomous weapons systems and a concept of limited action warfare beyond its borders. (Edmonds, 2021, 80, 115; McDermott, 2019).

Russia's strategy papers emphasize the goal of protecting Russia's world of values and ideas, as well as culture, and creating a unified character of Russia as a strong nation. In this context, the information confrontation and psychological warfare play a key role, highlighted by the effective use of AI and other advanced technologies (Edmonds, 2021, 82-83). According to Bērziņš (2019, 165-166): "The Russian view of modern warfare is based on the idea that the main battlespace is the mind. As a result, new-generation wars are to be dominated by information and psychological warfare in order to achieve superiority in troops and weapons control, morally and psychologically depressing an enemy's armed forces personnel and civilian population." As in his speech in 2017, Gerasimov (Komsomolskaya Pravda, 2017), seems to focus on the fight in the minds of the citizens, the sixth dimension of battlefields.

In summary, it can be said that the lines between war and peace and between defensive and offensive methods have been deliberately blurred in Russian military thinking (Foxall, 2021, 18). Unlike in the West, where peace is in principle considered a normal interstate situation, in Russian thinking wars and armed conflicts will continue uninterrupted. (Giles, 2021, 16-17). While Russia fears the use of asymmetric measures to influence its citizens, it also favours the use of these measures in creating a fog of war and influencing its adversaries. Information confrontation, information weapons (including cyber weapons utilizing AI) and informational-psychological operations has a centric position in Russian asymmetric warfare strategic planning.

3. Western perception of the character of cyberwar

The Strategic Concepts (2010, will be updated on 2022), the Emerging and Disruptive Technologies Coherent Implementation Strategy (2021), the NATO Warfighting Capstone Concept (NWCC, 2021) and the NATO Artificial Intelligence Strategy (2021) are key documents in mapping NATO's public stance on the character of war of the cyber era.

NWCC is adversary-centric and designed to acting across three operational contexts: shaping, contesting and fighting. It sets out five Warfare Development Imperatives: 1) Cognitive superiority; 2) Layered resilience; 3) Influence and power projection; 4) Integrated multi-domain defence; and 5) Cross-domain command (Tammen, 2021). The purpose of the NWCC is to create a vision for Alliance Warfare Development up to 2040 to allow the Alliance to protect NATO's core security interests in the future (Sweijs & all, 2020, 2).

A specialists are estimated, that Emerging and Disruptive Technologies (EDTs) are a challenge but also opportunity for NATO and the alliance to achieve dominance in key EDTs must be a strategic priority for the Alliance: "EDTs will disrupt, degrade and enable NATO military capabilities in the 2020-2040 timeframe. Such characteristics of modern technologies are drivers of the current evolution and revolution in data, AI, autonomy, space, quantum, hypersonics, biotechnologies and materials. Alone or in combination, they define the technological edge necessary for NATO's operational and organisational effectiveness." (Reflection Group, 2020, 13, 39).

The new technologies will change the nature of warfare, and enable new forms of attacks (Reflection Group, 2020, 16, 18, 31). NATO sees interoperability, a mix of old legacy systems and new weapon systems, the techno-

policy, legal and ethical issues as a challenge in the use and further development of AI and other EDTs (NSTO, 2020, 26, 39, 55). NATO's Artificial Intelligence Strategy (2021) identifies these challenges and sets out common principles to which the NATO and its Allies have committed themselves in developing and to use of AI and its applications: 1. Lawfulness; 2. Responsibility and Accountability; 3. Explainability and Traceability; 4. Reliability; 5. Governability; and 6. Bias Mitigation (NATO, 2021). These principles emphasize responsibility and, more broadly, Western values and norms in the theatre of future warfare, where AI offers immeasurable opportunities. As an example of these opportunities, AI can be used to enhance intelligence, surveillance and reconnaisance capacity in continuously ongoing Command and Control operations. The capability of advanced AI systems to collect, analyze, generate, and manipulate information opens new posibilities for information superiority at all levels of operational decision-making. Advanced artificial intelligence systems can create their own killing chains, mutable cyber weapons or even rewrite themselves, thus will be revolutionizing military operations at all levels. In the flexible defence or response cyber operations, the ability of AI to generate unique effects, randomly select locations for launching surprise operations and generate various responses, makes it difficult for an attacker to determine what kind of and where the countermeasures are needed. (Edmonds, 2021, 82; Chen, 2017, 104-105).

Artificial Intelligence and Robotic Autonomous Systems (RAS) are already here, but their advanced application for military purposes could give to states a significant military advantage, revolutionise military and strategic affairs and change the character of war (Tonin, 2019, 4). There are countless opportunities in the military sector to exploit AI and RAS. These can be used, for example, to facilitate autonomous and remote operations, operations both on physical and virtual A2/AD zones, to intensify the informed military decision making at all levels, to improve the situational awareness and resource management, and to increase the speed and scale of military action (Gray and Ertan, 2021, 22; Sayler and Hoadley, 2020; Tonin, 2019, 8).

But, there are also many challenges to development and use the advanced technologies to military purpose: Firstly, as with all weapons systems, there are ethical, political and legal issues, and especially when dealing with the international law and politics, issues of trust; secondly, there will be significant technological challenges: AI technology and systems use by that, need to be integrated into existing systems and ensure their interoperability. AI may be unpredictable or vulnerable to unique forms of manipulation or human based programming errors and cyber attacks. AI systems are brittle, opaque, and reliant on good data, and any failure in an AI enabled military system could have catastrophic consequences; and thirdly, challenges may be caused by a lack of financial and intellectual resources in the state seeking to exploit AI. The development of AI and related systems not only requires skilled and innovative people, but also changes the strategic military thinking and the allocation of available resources. (Gray and Ertan, 2021, 22; Sayler and Hoadley, 2020; Tonin, 2019, 6-8)

Western experts estimate that tomorrow's conflict will be characterized by the widening of the battlespace, the fusion of dimensions and the rise of borderless warfare. Future wars will include societal warfare (focused on disrupting and coercing societies) and cognitive warfare (focused on creating civilian disorder), alongside high-end conventional wars and wars fought by proxy (Sweijs & all, 2020, 3-4). However, the perception of the character of war is still hampered by traditional thinking about the distinction between peace, crisis and conflict paradigm (Tammen, 2021) An ability to project hard power will retain its place in the multi-domain warfare of the future. In multi-domain warfare, it is also necessary to develop A2/AD zones in order to reflect military strength in controversial areas based on technological advancements. Strategically future warfare is required for example cognitive superiority, full-spectrum engagement, and agile ways of adaptation. (Sweijs & all, 2020, 9, 11).

4. Conclusions

An examination of both Russia's and NATO's strategy documents shows that the use of AI and other EDTs in military activities and weapon systems, political will, and military strategic thinking, can been seen as key factors that will influence the formation of the character of war of the cyber era.

NATO and Russia both emphasize the importance of cyber methods, and in particular the development and exploitation of the opportunities offered by AI and RAS now and in a future warfare. Similarly, the creation of A2/AD zones and the ability to extend the battle to the adversary's A2/AD zones are seen as a significant factor in the military strategic thinking of both actors in cyber warfare. Both actors also recognize the key integration problem associated with the development of AI and RAS. New technology-based C5ISR and weapons systems should be able to be used integrated or at least in parallel with older systems, which are often expensive to be replaced and designed to have a long life cycle. The third common view is that there is already a war-like struggle in cyber domain and the permanent and effective ability to perform different and various levels of cyber operations is essential. This means that the use of advanced technology in surveillance and intelligence systems or in data collection and analysis alone is not enough. Operational efficiency also requires allowing the advanced technology to develop, to use and to real-time operate, for example, self-correcting and mutation-capable cyber weapon systems.

One of key difference between those actors is in the strategic communication, in the national narrative: Russia emphasizes the threat posed to the Russian people by the West and the readiness for sacrifices required to protect from it, while the West believes that citizens are safe and their information rights and freedoms must be protected and promoted. NATO and Russia share respect for human rights and security as common values in developing of AI and RAS, but otherwise the emphasis of values differs from emphasizing NATO's responsibility to promoting Russia's digital sovereignty goals. Both sides have their own values, their protection and the commitment of their citizens, as a basis for preparing for the war of the future. Indeed, the sixth generation war may be resolved by the minds of the people (soldiers, citizens and politicians): what kind of technology and what kind of weapon systems they are willing to accept and to use.

References

Abbott, D. (2010). The Handbook of Fifth-Generation Warfare. Nimble Books.

Ahvenainen, S. (1994). Sodankäynnistä, elektroniikasta ja elsosta. Tiede Ja Ase, 52(52), pp. 91–139. https://journal.fi/ta/article/view/47764.

Bērziņš J. (2014) Russia`s New Generation Warfare in Ukraine. Implications for Latvian Defence Forces, National Defence Academy of Latvia Center for Security and Strategic Research, 2014. http://www.naa.mil.lv/~/media/NAA/AZPC/Publikacijas/PP%2002-2014.ashx.

Bērziņš J. (2019) Not 'Hybrid' but New Generation Warfare. https://jamestown.org/wp-content/uploads/2019/02/Russias-Military-Strategy-and-Doctrine-web.pdf?x29008&x87069.

Chance, A. (2012) Motives Beyond Fear: Thucydides on Honor, Vengeance, and Liberty. https://dlib.bc.edu/islandora/object/bc-ir:101441/datastream/PDF/view.

Chen, J. (2017) Cyber Deterrence by Engagement and Surprise. PRISM 7, NO. 2, 2017. pp.101-107. https://ndupress.ndu.edu/Portals/68/Documents/prism/prism_7-2/prism_7-2.pdf.

Edmonds, J. and all (2021) Artificial Intelligence and Autonomy in Russia https://www.cna.org/CNA_files/centers/CNA/sppp/rsp/russia-ai/Russia-Artificial-Intelligence-Autonomy-Putin-Military.pdf

Foxall, A. (2021) Changing Character of Russia`s Understanding of War: Policy Implications for the UK and Its Allies. https://static1.squarespace.com/static/55faab67e4b0914105347194/t/60ad0d070095631c778111fe/1621953799713/How+Russia+Understands+War+2021.pdf.

Gerasimov, V. (2016): По опыту Сирии. 7.3.2016 Военно-промышленный курьер. https://vpk-news.ru/articles/29579.

Giles, K. (2021) What deters Russia. Enduring principles for responding to Moscow https://www.chathamhouse.org/sites/default/files/2021-10/21-09-23-what-deters-russia-giles.pdf.

Gray, C.S. (2010) War—Continuity in Change, and Change in Continuity. The US Army War College Quarterly: Parameters 40, 2. https://press.armywarcollege.edu/parameters/vol40/iss2/5.

Gray, M. & Ertan, A. (2021) Artificial Intelligence and Autonomy in the Military: An Overview of NATO Member States' Strategies and Deployment. NATO CCDCOE. https://ccdcoe.org/uploads/2021/12/Strategies_and_Deployment_A4.pdf.

Johnston, A. (1995). Cultural Realism: Strategic Culture and Grand Strategy in Chinese History. Princeton University Press 1995.

Kari, M. J (2019) Russian Strategic Culture in Cyberspace Theory of Strategic Culture – a tool to Explain Russia´s Cyber Threat Perception and Response to Cyber Threats. JYU DISSERTATIONS 122.

Kasapoglu, C. (2015) Russia`s Renewed Military Thinking: Non-Linear Warfare and Reflexive Control. Research Division – Nato Defence College, Rome – No. 121 – November 2015

Komsomolskaya Pravda, 26.12.2017: Начальник Генштаба Вооруженных сил России генерал армии Валерий Герасимов: «Мы переломили хребет ударным силам терроризма» http://archive.redstar.ru/index.php/component/k2/item/35551-my-perelomili-khrebet-udarnym-silam-terrorizma.

Krasnaya Zvezda (4.3.2019): Векторы развития военной стратегии. http://redstar.ru/vektory-razvitiya-voennoj-strategii/.

Kucharsky, L. (2018) Russian Multi-Domain Strategy against NATO: information confrontation and U.S. forward-deployed nuclear weapons in Europe. https://cgsr.llnl.gov/content/assets/docs/4Feb_IPb_against_NATO_nuclear_posture.pdf.

Lind, W.S and all (1989) The Changing Face of War: Into the Fourth Generation Marine Corps Gazette (pre-1994); Oct 1989; 73, 10; ProQuest Direct Complete pp. 22-26

McDermott, R. (2019): Russia's Military Scientists and Future Warfare. https://jamestown.org/program/russias-military-scientists-and-future-warfare/

NATO (2021) Summary of the NATO Artificial Intelligence Strategy. https://www.nato.int/cps/en/natohq/official_texts_187617.htm.

NATO Warfighting Capstone Concept (2021) https://www.act.nato.int/nwcc.

NATO Science & Technology Organization (2020): Science & Technology Trends 2020-2040 Exploring the S&T Edge. https://www.nato.int/nato_static_fl2014/assets/pdf/2020/4/pdf/190422-ST_Tech_Trends_Report_2020-2040.pdf.

Paronen, A. (2016) Onko Suomi sodassa? – Sodankäynnin viides sukupolvi. The Ulkopolitist. https://ulkopolitist.fi/2016/02/10/onko-suomi-sodassa-sodankaynnin-viides-sukupolvi/.

Reflection Group (2020) NATO 2030. United for a New Era. Analysis and Recommendations of the Reflection Group Appointed by the NATO Secretary General. https://www.nato.int/nato_static_fl2014/assets/pdf/2020/12/pdf/201201-Reflection-Group-Final-Report-Uni.pdf.

Sayler, K. M. and Hoadley, D. S. (2020) Artificial Intelligence and National Security. CRS Report R45178. https://sgp.fas.org/crs/natsec/R45178.pdf.

Slipchenko, V. (2013) "Information Resource and Information Confrontation: their Evolution, Role,and Place in Future War," Armeyskiy Sbornik (Army Journal), No. 10 2013, pp. 52-57,

Sukhankin, S. (2019) Russia's Offensive and Defensive Use of Information Security. In "Russia`s Military Strategy and Doctrine." Howard, G. and Czekaj, M (Eds.) The Jamestown Foundation, Washington, DC February 2019, pp. 302 – 342.

Sweijs, T. et all (2020) The NATO Warfighting Capstone Concept: Key Insights from the Global Expert Symposium Summer 2020. Hague Centre for Strategic Studies. https://www.jstor.org/stable/resrep26765.

Tammen, J.W. (2021) NATO Review (9.7.2021) NATO's Warfighting Capstone Concept: anticipating the changing character of war. https://www.nato.int/docu/review/articles/2021/07/09/natos-warfighting-capstone-concept-anticipating-the-changing-character-of-war/index.html.

The Doctrine of Information Security of the Russian Federation (2016). http://www.mid.ru/en/foreign_policy/official_documents/asset_publisher/CptICkB6BZ29/content/id/2563163

The Military Doctrine of the Russian Federation (2014). https://rusemb.org.uk/press/2029.

The National Security Strategy of the Russian Federation (2021). https://www.academia.edu/49526773/National_Security_Strategy_of_the_Russian_Federation_2021.

The National Strategy for the Development of Artificial Intelligence for the period until 2030 (2019). Center for Security and Emerging Technology, Trans.), October 10, 2019. https://cset.georgetown.edu/wp-content/uploads/Decree-of-the-President-of-the-RussianFederation-on-the-Development-of-Artificial-Intelligence-in-the-Russian-Federation-.pdf.

Thomas, T. (2016) Thinking Like A Russian Officer: Basic Factors And Contemporary Thinking On The Nature Of War. Foreign Military Studies Office. https://community.apan.org/wg/tradoc-g2/fmso/m/fmso-monographs/194971.

Tonin, M. (2019) 'Artificial Intelligence: Implications for NATO's Armed Forces'. NATO Science and Technology Committee Sub-Committee on Technology Trends and Security. 13 October 2019. https://www.nato-pa.int/document/2019-stcttc-2019-report-artificial-intelligence-tonin-149-stctts-19-e-rev1-fin.

U.S Joint Chief of Staff (2018) Joint Doctrine Note 1-18. Strategy. (JDN 1-8) https://fas.org/irp/doddir/dod/jdn1_18.pdf.

Vego, M. (2011) On Military Theory. Issue 62, 3 d quarter 2011 / JFQ. https://apps.dtic.mil/dtic/tr/fulltext/u2/a546600.pdf pp.60-67.

Enhancing the STIX Representation of MITRE ATT&CK for Group Filtering and Technique Prioritization

Mateusz Zych and Vasileios Mavroeidis
University of Oslo, Norway
mateusdz@ifi.uio.no
vasileim@ifi.uio.no

Abstract: In this paper, we enhance the machine-readable representation of the ATT&CK Groups knowledge base provided by MITRE in STIX 2.1 format to make available and queryable additional types of contextual information. Such information includes the motivations of activity groups, the countries they have originated from, and the sectors and countries they have targeted. We demonstrate how to utilize the enhanced model to construct intelligible queries to filter activity groups of interest and retrieve relevant tactical intelligence.

Keywords: cyber threat intelligence, mitre att&ck, stix, ttps, threat actor, knowledge representation

1. Introduction

The exponential increase in cyberattacks and sophisticated attack behavior push organizations to continuously invest in strengthening their cybersecurity posture (Brown & Lee 2021). Defenders to better respond to the current cyber threat landscape, improve their threat situational awareness, and stay resilient against cybersecurity threats have recognized the need to generate and utilize cyber threat intelligence and collaborate through information exchange to use others' experiences as their own organization's defense (Johnson et al. 2016).

There is a common understanding that for detection purposes, the value of atomic indicators like hash values, IP addresses, domain names, and host and network artifacts can vary due to their possible short lifespan and their vast amount that often leads to indicator and alert fatigue. David Bianco depicts that through a "pyramid of pain" (Bianco 2014). Bianco's pyramid of pain is shown in Figure 1 and represents the pain an adversary will suffer when a defender has denied the adversary the use of those indicators. At the apex of the pyramid of pain are Tactics, Techniques, and Procedures (TTPs) as a single grouping. When defenders detect and respond at the TTP level, they operate directly on adversary behavior. Detection at the TTP level can identify malicious activity that may not rely on prior knowledge of adversary tools and atomic indicators and can have a lasting impact on the attackers as they would have to change nearly every aspect of how they operate to avoid detection (Lee & Brown 2021).

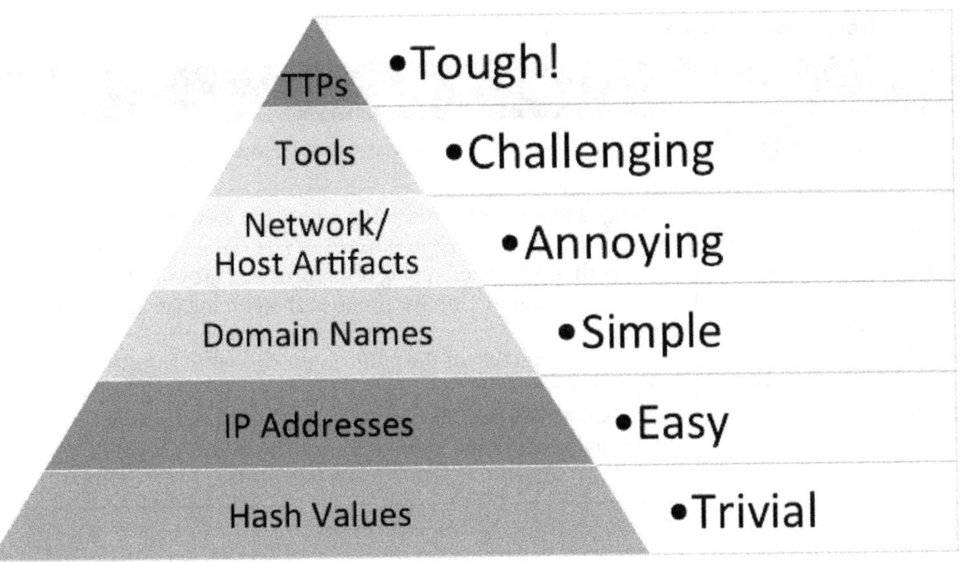

Figure 1: The Pyramid of Pain (Bianco 2014)

A globally-accessible effort that curates knowledge concerning adversary behavior is ATT&CK (Adversarial Tactics, Techniques, and Common Knowledge), led by MITRE and supported by the broader cybersecurity community via contributions. ATT&CK focuses on how adversaries compromise and operate within computer information networks. ATT&CK fundamentally is a set of taxonomies that provides defenders with a common language to map and communicate their findings pertinent to adversary behavior. It can support different areas of cyberspace defense, such as adversary emulation, behavioral analytics, cyber threat intelligence enrichment, defense gap assessment, red teaming, and SOC maturity assessment (Alexander, Belisle & Steele 2020). Furthermore, in the same project family, MITRE maintains a knowledge base of activity groups1 (referred to as Groups) and their technique use. For programmatic utilization, the entire ATT&CK knowledge base can be accessed through STIX interfaces that MITRE provides on their GitHub2.

This paper deals with enhancing the existing representation approach (model) of ATT&CK Groups in STIX 2.1 as a means of improving the amount and types of contextual information made available in a fully structured way. In particular, we focus on information that currently lies dormant in a semi-structured way in the description text field (see Table 1) of activity groups, such as group motivations, the countries they have originated from, and sectors and countries they have targeted. Using the proposed enhanced STIX 2.1 model, we demonstrate how to construct intelligible queries to filter activity groups of interest and retrieve relevant tactical intelligence.

The rest of the paper is structured as follows. Section 2 introduces the reader to ATT&CK Groups and discusses how the knowledge base is currently represented in STIX 2.1. Section 3 presents our proposed STIX 2.1 model that extends the current representation of the Groups knowledge base and discusses the set of new objects we utilize to make new types of information available. Section 4 demonstrates a use case where we use the proposed model to construct intelligible queries to filter activity groups of interest and then prioritize ATT&CK techniques that we should establish or validate controls against. Finally, section 5 concludes the paper.

2. ATT&CK Groups and their structure

Groups[3] are sets of related intrusion activity that are tracked by a common name in the security community and are aimed to collect ATT&CK techniques and software that have been reported to use. Some groups have multiple names associated with similar activities due to various organizations tracking similar activities by different names. Table 1 shows the set of properties (characterized as data items in the first column of Table 1) that formulate a single group. Briefly, a property of type "tag" points to an informational reference of a group, a property of type "field" includes information as free text, and a property of type "relationship" references an object that is directly related to the group such as ATT&CK techniques and software. Of particular importance for this research is the "description" property that, in many cases, carries information about a group's motivation, the country it originates from, as well as sectors and countries that have been targeted. As we will see in the following sections, we aim to make such information directly accessible and, consequently, queryable by extending the STIX 2.1 representation of the knowledge base.

Table 1: ATT&CK Group model (Strom et al. 2020)

Data Item	Type	Description
Name	Field	The name of the adversary group.
ID	Tag	Unique identifier for the group within the knowledgebase. Format: G####.
Associated Groups	Tag	Names that have overlapping references to a group entry and may refer to the same or similar group in threat intelligence reporting.
Version	Field	Version of the group in the format of MAJOR.MINOR.
Contributors	Tag	List of non-MITRE contributors (individual and/or organization) from first to most recent that contributed information on, about, or supporting the development of a group profile.

[1] https://attack.mitre.org/groups/
[2] https://github.com/mitre-attack/attack-stix-data
[3] https://attack.mitre.org/groups/

Data Item	Type	Description
Description	Field	A description of the group based on public threat reporting. It may contain dates of activity, suspected attribution details, targeted industries, and notable events that are attributed to the group's activities.
Associated Group Descriptions	Field	Section that can be used to describe the associated group names with references to the report used to tie the associated group to the primary group name.
Techniques / Sub-Techniques Used	Relationship / Field	List of (sub-)techniques that are used by the group with a field to describe details on how the technique is used. This represents the group's procedure (in the context of TTPs) for using a technique. Each technique should include a reference.
Software	Relationship / Field	List of software that the group has been reported to use with a field to describe details on how the software is used.

2.1 MITRE's STIX 2.1 representation of ATT&CK Groups

The STIX standard developed by the OASIS Cyber Threat Intelligence Technical Committee is a common language and serialization format used to represent and exchange cyber threat intelligence (Jordan, Piazza & Darley 2021). MITRE provides access to the ATT&CK knowledge base in STIX 2.0 and STIX 2.1. For the purpose of this research, we processed and manipulated version[4] 9 of the knowledge base in STIX 2.1. It is worth noting that the official web interface that allows users to interact with ATT&CK is built from the STIX data.

Figure 2 illustrates the set of STIX 2.1 object and relationship types leveraged to define groups that altogether comprise the ATT&CK Groups knowledge base. An "intrusion set" object (activity group or Group) connects with "attack pattern" objects (techniques/sub-techniques), "malware" objects (software), and "tool" objects (software).

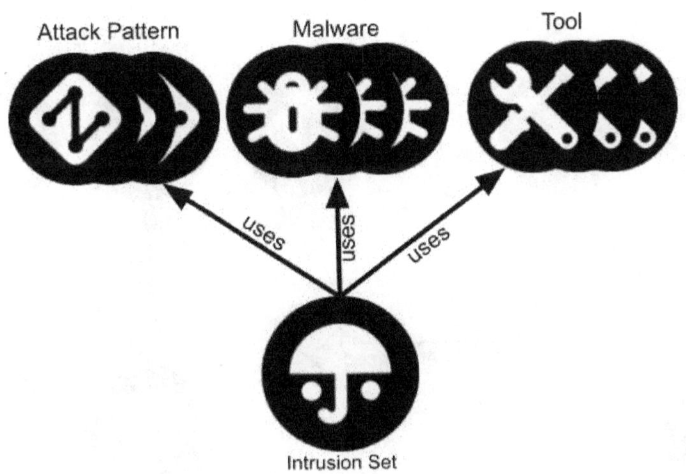

Figure 2: High-level STIX 2.1 representation of the Groups knowledge base

The intrusion set object type is used to represent activity groups. An intrusion set is a grouped set of adversarial behaviors and resources with common properties that is believed to be orchestrated by a single organization (Jordan, Piazza & Darley 2021). An attack pattern object describes the ATT&CK techniques and sub-techniques used in the intrusion set. A malware object represents commercial, custom closed source, or open-source software used for malicious purposes by adversaries (MITRE ATT&CK 2022). A tool object represents commercial, open-source, built-in, or publicly available software that could be used by a defender, pentester, red teamer, or adversary, and can include both software that generally is not found on an enterprise system as well as software generally available as part of an operating system that is already present in an environment (MITRE ATT&CK 2022).

[4] https://github.com/mitre/cti/tree/eb1b9385d44340ce867a77358c5f5aaed666e54c/enterprise-attack/intrusion-set

3. A proposed enhanced STIX 2.1 representation for the ATT&CK Groups knowledge base

As previously mentioned, this research proposes enhancements to MITRE's STIX 2.1 representation of the ATT&CK Groups knowledge base. We introduce additional types of contextual information that enable us to construct more intelligible queries and exploit the knowledge base in a more granular manner. Our proposed enhancements rely on information already part of the knowledge base but represented in a semi-structured form. In particular, information from the descriptions of activity groups, which in many cases, identify group motivations, the countries they originate from, as well as sectors and countries they have targeted. Example 1 presents such a case.

Example 1: Description of APT29 from the MITRE ATT&CK (version 9) Groups knowledge base[5].

> "APT29 is a threat group that has been attributed to Russia's Foreign Intelligence Service (SVR). They have operated since at least 2008, often targeting government networks in Europe and NATO member countries, research institutes, and think tanks. APT29 reportedly compromised the Democratic National Committee starting in the summer of 2015.
> In April 2021, the US and UK governments attributed the SolarWinds supply chain compromise cyber operation to the SVR; public statements included citations to APT29, Cozy Bear, and The Dukes. Victims of this campaign included government, consulting, technology, telecom, and other organizations in North America, Europe, Asia, and the Middle East. Industry reporting referred to the actors involved in this campaign as UNC2452, NOBELIUM, StellarParticle, and Dark Halo."

We extract the identified information and represent it in a fully structured manner using a set of distinct STIX 2.1 objects. In particular, the "location" object type is utilized to represent the suspected or confirmed country of origin of an activity group as well as targeted countries and regions. The "identity" object type is used to represent targeted sectors. Moreover, when feasible, the STIX 2.1 "intrusion set" objects are enriched with primary and secondary motivations (using the available primary and secondary motivations properties of the intrusion set object type). Figure 3 depicts the enhanced STIX 2.1 representation of the Groups knowledge base (version 9).

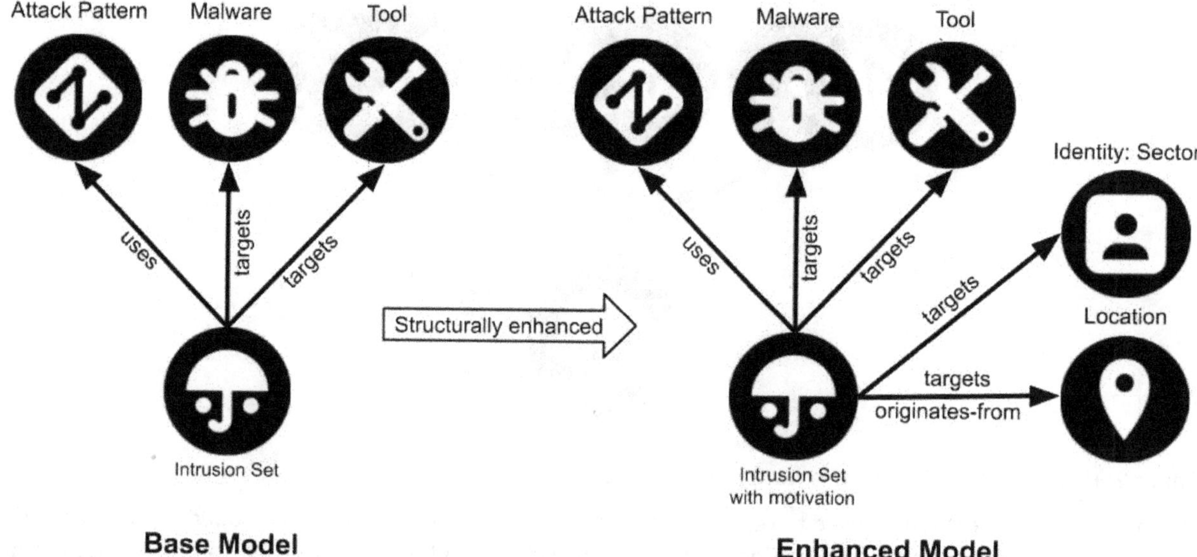

Figure 3: Enhanced STIX 2.1 representation of the ATT&CK Groups knowledge base

Similarly, Figure 4, based on our proposed model, depicts an enhanced STIX 2.1 representation of APT29 with the contextual prose presented in Example 1 transformed into structured queryable data elements. We have made the entire enhanced Groups knowledge base available on GitHub[6].

[5] https://attack.mitre.org/groups/G0016/

[6] https://github.com/fovea-research/SAG

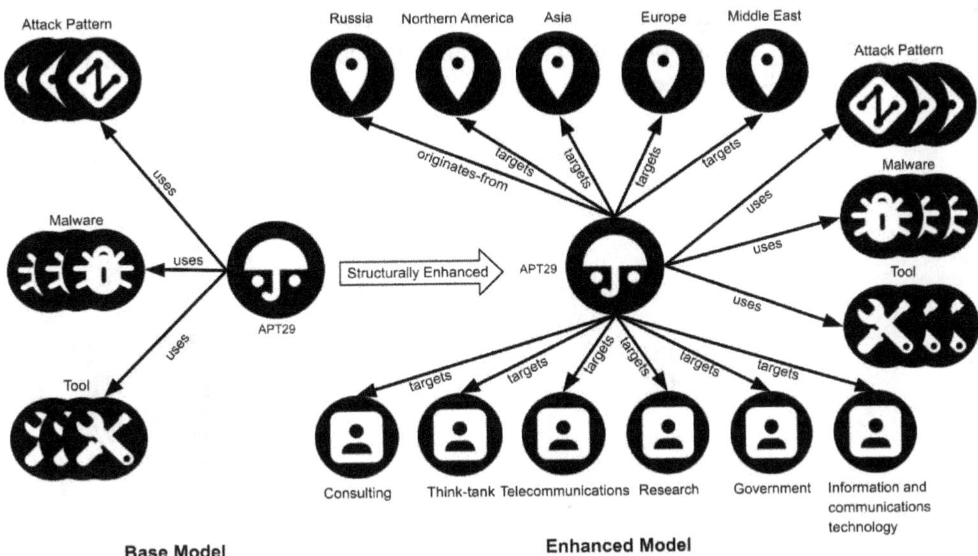

Figure 4: Graphical representation of APT29 using our enhanced model

4. Utilizing the enhanced STIX 2.1 Groups knowledge base for group filtering and technique prioritization

This section demonstrates how to use the new types of contextual information introduced in the enhanced STIX 2.1 representation of the Groups knowledge base to filter activity groups of interest. It is out of the scope of this paper to suggest specific data query- or programming languages to interact with the knowledge base. However, any programming language that can interact with JSON (JavaScript Object Notation) should be sufficient. It is also important to remark that the accuracy of the results retrieved might be questionable due to the use of non-standardized vocabularies in providing context that often leads to semantic ambiguity. For example, the number of sectors and sectors considered critical infrastructure might be dissimilar in different countries. Listing 1 presents an SQL-based query that aims to retrieve activity clusters that are believed to originate from the Russian Federation and have targeted establishments within the Government sector in the United States. The results indicate three activity groups to be of relevance.

```
SELECT * FROM GroupsKnowledgeBase
WHERE OriginatesFrom == "Russian Federation"
AND TargetSector == "Government"
AND TargetCountry == "United States"

Results:
APT28, APT29, Dragonfly 2.0
```

Listing 1: SQL query leveraging our model's contextual information types to retrieve activity groups of interest.

Awareness of the techniques and their frequency among relevant activity groups allows organizations to support decision-making to protect their assets. From that point, defenders can leverage the associated information for their use cases, such as prioritizing adversary techniques to conduct defense gap- or SOC (security operations center) maturity assessments. We retrieve the ATT&CK technique use of the groups and provide it as input to MITRE's ATT&CK Navigator[7] to identify and explore any commonalities. The technique overlap is presented in Figure 5. The light red color techniques do not overlap among the three groups. The red color techniques indicate use by two groups, and the dark red color techniques indicate use by all three groups.

5. Conclusion

In this paper, we extended (enhanced) the STIX 2.1 model of ATT&CK Groups with additional types of contextual information. Such information includes the groups' motivations, countries of origin, and targeted sectors and countries. We demonstrated how to use the proposed model to filter activity groups of interest and

[7] https://mitre-attack.github.io/attack-navigator/

consequently retrieve relevant tactical intelligence (behaviors/techniques). We used the ATT&CK techniques we extracted to identify and explore commonalities in technique use among different activity groups.

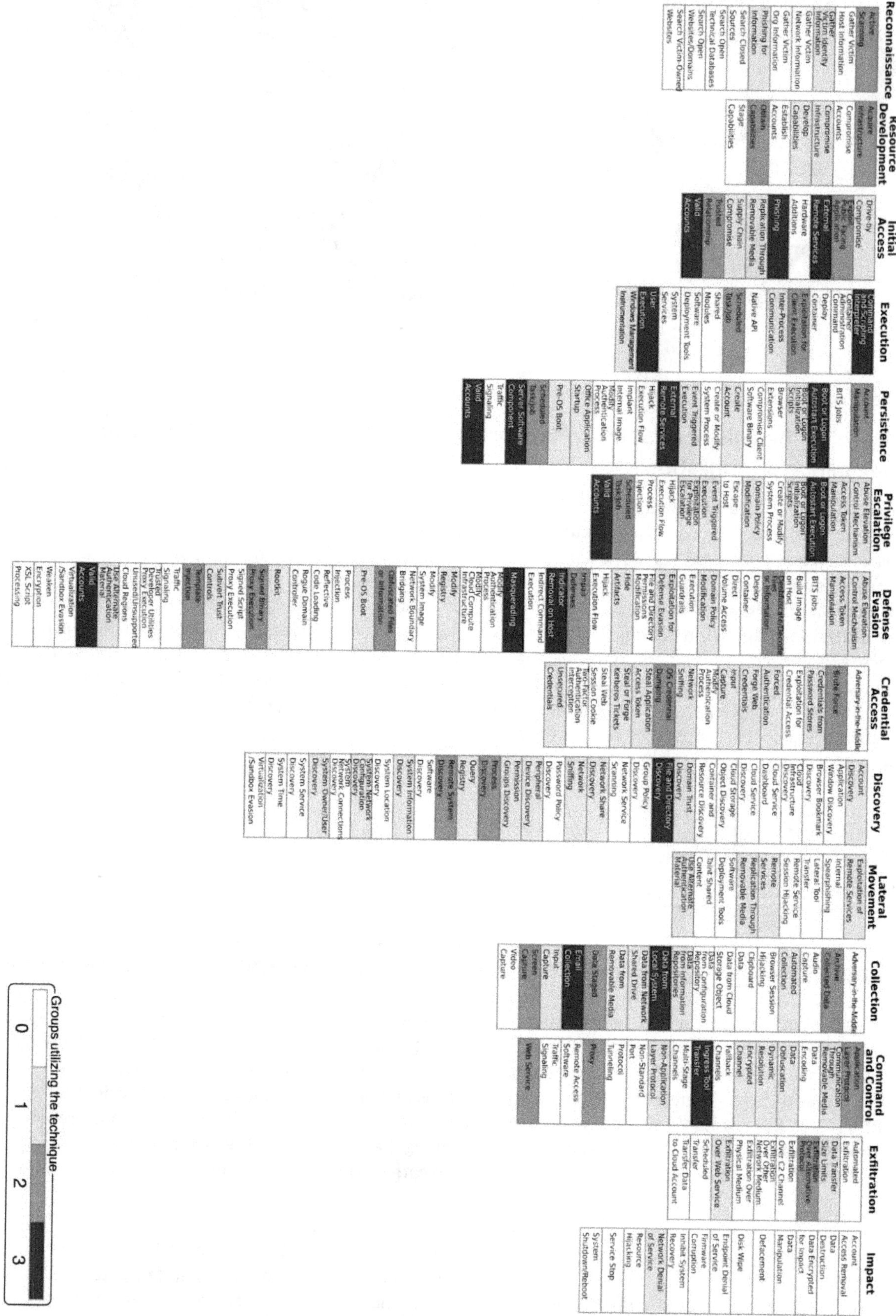

Figure 5: Heatmap showing commonalities in technique use among APT28, APT29, and Dragonfly 2.0

Acknowledgements

This research has received funding from the Research Council of Norway (forskningsrådet) under Grant Agreement No. 303585. In addition, this research work was supported by the European Health and Digital Executive Agency (HaDEA) under Grant Agreement No. INEA/CEF/ICT/A2020/2373266. The authors would like to express their appreciation to EclecticIQ for designing and making available icons to visualize the STIX language[8].

References

Alexander, O., Belisle, M. and Steele, J., (2020). MITRE ATT&CK® for Industrial Control Systems: Design and Philosophy. The MITRE Corporation: Bedford, MA, USA.

Bianco, D., (2019). The Pyramid of Pain (2014). Available at: http://detect-respond.blogspot.nl/2013/03/the-pyramid-of-pain.html.

Brown, R. and Lee, R.M., (2021). 2021 SANS Cyber Threat Intelligence (CTI) Survey. In Tech. Rep. SANS Institute.

Johnson, C., Badger, L., Waltermire, D., Snyder, J. and Skorupka, C., (2016). Guide to Cyber Threat Information Sharing. NIST special publication, 800(150).

Jordan, B., Piazza, R., and Darley T., (2021), July, STIX™ Standard, version2.1 ed., [Online] https://docs.oasis-open.org/cti/stix/v2.1/stix-v2.1.html, OASIS Cyber Threat Intelligence Technical Committee.

Lee, R.M., and Brown, R., (2021). FOR578 Cyber Threat Intelligence: 578.1 Cyber Threat Intelligence and Requirements. SANS.

MITRE ATT&CK, (2022), MITRE ATT&CK: Software. Available at: https://attack.mitre.org/software/

Strom, B.E., Applebaum, A., Miller, D.P., Nickels, K.C., Pennington, A.G. and Thomas, C.B., (2018). MITRE ATT&CK®: Design and philosophy. MITRE Product Mp, pp.18-0944. Revisited March 2020.

---

[8] https://github.com/eclecticiq/stix-icons

Masters Research Papers

Cyber Concerns With Cloud Computing

Jacob Chan and Mark Reith
Air Force Institute of Technology, Dayton OH, United States
jacob.chan@afit.edu
mark.reith@afit.edu

Abstract: The last two decades has seen a paradigm shift towards cloud-based services, cloud-central storage, and cloud computing. The benefits of this shift has been undeniable, including minimal user infrastructure needed to achieve what appears to be limitless data storage, powerful processing capabilities, and services that scale according to demand. However, when there is an upside there usually exists a downside. Cloud computing brings many security and cyber concerns that stems from the inherent insecurity of having large concentrations of data and assets in the cloud, making it a priority target for malicious actors. This survey paper will provide a review of the existing cyber concerns with cloud computing from a military perspective and point out future cyber concerns that may populate due to emerging technological advancements on the horizon.

Keywords: cloud computing, cloud-based services, cloud solutions, cloud computing security, cloud security concerns

1. Introduction

The continued paradigm shift in the last two decades towards cloud-based services and cloud storage solutions has provided not only a myriad of benefits for companies and individuals alike but has unveiled cyber concerns that stems from the inherent insecurity of having large concentrations of data and assets in the cloud, making it a priority target for malicious actors. This survey paper will provide a review of the existing cyber concerns with cloud computing that have been identified insofar from a military perspective and point out future cyber concerns that may populate due to emerging technological advancements on the horizon. To prepare readers in understanding relevant cyber concerns with cloud computing, this paper first provides background context on what cloud computing is, types of clouds and service models, and the benefits of cloud computing. Next, the paper goes into the depths of various cyber concerns related to cloud computing, ranging from confidentiality/privacy issues to insider threats, before highlighting potential cyber concerns for the future.

1.1 What is cloud computing?

The fundamental concept of cloud computing can be generalized to centralizing data storage, applications, and computing capacity to the network core (the Cloud) and providing end-user access to the Cloud as a service (Hayes, 2008). The idea of the Cloud is not a novel one and has been conceptualized as early as the 1950s, in the form of time-sharing systems (Earnest, 2016). During that time, computers were large mainframes that required an entire room to store and an equally large amount of power and electricity to host. The benefits provided by a mainframe were undeniable but not everyone had the money or room for one. Time-sharing systems connected smaller user terminals to mainframes and allowed people to leverage the benefits of computers without physical access to a mainframe, which was a predecessor to the client-server architecture of today. By the 1980s, personal computers had their own storage capacity and provided an array of applications that could be personalized to the individual user, making personal computers the ideal choice for computing and storage needs. In the past two decades, the tides of computing have once again shifted, with the increasing demand for digital storage and processing power for both individuals and organizations alike pushing the responsibilities and capabilities of digital storage and computing into the Cloud (Hayes, 2008). In particular, the Cloud offers an intriguing outlook for the U.S. military's growing demand for technological solutions (the U.S. was the country with the second highest R&D spending in 2021) (Szmigiera, 2021).

1.2 Three types of clouds

- 1) Public Clouds: The most common and vulnerable type of cloud are public clouds (Almorsy, Grundy and Müller, 2016). They include the free or paid cloud services and resources offered to the public. Companies that offer public clouds include Google, Amazon, Microsoft, IBM, Oracle, and more. Just as normal civilians do, U.S. military service members utilize public cloud services such as Google Docs or Dropbox all the time, in both their personal lives and at work.
- 2) Private Clouds: Private cloud resources and services are usually not free, and a private cloud is only accessible to private organizations or businesses that have permissions to access that specific cloud

(Srivastava and Khan, 2018). It is worthy to note that a public cloud provider can also offer private cloud services, through virtual partitioning of their cloud (Armbrust, 2010). An example is the Commercial Cloud Services (C2S) contract that the U.S. Intelligence Community (IC) has had with the one of the leading public cloud providers of the world, Amazon Web Services (AWS) since 2013 (Miller, 2021). Part of this $600 million contract is that AWS must service the private cloud needs of intelligence organizations such as the CIA, the FBI, and the NSA. In 2017, Amazon came out with the AWS Secret Region, which offers cloud services and storage for U.S. secret-level data and follows "the Director of National Intelligence (DNI) Intelligence Community Directive (ICD 503) and National Institute of Standards and Technology (NIST) Special Publication (SP) 800-53 Revision 4" (AWS Public Sector Blog Team, 2017). AWS Secret Region is offered to the U.S. IC as part of the C2S contract but non-IC U.S. government customers with secret-level clearance may also utilize AWS Secret Region with the proper contracts.

- 3) Hybrid Clouds: Hybrid clouds are a combination of multiple clouds, which can include public and private clouds (Srivastava and Khan, 2018). These types of clouds have become increasingly popular in the last couple years as cloud providers have become more competitive, with some offering distinct services. For example, in 2020, the U.S. IC awarded their Commercial Cloud Enterprise (C2E) contract to five different cloud providers: AWS, Microsoft, Google, IBM, and Oracle (Moss, 2020). The C2E contract, estimated to be worth north of $10 billion allows the IC to select among the five cloud providers that best fulfills their cloud needs. Back in July of this year, the DoD cancelled their Joint Enterprise Defense Infrastructure (JEDI) contract, worth $10 billion over 10 years with Microsoft in favor of the Joint Warfighting Cloud Capability (JWCC) contract, which seeks solutions from multiple cloud vendors (Feiner, 2021). According to an official statement released by the DoD, the JWCC contract will better fulfill the DoD's increasing need for cloud capabilities as posed by newer requirements such as the Joint All Domain Command and Control (JADC2) and the Artificial Intelligence and Data Acceleration (ADA) initiatives (Department of Defense, 2021).

1.3 Cloud service models

Cloud computing goes by many different names such as utility computing, on-demand computing, Software-as-a-Service (SaaS), internet as a platform, and many others (Hayes, 2008). Regardless of the name, most forms of cloud computing usually offer the following services.

- 1) Software as a Service (Saas): SaaS provides rich and practical applications to users that are easily accessible through a web browser without needing to install, update, and maintain software on their own personal computer (Srivastava and Khan, 2018). Examples of this include Google Workspace, which encompasses Google Docs, Google Spreadsheets, Google Drive, and Office 365, which boasts the entire Microsoft Office suite, all easily accessible online. As mentioned before, military service members use these cloud applications all the time. There are no specific regulations on cloud application usage for individual service members in the military. Instead, their usage is governed through the individual branch's regulations on information systems usage and classification guides.
- 2) Platform as a Service (PaaS): PaaS gives individual users and organizations an existing digital platform and frameworks to develop and customize applications, manage their business, or to carry out organizational functions irrespective of their own hardware setup (Srivastava and Khan, 2018). An example of this is Platform One, which is a secure DoD cloud platform that provides developers with existing tools, software, frameworks, and DevSecOps pipelines that allow them to solve complex problems using a 90% ready software solution from day one instead of starting from the ground up (Office of the Chief Software Officer, 2021). Simply put, Platform One allows DoD developers to focus on developing software solutions without having to worry about building a platform to develop software solutions.
- 3) Infrastructure as a Service (IaaS): IaaS refers to the physical and digital infrastructure needed to enable SaaS and PaaS on demand (Srivastava and Khan, 2018). This includes the servers, server racks, data centers, operating systems, communication hardware, and everything provided by cloud providers that cloud users interact with daily but never get to see. With IaaS, cloud customers do not need to set up their own infrastructure and can instead, rely on the existing infrastructure set up by the cloud provider. In 2010, the DoD started the Federal Data Center Consolidation Initiative (FDCCI) with goals of reducing the number of DoD-operated data centers by 60% by the end of 2018 (DoD Inspector General, 2016). The purpose was to save on costs associated with acquisition, sustainment, and manpower, reduce the DoD's real-estate and digital footprint which in turn reduces cyber-connected attack surfaces, and focus resources on necessary

war-fighting capability. Since 2016, the DoD has saved almost $2 billion from closing 210 tiered data centers and 3005 non-tiered data centers (Office of the Federal Chief Information Officer, 2021).

1.4 Benefits of cloud computing

The list of benefits for cloud computing is long and ever-growing. This paper will not extensively list all the benefits of cloud computing but will rather focus on the ones that will line the talk for cyber concerns mentioned later in the paper.

- 1) Minimal User Infrastructure: One of the biggest benefits about cloud computing that captivates customers is that there is minimal user infrastructure required. Cloud service providers supply the hardware and software infrastructure necessary for cloud applications, storage, and computing. Businesses and organizations can leverage this existing infrastructure for their own use and can save on costs of purchasing servers and setting up server racks, purchasing or renting space to host said servers, buying and configuring operating systems and server-side security such as antivirus, firewalls, or intrusion detection systems (IDS), and hiring/training employees to use, maintain, and upgrade this infrastructure. As mentioned previously, part of the DoD Cloud Strategy is to consolidate data centers and drive IT reform at the DoD (Department of Defense, 2018). By reducing unneeded infrastructure, the DoD hopes to reduce security risks/attack surface and redirect resources to war-fighters and workforces from different mission areas. Aligned with this initiative is the $62 million Private Cloud II contract that the U.S. Army awarded to IBM in 2016 (Serbu, 2016). The agreement states that IBM must set up, maintain, and operate a data center in Redstone Arsenal, an army post near Huntsville, Alabama, serving as a pilot initiative for a private company to operate a data center on U.S. military installation grounds. Although the Army added a data center through the contract, it is part of the consolidation effort to remove unnecessary data centers and push data center management responsibilities to private companies.
- 2) Amortized Spending: With cloud computing, there is no big upfront cost associated with purchasing expensive equipment and software (Almorsy, Grundy and Müller, 2016). Instead, there is an amortized cost in which cloud customers pay monthly (there exist options for yearly payments) based off how much computing power they used, the amount of storage they required, the type of services they needed, etc. This is especially relevant to the DoD, because a lot of organizations within the DoD are allotted funds per year. Some Air Force squadrons, or wings will spend the rest of their funds toward the end of the year because they may lose it if they don't use it, or it may affect how much they are allocated the following year. With cloud computing, spending is more streamlined and efficient because organizations are spending funds based off their usage and needs, which may change throughout the year as opposed to end-of-year spending to avoid wasting funds. It is important to note that while most military cloud utilization instances mentioned so far have been large ventures on the DoD (JWCC) or department level (Army Private Cloud II), smaller organizations such as Air Force squadron or wings can still utilize cloud computing and reap the benefits from amortized spending.
- 3) Elasticity: As mentioned just now, cloud computing provides customers with a flexibility in spending based off evolving needs, commonly referred to as elasticity. With elasticity, cloud customers are protected from over or under-provisioning (Armbrust, 2010). Businesses and organizations tend to provision service requirements according to peak hours. This may result in an over-provisioning of resources. For example, if the Air Force portal experiences peak hours at noon on Mondays and Fridays, the Air Force may add extra servers to meet that peak demand. However, the extra servers are idle and wasted during non-peak hours (which in this case, is most of the week). On the flip side, under-provisioning can be just as problematic. Imagine in the previous example that the Air Force decides to provision just enough resources to handle normal usage. There is an unexpected event (i.e., COVID-19 or a natural disaster) and the Air Force portal is receiving an influx in activity. In this instance, it is paramount for the Air Force to increase resources to meet demand. However, this demand may soon die down and the Air Force is left with even more wasted resources. Cloud computing offers a flexible and scalable solution for businesses and organizations to meet elastic and changing demands (Zissis and Lekkas, 2012). Amazon Web Services (AWS) allows customers to "pay as you go" (Bishe, 2018). They can pay for storage by the gigabyte and processing power by the hour, allowing businesses and organizations to scale up during peak-hours and scale back down afterwards.

1.5 Prior cloud security research

Chen, Paxson and Katz (2010) attempts to separate security problems unique to cloud computing from the legacy security problems that existed since the development of time-sharing systems. The authors argue that most cloud security concerns are not new but rather variations of historical web application problems. There are, however, two new security concerns specific to and brought on by cloud computing: multi-party trust considerations and mutual auditability. Trust considerations are different for the cloud because multiple parties are sharing resources. In particular, concerns exist with cloud customers being able to view each other's activity, businesses accessing proprietary data from their competition (cloud providers can also be the ones accessing competitive data), or cloud customers using the Cloud to run botnets, spam campaigns, or password brute-forcers. These issues result in the need of mutual auditability between cloud customers and cloud providers. Zissis and Lekkas (2012) examines unique security requirements related to the cloud and proposes a solution using a Trusted Third Party alongside Public Key Infrastructure (PKI) to ensure confidentiality, integrity, and authenticity of data. They highlight the notion of trust in traditional computing, which is based on security policy vs. cloud computing, in which users have to trust the processes and security set in place by the cloud provider. Furthermore, trust is delegated to cloud providers in public clouds while trust remains within the organization for a private cloud. The authors proposes a trusted authority (Trusted Third Party) that addresses security concerns within the cloud and facilitating trust between cloud providers and cloud customers. Almorsy, Grundy and Müller (2016) explores the cloud security problem from multiple perspectives including the cloud stakeholder's perspective, the architecture-level perspective, cloud characteristics perspective, and cloud service model delivery perspective. Cloud stakeholders include cloud providers, service providers to deliver content and services using the Cloud, and service users. Each stakeholder has different security needs and expectations. The authors propose service-level agreements (SLAs) and security transparency as a solution to resolve such conflicts. They divide cloud issues into subsections related to the different SaaS, PaaS, and IaaS service models and provides a cloud computing dependencies stack containing VMs, APIs, Services and Applications that the cloud service models rely on, which reveals vulnerabilities inherent in the cloud architecture. While the aforementioned authors do a great job highlighting cloud-related security issues, they do not touch upon the issues from a military perspective. Cloud issues may have differing levels of effects, likelihood, and tolerance for the military than civilian customers. For example, it can be argued that most if not all cloud customers likely value mutual trust with cloud providers. From the perspective of the military, trust may be the single most important requirement within cloud computing and a lack of data confidentiality is often intolerable (i.e. recent discussions on supply chain security). While many cloud customers, especially big companies, also have a lot to lose in the event of a data breach, it usually does not affect national security. Military and defense contractor customers often deal with classified data, which may bring upon unique issues or require specialized solutions, and it may be valuable to consider cloud security from their perspective. Ďulík (2016) looks at the benefits cloud computing can provide to the military and potential risks the military should consider. The author considers private clouds to be most appropriate for the military and highlights the different security mechanisms provided by the private cloud reference model. The author does a great job explaining the benefits, risks, and security issues special to the military at the time but does not look at how upcoming technology can impact cloud security for the military.

2. Cyber concerns

While many advantages of cloud computing come from centralizing data and services within the Cloud, it is also this inherent concentration of data and assets that brings about cyber concerns in the first place. Although, it is true that some of the security concerns seen in cloud computing can also be related back to traditional decentralized computing, this paper will not compare the security concerns of cloud computing and traditional computing, nor will it assume on which is more secure. Whether the Cloud is perceived to be secure or not, it will, no doubt, continue to proliferate and be used globally. The number of internet users utilizing some type of cloud service has grown from 2.4 billion in 2013 to 3.6 billion in 2018 (Statista, 2014). Since the Cloud will continue to grow regardless, it is best to focus our attention on security issues with cloud computing and hopefully spark some innovative solutions to fix them.

2.1 Centralized target

The top three industries targeted by cyber espionage in 2020 were finance, information, and healthcare. Cloud computing makes it easier for malicious actors by congregating the data and operations of multiple sectors in one place, making the Cloud a centralized target. Beyond that, cloud computing generated over $300 billion dollars of revenue in 2020. This makes cloud providers themselves an attractive target for cyber espionage, ransomware, denial-of-service attacks, phishing attempts, and more. With the emergence of private clouds for the military and the DoD's increasing reliance on cloud services, the Cloud has become a viable attack vector for both state and non-state adversaries that hope to steal sensitive data from the U.S. government.

2.2 Single-point-of-failure

Beyond client-side systems and the actual data, everything else in cloud computing are provided by cloud providers including servers, buildings that house the servers, software, data storage, processors, and services. This makes the Cloud, a single-point-of-failure. It can be argued that most cloud providers, being multi-billion-dollar businesses, have redundancy built-in such as multiple data centers, emergency power, backup storage, etc. However, their redundant systems and infrastructure could be vulnerable to the same attacks or issues that caused the original systems to go down if they use similar software, servers, or if their infrastructure is set up the same way (Armbrust, 2010). Some events also cannot be planned, such as if the cloud provider goes out of business. If a cloud provider goes out of business or gets knocked off the net, can a customer's data be transferred to another cloud provider? Can it be transferred in time before the original cloud provider's systems get decommissioned? If the cloud goes down, can businesses or organizations still function, or will they have to shut down? These questions are especially concerning for the military because the military is responsible for a wide array of critical functions including national security. If the cloud goes down, the military cannot simply go home for the day. Thus, the DoD is especially concerned with redundancy and recovery options that exist with cloud computing. One solution to the single-point-of-failure problem is to utilize multiple cloud providers, but the downside is that there may be hefty additional costs associated with maintaining the same instances of data and applications with multiple vendors. The DoD is headed towards this direction of leveraging multiple cloud providers as evident in the 2018 DoD Cloud Strategy and the JWCC contract (Department of Defense, 2018).

2.3 Insider threat

A lot of security concerns are focused on the security of the Cloud and how to prevent intruders from getting in. Other threats come from within the Cloud. In the Trojan War, instead of breaking into the impenetrable walls of Troy, the Greeks decided to hide in the Trojan Horse and win the war from the inside. In the story of cloud computing, Troy represents cloud providers such as Amazon, Google, or Facebook and the Greeks can be anybody. Of course, any business or organization can be a target of an insider threat, but this threat is amplified for cloud providers because of the value of the data contained within the Cloud and the fact that attacking the Cloud is like an attack on thousands of valuable business or organizations simultaneously. Part of this problem can be mitigated by using application-level encryption as an added layer of security (Zissis and Lekkas, 2012). However, the threat posed to the Cloud isn't an insider breaking through encryption as much as the insider gaining access to this information. There exist tools that can break through encryption but there is no tool that can gather the encrypted data of thousands of valuable companies in one place. Correction, the Cloud is that tool. On the government side, military and government private clouds often contain sensitive information that may have ties to national security, policy, and objectives. Some of this information could be classified as well as seen in the C2S and Army Private Cloud II contract. From a DoD perspective, it is paramount for cloud providers that service government needs to have insider threat training for employees, an intensive background check and classification process for new employees and re-evaluations every few years, information sharing guidelines based off need-to-know, detailed logs of employee actions and regular auditing of logs, strong authentication and access control to the Cloud, heavy encryption of customer data that is not readable by employees, and limitation of employee or administrator power.

2.4 Confidentiality/privacy

It is common practice for cloud providers to service multiple clients using the same physical systems and storage devices (Almorsy, Grundy and Müller, 2016). This type of service model can be vulnerable to data spillage, which can result in catastrophic damage to confidential or proprietary information. According to Statista.com, most data stored in the Cloud is considered sensitive data. Likewise, military and government clouds contain sensitive and classified data. Military service members can also accidentally store sensitive data on public clouds. Once again, the DoD is concerned with whether the cloud provider has systems to place to prevent employees and other customers from accessing government data. A partial solution to this problem could be having cloud providers operate private data centers solely for government use as seen in the Army's Private Cloud II contract with IBM. The latter concern would have to be fixed through stricter and more definitive guidelines on service member usage of the Cloud.

2.5 Availability

Availability of data or service is important in almost any internet-based application, but is scaled up for the Cloud because more customers can be affected by loss of availability. An investment firm could lose millions of dollars if they lost access to proprietary market data that allowed them to move in or out of stocks at a moment's notice. A patient's life could be at risk if a hospital loss access to the Cloud that they used to store patient data. To be fair, a lot of this important data is typically stored locally too but as the Cloud continues to grow in size and application, the consequences of availability issues could grow exponentially. As mentioned before, the military is responsible for national security and numerous missions that span the globe. A lack of availability for military clouds could have a detrimental effect to national security, the economy, diplomacy, and policy, and much more. Using multiple cloud providers certainly helps because it less likely for multiple cloud providers to be unavailable at the same time. However, it is important for the DoD and cloud providers to have disaster recovery plans, continuity of operations procedures, and redundant non-cloud systems in place to mitigate loss of cloud availability.

3. Future cyber concerns

Gordon Moore, the co-founder of Intel stated in 1965 that the number of transistors on a circuit will double every two years, hence doubling computing power every two years. Although the effects of Moore's Law have diminished in the past few years, technological advancements have continued to emerge and will likely continue to prosper for the foreseeable future. Some of these technologies may benefit cloud computing while others may create concerns associated with cloud computing. This section will highlight some of these concerns.

3.1 Quantum computing

Quantum computing is often touted as the next big thing in computers, with the common belief that a sophisticated quantum computer could solve traditional computer problems in a fraction of the time. In theory, RSA encryption using a 2048-bit key would take trillions of years to break using traditional computers (Baumhof, 2019). A quantum computer with 4099 perfectly stable qubits (quantum bits) would break the same algorithm in about 10 seconds. Current quantum technology is still quite a bit off from reaching 4099 qubits, the standard of data representation in quantum computing, like the bit in traditional computing. The most powerful quantum computer today, the IBM Eagle only has 127 qubits (Sparkes, 2021). While this is an impressive accomplishment nonetheless, the disparity between the current and the theoretical implies that quantum computing is still in its infancy and much more needs to be done before it can reach its potential. However, the largest quantum computer in 2019 was the Google Bristone with just 72 qubits. The number of qubits in the leading quantum computers essentially doubled over 2 years. This is both good and bad news. The good news is obvious because of the benefits quantum computing touts. The bad news is that advancements with quantum computing would likely make brute-forcing advanced encryption schemes possible and render traditional encryption useless. The country that wins the quantum computing race will likely gain a major boost in global influence and national capability. In the hypothetical situation that a foreign nation develops a quantum computer powerful enough to crack encryption or authentication, U.S. Fortune 500 companies, many of which are cloud providers, would be likely targets. U.S. government assets would also be targets, including military and DoD clouds. These collective clouds likely provide the most valuable data to an adversary and have the most impact to the U.S.. From a cloud perspective, it is vital for the U.S. be the front-runners in quantum technology and also come out with the next-generation of encryption algorithms before quantum computing makes traditional encryption schemes obsolete.

3.2 Internet of things

As the "Internet of Things" (IoT) continues to grow to encompass more devices, the threat towards cloud computing also increases. Only a few decades ago, airman had to physically go on base to access a Non-classified Internet Protocol Router (NIPR) computer. Now, it is possible to use technologies such as Virtual Desktop Infrastructure (VDI) or VMWare to access NIPR workstations from at home. Apps also exist to access their emails from their phone. In 2019, the 445th Airlift Wing at Buckley AFB came up with the "Desktop Anywhere" program, which allows Air Force reservists to run a stand-alone Air Force desktop over their personal computers and phones (Amidon, 2019). More recently, the U.S. Army Futures Command has implemented ways to allow soldiers to access Secret Internet Protocol Router (SIPR) computers from remote locations (DEVCOM C5ISR Center Public Affairs, 2021). The evolution of IoT introduces more devices to military and government clouds, including personal devices. These IoT devices, if not configured properly with security in mind, can lead to increased attack vectors for adversaries.

3.3 Augmented reality

In the past decade or so, life has slowly trended towards a virtual side, with many people opting to shop online, meet people through dating apps, and even buy cars online instead of going to a dealership. COVID-19 further accelerated this change, making virtual concerts, virtual hangouts, distance learning, and teleworking the norm. Adjacent to this is augmented reality and virtual reality, which include entire industries built on making virtual representations of the real world. Just recently, Facebook publicized the imminent release of the Metaverse, which is a virtual extension of the real world that people can plug into and access (Clark, 2021). Many service members will, no doubt, plug into the Metaverse and its applications will eventually be adopted by the DoD to enhance national objectives. A project of this size will likely be stored in the Cloud. Let's push realism aside for a bit and discuss a hypothetical situation. What if augmented reality allowed soldiers to display a map that showed locations of friendly forces? What happens if the soldier gets plugged out mid-battle? They just lost situational awareness of where their team is. Or what happens if adversaries somehow gained access to the soldier's map? They now know the soldier's location and their entire team's location. The bottom line is augmented reality has the potential to bring in new capabilities for the DoD. However, if augmented reality data is stored in the Cloud, it will be subjected to the same vulnerabilities that anything else in the Cloud faces because the Cloud is a single-point-of-failure.

4. Conclusion

Cloud computing has seen rapid deployment and adoption in the past two decades thanks to its myriad of benefits including minimal user infrastructure needed, amortized spending, elasticity, limitless storage, and more. Cloud computing does not come without concerns, many of which are directly tied to the centralization of data, applications, services, and functionality in one place, which is a key essence of the Cloud. Despite these concerns, the DoD and other U.S. government entities such as the Intelligence Community has embraced cloud computing as means of accomplishing and enhancing national objectives. Many cyber concerns with cloud computing such as concerns with privacy and confidentiality of data, availability of data and services, insider threats, and vendor lock-in have magnified effects and consequences for the U.S. government due to the responsibilities that the government has in relation to national functions and the often-sensitive nature of the data dealt with by government entities. To mitigate some of these issues, the DoD and U.S. government is moving towards leveraging multiple providers to provide redundancy and added benefits. However, the emergence of newer technologies such as quantum computing, IoT, autonomous vehicles, and augmented reality, may lead to more cyber concerns in the future.

References

Almorsy, M., Grundy, J. and Müller, I., 2016. An analysis of the cloud computing security problem. *arXiv preprint arXiv:1609.01107*.

Amidon, E. (2019) 'Desktop Anywhere: Innovation in action' [online]. Available at: https://www.445aw.afrc.af.mil/News/Article-Display/Article/1866787/desktop-anywhere-innovation-in-action/.

Armbrust, M., Fox, A., Griffith, R., Joseph, A.D., Katz, R., Konwinski, A., Lee, G., Patterson, D., Rabkin, A., Stoica, I. and Zaharia, M., 2010. A view of cloud computing. *Communications of the ACM, 53*(4), pp.50-58.

AWS Public Sector Blog Team (2017) 'Announcing the New AWS Secret Region' [online]. Available at: https://aws.amazon.com/blogs/publicsector/announcing-the-new-aws-secret-region/.

AWS (2021) 'Amazon S3 FAQs' [online]. Available at: https://aws.amazon.com/s3/faqs/.

Baumhof, A. (2019) 'Breaking RSA Encryption – an Update on the State-of-the-Art' [online]. Available at: https://www.quintessencelabs.com/blog/breaking-rsa-encryption-update-state-art/.

Bishe, A. (2018) 'How to optimize cost savings in AWS Marketplace' [online]. Available at: https://aws.amazon.com/blogs/awsmarketplace/how-to-optimize-cost-savings-in-aws-marketplace/.

Chen, Y., Paxson, V. and Katz, R.H., 2010. What's new about cloud computing security. *University of California, Berkeley Report No. UCB/EECS-2010-5 January, 20*(2010), pp.2010-5.

Clark, P. (2021) 'The Metaverse Has Already Arrived. Here's What That Actually Means' [online]. Available at: https://time.com/6116826/what-is-the-metaverse/.

Department of Defense (2018) 'DoD Cloud Strategy' [online]. Επιμέλεια D. of Defense. Available at: https://media.defense.gov/2019/Feb/04/2002085866/-1/-1/1/DOD-CLOUD-STRATEGY.PDF.

Department of Defense (2021) 'Future of the Joint Enterprise Defense Infrastructure Cloud Contract' [online]. Available at: https://www.defense.gov/News/Releases/release/article/2682992/future-of-the-joint-enterprise-defense-infrastructure-cloud-contract/.

DEVCOM C5ISR Center Public Affairs (2021) 'Army Futures Command enables classified work from remote locations' [online]. Available at: https://www.army.mil/article/244545/army_futures_command_enables_classified_work_from_remote_locations.

Ďulík, M. and Junior, M.Ď., 2016. Security in military cloud computing applications.

Earnest, L. (2016) 'Who invented Timesharing?' [online]. Available at: https://web.stanford.edu/~learnest/nets/timesharing.htm#:~:text=The%20first%20commercial%20timesharing%20system,and%20began%20working%20in%201965.

Office of the Inspector General (2016) 'DoD's Efforts to Consolidate Data Centers Need Improvement' [online]. Available at: https://media.defense.gov/2016/Mar/29/2001714226/-1/-1/1/DODIG-2016-068.pdf.

Google (2021) 'Discover our data center locations' [online]. Available at: https://www.google.com/about/datacenters/locations/.

Hayes, B., 2008. Cloud computing. Vol. 51 No. 7, Pages 9-11.

Kay, G. (2022) 'A 19-year-old security researcher describes how he remotely hacked into over 25 Teslas' [online].

Khalil, I.M., Khreishah, A. and Azeem, M., 2014. Cloud computing security: A survey. *Computers*, *3*(1), pp.1-35.

Miller, J. (2021) 'As C2E gets going, DIA sets its strategy for more cloud services' [online]. Available at: https://federalnewsnetwork.com/ask-the-cio/2021/04/as-c2e-gets-going-dia-sets-its-strategy-for-more-cloud-services/.

Moss, S. (2020) 'CIA awards multibillion C2E cloud contract to AWS, Microsoft, Google, Oracle, and IBM' [online]. Available at: https://www.datacenterdynamics.com/en/news/cia-awards-multibillion-c2e-cloud-contract-aws-microsoft-google-oracle-and-ibm/.

Platform One (2021) 'WHAT CAN PLATFORM ONE DO FOR YOU?' [online] Available at: https://p1.dso.mil/#/.

Serbu, J. (2016) 'IBM wins $62 million contract to run private cloud pilot at Army's Redstone Arsenal' [online]. Available at: https://federalnewsnetwork.com/army/2016/10/ibm-wins-62-million-contract-run-private-cloud-pilot-armys-redstone-arsenal/.

Seredynski, P. (2021) 'Gathering clouds will form autonomy's computing backbone' [online]. Available at: https://www.sae.org/news/2021/04/autonomous-vehicles-and-their-cloud-computing-networks.

Sparkes, M. (2021) 'IBM creates largest ever superconducting quantum computer' [online]. Available at: https://www.newscientist.com/article/2297583-ibm-creates-largest-ever-superconducting-quantum-computer/.

Srivastava, P. and Khan, R., 2018. A review paper on cloud computing. *International Journal of Advanced Research in Computer Science and Software Engineering*, *8*(6), pp.17-20.

Statista Research Department (2014) 'Number of consumer cloud-based service users worldwide in 2013 and 2018(in billions)' [online]. Available at: https://www.statista.com/statistics/321215/global-consumer-cloud-computing-users/.

Szmigiera, M. (2021) 'Leading countries by gross research and development (R&D) expenditure worldwide in 2021' [online]. Available at: https://www.statista.com/statistics/732247/worldwide-research-and-development-gross-expenditure-top-countries/.

Office of the Federal Chief Information Officer (2021) 'Data Center Optimization Initiative' [online]. Available at: https://datacenters.cio.gov/.

Zissis, D. and Lekkas, D., 2012. Addressing cloud computing security issues. *Future Generation computer systems*, *28*(3), pp.583-592.

Impact of Information Security Threats on Small Businesses During the Covid-19 Pandemic

Inga Mzileni and Tabisa Ncubukezi
Information Technology Department, Faculty of Informatics and Design, Cape Peninsula University of Technology, Cape Town, South Africa
nizoinga6@gmail.com
ncubukezit@cput.ac.za

Abstract: Information is a significant asset of any organization. The increased information demand by all parties has gained attention and raised security concerns – especially in this digital era where everyone depends heavily on the Internet. The Internet and online platforms expose valuable information to various information threats. These pervasive threats compromise information privacy, safety, and security. Legitimate people and criminals compete to access information. Criminals use innovative ways to gradually increase information security threats, especially in the small business sector with only a minimal budget for proactive security measures. Due to the scarcity of academic research on information security threats for small businesses, this study presents the impact of security threats on businesses during the global Covid-19 pandemic. A qualitative survey within the interpretive approach was used to gather data from 20 small businesses in Western Cape, South Africa, to fill this gap. The study used judgmental sampling to select research participants who are business owners. Data were analyzed using thematic analysis. The results indicated the knowledge gap relating to information threats, even though most businesses are familiar with the costly and negative impact of threats on business operations, resulting in business discontinuity. However, some small business sectors showed minimal awareness a6nd understanding of information security threats, their impact, and proactive mitigation strategies. The study concluded with recommendations to protect against information security threats.

Keywords: cyber-attacks, cyber security, information security, information security threats, internet, small businesses

1. Introduction

Globally, all institutions possess valuable information related to assets, financial payments, customers, personal identification, and payment cards. The global Covid-19 pandemic forced all institutions to exchange information on the Internet, granting access to legitimate users and criminals. This process exposes information to threats that render the small business sector vulnerable (Ncubukezi & Mwansa, 2021). Small business exposure to information threats negatively impacts their daily activities, resulting in numerous business risks (Ncubukezi, Mwansa & Rocaries, 2020). The assets, information, and people associated with the small business sector become vulnerable and exposed to malicious attacks, compromising the state of a business. Examples of information security threats include software attacks, intellectual property theft, and sabotage (Holovkin, Tavolzhanskyi & Lysodyed, 2021).

Information security threats gradually increase, leading to data loss, corruption, and disruption of normal business operations (Whitman & Mattord, 2021). Consequently, the safety of business information and other assets is a substantial concern, especially when a business relies on the Internet. The presence of information security threats among small businesses can result in a significant data breach: loss, manipulation, or deletion (Biener, Eling & Wirfs, 2015).

Cyber security has escalated in the challenge, especially during the global pandemic, as opportunities for hackers, attackers, and scammers to take advantage of emergencies are prevalent, especially when people are frightened (Khan, Brohi, Zaman & 2020:2). This work, then, focuses on the impact of information security threats among small businesses during the Covid-19 pandemic. To achieve this, the paper:

- identifies the persuasive information security threats;
- determines the impact of information security threats on small businesses; and
- determines measures to protect against information security threats.

The entire paper is laid out as follows: the subsequent sections address the literature review, research methodology, results, and discussions, followed by recommendations and concluding remarks.

2. Literature review

The coronavirus pandemic (Covid-19) started in 2019 and quickly escalated into a global crisis, resulting in the mass quarantine of citizens worldwide. Global restrictions forced all institutions, including small businesses, to operate from homes (Georgiadou, Mouzakitis & Askounis, 2021), which increased Internet dependency on routine business activities (Ncubukezi, Mwansa & Rocaries, 2021). In recent times, as businesses globally face transformation, most have become increasingly innovative, competitive, and challenging, while concomitantly, security risks targeting information systems have also increased (Gerić & Hutinski, 2007). Small businesses are frequently receiving threats and attacks (Akpan, Udoh & Adebisi, 2020). Small businesses are presented below.

2.1 Overview of small business

'Small' in terms of qualifying for government support and preferential tax policy varies by country and industry. The business sector is determined by size, with small to medium-sized (SMEs) businesses employing fewer than 200 employees. Small businesses can be privately owned corporations, partnerships, or sole proprietors with fewer employees and lower annual revenue than a corporation or regular-sized company (Itliong, 2020). Small businesses generally employ a range of 50 to 60% of South Africa's workforce and contribute significantly to around 34% of the gross domestic product (GDP) (IFC World Bank Group, 2021). Small businesses contribute to GDP (Paulsen & Toth, 2016).

Small businesses employ fewer than 50 employees in comparison to medium businesses, with some employing fewer than 20 depending on the industry and micro sizes. This study focuses on SMEs that operate as Internet cafés with 3 to 10 employees, depending on the number of services a manager undertakes. For example, Internet cafés typically provide Internet services, information and communication technology (ICT), and document services such as faxing, copying, and printing.

2.2 Impact of Covid-19 among businesses

In recent times, global businesses have grown increasingly competitive, with small businesses often on the receiving end. When the coronavirus pandemic (Covid-19) started in 2019 and quickly became a global crisis, the world saw the mass quarantine of hundreds of millions of citizens worldwide. Covid-19 pandemic-imposed lockdowns in most nations affected all industries and service sectors. The pandemic is anticipated to permanently normalize the Internet use and continue to force businesses to adopt online service strategies (Herath & Herath, 2020). The rapid and massive shift to online revealed a concern for small businesses. For the most part, they do not have sufficiently trained staff for the new and unfamiliar tools now necessary to remain competitive in these trying times. This inexperienced sector with minimal skills in technology usage has become easy target (Monteith et al., 2021), especially as many people, including businesses, perform their daily activities at home relying on a connection to the Internet, even though the small business sector does not have diverse IT and cyber security personnel compared to large enterprises (Ikpe et al., 2020). As a result, most small business managers have little understanding of information security and security threats.

2.3 Why information security?

As one of the main business assets, information should always be protected and guarded against pervasive, disruptive, intruding threats (Lundgren & Möller, 2019). Information security sometimes referred to as InfoSec, defends information from unauthorized access, use, disclosure, alteration, recording, inspection, or recording (Alhassan & Adjei-Quaye, 2017). Consequently, information security should be prioritized in all institutions. Information security practice prevents unauthorized access to information, unauthorized data modification, and deletion of data. The network and the computer are the main tools used to deploy data breaches, commonly coming as viruses, spam, phishing, and identity theft. These data breaches compromise the confidentiality, integrity, and availability (CIA) of business information, the fundamental information security principles (Zafar, Ko & Osei-Bryson, 2016).

Figure 1 shows the three primary components of the CIA triad – confidentiality, integrity, and availability – for guiding the sector's security procedures and policies (Moura & Serrao, 2016). The confidentiality component promotes data availability to those with the proper authorization, while the integrity principle focuses on information consistency, accuracy, and trustworthiness. Finally, the availability principle ensures easy access to authorized recipients (Thapa & Camtepe, 2021). These principles protect a business's digital assets against ever-looming cyber-attacks.

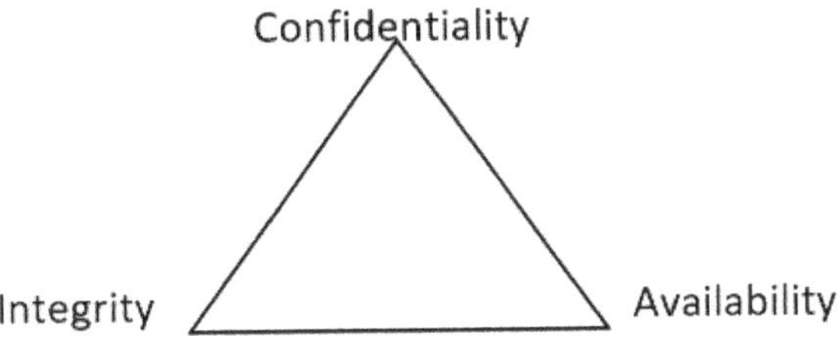

Figure 1: CIA triad security principles (Moura & Serrao, 2016)

As the key aspect of every business, information security aims to ensure business continuity and minimize unforeseen damages by limiting the impact of security incidents (Von Solms & Van Niekerk, 2013). Information security measures are required because the technology applied to information creates risk and is important to an organization's dependence on information technology. This is significant when an organization's information is exposed to risk. As failures of information security measures are adverse events that cause business losses, information security risk management is essential for every business (Blakley, McDermott & Geer, 2001).

2.4 Security threats

The global pandemic has revealed the state of safety and security of businesses. A successful business prioritises the security of the business at all levels, namely: personnel, network, systems, hardware and information. Improved safety and security includes the protection of sensitive data, personally identifiable information, protected health information, and governmental and industrial information systems from theft and damage attempted by criminals and adversaries. Increased Internet usage by businesses necessary for visibility in the market to attract new prospects, increase profit and improve communication has exposed all sectors to a diverse range of security threats, such that security threats are presently a major challenge within the small business sector.

But in these trying times, the overall security of small businesses is undeniably challenging. Hackers, attackers, and scammers take advantage of emergencies, primarily when people are frightened (Khan, Brohi & Zaman, 2020). Cyber-attacks result in a leakage of financial information, among other losses, which damage businesses. Therefore, the safety of business information is a critical and significant concern. Protecting computer operating systems, networks, and data from cyber-attacks requires special skills and knowledge. Heidt, Gerlach, and Buxmann (2019) confirm that most small institutions do not have a budget to hire IT specialists to secure their systems. This opens the door for crimes within the business sector that threaten security. Cybercrimes are human-generated and can be planned or unplanned (Ncubukezi, 2022).

2.5 Cybercrime

Cybercrimes are typically regarded as computer crimes because criminals use a computer as a tool to commit illegal actions such as fraud, identity theft, or privacy violation (Monteith et al., 2021). Cybercrime is often described as a crime triangle, which specifies that for cybercrime to occur, three factors must exist: a victim, a motive, and an opportunity (Lallie et al., 2021). In such a scenario, the victim is the target of the attack while the motive forces information threats and attack, and the opportunity presents a chance for the crime to occur (Lallie et al., 2021). Due to the valuable information possessed by small businesses, they are ready targets of cyber-attacks and crimes.

Cybercrimes corrupt the state of security of the business's hygiene at all levels of information. The Covid-19 pandemic gave both legit users and criminals equal chances to perform activities in the business space. As victims of criminal intentions, businesses are easy cyberspace victims (Ncubukezi, Mwansa & Rocaries, 2020). Some attackers select victims based on susceptibility to an attack; these attacks are called opportunistic attacks. Opportunistic attackers seek to maximize their gain and, therefore, wait for the best time to launch an attack where conditions fit.

3. Related work

The Covid19 global pandemic has caused a panic to many institutions. The global pandemic raised cybersecurity related issues especially during the "new normal." As a results there are a number of studies that have been conducted in relation to the impact of the global pandemic. A study conducted by Georgiadou, Mouzakitis and Askounis (2021) evaluated cyber security culture readiness among the organizations that are based on various countries. In year 2020, their study used online surveys with 23 questions to collect data from 264 employees working at home during the sudden Covid19 pandemic. The results of their study were discussed and recommendations addressed both vulnerabilities and needs for security culture.

Another study analysed the cyber-crime during COVID-19 pandemic focusing on the variety of cyber attacks. The study revealed cyber-attack modus-operandi of campaigns and steadily attacks which ultimately became prevalent resulting to a maximum of three attacks being reported a day (Lallie et al., 2021). In addition, Pranggono and Arabo (2021) studied cybersecurity issues that have occurred during the coronavirus (COVID-19) pandemic. Their paper addressed the correlation between the cyber attacks, their risks caused and the pandemic which increased anxiety and fear. In their study it came out that the healthecare sectros are mostly vulnerable to cyber attacks. The study recommended the practical approaches that proactively reduce cyber-attacks related risks among the sector. the impact of the information security threats among the small business sectors during the hitting global pandemic. The study is conducted in South Africa, Cape Town. A total of twenty small businesses were sampled and the results were reported and discussed.

4. Research method

While there are numerous research designs, the most prominent five are as follows: narrative research, phenomenology research, grounded theory research, ethnology research, and case study research. Phenomenology provides the best approach for this study. In the phenomenological approach, the focus describes the meaning for individuals of their lived experience of a concept. The study describes what all participants have in common as they experience the phenomenon. Phenomenology has two methods: hermeneutic phenomenology – the theory and methodology of interpretation; and empirical phenomenology – minimal focus on the interpretations of the research and more on describing participant experiences.

Qualitative approach: The study adopts a qualitative interpretation research paradigm to study the research participants in their natural setting and understand opinions and perceptions. The approach provides great value in relation to diverse environmentns and phenomena by describing events or experiences of the research participants from a wide range of sectors and different roles. So this study reports the information security threats among small businesses during the Covid-19 pandemic. This qualitative interpretive research gathers data from small businesses. The approach helps to develop concepts and visions for problems. For this study, the researchers were interested in the meaning and meaning-making process because qualitative data never speaks for itself but needs to be given meaning.

Research participants: The study focused on male and female small business managers and owners, between 23 and 30 years old, as research participants. Business owners and managers, knowledgeable about business systems, are responsible for overall business management. These participants were invited to participate in the study through emails. Only those who voluntarily responded to the communication request were part of the study. None of the participants were reminded or persuaded to participate in the study. The selection of the sampling method used in the study is presented below.

Sampling and data collection method: Twenty (20) participants were selected using judgemental sampling within the qualitative interpreting approach. The chosen sampling method helps researchers get more information on the collected data. In addition, it helps to describe the findings of the main impact based on the population. Information gathered will be used for research purposes to achieve the aim of the study. Due to national and global restrictions, this work used an online survey, created on Google Forms, as the data collection method. This was the most convenient data collection method available. The researcher sent email invitations to small business managers. Those willing to participate responded and were sent a link to complete a qualitative online survey. The study received an 85% response rate.

Data analysis: Data analysis provides theorized and interpretative accounts, socially located explorations of experiences (Braun et al., 2021), and sense-making of information security threats to small businesses. Data gathered from the qualitative survey were analyzed using thematic analysis. The researcher examined the collected data to identify common themes relating to repeated topics, ideas, and patterns. The researcher further worked with the data collected during the analysis and assigned preliminary codes to describe the content. The researcher searched for themes in the codes across different responses. The themes analyzed in the study were as follows:

- participant background;
- common threats experienced by small businesses;
- impact of information security threats, and
- security measures to improve small business security.

Themes were reviewed, defined, and named to produce a report. These themes are clearly presented in the results and discussion section.

Ethical considerations: The researcher, having received ethical clearance from the departmental ethics committee, assured respondents of confidentiality, privacy, and anonymity. The information collected in this study is only used for research purposes. In addition, research participants were not forced to participate in the study and were given sufficient time to decide. The participation was voluntary, and the participants were made aware to withdraw at any time without fear of consequences.

5. Results and discussions

This section reports the experiences of the small business with the impact of information security threats. Results are presented according to the pervasive information security threats, their impact, and the measures used to protect against these information security threats in the small business sector.

5.1 Participant background

Data were collected from Internet café businesses in South Africa in Western Cape. The male and female participants were adults aged between 23 to 30 years. The participants were selected from the townships in the Western Cape. Most businesses have been operating for over eight (8) years, while others are still in the early years, just finding their feet in the business world. All participants use cyberspace to attract new prospects and engage in daily business activities, making them vulnerable to various information threats. Figures 2 to Figure 4 illustrate participant gender, participant age, and the number of years of each business.

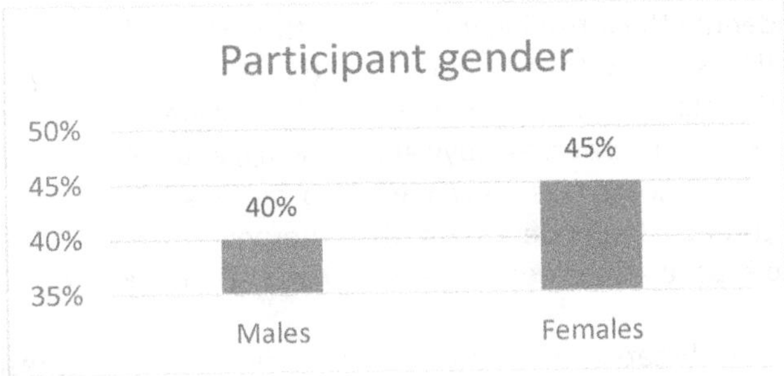

Figure 2: Participant gender

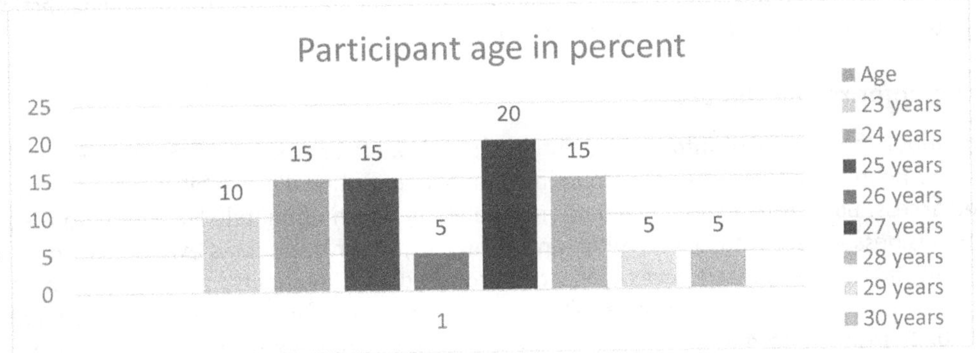

Figure 3: Participant age

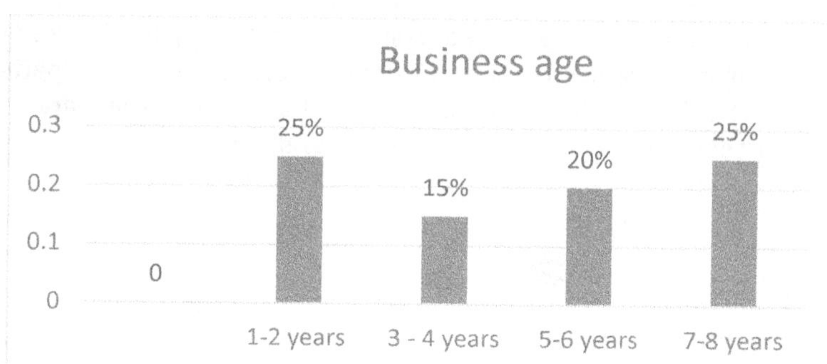

Figure 4: Business age

5.2 Common information threats

When asked about pervasive information security threats that the small business sector experiences, participants indicated that they are exposed to phishing, unauthorized access, modification of data, spyware, Trojan horses, and malware.

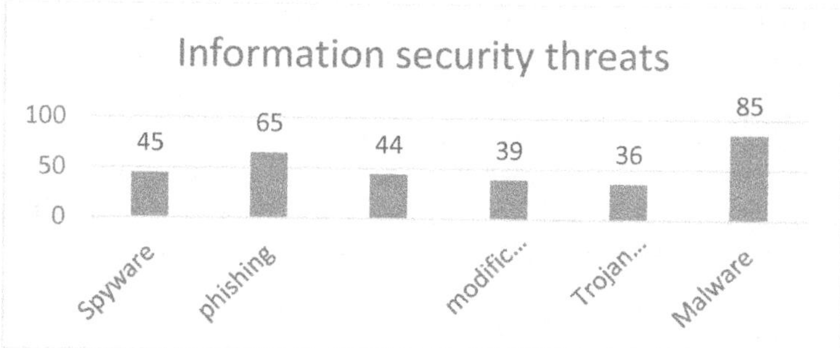

Figure 5: Information threats

Figure 5 shows information security threats that businesses experienced: 45% of the businesses experienced spyware, 65% phishing, 44% unauthorized access, 39% modification of data, 36% Trojan horses, and 85% malware attacks. The results show that all the selected businesses were victims of information threats. The impact of the information threats weakens business information confidentiality, integrity, availability, and authorized access. Information threats are intrusive and disconcerting attacks within the small business sector. All businesses need to identify the intruding information threats which increased during the global pandemic. Identifying the information security threats influences the strategies that can be used to mitigate the risks caused by information threats (Fedushko & Benova, 2019). Businesses experience these information threats through the use and access to ICT resources. Spyware, phishing, and Trojan horses result from connection to the Internet. Even though cyberspace brings convenience to businesses, equally this access opens channels for increased information threats. Dependency on the Internet and other information and communication technologies continue to expose all business sectors to threats (Ncubukezi & Mwansa, 2021).

In addition, the information threats can be caused by employee ignorance, resistance to change, mischievous behaviors, minimal awareness levels, and access to unknown websites (Safa et al., 2018). Likewise, other information threats are often caused by memory sticks and external hard drives. The following section presents the impact of these information threats (Ncubukezi, 2022).

5.3 Impact of information threats

When asked about the impact of information threats, participants shared their experiences that affect the business assets, people, systems, and information. Criminals intentionally exploit business systems to gain unauthorized access. Businesses experience data manipulation, which ultimately leads to data breaches. Even though some businesses have established mitigation strategies, other businesses experience denial of service, unavailability of data; unauthorized access; and compromised confidentiality, privacy, and integrity, which ultimately cause data breaches. These information security breaches are caused by insiders and outsiders (Ncubukezi, 2022). Insider information threats are often caused by employee ignorance, poor decision making, lack of skills, poorly enforced security strategies, understaffing, poor security guidelines, and a technological knowledge gap (Kluge, Sasse & Verret, 2022; Singh & Singh, 2022).

When security for business resources has not been appropriately implemented, businesses are vulnerable to various threats. Information security threats – software attacks, intellectual property theft, identity theft, equipment or information, sabotage, and information extortion (Deeks, 2020) – severely impact business information. The results of information security can diminish a business's reputation and hinder financial groups (Ncubukezi, 2022). Figure 6 illustrates the impact of information threats.

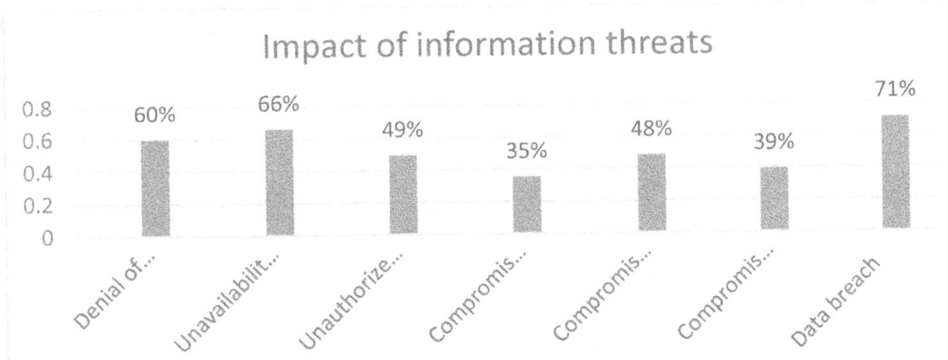

Figure 6: Impact of information threats

5.4 Measures to protect against the information security threats

When businesses were asked about security measures, participants indicated that they use passwords to protect sensitive information and resources. However, passwords are ineffective if they are not regularly updated or meet acceptable password criteria. Weak passwords cause a major loophole in the system (Jiow, Mwagwabi & Low-Lim, 2021). Firewalls detect information security threats penetrating the system (Emm, 2013). The lack of firewalls within the small business space increases the chances of unauthorized access to the system. An effective firewall filters both the incoming and outgoing traffic on the network. According to Okoye (2017), most managers are concerned about the confidentiality of an organization's operational data rather than protecting the firm's knowledge and information assets. Hutchings, Smith, and James (2013) suggest that "small businesses face several computer security threats and may lack the time and technological resources that enable software updates and patches that improve security." Due to the severity of risks caused by the global pandemic, software manufacturers improve their security by updating and edifying software patches; failure to run software updates exposes systems to a diverse range of threats.

The small business sector should invest in securing overall resources and information. They should adopt alternative and advanced strategies to detect early warning signs. In addition, businesses should adopt cloud storages that will serve as a secure backup system, as cloud storage has a high level of security which will heighten the safety and security of information. The idea is to back up data according to each business's need, especially free cloud storage, likely meeting small business needs (Hutchings, Smith & James, 2013). Businesses should also always have up-to-date antivirus and antispyware. When business devices are connected to the network, the devices with antivirus and antispyware will automatically run updates to detect abnormalities on the business network and system. In addition, device encryption is used to improve security.

Data can be protected in multiple ways. Small businesses need a security strategy to protect their own business, customers, and data. Information security threats always scope for ways to gain access into a system, detrimentally impacting businesses, resulting in data loss, device theft, breaches, and financial loss. All businesses encounter different information risks based on their services, but all are impacted negatively by information threats. Security threats include everything that protects sensitive data, personally identifiable information, protected health information, and governmental and industrial information systems from theft and damage attempted by criminals and adversaries. Protecting computer operating systems, networks, and data from cyber-attacks requires special skills and knowledge. It may be expensive for growing businesses to hire an IT specialist to secure their systems.

Table 1: Summary of research questions about mitigation strategies

Questions	Measures used	Business responses
What security measures do businesses implement?	Passwords	100% of the respondents indicated that they use passwords despite not complying with the accepted password criteria or regularly updating them.
	Firewall	60% of the respondents indicated that they have a firewall, and another 40% do not use a firewall to filter incoming traffic.
	Backup system and cloud storage	68% of the businesses back up their information on the cloud, while 32% do not back up their information. Instead, they trust their hard drives.
	Software updates and patches	56% of the businesses automatically run software updates which increase the safety and security of information. Some businesses do not run software updates, which creates a loophole for various information threats.
	Antivirus and antispyware	40% of the respondents operate on stand-alone devices which do not allow automatic antivirus updates and antispyware.

5.5 Recommendations and future research

The evident increase in cyber-attacks within small businesses in South Africa requires investment in proactive defensive technologies to reduce information risks and threats. Approximately 43% of cyber-attacks are aimed particularly at small businesses in the health, insurance, retail, financial and legal sectors (Ginindza, 2021). The research participants shared their experiences with information security threats and the impact on small businesses, especially Internet cafés. The study revealed a need to equip small businesses with ways to combat information security threats. Effective information security requires enforced policies, training management, and established security controls (Cheung, 2014). The small business sector should be aware of the information threats to effectively protect against their assets and infrastructure before deploying security measures (Anderson, 2003). Pranggono and Arabo (2021) suggest using a virtual private network continuous training for system end-users. Due to high exposure to information threats during the Covid-19 pandemic, the following are recommended for businesses:

- Implement security on information, systems, network, devices, and personnel.
- Always use firewalls to filter incoming and outgoing traffic.
- Use virtual private networks (VPNs) for secure Internet connections.
- Enforce strong password criteria.
- Regularly update software and restrict automatic installations, especially from unknown websites.
- Provide security training throughout the year, updated regularly.
- Use encryption and multi-factor authentication.

This study should include other business sectors in the future.

5.5.1 Contribution and significance

The global Covid-19 pandemic became a fertile season for information threats. It influenced the use of technology for reaching out to loved ones, other businesses, suppliers, and new business partners. The study's findings showed small businesses' current state of information security. First, this study gave insight into the current state of information security within the small business sector. Secondly, this study generated awareness about the current information security, its impact, and security measures in the small business sector. The more information available about information security, the better strategies can be established. And finally, this study challenged businesses to explore lasting and robust security strategies to strengthen information security.

6. Conclusion

The Covid-19 pandemic affected everyone and introduced new changes and rules, resulting in thousands of deaths, self-isolation, lockdown, and increased Internet dependency. Small businesses, in particular, have had to adapt and adopt new technologies, which have increased exposure and vulnerability to cyber-attacks and information threats. The study determined the nature of information security threats in the small business sector, threats that steal private and sensitive data, finances, and client information. The research revealed that the small business sector is exposed to various information threats that negatively affect businesses. No business can avoid information threats. In addition, the study revealed the significant information security risks and the impact of intrusive information threats on the small business sector. In response, businesses shared their current security strategies for reducing business risks.

References

Akpan, I.J., Udoh, E.A.P. & Adebisi, B. (2020). "Small business awareness and adoption of state-of-the-art technologies in emerging and developing markets, and lessons from the COVID-19 pandemic". *Journal of Small Business & Entrepreneurship*, pp. 1-18.

Alhassan, M.M. & Adjei-Quaye, A. (2017). "Information security in an organization." *International Journal of Computer (IJC), 24*(1), pp. 100-116.

Anderson, J.M. (2003). "Why do we need a new definition of information security." *Computers & Security, 22*(4), pp. 308-313.

Biener, C., Eling, M. & Wirfs, J.H. (2015). "Insurability of cyber risk: An empirical analysis." *The Geneva Papers on Risk and Insurance-Issues and Practice, 40*(1), pp. 131-158.

Blakley, B., McDermott, E. & Geer, D. (2001). "Information security is information risk management." In *Proceedings of the 2001 workshop on New security paradigms* (pp. 97-104).

Braun, V., Clarke, V., Boulton, E., Davey, L. & McEvoy, C. (2021). "The online survey as a qualitative research tool." *International Journal of Social Research Methodology, 24*(6), pp. 641-654.

Cheung, S.K. (2014). "Information security management for higher education institutions." In *Intelligent Data analysis and its Applications, Volume I* (pp. 11-19). Springer, Cham.

Deeks, A. (2020). "Secrecy Surrogates." *Virginia Law Review, 106*(7), pp. 1395-1477.

Emm, D. (2013). "Security for SMBs: Why it's not just big businesses that should be concerned." *Computer Fraud & Security*, 2013(4), pp. 5-8.

Fedushko, S. & Benova, E. (2019). "Semantic analysis for information and communication threats detection of online service users." *Procedia Computer Science, 160*, pp. 254-259.

Georgiadou, A., Mouzakitis, S. & Askounis, D. (2021). "Working from home during COVID-19 crisis: a cyber security culture assessment survey." *Security Journal*, pp. 1-20.

Gerić, S. & Hutinski, Ž. (2007). "Information system security threats classifications." *Journal of Information and organizational sciences, 31*(1), pp. 51-61.

Ginindza, B. (2021). "Work from home increases cyber-attack risks for SMEs" [Accessed on 13 March 2022] available from https://www.iol.co.za/business-report/companies/work-from-home-increases-cyber-attack-risks-for-smes-8991e5de-5bdc-48d5-afa0-3c8c56ff4a2a

Heidt, M., Gerlach, J.P. & Buxmann, P. (2019). "Investigating the security divide between SME and large companies: How SME characteristics influence organizational IT security investments." *Information Systems Frontiers, 21*(6), pp. 1285-1305.

Herath, T. & Herath, HS (2020). "Coping with the new normal imposed by the COVID-19 pandemic: Lessons for technology management and governance." *Information Systems Management, 37*(4), pp. 277-283.

Holovkin, B.M., Tavolzhanskyi, O.V. & Lysodyed, O.V. (2021). "Corruption as a cybersecurity threat in conditions of the new world's order." *Linguistics and Culture Review, 5*(S3), pp. 499-512.

Hutchings, A., Smith, R.G. & James, L. (2013). "Cloud computing for small business: Criminal and Security threats and prevention measures." *Trends and Issues in Crime and Criminal Justice*, (456), pp. 1-8.

Itliong, J. (2020). Online Strategies For Small Businesses Affected By Covid-19: A Social Media And Social Commerce Approach. Masters Thesis California State University, San Bernardino.

Jiow, H.J., Mwagwabi, F. & Low-Lim, A. (2021). "Effectiveness of protection motivation theory based: Password hygiene training programme for youth media literacy education." *Journal of Media Literacy Education*, 13(1), pp. 67-78.

Kluge, A., Sasse, M.A. & Verret, I. (2022). "Why IT Security Needs Therapy." In *Computer Security. ESORICS 2021 International Workshops: CyberICPS, SECPRE, ADIoT, SPOSE, CPS4CIP, and CDT&SECOMANE, Darmstadt, Germany, October 4–8, 2021, Revised Selected Papers* (Vol. 13106, p. 335). Springer Nature.

Lallie, H.S., Shepherd, L.A., Nurse, J.R., Erola, A., Epiphaniou, G., Maple, C. & Bellekens, X. (2021). Cyber security in the age of COVID-19: A timeline and analysis of cyber-crime and cyber-attacks during the pandemic. *Computers & Security*, 105, p. 102248.

Lundgren, B. & Möller, N. (2019). "Defining information security." *Science and engineering ethics, 25*(2), pp. 419-441.

Monteith, S., Bauer, M., Alda, M., Geddes, J., Whybrow, P.C. & Glenn, T. (2021). "Increasing cybercrime since the pandemic: Concerns for psychiatry." *Current psychiatry reports, 23*(4), pp. 1-9.

Moura, J. & Serrão, C. (2015). "Security and privacy issues of big data." In *Handbook of research on trends and future directions in big data and web intelligence* (pp. 20-52). IGI Global.

Ncubukezi, T. (2021). "Human errors: A cybersecurity concern and the weakest link to small businesses." *International Conference on Cyber Warfare and Security*, 17, pp. 395-403.

Ncubukezi, T. & Mwansa, L. (2021). "Best Practices Used by Businesses to Maintain Good Cyber Hygiene During Covid19 Pandemic." *International Conference for Internet Technology and Secured Transactions* , 9, pp. 714–721.

Ncubukezi, T., Mwansa, L. & Rocaries, F. (2020). "Review of the current cyber hygiene in small and medium-sized businesses." *International Conference for Internet Technology and Secured Transactions,* 15, pp. 283–288.

Ncubukezi, T., Mwansa, L. & Rocaries, F. (2021). "An analysis of the cybercrimes within the Western Cape small and medium-sized enterprises." *International Conference on Cyber Warfare and Security*, 16, pp. 425-435.

Okoye, SI (2017). "*Strategies to minimize the effects of information security threats on business performance.*" (Doctoral dissertation, Walden University).

Paulsen, C. & Toth, P. (2016). "*Small business information security: The fundamentals*." (No. NIST Internal or Interagency Reports (NISTIR) 7621 Rev. 1). National Institute of Standards and Technology.

Pranggono, B. & Arabo, A. (2021). "COVID-19 pandemic cybersecurity issues." *Internet Technology Letters*, 4(2), e247.

Safa, N.S., Maple, C., Watson, T. & Von Solms, R. (2018). "Motivation and opportunity-based model to reduce information security insider threats in organizations." *Journal of information security and applications*, 40, pp. 247-257.

Singh, I. & Singh, Y. (2022). "Cyber-Security Knowledge and Practice of Nurses in Private Hospitals in Northern Durban, Kwazulu-Natal. " *Journal of Theoretical and Applied Information Technology, 100*(1).

Thapa, C. & Camtepe, S. (2021). "Precision health data: Requirements, challenges and existing data security and privacy techniques." *Computers in biology and medicine*, 129, p. 104130.

Von Solms, R. & Van Niekerk, J. (2013). "From information security to cyber security." *Computers & Security*, 38, pp. 97-102.

Whitman, M.E. & Mattord, H.J. (2021). "*Principles of information security*." Cengage Learning.

Zafar, H., Ko, M.S. & Osei-Bryson, K.M. (2016). "The value of the CIO in the top management team on performance in the case of information security breaches." *Information Systems Frontiers*, 18(6), pp. 1205-1215.

Analysis of Sexual Abuse of Children Online and CAM Investigations in Europe

Johanna Parviainen and Jyri Rajamäki
Laurea University of Applied Sciences, Espoo, Finland
Johanna.Parviainen@student.laurea.fi
Jyri.Rajamaki@laurea.fi

Abstract: Child sexual abuse or child´s exploitation online as sexual violence including Child Abuse Materials (CAM/CSAM) is a global phenomenon. This case study aims to get information on the current nature of crimes by online published surveys, reports, articles, and documents as an international and cross-border cybercrime in Europe. To get information of children´s own experiences of some European countries, information on how they react to sexual messages or sexual harassment online or how they recognize a threat to be a victim of sexual abuse online are important aspects to understand the phenomenon at all. The sexually motivated offenders and their behavior online conversations are also important to recognize to get more information of this criminal activity at all. If sexual abuse has been done only online, the knowledge of the current events helps law enforcement authorities (LEAs) to understand how they could find reliably the needed digital evidence for pre-trial investigations and judicial processes. The authorities' workload can be high in CAM/CSAM cases first with handling enormous digital data, but also with nature of cases which has seen widely causing different forms of stress also to professionals. From this point of view, this study also aims to describe how the different forensic tools and technological solutions would help LEAs with their jobs, for example, by classifying different materials into different categories, recognizing better victims and suspects, or winning time to investigate other crimes.

Keywords: cybercrime, child abuse, child abuse materials, sexual harassment, CAM investigation

1. Introduction

Digitalization and using the internet online is a part of our everyday life. Also, children take part in a vibrant online environment by using new tools, technology advances, infinity novelties, and opportunities. Comparing earlier generations, children as a representant of the so-called digital native generation can have already completely different skills to gather information or excellent learned multitasking skills (Pongrácz, 2019). Children´s rights and their fundamental freedom as digital citizenship in free societies are important. To their fullest development, potential, and participation in the digital environment and infrastructure with communication technology devices, connectivity, services, and contents, everyone should have affordable access and the possibility to use the digital environment. It has been seen that holistic participation provides resilience and response to life both online and offline. However, it is important to ensure that digital content fits children's age and maturity as age-appropriate and child-friendly content in a different area of living also providing a plurality of high-quality information sources with different services online. Taking care of family and other social relationships online is important also to children. Council of Europe (2018) has seen that all member states must protect children in the digital environment from violence, exploitation, or abusing avoiding risks. Risks can be associated with harmful or illegal activities or events in which children should have competence wisely commitment to the digital environment. These dangers can be multiple such as extremely online recruitment to criminal offenses, extremist political or religious movements. Online can also include risks for bullying, harassment, hate speech, discrimination, racism, stalking, grooming, adult pornography, or child sexual materials. Children´s needs and vulnerable settings should be able to guarantee from all these (Council of Europe, 2018).

Every environment can include harmful or potential risks because it is natural that we all can meet at the same time with good things different bad things, threats, dangers, and harmful situations there we are – in this case in the digital world online. That is why we need knowledge of individual risk online. When we are talking about children, it is necessary to notice that they can naturally have a low sense of danger or they cannot interpret something in their ages right or as a risk at all (Pongrácz, 2019). The growing concern has seen that different platforms on the internet may include individuals with a sexual interest in children (Kloess, et al., 2017), who are using the widely opened possibilities of the internet to possess, product and distribute illegal materials such as CAM/CSAM (Brown, 2017). Recognizing the current nature of these crimes can give better possibilities to understand and react to this multidisciplinary and comprehensive phenomenon (Europol, 2017).

Children have a risk of being a victim of sexual harassment or sexual abuse with sexual messages, asked for or send nude photos and CAM/CSAM materials photos or videos. The offender's intention, criminal activities, and illegal material have harmful reactions or distress to child victims internationally and cross-border. This case study, focusing on the general European context, investigates child sexual abuse or child´s exploitation online including CAM/CSAM as sexual violence. The research material contains different reports, articles, documents, and common conventions behind different legislations of child sexual abuse online or child exploitation as an international cybercrime. These sources were chosen mainly from the latest year as a sample to get updated knowledge of victims, offenders, technological issues, and point of view of judicial processes as digital evidence and investigators' workload in pre-trial investigations. The study explores this complexity and widespread phenomenon as a crime against children and the dissemination of CAM/CSAM material.

2. Children's online behavior

The EU Kids Online 2020 survey, summarized in Table 1, gained results from 25 000 children aged 9-16 in 19 European countries between autumn 2017 and summer 2019. The survey mapped children's internet access, online practices, skills, risks, and opportunities by summarizing the main findings as a key topic from each area. The risks and opportunities mentioned as specific activities or experiences that could lead to harm or a positive outcome, including overall negative experiences, online aggression and cyberbullying, encountering potentially harmful content, experiencing data misuse, excessive internet use, sexting, seeing sexual images, meeting new people online and preference for online communication. In most countries, children had happened something bothered or upset in the past year online, not frequently, but a few times. Harmful content was named including, for example, cyberbullying, hate messages, requires to be very thin, personal data misuse, or get viruses or spyware. Also spending time daily or weekly online or other excessive internet use has reported in some cases to be harmful impacting to children eating, sleeping or leaving them less time to spend with friends, family or doing schoolwork but the majority of the children in all of the countries had not shared this experience. To clarify sexual messages by this survey it was first noticed that many children exchanged sexual messages with a peer or girl/boyfriend, which were seen quite normal development to be using digital environments online context day-to-day lives in every life´s area considering also flirting, exploration of sexuality and the establishment. Maintenance of intimate relationships which also could be mediated via technology did not constitute a risk of serious harm. Commonly it was also reported that many can feel it be easier themselves online than meeting people face-to-face, but the majority of the children in all countries had told that they never talk about personal things online that they did not talk about face-to-face. However, eight countries reported that most of the children were talking about different things online than offline at least sometimes or more often. (Smahel, et al., 2020)

The EU Kids Online 2020 survey studied how to ensure a balanced approach to sexual images online or other media considering all kinds of media-related exposure to sexual images, not only online materials. Many children could seek sexual content by curiosity, trying to find answers for questions they have about puberty or knowledge of their own body and sexual identity. To clarify more deeply this part of images which can cause harm or potential risks as a bothering or upsetting material, children had named for example being exposed to online sexual content, aggressive content, and other types of unwanted sexual content. An average of 22 percent of children, commonly older children, reported that they had received sexual messages in the past year. Also, an average of 6 percent of children reported that they had sent sexual messages. If children were asked or to him/her had received sexual requests to exchange sexual messages or material online, they had mainly been girls between 12 to 16 years. Especially girls reported that they had felt upset, distressed, and potential harm after this. Sending sexual messages, images and videos could always have the potential to be distributed and made public without original participations control. Commonly, several children had not told harmful experiences to anyone and if, they had told to a parent, friend, teacher or to professionals who helps children. Most children say they usually know how to react to the online behaviors of others which they don't like for example coping with blocking the person, talking to other people, ignoring the problem, using technical measures, or confronting the stressor or aggressor. Complex sexting using different electronic devices could lead to grooming or sexually abusive behavior efforts by adults. These situations included also risks for sending or receiving intimate digital material (Smahel, et al., 2020).

Table 1: Examples from different European countries by a survey to get a wider approach to the experience of harmful content or risk for abuse children online (Smahel, et al., 2020)

Country	Findings
Croatia	30 percent of children aged 9-17 had seen sexual content online in the past year almost two-thirds of children had seen sexual photos or films with nudity intentionally without any intention to see these contents
Portugal	33 percent of children and youths have reported seen sexual content online in the past year
Czech Republic	girls have more often specially asked to share intimate information boys have received sexually charged messages girls and younger children have reported being upset being attacked online
Germany	bullying, receiving, or sending sexual messages has seen being quite normal with digitally engaged children children have been also exposed to various cyber risks by many opportunities of being online, such as exposing also sexual, violent, or hateful content 63 percent of children have reported that they feel pleasant experience to meet an online contact in real life
Switzerland	many young children have online social media profiles, and they see social media useful tool to find new friends older children have seen to be vulnerable to more risky experiences by having more skills and owning smartphones many meets with online known in real life face-to-face and mostly they have felt experiences enjoyed and positive

Comparing these findings to earlier studies, the wide view of point can have still more specific details to take consider. For example, in the Czech has been recognized that specific social networks´ online blackmailing of intimate images has affected 6 to 8 percent of Czech children, which has meant more than 21,000 children (Kopecký, 2016). German survey has pointed out that parents or carers may thoughtlessly add their children´s risk online publishing children´s or adolescents` pictures online without their permission. It has been seen as important that parents or carers are aware of possible negative consequences, hurtful comments, or their children´s right to privacy at all (Smahel, et al., 2020). In spring 2021, Save the Children Finland has surveyed grooming by children´s self-reported online survey. Children mainly recognize different conversations, messages, or images which can lead to grooming quite well, but still, there can be a gap between their knowledge and real-life actions in real situations how they react to contact-making or how ready they are to tell parents at least of harmful situations online (Juusola, et al., 2021). One meaningful question is also that in general, meeting a stranger in real life after online knowing in the theme of child sexual abuse has depicted as especially risky ordering to tell someone about the meeting or meet unknown only in a public place, but a Swizz's survey also wants to consider that average 70 percent of all children in the survey reported that they were happy after the meetings which could be an opportunity and benefits for children (Smahel, et al., 2020).

3. Grooming, child sexual abuse online and CSAM

Figure 1 summarizes the phenomenon of child sexual abuse online.

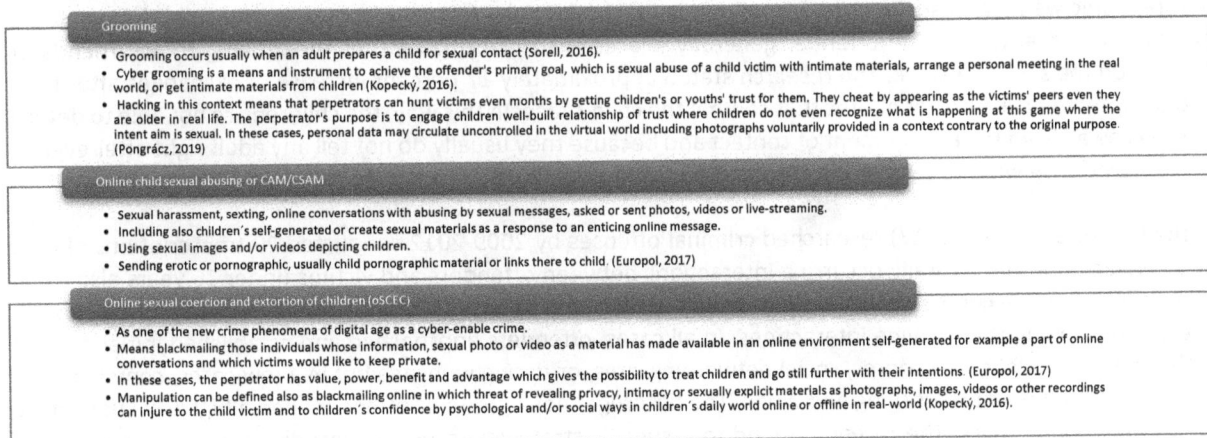

Figure 1: Terminology definitions

Contact making to a child normally starts with a conversation on media platforms, chat applications, or online games or gaming platforms. It is quite common that conversations move from one online platform to another including social engineering methods with manipulative or threatening tactics to create an individual relationship. The perpetrators can be peer-perpetrated with harmful and bullying consequences to the victim, but most of the cases are made by sexually motivated adults who use technology to engage in sexual contact with children through processes of sexual grooming to get CSAM or meet victims for physical sexual activity. The perpetrators may also be organized groups with sexual and/or financial motivations. Financially motivated offenders' main goal is to obtain money and procurement sexual material for commercial distribution. Usually, they had been located to organized criminal groups which generally operate in teams and with contacts between countries with a common language outside of the EU both national and international levels. In these cases, perpetrators can include both genders, they do not know the victim personally and they can target widely potential victims. Usually, they have used pre-recorded footage by enticing online messages or manipulative techniques obtained from pornography and live-sex camera sites. Mainly manipulations can occur, for example, by pretending to be younger (peer), with sexual reciprocation ("I do, if you do" meanings by shared images or live streaming), assuring friendship or romantic relationship online, using multiple fake online identities (pretending to be a supportive friend or a sympathetic victim of the same offender) or offering something (e.g., money, drugs) to the child. It can be enough to threaten with messages for example to post images to a place where family and friends could see, for example, victims naked photos or threat to create sexual content of child by using digital editing tools or record child unknowingly, but in the worst cases, threats can be threatening to hurt a child or family members physically or commit suicide (Europol, 2017). However, it is quite usual in real life that those who have picked up an online investigation into child abuse images have not previously been known to the police or other authorities (Brown, 2017). In the next two chapters, couple examples of from earlier research in Europe describe closer offenders' motivation as a criminal activity.

In the Czech Republic, Kopecký (2016) surveyed children's prevalence of cyberbullying and other risky forms of communication in the years 2014-2016. He focused on specific cases of children's extortion that had been captured and dealt with by the Centre for the Prevention of Risky Virtual Communication and Police for the research. He found victims ages were 11-17 years by whom 73 percent were girls. He also stated that 15-17 years old children were in a worse situation regarding blackmail even younger children would have a lack of life experiences. Recognized techniques had focused on gaining children's confidence, luring out intimate material, and subsequent blackmailing. Kopecký identified a simplified model of blackmailing of children in five stages through the usage of ICT by specific, repetitive forms of communication within the behaviors of the offenders. First, the offender gets in touch with the child by a conversation which can typically be initiated sexually or reflect straight to exchange or have photos. After it, the offender begins to positively evaluate by flattering or praising the victim all materials sent by victims and ask smoothly for more photos or other materials to achieve children's admiration and respect. In the third stage, the offender attempts to confirm that the victim´s identity leads to a real child and that the photos are authentic by commonly identifying verification photos with the note or date. After this, photos get more increasing and intensifying nature between the victim and the offender where children's participation can also imitate offenders´ behavior in exchanging naked photos or sexual acts when a child can send to the offender almost anything. The offender can also send photos from erotic or pornographic websites, usually child pornographic material, to the child. At last, when the child decides to be not to respond anymore, the offender generally proceeds to blackmail either threatening to tell friends or parents on the social network. The research stated approximately 17 percent increase of exchanges after that, and the offender can also get photos that contain the child´s face. In this way, the children are not able to detach or break away from the engagement of contact and because they usually do not tell any adult right after events, there are no easy ways to stop or interrupt actions. (Kopecký, 2016)

In the UK, Kloess, et al. (2017) researched criminal offenses by 2009-2012 conducted 29 transcriptions of chat logs conversations in naturally occurring interactions between offenders and victims under 16 years also using each case the whole contextual police reports had ascertained offenders underlying motivation and function to get children to engage in online interactions. In all cases, offenders approached and established initial contact with victims via public chat rooms, social networking sites, or dating websites but in a few cases, conversations were moved to instant messaging and email. The heterogeneous group of offenders had a highly sexual nature of conversational topics. The research found four specific strategies to engage victims: directness in initiating online sexual activity, pursuing sexual information, the next step, and fantasy rehearsal even if by manipulative strategies. Commonly offenders could design their communication in such a way that frequently achieved their goal of sexual arousal and gratification trying to achieve victims' compliance by overcoming victims' resistance to incite victims into performing sexual acts on online sexual activity. Offenders can be suggestive or flattering or enquire about the victim´s sexual likings, preferences, practices, or previous sexual experiences. The offenders can deal with their own experiences or use emotions such as normalizing the nature of behavior or performing playful to belittling situations. For example, repeating initial offenders' questions, requirements, or compliance, insinuating through offender´s disappointment, sadness, or other feelings offenders try to find sexual arousing. It is also possible to use other victims recorded sexual behavior online encouraging victims to respond by likelihood way. Also, direct attempts to initiate sexual activity via webcam to show parts of body or genitalia, masturbation, use foreign objects to perform penetrative sexual acts or ask for arrangements for physical meetings or conduct sexual contact are usual. Mainly offenders ensure contacts continue to the future respectively even if they also consider minimizing the risk of disclosure and detection, but there can also be highly manipulative, sexual of fantasy fulfillment motivated and even aggressive directness to the victim after a few minutes´ conversations. (Kloess, et al., 2017)

4. CAM/CSAM investigation methods

The purpose of digital forensic investigations is to get digital evidence of a specific criminal event. In these cases, in which offenders have operated in illegal activities through their digital devices, by seized devices appropriate analysis of digital examination is a possibility to get needed information for this evidence, find information of victims, other offenders, or CSAM. These digital traces with every single item of digital evidence of different events will guarantee both victim´s and suspect´s rights in the judicial process by presenting admissibility and incriminating evidence to a court and other legal liabilities. Forensics needs are challenged also by the development of technology. It is not only a question of new possibilities or threats, even if it is a question about day-to-day practices. Data per person on average has increased enormously and it leads to the continuous growth of data scales in examinations which has quite commonly used for data storage terabyte-sized hard disks and cloud services by high capacity and low costs. It is quite normal that almost everyone has a large amount of digital information spread throughout separate personal devices and cloud-based personal accounts. This also means an ever-growing workload to online child abuse investigators and criminal technicians. All of this can cause also delays first for investigations and all trial processes. (Acar, 2018)

In contexts of cybercrimes, digital pieces of evidence of the crime can even itself be enough for prosecution and conviction in the trial (Denk-Florea, et al., 2020) or they are important with other pieces of evidence by side with child´s interview or other´s interrogations. For example, two studies in the last years have examined law enforcement professionals' workload and distress with CSAM investigation. A survey from the UK has shared the professionals' negative emotions like shock, anger, desensitization to material, desire not to look at the material, physical responses of sickness or empathy feelings for the victims, or feelings of inadequacy. Many events of cases have been named as a highly disturbing material, but also the duration and frequency of viewing materials and the uniqueness of each criminal case can affect distress. For example, especially audio or written information included videos were recognized to be highly distressing by their way to witness and recollect victims' traumatic event experience again. In the CSAM context, all indications include abuse and contents can at least be so raw with intrusive images, unwanted thoughts, or flashbacks that there can be hard to find resilience or psychological safety to professionals heightened rationale to do their job even they have commitment and motivation to important jobs to seek norm violations or illegal practices as their job (Denk-Florea, et al., 2020). A survey in the US has researched that there is an increased risk for experiencing psychological distress, different transforms or stress symptoms with can occur as emotional stress by indirectly stressful experiences without non-specific traumatic events; such as compassion fatigue. Especially situations can be stressful for those investigators who have the overlap dual roles as both duties in smaller units; duties as a digital forensic examiner and a case-investigators even in each case meeting also in face-to-face interaction with interrogations the child victims, parents or family members and offender or offenders. (Seigfried-Spellar, 2017)

Images depicting children in sexual acts are under criminal offense mainly around the world their possession, production, or distribution, which means that law´s everywhere should intervene, reinforce and reflect unwanted events. By continuing growth of digital materials has indicated that also image-related offenses would come to be increasingly challenging technological solutions to identify illegal materials and that still demanding also more forensic examinations of seized equipment or devices and setting new requirements to all of them who are working with these cases (Brown, 2017). As an international and cross-border crime, child sexual abuse online and CAM/CSAM means that in many cases the offender, victim, and data can all relate to different countries. Usually challenges to forensic issues has associated with the locality with legal frames to get data from countries where the some company´s headquarter or operational center is located to technological possibilities and capabilities particularly or to anonymization technologies, encryptions or peer-to-peer structures, real possibilities to victim and offender identifications, particularly to the actual current legal issues or all these issues limitations together (Acar, 2017). It has also been recognized that in a few cases, the offenders have used sophisticated computer programs which have allowed the users to remotely control a PC or laptop webcam to record or acquire intimate materials without the victim´s knowledge (Kopecký, 2016). Internet anonymity has been seen as one of the key risk factors for harmful harassment, abuses or uncontrolled behavior at all no matter if there is a question of individual or organized crimes because of hiding your real name and identity that perpetrator who has a sexual desire for children as a personality disorder can cause many risks to keep their experiences and activities secret by transferring child pornographic or child sexual abuse materials or images anonymously through the network or seeking for victims (Pongrácz, 2019). In Dark Web´s different platforms, anonymity can protect perpetrators online without risks of being caught and they can form groups of hundreds or thousands of members scattering all over the planet to deal individual´s sexual interest in children or an interest in child abuse images production, distribution or exchange (Brown, 2017).

Recognizing illegal material by technological ways would get advice and help for all actors. Attempts to reduce or eliminate the traffic images online can start from to use of known URLs with illegal contains to block them by lists or restrict access to certain pages. The hashes can help to develop PhotoDNA which is meant for image-matching technology as a unique signature for a digital image. By this signature, it would be easier to identify, report, eliminate and find copies of images or material of CSAM online (Brown, 2017). In the latest years, the discussion of different possibilities which technologies can occur to the offender, but also current technologies or future´s new technologies which can or could be helpful or advancing to authorities' jobs in prevention or detection. For example, Acar (2017) has seen that applying web cameras and video streaming through Voice-over-IP (VoIP) applications with audio and video communications can be one of the emerging forms of online child sexual abusing between the offenders and victims as an efficient, fast, cheap and private technology without no clue to the real world. Acar has mentioned that discussion of new or better practical solutions needs to solve cases with a technical and legal debate together, also with private sectors entities or as a public-private partnership. Video chats, for example, could enable access to the third part in real-time communications considering issues such as privacy, needed information to protect children but also to get solid evidence of recordings of criminal activities for the legal processes. According to Acar (2017), fully automated chatbots with text-to-speech recognition and big data analysis of metadata (attributes of a resource such as a date, creator, and location of IP addresses) including content data (communications texts, audio and video files) by VoIP companies could be helpful items to get the situation to better. In particular, crime prevention needs technology considering effective but also cost-efficient strategies. Technological capabilities should be taken into consideration proactively, considering that offenders will always apply new technologies for criminal activities too. (Acar, 2017) Maybe, automated methods are not adequate alone for current practical needs and measures of digital forensic investigations, but for example, automated triage system including time-consuming report writing could be quicker and easier for investigators workload as a job but also by a psychological way without seeing distressing materials. They could help to allocate LEAs' limited resources, for example, for quicker identification of victims. (Acar, 2018)

5. Discussion

Child sexual exploitation is one of the EU´s priorities in the fight against serious and organized crime (Europol, 2022). Commonly, cybercrimes are expected to come agile exploiting new technologies fast and tailoring their attacks using new methods and cooperating in new ways. This requires LEAs to have capabilities and possibilities for how to fight cybercrime (Interpol, 2022). The phenomenon was recognized already at the beginning of century 2000. There are different conventions or recommendations which have had impacts on domestic legislation in European countries or impacts to different measures or interventions as a co-operation between each other.

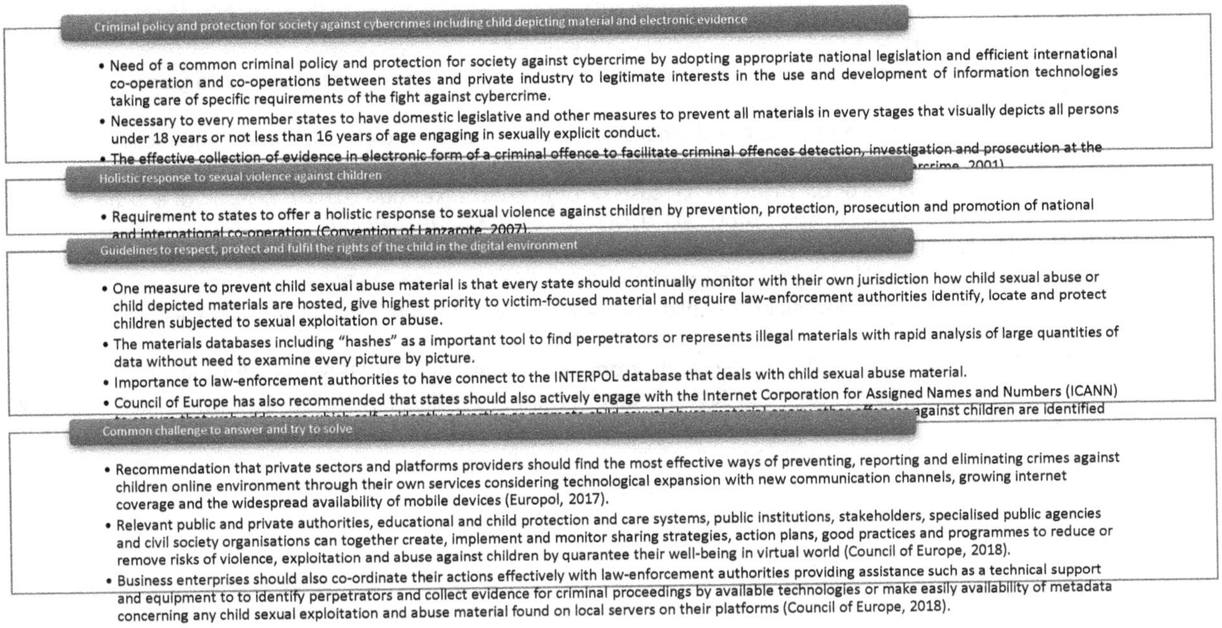

Criminal policy and protection for society against cybercrimes including child depicting material and electronic evidence

- Need of a common criminal policy and protection for society against cybercrime by adopting appropriate national legislation and efficient international co-operation and co-operations between states and private industry to legitimate interests in the use and development of information technologies taking care of specific requirements of the fight against cybercrime.
- Necessary to every member states to have domestic legislative and other measures to prevent all materials in every stages that visually depicts all persons under 18 years or not less than 16 years of age engaging in sexually explicit conduct.
- The effective collection of evidence in electronic form of a criminal offence to facilitate criminal offences detection, investigation and prosecution at the … rcrime, 2001).

Holistic response to sexual violence against children

- Requirement to states to offer a holistic response to sexual violence against children by prevention, protection, prosecution and promotion of national and international co-operation (Convention of Lanzarote, 2007).

Guidelines to respect, protect and fulfil the rights of the child in the digital environment

- One measure to prevent child sexual abuse material is that every state should continually monitor with their own jurisdiction how child sexual abuse or child depicted materials are hosted, give highest priority to victim-focused material and require law-enforcement authorities identify, locate and protect children subjected to sexual exploitation or abuse.
- The materials databases including "hashes" as a important tool to find perpetrators or represents illegal materials with rapid analysis of large quantities of data without need to examine every picture by picture.
- Importance to law-enforcement authorities to have connect to the INTERPOL database that deals with child sexual abuse material.
- Council of Europe has also recommended that states should also actively engage with the Internet Corporation for Assigned Names and Numbers (ICANN) … against children are identified

Common challenge to answer and try to solve

- Recommendation that private sectors and platforms providers should find the most effective ways of preventing, reporting and eliminating crimes against children online environment through their own services considering technological expansion with new communication channels, growing internet coverage and the widespread availability of mobile devices (Europol, 2017).
- Relevant public and private authorities, educational and child protection and care systems, public institutions, stakeholders, specialised public agencies and civil society organisations can together create, implement and monitor sharing strategies, action plans, good practices and programmes to reduce or remove risks of violence, exploitation and abuse against children by quarantee their well-being in virtual world (Council of Europe, 2018).
- Business enterprises should also co-ordinate their actions effectively with law-enforcement authorities providing assistance such as a technical support and equipment to to identify perpetrators and collect evidence for criminal proceedings by available technologies or make easily availability of metadata concerning any child sexual exploitation and abuse material found on local servers on their platforms (Council of Europe, 2018).

Figure 2: Approaches for preventing children online sexual abusing

Many sources consider the phenomenon is a collective problem. As Figure 2 presents, many challenges exist for dealing with the problem with significant technological players. Some players have already reunited for developing systems to locate contents of CSAM by applying publicly available information, announcements, and warnings to commit a criminal offense or to help sites visitors to find potential sources to get help if they were worried about their behavior (Brown, 2017). Former legislations and authorities' jurisdictions have placed requirements to different actors use their possibilities by side authorities work in societies. Combining earlier and future approaches together, a good possibility exists for handling the phenomenon better at the European level considering also different actors' possible interventions widely in every stage of every society.

Recent studies recognize offenders' manipulative ways of finding child victims which it is important to take also consider these children´s self-reported information of various situations online. Possible harmful situations also require awareness of possible new developing ways to these contact makings and how different actors can follow and prevent these situations even if different obstacles for getting better results to prevent, detect or solve cases are quite well known. New legislation and wide co-operation at all societies levels are needed for offering better safety in the digital environment also for children. Technological issues are both possibilities that the offenders can apply, but technology also provides possibilities to protect children online. The development should take consider also possible forensic tools and these technological solutions which could help LEAs´ workload to clarify possible crimes effectively. Further research is needed to focus on almost every aspect of this phenomenon considering time preference; for example, possible new technological developmental issues, by following possible changes in victims and offenders' behavior or different interventions impacts to phenomenon would be important examine.

References

Acar, K. V., 2017. "Webcam Child Prostitution: An Exploration of Current and Futuristic Methods of Detection", International Journal of Cyber Criminology 11(1):98-109. DOI:10.5281/zenodo.495775

Acar, K. V., 2018. "OSINT by Crowdsourcing: A Theoretical Model for Online Child Abuse Investigations", International Journal of Cyber Criminology 12(1):206-229. DOI:10.5281/zenodo.1467897.

Brown, J., 2017. *Online Risk to Children: Impact, Protection and Prevention*, John Wiley & Sons, Ltd.

Council of Europe, 2001. Convention on Cybercrime.

Council of Europe, 2007. Lanzarote Convention.

Council of Europe, 2018. Guidelines to respect, protect and fulfil the rights of the child in the digital environment, Recommendation CM/Rec (2018) of the Committee of Ministers, Council of Europe.

Denk-Florea, C.-B. et al., 2020. *Understanding and supporting law enforcement professionals working with distressing material: Findings from a qualitative study*, San Francisco: PLoS One; San Francisco.

Europol, 2017. Online sexual coercion and extortion as a form of crime affecting children - Law enforcement perspective, Europol, EC3, European Cybercrime Center.

Europol, 2022. Crime Areas; Child Sexual Exploitation.

Interpol, 2022. Crimes; Crimes against children.

Interpol, 2022. Crimes; Cybercrime.

Juusola, A. et al., 2021. "Grooming lasten silmin – selvitys 11-17-vuotiaiden lasten ja nuorten kokemasta groomingista netissä", Pelastakaa Lapset ry.

Kloess, J. A. et al., 2017. "A Qualitative Analysis of Offender´s Modus Operandi in Sexually Exploitative Interactions with Children Online", Criminology & Criminal Justice, 29(6): 563-591. DOI:10.1177/1079063215612442

Kopecký, K., 2016. "Online blackmail of Czech children focused on so-called "sextortion" (analysis of culprit and victim behaviors)", Telematics and Informatics, 34(1): 11-19. DOI:10.1016/j.tele.2016.04.004

Pongrácz, I., 2019. Children, Online Dangers and Solutions.

Seigfried-Spellar, K. C., 2017. "Assessing the Psychological Well-being and Coping Mechanisms of Law Enforcement Investigators vs. Digital Forensic Examiners of Child Pornography Investigations", Society for Police and Criminal Psychology.

Smahel, D. et al., 2020. EU Kids Online 2020: Survey results from 19 countries. DOI: 10.21953/lse.47fdeqj01ofo

Sorell, T., 2016. "Online Grooming and Preventive Justice", Criminal Law and Philosophy 11:705–724. DOI: 10.1007/s11572-016-9401-x.

Forensic Trails Obfuscation and Preservation via Hard Drive Firmware

Paul Underhill[1], Toyosi Oyinloye[1], Lee Speakman[2], and Thaddeus Eze[1]
[1]The Department of Computer Science, University of Chester, UK
[2]School of Science, Engineering and Environment, University of Salford, UK
p.underhill@chester.ac.uk
t.oyinloye@chester.ac.uk
l.speakman@salford.ac.uk
t.eze@chester.ac.uk

Abstract: The hard disk drive stores data the user is creating, modifying, and deleting while a firmware facilitates communication between the drive and the operating system. The firmware tells the device and machine how to communicate with each other and will share useful information such as, disk size and information on any bad sectors. Current research shows that exploits exist that can manipulate these outputs. As an attacker, you can change the size of the disk displayed to the operating system to hide data in, likewise by marking an area of the disk as bad. Users may not be aware of these changes as the operating system will accept the readings from the firmware. However, although the data is not reachable via the operating system this paper looks at the traceability of manipulated data using data recovery software FTK Imager, Recuva, EaseUS and FEX Imager. This report examines the use of malicious techniques to thwart digital forensic procedures by manipulating the firmware. It is shown how this is possible and current forensic techniques or software does not easily detect a change within the firmware. However, with the use of various forensic tools, obfuscated trails are detectable. This report follows a black box testing methodology to show the validation of forensic tools or software against anti-forensic techniques. The analysis of the results showed that most tools can find the firmware changes, however, it requires an analyst to spot the subtle differences between standard and manipulated devices. The use of multiple software tools can help an analyst spot the inconsistencies.

Keywords: hard drive firmware, digital forensics, data recovery, data manipulation, security analysis

1. Introduction

The use of forensic analysis to examine devices has become a standardised way to collect evidence or help provide insight into cyber attacks. The use of these methods are relied upon in examinations and are expected to show a true, un-modified record of the data on the device. However, what if the firmware of the device could be changed? This could hide, change, or manipulate what an analyst sees when first mounting the drive to their machine. This could prevent an accurate and comprehensive analysis of the device, especially if the analysts do not probe further into the reliability of the devices recorded details such as its true size and features.

The process of compromising or changing the integrity of evidence in this way is known as anti-forensics (Harris, 2006). The Parliamentary Office of Science and Technology (2015) explain that having such a wide range of devices and firmware will increase difficultly for an analyst to spot maliciously changed storge media. The National Police Chiefs Council (Hewitt, 2021) suggest that the biggest issue with the collection of evidence will be ensuring analysts have the knowledge and capability to respond to these more sophisticated attacks. The last point the NPCC make is the issue of completing investigations promptly while sticking to strict budgets. Here a challenge arises, to perform a thorough review of the device in a forensically sound manner. To overcome this the Attorney General's Guidelines (2013) allows tools to be employed for imaging, searching and collection of evidence to increase efficiency. This paper will show how the use of malicious firmware could thwart detection when analysed with these tools, and how a manual inspection is a more reliable way to find the data. The main test will be to examine a drive where the data has been saved to a hidden area on the disk that has been changed at a firmware level. The success of each tool will be recorded, any further work that can build upon this and any solutions that can be offered to identity compromised firmware will also be discussed.

1.1 Research question / aim

Etow (2020) found little research has been published assessing the reliability of forensic tools against anti-forensics. This lack of findings means methodologies for anti-forensic attacks have not been securitised for their effectiveness. Therefore, this research will complete an anti-forensic strategy to test forensic and recovery software to find its effectiveness.

The aim of the research is to manipulate displayed data on a disk by editing the firmware. This leads to the research questions, 'Can the change be implemented without current forensic software detecting it and can the data be extracted or recovered in a forensically sound manner'? This will allow the findings to conclude any forensic failings during the testing stages.

2. Literature review

2.1 Firmware overview

All devices that communicate between an Operating System (OS)(software) and a storage medium (hardware) will need some sort of translation to be able to interpret the transferred data. This is the principle of firmware, it acts as a bridge between the hardware and software, by translating data from one another. The firmware has the highest privilege on a system (Hassan, Markantonakis and Akram, 2016). This allows the OS to tell the user details of the drive, such as the size or prevent the OS from saving data to any areas on the disk that may be marked as "bad".

When the device is turned on the firmware will begin a self-check, this ensures correct device operation and will mark unreadable areas of the disk or hide any areas that are not used by the OS. Each device has specific firmware and unique checks (Sutherland, Davies and Blyth, 2011). However, each firmware has a similar layout. For example, part A of the firmware will be on the printed circuit board and the drive will contain a service zone to store defect lists and self-monitoring analysis and reporting technology attributes which monitor the health of the drive. For example, read error rate, performance, temperatures, and the total number of hours the device has been used. This is a small selection of recordable attributes and encompasses numerous different tests (NTFS.com, 2021). Firmware part B will hold the Logical bock Addressing (LBA) translator table. If all the checks are successful it will begin to load the OS.

2.2 Forensic operations

Digital forensics is the science of dealing with digital information created, stored, and transmitted by electronic devices (Shanmugam, 2011). The data uncovered can be used to inform investigations and legal proceedings. The data contained on devices must be extracted, analysed, and presented using methods and software that is validated by the UK Forensic Science Regulator. This means the use of a scientifically proven method aimed at preservation, validation, identification, analysis, documentation, and presentation of such evidence (UK Forensic Science Regulator, 2015).

The Forensic Science Regulator (2015) has a strict framework so that investigative processes can be repeatable and therefore relied upon at court. This is also true for the tools used, as both the regulator and investigating analyst will need to validate software before it can be approved for use (UK Forensic Science Regulator, 2015). However, Becket and Slay (2007) explain these tests may not find all errors or limitations of the software, as this would require extensive resources which may not be readily available. The validation criteria are set out in standards such as ISO 17025:2005, which is balanced between cost, risk, and technical possibilities. Therefore, the limitations of software, and loopholes in standards, allows malicious users to exploit issues that may not have been explored. Most forensic analysis is focused on the operating system level. They will take direct copies of items such as storage information or the registry (Jeong and Lee, 2019; Wundram et al., 2013). This relies on the firmware being correct as the operating system will display what it receives.

2.3 Anti-forensic techniques

It is difficult to pinpoint the first firmware level attack. However, "Vault 7", a collection of documents found on Wikki Leaks shows that the CIA had capabilities in 2013 of snooping via firmware vulnerabilities, such as rootkits and injecting malicious code (Vording, 2018). It is suggested that four categories exist for firmware level attacks against forensic processes. Artefact wiping, data hiding, trail obfuscation and attacks against the tools or the processes (Rogers and Lockheed, 2005; Chandran, 2013; Majed, Noura and Chehab, 2020).

2.3.1 *Artifact wiping*

Stewin and Bystrov (2012) discovered it is possible to extract or delete cryptographic keys from a SSD device. This can be implemented by placing malware within the firmware, as the high privilege level could allow access to the Direct Memory Access (DMA) area. The DMA is used by the OS and legitimate software to securely communicate with hardware devices. The research by O' 'Hara and Malisow (2016) showed that by removing these keys it would be impossible for an analyst to access any data. Additionally, if the use of crypto shredding were employed it would render the whole disk useless and completely unrecoverable (O' 'Hara and Malisow, 2016).

Harris (2006) found that using hard drive ATA secure erase features can make it almost impossible for data to be recovered. The ATA erase feature is a disk scrubber built into the device at a firmware level and will destroy all data on all areas of the disk, regardless of whether it is protected by the OS or other such features. Marupudi (2017) found that using the secure erase setting to permanently write and erase can cause wear levelling in SSD drives. This is when areas of the device have been written to, deleted and re-written until the area is damaged by overuse, becoming unreliable to store data. By allowing this to run over the whole disk it would render the storage device useless to any forensic analysis.

2.3.2 Data hiding

As the OS relies on firmware, data can be manipulated so rogue information about the hard drive is sent to the OS. These changes may not be apparent to a user, as performance may not seem to be affected during typical use (Haswell, 2006). Haswell (2006) shows that the firmware can tell the OS that the size of the internal storage is different to its physical. He found that a malicious user can add data to these hidden areas, and the OS will not display them. Brezinski and Killalea (2002) showed similar findings and found that malicious software can also be obfuscated in these hidden areas. This can then run during the imaging process making changes to the disk and invalidating the results. This poses an issue for law enforcement, as some forensic tools have been identified as not discovering the hidden areas on the internal storage device (Surtherland et al., 2009). Their research found that if the malicious user marks sectors as bad on the drives error lists within the firmware, the tools would not display the bad areas as they rely on the firmware's translator operating correctly.

Gruhn (2017) conducted similar research, showing that the size of the drive can be changed by changing the firmware overlays, partition tables and marking sectors as bad. However, went on to show that the EEPROM can be used as standalone hidden storage. Gruhn (2017) and Jeong and Lee (2019) both suggest that this area is not scrutinised by forensic tools or processes.

2.3.3 Trail obfuscation

Trail obfuscation is to mislead or confuse a forensic technique or analyst and ultimately the investigation. This can be achieved in several ways, deleting logs, IP/MAC spoofing, proxy servers, the use of zombie accounts and misinformation (Shanmugam (2011). Shanmugam (2011) shows this is possible with the use of the Metasploit framework "transmogrify". He showed that you can change the header file information, allowing the malicious user to change an image to a document at a header level. When the forensic tools scan the device for images this malicious file may not show as it is declared as a document.

Cho (2016) found several tools that can be used to change the creation, last modified and recent timestamps for files on the OS. Once this change has been made the forensic tools would display the modified date and time to the analyst. Schicht (2014) backs this statement saying the changes will not be detected unless the analysts look in shadow copies and logfiles. The recommendation is to also change the data for the shadow copy. Schicht (2014) suggests the logs are only short so the evidence will not be retained for long. He suggests that finding this information is "not trivial work". Gul and Kugu (2017) found that similar software "Timestomp" successfully changed the dates and times. But the forensics tools could easily spot the changes. The research found that copying the file to a new location after using the tool then prevented the original data from being found.

2.3.4 Attacks against forensics tools and processes

Firmware anti-forensics can be integrated at the time of imaging, this could be to destroy, hide or manipulate data. All these examples would damage the integrity of the evidence. Laurenson (2016) suggests current forensic software tools do not interrogate these damaged areas, leading to a gap for malicious attackers to manipulate evidence. Du, Ledwith and Scanlon (2018) further back up Laurenson (2016) and compared several forensic tools, finding that they did not scan the firmware of disks. Sutherland et al. (2009) used well-known forensic software Encase, FTK and AccessData. They found that changing the error lists in the firmware was not detected by any of the software tested. Further analysis also found that changing the Logical Block Addressing (LBA) table to the actual Cylinder Head-Sector (CHS) location prevented the software from making a full image. These issues are caused by the forensic software relying on the LBA provided by the firmware.

Harris (2006) argues that the above research focuses too much on software ability and suggests that research should focus on the human element of forensic analysis. He explains that as humans we can be easily deceived by invalid dates or deleted logs. Therefore Harris (2006) and Rogers and Lockheed (2005) both suggest not to automate the whole procedure. Harris (2006) further explains that the software could work on a "what should be" rather than a "what is" functionality. This would allow the software to scan the master file table and notify the analysis of any non-standard code, such as NTFS flags. This would allow the analysts to review the change and allow the possibility of spotting anti-forensic techniques before they cause further issues. Lieberman (2000), found that less automation allowed analysts to follow impulses or intuitions leading to the examination being impaired. Adderley (2019) also noted that manual examinations would become very timely and costly, impacting digital forensics within law enforcement. He suggests that a mix of both automated and human manual exploration would be the best of both approaches.

3. Research methodology

This research is based on black box testing by Wilsdon and Slays (2006). They suggest following the six steps shown below:

- 1. Acquisition of software
- a) The software used to forensically analyse the procedure should be listed with version numbers, installed patches or any modifications that have been made to the tool.
- 2. Identification of software functionalities
- a) The functions of the tool are documented, via user manuals or use of the software etc. Only identified functionality of the software should be evaluated.
- 3. Development of test cases and reference sets

All tests must be based on black box testing methods, examining previously listed functions.

- a).Tests are prepared for each function of each tool. Some tests can be used on multiple tools, but all tests should be based on real-world scenarios.
- 4. Development of result acceptance spectrum
- a). The use of black-box testing allows the expected result to be expressed in advance. This allows any deviation from this result to be easily identified.
- 5. Execution of tests and evaluation of results
- a). All results will be compared against the acceptance spectrum compiled in section four.
- b). Any function that is not an expected result will be evaluated as "Failed."
- c). Any functions that are expected will be evaluated as "Passed."
- 6. Release of evaluation results
- a) Upon conclusion of all previous phases, the results should be made available to the community.

(Bhat, AlZahrani and Wani, 2021; Flandrin et al., (2014).

The results within this report will be presented on a scale from 0-3.

0 – No trace of data at all.
1 - Location or name of file found but no data within.
2 - Location and name of file found, some data recovered from within.
3 - Fully recovered document with all contents intact.

The use of this scale will allow a range of results to be shown, as data recovery can be partially fulfilled the scale will help show the true reliability of the tests and results.

3.1 Tools and software

The below tools have been selected for the research implementation tasks. This includes software to image, check, analyse the drive, and data. In line with the methodology, the tools functions have been listed, version numbers provided and a brief description. To further add to this the hardware used has been listed with additional relevant information.

Table 1: Information on tools, software, and hardware

Tool Name	Version	Functions
Md5Checker	3.3	Calculate and display MD5 Checksums. Verify if files have been changed and integrity of files.
FTK Imager	4.5	Create forensic images of various media. Preview and recover files, folders, and contents of files.
Recuva	1.53.1087	Advanced file recovery.
EaseUS	14.2	Recover from lost or deleted partitions.
FEX Imager	2.2.0(263)	Acquire disk images, files and hash information.
Hex Editor HxD	2.5.0.0	Hex Editor.
HDDHackr	1.40	Flashes the firmware on hard drives.
Bootable USB Creator	4.0	Creates bootable disk on USB to install Dos.
Western Digital 320GB	WD3200BEVT	The hard drive that will be manipulated.

The below files were added to the disk. The table below shows the data in the original state, once the tests have been concluded the original hash values will be compared against the extracted data to identify any differences.

Table 2: Original data on the device

File Type	File Name	Md5	Content
.docx	doc1	8A6AF8B3B74C2D9A44868F48B82EBB9A	Text
.txt	text doc	0C46BDBA0B995AEAF4A42BBE527D8B25	Text
.png	image 1	C37EE8102ADE5A30A390FE81F3D75014	Image
.jpg	image 2	066F76801FBE38BB54015AFA60ED51CB	Image

4. Implementation

When first connecting the device to the machine (in its original state) we can see Windows recognises it as a 298.09GB disk. This is close to 320GB and would be expected as other space is used for configuration, shadow copy services etc. Before editing the firmware, a copy of the original is taken from the disk. The Hex editor is then used to examine sector 16, the security sector for the drive containing information on the firmware and disk. Figure 1 shows the first bytes of data in that sector. The hex editor displays a series of "20's" this is hex for white space or empty areas. After this, a list of "decoded text" is shown. Each of these are explained in table 3 and shown in the corresponding figures.

Table 3: The layout of sector 16

Offset	Data Type	Decoded Text	Hex Value
00 – 0B	Empty	Space	
0C - 13	Serial Number	6VDC44HD	36 56 44 43 34 48 44
14 – 1B	Firmware Revision	0002CE02	30 30 30 32 43 30 32
1C - 26	Manufacturer/Model Number	ST9320325AS	53 54 39 33 32 3033 32 35 41 53
58 – 5B	LBA Size		B0 EA 42 25

```
Offset(h) 00 01 02 03 04 05 06 07 08 09 0A 0B 0C 0D 0E 0F  Decoded text
00000000  20 20 20 20 20 20 20 20 20 20 20 20 36 56 44 43              6VDC
00000010  34 34 48 44 30 30 30 32 43 45 30 32 53 54 39 33  44HD0002CE02ST93
00000020  32 30 33 32 35 41 53 20 20 20 20 20 20 20 20 20  20325AS
00000030  20 20 20 20 20 20 20 20 20 20 20 20 20 20 20 20
```

Figure 1: The serial number location

Understanding the layout allows the correct area to be changed. For instance, if a user intended to update the serial or model name, the numbers or letters can be converted to hex and entered here. This will produce a .bin file that can be uploaded to the device.

4.1 Changing the device size

Only four bits need to be changed to show the disk drive size at offset 58 to 5B. The disk size is displayed in little-endian, so B0 EA 42 25 becomes 25 42 EA B0. By multiplying this by 512 (the standard sector size) it will give the total disk size. However, this is in Hex, converting the hex to decimal can be seen in table 4.

Table 4: Converting hex to decimal original hard drive size

Hex	25 42 EA B0	X	200	=	4A 85D5 6000
Decimal	625,142,448	X	512	=	320,072,933,376

Table 4 shows that the original size of this drive is 320GB. To update the firmware to show the disk size as 120GB the following hex would need to be used: B0 4B F9 0D converted to little-endian 0D F9 4B B0.

Table 5: Converting hex to decimal for new firmware

Hex	0D F9 4B B0	x	200	=	1B F297 6000
Decimal	234,441,648	x	512	=	120,034,123,776

Therefore, giving us a total space of 120GB. At this point, the .bin file is ready to be uploaded to the device. This is done using HDDHACKR. The software works by sending vendor-specific commands to modify the firmware. It allows the user to enter specific commands.

5. Results

When connecting the device to a Windows operating system the drive is showing as 111.79GB. At this point the recovery is taken pre and post initialisation.

5.1 Unallocated volume tests

The first set of results are taken before the disc is not initialised. When scanning the drive with FTK imager it shows the device as 120GB. The forensic software showed that it has unpartitioned space [basic disk]. Within this unpartitioned space, it shows some data exists, when searching more in-depth you can find the data fully recovered. EaseUS data recovery shows the drive as 111.79GB, as expected with the changed firmware. However, lists it as a 'device' rather than a hard drive, as previously shown. All data was successfully recovered.

Before scanning the drive FEX shows the user an overview screen. Here the software displayed the serial number as 000000000. However, the sectors (234441648) and sector size (512) displayed shows that the drive is 120GB as programmed within the firmware. All data was successfully recovered.

Recuva was unable to recover any of the data on the disk.

Table 6: Results from the unallocated disk drive

File type	Windows Explorer	Recuva	EaseUS	FTK imager	FEX Imager
Docx	0	0	3	3	3
Text	0	0	3	3	3
.png	0	0	3	3	3
.jpg	0	0	3	3	3
Md5	0	0	3	3	3

5.2 Allocated volume tests

For these series of tests, the drive was assigned a drive letter to simulate a normal drive on Windows operating system. The same recovery and forensic analysis methods are employed.

FTK and FEX did find traces of the data but was not able to display it fully. Although Recuva and EaseUS used advanced or "deep" scans fully recovering the images and the .docx file. However, they did not find any trace of the text document. All other data was found. Recuva found the two images and the document however the .jpg md5 had changed from the original file. All other MD5's stayed the same.

The forensic tools were able to detect these files but could not display them. This may be due to the software not being able to recover these files in a forensically sound way. This can be proposed as the recovery software changing the MD5 hash value.

Table 7: Results from the allocated volume

File type	Windows Explorer	Recuva	EaseUS	FTK imager	FEX Imager
Docx	0	3	3	1	1
Text	0	0	0	1	1
.png	0	3	3	1	1
.jpg	0	3	3	1	1
Md5	0	0	0	0	0

6. Analysis

6.1 Drive unallocated

When scanning the drive FEX imager showed the serial number as 0000000. If the serial was explicitly entered into the firmware, it may not have displayed this. However, the software still displayed the size as 120GB. This mismatch in the firmware may allow analysists to pick up on this.

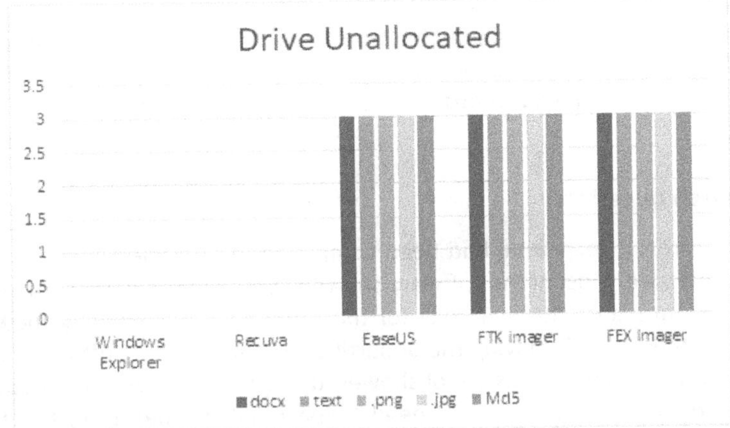

Figure 2: Results from the unallocated drive

Once the firmware was uploaded the drive was shown to the host machine as one drive with a total of 120GB. This new total is the unallocated size, and the first examinations were taken when the drive was in the uninitialized state. As the disk cannot be mounted by the operating system Windows and Recuva cannot begin to extract any data. This may suggest that Recuva is relying on the OS to display and scan the drive. EaseUS, FTK and FEX imager all managed to extract the data in a forensically sound manner (figure 2). All the data was clearly shown in "unpartitioned space". This could suggest that these software tools do not rely on the firmware to give an accurate reading from the disk.

6.2 Drive allocated

The drive was then allocated a drive letter and the tools used to scan the disk again. Windows relied on the firmware to give all details to display. Therefore, no data was found as the new allocated drive "D" was an empty drive.

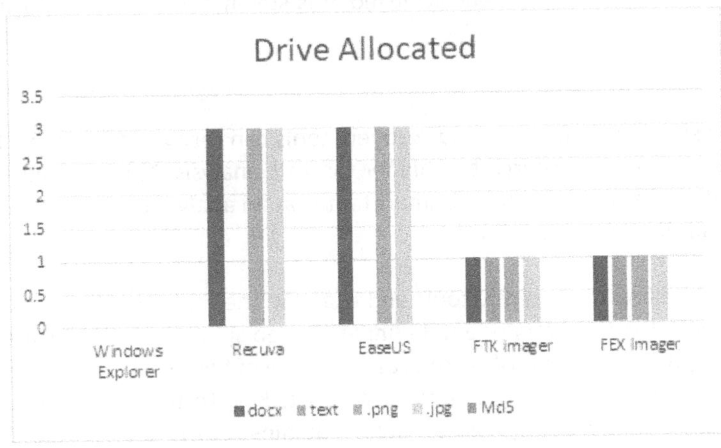

Figure 3: Results from the allocated drive

Both recovery software tools managed to recover a readable version of docx, png and jpg. However, although the data was fully recovered and looked identical to the originals, the MD5 hashes did not match. The recovered files also did not have the same title as originally given; this may suggest the header information has been lost or removed for these files, or the forensic tools have renamed the found files. It is unsure why this has happened; it may be that this change occurred when the drive was allocated a drive letter. As the previous test provides all MD5 hashes are maintained (figure 3).

FTK and FEX imager did find traces of the data but not fully recovered within the software itself. The tools may have prevented the automatic recovery of these files as it would cause the MD5 to be changed. This would then prevent the data from being used in law enforcement or as evidence in cases. As both forensic tools did not provide a full recovery of the files the hashes could not be compared.

The lack of recovery of the '.txt' file from both data recovery programmes may be from the tools not displaying unrecoverable data to be more user friendly. Another approach to finding the text document could be data carving, where the user can search the HEX for specific details within the document. File carving can be completed using tools that search the files header information, where contains the start of data points and 'the end of file' in the unallocated data. The data between the two points can be extracted and examined further. Processes such as data carving could lead to further data being recovered; however, this was out of scope for this experiment.

6.3 Answers to research question

To determine if the anti-forensic technique had been completed in an undetectable way, the disk drive was examined with forensic software. Most of the software showed maliciously changed firmware values, and found the unpartitioned space from scans. The main test for the tools was to see if they showed any traces that a change had taken place. Other than displaying the unpartitioned space, most software packages did not show any tampering of the drive. Only one forensic tool showed that the disk drive serial number was incorrect. In most investigations the recovery of data would need to be reliable. Therefore, the firmware change has obstructed some of the forensic tools from acquiring the data in a forensically sound manner. This was shown by the altered files, backed up by the MD5 hash not being the same as the original file.

A1 Based on the research findings within this study, the data shows that the firmware can be manipulated to change, hide, or delete data. Although each of these had various outcomes overall the idea to prevent investigations from proceeding was seen as achieved.

A2 The findings have shown that the technique can be hidden from most software used here. All but one has shown that the drive was not tampered with. Although the results did find the unpartitioned space, the software did not suggest any tampering or changes to the disk had been made or found. The only software that did was FTK imager, with the unusual serial number.

A3 The results from this experiment have shown that most data could be found but not necessarily read, displayed, or recovered in a forensically sound manner after the firmware change. This shows that it would hinder an investigation. Although this goal was achieved, it is still important to point out that recovery of the data is still possible.

7. Conclusion

To conclude, it has been shown that forensic and recovery tools can recover data in most situations. However, some of the techniques within this research can hinder forensic analysis. This is mainly recovering deleted or hidden files, impacting the reliability of data. It is important that an analyst can spot signs of malicious changes before data is extracted for use in an investigation.

Further research on decompiling the forensic tools and examine the inner workings could be conducted. This may give a clearer answer to the discovery of empty unpartitioned space and why it exists on a disk that shows it as different size. This might show if the software uses a mixture of hardware and software-based firmware to scan the drive, or if the software searches the drive without talking to the firmware. Research on the use of hardware modified firmware would be a good next step. Hutchins (2015), Cipriani (2016) and Yliluoma (2018) all report similar attacks outlined in this paper but with tougher to recover artefacts.

Overall, all the signs of tampering will most likely be picked up by some tools. Therefore, the analyst should not rely on any one tool but a library. The examination should closely note the serial numbers and the size displayed by the software compared to the drive itself, although the label could easily be changed. If this is the case, comparing the model number and name is important. Most manufactures will have white space before or after the manufacturing name. This again can be an oversite by an attacker who may not have observed this when modifying the firmware.

References

Adderley, A., 2019. Graph-based Temporal Analysis in Digital Forensics. Masters. Harvard University.

Attorney General's Office, 2022. Attorney General's Guidelines on Disclosure For investigators, prosecutors and defence practitioners. [online] Assets Publishing Service. Available at: < shorturl.at/jyHJ7 > [Accessed 26 August 2021].

Bhat, W., AlZahrani, A. and Wani, M., 2021. Can computer forensic tools be trusted in digital investigations? Science & Justice, 61(2), pp.198-203.

Brezinski and Killalea, 2002. Guidelines for Evidence Collection and Archiving. [online] Rfc-editor. Available at: <https://www.rfc-editor.org/pdfrfc/rfc3227.txt.pdf> [Accessed 26 July 2021].

Chandran, R., 2022. Overview of Digital Forensics and Anti-Forensics Techniques. Masters. Auckland University of Technology.

Cho, G., 2016. Data Hiding in NTFS Timestamps for Anti-Forensics. International Journal of Internet, Broadcasting and Communication, 8(3), pp.31-40.

Cipriani, T., 2016. Coreboot on the ThinkPad X220 with a Raspberry Pi. [online] Tylercipriani.com. Available at: <https://tylercipriani.com/blog/2016/11/13/coreboot-on-the-thinkpad-x220-with-a-raspberry-pi/> [Accessed 12 June 2021].

Du, X., Ledwith, P. and Scanlon, M., 2018. Deduplicated Disk Image Evidence Acquisition and Forensically-Sound Reconstruction. 2018 17th IEEE International Conference On Trust, Security And Privacy In Computing And Communications/ 12th IEEE International Conference On Big Data Science And Engineering (TrustCom/BigDataSE), pp.1674-1679.

Etow, T., n.d. Impact Of Anti-Forensics Techniques On Digital Forensics Investigation. Ph.D. Linnaeus University.

Gruhn, M., 2017. Forensic limbo: Towards subverting hard disk firmware bootkits. Digital Investigation, 23, pp.138-150.

Gul, M. and Kugu, E., 2017. A survey on anti-forensics techniques. 2017 International Artificial Intelligence and Data Processing Symposium (IDAP), pp.1-6.

Harris, R., 2006. Arriving at an anti-forensics consensus: Examining how to define and control the anti-forensics problem. Digital Investigation, 3, pp.44-49.

Hassan, R., Markantonakis, K. and Akram, R., 2016. Can You Call the Software in Your Device be Firmware?. 2016 IEEE 13th International Conference on e-Business Engineering (ICEBE), pp.188-195.

Haswell, J., 2016. SSD Architectures to Ensure Security and Performance. In: Flash Memory Summit.

Hewitt, M., 2021. Npcc. [online] Npcc Police UK. Available at: <https://www.npcc.police.uk/ThePoliceChiefsBlog/Default.aspx> [Accessed 10 March 2021].

Hutchins, M., 2016. Hard Disk Firmware Hacking (Part 1). [online] MalwareTech. Available at: < shorturl.at/fvHR0 > [Accessed 29 June 2021].

Jeong, D. and Lee, S., 2019. Forensic signature for tracking storage devices: Analysis of UEFI firmware image, disk signature and windows artifacts. Digital Investigation, 29, pp.21-27.

Laurenson, T., 2016. Automated Digital Forensic Triage: Rapid Detection of Anti-Forensic Tools. Ph.D. University of Otago.

Lieberman, M., 2000. Intuition: A social cognitive neuroscience approach. Psychological Bulletin, 126(1), pp.109-137.

Majed, H., Noura, H. and Chehab, A., 2020. Overview of Digital Forensics and Anti-Forensics Techniques. 2020 8th International Symposium on Digital Forensics and Security (ISDFS),.

Marupudi, S., 2017. Solid State Drive: New Challenge for Forensic Investigation. [online] St. Cloud State University. Available at: <https://repository.stcloudstate.edu/msia_etds?> [Accessed 16 July 2021].

NTFS.com, 2021. S.M.A.R.T. Attributes. [online] NTFS. Available at: <https://ntfs.com/disk-monitor-smart-attributes.htm> [Accessed 17 June 2021].

O'Hara, B. and Malisow, B., 2017. Ccsp (ISC)2 certified cloud security professional official study guide. CA: John Wiley & Sons.

Rogers, M. and Lockheed, M., 2005. Anti-forensics - Lockheed Martin. CA.

Schicht, J., 2014. GitHub - jschicht/SetMace: Manipulate timestamps on NTFS. [online] GitHub. Available at: <https://github.com/jschicht/SetMace> [Accessed 17 June 2021].

The Parliamentary Office of Science and Technology, 2016. Digital Forensics and Crime. [online] Parliament UK. Available at: <https://post.parliament.uk/research-briefings/post-pn-0520/> [Accessed 19 June 2021].

Shanmugam, K., 2011. Validating digital forensic evidence. Ph.D. Brunel University Uxbridge.

Stewin, P. and Bystrov, I., 2013. Understanding DMA Malware. Detection of Intrusions and Malware, and Vulnerability Assessment, pp.21-41.

Sutherland, I., Davies, G. and Blyth, A., 2011. Malware and steganography in hard disk firmware. Journal in Computer Virology, 7(3), pp.215-219.

Sutherland, I., Davies, G., Pringle, N. and Blyth, A., 2009. The Impact of Hard Disk Firmware Steganography on Computer Forensics. Journal of Digital Forensics, Security and Law, 4(2), pp.73-84.

UK Forensic Science Regulator, 2015. Overseeing Quality. [online] publishing Service UK. Available at: <https://assets.publishing.service.gov.uk/government/uploads/system/uploads/attachment_data/file/470526/FSR_Newsletter_26__October_2015.pdf> [Accessed 20 July 2021].

Vording, J., 2018. Vault 7 and the Paradox of Democratic Society. Masters. Leiden University.

Wilsdon, T. and Slay, J., 2006. Validation of forensic computing software utilizing black box testing techniques. Edith Cowan University.

Wundram, M., Freiling, F. and Moch, C., 2013. Anti-forensics: The Next Step in Digital Forensics Tool Testing. 2013 Seventh International Conference on IT Security Incident Management and IT Forensics, pp.83-97.

Yliluoma, J., 2018. Replacing BIOS in a ThinkPad with GPL CoreBoot. [online] Youtube. Available at: < shorturl.at/cuvyS > [Accessed 12 August 2020].

Work in Progress Papers

DRAM-Based Physically Unclonable Functions and the Need for Proper Evaluation

Pascal Ahr, Christoph Lipps and Hans Dieter Schotten
German Research Center for Artificial Intelligence, Kaiserslautern, Germany
Pascal.Ahr@dfki.de
Christoph.Lipps@dfki.de
Hans_Dieter.Schotten@dfki.de

Abstract: Dynamic Random-Access Memory (DRAM)-based Physically Unclonable Functions (PUFs) are a part of the Physical Layer Security (PhySec) domain. Those electrical PUFs are memory based and exhibit a high availability, Shannon Entropy, low energy consumption and high amount of Challenge Response Pairs (CRPs). Because of those properties, the DRAM PUF is a promising approach for security applications in the Industrial Internet of Things (IIoT) context as well as securing the Sixth-Generation (6G) Wireless Systems and edge computing. DRAM, with its most common one-Transistor one-Capacitor (1T1C) architecture, and as a volatile memory is embedded in almost every modern computing unit. Regarding the PUF security applications, four main types of applications are currently distinguished in the scientific community: *Retention Error, Row Hammer, Startup* and *Latency* PUFs. Thereby these differ in their procedure in how responses are generated as well as by the physical mechanisms. Each of them with varying properties in terms of availability, reliability, uniqueness and uniformity. To examine this, and to obtain comparable results, this work proposes to compare the four different DRAM-PUF types i) with the same metrics of evaluation and ii) implemented on the same DRAM cells. This represents both the difference with regard to the work done in the literature and the added value of this work presented. As far as known, there is no work to date that performs the intended evaluations using the same evaluation platform under the identical conditions. However, this is required for comparable results. This consistent comparison is ensured by a self-developed and implemented evaluation platform, which is accordingly equipped with a significant number of DRAMs. By an appropriate high volume of measurements, a corresponding resolution will be given. Monitoring the environmental conditions prevents from wrong interpretations caused by environmental influences but also provides useful context information. Furthermore, a detailed technical and physical background will be described. The results of this approach will assist by the consideration of which DRAM-PUF is appropriate in which (environmental) conditions and thereby provide a guideline for practitioners.

Keywords: physical layer security, physically unclonable functions, dynamic random-access memory, DRAM PUFs, secure 6G, secure edge computing

1. Physically Unclonable Functions as hardware security anchors

Interconnectivity, as one of the drivers and core technologies of the fourth industrial revolution, overcomes the limitations of local computers. Concepts like the Industrial Internet of Things (IIoT), edge computing and Next Generation Wireless Systems, such as in particular the development of the Sixth-Generation (6G) are leading to a high increase of low complexity and Reduced Capacity (RedCap) devices (Lipps, et al., 2021). The challenge facing the systems, however, is secure communication and the doubtless identification and authentication of the participants. But, according to the computing power, size and energy constrains of IIoT devices, "traditional" cryptographic methods like the Rivest-Shamir-Adleman (RSA) algorithm are not sufficient and therefore new and appropriate approaches are required.

One promising approach are Physically Unclonable Functions (PUFs), a "physical entity whose behaviour is a function of its structure and the intrinsic variation of its manufacturing process" (Halak, 2018). Due to manufacturing related and uninfluenceable influences are those variations unique for every physical entity and can be considered as a technical fingerprint of the semiconductor device. An additional benefit is their Non-Volatile Memory (NVM): A PUF response can be regenerated the time needed (Lipps, et al., 2018). Furthermore, Random-Access Memory (RAM) is intergraded in almost every computing unit and thus inherently existing in many systems. Now that the Static RAM (SRAM) variant has been discussed in detail in the scientific community, the focus is currently shifting to the Dynamic RAM (DRAM) one (Ahr, Noushinfar & Lipps, 2021). But, although there is work available (Anagnostopoulos, et al., 2018), these merely compare different approaches, whereas the present work calls for a unified and standardized approach.

Therefore, this work in progress paper focusses to close the lack of consistency over the four DRAM PUF types: *Retention Error, Row Hammer, Startup* and *Latency* PUFs; and to provide a sound comparison. This is structured as follows: Section 2 explains the DRAM and its PUF types while Section 3 describes the test setup. Afterwards in Section 4 metrics to compare the different types are introduced. Finally, Section 5 summarizes the work.

2. Dynamic-RAM based Physically Unclonable Functions

Dynamic Random-Access Memory is a volatile memory with its most common architecture, the one-transistor one-capacitor (1T1C) structure. As depicted in **Figure** 1, the storage capacitor (C_S) is connected to the bitline (BL) via a Metal-Oxide-Semiconductor Field Effect Transistor (MOSFET) (M1) and "stores" the content of the cell. A DRAM is built of many such cells, while a specific word can be accessed via the wordline (WL). This in turn is connected to the gate of the MOSFET (M1). Due to its dimensions, the BL has a capacitance much higher than C_S, which leads to a very small exchange of electrons and therefore a very small change of the bitline voltage V_{BL}, while connected. To "detect" the voltage change, a sense amplifier (SensAmp) is used. These are two cross coupled inverters in a feedback loop and thereby build a bistable system with the two states *logical zero* and *logical one*.

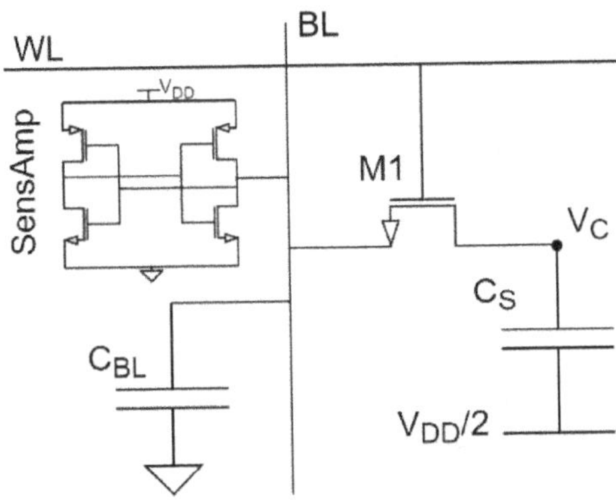

Figure 1: 1T1C DRAM Cell

During a read out process the BL is pre-charged to the tipping point of the SensAmp. Afterwards the C_S is connected and according to its content, the bistable system will adopt one of the two possible states. To reduce the stress caused by the electrical field, one plate of the C_S is biased to $V_{DD}/2$ (Tehranipoor , et al., 2017). But, as the storage capacitor and MOSFET are not ideal they exhibit leakage effects leading to discharging of C_S and therefore to the loss of the stored data. To avoid this, a refreshing of the system is done periodically.

The semiconductors are influenced by manufacturing process fluctuations and can be used to build four different DRAM PUF types: *Retention Error, Row Hammer, Startup* and *Latency* PUFs.

2.1 Retention Error

The Retention Error PUF uses the leakage effect of the non-ideal components of the DRAM cell. This leakage effect is individual for every cell and influenced by the manufacturing fluctuations. C_S is charged to V_{DD} by writing and reading out after a predefined time while disabling the refreshing mechanism. Either the cell will remain enough electrons to charge the BL over the tipping point of the SensAmp and will read out as logical one or not and will read out as logical zero. This pattern is used to form the PUF response (Keller, et al., 2014).

2.2 Row Hammer

The hammer procedure, a rapidly access to a DRAM row, so called Hammer Rows, causes a quick leakage in cells of adjacent rows, called PUF rows. The quantity of this effect is influenced by manufacturing related fluctuations of the Integrated Circuit (IC). The response is generated by pre-charging C_S, performing the hammer procedure and read the remaining content of the DRAM (Schaller, et al., 2017).

2.3 Latency

A DRAM cell needs some time to store or read data correctly. If this time is violated the operation failed. Manufacturing fluctuations causes an individual time for each cell. The procedure is to write known data in a specific memory region normally and read those with reduced time. Some cells will read correctly while others not. This pattern forms the response of the latency PUF and is possible with reduced write time as well (Hashemian , et al., 2015) (Kim , et al., 2018).

2.4 Startup

After the Startup of a discharged DRAM, all cells are in an undefined state. Due to the pre-charging of one plate of the C_S, point V_C in Figure 1 has got a potential different from GND. By reading out the Sartup state, either point V_C is charged enough to cause the SensAmp to adopt a logical one or not and a logical zero is read (Tehranipoor , et al., 2017).

3. Evaluation platform

To provide the required consistency, all four different DRAM-PUF types shall be integrated successively on the same DRAM modules. Only this enables a DRAM independent proper comparison. In order to consider the environmental conditions as influencing factors in the evaluation, the analysis will be performed in a climatic chamber in which, among other things, the temperature as well as the humidity can be adjusted. The respective conditions are recorded via a corresponding sensor system and taken into account in the evaluation with respect to possible correlations of the PUF responses.

As basis for the evaluation platform a Xilinx EK-U1-ZCU 104G Field Programmable Gate Array (FPGA) development board with a connection from FPGA to embedded DRAM Small Outline Dual Inline Memory Module (SODIMM) is considered. This enables the usage of the Double Data Rate (DDR) interface of the DRAM modules. Besides, by using the interface, it is possible to control the DRAM sufficiently. The read out of the PUF responses will be send from the FPGA board via ethernet or via Peripheral Component Interconnect (PCI) BUS to a conventional computer. This is enabled by an onboard processor and installed LINUX Operating System (OS) and provide a sufficient transfer of the measured data.

4. Metrices

As already mentioned, a uniformity of evaluation is mandatory for an adequate scientific exploitation of the results. This also includes working with appropriately standardized matrices. According to *Basel Halak* the uniqueness, reliability and uniformity are the most important metrics to classify the quality of a PUF (Halak, 2018). The calculations are based on the Hamming Distance (HD) and the Hamming Weight (HW).

In addition, factors and parameters such as the implement-ability in already existing devices, temperature dependency, time to generate PUF response, time when the PUF is available and the correlation with environmental conditions are considered as well.

4.1 Uniqueness

This metric indicates how unique a specific PUF chip is and its distinguishability form other chips. If two chips are completely nonidentical the value is 50%. The uniqueness is calculated by

$$\mathrm{HD_{inter}} = \frac{2}{k \cdot (k-1)} \sum_{i=1}^{k-1} \sum_{j=i+1}^{k} \frac{\mathrm{HD}\left(R_i(n), R_j(n)\right)}{n} \cdot 100\% \qquad (1)$$

With the same challenge and $\mathrm{R_i(n)}$ as the response of chip i and $\mathrm{R_j(n)}$ as the response of chip j out of k chips.

4.2 Reliability

The ability of a PUF to generate a consistent response to a given challenge is given by the Reliability and the best possible value is 100%. The reliability is calculated as follows

$$\mathrm{HD_{intra}} = \frac{1}{k} \sum_{i=1}^{k} \frac{\mathrm{HD}\left(R_i(n), R_i'(n)\right)}{n} \cdot 100\% \qquad (2)$$

$$\mathrm{reliability} = 100\% - \mathrm{HD_{intra}} \qquad (3)$$

With different condition for the same challenge and $R_i'(n)$ as the response of the same chip.

4.3 Uniformity

How unpredictable a specific response of a PUF is, is given by the uniformity, whereby a value of 50% indicates true randomness.

$$\text{Uniformity} = \frac{1}{k}\sum_{i=1}^{k} r_i \cdot 100\% \tag{4}$$

With the same chip and r_i as the HW of the ith out of k PUF responses.

5. Conclusion

Physically Unclonable Functions, especially in their application as arbiter, flip-flop, and SRAM-based approaches are well researched applications. However, there is a growing trend towards DRAM-based approaches as well. Unfortunately, the four common derivatives *Retention Error, Row Hammer, Latency and Startup* have not yet been examined uniformly or have merely been considered separately from one another, leading to not fully comparable results.

This work in progress proposes to examine all methods uniformly on a common evaluation platform using the same metrics. By considering additional information like implement-ability in already existing devices, temperature dependency, time to generate the PUF response, time when the PUF is available and correlation with the environmental conditions, a database is created which allows to select a suitable DRAM for a specific application and thus to give the reader a concrete assistance. The next step will be the implementation of the proposed evaluation platform and the collection of the corresponding data in order to make them available in a clearly structured way.

Acknowledgements

This work has been supported by the Federal Ministry of Education and Research of the Federal Republic of Germany (Förderkennzeichen 16KIS1283, AI-NET PROTECT). The authors alone are responsible for the content of the paper.

References

Ahr, P., Noushinfar, M. & Lipps, C., "RAM-Based PUFs: Comparing Static- and Dynamic Random Access Memory", *Workshop on Next Generation Networks and Application*, Kaiserslautern, Germany, 2021.

Anagnostopoulos, N., Katzenbeisser, S., Chandy, J. & Tehranipoor, F., "An Overview of DRAM-Based Security Primitives", *Cryptography*, DOI:10.3390/cryptography202007, 2018.

Halak, B., "Physically Unclonable Functions - From Basic Design Principles to Advanced Hardware Security Applicaitons", Springer International Publishing, ISBN: 978-3-319-76804-5, 2018.

Hashemian, M.S., Singh, B., Wolff, F., Weyer, D., Clay, S & Papachristou, C., "A Robust Authentication Methodology using Physically Unclonable Functions in DRAM Arrays", *Design, Automation Test in Europe Conference Exhibition (DATE)*, Grenoble, France, DOI: 10.7873/DATE.2015.0308, 2015.

Keller , C., Gurkaynak , F., Kaeslin , H. & Felber, N., "Dynamic memory-based physically unclonable function for the generation of unique identifiers and true random numbers", *IEEE International Symposium on Circuits and Systems (ISCAS)*, Melbourne, VIC, Australia, DOI: 10.1109/ISCAS.2014.6865740, 2014.

Kim , J. S., Patel , M., Hassan , H. & Mutlu, O., "The DRAM Latency PUF: Quickly Evaluating Physical Unclonable Functions by Exploiting the Latency-Reliability Tradeoff in Modern Commodity DRAM Devices", *IEEE International Symposium on High Performance Computer Architecture (HPCA)*, Vienna, Austria, DOI: 10.1109/hpca.2018.00026, 2018.

Lipps, C., Weinand, A., Krummacker, D., Fischer, C. & Schotten, H.D., "Proof of Concept for IoT Device Authentication Based on SRAM PUFs Using ATMEGA 2560-MCU", *1st International Conference on Data Intelligence and Security (ICDIS)*, South Padre Island, TX, USA, DOI: 10.1109/ICDIS.2018.00013, 2018.

Lipps, C., Baradie, S., Noushinfar, M., Herbst, J., Weinand, A. & Schotten, H.D., "Towards the Sixth Generation (6G) Wireless Systems: Thoughts on Physical Layer Security", *Mobile Communication - Technologies and Applications - 25. VDE/ITG Fachtagung Mobilkommunikation*, Osnabrück, Germany, 2021.

Schaller, A., Xiong, W., Anagnostopoulos, N.A., Saleem, M.U., Gabmeyer, S., Katzenbeiser, S. & Szefer, J., "Intrinsic Rowhammer PUFs: Leveraging the Rowhammer effect for improved security", *IEEE International Symposium on Hardware Oriented Security and Trust (HOST)*, Mclean, VA, USA, DOI: 10.1109/HST.2017.7951729, 2017.

Tehranipoor , F., Karimian , N., Yan , W. & Chandy, J.A., "DRAM-Based Intrinsic Physically Unclonable Functions for System-Level Security and Authentication", *IEEE Transactions on Very Large Scale Integration (VLSI) Systems*, vol. 25, no. 3, DOI:10.1109/TVLSI.2016.2606658, 2017.

ECHO Cyber-Skills Framework as a Cyber-Skills Education and Training Tool in Health and Medical Tourism

Eleonora Beltempo, Jussi Karvonen and Jyri Rajamäki
Laurea University of Applied Sciences, Finland
eleonora.beltempo@student.laurea.fi
jussi.karvonen@student.laurea.fi
jyri.rajamaki@laurea.fi

Abstract: The ECHO Horizon 2020 Project develops a European cybersecurity ecosystem. One of its assets is the ECHO Cyber-Skills Framework (ECSF). This work in progress paper aims to improve cybersecurity education and training in the healthcare industry including health and medical tourism. First, this paper finds out how ECSF will benefit the healthcare sector regarding cyber-skills and awareness in order to create a more secure information technology (IT) environment when it comes to healthcare. Based on these findings, the paper proposes a strategy to adopt ECSF in order to improve the existing state of IT security and increase worker and management awareness and understanding. Finally, the paper looks at ECSF's possibilities to be a tool for education and training in health and medical tourism.

Keywords: ECHO project, cyber-skills framework, cybersecurity, health and medical tourism

1. Introduction

ECHO (the European network of Cybersecurity centres and competence Hub for innovation and Operations) H2020 Project develops a European Cybersecurity ecosystem, to support secure cooperation and development of the European market, as well as to protect the citizens of the European Union against cyber threats and incidents (ECHO, 2021). The ECHO Cyber-Skills Framework (ECSF) aims at providing a foundation and practical guidelines for better defining the knowledge and skill gaps in the healthcare, transport and energy industries as well as for the development of cybersecurity education and training programs that address those gaps. The ECSF serves as an inventory tool, providing methodological guidelines for the design, update and development of training programs and curricula, both within the framework of the ECHO project, as well as within the scope of relevant EU initiatives, as a common reference model for capacity building (Varbanov, 2021). Figure 1 presents the main components of the ECSF.

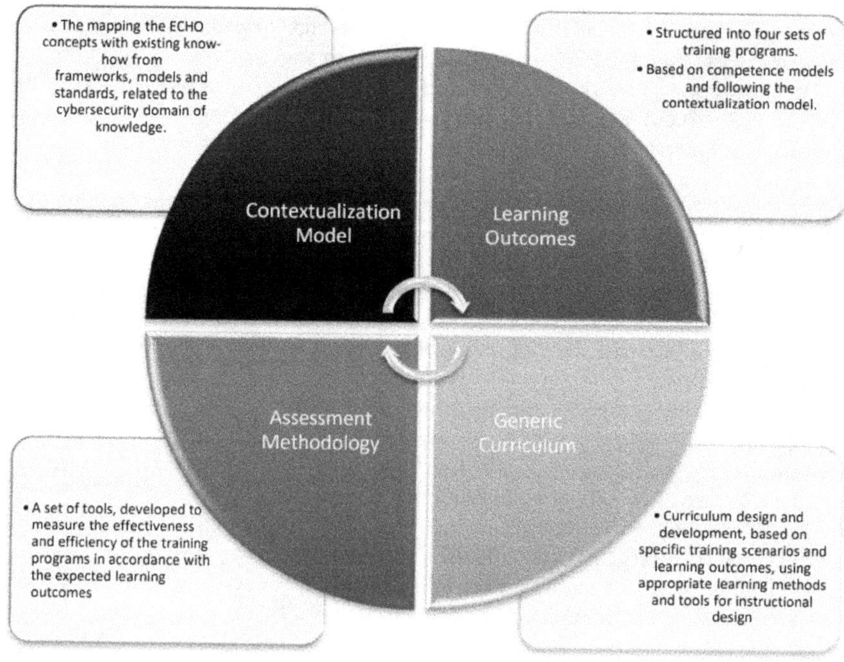
• The mapping the ECHO concepts with existing know-how from frameworks, models and standards, related to the cybersecurity domain of knowledge.
• Structured into four sets of training programs.
• Based on competence models and following the contextualization model.
Contextualization Model
Learning Outcomes
Assessment Methodology
Generic Curriculum
• A set of tools, developed to measure the effectiveness and efficiency of the training programs in accordance with the expected learning outcomes
• Curriculum design and development, based on specific training scenarios and learning outcomes, using appropriate learning methods and tools for instructional design

Figure 1: Main Components of the ECSF (Varbanov, 2021)

This work in progress paper finds out the possibilities of utilizing the ECSF in the healthcare industry including health and medical tourism. Although the concept of health and medical tourism is widespread, also the following terms are used when speaking of travel-based health-related activities: health tourism, medical tourism, wellness tourism, spa tourism and medical travel (Romanova, Vetitnev & Dimanche, 2015). Figure 2 illustrates different types of health and medical tourism.

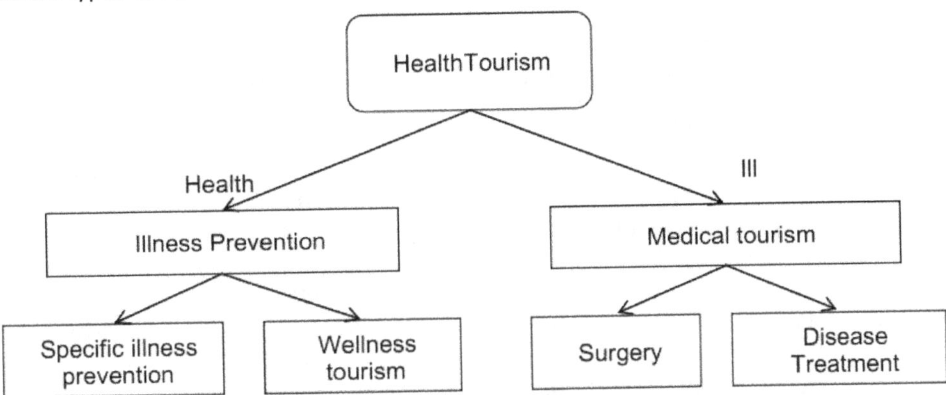

Figure 2: Typology of Health and Medical Tourism (adapted from Romanova et. al, 2015)

This study starts by answering three research questions to see how ECHO and the healthcare sector would both benefit from the possible implementation of ECSF in the healthcare sector:

- How ECHO cyber-skills framework would be beneficial for healthcare?
- What is the best way to implement the ECHO cyber-skills framework into the healthcare sector?
- How the ECHO network will benefit from this implementation.

After answering these questions, this paper combines the findings and proposes how to use ECSF in health and medical tourism.

2. Findings

From our study, we combine the information that has been gathered with the help of our research questions. We figure out what are the problems within healthcare sectors organizations, what is the best way to implement the ECSF, and how can ECHO benefit from the possible implementation of ECSF into the healthcare sector. We also search that if there are any public cyber-skills frameworks that we could possibly reference to see if there is an increase in employees' skills and awareness.

2.1 ECSF benefits for healthcare

To find out about how a cyber-skills framework could be beneficial for healthcare we must first find out about the gaps in healthcare sectors' cyber-skills, knowledge, and the state of awareness about possible threats. Secondly, we have to find out is there any kind of basic training for healthcare sectors staff regarding cyber-skills and knowledge.

The current gaps in the healthcare sector's cyber-skills, knowledge, and awareness are quite huge. According to IONOS Cloud, 37% of healthcare IT employees say their organization is at risk of security because of skills gaps in the field (Mageit, 2021). IONOS Cloud's report continues that 39% of IT professionals in the healthcare sector state that they have gaps in data protection, 25% say they are not adhering to the legislation, and 21% do not follow proper data protection measures (Mageit, 2021). These gaps and threats caused by them can lead to attackers leaking sensitive information about patients which can cause risks for health and safety, and not using critical medical services because they can be affected by unauthorized control (Varbanov, 2021).

The healthcare sector includes private and public hospitals, as well as the companies that manufacture medical devices and the pharmaceutical industry (Varbanov, 2021). The assets that ECHO Cyber-skills Framework focuses on an organizational level are the ICT assets of the company and the professionals who are responsible for making sure the ICT assets are secured. Data and information are the most important assets in a healthcare organization. From the operational technology standpoint for the healthcare sector, devices and equipment that produce data that can be exchanged in and outside the organization are key assets.

From these findings, we can determine that using ECSF would increase cyber-skills and awareness within healthcare sectors employees and lower the risk of the organizations within the sector being attacked or having a close call. Organizations can also adapt to the situations from learning from previous incidents.

2.2 Implementation of ECHO cyber-skills framework to the healthcare sector

The ECHO Multi-sector Assessment Framework (E-MAF) supports risk management decision-making, i.e. it provides a framework for understanding cyber risks and, on that basis, supports decisions on where to invest human, technological and financial resources to reduce those risks to an acceptable degree (Tagarev, Pappalardo & Stoianov, 2020). To properly implement ECSF into the healthcare sector, E-MAF has to be used to address the needs and gaps. E-MAF provides the ECSF the ways to analyze challenges and opportunities, and how to assist the development of cybersecurity technology roadmaps. Using E-MAF provides educational portfolios and training programs, and a unified definition of what skills and qualifications are needed. E-MAF includes the ECHO Security Control, which provides more specific measures divided into four different levels: Organizational, Technical, Functional, and Non-Functional. Organizational implies what the cybersecurity professionals should be able to perform, including the application of security controls, mitigation, and countermeasures. Technical implies the skills and competencies professionals must have in able to demonstrate, which includes security controls. Functional implies sector-specific knowledge the professionals must demonstrate to complete modules or programs. Non-Functional implies knowledge that carries to achieving organizational, technical, and learning objectives within the organization.

There is a lack of publicly used frameworks that focus on cyber-skills training. Hübner et al. (2019) presents the TIGER International Recommendation Framework of Core Competencies in Health Informatics 2.0. Their framework is meant to augment the scope from nursing towards a series of six other professional roles, i.e. direct patient care, health information management, executives, chief information officers, engineers and health IT specialists and researchers and educators. Another example of existing frameworks is the NICE Framework (National Institute of Standards and Technology, 2021), but it is not entirely focused on training cyber-skills and increasing awareness. It does enable cybersecurity education and training, but it is more focused on developing and supporting the workforce so they are capable of meeting cybersecurity needs. The NICE Workforce Framework for Cybersecurity is a good example when providing information about what the employees need to and how to continuously describe learner capabilities. The Frameworks benefits include, for example, enhancing employee skills, understanding needs and skills gaps in the workforce, and hiring the right people for the job, when ECSF wants to train the current employees of the organization to identify and act accordingly in situations where the risk of incidents happening regarding cybersecurity are possible.

Considering how various it is the sector regarding healthcare, a slow implementation based on staff training is essential to guarantee the success of the framework. The continuous improvement of the courses and the new tools that ECHO will develop accordingly will ensure a safe and secure environment for both the staff and the customers.

2.3 How the ECHO network will benefit from this implementation?

This study aims to find out how the ECHO network would benefit from the possible implementation of ECSF in the healthcare sector. From the possible implementation of ECSF in the healthcare sector, ECHO would form a possibly long-lasting partnership with healthcare sectors organizations. Also, from this implementation, ECHO gives themselves information and data to develop more not just with ECSF, but in many more subjects. From the implementation of ECSF into the healthcare sector, ECHO could open doors to even more sectors to produce and implement different frameworks and tools. All of these things mentioned before would let ECHO grow more itself. ECHO would also be a big part of the digitalization of the healthcare sector within the EU.

3. ECSF as a tool in education and training in health and medical tourism

The purpose of medical tourism is to go to another country for medical procedures, for example, receive treatment for a condition, or to seek enhancement. The motivation for this tourism usually is a lower cost of care or higher quality care. These activities usually are reactive to illnesses that are medically necessary or overseen by a doctor (Global Wellness Institute, 2021).

Using ECSF as a tool in education and training in health and wellness tourism would be limited to EU member states. The European Commission (2021) adopted a Recommendation on a European electronic health record exchange to make the flow of Protected Health Information (PHI) of European citizens more quickly to access and share. European Commission also mentions in the article, that ensuring citizens secure access to their data develops the transformation of health and care even more digital. Using ECSF as a part of making sure the secure access and transformation of PHI of European citizens would be a tremendous thing for both EU and ECHO. Implementing ECSF as a part of the EU member states healthcare framework would secure the availability and transformation of PHI, while also ensuring the constant training and development of cyber-skills and awareness of healthcare sectors employees.

4. Discussion

Free movement of people is one of the cornerstones of the European Union. According to the Directive on Cross-Border Healthcare, which has been implemented in the entirety of the EU since 2013 for European citizens, no matter where they live, they have the right to choose where to receive medical treatment across the EU and to be compensated for it. However, in order to secure the above-mentioned rights and unleash the potential of cross-border healthcare exchange, new solutions are needed to secure the storage and cross-border exchange of health data. After the revelations of Edgar Snowden, it is more probable that widely used closed-source security solutions have serious defects and intentionally planted backdoors. It is widely accepted that real information security can increasingly be based on the openness and transparency of the security solution and the secrecy of its encryption keys.

The healthcare sector benefits from using ECSF could be increasing awareness about possible threats, their capabilities on how to work with IT devices without causing possible incidents, and constantly adapting and learning from possible threats. However, due to differences between countries, implementing a unified cyber-skills framework might be incompatible with different nations, but as Nurse, Adamos & Di Franco (2021) mentions in their report about the European cybersecurity skills framework, forming a unified framework that would take into account the needs of EU and their member states is vital for going even further in Europe's digital future.

Acknowledgements

This work was supported by the ECHO project which has received funding from the European Union's Horizon 2020 research and innovation programme under the grant agreement no.830943.

References

ECHO (2021). *Project summary*, [online], https://echonetwork.eu/project-summary/

European Commission (2021). *Exchange of electronic health records across the EU*, [online], 31st August 2021, https://digital-strategy.ec.europa.eu/en/policies/electronic-health-records

Global Wellness Institute (2021). *Wellness Tourism,* [online], https://globalwellnessinstitute.org/what-is-wellness/what-is-wellness-tourism/

Hübner, U., Thye, J., Shaw, T., Elias, B., Egbert, N., Saranto, K., Babitsch, B., Procter, P. and Ball, M. (2019). 'Towards the TIGER International Framework for Recommendations of Core Competencies in Health Informatics 2.0: Extending the Scope and the Roles', *Studies in Health Technology and Informatics*, Volume 264: MEDINFO 2019: Health and Wellbeing e-Networks for All

Mageit, S. (2021). 'Skills gap in healthcare IT industry causes security threats, according to new report', *Healthcare IT News*, 16th September 2021. https://www.healthcareitnews.com/news/emea/skills-gap-healthcare-it-industry-cause-security-threats-according-new-report

National Institute of Standards and Technology (2021). *Workforce Framework for Cybersecurity (NICE Framework),* [online], NIST Special Publication 800-181. https://www.nist.gov/system/files/documents/2021/05/05/NICE%20Framework%20%28NIST%20SP%20800-181%29_one-pager_508Compliant.pdf

Nurse, J., Adamos, K. and Di Franco, F. (2021). *Addressing the EU Cybersecurity Skills Shortage and Gap Through Higher Education*, European Union Agency for Cybersecurity (ENISA). DOI: 10.2824/033355

Romanova, G., Vetitnev, A. and Dimanche, F. (2015). 'Health and Wellness Tourism', In F. Dimanche and L. Andrades (Eds.) *Tourism in Russia: A Management Handbook*, Emerald, pp.231-287

Tagarev, T., Pappalardo, M, and Stoianov. N. (2020). 'A Logical Model for Multi-Sector Cyber Risk Management', *Information & Security: An International Journal* 47, no. 1, pp. 13-26. https://doi.org/10.11610/isij.4701

Varbanov, P. (2021). *D2.6 ECHO Cyberskills Framework,* [online], https://echonetwork.eu/wp-content/uploads/2021/03/ECHO_D2.6_Cyberskills-Framework.pdf

How to Utilize E-EWS as a Tool in Healthcare

Janne Lahdenperä, Joonas Muhonen and Jyri Rajamäki
Laurea University of Applied Sciences, Espoo, Finland
janne.lahdenpera@student.laurea.fi
joonas.muhonen@student.laurea.fi
jyri.rajamaki@laurea.fi

Abstract: ECHO (the European network of Cybersecurity centres and competence Hub for innovation and Operations) is one of the four pilot projects under the European Commission's H2020 Program. This work-in-progress paper relates to the project's task "ECHO Early-Warning Systems (E-EWS) / ECHO Federated Cyber Range (E-FCR) Demonstration Workshops" that will be implemented during 2021 and 2022. As the healthcare industry becomes more connected to the Internet, the possibilities for disastrous cyber-attacks rise accordingly. Well-performing warning systems and robust information sharing between different parties are essential tools to help prevent these attacks. The aim of this paper is to find out how to utilize E-EWS as a tool in the healthcare sector. We started by mapping out the existing Early Warning Systems related to healthcare. At the same time, we researched the different implementations of E-EWS into already existing national systems and how possible information sharing could be done. As a result, we found that there does not seem to be any widely used international Early Warning Systems in use in the healthcare field and can conclude that implementing E-EWS could have significant benefits for the whole industry. A working Early Warning System can help to prevent cyber threats and save lives. However, there are many challenges involved in the implementation. First, the healthcare field is very fragmented with many different private and national actors, and second, the structural differences between different EU countries bring their own problems. Thus, the successful implementation of E-EWS in healthcare depends mainly on how all the different actors can cooperate.

Keywords: healthcare, ECHO project, early warning, cybersecurity, information sharing

1. Introduction

Several countries classify healthcare as one of the most critical infrastructures (Pappalardo et al., 2020). In the event of damage, the consequences will be reflected throughout the community. Digital infrastructure is part of modern healthcare and new technologies are also being used in various sectors of healthcare. This increases the number of threats related to information and cyber security. The most serious cyber threats in the near future can be considered to be cyber-attacks on healthcare (O'Brien et al., 2020).

To develop a common cybersecurity strategy for Europe, the European Commission has, under the H2020 Program, formed the European network of Cybersecurity centres and competence Hub for innovation and Operations (ECHO) project. One of the aims of the ECHO is to strengthen the cyber defence of the European Union by supporting secure collaboration between EU members.

Healthcare is made up of many different actors, including hospital environments, the pharmaceutical industry and various healthcare facilities. One of ECHO's objectives is to protect these interests. The ECHO Early Warning System (E-EWS) can be used to improve the defence against cyber threats to the healthcare system throughout the European Union. A well-designed E-EWS enables real-time data sharing between different countries and actors, thus improving the ability to defend against various cyber threats. ECHO's Federated Cyber Range (E-FCR) supports the shortage of health threats and can be used to learn how to use the E-EWS environment and to create different scenarios (ECHO, 2021).

The main goal of this work-in-progress paper is to find out how to utilize the ECHO Early Warning System as a tool in healthcare. After the introduction, section 2 familiarizes us with the different cyber threats the healthcare industry faces, and section 3 deals with early warning systems in operational use. Section 4 proposes how E-EWS can be implemented in the healthcare sector. Finally, section 5 concludes the paper and suggests future actions.

2. Cyber threats in healthcare

Healthcare faces a wide range of cyber threats, and it is highly vulnerable to these factors. According to the European Union Agency for Network and Information Security (ENISA), the most important cybersecurity challenges of healthcare infrastructures and systems are: 1) systems availability; 2) lack of interoperability; 3) access control and authentication; 4) data integrity; 5) network security; 6) security expertise and awareness; 7) data loss; 8) standardization, compliance, and trust; 9) cross-border incidents; and 10) incidents management

(Liveri, Sarri & Skouloudi, 2015). The number of cyber-attacks has increased significantly in recent years. This has been driven by the COVID-19 pandemic and the increased use of technology in the pharmaceutical industry and patient care. The operating environment for healthcare is diverse and includes many systems, data and equipment that are critical to operations. The impact of cyber-attacks on health care can be very harmful and, at worst, totally devastating. They have a direct impact on patient safety and thus on human lives. In a survey conducted by O'Brien et al (2020), none of the reported impacts of serious cyberattacks (N=9) include direct loss of life (Table 1), but as the number of attacks increases, so does the likelihood of fatalities.

Table 1: Reported impact of the most serious cyber attack

Impact	Amount
No effect	44%
Patient appointment cancelled	11%
Impact on internal project progress	11%
Opportunity costs (remediation work)	11%
Work systems down	11%
Data lost	11%

According to O'Brien et al (2020), the most common cyber-attacks on the healthcare system have been ransomware, denial-of-service attacks, data breaches and different phishing methods. Different pandemics will most likely increase the likelihood of these threats.

The effects of cyber threats can be mitigated by security training of active health care personnel using ECHO Cybersecurity Skills Framework (E-CSF). Effective and versatile training has a big effect in reducing human error. Phishing scams are less likely to be successful, as trained personnel are able to handle information, programs, and networks with caution (Pappalardo et al., 2020). However, this does not prevent the attacks from taking place.

Cyber-attacks on healthcare are mostly driven by financial goals. However, these goals may change in the future to other goals that appropriately seek to compromise patient safety. Such threats are caused by cyber warfare as well as terrorism, which have different motivations. The damage they cause to healthcare can be immeasurable. From the point of view of the cyberwar, healthcare is a strategically interesting target, as its effects extend to the army as well as to civilians (Wairimu, 2021).

3. Existing early warning systems

One example of operational early warning systems is Havaro, with Havaro 2.0 under development. Havaro operates in Finland under the auspices of the national cyber security centre Traficom (Rajamäki et al., 2019). Simola and Lehto (2020) have studied Havaro as a part of the E-EWS. It can be assumed that similar early warning systems are also in use in several European Union countries. Unfortunately, there is very little literature on the use of early warning systems in healthcare related to cybersecurity. This may be because these are individual players whose customers are in the private sector.

For non-EU actors such as the UK Cyber Security Centre (NCSC), the Early Warning System is provided in its own domain (National Cyber Security Centre, 2021). This is a similar concept to Havaro. Regarding North Atlantic Treaty Organization (NATO) member states, Dupuy et al (2020) stress the importance of creating operating models that provide early warning systems for attacks against NATO's energy sector, but there is no information if these systems have been built and if they could also be used in the healthcare sector.

4. E-EWS implementation proposal for the healthcare sector

At its core, E-EWS is an information-sharing platform. Communication between different organizations is based on mutual trust (Kirkov et al., 2020). If this trust is abused and false data entered into the system, the results can be disastrous. As the healthcare industry field is very fractured into multiple national and private actors, adding new organizations to E-EWS is not simple. One way to implement this would be to have primarily national warning systems implementing E-EWS and have different private and national health care providers get their information from these national systems.

There has already been some research into how the Finnish EWS Havaro could share information with E-EWS (Rajamäki, 2019). There would be one centralised hub for the system and many national sub-hubs. The participants do not exchange information directly with each other, because all of the information exchange is done between hubs. In this way, all of the participants can get their information from a trusted source.

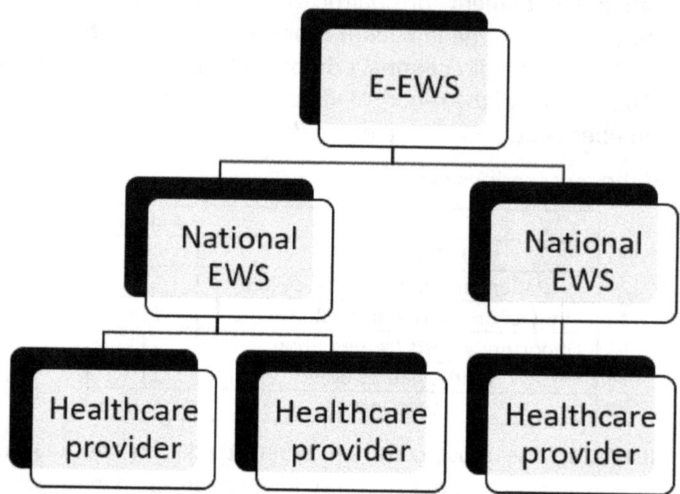

Figure 1: Proposed implementation hierarchy of E-EWS in the healthcare sector

With healthcare, this would mean that any provider that wants to be a part of the E-EWS would get their security information primarily from their national system and would not interact directly with E-EWS (Figure 1). This can also provide elasticity to the system as a whole, as it is much easier to add or remove organizations from the system. As the European Union consists of many diverse participants with their own laws and regulations, this implementation gives room for various practical ways of execution.

5. Conclusions

This work-in-progress paper starts with two questions in mind: 1) What early warning systems are there already in use in the healthcare field? 2) How could E-EWS be implemented into healthcare?

The lack of information about existing early warning systems was very surprising. It may be that the operators of these systems do not want to publicly share information about their inner workings in fear of potential security breaches. One more thing to note is that as many of the cyber threats in the healthcare industry can be best mitigated by increased training of medical personnel, the importance of EWSs may not be very high. But as the amount of networked health care equipment increases, so does the need for efficient EWSs.

The implementation of E-EWS into healthcare systems may be best carried out via national systems. The industry is very diverse and adding both national and private organizations from many different areas might prove to be very slow, difficult, and costly. By having health care providers connect primarily to their own national warning systems and only through them to the E-EWS, the operation of an EU-spanning warning system could be very manageable.

The practical implementations of E-EWS are an area that can benefit very much from further research. As the system is still primarily in the development phase, having clear and ready-made implementations could help adapt the technology for use in the healthcare field.

References

Dupuy, A., Iftimie, I., Nussbaum, D. and Pickl, S. 2020. Cyber as a Hybrid Threat to NATO's Operational Energy Security. Proceedings of the 19th European Conference on Cyber Warfare and Security (ECCWS20), University of Chester, UK, pp. 98-106.

ECHO, 2021. ECHO Federated Cyber Range, [online], https://echonetwork.eu/echo-federated-cyber-range/

Liveri, D., Sarri, A. and Skouloudi, C. 2015. Security and Resilience in eHealth: Security Challenges and Risks, ENISA.

National Cyber Security Centre, 2021. Early Warning, [online], NCSC.GOV.UK, https://www.ncsc.gov.uk/information/early-warning-service

O'Brien, N., Graß, E., Martin, G. Durkin, M. Darzi, A and Ghafur, S. 2020. Safeguarding our healthcare systems: A global framework for cybersecurity. World Innovation Summit for Health, Doha, Qatar.

Pappalardo, M. et al. 2020. D2.2 ECHO Multi-Sector Assessment Framework, [online], ECHO Network, https://echonetwork.eu/wp-content/uploads/2020/11/ECHO_D2.2-Derivation-of-ECHO-Multi-sector-Assessment-Framework_v2.4.pdf

Rajamäki, J. et al. 2019. D3.6 ECHO Information Sharing Models, [online], ECHO Network, https://echonetwork.eu/wp-content/uploads/2020/02/ECHO_D3.6-ECHO-Information-Sharing-Models-v1.0.pdf

Kirkov, P. et al. 2020. D4.3 Inter-Sector Cybersecurity Technology Roadmap, [online], ECHO Network, https://echonetwork.eu/wp-content/uploads/2020/11/ECHO_D4.3-INTER-SECTOR-CYBERSECURITY-TECHNOLOGY-ROADMAP-v1.0.pdf

Simola, J., & Lehto, M. (2020). National cyber threat prevention mechanism as a part of the E-EWS. Proceedings of the 15th International Conference on Cyber Warfare and Security (ICCWS 2020). Academic Conferences International, pp. 539-548.

Wairimu, S. 2021, e-Health as a Target in Cyberwar: Expecting the Worst. Proceedings of the 20th European Conference on Cyber Warfare and Security (ECCWS21), University of Chester, UK, pp. 549-557.

www.ingramcontent.com/pod-product-compliance
Lightning Source LLC
Chambersburg PA
CBHW082304210326
41598CB00028B/4436
* 9 7 8 1 9 1 4 5 8 7 4 0 5 *